Aspects of Harmonic Analysis on Locally Compact Abelian Groups

Mathematical Foundations for Computer Vision, Machine Learning, and Robotics

Series Editors: Jean Gallier (*University of Pennsylvania, USA*)
Jocelyn Quaintance (*University of Pennsylvania, USA*)

Published

Vol. 1 *Aspects of Harmonic Analysis on Locally Compact Abelian Groups*
by Jean Gallier and Jocelyn Quaintance

Mathematical Foundations for Computer Vision, Machine Learning, and Robotics: Vol. 1

Aspects of Harmonic Analysis on Locally Compact Abelian Groups

Jean Gallier
University of Pennsylvania, USA

Jocelyn Quaintance
University of Pennsylvania, USA

World Scientific

NEW JERSEY • LONDON • SINGAPORE • BEIJING • SHANGHAI • HONG KONG • TAIPEI • CHENNAI • TOKYO

Published by

World Scientific Publishing Co. Pte. Ltd.

5 Toh Tuck Link, Singapore 596224

USA office: 27 Warren Street, Suite 401-402, Hackensack, NJ 07601

UK office: 57 Shelton Street, Covent Garden, London WC2H 9HE

Library of Congress Cataloging-in-Publication Data

Names: Gallier, Jean H., author. | Quaintance, Jocelyn, author.

Title: Aspects of harmonic analysis on locally compact Abelian groups / Jean Gallier, University of Pennsylvania, USA, Jocelyn Quaintance, University of Pennsylvania, USA.

Description: New Jersey : World Scientific, [2024] | Series: Mathematical Foundations for Computer Vision, Machine Learning, and Robotics ; volume 1 | Includes bibliographical references and index.

Identifiers: LCCN 2024008416 | ISBN 9789811291715 (hardcover) | ISBN 9789811291722 (ebook for institutions) | ISBN 9789811291739 (ebook for individuals)

Subjects: LCSH: Locally compact Abelian groups. | Harmonic analysis.

Classification: LCC QA387 .G35 2024 | DDC 512/.25--dc23/eng/20240314

LC record available at https://lccn.loc.gov/2024008416

British Library Cataloguing-in-Publication Data

A catalogue record for this book is available from the British Library.

For any available supplementary material, please visit
https://www.worldscientific.com/worldscibooks/10.1142/13799#t=suppl

Preface

The question that motivated writing this book is:

What is the Fourier transform?

We were quite surprised by how involved the answer is, and how much mathematics is needed to answer it, from measure theory, integration theory, some functional analysis, to some representation theory.

First we should be a little more precise about our question. We should ask two questions:

(1) What is the *input domain* of the Fourier transform?
(2) What is the *output domain* of the Fourier transform?

The answer to (1) is that the domain of the Fourier transform, denoted by $\mathcal{F}$, is a set of functions on a *group* G. Now in order for the Fourier transform to be useful, it should behave well with respect to *convolution* (denoted $f * g$) on the set of functions on G, which implies that these functions should be *integrable*.

This leads to the first subtopic, which is *what is integration on a group*? The technical answer involves the *Haar measure* on a locally compact group. Thus, any serious effort to understand what the Fourier transform is entails learning a certain amount of measure theory and integration theory, passing through versions of the Radon–Riesz theorem relating Radon functionals and Borel measures, and culminating with the construction of the Haar measure. The two candidates for the domain of the Fourier transform are the spaces $\mathrm{L}^1(G)$ and $\mathrm{L}^2(G)$. Unfortunately, convolution and the Fourier transform are not necessarily defined for functions in $\mathrm{L}^2(G)$, so the domain of the Fourier transform is $L^1(G)$. Then the equation $\mathcal{F}(f * g) = \mathcal{F}(f)\mathcal{F}(g)$ holds, as desired. If G is a compact group, $L^2(G)$ is a suitable (and better) domain.

The answer to Question (2) is more complicated, and depends heavily on whether the group G is commutative or not. The answer is much simpler if G is commutative. In both cases, the output domain of the Fourier transform should be a set of functions from a space Y to a space Z.

If G is commutative, then we can pick $Z = \mathbb{C}$. However, the space Y is rarely equal to G (except when $G = \mathbb{R}$). It turns out that a good theory (which means that it covers all cases already known) is obtained by picking Y to be the group $\widehat{G}$, the *Pontrjagin dual* of G, which consists of the *characters* of the group G. A character of G is a continuous homomorphism $\chi\colon G \to \mathbf{U}(1)$ from G to the group of complex numbers of absolute value 1. For any function $f \in L^1(G)$, the Fourier transform $\mathcal{F}(f)$ of f is then a function

$$\mathcal{F}(f)\colon \widehat{G} \to \mathbb{C}.$$

In general, $\widehat{G}$ is completely different from G, and this creates problems. For the familiar cases, $G = \mathbb{T} = \mathbf{U}(1) = \{z \in \mathbb{C} \mid |z| = 1\}$, $G = \mathbb{Z}$, $G = \mathbb{R}$, and $G = \mathbb{Z}/n\mathbb{Z}$, the characters are well known, namely $\widehat{\mathbf{T}} = \mathbb{Z}$, $\widehat{\mathbb{Z}} = \mathbf{T}$, $\widehat{\mathbb{R}} = \mathbb{R}$, and $\widehat{\mathbb{Z}/n\mathbb{Z}} = \mathbb{Z}/n\mathbb{Z}$. The case $G = \mathbb{Z}/n\mathbb{Z}$ corresponds to the *discrete Fourier transform*.

For the groups listed above, we know that under some suitable restriction, we have *Fourier inversion*, which means that there is some transform $\overline{\mathcal{F}}$ (called *Fourier cotransform*) such that

$$f = \overline{\mathcal{F}}(\mathcal{F}(f)). \tag{$*$}$$

We have to be a bit careful because the domain of $\overline{\mathcal{F}}$ is $L^1(\widehat{G})$, and not $L^1(G)$, are they are usually very different because in general G and $\widehat{G}$ are *not* isomorphic. Then (assuming that it makes sense), $\overline{\mathcal{F}}(\mathcal{F}(f))$ is a function with domain $\widehat{\widehat{G}}$, so there seems no hope, except in very special cases such as $G = \mathbb{R}$, that $(*)$ could hold. Fortunately, *Pontrjagin duality* asserts that G and $\widehat{\widehat{G}}$ are isomorphic, so $(*)$ holds (under suitable conditions) in the form

$$f = \overline{\mathcal{F}}(\mathcal{F}(f)) \circ \eta,$$

where $\eta\colon G \to \widehat{\widehat{G}}$ is a canonical isomorphism.

If G is a commutative locally compact group, there is a beautiful and well understood theory of the Fourier transform based on results of Gelfand,

Pontrjagin, and André Weil. In particular, even though the Fourier transform is not defined on $\mathrm{L}^2(G)$ in general, for any function $f \in \mathrm{L}^1(G) \cap \mathrm{L}^2(G)$, we have $\mathcal{F}(f) \in \mathrm{L}^2(\widehat{G})$, and by Plancherel's theorem, the Fourier transform extends in a unique way to an isometric isomorphism between $\mathrm{L}^2(G)$ and $\mathrm{L}^2(\widehat{G})$ (see Section 10.8). Furthermore, if we identify G and $\widehat{\widehat{G}}$ by Pontrjagin duality, then $\mathcal{F}$ and $\overline{\mathcal{F}}$ are mutual inverses (see Section 10.9). Harmonic analysis on locally compact abelian groups is covered quite thoroughly in this book (Volume I).

If G is *not* commutative, things are a lot tougher. Characters no longer provide a good input domain, and instead one has to turn to *unitary representations*. A unitary representation is a homomorphism $U\colon G \to \mathbf{U}(H)$ satisfying a certain continuity property, where $\mathbf{U}(H)$ is the group of unitary operators on the Hilbert space H. Then $\widehat{G}$ is the set of equivalence classes of irreducible unitary representations of G, but it is no longer a group. Aspects of harmonic analysis on noncommutative locally compact groups (based on representation theory) are presented in a second book (Volume II).

In particular, we discuss quite extensively the case where G is compact. In this case, an important theorem due to Peter and Weyl gives a nice decomposition of $\mathrm{L}^2(G)$ as a Hilbert sum of finite-dimensional matrix algebras corresponding to the irreducible unitary representations of G.

Acknowledgement: Many thanks to the participants of the "underground" Tuesday meetings, Christine Allen-Blanchette, Carlos Esteves, Stephen Phillips, and João Sedoc, for catching mistakes and for many helpful comments. We also thank Kostas Daniilidis for being a source of inspiration. Our debt to J. Dieudonné, G. Folland, E. Hewitt and K.A. Ross, A. Knapp, S. Lang, A.A. Kirillov, Laurent Schwartz, E. Stein and R. Shakarchi, and W. Rudin, is enormous. Every result in this manuscript is found in one form or another in their seminal books.

Contents

Chapter 1

Introduction

The main topic of this two volume book is the Fourier transform and Fourier series, both understood in a broad sense.

Historically, trigonometric series were first used to solve equations arising in physics, such as the wave equation or the heat equation. D'Alembert used trigonometric series (1747) to solve the equation of a vibrating string, elaborated by Euler a year later, and then solved in a different way essentially using Fourier series by D. Bernoulli (1753). However it was Fourier who introduced and developed Fourier series in order to solve the heat equation, in a sequence of works on heat diffusion, starting in 1807, and culminating with his famous book, *Théorie analytique de la chaleur*, published in 1822.

Originally, the theory of Fourier series is meant to deal with periodic functions on the circle $\mathbb{T} = \mathbf{U}(1) = \{z \in \mathbb{C} \mid |z| = 1\}$, say functions with period 2π. Remarkably the theory of Fourier series is captured by the following two equations:

$$f(\theta) = \sum_{m \in \mathbb{Z}} c_m e^{im\theta}. \tag{1}$$

$$c_m = \int_{-\pi}^{\pi} f(\theta) e^{-im\theta} \frac{d\theta}{2\pi}. \tag{2}$$

Equation (1) involves a series, and Equation (2) involves an integral. There are two ways of interpreting these equations.

The first way consists of starting with a convergent series as given by the right-hand side of (1) (of course $c_n \in \mathbb{C}$), and to ask what kind of function is obtained. A second question is the following: Are the coefficients in (1) computable in terms of the formulae given by (2)?

The second way is to start with a periodic function f, apply Equation (2) to obtain the c_m, called *Fourier coefficients*, and then to consider Equation (1). Does the series $\sum_{m\in\mathbb{Z}} c_m e^{im\theta}$ (called *Fourier series*) converge at all? Does it converge to f?

Observe that the expression $f(\theta) = \sum_{m\in\mathbb{Z}} c_m e^{im\theta}$ may be interpreted as a countably infinite superposition of elementary periodic functions (the harmonics), intuitively representing simple wave functions, the functions $\theta \mapsto e^{im\theta}$. We can think of m as the frequency of this wave function.

The above questions were first considered by Fourier. Fourier boldly claimed that *every* function can be represented by a Fourier series. Of course, this is false, and for several reasons. First, one needs to define what is an integrable function. Second, it depends on the kind of convergence that are we dealing with. Remarkably, Fourier was almost right, because for every function f in $\mathrm{L}^2(\mathbb{T})$, a famous and deep theorem of Carleson states that its Fourier series converges to f almost everywhere in the L^2-norm.

Given a periodic function f, the problem of determining when f can be reconstructed as the Fourier series (Equation (1)) given by its Fourier coefficients c_m (Equation (2)) is called the problem of *Fourier inversion*. To discuss this problem, it is useful to adopt a more general point of view of the correspondence between functions and Fourier coefficients, and Fourier coefficients and Fourier series.

Given a function $f \in \mathrm{L}^1(\mathbb{T})$, Equation (2) yields the $\mathbb{Z}$-indexed sequence $(c_m)_{m\in\mathbb{Z}}$ of Fourier coefficients of f, with

$$c_m = \int_{-\pi}^{\pi} f(\theta)e^{-im\theta}\,\frac{d\theta}{2\pi},$$

which we call the *Fourier transform* of f, and denote by $\widehat{f}$, or $\mathcal{F}(f)$. We can view the Fourier transform $\mathcal{F}(f)$ of f as a function $\mathcal{F}(f)\colon \mathbb{Z} \to \mathbb{C}$ with domain $\mathbb{Z}$.

On the other hand, given a $\mathbb{Z}$-indexed sequence $c = (c_m)_{m\in\mathbb{Z}}$ of complex numbers c_m, we can define the Fourier series $\overline{\mathcal{F}}(c)$ associated with c, or *Fourier cotransform* of c, given by

$$\overline{\mathcal{F}}(c)(\theta) = \sum_{m\in\mathbb{Z}} c_m e^{im\theta}.$$

This time, $\overline{\mathcal{F}}(c)$ is a function $\overline{\mathcal{F}}(c)\colon \mathbb{T} \to \mathbb{C}$ with domain $\mathbb{T}$. Fourier inversion can be stated as the equation

$$f(\theta) = ((\overline{\mathcal{F}} \circ \mathcal{F})(f))(\theta).$$

Of course, there is an issue of convergence. Namely, in general, $\widehat{f} = \mathcal{F}(f)$ does not belong to $\ell^1(\mathbb{Z})$. There are special cases for which Fourier inversion holds, in particular, if $f \in \mathrm{L}^2(\mathbb{T})$.

Let us now consider the Fourier transform of (not necessarily periodic) functions defined on $\mathbb{R}$. For any function $f \in \mathrm{L}^1(\mathbb{R})$, the *Fourier transform* $\widehat{f} = \mathcal{F}(f)$ of f is the function $\mathcal{F}(f) \colon \mathbb{R} \to \mathbb{C}$ defined on $\mathbb{R}$ given by

$$\widehat{f}(x) = \mathcal{F}(f)(x) = \int_{\mathbb{R}} f(y) e^{-iyx} \frac{dx(y)}{\sqrt{2\pi}},$$

and the *Fourier cotransform* $\overline{\mathcal{F}}(f)$ of f is the function $\overline{\mathcal{F}}(f) \colon \mathbb{R} \to \mathbb{C}$ defined on $\mathbb{R}$ given by

$$\overline{\mathcal{F}} f(x) = \int_{\mathbb{R}} f(y) e^{iyx} \frac{dx(y)}{\sqrt{2\pi}}.$$

This time, the domain of the Fourier transform is the same as the domain of the Fourier cotransform, but this is an exceptional situation. Also, in general the Fourier transform $\widehat{f}$ is not integrable, so Fourier inversion only holds in special cases.

The preceding examples suggest two questions:

(1) What is the *input domain* of the Fourier transform?
(2) What is the *output domain* of the Fourier transform?

The answer to (1) is that the domain of the Fourier transform, denoted by $\mathcal{F}$, is a set of functions on a *group* G. In order for the Fourier transform to be useful, it should behave well with respect to an operation on the set of functions on G called *convolution* (denoted $f * g$), which implies that these functions should be *integrable*.

This leads to the first subtopic, which is: *what is integration on a group?* The technical answer involves the *Haar measure* on a locally compact group. Thus, any serious effort to understand what the Fourier transform is entails learning a certain amount of measure theory and integration theory, passing through versions of the Radon–Riesz theorem relating Radon functionals and Borel measures, and culminating with the construction of the Haar measure. This preliminary material is discussed in Chapters 2, 3, 4, 5, 7, and 8.

Chapter 2 gathers some basic results about function spaces, in particular, about different types of convergence (pointwise, uniform, compact).

Some sophisticated notions cannot be avoided, such as equicontinuity, filters, topologies defined by semi-norms, and Fréchet spaces.

Chapter 3 provides a quick review of the Riemann integral and its generalization to regulated functions.

Chapter 4 is devoted to basics of measure theory: σ-algebras, semi-algebras, measurable spaces, monotone classes, (positive) measures, measure spaces, null sets, and properties holding almost everywhere. We also define outer measures and prove Carathéodory's theorem which gives a method for constructing a measure from an outer measure. We conclude by using Carathéodory's theorem to define the Lebesgue measure on $\mathbb{R}$ and $\mathbb{R}^n$ from the Lebesgue outer measure. Our presentation relies on Halmos [Halmos (1978)], Rudin [Rudin (1987)], Lang [Lang (1996)], and Schwartz [Schwartz (1993a)].

Chapter 5 develops the theory of Lebesgue integration in a fairly general context, namely functions defined on a measure space taking values in a Banach space. This integral is usually known as the Bochner integral (developed independently by Dunford). We agree with Lang ([Lang (1996)]) that the investment needed to deal with functions taking values in a Banach space rather than in $\mathbb{R}$ is minor, and that the reward is worthwhile. This approach is presented in detail in Dunford and Schwartz [Dunford and Schwartz (1988)], and more recent (and easier to read) expositions of this method are given in Lang [Lang (1996)] and Marle [Marle (1974)].

After reading this chapter, the reader will know what are the spaces $\mathrm{L}^1(X)$, $\mathrm{L}^2(X)$, and $\mathrm{L}^\infty(X)$, which is essential to move on to the study of harmonic analysis on locally compact abelian groups (abbreviated as LCA groups), which is the subject of this book (Volume I). In Chapter 5 we provide some proofs.

Chapter 7 presents the theory of integration on locally compact spaces due to Radon and Riesz based on linear functionals on the space of continuous functionals with compact support. Although this material is well-known to analysts, it may be less familiar to other mathematicians, and most computer scientists have not been exposed to it. Our presentation relies heavily on Rudin [Rudin (1987)] (Chapter 2), Lang [Lang (1996)] (Chapter IX), Folland [Folland (1999)] (Chapter 7), Marle [Marle (1974)], and Schwartz [Schwartz (1993a)]. We also borrowed much from Dieudonné [Dieudonné (2005)] (Chapter XIII).

We state the famous representation theorem of Radon and Riesz for positive linear functionals and certain types of positive Borel measures (Theorem 7.8 and Theorem 7.15). Here, inspired by Folland and Lang, we define a σ- *Radon measure* as a Borel measure which is outer regular, σ-inner regular, and finite on compact subsets. A *Radon measure* is a σ-Radon measure which is also inner regular. Linear functionals which are bounded on the space of continuous functions with support contained in a fixed compact support are called *Radon functionals*. We have avoided Bourbaki and Dieudonné's use of the term Radon measure for a Radon functional, which is just too confusing.

We define complex measures, and following Rudin, we present the Radon–Riesz correspondence between bounded Radon functionals and complex (regular) measures (Theorem 7.30). This theorem is absolutely crucial to the construction of the Haar measure and to the definition of the convolution of complex measures and of functions.

Chapter 8 contains a rather complete discussion of the Haar measure on a locally compact group, convolution, and the application of convolution to regularization. After some preliminaries about topological groups (Section 8.1), we describe the method for constructing a left Haar measure from a left Haar functional, following essentially Weil's proof as presented in Folland [Folland (1995)] (see Sections 8.2 and 8.3). We prove almost everything, except for a technical lemma. Then we prove the uniqueness of the left Haar measure up to a positive constant, using Dieudonné's method [Dieudonné (2005)] (Section 8.4). We introduce the modular function and the modulus of an automorphism. We show how to use the Haar measure to construct a hermitian inner product invariant under the representation of a compact group. We discuss G-invariant measures on homogeneous spaces.

One of the main applications of the Haar measure is the definition of the convolution $\mu * \nu$ of (complex) measures and the convolution $f * g$ of functions; see Section 8.11. Under convolution, the set $\mathcal{M}^1(G)$ of complex regular measures is a Banach algebra with an involution, and a multiplicative unit element. This algebra contains the Banach subalgebra $\mathrm{L}^1(G)$, which doesn't have a multiplicative unit in general. In Section 8.14, we show that by convolving a function f with functions g_n from a "well-behaved" family we obtain a sequence $(f * g_n)$ of functions more regular that f that converge to f. This technique is known as *regularization*.

Chapter 8 is the last of the chapters dealing with background material. Similar material is coved in Folland [Folland (1995)], and very extensively in Hewitt and Ross [Hewitt and Ross (1979)] (over 400 pages).

The main chapters presenting some elements of harmonic analysis on locally compact abelian groups, in particular the Fourier transform, are:

(1) Chapter 6, in which the classical theory of the Fourier transform (and cotransform) on $\mathbb{T}$, $\mathbb{R}$, and then $\mathbb{T}^n$ and $\mathbb{R}^n$, is presented. We also present the sampling theorem due to Shannon, and discuss the Heisenberg uncertainty principle. Our presentation is inspired by Rudin [Rudin (1987)], Folland [Folland (1992, 1999)], Stein and Shakarchi [Stein and Shakarchi (2003)], and Malliavin [Malliavian *et al.* (1995)].
(2) Chapter 10, which is devoted to harmonic analysis on *locally compact abelian groups*, based on the seminal work of A. Weil, Gelfand, and Pontrjagin. Our presentation is based on Folland [Folland (1995)] and Bourbaki [Bourbaki (1967b)].

Chapter 10 requires more preparatory material.

If G is a commutative locally compact group, then the domain of the Fourier transform on $\mathrm{L}^1(G)$ is the group $\widehat{G}$ of characters of G, the homomorphisms $\chi\colon G \to \mathbb{C}$ such that $|\chi(g)| = 1$ for all $g \in G$. The group $\widehat{G}$ is called the *Pontrjagin dual* of G. It turns out that $\widehat{G}$ is homeomorphic to the space $\mathsf{X}(\mathrm{L}^1(G))$ of characters of the Banach algebra $\mathrm{L}^1(G)$. Thus we need some knowledge about normed algebras. Chapter 9 presents the basic theory of normed algebras and their spectral theory needed for Chapter 10. The study of algebras and normed algebras focuses on three concepts:

(1) The notion of *spectrum* $\sigma(a)$ of an element a of an algebra A.
(2) If A is a commutative algebra, the notion of *character*, and the space $\mathsf{X}(A)$ of characters of A.
(3) If A is a commutative algebra, the notion of *Gelfand transform*, $\mathcal{G}\colon A \to \mathcal{C}(\mathsf{X}(A); \mathbb{C})$.

The Gelfand transform from $\mathrm{L}^1(G)$ to $\mathsf{X}(\mathrm{L}^1(G))$ is the Fourier cotransform on $\mathrm{L}^1(G)$. Our presentation is inspired by Dieudonné [Dieudonné (2005)], Bourbaki [Bourbaki (1967b)], and Rudin [Rudin (1991)].

If G is a locally compact abelian group, then for any function $f \in \mathrm{L}^1(G)$, the Fourier transform $\mathcal{F}(f)$ of f is then a function

$$\mathcal{F}(f)\colon \widehat{G} \to \mathbb{C}.$$

In general, $\widehat{G}$ is completely different from G, and this creates problems. For the familiar cases, $G = \mathbb{T} \cong \mathbf{U}(1)$, $G = \mathbb{Z}$, $G = \mathbb{R}$, and $G = \mathbb{Z}/n\mathbb{Z}$, the characters are well known. The case $G = \mathbb{Z}/n\mathbb{Z}$ corresponds to the *discrete Fourier transform*.

For the groups listed above, we know that under some suitable restriction, we have *Fourier inversion*, which means that there is some transform $\overline{\mathcal{F}}$ (called *Fourier cotransform*) such that

$$f = \overline{\mathcal{F}}(\mathcal{F}(f)). \tag{$*$}$$

We have to be a bit careful because the domain of $\overline{\mathcal{F}}$ is $\mathrm{L}^1(\widehat{G})$, and not $\mathrm{L}^1(G)$, are they are usually very different because in general G and $\widehat{G}$ are *not* isomorphic. Then (assuming that it makes sense), $\overline{\mathcal{F}}(\mathcal{F}(f))$ is a function with domain $\widehat{\widehat{G}}$, so there seems no hope, except in very special cases such as $G = \mathbb{R}$, that $(*)$ could hold. Fortunately, *Pontrjagin duality* asserts that G and $\widehat{\widehat{G}}$ are isomorphic, so $(*)$ holds (under suitable conditions) in the form

$$f = \overline{\mathcal{F}}(\mathcal{F}(f)) \circ \eta,$$

where $\eta\colon G \to \widehat{\widehat{G}}$ is a canonical isomorphism.

If G is a commutative locally compact group, there is a beautiful and well understood theory of the Fourier transform based on results of Gelfand, Pontrjagin, and André Weil presented in Chapter 10. In particular, even though the Fourier transform is not defined on $\mathrm{L}^2(G)$ in general, for any function $f \in \mathrm{L}^1(G) \cap \mathrm{L}^2(G)$, we have $\mathcal{F}(f) \in \mathrm{L}^2(\widehat{G})$, and by Plancherel's theorem, the Fourier transform extends in a unique way to an isometric isomorphism between $\mathrm{L}^2(G)$ and $\mathrm{L}^2(\widehat{G})$. Furthermore, if we identify G and $\widehat{\widehat{G}}$ by Pontrjagin duality, then $\mathcal{F}$ and $\overline{\mathcal{F}}$ are mutual inverses.

If G is *not* commutative, things are a lot tougher. Characters no longer provide a good input domain, and instead one has to turn to *unitary representations*. Some aspects of harmonic analysis on noncommutative locally compact groups are presented in a second book (Volume II).

More basic background material dealing with elementary topology, matrix norms, groups and group actions, and Hilbert spaces is found in Appendices A, B, C, D and E. These chapters should be considered as appendices and should be consulted by need.

To keep the length of this book under control, we resigned ourselves to omit many proofs. This is unfortunate because some beautiful proofs (such as the proof of the Radon–Riesz theorem for bounded Radon functional) had to be omitted. However, whenever a proof is omitted, we provide precise pointers to sources where such a proof is given.

After Chapter 1 the logical starting point of this book is Chapter 2, followed by the other chapters in consecutive order. However, some readers might find it more illuminating to proceed directly to Chapter 6 which provides a less abstract view of Fourier analysis and harmonic analysis. Readers not familiar with the Lebesgue theory of integration should not be concerned, and they should replace this fancy notion with the notion of integral that they are familiar with. The consequence of such a simplifying assumption is that some of the results may not be quite correct, but this should be a good motivation to return to the chapters dealing with measure theory and integration.

Chapter 2

Function Spaces Often Encountered

Various spaces of functions $f\colon E \to F$ from a topological space E to a metric space or a normed vector space F come up all the time. The most frequently encountered are the spaces $(F^E)_b$ of bounded functions, the spaces $\mathcal{K}(E;F)$ of continuous functions with compact support, the spaces $\mathcal{C}_0(E;F)$ of continuous functions which tend to zero at infinity, and the spaces $\mathcal{C}_b(E;F)$ of continuous bounded functions. When F is a normed vector space, all these spaces are normed vector spaces with the sup norm. An important issue about function spaces is the convergence of sequences of functions. We review the main three notions, pointwise convergence (also known as simple convergence), uniform convergence, and compact convergence. A sequence of continuous functions may converge pointwise to a function which is not continuous. Uniform convergence has a better behavior. If F is a complete normed vector space, then both spaces $\mathcal{C}_b(E;F)$ and $(F^E)_b$ are also complete under uniform convergence. An interesting family of functions in $(F^{[a,b]})_b$ is the space $\mathrm{Reg}([a,b];F)$ of regulated functions. These functions have at most only countably many simple kinds of discontinuities called discontinuities of the first kind. If F is a complete normed vector space, then the space $\mathrm{Reg}([a,b];F)$ is complete. It contains a subspace $\mathrm{Step}([a,b];F)$ consisting of very simple functions called step functions, which take finitely many different values on consecutive open intervals. The space $\mathrm{Step}([a,b];F)$ is dense in $\mathrm{Reg}([a,b];F)$. If E is a locally compact space, then the space $\mathcal{C}_0(E;\mathbb{C})$ is the closure of $\mathcal{K}(E;\mathbb{C})$ in $\mathcal{C}_b(E;\mathbb{C})$. This chapter relies heavily on the material discussed Appendix A so the reader may want to refer to this appendix whenever the need arises.

2.1 The Function Space F^E and Pointwise Convergence

In this section we study the space of functions $f\colon E \to F$, where E and F are arbitrary topological spaces. We denote the set of all functions from E to F by F^E.

Our first goal is to make F^E into a topological space in its own right. Surprisingly, one of the easiest ways to describe a topology on F^E is to follow Tychonoff and observe that

$$F^E \cong \prod_{x\in E} F_x, \qquad F_x = F.$$

Since F^E is isomorphic to an E-indexed product space, we may give it a product topology as follows: a subset of functions in F^E is open if it is the union of subsets U_A of functions $f\colon E \to F$ for which there is some *finite* subset A of E such that $f(x) \in U_x$ for all $x \in A$, where U_x is an open subset of F, and $f(x) \in F$ is arbitrary for all $x \in E - A$; see Figure 2.1.

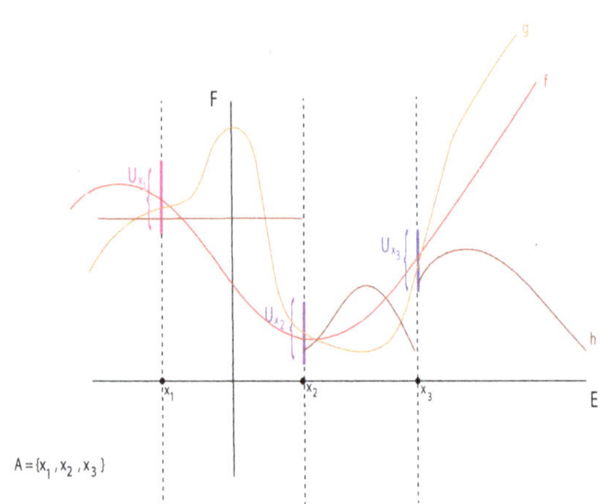

Fig. 2.1 A schematic illustration of an open set U_A of F^E, (where the reader may assume $E = F = \mathbb{R}$). The three functions $f, g, h \in U_A$ since they pass "through" the open sets U_{x_i}, for $1 \le i \le 3$.

Equivalently, for any $x \in E$ and any open subset U of F, let $S(x, U)$ be the set

$$S(x,U) = \{f \mid f \in F^E,\ f(x) \in U\};$$

see Figure 2.2. Then observe that

$$U_A = \bigcap_{x \in A} S(x, U_x), \quad A \text{ finite},$$

that is, the sets $S(x, U)$ form a subbasis of the product topology on F^E.

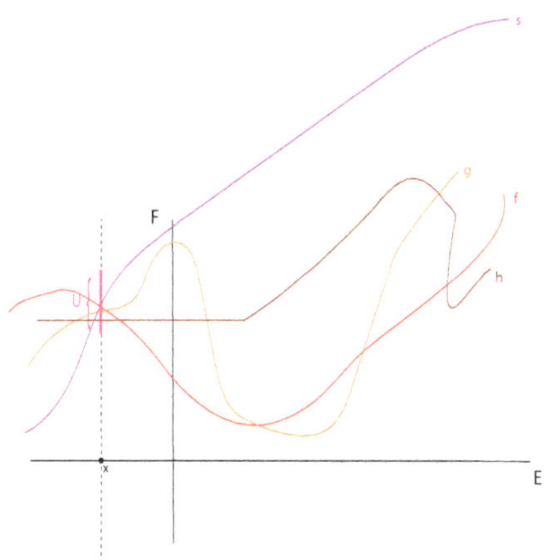

Fig. 2.2 A schematic illustration of an open set $S(x, U)$ of F^E, (where the reader may assume $E = F = \mathbb{R}$). The four functions $f, g, h, s \in S(x, U)$ since they pass "through" the open set U.

For every $x \in E$, if $\pi_x \colon F^E \to F$ is the projection map given by

$$\pi_x(f) = f(x), \qquad f \in F^E,$$

(evaluation at x), then the product topology on F^E is the weakest topology that makes all the π_x continuous. Indeed, the weakest topology on F^E making all the π_x continuous consists of all unions of finite intersections of subsets of F^E of the form $\pi_x^{-1}(U_x)$, for any open subset U_x of F, but

$$\pi_x^{-1}(U_x) = S(x, U_x),$$

is one of the sets in the subbasis defined above. For this reason, the product topology on F^E is also called the *weak topology* induced by the family of functions $(\pi_x)_{x \in E}$; see Rudin [Rudin (1991)] (Chapter 3, Section 3.8).

Now that we have made F^E into a topological space, we can ask ourselves what it means for a sequence $(f_n)_{n \geq 1}$ of functions $f_n \colon E \to F$ to converge to f. By definition of the product topology, $(f_n)_{n \geq 1}$ converges to f if and only if given any subbasic open set $S(x, U)$ containing f, there exists $n_0 \geq 0$ such that $f_n \in S(x, U)$ whenever $n \geq n_0$. A moment of reflection shows that we may reinterpret the previous statement as saying for a *fixed* point $x \in E$, $f_n(x)$ becomes "arbitrarily" close to $f(x)$. This reinterpretation is rigorously stated in terms of pointwise convergence, namely that for *fixed* $x \in E$, $\lim_{n\to\infty} f_n(x) = f(x)$. The notion of pointwise convergence does not require F to be a metric space, but since this is the situation we most often encounter, we give the definition assuming that (F, d) is a metric space,

Definition 2.1. Let (F, d) be a metric space. A sequence $(f_n)_{n\geq 1}$ of functions $f_n : E \to F$ *converges pointwise* (or *converges simply*) to a function $f : E \to F$ if for every $x \in E$, for every $\epsilon > 0$, there is some $N > 0$ such that

$$d(f_n(x), f(x)) < \epsilon \quad \text{for all } n \geq N.$$

Definition 2.1 is illustrated in Figure 2.3.

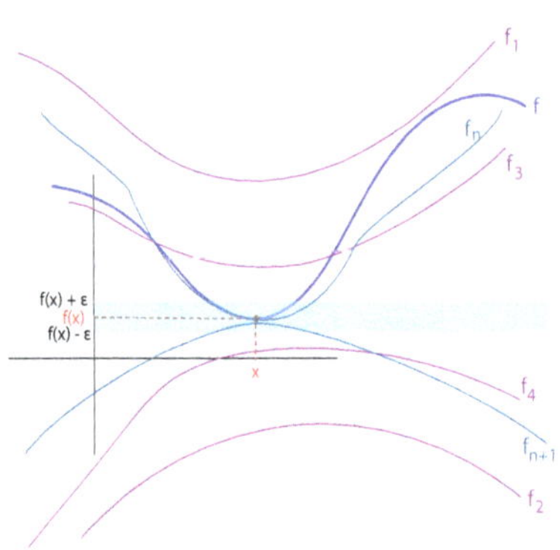

Fig. 2.3 A schematic illustration of $f_n(x)$ converging pointwise $f(x)$, where $E = F = \mathbb{R}$. As n increases, the graph of $f_n(x)$ near x must be in the band determined by the graphs of $f(x) - \epsilon$ and $f(x) + \epsilon$.

To reiterate, Definition 2.1 says that for every $x \in E$, the sequence $(f_n(x))_{n\geq 1}$ converges to $f(x)$. Observe that the above ϵ depends on x.

A sequence $(f_n)_{n\geq 1}$ of elements of F^E converges pointwise to $f \in F^E$ iff the sequence $(f_n)_{n\geq 1}$ converges to f in the product topology; see Munkres [Munkres (2000)] (Chapter 7, Section 46, Theorem 46.1), or Folland [Folland (1999)] (Chapter 4, Proposition 4.12). Consequently, the product (weak) topology is also called the topology of pointwise convergence and pointwise convergence is also known as *weak convergence*. We summarize the previous discussion in the following definition.

Definition 2.2. If F is any topological space and E is any set, the topology on F^E having the sets

$$S(x, U) = \{f \mid f \in F^E,\ f(x) \in U\}, \qquad x \in E,\ U \text{ open in } F,$$

as a subbasis is the *topology of pointwise convergence*. An open subset of F^E in this topology is any union (possibly infinite) of finite intersections of subsets of the form $S(x, U)$ as above.

If F is Hausdorff, so is the topology of pointwise convergence. Indeed, if $f, g \in F^E$ and $f \neq g$, then there is some $x \in E$ such that $f(x) \neq g(x)$, and since F is Hausdorff, there exist two disjoint open subsets $U_{f(x)}$ and $U_{g(x)}$ with $f(x) \in U_{f(x)}$ and $g(x) \in U_{g(x)}$. Then $\pi_x^{-1}(U_{f(x)})$ and $\pi_x^{-1}(U_{g(x)})$ are disjoint open subsets with $f \in \pi_x^{-1}(U_{f(x)})$ and $g \in \pi_x^{-1}(U_{g(x)})$; see Figure 2.4.

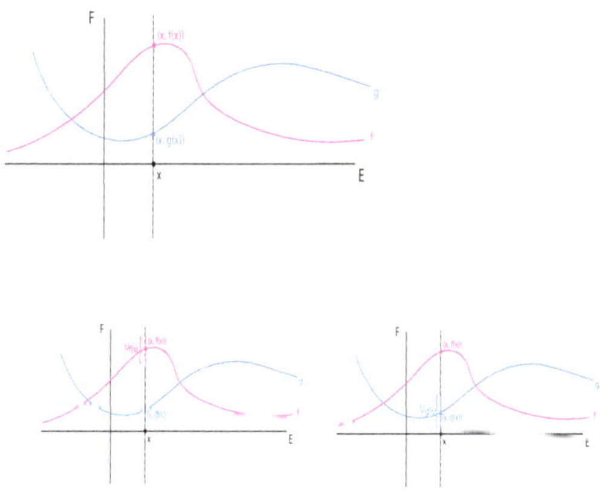

Fig. 2.4 If F is Hausdorff, so is the topology of pointwise convergence. For convenience, let $E = F = \mathbb{R}$. The top figure illustrates two distinct elements of F^E. The bottom left figure illustrates the open set $\pi_x^{-1}(U_{f(x)})$, while the bottom right figure illustrates the open set $\pi_x^{-1}(U_{g(x)})$. These two sets separate f and g within F^E.

When F is a metric space there are two important subsets within F^E, the subspace of *continuous* functions $\mathcal{C}(E; F)$, and the subspace of *bounded* functions $(F^E)_b$. As shown in Figure 2.5, both $\mathcal{C}(E; F)$ and $(F^E)_b$ inherit subspace topologies from the product topology of F^E. But if F is either a metric or a normed vector space, we can place "finer" topologies on both $\mathcal{C}(E;F)$ and $(F^E)_b$. In the next section we discuss how such a topology makes $(F^E)_b$ into its own metric space by considering it an independent space in its own right, not necessarily embedded in F^E.

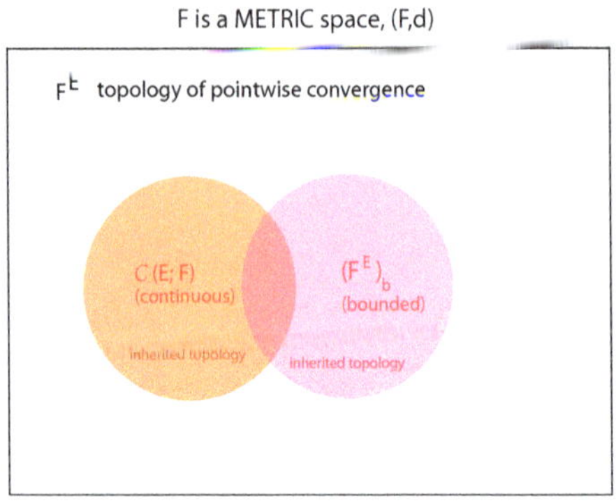

Fig. 2.5 A Venn diagram of illustration of F^E and the subsets $\mathcal{C}(E;F)$ and $(F^E)_b$ with the inherited topology of pointwise convergence.

2.2 Spaces of Bounded Functions

In this section we are dealing with functions $f\colon E \to F$, where F is either a metric space or a normed vector space.

First assume that F is a metric space with metric d. We would like to make F^E into a metric space. It is natural to define a metric on F^E by setting

$$d_\infty(f,g) = \sup_{x\in E} d(f(x),g(x))$$

for any two functions $f,g\colon E \to F$, but if $d(f(x),g(x))$ is unbounded as x ranges over E, the expression $\sup_{x\in E} d(f(x),g(x))$ is undefined. Therefore, we consider the space of bounded functions defined as follows.

Definition 2.3. If (F,d) is a metric space, a function $f\colon E \to F$ is *bounded* if its image $f(E)$ is bounded in F, which means that $f(E) \subseteq B(a,\alpha)$, for some closed ball $B(a,\alpha)$ of center a and radius $\alpha > 0$. See Figure 2.6. The space of bounded functions $f\colon E \to F$ is denoted by $(F^E)_b$.

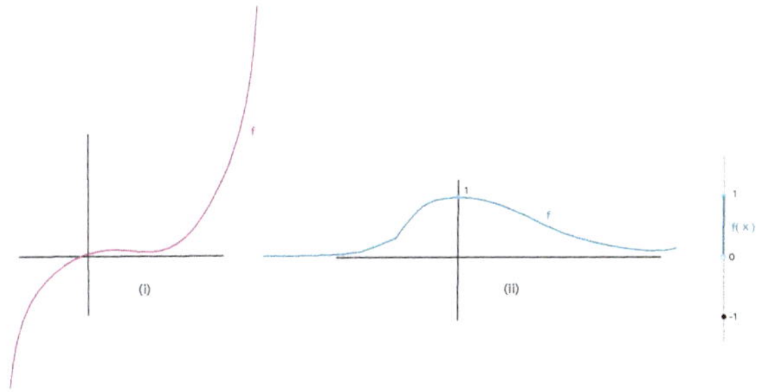

Fig. 2.6 Let $E = F = \mathbb{R}$ with the Euclidean metric. In Figure (i), f is unbounded since $f(E) = \mathbb{R}$. In Figure (ii), $f \in (F^E)_b$ since $f(E) = (0, 1]$ and $(0, 1] \subset B(0, 1) = [-1, 1]$.

If $f\colon E \to F$ and $g\colon E \to F$ are bounded functions, then it is easy to see that if $f(E) \subseteq B(a, \alpha)$ and if $g(E) \subseteq B(b, \beta)$, then

$$d(f(x), g(x)) \leq \alpha + \beta + d(a, b) \quad \text{for all } x \in E;$$

see Figure 2.7. Therefore, $\sup_{x \in E} d(f(x), g(x))$ is well defined. It is easy to check that if we define

$$d_\infty(f, g) = \sup_{x \in E} d(f(x), g(x))$$

for any two bounded functions f, g, then d is indeed a metric on $(F^E)_b$.

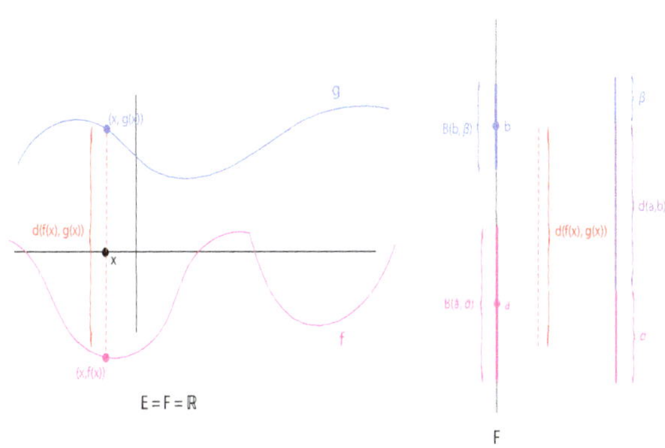

Fig. 2.7 An illustration of $d(f(x), g(x)) \leq \alpha + \beta + d(a, b)$, when $E = F = \mathbb{R}$ with the Euclidean metric.

Definition 2.4. If (F, d) is a metric space, then for any two bounded functions $f, g \in (F^E)_b$, the quantity

$$d_\infty(f,g) = \sup_{x \in E} d(f(x), g(x))$$

is a metric on $(F^E)_b$. See Figure 2.8.

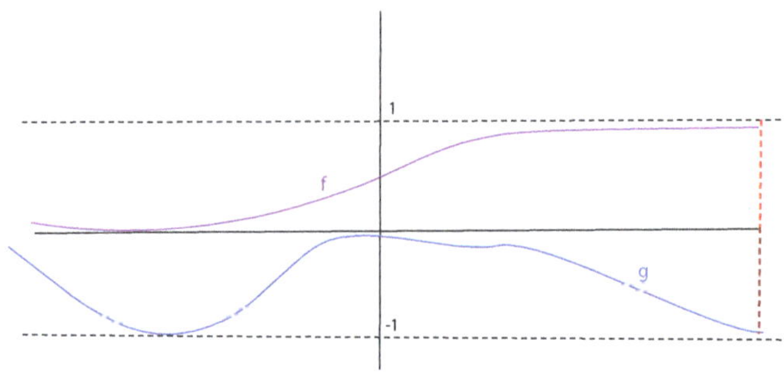

Fig. 2.8 Let $E = F = \mathbb{R}$ with the Euclidean metric. Both $f, g \in (F^E)_b$ since $f(E) = (0,1)$, while $g(E) = [-1, 0)$. The concatenation of the vertical dashed red lines is $d_\infty(f,g) = \sup_{x \in E} d(f(x), g(x)) = 1 - (-1) = 2$.

If $(F, \| \, \|)$ is normed metric space, then F^E is a vector space, and it is easy to check that $(F^E)_b$ is also a vector space. For any bounded function $f \colon E \to F$ (which means that $f(E) \subseteq B(0, \alpha)$, for some closed ball $B(0, \alpha)$), then

$$\|f\|_\infty = \sup_{x \in E} \|f(x)\|$$

is a norm on the vector space $(F^E)_b$.

Definition 2.5. If $(F, \| \, \|)$ is a normed vector space, then for any bounded function $f \in (F^E)_b$, the quantity

$$\|f\|_\infty = \sup_{x \in E} \|f(x)\|$$

is a norm on $(F^E)_b$, often called the *sup norm*; see Figure 2.9.

The following important theorem can be shown; see Schwartz [Schwartz (1980)] (Chapter XV, Section 1, Theorem 1).

Theorem 2.1. *(1) If (F, d) is a complete metric space, then $((F^E)_b, d_\infty)$ is also a complete metric space.*

(2) If $(F, \| \, \|)$ is a complete normed vector space, then $((F^E)_b, \| \, \|_\infty)$ is also a complete normed vector space.

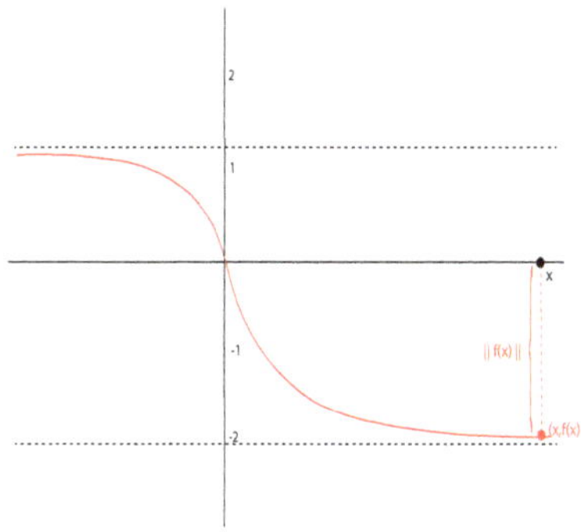

E = F = ℝ

Fig. 2.9 Let $f \in (F^E)_b$, where $E = F = \mathbb{R}$ with norm given by the absolute value. Then $\|f\|_\infty = 2$.

2.3 Uniform Convergence of Functions

When dealing with spaces of functions, a crucial issue is to identify notions of limit that preserve certain desirable properties, such as continuity.

Unfortunately the notion of pointwise convergence within F^E does not have such a property. If a sequence $(f_n)_{n\geq 1}$ of continuous functions converges pointwise to a function f, this f is not necessarily continuous. For example, the functions $f_n \colon [0,1] \to \mathbb{R}$ given by $f_n(x) = x^n$ are continuous, and the sequence $(f_n)_{n\geq 1}$ converges pointwise to the discontinuous function $f \colon [0,1] \to \mathbb{R}$ given by

$$f(x) = \begin{cases} 0 & \text{if } 0 \leq x < 1 \\ 1 & \text{if } x = 1, \end{cases}$$

as evidenced by Figure 2.10.

However, if F is a metric space there is a stronger notion of convergence, uniform convergence, which ensure that continuity *is preserved* in the limit.

Definition 2.6. Let (F, d) be a metric space. A sequence $(f_n)_{n\geq 1}$ of functions $f_n \colon E \to F$ *converges uniformly* to a function $f \colon E \to F$ if for every $\epsilon > 0$, there is some $N > 0$ such that

$$d(f_n(x), f(x)) < \epsilon \quad \text{for all } n \geq N \text{ and for all } x \in E.$$

See Figure 2.3.

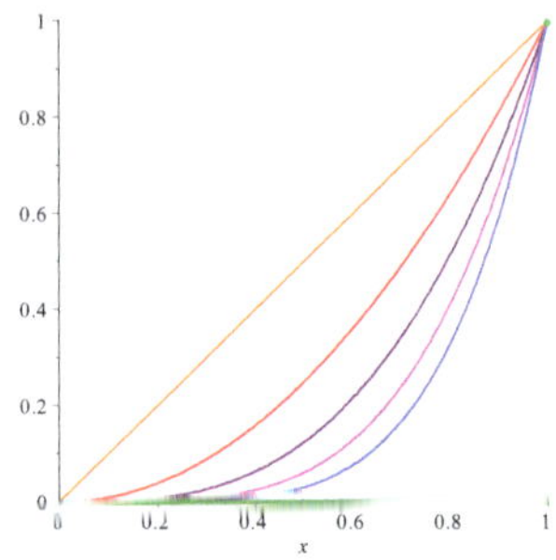

Fig. 2.10 The sequence of functions $f_n(x) = x^n$ over $[0,1]$ converges pointwise to the discontinuous green graph.

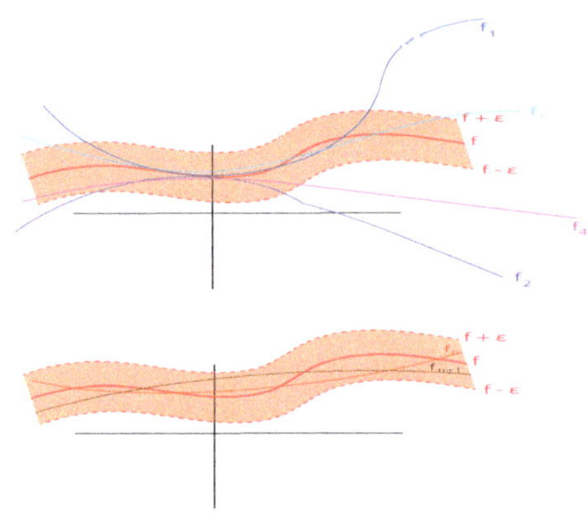

Fig. 2.11 A schematic illustration of f_n converging uniformly to f, where $E = F = \mathbb{R}$. As n increases, the graph of f_n must lie entirely in the band determined by the graphs of $f - \epsilon$ and $f + \epsilon$.

Observe that convergence in the metric space of bounded functions $((F^E)_b, d_\infty)$ is the uniform convergence of sequences of functions. Similarly, convergence in the normed vector space of bounded functions $((F^E)_b, \|\ \|_\infty)$ is the uniform convergence of sequences of functions. For this reason, the topology on $(F^E)_b$ induced by the metric d_∞ (or the norm $\|\ \|_\infty$) is sometimes called the *topology of uniform convergence*. Figure 2.12 illustrates how the intrinsic metric based topology of uniform convergence is the finer topology which replaces the inherited topology of pointwise convergence.

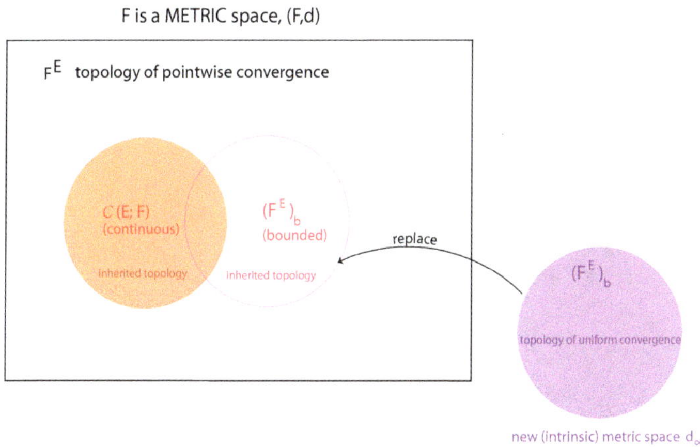

Fig. 2.12 A Venn diagram illustration of F^E and two of its subspaces; $\mathcal{C}(E;F)$, which has the inherited topology of pointwise convergence, and $(F^E)_b$, which has the inherited topology of pointwise convergence replaced with the topology of uniform convergence.

The difference between simple (pointwise) and uniform convergence is that in uniform convergence, ϵ is independent of x. For example the functions $f_n\colon [0,2\pi] \to \mathbb{R}$ defined by $f_n(x) = n\sin\left(\frac{x}{n}\right)$ converges uniformly to $f(x) = x$, as evidenced by Figure 2.13. Consequently, uniform convergence implies simple convergence, but the converse is false, as the following examples illustrate.

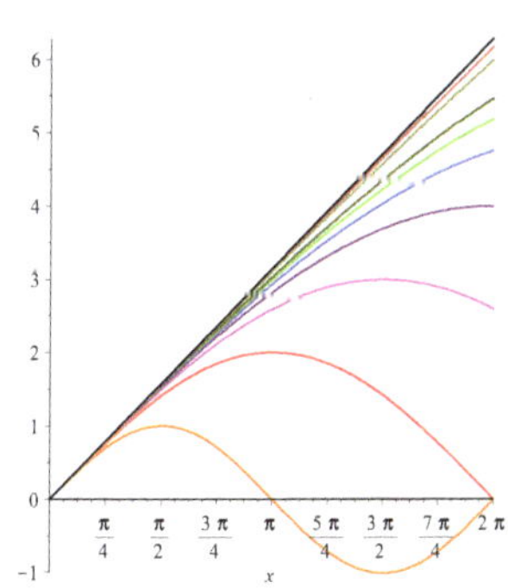

Fig. 2.13 The colored functions $f_n(x) = n\sin\left(\frac{x}{n}\right)$, over the domain $[0,2\pi]$, converge uniformly to the black line $f(x) = x$.

Example 2.1. Let $g\colon \mathbb{R} \to \mathbb{R}$ be the function given by

$$g(x) = \frac{1}{1+x^2},$$

and for every $n \geq 1$, let $f_n\colon \mathbb{R} \to \mathbb{R}$ be the function given by

$$f_n(x) = \frac{1}{1+(x-n)^2}.$$

The function f_n is obtained by translating g to the right using the translation $x \mapsto x+n$; see Figure 2.14.

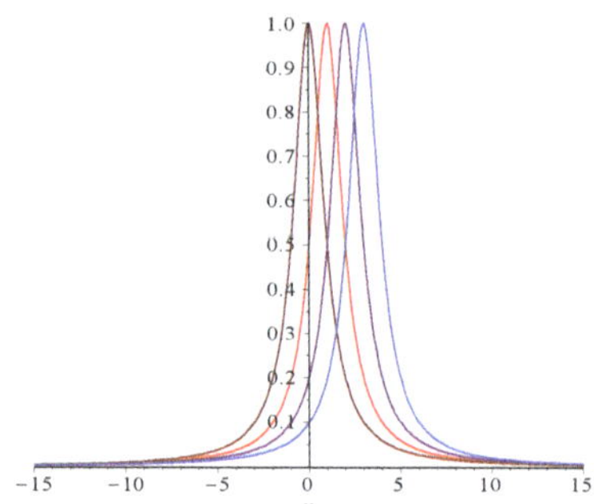

Fig. 2.14 The bell curve graphs of Example 2.1; $g(x)$ in brown; $f_1(x)$ in red; $f_2(x)$ in purple; $f_3(x)$ in blue.

Since

$$\lim_{n\mapsto\infty} \frac{1}{1+(x-n)^2} = 0,$$

the sequence $(f_n)_{n\geq 1}$ converges pointwise to the zero function f given by $f(x) = 0$ for all $x \in \mathbb{R}$. However, since the maximum of each f_n is 1, we have

$$d_\infty(f_n, f) = 1 \quad \text{for all } n \geq 1,$$

so the sequence $(f_n)_{n\geq 1}$ does not converge uniformly to the zero function.

Example 2.2. Pick any positive real $\alpha > 0$. For each $n \geq 1$, let $f_n\colon \mathbb{R} \to \mathbb{R}$ be the piecewise affine function defined as follows (see Figure 2.15):

$$f_n(x) = \begin{cases} 0 & \text{if } x \leq 0 \text{ or } x \geq 1/n \\ (2n)n^\alpha x & \text{if } 0 \leq x \leq 1/(2n) \\ 2n^\alpha(1-nx) & \text{if } 1/(2n) \leq x \leq 1/n. \end{cases}$$

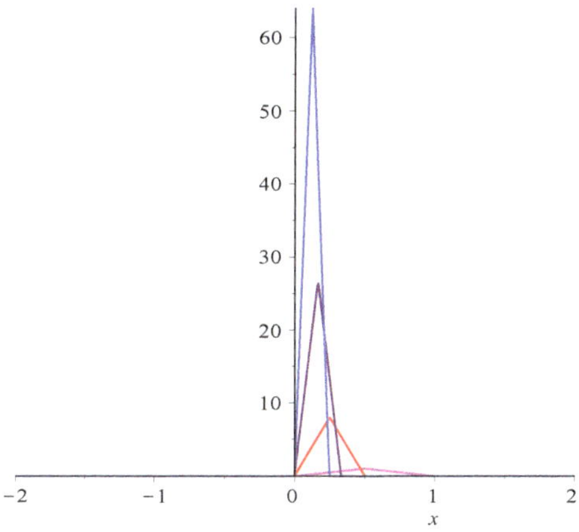

Fig. 2.15 The piecewise affine functions of Example 2.2 with $\alpha = 3$; $f_1(x)$ in magenta; $f_2(x)$ in red; $f_3(x)$ in purple; $f_4(x)$ in blue. Each $f_n(x)$ has a symmetrical triangular peak. As n increases, the peak becomes taller and thinner.

For every $x > 0$, there is some n such that $1/n < x$, so $\lim_{n\mapsto\infty} f_n(x) = 0$ for $x > 0$, and since $f_n(x) = 0$ for $x \leq 0$, we see that the sequence $(f_n)_{n\geq 1}$ converges pointwise to the zero function f. However, the maximum of f_n is n^α (for $x = 1/(2n)$) so

$$d_\infty(f_n, f) = n^\alpha,$$

and $\lim_{n\mapsto\infty} d_\infty(f_n, f) = \infty$, so the sequence $(f_n)_{n\geq 1}$ does not converge uniformly to the zero function.

If E is a topological space, it is useful to define the following local notion of uniform convergence.

Definition 2.7. Let E be a topological space and let (F, d) be a metric space. A sequence $(f_n)_{n\geq 1}$ of functions $f_n\colon E \to F$ *converges locally uniformly* to a function $f\colon E \to F$ if for every $x \in E$, there is some open subset U of E containing x such that for every $\epsilon > 0$, there is some $N > 0$ such that

$$d(f_n(x), f(x)) < \epsilon \quad \text{for all } n \geq N \text{ and for all } x \in U;$$

see Figure 2.16.

If E is locally compact, it is easy to see that a sequence $(f_n)_{n\geq 1}$ converges locally uniformly iff it converges uniformly on every compact subset of E.

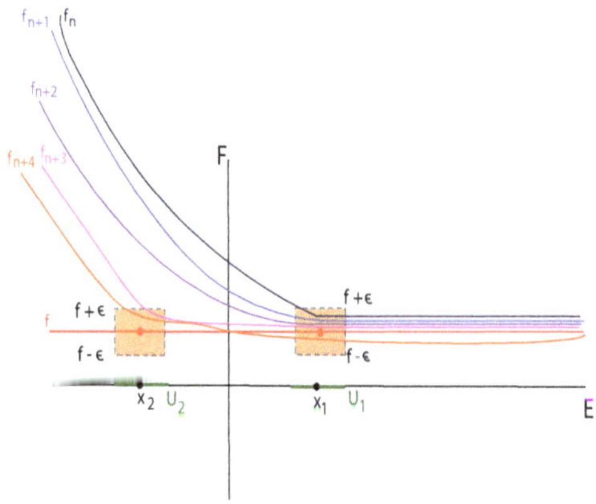

Fig. 2.16 Let $E = F = \mathbb{R}$ with the Euclidean metric, and let f be red horizontal line. The sequence $(f_n)_{n>1}$ converges locally uniformly to f. Note that for a given x and a given ϵ, the N will vary. For example, for x_1, $N = n$, while for x_2, $N = n + 4$.

As we saw at the beginning of this section, the pointwise limit of a sequence $(f_n)_{n\geq 1}$ of continuous functions needs not be continuous. However, if the convergence is locally uniform, then the limit is continuous. The following theorem gives sufficient conditions for the limit of a sequence of continuous functions to be continuous.

Theorem 2.2. *Let E be a topological space, (F, d) be a metric space, and let $(f_n)_{n\geq 1}$ be a sequence of functions $f_n \colon E \to F$ converging locally uniformly to a function $f \colon E \to F$. Then the following properties hold:*

(1) If the functions f_n are continuous at some point $a \in E$, then the limit f is also continuous at a.
(2) If the functions f_n are continuous (on the whole of E), then the limit f is also continuous (on the whole of E).
(3) If E is a metric space, the sequence $(f_n)_{n\geq 1}$ converges uniformly to f, and the f_n are uniformly continuous on E, then the limit f is also uniformly continuous on E.

The proof of Theorem 2.2 can be found in Schwartz [Schwartz (1980)] (Chapter XV, Section 4, Theorem 1).

Here are a few applications of Theorem 2.2.

Definition 2.8. Let E be a topological space, and let (F, d) be a metric space. The metric subspace of $((F^E)_b, d_\infty)$ consisting of all continuous bounded functions $f: E \to F$ is denoted $\mathcal{C}_b(E; F)$. If $(E, \| \|)$ is a normed vector space, the normed subspace of $((F^E)_b, \| \|_\infty)$ consisting of all continuous bounded functions $f: E \to F$ is also denoted $\mathcal{C}_b(E; F)$.

Proposition 2.3. *Let E be a topological space, and let (F, d) be a metric space. The metric subspace $\mathcal{C}_b(E; F)$ of $((F^E)_b, d_\infty)$ is closed. If (F, d) is a complete metric space, then $(\mathcal{C}_b(E; F), d_\infty)$ is also complete.*

Proposition 2.4. *Let E be a topological space, and let $(F, \| \|)$ be a normed vector space. The normed subspace $\mathcal{C}_b(E; F)$ of $((F^E)_b, \| \|_\infty)$ is closed. If $(F, \| \|)$ is a complete normed vector space, then $(\mathcal{C}_b(E; F), \| \|_\infty)$ is also complete.*

An important special case of Proposition 2.4 is the case where $F = \mathbb{R}$ or $F = \mathbb{C}$, namely, our functions are real-valued continuous and bounded functions $f: E \to \mathbb{R}$, or complex-valued continuous and bounded functions $f: E \to \mathbb{C}$. The spaces of functions $(\mathcal{C}_b(E; \mathbb{R}), d_\infty)$ and $(\mathcal{C}_b(E; \mathbb{C}), \| \|_\infty)$ are complete.

If E is compact and if $(F, \| \|)$ is a complete normed vector space, then every continuous function $f: E \to F$ is bounded. As a consequence, the space $\mathcal{C}(E; F)$ of continuous functions $f: E \to F$ is complete.

2.4 Compact Convergence and the Space of Continuous Functions

In the past two sections, for the case of a metric space (F, d), we investigated $(F^E)_b$. The topology we placed on $(F^E)_b$, that of uniform convergence, was intrinsic in nature and *was not* induced by the topology of pointwise convergence on F^E, but in fact finer than the induced topology. Still assuming that F is a metric space, we now want to investigate $\mathcal{C}(E; F)$. Unlike the case of $(F^E)_b$, we *can* use F^E to induce an appropriate topology on $\mathcal{C}(E; F)$, but the key is to create a *new* finer topology on F^E, namely that of compact convergence. This is not an arbitrary choice, but one based on experience, since the topology of compact convergence occurs in the definition of the dual of an abelian locally compact group.

Definition 2.9. Let E be any topological space and let (F, d) be a metric space. For any $\epsilon > 0$, any $f \in F^E$, and any compact subset K of E, define the set $B_K(f, \epsilon)$ by

$$B_K(f, \epsilon) = \left\{ g \in F^E \mid \sup_{x \in K} d(f(x), g(x)) < \epsilon \right\};$$

see Figure 2.17. The family of sets $B_K(f,\epsilon)$ is a subbasis of the *topology of compact convergence*; that is, an open set of F^E in this topology is any union (possibly infinite) of finite intersections of subsets of the form $B_K(f,\epsilon)$. The space of continuous functions from E to F with the topology of compact convergence is denoted by $(F^E)_c$.

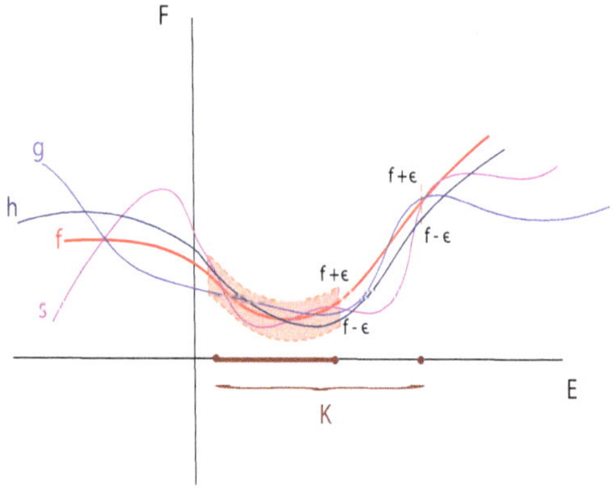

Fig. 2.17 Let $E = F = \mathbb{R}$ with the Euclidean metric, and let K be the disjoint union of the brown closed interval and single point. Then $g, h, s \in B_K(f,\epsilon)$.

The difference between this topology and the topology of pointwise convergence is that a general basis subset containing a function f contains functions that are close to f not just at finitely many points, but at all points of some compact subset. Thus the topology of pointwise convergence is weaker than the topology of compact convergence, which itself is weaker than the topology of uniform convergence. It is easy to see the sets $B_K(f,\epsilon)$ actually form a basis of the topology of compact convergence (they are closed under finite intersections).

It is easy to show that a sequence (f_n) of functions in F^E converges to a function f in the topology of compact convergence iff for every compact subset K of E, the sequence (f_n) converges uniformly to f on K.

If the space E is compactly generated, then the topology of compact convergence is even better behaved.

Definition 2.10. A topological space E is *compactly generated* if any subset U of E is open if and only if $U \cap K$ is open in K for every compact subset K.

The following result is shown in Munkres [Munkres (2000)] (Chapter 7, Section 46, Lemma 46.3).

Proposition 2.5. *If a topological space E is locally compact or first countable, then it is compactly generated.*

A nice consequence of E being compactly generated is that, as in the case of uniform convergence, the limit of a sequence of continuous functions that converges to a function f in the topology of compact convergence is continuous.

Proposition 2.6. *Let E be a compactly generated topological space and let (F, d) be a metric space. Then the space $\mathcal{C}(E; F)$ of continuous functions from E to F is closed in F^E in the topology of compact convergence.*

Proposition 2.6 is proven in Munkres [Munkres (2000)] (Chapter 7, Section 46, Theorem 46.5).

In many applications we are interested in considering the space $\mathcal{C}(E; F)$ of continuous functions from E to F as an independent space in its own right, not as a subspace embedded in F^E. As such there is an *intrinsic* way to define a topology on $\mathcal{C}(E; F)$ which has the advantage of *not* requiring F to be a metric space. Fortunately, as we will discover, and as illustrated in Figure 2.18, if F is a metric space, this intrinsic methodology corresponds to the inherent topology of compact convergence.

Definition 2.11. Let E and F be two topological spaces. For any compact subset K of E and any open subset U of F, let $S(K, U)$ be the set of continuous functions

$$S(K, U) = \{f \mid f \in \mathcal{C}(E; F),\ f(K) \subseteq U\};$$

see Figure 2.19. The sets $S(K, U)$ form a subbasis for a topology on $\mathcal{C}(E; F)$ called the *compact-open topology*. An open set in the topology is any union (possibly infinite) of finite intersections of subsets of the form $S(K, U)$.

It is immediately verified that if F is Hausdorff, then the compact-open topology on $\mathcal{C}(E; F)$ is Hausdorff.

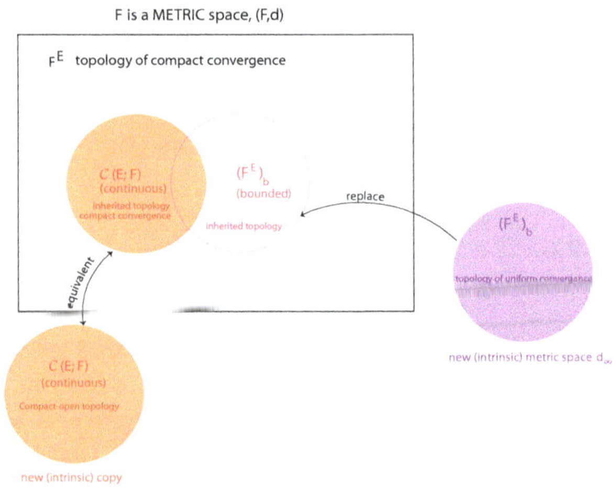

Fig. 2.18 A Venn diagram illustration of F^E with the finer topology of compact convergence, along with the subsets $\mathcal{C}(E;F)$ and $(F^E)_b$. There are two equivalent approaches for placing a topology on $\mathcal{C}(E;F)$, an inherited subspace approach and an intrinsic approach. The metric topology on $(F^E)_b$ still requires the intrinsic approach.

Remark: Observe that the open subsets $S(x,U)$ of the topology of pointwise convergence can be viewed as the result of restricting K to be a single point but relaxing f to belong to F^E.

The compact-open topology is interesting in its own right and coincides with the topology of compact convergence when F is a metric space. The following result is proven in Munkres [Munkres (2000)] (Chapter 7, Section 46, Theorem 46.8).

Proposition 2.7. *If E is a topological space and if (F,d) is a metric space, then on the space $\mathcal{C}(E;F)$ of continuous functions from E to F, the compact-open topology and the topology of compact convergence coincide.*

2.5 Equicontinuous Sets of Continuous Functions

Recall that in uniform convergence the limit of a sequence of continuous function is continuous. Another notion that is often useful to show that a sequence of continuous functions converges pointwise to a continuous function is the notion of an equicontinuous set of functions. Intuitively speaking equicontinuity is of sort of uniform continuity for sets of functions.

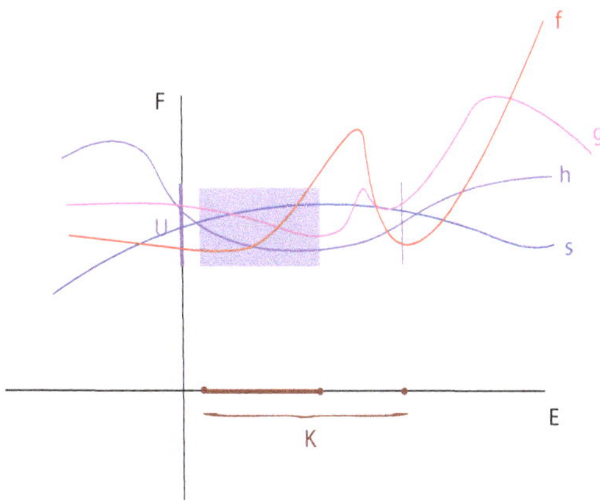

Fig. 2.19 Let $E = F = \mathbb{R}$ with the Euclidean metric, and let K be the disjoint union of the brown closed interval and single point. Let U be the purple open interval. Then $f, g, h, s \in S(K, U)$ since each function passes through the light purple region.

Definition 2.12. Let E be a topological space and let (F, d_F) be a metric space. A subset $S \subseteq \mathcal{C}(E; F)$ of the set of continuous functions from E to F is *equicontinuous at a point* $x_0 \in E$ if for every $\epsilon > 0$, there is some open subset $U \subseteq E$ containing x_0 such that $d_F(f(x), f(x_0)) \leq \epsilon$ for all $x \in U$ and for *all* $f \in S$; see Figure 2.20. If E is also a metric space with metric d_E, then the above condition says that for every $\epsilon > 0$ and for *all* $f \in S$, there is some $\eta > 0$ such that $d_F(f(x), f(x_0)) \leq \epsilon$ whenever $d_E(x, x_0) \leq \eta$. The set of functions S is *equicontinuous* if it is equicontinuous at every point $x \in E$.

For example, if E is a metric space and if there exists two constants $c, \alpha > 0$ such that we have the Lipschitz condition

$$d_F(f(x), f(y)) \leq c(d_E(x, y))^\alpha, \qquad \text{for all } f \in S \text{ and all } x, y \in E,$$

then S is equicontinuous.

Proposition 2.8. *Let (f_n) be a sequence of functions $f_n \in \mathcal{C}(E; F)$, and let (x_n) be a sequence of points $x_n \in E$. If the set $\{f_n\}$ is equicontinuous, the sequence (x_n) converges to $x \in E$, and the sequence (f_n) converges pointwise to some function $f\colon E \to F$, then the sequence $(f_n(x_n))$ converges to $f(x) \in F$.*

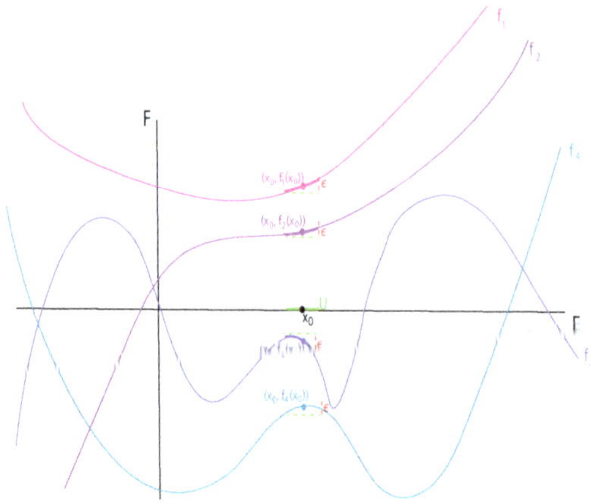

Fig. 2.20 Let $E = F = \mathbb{R}$ with the Euclidean metric, and let U be the green open interval containing x_0. The set $S = \{f_1, f_2, f_3, f_4\}$ is equicontinuous at x_0.

Proof. We have, as shown in Figure 2.21, the inequality

$$d_F(f_n(x_n), f(x)) \leq d_F(f_n(x_n), f_n(x)) + d_F(f_n(x), f(x)).$$

For every $\epsilon > 0$, since the sequence (f_n) converges pointwise to f, there is

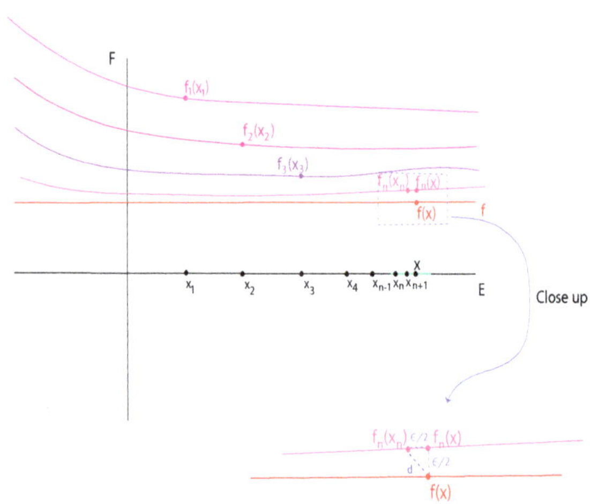

Fig. 2.21 An illustration of $d_F(f_n(x_n), f(x)) \leq d_F(f_n(x_n), f_n(x)) + d_F(f_n(x), f(x))$, where $E = F = \mathbb{R}$. For simplicity we suppressed the first coordinate of the ordered pair.

some $N_2 > 0$ such that $d_F(f_n(x), f(x)) \leq \epsilon/2$ for all $n \geq N_2$. Since $\{f_n\}$ is equicontinuous, there is some open subset $U \subseteq E$ containing x such that

$$d_F(f_n(y), f_n(x)) \leq \epsilon/2 \qquad \text{for all } n \geq 1 \text{ and all } y \in U.$$

Since (x_n) converges to x, there is some $N_1 > 0$ such that $x_n \in U$ for all $n \geq N_1$, so

$$d_F(f_n(x_n), f_n(x)) \leq \epsilon/2 \qquad \text{for all } n \geq N_1,$$

and for all $n \geq \max\{N_1, N_2\}$, we have $d_F(f_n(x_n), f(x)) \leq \epsilon$, which proves that $(f_n(x_n))$ converges to $f(x)$. □

There are various results about equicontinuous sets of functions usually known as variants of *Ascoli's theorem*. Schwartz [Schwartz (1980)] (Chapter XX) gives one of the most complete expositions we are aware of. We only consider three variants of Ascoli's theorem that we will need.

Theorem 2.9. *(Ascoli I) Let E be a topological space, let (F, d_F) be a metric space, and let $S \subseteq \mathcal{C}(E; F)$ be a set of equicontinuous functions at some $x_0 \in E$. Then the closure $\overline{S}$ of S in F^E with the topology of pointwise convergence is also equicontinuous at x_0. As a corollary, if $S \subseteq \mathcal{C}(E; F)$ is a set of equicontinuous functions, then every function $f \in \overline{S}$ is continuous, and for every sequence (f_n) of functions $f_n \in S$, if (f_n) converges pointwise to a function $f \in F^E$, then f is continuous.*

Proof. Since S is equicontinuous at x_0, for every $\epsilon > 0$, there is some open subset $U \subseteq E$ containing x_0 such that

$$d_F(f(x_0), f(x)) \leq \epsilon, \qquad \text{for all } f \in S \text{ and all } x \in U.$$

But for $x \in U$ fixed, the map $f \mapsto (f(x_0), f(x))$ from F^E to $F^2 = F \times F$ is continuous (this is a projection onto a product), and d_F is continuous on F^2. As a consequence, the set

$$\{f \in F^E \mid d_F(f(x_0), f(x)) \leq \epsilon\}$$

is closed in F^E, and since it contains S, it also contains $\overline{S}$. Thus, for every $\epsilon > 0$, we found an open subset U containing x_0 such that $d_F(f(x_0), f(x)) \leq \epsilon$ for all $x \in U$ and all $f \in \overline{S}$, which means that $\overline{S}$ is equicontinuous.

Since every function in an equicontinuous set of functions is continuous, every function $f \in \overline{S}$ is continuous. By definition of the pointwise topology, if a sequence (f_n) of functions $f_n \in S$ converges pointwise to a function $f \in F^E$, then $f \in \overline{S}$, so f is continuous. □

Dieudonné proves a weaker version of Theorem 2.9, namely that for every subset S of the space of bounded continuous functions $\mathcal{C}_b(E;F)$, if S is equicontinuous, then its closure $\overline{S}$ is also equicontinuous. This is Proposition 7.5.4 in Dieudonné [Dieudonné (2010)] (Chapter 7, Section 5).

The second version of Ascoli's theorem involves a dense subset E_0 of E. We need the following variant of Definition 2.2.

Definition 2.13. The topology of *pointwise convergence in E_0* is the topology on F^E having the sets

$$\mathcal{U}(x,U) = \{f \mid f \in F^E,\ f(x) \in U\}, \qquad x \in E_0,\ U \text{ open in } F,$$

as a subbasis.

Theorem 2.10. *(Ascoli II) Let E be a topological space, let (F, d_F) be a metric space, E_0 be a dense subset of E, and $S \subseteq \mathcal{C}(E;F)$ be a set of equicontinuous functions. Then the topology of pointwise convergence in E_0, the topology of pointwise convergence, and the topology of compact convergence (all three topologies being defined in F^E), induce identical topologies on S.*

Theorem 2.10 is proven in Schwartz [Schwartz (1980)] (Chapter XX, Theorem XX.3.1). The following corollaries of Theorem 2.10 are particularly useful. The first of these two propositions is an immediate consequence of Theorem 2.10.

Proposition 2.11. *Let E be a topological space and let (F, d_F) be a metric space. If a sequence (f_n) of continuous functions $f_n \in \mathcal{C}(E;F)$ converges pointwise to a function $f \in F^E$ and if $\{f_n\}$ is equicontinuous, then f is continuous and the sequence (f_n) converges uniformly to f on every compact subset.*

Proposition 2.12. *Let E be a topological space, E_0 a dense subset of E, and let (F, d_F) be a metric space. If the following properties hold:*

(1) The sequence (f_n) of continuous functions $f_n \in \mathcal{C}(E;F)$ converges pointwise for every $x \in E_0$;
(2) The set $\{f_n\}$ is equicontinuous;
(3) The set $\{f_n(x) \mid n \geq 1\}$ is contained in a complete subset of F for every $x \in E$;

then the sequence (f_n) converges pointwise (for all $x \in E$) to a continuous function f, and (f_n) converges uniformly to f on every compact subset. If F complete, then Condition (3) is automatically satisfied and can be omitted.

Proof. Since by Theorem 2.10, the topology of pointwise convergence on E_0 is identical to the topology of pointwise convergence on E, as the sequence (f_n) converges pointwise for every $x \in E_0$, it also converges pointwise for every $x \in E$. This implies that for every x, the sequence $(f_n(x))$ is a Cauchy sequence in F, but since by (3) the set $\{f_n(x) \mid n \geq 1\}$ is contained in a complete subset of F, the sequence $(f_n(x))$ converges. Thus (f_n) converges pointwise to a function $f \in F^E$, and since $\{f_n\}$ is equicontinuous, by Proposition 2.11, the function f is continuous, and (f_n) converges uniformly to f on every compact subset. □

Dieudonné proves a special case of Proposition 2.12 where E is a metric space, F is a complete normed vector space (a Banach space), the functions f_n are continuous and bounded, and $\{f_n\}$ is equicontinuous; see Proposition 7.5.5 and Proposition 7.5.6 in [Dieudonné (2010)] (Chapter 7, Section 5).

In most applications of Ascoli I and II, E is a metric space and F is a (complete) normed vector space. The following result about sets of continuous linear maps will be needed.

Proposition 2.13. *Let E be a metrizable vector space and F be a normed vector space. A subset of continuous linear maps $S \subseteq \mathcal{L}(E; F)$ is equicontinuous if and only if there is some open subset $V \subseteq E$ containing 0 and some real $c > 0$ such that $\|f(x)\| \leq c$ for all $x \in V$ and all $f \in S$.*

Proof. If S is equicontinuous, then obviously the property of the proposition holds. Conversely, for any $\epsilon > 0$, the condition $\|f(x)\| \leq c$ for all $x \in V$ and all $f \in S$ implies that $\|f(x)\| \leq \epsilon$ for all $x \in (\epsilon/c)V$ and all $f \in S$, so S is equicontinuous at 0. For any $x_0 \in E$, and for all $x \in x_0 + (\epsilon/c)V$, we have

$$\|f(x) - f(x_0)\| = \|f(x - x_0)\| \leq \epsilon$$

for all $f \in S$, that is, S is equicontinuous at x_0. □

A third version of Ascoli's theorem involving relative compactness will be needed in Vol II, Section 9.1. Recall that a subset A of a Hausforff space X is relatively compact if its closure $\overline{A}$ is compact in X.

Theorem 2.14. *(Ascoli III) Let E be a topological space, let (F, d_F) be a metric space, and let $S \subseteq \mathcal{C}(E; F)$ be a set of continuous functions. Assume the following two conditions hold:*

(1) The set S is equicontinuous.

(2) For every $x \in E$, the set $S(x) = \{f(x) \mid f \in S\}$ is relatively compact in F.

Then the set S is relatively compact in the space $(F^E)_c$ of continuous functions from E to F with the topology of compact convergence. Conversely, if E is locally compact and if the set S is relatively compact in the space $(F^E)_c$, then Conditions (1) and (2) hold.

Proof. A complete proof is given in Schwartz [Schwartz (1980)] (Chapter XX, Theorem XX.4.1). We only prove the first part of the theorem. The proof uses Tychonoff's powerful product theorem. By hypothesis, for every $x \in E$, the closure $\overline{S(x)}$ of $S(x)$ is compact in F, so by Tychonoff's theorem, the product $\prod_{x\in E} \overline{S(x)}$ is compact in F. By definition of the above product, this means that the set $\widehat{S}$ of functions $f \in F^E$ such that $f(x) \in \overline{S(x)}$ for all $x \in E$ is compact in F^E with the topology of pointwise convergence. Since S is contained in the compact set $\widehat{S}$, we deduce that its closure $\overline{S}$ is compact in F^E (with the topology of pointwise convergence). By Ascoli I (Theorem 2.9), since S is equicontinuous, the set $\overline{S}$ is also equicontinuous. By Ascoli II (Theorem 2.10), since the restriction to $\overline{S}$ of the topology of pointwise convergence on F^E coincides with the restriction to $\overline{S}$ of the topology of compact convergence on F^E, the set $\overline{S}$ is compact in $(F^E)_c$, and thus S is relatively compact in $(F^E)_c$. □

The special case of Theorem 2.14 in which E is compact and F is a Banach space is proven in Dieudonné [Dieudonné (2010)] (Chapter 7, Section 5, Theorem 5.7.5). Because F is complete the proof is simpler and does not use Tychonoff's theorem.

2.6 Continuous Functions of Compact Support

In this section we consider F to be a normed vector space. We know that two important subspaces of F^E are $(F^E)_b$, the space of bounded functions, and $\mathcal{C}(E; F)$, the space of continuous functions. The intersection of these subspaces, with the inherited sup norm of $(F^E)_b$, is the space of continuous bounded functions $\mathcal{C}_b(E; F)$. Within $\mathcal{C}_b(E; F)$ there is another interesting subspace, namely $\mathcal{C}_c(E; F)$, the space of continuous functions of compact support. In this section we investigate $\mathcal{C}_c(E; F)$ and describe its closure within $\mathcal{C}_b(E; F)$. So first we recall what is the support of a function.

Definition 2.14. Given any function $f\colon E \to F$, where E is a topological space and F is a vector space, the support $\mathrm{supp}(f)$ of f is the closure of the subset of E where f is nonzero, that is, $\mathrm{supp}(f) = \overline{\{x \in E \mid f(x) \neq 0\}}$. The function f has *compact support* if its support $\mathrm{supp}(f)$ is compact. If E is Hausdorff, this is equivalent to saying that f vanishes outside some compact subset K of E. See Figure 2.22.

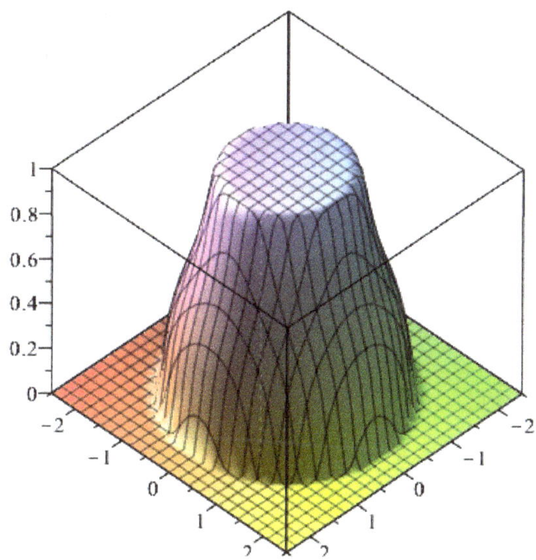

Fig. 2.22 The graph of $f\colon \mathbb{R}^2 \to \mathbb{R}$ with compact support $\mathrm{supp} = \overline{B(0,2)} = \{(x,y) \in \mathbb{R}^2 \mid x^2 + y^2 \leq 2\}$.

It is easy to see that the set of continuous functions $f\colon E \to F$ with compact support is a vector space.

Definition 2.15. The vector space of continuous functions $f\colon E \to F$ with compact support is denoted by $\mathcal{C}_c(E;F)$, or $\mathcal{K}(E;F)$. For every compact subset K of E, we denote by $\mathcal{K}(K;F)$ the space of continuous functions whose support is contained in K. Then

$$\mathcal{K}(E;F) = \bigcup_{K \subseteq E,\, K \text{ compact}} \mathcal{K}(K;F).$$

Observe that every function in $\mathcal{K}(E;F)$ is bounded, that is, $\mathcal{K}(E;F) \subseteq \mathcal{C}_b(E;F)$.

If $F = \mathbb{R}$ or $F = \mathbb{C}$, then we write $\mathcal{K}_{\mathbb{R}}(E)$ or $\mathcal{K}_{\mathbb{C}}(E)$ for $\mathcal{K}(E; F)$. Radon functionals are certain kinds of linear forms on $\mathcal{K}_{\mathbb{C}}(E)$.

The following results will be needed in Vol II, Chapter 3.

Proposition 2.15. *If E is a compact metric space, then the spaces $\mathcal{C}_{\mathbb{R}}(E)$ and $\mathcal{C}_{\mathbb{C}}(E)$ are separable.*

Proof sketch. The proof is nontrivial and can be found in Dieudonné [Dieudonné (2010)] (Chapter 7, Theorem 7.4.4). The proof makes a crucial use of the Stone–Weierstrass Theorem (Theorem 9.36) The first step is to observe that it suffices to prove that $\mathcal{C}_{\mathbb{R}}(E)$ is separable because $\mathcal{C}_{\mathbb{C}}(E)$ is the direct (topological) sum of $\mathcal{C}_{\mathbb{R}}(E)$ and $i\mathcal{C}_{\mathbb{R}}(E)$. As a second step, we observe that by Proposition A.47, since E is a compact metric space, it is separable, and by Proposition A.46, a separable metric space is second countable. Thus there is a countable base (U_n) for the topology. Then the trick is to define the family of continuous functions $g_n(t) = d(t, E - U_n)$ (see Definition A.5 for the definition of the distance to a subset). The next step is to define the subalgebra B of $\mathcal{C}_{\mathbb{R}}(E)$ generated by the monomials $g_{i_1}^{m_1}(t) \cdots g_{i_k}^{m_k}(t)$ and to check that B satisfies the hypotheses of the Stone–Weierstrass Theorem (Theorem 9.36). The final step is to show that by using rational linear combinations of the monomials $g_{i_1}^{m_1}(t) \cdots g_{i_k}^{m_k}(t)$ we obtain a countable dense subset of $\mathcal{C}_{\mathbb{R}}(E)$ (see Dieudonné [Dieudonné (2010)] (Chapter 5, Theorem 5.10.1)). □

Proposition 2.16. *If E is a locally compact separable metric space, then the spaces $\mathcal{K}_{\mathbb{R}}(E)$ and $\mathcal{K}_{\mathbb{C}}(E)$ are separable.*

Proof sketch. A proof is implicitly given in Dieudonné [Dieudonné (2005)] (Chapter XIII, Theorem 13.11.6). As in the proof of Proposition 2.15 it suffices to prove our result for $\mathcal{K}_{\mathbb{R}}(E)$. By Proposition A.49(1), since E is locally compact, metric, and separable, there is a countable sequence (K_n) of compact subsets of E such that $K_n \subseteq K_{n+1}$ and $E = \bigcup_{n \geq 1} K_n$. Then

$$\mathcal{K}_{\mathbb{R}}(E) = \bigcup_{n \geq 1} \mathcal{K}_{\mathbb{R}}(K_n).$$

By Proposition 2.15, for each $n \geq 1$, there is a dense sequence $(g_{m,n})_{n \geq 1}$ in $\mathcal{K}_{\mathbb{R}}(K_n)$. Then the countable double sequence $(g_{m,n})$ is dense in $\mathcal{K}_{\mathbb{R}}(E)$. □

If $(F, \| \, \|)$ is a Banach space and K is a fixed compact subset of E, then so is $\mathcal{K}(K; F)$ (for the sup norm $\| \, \|_\infty$), because it is closed in $\mathcal{C}_b(E; F)$. However, the normed vector space $(\mathcal{K}(E; F), \| \, \|_\infty)$ is *not* complete!

Example 2.3. For every $n \geq 1$, consider the function $u_n \colon \mathbb{R} \to \mathbb{R}$ defined as follows:

$$u_n(x) = \begin{cases} 1 & \text{if } -n \leq x \leq n \\ x + n + 1 & \text{if } -(n+1) \leq x \leq -n \\ -x + n + 1 & \text{if } n \leq x \leq n + 1 \\ 0 & \text{if } |x| \geq n + 1. \end{cases}$$

Now consider the sequence of functions (f_n) given by

$$f_n(x) = u_n e^{-|x|}.$$

Each function f_n is continuous and has compact support $[-(n+1), n+1]$, and it is easy to show that the sequence (f_n) converges uniformly to the function f given by $f(x) = e^{-|x|}$, but f does not have compact support. The problem is that the domains of the functions f_n, although compact, keep growing as n goes to infinity. See Figure 2.23.

Example 2.3 shows that the normed vector space $(\mathcal{K}(E; F), \| \, \|_\infty)$ is *not closed* in the complete normed vector space $(\mathcal{C}_b(E; F), \| \, \|_\infty)$. It would be useful to identify the closure $\overline{\mathcal{K}(E; F)}$ of $\mathcal{K}(E; F)$ in $\mathcal{C}_b(E; F)$, and this can indeed be done when E is locally compact.

Assume that f belongs to the closure $\overline{\mathcal{K}(E; F)}$ of $\mathcal{K}(E; F)$. This means that there is a sequence (f_n) of functions $f_n \in \mathcal{K}(E; F)$ such that $\lim_{n \mapsto \infty} \|f - f_n\|_\infty = 0$, so for every $\epsilon > 0$, there is some $n \geq 1$ such that $\|f(x) - f_n(x)\| \leq \epsilon$ for all $x \in E$, and since f_n has compact support, there is some compact subset K of E such that $\|f(x)\| \leq \epsilon$ for all $x \in E - K$. This suggests the following definition.

Definition 2.16. The subspace of $\mathcal{C}_b(E; F)$, denoted $\mathcal{C}_0(E; F)$, consisting of the continuous functions f such that for every $\epsilon > 0$, there is some compact subset K of E such that $\|f(x)\| \leq \epsilon$ for all $x \in E - K$, is called the space of *continuous functions which tend to* 0 *at infinity*; see Figure 2.24.

If E is compact, we can pick $K = E$, in which case $E - K = \emptyset$. This shows that Definition 2.16 has been designed so that if E is compact, then $\mathcal{C}_0(E; F) = \mathcal{C}(E; F) = \mathcal{K}(E; F)$.

Observe that if $E = \mathbb{R}$, then a function $f \in \mathcal{C}_0(\mathbb{R}; F)$ does indeed have the property that $\lim_{x \mapsto -\infty} f(x) = \lim_{x \mapsto +\infty} f(x) = 0$; see Figure 2.25.

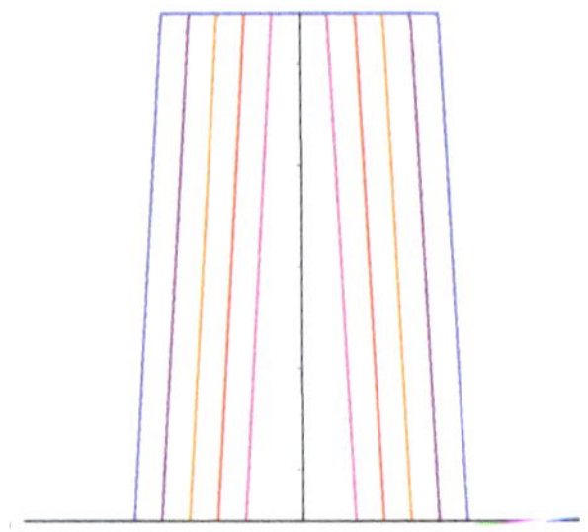

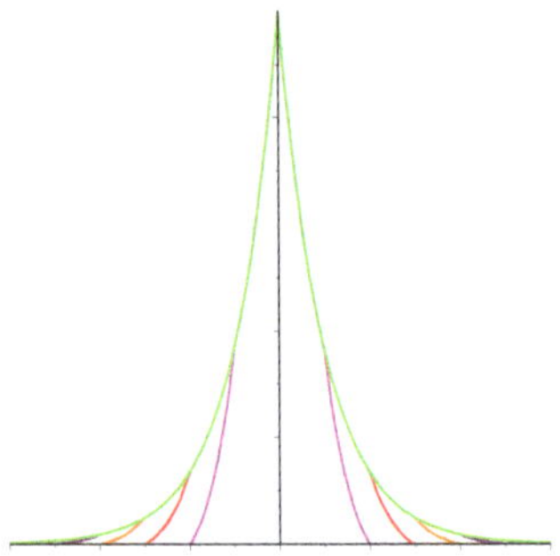

Fig. 2.23 The functions of Example 2.3. Figure (i) illustrates the $u_1(x)$ in magenta; $u_2(x)$ in red, $u_3(x)$ in orange, $u_4(x)$ in purple, and $u_5(x)$ in blue. Figure (ii) uses the same color scheme to illustrate the corresponding $f_n(x)$. Note these $f_n(x)$ converge uniformly to green $f(x) = e^{-|x|}$.

We showed that $\mathcal{K}(E; F) \subseteq \mathcal{C}_0(E; F)$. If E is locally compact, then we have the following result from Dieudonné [Dieudonné (2005)] (Chapter XIII, Section 20) and Rudin [Rudin (1987)] (Chapter 3, Theorem 3.17).

Proposition 2.17. *If E is locally compact, then $\mathcal{C}_0(E; \mathbb{C})$ is the closure of $\mathcal{K}(E; \mathbb{C})$ in $\mathcal{C}_b(E; \mathbb{C})$. Consequently, $\mathcal{C}_0(E; \mathbb{C})$ is complete.*

Proof. We already showed just before Definition 2.16 that if a function f belongs to the closure of $\mathcal{K}(E; \mathbb{C})$, then it tends to zero at infinity. Conversely, pick any f in $\mathcal{C}_0(E; \mathbb{C})$. For every $\epsilon > 0$, there is a compact subset K of E such that $|f(x)| < \epsilon$ outside of K. By Proposition A.39, there is continuous function $g \colon E \to [0, 1]$ with compact support such that $g(x) = 1$ for all $x \in K$. Clearly $fg \in \mathcal{K}_{\mathbb{C}}(E)$, and $\|fg - f\|_\infty < \epsilon$. This shows that $\mathcal{K}(E; \mathbb{C})$ is dense in $\mathcal{C}_0(E; \mathbb{C})$. $\square$

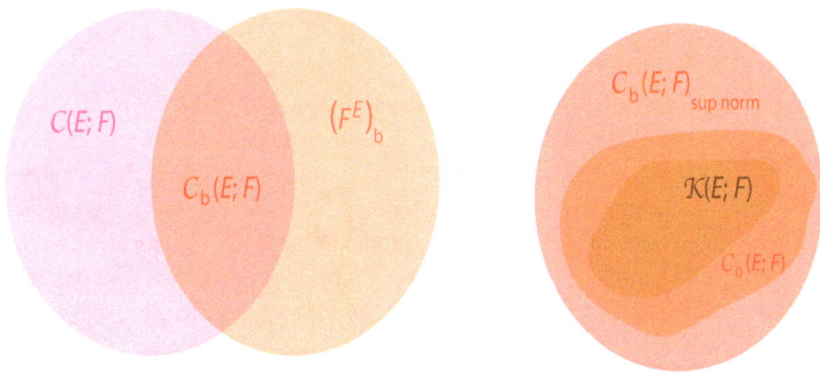

Fig. 2.24 The Venn diagram relationships between $\mathcal{C}_b(E;F) = (F^E)_b \cap \mathcal{C}(E;F)$, and the subspaces $\mathcal{K}(E;F)$ and $\mathcal{C}_0(E;F)$, where $\mathcal{K}(E;F) \subseteq \mathcal{C}_0(E;F) \subseteq \mathcal{C}_b(E;F)$.

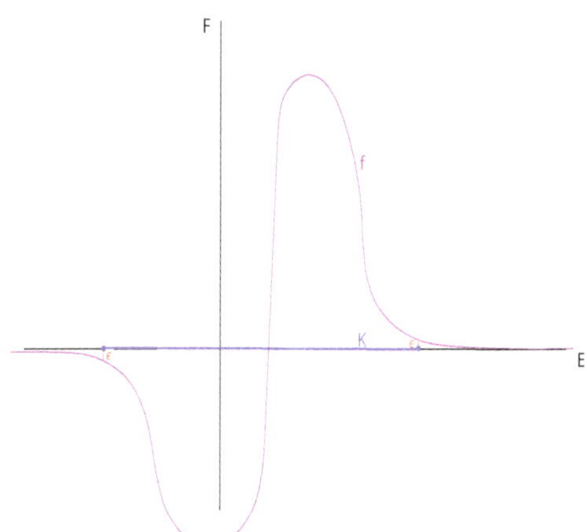

Fig. 2.25 A schematic illustration of $f \in \mathcal{C}_0(\mathbb{R};F)$, where the reader may consider $F = \mathbb{R}$.

In summary, if E is locally compact, then we have the inclusions

$$\mathcal{K}(E;\mathbb{C}) \subseteq \mathcal{C}_0(E;\mathbb{C}) \subseteq \mathcal{C}_b(E;\mathbb{C}),$$

with $\mathcal{C}_0(E;\mathbb{C})$ and $\mathcal{C}_b(E;\mathbb{C})$ complete, and $\mathcal{K}(E;\mathbb{C})$ dense in $\mathcal{C}_0(E;\mathbb{C})$. If E is not compact, these inclusions are strict in general. It turns out that the space of continuous linear forms on $\mathcal{C}_0(E;\mathbb{C})$ is isomorphic to the space of bounded Radon functionals.

2.7 Topologies Defined by Semi-Norms; Fréchet Spaces

Certain function spaces, such as the space $\mathcal{C}(X;\mathbb{C})$ of continuous functions on a topological space X, do not come with "natural" topologies defined by a norm or a metric for which they are complete. However, the weaker notion of semi-norm can be used to define a topology, and under certain conditions, although such topologies are not defined by any norm, they are metrizable and complete. In this section we briefly discuss the use of semi-norms to define topologies. It turns out that the corresponding spaces are locally convex.

Recall from Definition B.1 that a semi-norm satisfies Properties (N2) and (N3) of a norm, but in general does not satisfy Condition (N1), so $\|x\| = 0$ does not necessarily imply that $x = 0$. Here is a method for defining a topology on a vector space using a family of semi-norms.

Definition 2.17. Let X be a vector space and let $(p_\alpha)_{\alpha\in I}$ be a family of semi-norms on X. For every $x \in X$, every $\epsilon > 0$, and every $\alpha \in I$, let

$$U_{x,\alpha,\epsilon} = \{y \in X \mid p_\alpha(y - x) < \epsilon\}.$$

The *topology induced by the family of semi-norms* $(p_\alpha)_{\alpha\in I}$ is the weakest (coarsest) topology whose open sets are arbitrary unions of finite intersections of subsets of the form $U_{x,\alpha,\epsilon}$.

We can think of the subset $U_{x,\alpha,\epsilon}$ as an open ball of center x and radius ϵ in X, determined by the semi-norm p_α.

Two good examples of topologies induced by families of semi-norms are the topology of pointwise convergence and the topology of compact convergence on a normed vector space F.

Example 2.4. Let E be any set and let F be a normed vector space. If we define the family of semi-norms $(p_x)_{x\in E}$ by

$$p_x(f) = \|f(x)\|, \quad f \in F^E,\ x \in E,$$

then it is easy to see that the topology defined by the family $(p_x)_{x\in E}$ is the topology of pointwise convergence on F^E, which has the subsets

$$S(x, U) = \{f \mid f \in F^E,\ f(x) \in U\}, \qquad x \in E,\ U \text{ open in } F,$$

as a subbasis.

Example 2.5. Let E be a topological space and let F be a normed vector space. If we define the family of semi-norms $\{p_K \mid K \text{ compact in } E\}$, by

$$p_K(f) = \sup_{x \in K} \|f(x)\|, \quad f \in F^E,\ K \text{ compact in } E,$$

then it is easy to see that the topology defined by the family (p_K) is the topology of compact convergence on F^E, which has the subsets

$$B_K(f, \epsilon) = \left\{ g \in F^E \mid \sup_{x \in K} d(f(x), g(x)) < \epsilon \right\}$$

as a subbasis.

We have made our vector space X into a topological space but it is not clear that the operations (addition and scalar multiplication) are continuous. Also, in general, this topology is not Hausdorff. The following proposition addresses these issues.

Proposition 2.18. *Let X be a vector space and let $(p_\alpha)_{\alpha \in I}$ be a family of semi-norms on X.*

(1) With the topology induced by the family of semi-norms $(p_\alpha)_{\alpha \in I}$, addition and scalar multiplication are continuous, so X is a topological vector space.
(2) For every $x \in X$, the finite intersections of subsets of the form $U_{x,\alpha,\epsilon}$ is a neighborhood base of x.
(3) Every open set $U_{x,\alpha,\epsilon}$ is convex.
(4) Every p_α is continuous.
(5) The topology induced by the family of semi-norms is Hausdorff if and only if, for every $x \neq 0$, there is some $\alpha \in I$ such that $p_\alpha(x) \neq 0$.

Proposition 2.18 is proven in Folland [Folland (1999)] (Chapter 5, Section 5.4, Theorem 5.14), or Rudin [Rudin (1991)] (Chapter 1, Theorem 1.37). In view of Property (3), the topological space X is said to be *locally convex*.

In a vector space X whose topology defined by a family of semi-norms $(p_\alpha)_{\alpha \in I}$ is Hausdorff, it is easy to see that the convergence of a sequence (x_n) to a limit x is expressed conveniently as follows.

Proposition 2.19. *Let X be a space whose topology is defined by a family of semi-norms $(p_\alpha)_{\alpha \in I}$. If X is Hausdorff, then a sequence (x_n) converges to a limit x iff for every $\alpha \in I$, for every $\epsilon > 0$, there is some $N_\alpha > 0$ such that $p_\alpha(x - x_n) \leq \epsilon$ for all $n \geq N_\alpha$; equivalently,* $\lim_{n \mapsto \infty} p_\alpha(x - x_n) = 0$, *for every $\alpha \in I$.*

When the index family I is countable and the topology induced by a family of semi-norm is Hausdorff, then X is actually metrizable.

Proposition 2.20. *Let X be a vector space and let $(p_\alpha)_{\alpha \in I}$ be a family of semi-norms on X. If the topology induced by $(p_\alpha)_{\alpha \in I}$ is Hausdorff and if I is countable, then X is metrizable with a translation-invariant metric d (this means that $d(a, b) = d(a + u, b + u)$ for all $a, b, u \in X$). In fact, we can use the metric d given by*

$$d(x, y) = \sum_{m=0}^{\infty} \frac{1}{2^m} \frac{p_m(y - x)}{1 + p_m(y - x)}.$$

Proposition 2.18 is proven in Dieudonné [Dieudonné (2005)] (Chapter 12, Section 4, Theorem 12.4.6), and in Rudin [Rudin (1991)] (Chapter 1, Page 29, with an equivalent metric).

Definition 2.18. A vector space X whose topology is defined by a countable family of semi-norms, and which is Hausdorff and complete for some translation-invariant metric defining the topology of X is called a *Fréchet space*.

A prime example of a Fréchet space is the space $\mathcal{C}(X; \mathbb{C})$ of continuous functions on a separable, locally compact, metrizable space X. This will be proven shortly.

The following technical result is needed to prove Proposition 2.22.

Proposition 2.21. *Let X be a metrizable Hausdorff topological vector space. For any translation-invariant metric d defining the topology of X, a sequence (x_n) is a Cauchy sequence if and only if for every neighborhood V of 0, there is some $N > 0$ such that $x_m - x_n \in V$ for all m, n such that $m \geq N$ and $n \geq N$.*

Proof. A slightly more general result is proven for topological groups in Dieudonné [Dieudonné (2005)] (Chapter 12, Section 9, Theorem 12.9.2) and Rudin [Rudin (1991)] (Chapter 1, Page 21). If a metric d defining the topology of X is translation-invariant, then

$$d(x_n, x_m) = d(0, x_m - x_n),$$

and the sequence (x_n) is a Cauchy sequence iff for every $\epsilon > 0$, there is some $N > 0$ such that $d(0, x_m - x_n) < \epsilon$ for all $m \geq N$ and $n \geq N$, which is equivalent to saying that $x_m - x_n \in V$, where V is the open ball of center 0 and radius ϵ, which is an open subset of X, by definition of the metric topology. Conversely, since the topology of X is defined by the metric d, every open ball of center 0 is an open set, so the condition of the proposition implies that (x_n) is a Cauchy sequence for every translation-invariant metric defining the topology of X. $\square$

Proposition 2.22. *If a metrizable topological vector space X is Hausdorff and complete for some translation-invariant metric d defining the topology of X, then it is also complete for every translation-invariant metric d' defining the topology of X,*

We now prove that the space $\mathcal{C}(X;\mathbb{C})$ of continuous functions on a separable, locally compact, metrizable space X is a Fréchet space.

Recall from Proposition A.49 that since X is metrizable, there is a sequence $(U_n)_{n\geq 0}$ of open subsets such that for all $n\in\mathbb{N}$, $U_n\subseteq U_{n+1}$, $\overline{U_n}$ is compact, $\overline{U_n}\subseteq U_{n+1}$, and $X=\bigcup_{n\geq 0}U_n=\bigcup_{n\geq 0}\overline{U_n}$. For every $n\in\mathbb{N}$, define the function $p_n\colon\mathcal{C}(X;\mathbb{C})\to\mathbb{R}$ by

$$p_n(f)=\sup_{x\in U_n}|f(x)|,\qquad f\in\mathcal{C}(X;\mathbb{C}).$$

It is immediately verified that the p_n are semi-norms (but none of the p_n are norms if X is not compact). For each $f\in\mathcal{C}(X;\mathbb{C})$, if $f\neq 0$, then there is some n such that $x\in U_n$, hence $p_n(f)\neq 0$. Thus, by Proposition 2.18(5), the space $\mathcal{C}(X;\mathbb{C})$ with the topology induced by the family of semi-norms (p_n) is Hausdorff. By Proposition 2.20, this topology is metrizable. Note that the restriction of p_{n+1} to the compact subset $\overline{U_n}$ is actually a norm, and by definition of the metric d given by Proposition 2.20, the restriction of d to $\overline{U_n}$ is equivalent to p_{n+1}.

Proposition 2.23. *Let X be a separable, locally compact, metrizable space. The space $\mathcal{C}(X;\mathbb{C})$ with the topology induced by the family of semi-norms (p_n) is complete. Therefore, it is a Fréchet space.*

Proof. Since the restriction of the metric d to $\overline{U_n}$ is equivalent to p_{n+1}, by Proposition 2.22, a sequence (f_k) of functions in $\mathcal{C}(X;\mathbb{C})$ is a Cauchy sequence if for every n, the sequence of restrictions $f_k|\overline{U_n}$ is a Cauchy sequence in the Banach space $\mathcal{C}(\overline{U_n};\mathbb{C})$, hence converges uniformly in $\overline{U_n}$ to a continuous function $g_n\in\mathcal{C}(\overline{U_n};\mathbb{C})$. Since $g_{n+1}|\overline{U_n}=g_n$, there exists a continuous function $f\in\mathcal{C}(X;\mathbb{C})$ whose restriction to each U_n agrees with the restriction of g_n to U_n; see Figure 2.26.

It is clear that $\lim_{m\mapsto\infty} p_n(f - f_m) = 0$ for all $n \geq 0$, hence by Proposition 2.19, f is the limit of the Cauchy sequence (f_k), and $\mathcal{C}(X;\mathbb{C})$ is complete. □

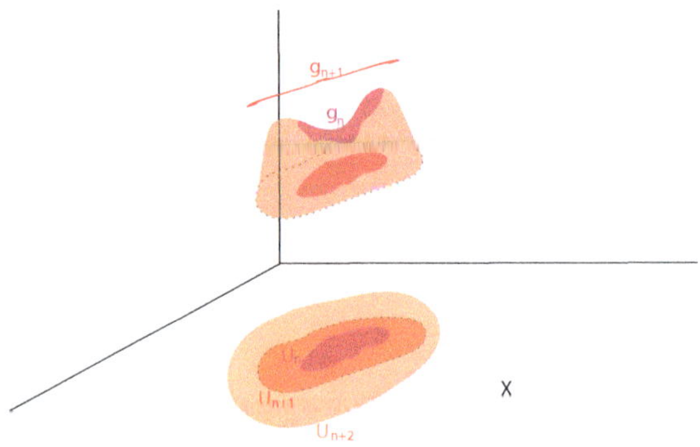

Fig. 2.26 A schematic illustration of the function g_n and its continuous extension g_{n+1}. In this figure X is represented by the horizontal plane and $\mathbb{C}$ is the vertical axis. The graph of g_{n+1} is the dusty rose surface while the graph of g_n is the plum surface patch inside of that surface.

It is shown in Rudin [Rudin (1991)] (Chapter 1, Example 1.44) that the Fréchet space $\mathcal{C}(X;\mathbb{C})$ is not normable.

The following result is shown in Dieudonné [Dieudonné (2005)] (Chapter 12, Section 14, Theorem 12.14.6.2).

Proposition 2.24. *Let X be a separable, locally compact, metrizable space. The Fréchet space $\mathcal{C}(X;\mathbb{C})$ is separable. In fact, there is a countable dense set consisting of continuous functions with compact support.*

Another good example of a Fréchet space is the Schwartz space; see Section 6.8.

2.8 Regulated Functions

In the last two sections we focused on $\mathcal{C}(E;F)$ where F is a normed vector space. We return to $(F^E)_b$ and in preparation for the next chapter on the Riemann integral investigate two important subspaces of $(F^{\mathbb{R}})_b$, the space of regulated functions and then the space of step functions, both of which inherit the sup norm from $(F^E)_b$. Since the space of regulated functions contains the space of step functions we begin with the definition of the larger subspace. Recall that there are four kinds of intervals of $\mathbb{R}$: (a, b), $[a, b)$, $(a, b]$, and $[a, b]$, with $a < b$. By convention, $(a, b) = [a, b)$ if $a = -\infty$, and $(a, b) = (a, b]$ if $b = \infty$.

Definition 2.19. Let I be an interval of $\mathbb{R}$, and let F be a metric space (or a normed vector space). Given a function $f\colon I \to F$, for any $x \in I$ with $x \neq b$, we say that f *has a limit to the right in* x if $\lim_{y\in I,\, y>x} f(y)$ exists as $y \in I$ tends to x from above. This limit is denoted by $f(x+)$. For any $x \in I$ with $x \neq a$, we say that f *has a limit to the left in* x if $\lim_{y\in I,\, y<x} f(y)$ exists as $y \in I$ tends to x from below. This limit is denoted by $f(x-)$. Given any interval I, a function $f\colon I \to F$ is a *regulated function* (or *ruled function*) if it has a left limit and a right limit for every $x \in I$. If F is a metric space (or a normed vector space), a function $f\colon \mathbb{R} \to F$ is a *regulated function* (or *ruled function*) if there is some interval I such that f vanishes outside I, and the restriction $f\colon I \to F$ of f to I is regulated. See Figure 2.27.

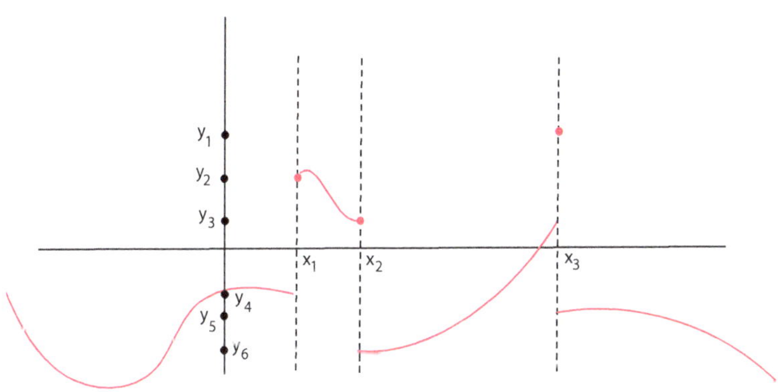

Fig. 2.27 An illustration of a regulated function $f\colon \mathbb{R} \to \mathbb{R}$. This function has three discontinuities x_1, x_2, and x_3, each of the first kind. Note that $f(x_1-) = y_4$, $f(x_1+) = f(x_1) = y_2$, $f(x_2-) = f(x_2) = y_3$, $f(x_2+) = y_6$, $f(x_3-) = y_3$, $f(x_3+) = y_5$, yet $f(x_3) = y_1$.

The notion of a regulated function can also be defined in terms of certain kinds of discontinuities.

Definition 2.20. Let I be an interval of $\mathbb{R}$, and let F be a metric space (or a normed vector space). Given a function $f: I \to F$, we say that a point $x \in I$ is a *discontinuity of the first kind* if the left limit $f(x-)$ and the right limit $f(x+)$ both exist, but $f(x-) \neq f(x)$ or $f(x+) \neq f(x)$.

It is clear that a function $f: I \to F$ is regulated iff for every $x \in I$, either f is continuous or x is a discontinuity of the first kind. Thus every continuous function is a regulated function. It is also easy to see that a monotonic function $f: I \to \mathbb{R}$ is a regulated function.

The function $f: \mathbb{R} \to \mathbb{R}$ defined by

$$f(x) = \begin{cases} \sin\left(\frac{1}{x}\right) & \text{if } x \neq 0 \\ 0 & \text{if } x = 0 \end{cases}$$

is discontinuous at $x = 0$, but this is not a discontinuity of the first kind. See Figure 2.28.

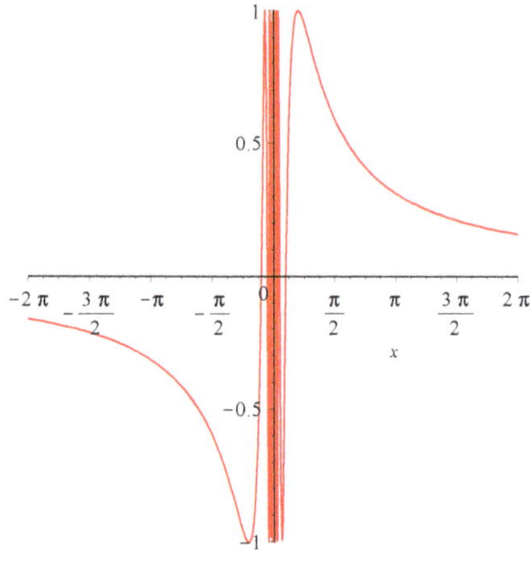

Fig. 2.28 The graph of $f(x) = \sin\left(\frac{1}{x}\right)$, $x \neq 0$.

The following result is shown in Schwartz [Schwartz (1992)] (Chapter III, Section 2, Theorem 3.2.3).

Proposition 2.25. *If $f: I \to F$ is a regulated function (where F is a metric space), then f has at most countably many discontinuities of the first kind.*

Regulated functions on a closed and bounded interval $[a, b]$ must be bounded. As a consequence, they arise as limits of uniformly convergent sequences of step functions.

Definition 2.21. A function $f\colon \mathbb{R} \to F$ (where F is any set) is a *step function* if there is a finite sequence $(a_0, a_1, \ldots, a_n)$ of real numbers such that $a_k < a_{k+1}$ for $k = 0, \ldots, n-1$, and f is constant on each of the open intervals $(-\infty, a_0)$, (a_k, a_{k+1}) for $k = 0, \ldots, n-1$, and $(a_n, +\infty)$. The sequence $(a_0, a_1, \ldots, a_n)$ is called an *admissible subdivision* for f. See Figure 2.29. If a step function f has compact support, then we assume that f vanishes on $(-\infty, a_0)$ and on $(a_n, +\infty)$ for any admissible subdivision $(a_0, a_1, \ldots, a_n)$ for f. By a step function $f\colon [a, b] \to F$, we mean a step function such that $f(x) = 0$ for all $x \leq a$ and for all $x \geq b$.

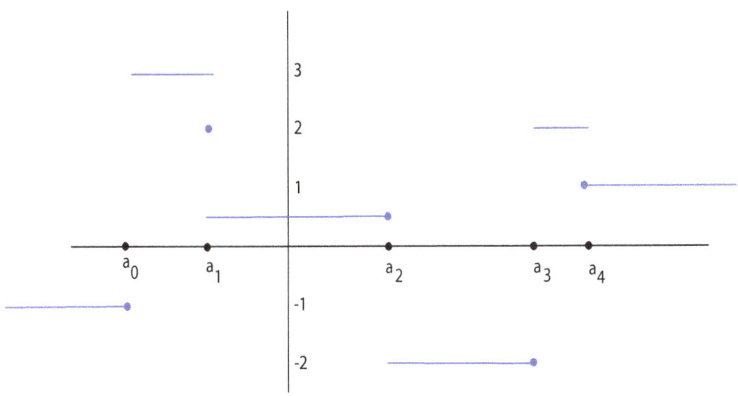

Fig. 2.29 An illustration of a step $f\colon \mathbb{R} \to \mathbb{R}$ with admissible subdivision $(a_0, a_1, a_2, a_3, a_4)$.

Observe that Definition 2.21 does not make any restriction on the values $f(a_k)$, but a step function is regulated. Also, by refining a given subdivision, a given step function admits infinitely many admissible subdivisions.

The following result is easy to prove.

Proposition 2.26. *If F is a vector space, then the set of step functions $f\colon \mathbb{R} \to F$ is a vector space denoted by* $\mathrm{Step}(\mathbb{R}; F)$. *The set of step functions $f\colon [a, b] \to F$ is also vector space denoted by* $\mathrm{Step}([a, b]; F)$.

The following proposition is much more interesting.

Proposition 2.27. *Let F be a metric space and let $[a, b]$ be a closed and bounded interval. Then every regulated function $f\colon [a, b] \to F$ is the limit of a uniformly convergent sequence $(f_n)_{n\geq 1}$ of step functions $f_n\colon [a, b] \to F$. Furthermore, if F is a complete metric space, then the limit of any uniformly convergent sequence $(f_n)_{n\geq 1}$ of step functions is a regulated function.*

The proof of Proposition 2.27 is given in Schwartz [Schwartz (1992)] (Chapter III, Section 2, Theorem 3.2.9).

As a corollary of Proposition 2.27 we have the following result.

Proposition 2.28. *If F is a complete metric space, then the space of regulated functions on $[a, b]$ is closed in $(F^{[a,b]})_b$, and the space of step functions on $[a, b]$ is dense in the space of regulated functions on $[a, b]$. Thus if F is complete, since $(F^{[a,b]})_b$ is complete, the space of regulated function on $[a, b]$ is also complete.*

Another corollary of Proposition 2.27 is that every continuous function $f\colon [a, b] \to F$ to a metric space F is the limit of a uniformly convergent sequence $(f_n)_{n\geq 1}$ of step functions $f_n\colon [a, b] \to F$.

If F is a vector space, the set of regulated functions defined on the closed and bounded interval $[a, b]$ is a vector space denoted by $\mathrm{Reg}([a, b]; F)$. Then Proposition 2.27 implies the following result.

Proposition 2.29. *Let F be a complete normed vector space. The space $\mathrm{Reg}([a, b]; F)$ of regulated functions on $[a, b]$ is complete, and the space $\mathrm{Step}([a, b]; F)$ is dense in $\mathrm{Reg}([a, b]; F)$.*

Step functions can be used to define the Riemann integral. To do so it is convenient to consider functions of finite support. Furthermore, a modified version of step functions involving a measure will be used to define the integral on a measure space.

2.9 Problems

Problem 2.1. Prove Theorem 2.1. Hint: See Schwartz [Schwartz (1980)] (Chapter XV, Section 1, Theorem 1).

Problem 2.2. Prove Theorem 2.2. Hint: See Schwartz [Schwartz (1980)] (Chapter XV, Section 4, Theorem 1).

Problem 2.3. Prove that the sets $B_K(f, \epsilon)$ form a basis for the topology of compact convergence.

Problem 2.4. Prove Proposition 2.6. Hint: See Munkres [Munkres (2000)] (Chapter 7, Section 46, Theorem 46.5).

Problem 2.5. Prove that if F is Hausdorff, then $\mathcal{C}(E, F)$ is also Hausdorff with respect to the compact-open topology.

Problem 2.6. Prove Proposition 2.7. Hint: See Munkres [Munkres (2000)] (Chapter 7, Section 46, Theorem 46.8).

Problem 2.7. Advanced Exercise: Prove Theorem 2.10. Hint: See Schwartz [Schwartz (1980)] (Chapter XX, Theorem XX.3.1).

Problem 2.8. Prove Proposition 2.18. Hint: See Folland [Folland (1999)] (Chapter 5, Section 5.4, Theorem 5.14) or Rudin [Rudin (1991)] (Chapter 1, Theorem 1.37).

Problem 2.9. Prove Proposition 2.19.

Problem 2.10. Prove Proposition 2.27. Hint: See Schwartz [Schwartz (1992)] (Chapter III, Section 2, Theorem 3.2.9).

Chapter 3

The Riemann Integral

Let $f\colon [a,b] \to \mathbb{R}$ be a continuous function. Intuitively, the Riemann integral $\int_a^b f(t)dt$ is the area of the surface "under the curve" $t \mapsto f(t)$ from $x = a$ to $x = b$. It can be approximated by the sum $s_T(f)$ (called *Cauchy-Riemann sum*) of the areas $(t_{k+1} - t_k)f(t_k)$ of $n \geq 1$ narrow rectangles, where $T = (t_0, t_1, \ldots, t_n)$ is any sequence of reals such that $t_0 = a$, $t_n = b$ and $t_k < t_{k+1}$, for $k = 0, \ldots, n-1$; see Figure 3.1. The fact that the function f is continuous on the compact interval $[a,b]$ implies that the sums $s_T(f)$ have a limit when the diameter of the subdivision tends to zero (see Definition 3.1), which means the maximum of the distances $t_{k+1} - t_k$ tends to zero (as n goes to infinity), and this limit is independent of the subdivision. Thus we can define the Riemann integral $\int_a^b f(t)dt$ as this common limit. The mapping $f \mapsto \int_a^b f(t)dt$ is a positive linear form on the space of continuous functions on $[a,b]$. This procedure applies unchanged to continuous functions $f\colon [a,b] \to F$, where F is a complete normed vector space. The method can be adapted to define the integral of regulated functions (see Definition 2.19). We proceed in two steps:

(1) The method of Cauchy-Riemann sums is easily adapted to define the notion of integral for a step function (see Definition 2.21). This yields a mapping $\int\colon \mathrm{Step}([a,b];F) \to F$ which is easily seen to be linear and continuous.

(2) By Proposition 2.29, the vector space $\mathrm{Step}([a,b];F)$ of step functions over $[a,b]$ is dense in $\mathrm{Reg}([a,b];F)$, the space of regulated functions over $[a,b]$, and $\mathrm{Reg}([a,b];F)$ is complete. By Theorem A.73, the continuous linear map $\int\colon \mathrm{Step}([a,b];F) \to F$ has a unique extension $\int\colon \mathrm{Reg}([a,b];F) \to F$ to $\mathrm{Reg}([a,b];F)$, which is also continuous and linear. This is how the integral of a regulated function is defined.

In summary, we define an "obvious" notion of integral on the simple set Step$([a,b];F)$. It is a linear and continuous mapping, so we extend it by continuity to the bigger space Reg$([a,b];F)$ in which Step$([a,b];F)$ is dense.

3.1 Riemann Integral of a Continuous Function

In this section we define the Riemann integral of a real-valued continuous function.

Definition 3.1. Let $a < b$ be any two reals. A set $T = \{t_0, t_1, \ldots, t_n\}$ of reals such that $t_0 = a$, $t_n = b$ and $t_k < t_{k+1}$, for $k = 0, \ldots, n-1$, is called a *subdivision* of $[a,b]$. The *diameter* $\delta(T)$ of T is defined by

$$\delta(T) = \max_{0 \leq k \leq n-1} (t_{k+1} - t_k).$$

Given a continuous function $f\colon [a,b] \to \mathbb{R}$, define the *Cauchy–Riemann sum* $s_T(f)$ by

$$s_T(f) = \sum_{k=0}^{n-1} (t_{k+1} - t_k) f(t_k).$$

See Figure 3.1.

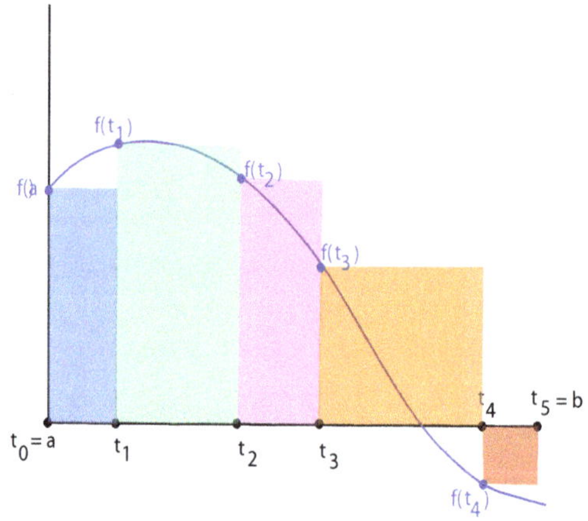

Fig. 3.1 The Cauchy–Riemann sum $s_T(f) = \sum_{k=0}^{4} (t_{k+1} - t_k) f(t_k)$ is the "signed" area represented by the pastel shaded boxes.

Observe that

$$\sum_{k=0}^{n-1}(t_{k+1} - t_k) = b - a.$$

We immediately check that s_T is a linear form on the set of continuous functions on $[a, b]$. Furthermore, if $f \geq 0$, which means that $f(t) \geq 0$ for all $t \in [a, b]$, then $s_T(f) \geq 0$.

The question is, as the subdivision T becomes finer and finer, in the sense that $\delta(T)$ becomes smaller and smaller (which means that n gets bigger and bigger), do the sums $s_T(f)$ have a limit?

The answer is *yes*.

The reason is that a continuous function on a compact interval $[a, b]$ is uniformly continuous, and this implies that for any sequence (T_m) of subdivisions such that $\delta(T_m) \to 0$ as m goes to infinity, the sums $s_{T_m}(f)$ form a Cauchy sequence, as we now explain.

Proposition 3.1. *Let $f\colon [a, b] \to \mathbb{R}$ be a continuous function defined on a closed and bounded (compact) interval $[a, b]$. For every $\epsilon > 0$, there is some $\eta > 0$ such that for any two subdivisions T and T' of $[a, b]$ such that $\delta(T) < \eta$ and $\delta(T') < \eta$, we have*

$$|s_T(f) - s_{T'}(f)| < \epsilon.$$

Proof. Since a continuous function on $[a, b]$ is actually uniformly continuous, for any $\epsilon > 0$, we can find some $\eta > 0$ such that

$$|f(x) - f(x')| < \epsilon/2(b - a) \quad \text{for all } x, x' \in [a, b] \text{ such that } |x - x'| < \eta.$$

If $T = \{t_0, t_1, \ldots, t_n\}$ and $T' = \{t'_0, t'_1, \ldots, t'_{n'}\}$, let $T'' = T \cup T'$ and let T''_k be the subdivision $T''_k = T' \cap [t_k, t_{k+1}]$, more precisely, $T''_k = \{s_0, s_1, \ldots, s_r\}$, with $s_0 = t_k$, $s_r = t_{k+1}$, and

$$\{s_1, \ldots, s_{r-1}\} = \{t'_j \mid t_k < t'_j < t_{k+1}\},$$

for $k = 0, \ldots, n - 1$, with $r = 0$ if the above set on the right-hand side is empty. Then we immediately check that

$$T'' = \bigcup_{k=0}^{n-1} T''_k, \quad \text{and} \quad s_{T''}(f) = \sum_{k=0}^{n-1} s_{T''_k}(f).$$

See Figure 3.2.

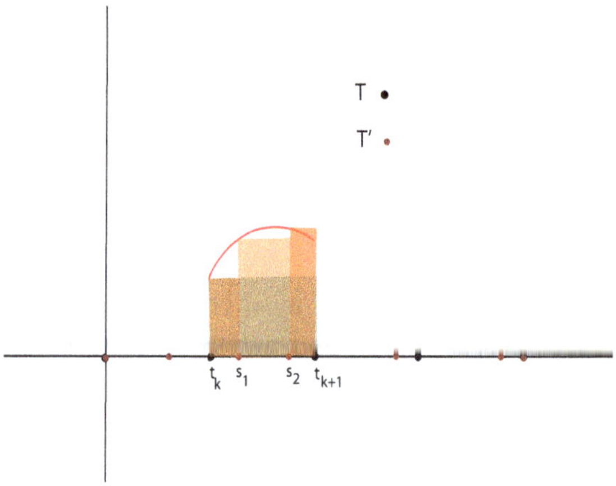

Fig. 3.2 An illustration of the refinement $s_{T_k''}(f)$ utilized in the proof of Proposition 3.1. Note that T is given by the black dots while T' is given by the brown dots.

Since $s_{T_k''}(f)$ is of the form

$$s_{T_k''}(f) = \sum_{i=0}^{r-1}(s_{i+1} - s_i)f(s_i),$$

where $t_k \leq s_i \leq t_{k+1}$ for $i = 0, \ldots, r$, and since $\sum_{i=0}^{r-1}(s_{i+1} - s_i) = s_r - s_0 = t_{k+1} - t_k$, we have

$$\begin{aligned}
|s_{T_k''}(f) - (t_{k+1} - t_k)f(t_k)| &= \left|\sum_{i=0}^{r-1}(s_{i+1} - s_i)f(s_i) - (t_{k+1} - t_k)f(t_k)\right| \\
&= \left|\sum_{i=0}^{r-1}(s_{i+1} - s_i)(f(s_i) - f(t_k))\right| \\
&\leq \sum_{i=0}^{r-1}|(s_{i+1} - s_i)||(f(s_i) - f(t_k))| \\
&< \sum_{i=0}^{r-1}|(s_{i+1} - s_i)|\frac{\epsilon}{2(b-a)} \\
&= (t_{k+1} - t_k)\frac{\epsilon}{2(b-a)}.
\end{aligned}$$

As a consequence, we obtain

$$\begin{aligned}
|s_T(f) - s_{T''}(f)| &= \left|\sum_{k=0}^{n-1}(t_{k+1} - t_k)f(t_k) - \sum_{k=0}^{n-1} s_{T''_k}(f)\right| \\
&\leq \sum_{k=0}^{n-1} |s_{T''_k}(f) - (t_{k+1} - t_k)f(t_k)| \\
&< \sum_{k=0}^{n-1}(t_{k+1} - t_k)\frac{\epsilon}{2(b-a)} \\
&\leq \frac{\epsilon}{2},
\end{aligned}$$

that is,

$$|s_T(f) - s_{T''}(f)| < \frac{\epsilon}{2}.$$

By a similar argument applied to T', we obtain

$$|s_{T'}(f) - s_{T''}(f)| < \frac{\epsilon}{2}.$$

But then we obtain

$$\begin{aligned}
|s_T(f) - s_{T'}(f)| &= |s_T(f) - s_{T''}(f) + s_{T''}(f) - s_{T'}(f)| \\
&\leq |s_T(f) - s_{T''}(f)| + |s_{T''}(f) - s_{T'}(f)| \\
&< \frac{\epsilon}{2} + \frac{\epsilon}{2} = \epsilon,
\end{aligned}$$

as claimed. □

Remark: It is easy to check that the proof of Proposition 3.1 is still valid if we use more general Cauchy-Riemann sums. Namely, given a subdivision $T = \{t_0, t_1, \ldots, t_n\}$ of $[a, b]$, and any choice of reals $\theta_1, \ldots, \theta_n$ such that $t_k \leq \theta_{k+1} \leq t_{k+1}$ for $k = 0, \ldots, n-1$, define $s_T(f)$ as

$$s_T(f) = \sum_{k=0}^{n-1}(t_{k+1} - t_k)f(\theta_{k+1});$$

see Figure 3.3.

Proposition 3.1 implies the following result, which establishes the existence of the Riemann integral of a continuous function defined on a closed and bounded (compact) interval $[a, b]$.

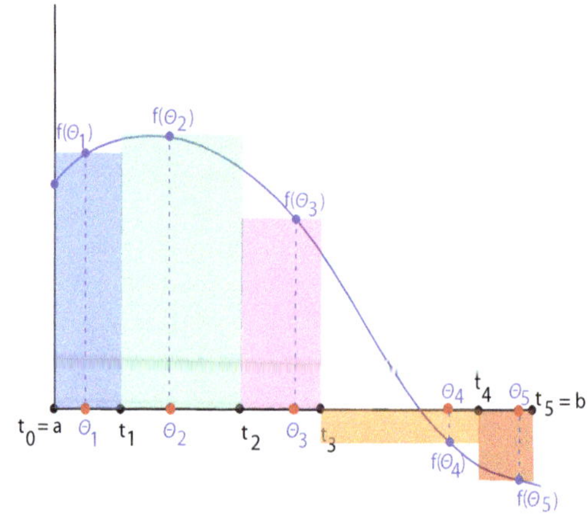

Fig. 3.3 The general Cauchy–Riemann sum $s_T(f) = \sum_{k=0}^{4}(t_{k+1} - t_k)f(\theta_{k+1})$ is the "signed" area represented by the pastel shaded boxes.

Theorem 3.2. *Let $f\colon [a,b] \to \mathbb{R}$ be a continuous function defined on a closed and bounded (compact) interval $[a,b]$. For every sequence $\mathcal{T} = (T_m)_{m\geq 1}$ of subdivisions of $[a,b]$ such that $\lim_{m\mapsto\infty} \delta(T_m) = 0$, the sequence $(s_{T_m}(f))_{m\geq 1}$ is a Cauchy sequence, and thus has a limit $S_{\mathcal{T}}(f)$. For any two sequences $\mathcal{T} = (T_m)_{m\geq 1}$ and $\mathcal{T}' = (T'_m)_{m\geq 1}$ of subdivisions of $[a,b]$, if $\lim_{m\mapsto\infty} \delta(T_m) = 0$ and $\lim_{m\mapsto\infty} \delta(T'_m) = 0$, then $S_{\mathcal{T}}(f) = S_{\mathcal{T}'}(f)$, that is, the limit of the sequence $(s_{T_m}(f))$ is independent of the sequence $\mathcal{T} = (T_m)_{m\geq 1}$ such that $\lim_{m\mapsto\infty} \delta(T_m) = 0$.*

Proof. Pick any $\epsilon > 0$, and let $\eta > 0$ be some number given by Proposition 3.1, such that for any two subdivisions T and T' of $[a,b]$ such that $\delta(T) < \eta$ and $\delta(T') < \eta$, we have

$$|s_T(f) - s_{T'}(f)| < \epsilon.$$

Since $\lim_{m\mapsto\infty} \delta(T_m) = 0$, there is some $N > 0$ such that for all $m, n \geq N$, we have $\delta(T_m) < \eta$ and $\delta(T_n) < \eta$, which by the definition of η, implies that

$$|s_{T_m}(f) - s_{T_n}(f)| < \epsilon \quad \text{for all } m, n \geq N.$$

Therefore, $(s_{T_m}(f))$ is a Cauchy sequence. Since $\mathbb{R}$ is a complete metric space, this sequence has a limit $S_{\mathcal{T}}(f)$. The same argument shows that $(s_{T'_m}(f))$ is a Cauchy sequence which has a limit $S_{\mathcal{T}'}(f)$.

Since by hypothesis $\lim_{m\mapsto\infty}\delta(T_m)=0$ and $\lim_{m\mapsto\infty}\delta(T'_m)=0$, there is some $N>0$ such that for all $m\geq N$, we have $\delta(T_m)<\eta$ and $\delta(T'_m)<\eta$, so by Proposition 3.1,

$$|s_{T_m}(f)-s_{T'_m}(f)|<\epsilon \quad \text{for all } m\geq N. \tag{eq1}$$

By the triangle inequality

$$|S_{\mathcal{T}'}(f)-S_{\mathcal{T}}(f)|\leq|S_{\mathcal{T}'}(f)-s_{T'_m}(f)|+|s_{T'_m}(f)-s_{T_m}(f)|+|s_{T_m}(f)-S_{\mathcal{T}}(f)|,$$

since the Cauchy sequences $(s_{T_m}(f))$ and $(s_{T'_m}(f))$ converge and (eq1) holds, we deduce that $S_{\mathcal{T}'}(f)=S_{\mathcal{T}}(f)$, that is, the sequences $(s_{T_m}(f))$ and $(s_{T'_m}(f))$ have the same limit. ☐

Theorem 3.2 also holds for the more general Cauchy-Riemann sums defined in the Remark after Proposition 3.1.

Theorem 3.2 justifies the following definition.

Definition 3.2. Let $f\colon[a,b]\to\mathbb{R}$ be a continuous function defined on a closed and bounded (compact) interval $[a,b]$. The common limit $S_{\mathcal{T}}(f)$ of the Cauchy sequences $(s_{T_m}(f))_{m\geq1}$, for all sequences $\mathcal{T}=(T_m)_{m\geq1}$ of subdivisions of $[a,b]$ such that $\lim_{m\mapsto\infty}\delta(T_m)=0$, is called the *Riemann integral* of f, and is denoted by $\int_a^b f(t)dt$.

The following are basic properties of the Riemann integral, which are easy to prove (using suitable subdivisions of $[a,b]$):

(1) The mapping $f\mapsto\int_a^b f(t)dt$ is a *linear form* on the space of continuous functions on $[a,b]$. This means that for any two continuous functions $f,g\colon[a,b]\to\mathbb{R}$ and any scalar $\lambda\in\mathbb{R}$,

$$\int_a^b(f+g)(t)dt=\int_a^b f(t)dt+\int_a^b g(t)dt$$

$$\int_a^b(\lambda f)(t)dt=\lambda\int_a^b f(t)dt,$$

where, as usual, $f+g$ is the function given by $(f+g)(t)=f(t)+g(t)$, and λf is the function given by $(\lambda f)(t)=\lambda f(t)$, for all $t\in[a,b]$. Furthermore, it is a positive linear form, which means that if $f\geq0$, then $\int_a^b f(t)dt\geq0$. These seemingly innocuous properties turn out to be very important. Indeed, we will see later how the notion of integral on a locally compact space can be defined in terms of such linear forms (Radon functionals).

(2)

$$\left|\int_a^b f(t)dt\right| \leq \int_a^b |f(t)|dt \leq (b-a) \max_{t\in[a,b]} |f(t)|;$$

see Figure 3.4.

(3) If $f \geq 0$ and $f(t) > 0$ for some $t \in [a,b]$, then $\int_a^b f(t)dt > 0$.

(4) If $a < b < c$, then

$$\int_a^b f(t)dt + \int_b^c f(t)dt = \int_a^c f(t)dt;$$

see Figure 3.5.

(5) If $H\colon [a,b] \to \mathbb{R}$ is the function given by

$$H(x) = \int_a^x f(t)dt,$$

then H is differentiable on $[a,b]$ and $H'(x) = f(x)$ (the so-called *first fundamental theorem of calculus*).

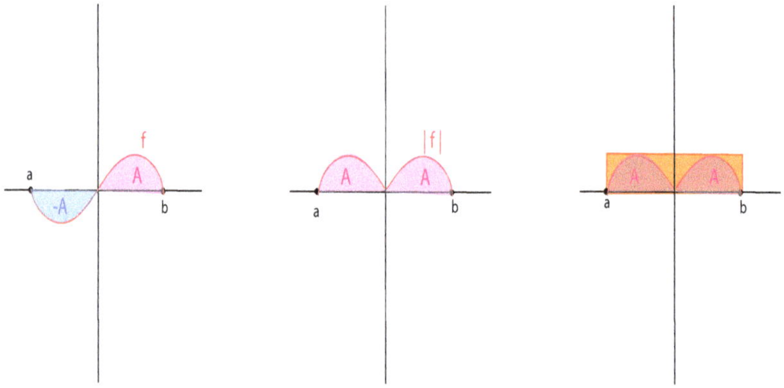

Fig. 3.4 The left figure illustrates $\int_a^b f(t)dt = A + (-A) = 0$, while the middle figure illustrates $\int_a^b |f(t)|dt = 2A$, so $\left|\int_a^b f(t)dt\right| \leq \int_a^b |f(t)|dt$. The right figure shows that $\int_a^b |f(t)|dt$ is contained within the orange rectangle of area $(b-a)\max_{t\in[a,b]} |f(t)|$.

The process that we just described only requires that the codomain be complete and that a continuous function $f\colon [a,b] \to F$ be uniformly continuous. We also need the linear combinations $\sum_{k=0}^{n-1}(t_{k+1} - t_k)f(t_k)$ to make sense, so F should be a vector space. If we assume that F is a

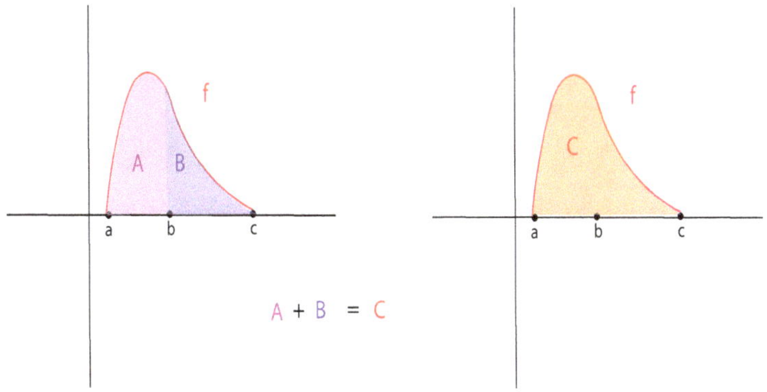

Fig. 3.5 In the left figure, $\int_a^b f(t)dt = A$, while $\int_b^c f(t)dt = B$, and this is equal to the peach area under the curve from a to c.

complete normed vector space (a Banach space), then the Riemann integral of a continuous vector-valued function $f\colon [a,b] \to F$ can be defined by using the method that we just presented.

In the next section we show how to define the integral of functions with discontinuities, provided that these discontinuities are "reasonable." For this, a new crucial idea is needed: to define the integral on a class of simple functions with a finite number of reasonable discontinuities, and then to extend the integral to a bigger class of functions by taking limits of simple functions. For this process to work, the bigger space of functions should be complete.

3.2 The Riemann Integral of Regulated Functions

In this section we show how to define the integral of regulated functions $f\colon [a,b] \to F$, where F is any complete normed vector space, in particular $\mathbb{R}$ or any finite-dimensional vector space (real or complex).

The first key ingredient is that the method of Cauchy-Riemann sums can be immediately adapted to define the notion of integral for a step function. The mapping $\int\colon \mathrm{Step}([a,b];F) \to F$ is seen to be linear and continuous.

The second key ingredient is that, by Proposition 2.29, the vector space $\mathrm{Step}([a,b];F)$ of step functions over $[a,b]$ is dense in $\mathrm{Reg}([a,b];F)$, the space of regulated functions over $[a,b]$, and $\mathrm{Reg}([a,b];F)$ is complete, where $[a,b]$ is a closed and bounded interval.

Because $\mathrm{Step}([a,b];F)$ is dense in $\mathrm{Reg}([a,b];F)$ and $\mathrm{Reg}([a,b];F)$ is complete, by Theorem A.73, the continuous linear map $\int\colon \mathrm{Step}([a,b];F) \to F$ has a unique extension $\int\colon \mathrm{Reg}([a,b];F) \to F$ to $\mathrm{Reg}([a,b];F)$, which is also continuous and linear. This is how the integral of a regulated function is defined.

Thus it remains to define the integral of a step function.

Definition 3.3. Let $f\colon [a,b] \to F$ be a step function. For any admissible subdivision $T = (a_0, a_1, \ldots, a_n)$ for f, for any sequence $\xi = (\xi_1, \ldots, \xi_n)$ of reals such that $\xi_{k+1} \in (a_k, a_{k+1})$ for $k = 0, \ldots, n-1$, define $s_{T,\xi}(f)$ by

$$s_{T,\xi}(f) = \sum_{k=0}^{n-1} (a_{k+1} - a_k) f(\xi_{k+1}).$$

See Figure 3.6.

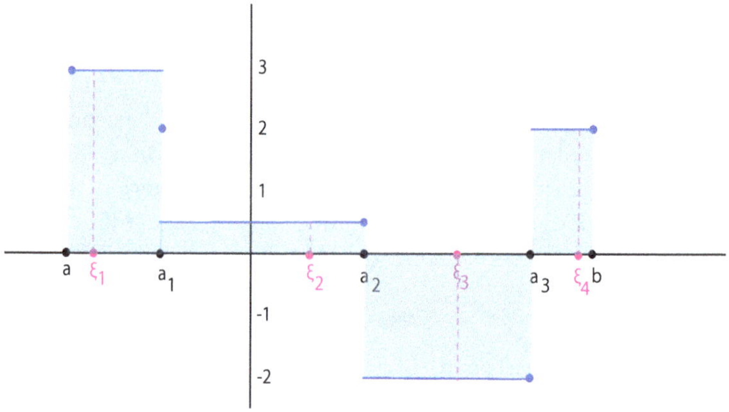

Fig. 3.6 An illustration of $s_{T,\xi}(f) = \sum_{k=0}^{3}(a_{k+1} - a_k)f(\xi_{k+1})$ for step function $f\colon [a,b] \to \mathbb{R}$.

The above is a linear combination of vectors in F, and since F is a vector space, it is well defined. Note that because $\xi_{k+1} \in (a_k, a_{k+1})$, $s_{T,\xi}(f)$ does not depend on the value of f at the a_k. For simplicity of language, we refer to a pair (T, ξ) as in Definition 3.3 as an *admissible pair* for f.

The problem with the above definition of $s_{T,\xi}(f)$ is that it depends on the admissible subdivision T, and on ξ, but because f is a step function, it is constant on each interval (a_k, a_{k+1}), so in fact $s_{T,\xi}(f)$ is independent of the admissible pair (T, ξ).

Proposition 3.3. *Given a step function $f\colon [a,b] \to F$, for any two admissible pairs (T, ξ) and (T', ξ') for f, we have $s_{T,\xi}(f) = s_{T',\xi'}(f)$.*

Proposition 3.3 is proved by using an admissible pair which is finer than both (T, ξ) and (T', ξ'). The details are left to the reader, or see Schwartz [Schwartz (1993a)] (Chapter V, Section §1).

Proposition 3.3 justifies the following definition.

Definition 3.4. Let $f\colon [a,b] \to F$ be a step function. The *integral* of f, denoted $\int_{[a,b]} f$, is the common value of the sum $s_{T,\xi}(f)$, for any any admissible pair (T, ξ) for f.

The following proposition follows almost immediately from the definitions.

Proposition 3.4. *The map $\int\colon \mathrm{Step}([a,b]; F) \to F$, where $\int f = \int_{[a,b]} f$ is the integral defined in Definition 3.4, is linear. Furthermore, we have*

$$\left\| \int_{[a,b]} f \right\| \leq \int_{[a,b]} \|f\| \quad \textit{and} \quad \left\| \int_{[a,b]} f \right\| \leq (b-a) \|f\|_\infty,$$

where $\|f\|$ means the real-valued function $x \mapsto \|f(x)\|$. If $f = \mathbb{R}$ and if $f \geq 0$, then $\int_{[a,b]} f \geq 0$.

Proposition 3.4 shows that the map $\int\colon \mathrm{Step}([a,b]; F) \to F$ is linear and continuous. As we explained earlier, by Theorem A.73, the map $\int\colon \mathrm{Step}([a,b]; F) \to F$ has a unique extension $\int\colon \mathrm{Reg}([a,b]; F) \to F$ to $\mathrm{Reg}([a,b]; F)$, which is also linear and continuous.

Definition 3.5. The *integral* $\int_{[a,b]} f$ of any regulated function $f \in \mathrm{Reg}([a,b]; F)$ is equal to $\int f$, where $\int\colon \mathrm{Reg}([a,b]; F) \to F$ is the unique linear and continuous extension of the linear and continuous map $\int\colon \mathrm{Step}([a,b]; F) \to F$. This integral is called the *Riemann integral* of the regulated function f.

Definition 3.5 is not very constructive. It turns out that the Riemann integral of a regulated function can be defined more directly in terms of generalized Riemann sums. This approach is presented in Schwartz [Schwartz (1993a)] (Chapter V, Section §1).

Note that we actually haven't defined the notion of Riemann-integrable function. What we did is to exhibit a family of functions, the regulated functions, which are Riemann-integrable function. The notion of Riemann-integrable function is defined in various books, including Schwartz [Schwartz (1993a)]. This can be done using the notion of upper integral $\int^* f$, which is defined for a positive function $f \in \mathcal{K}(\mathbb{R}, F)$ as the infimum of the integrals of the step functions that bound f from above.

The space of Riemann-integrable functions contains other functions besides the regulated functions. For example, functions with compact support which are continuous except at finitely many points, are Riemann-integrable. The function $x \mapsto \sin(1/x)$ is such a function (with value 0 at $x = 0$). It is Riemann-integrable on $[0, 1]$, even though 0 is not a discontinuity of the first kind.

The method of this section, which consists in defining the notion of integral for a "big" set of functions, such as $\mathrm{Reg}([a, b]; F)$, by first defining a notion of integral on a very simple set of functions for which the definition is obvious, such as $\mathrm{Step}([a, b]; F)$, and then to extend the integral on $\mathrm{Step}([a, b]; F)$ to a notion of integral on $\mathrm{Reg}([a, b]; F)$ using a completion process, is a key idea. In this situation we are lucky that $\mathrm{Reg}([a, b]; F)$ is complete.

In order to define a notion of integral for functions defined on a domain X which is more general than a compact interval $[a, b]$ of $\mathbb{R}$, we can proceed as above, but some additional structure on X is needed to define step functions and the notion of integral of step functions. This new ingredient is the notion of *measure*. The other technical difficulty is that the completion of the space of generalized step functions is not a space identifiable with a space of familiar functions. By Theorem A.72, the completion always exists, but its elements are equivalence classes of functions, so it will take some work to exhibit this space as a set of functions.

3.3 Problems

Problem 3.1. Let $f\colon [a, b] \to \mathbb{R}$ be a continuous function. Prove that f is uniformly continuous.

Problem 3.2. Check that the proof of Proposition 3.1 is valid for general Cauchy-Riemann sums defined as follows $s_T(f)$: given a subdivision $T = \{t_0, t_1, \ldots, t_n\}$ of $[a, b]$, and any choice of real numbers $\theta_1, \ldots, \theta_n$ such that

$t_k \leq \theta_{k+1} \leq t_{k+1}$ for $k = 0, \ldots, n-1$, define $s_T(f)$ as

$$s_T(f) = \sum_{k=0}^{n-1} (t_{k+1} - t_k) f(\theta_{k+1}).$$

Problem 3.3. Check that Theorem 3.2 holds for the more general Cauchy-Riemann sums $s_T(f)$ defined in Problem 3.2.

Problem 3.4. Let $f\colon [a,b] \to \mathbb{R}$ be a continuous function. Prove the following properties of the Riemann integral.

(1) $\left|\int_a^b f(t)dt\right| \leq \int_a^b |f(t)|dt \leq (b-a)\max_{t\in[a,b]} |f(t)|$.
(2) If $f \geq 0$ and $f(t) > 0$ for some $t \in [a,b]$, then $\int_a^b f(t)dt > 0$.
(3) If $a < b < c$, then

$$\int_a^b f(t)dt + \int_b^c f(t)dt = \int_a^c f(t)dt.$$

(4) If $H\colon [a,b] \to \mathbb{R}$ is the function given by

$$H(x) = \int_a^x f(t)dt,$$

then H is differentiable on $[a,b]$ and $H'(x) = f(x)$.

Problem 3.5. Prove Proposition 3.3. Hint: Use an admissible pair which is finer than both (T,ξ) and (T',ξ'). Alternatively, see Schwartz [Schwartz (1993a)] (Chapter V, Section §1).

Problem 3.6. Prove Proposition 3.4.

Chapter 4

Measure Theory; Basic Notions

Let X be a nonempty set. Intuitively, a measure on X is a function μ that assigns a nonnegative real number $\mu(A)$ to every subset A in some specified nonempty collection $\mathcal{A}$ of subsets of X, where $\mu(A)$ is a generalization of the notion of length, area, or volume. For example, any natural measure μ on $\mathbb{R}$ should have the property that $\mu((a,b)) = \mu([a,b]) = b - a$ for all $a \leq b$. It is natural to require that if a subset A is sliced into countably many pairwise disjoint small pieces A_i, then $\mu(A) = \mu\left(\bigcup_{n=1}^{\infty} A_i\right) = \sum_{n=1}^{\infty} \mu(A_i)$. This property is called *σ-additivity*. Then the family $\mathcal{A}$ of subsets on which μ is defined should be closed under countable unions. It is also natural to require $\mathcal{A}$ to be closed under complementation. This leads to the important notion of a *σ-algebra*, which is closed under complementation and countable unions. The weaker notion which only requires closure under complementation and closure under finite unions is that of an *algebra*. In general it is not easy to construct nontrivial σ-algebras, so it is useful to have tools to do so. A pair $(X, \mathcal{A})$ consisting of a nonempty set and a σ-algebra $\mathcal{A}$ is called a *measurable space*.

Given any nonempty family $\mathcal{S}$ of subsets of X, there is a smallest σ-algebra $\mathcal{A}(\mathcal{S})$ containing $\mathcal{S}$. If X is a topological space, then the σ-algebra $\mathcal{B}(X)$ containing the open subsets of X is an important σ-algebra called the *Borel σ-algebra*.

The notion of *monotone class* is also useful to construct σ-algebras. Given any nonempty family $\mathcal{S}$ of subsets of X, there is a smallest monotone class $\mathfrak{M}(\mathcal{S})$ containing $\mathcal{S}$. Given an algebra $\mathcal{B}$, the smallest σ-algebra $\mathcal{A}(\mathcal{B})$ containing $\mathcal{B}$ and the smallest monotone class $\mathfrak{M}(\mathcal{B})$ containing $\mathcal{B}$ are identical: $\mathcal{A}(\mathcal{B}) = \mathfrak{M}(\mathcal{B})$.

Next we define (positive) measures on a σ-algebra. A triple $(X, \mathcal{A}, \mu)$ consisting of a nonempty set, a σ-algebra $\mathcal{A}$, and a measure μ on $\mathcal{A}$ is called a *measure space*. We investigate a few properties of measures. In particular, we show that every measure can be extended to a *complete measure*, which means that all $A \in \mathcal{A}$, if $\mu(A) = 0$, then $B \in \mathcal{A}$ for all $B \subseteq A$.

As we said earlier, it is not easy to construct nontrivial measures. A very useful concept to achieve this is the notion of *outer measure*, introduced in Section 4.4. Outer measures are defined for all subsets of X, which makes them much easier to construct. In particular, we construct the *Lebesgue outer measure* μ_L^*.

A fundamental theorem due to Carathéodory shows that every outer measure induces a measure space; see Theorem 4.11.

By applying Theorem 4.11 to the outer measure μ_L^* we obtain the σ-algebra $\mathcal{L}(\mathbb{R})$ of *Lebesgue measurable sets* and the *Lebesgue measure* μ_L; see Section 4.5. The Borel σ-algebra $\mathcal{B}(\mathbb{R})$ is properly contained in the σ-algebra $\mathcal{L}(\mathbb{R})$ of Lebesgue measurable sets, and there are subsets of $\mathbb{R}$ that are not Lebesgue measurable sets (assuming the axiom of choice). We also discuss various regularity properties of the Lebesgue measure.

4.1 σ-Algebras

Let X be a nonempty set. We would like to define the notion of "measure" for the subsets of X in such a way that familiar properties of the notion of length, area, or volume of polyhedral objects in $\mathbb{R}$, $\mathbb{R}^2$ or $\mathbb{R}^3$ hold. The measure $m(A)$ of a subset of X should be nonnegative, but we have to allow "big" objects to have infinite measure so it is desirable to extend the nonnnegative real numbers by adding a new element corresponding to infinity.

Technically, we define $\overline{R}_+$ as the union

$$\overline{R}_+ = \{\alpha \in \mathbb{R} \mid \alpha \geq 0\} \cup \{+\infty\} = \mathbb{R}_+ \cup \{+\infty\},$$

where $+\infty$ is *not* in $\mathbb{R}_+$, and we assume that the following properties hold:

(a) $\alpha < +\infty$, for all $\alpha \in \mathbb{R}_+$,
(b) $\alpha + (+\infty) = (+\infty) + \alpha = +\infty$, for all $\alpha \in \overline{\mathbb{R}}_+$,
(c) $\alpha \cdot (+\infty) = (+\infty) \cdot \alpha = +\infty$, for all $\alpha \in \overline{\mathbb{R}}_+ - \{0\}$,
(d) $0 \cdot (+\infty) = (+\infty) \cdot 0 = 0$,

(e) If $(\alpha_i)_{i\geq 1}$ is a sequence with $\alpha_i \in \overline{\mathbb{R}}_+$, and if $\alpha_i = +\infty$ for some i, then $\sum_{i=1}^{\infty} \alpha_i = +\infty$.

The set $\overline{R}_+$ is also denoted by $[0, +\infty]$.

In this chapter we closely follow Halmos [Halmos (1978)] and some course notes given by Philippe G. Ciarlet in 1970–1971 at ENPC (Paris, France). Other nice (but concise) presentations can be found in Rudin [Rudin (1987)], Folland [Folland (1999)], and Lang [Lang (1996)]. A very detailed presentation is given in Schwartz [Schwartz (1993a)].

An "ideal measure" should be a function m satisfying the following properties:

(1) $m\colon 2^X \to [0, +\infty]$, that is, m is defined for *all* subsets of X.
(2) $m(\emptyset) = 0$.
(3) For any countable sequence $(A_i)_{i\geq 1}$ of subsets A_i of X such that $A_i \cap A_j = \emptyset$ for all $i \neq j$,

$$m\left(\bigcup_{i=1}^{\infty} A_i\right) = \sum_{i=1}^{\infty} m(A_i).$$

This property is called *σ-additivity.*

The intuition behind σ-additivity is that if we slice an object A into countably many pairwise disjoint small pieces A_i, then the measure $m(A)$ of A should be the sum of the measures $m(A_i)$ of the pieces A_i.

Observe that by choosing a sequence $(A_i)_{i\geq 1}$ such that $A_j = \emptyset$ for all $j > n$, and $A_i \cap A_j = \emptyset$ if $i \neq j$, we obtain the property

$$m\left(\bigcup_{i=1}^{n} A_i\right) = \sum_{i=1}^{n} m(A_i),$$

known as *additivity*; see Figure 4.1.

For any two subsets A and B, if $A \subseteq B$, we can write $B = A \cup (B - A)$, with $A \cap (B - A) = \emptyset$, so by additivity,

$$m(B) = m(A) + m(B - A),$$

and since $m(B - A) \geq 0$, we obtain

$$m(A) \leq m(B);$$

see Figure 4.2.

We claim that the following property holds.

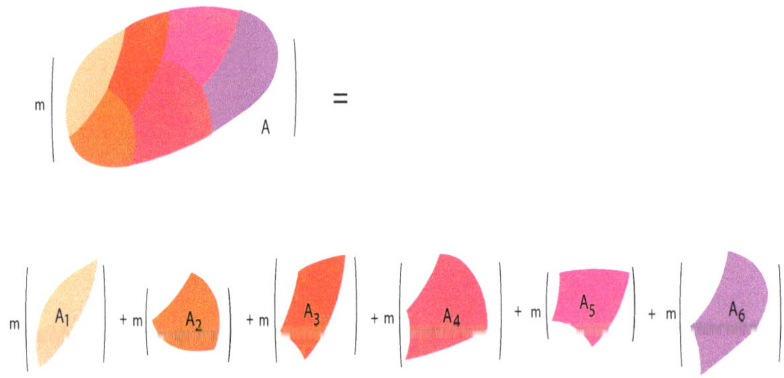

Fig. 4.1 A pictorial representation of the identity $m\left(\bigcup_{i=1}^{6} A_i\right) = \sum_{i=1}^{6} m(A_i)$.

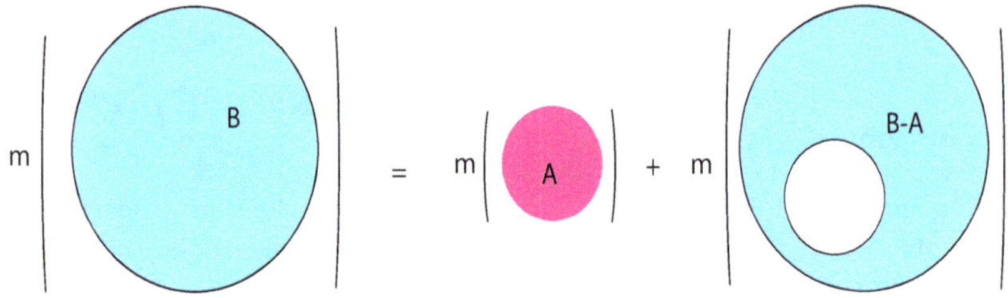

Fig. 4.2 A pictorial representation of the identities $m(B) = m(A) + m(B - A)$ and $m(A) \leq m(B)$.

Proposition 4.1. *If a function m satisfies Properties (1–3) above, then for any countable sequence $(A_i)_{i\geq 1}$ of (not necessarily pairwise disjoint) subsets A_i of X,*

$$m\left(\bigcup_{i=1}^{\infty} A_i\right) \leq \sum_{i=1}^{\infty} m(A_i).$$

Proof. Define the sequence (B_i) of subsets of X as follows: $B_1 = A_1$, $B_2 = A_2 - A_1, \ldots, B_i = A_i - \left(\bigcup_{j=1}^{i-1} A_j\right)$, for all $i \geq 2$. See Figure 4.3.

It is easy to check that $\bigcup_{i=1}^{\infty} B_i = \bigcup_{i=1}^{\infty} A_i$, $B_i \cap B_j = \emptyset$ for all $i \neq j$,

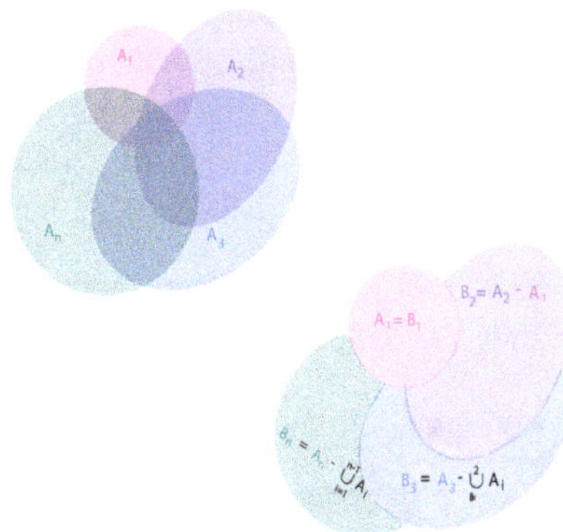

Fig. 4.3 A schematic illustration of the set construction (B_i).

and $m(B_i) \leq m(A_i)$ for all $i \geq 1$, so by σ-additivity,

$$m\left(\bigcup_{i=1}^{\infty} A_i\right) = m\left(\bigcup_{i=1}^{\infty} B_i\right) = \sum_{i=1}^{\infty} m(B_i) \leq \sum_{i=1}^{\infty} m(A_i),$$

as claimed. □

In general, for an arbitrary set X, there may be no function m satisfying Properties (1–3) for *all* subsets of X, as well as certain desirable properties. For example, there is no such translation invariant function on $2^{\mathbb{R}}$ such that $m([0,1)) \neq 0$ and $m([0,1)) \neq +\infty$, and no such translation invariant function on $2^{\mathbb{R}}$ such that $m([a,b]) = b-a$ for every interval $[a,b]$; see Section 4.5.

Thus we are led to relax some of these conditions. There are two options:

(1) The first option is to relax (3) by replacing it by the result of Proposition 4.1, namely

   (3') For any countable sequence $(A_i)_{i\geq 1}$ of subsets A_i of X,

$$m\left(\bigcup_{i=1}^{\infty} A_i\right) \leq \sum_{i=1}^{\infty} m(A_i).$$

   This approach leads to *outer measures*, and is discussed in Section 4.4.

(2) Condition (3) is highly desirable, so the second option is to restrict the domain of m to be a proper family of subsets of X; the right notion is that of a *σ-algebra*.

The notion of a σ-algebra is more important that the notion of outer measure, which is needed for technical reasons. Thus we now consider Option 2, and define σ-algebras. Once the notion of σ-algebra is defined, we will be able to define the abstract notion of a measure (see Definition 4.9), which is the crucial ingredient in defining a general notion of integral.

Definition 4.1. Let X be any nonempty set. A family $\mathcal{A}$ of subsets of X is a *σ-algebra* if it satisfies the following conditions:

(A1) $X \in \mathcal{A}$.
(A2) For every subset A of X, if $A \in \mathcal{A}$, then $X - A \in \mathcal{A}$ (closure under complementation).
(σ-A3) For every countable family $(A_i)_{i \geq 1}$ of subsets of X, if $A_i \in \mathcal{A}$ for all $i \geq 1$, then $\bigcup_{i=1}^{\infty} A_i \in \mathcal{A}$ (closure under countable unions).

From (A1) and (A2), we see that $\emptyset \in \mathcal{A}$. From (A2) and ($\sigma$-A3) and the fact that $A = X - (X - A)$ and $\bigcap_{i=1}^{\infty} A_i = X - (\bigcup_{i=1}^{\infty}(X - A_i))$, if $A_i \in \mathcal{A}$ for all $i \geq 1$, then $\bigcap_{i=1}^{\infty} A_i \in \mathcal{A}$ (closure under countable intersections). In particular, if we let $A_i = \emptyset$ for all $i \geq 3$, we see that if $A_1 \in \mathcal{A}$ and $A_2 \in \mathcal{A}$, then $A_1 \cup A_2 \in \mathcal{A}$ and $A_1 \cap A_2 \in \mathcal{A}$. Since $A_1 - A_2 = A_1 \cap (X - A_2)$, we also have $A_1 - A_2 \in \mathcal{A}$.

Axiom (σ-A3) is a strong condition, and this is the reason why it is not easy to construct nontrivial σ-algebras. There are two extreme σ-algebras:

(1) $\mathcal{A} = \{\emptyset, X\}$.
(2) $\mathcal{A} = 2^X$, the family of all subsets of X.

Interesting σ-algebra lie in-between.

Remarks:

(1) Some authors use the term *σ-field* instead of σ-algebra. This is a rather unfortunate terminology, because in algebra, a field is a set with two operations that have identity elements. Here the operations are union and intersection, but there is no identity element for intersection.
(2) If we weaken Condition σ-A3 to *finite* unions, then we obtain a structure called an *algebra* (or *boolean algebra* of sets).

Definition 4.2. Let X be any nonempty set. A family $\mathcal{B}$ of subsets of X is an *algebra* (or *boolean algebra* of sets) if it satisfies the following conditions:

(A1) $X \in \mathcal{B}$.
(A2) For every subset A of X, if $A \in \mathcal{B}$, then $X - A \in \mathcal{B}$ (closure under complementation).
(A3) For every finite family $(A_i)_{i=1}^n$ of subsets of X, if $A_i \in \mathcal{B}$ for all $i = 1, \ldots, n$, then $\bigcup_{i=1}^n A_i \in \mathcal{B}$ (closure under finite unions).

As in the case of σ-algebras, an algebra contains $\emptyset$ and is closed under (finite) unions and intersections. In the construction of a product of measurable spaces, another notion of algebra will come up. These are the semi-algebras.

Definition 4.3. Let X be any nonempty set. A family $\mathcal{S}$ of subsets of X is a *semi-algebra* if it satisfies the following conditions:

(S1) $X, \emptyset \in \mathcal{S}$.
(S2) For all $A, B \in \mathcal{S}$, we have $A \cap B \in \mathcal{S}$.
(S3) For all $A \in \mathcal{S}$, we have $X - A = X_1 \cup \cdots \cup X_n$, for finitely many pairwise disjoint subsets $X_i \in \mathcal{S}$.

Example 4.1. First consider the family of intervals of $\mathbb{R}$ of the form $[a, b)$, with $a \leq b$, where $a = -\infty$ or $b = \infty$ is allowed. By convention, let $[a, b) = \emptyset$ if $a > b$. This is a semi-algebra, because

$$[a_1, b_1) \cap [a_2, b_2) = [\max(a_1, a_2), \min(b_1, b_2)),$$

and

$$X - [a, b) = [-\infty, a) \cup [b, \infty);$$

see Figure 4.4.

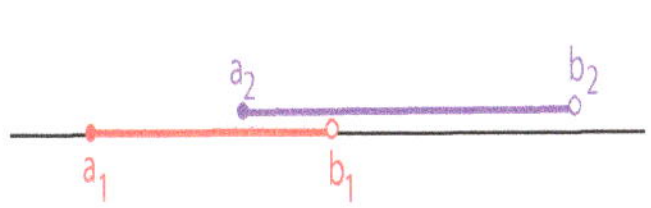

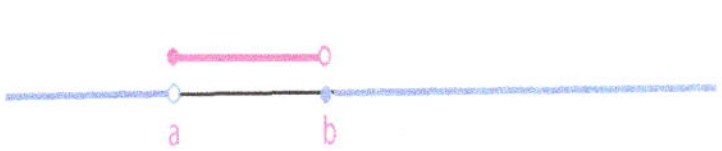

Fig. 4.4 The left figure illustrates $[a_1, b_1) \cap [a_2, b_2) = [\max(a_1, a_2), \min(b_1, b_2)) = [a_2, b_1)$, while the right figure illustrates $\mathbb{R} - [a, b) = [-\infty, a) \cup [b, \infty)$.

Example 4.2. Next, let X and Y be two nonempty sets, and let $\mathcal{A}$ be an algebra on X and let $\mathcal{B}$ be an algebra on Y. Define the set $\mathcal{R}$ of *rectangles* in $X \times Y$ as follows:

$$\mathcal{R} = \{A \times B \in X \times Y \mid A \in \mathcal{A},\ B \in \mathcal{B}\}.$$

It is easy to check that $\mathcal{R}$ is a semi-algebra. For example,

$$(A_1 \times B_1) \cap (A_2 \times B_2) = (A_1 \cap A_2) \times (B_1 \cap B_2),$$

and

$$(X \times Y) - (A \times B) = ((X - A) \times (Y - B)) \cup ((X - A) \times B) \cup (A \times (Y - B));$$

see Figure 4.5.

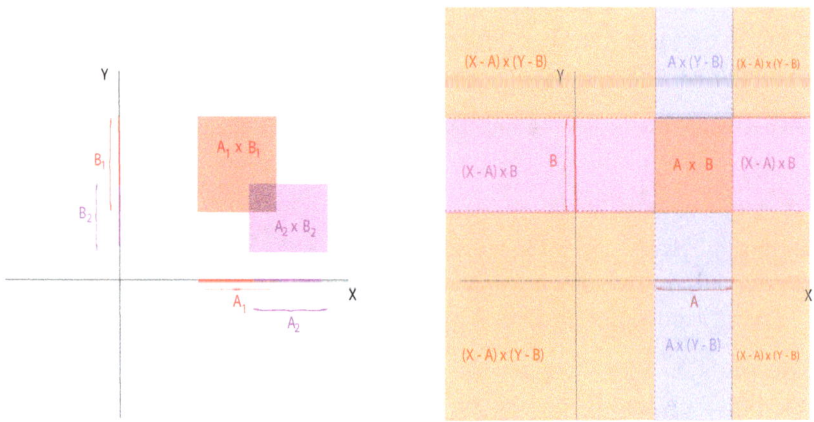

Fig. 4.5 Let X and Y be arbitrary topological spaces (for example $\mathbb{R}$). The left figure illustrates $(A_1 \times B_1) \cap (A_2 \times B_2) = (A_1 \cap A_2) \times (B_1 \cap B_2)$ as the overlap of the red and lilac rectangles while the right figure illustrates $(X \times Y) - (A \times B) = ((X - A) \times (Y - B)) \cup ((X - A) \times B) \cup (A \times (Y - B))$.

Then it can be shown that the set $\mathcal{B}(\mathcal{R})$ of finite unions of pairwise disjoint sets in $\mathcal{R}$ is the smallest algebra containing the semi-algebra $\mathcal{R}$. This algebra will be used to construct the product of measurable spaces.

The following result can be shown.

Proposition 4.2. *Given a semi-algebra $\mathcal{S}$, the smallest algebra $\mathcal{B}(\mathcal{S})$ containing $\mathcal{S}$ is the family of finite unions of pairwise disjoint subsets in $\mathcal{S}$.*

Definition 4.4. Let X be any nonempty set. A pair $(X, \mathcal{A})$ where $\mathcal{A}$ is a σ-algebra of subsets of X is called a *measurable space*. The subsets of X that belong to $\mathcal{A}$ are called the *measurable subsets* of X.

Proposition 4.3. *Let X be any nonempty set, and let $\mathcal{S}$ be any nonempty family of subsets of X. Then there is a σ-algebra $\mathcal{A}(\mathcal{S})$ with the following properties:*

(a) $\mathcal{S} \subseteq \mathcal{A}(\mathcal{S})$.
(b) If $\mathcal{A}'$ *is any* σ*-algebra such that* $\mathcal{S} \subseteq \mathcal{A}'$*, then* $\mathcal{A}(\mathcal{S}) \subseteq \mathcal{A}'$.

This means that $\mathcal{A}(\mathcal{S})$ *is the smallest* σ*-algebra containing* $\mathcal{S}$.

Definition 4.5. Let X be any nonempty set, and let $\mathcal{S}$ be any nonempty family of subsets of X. The smallest σ-algebra $\mathcal{A}(\mathcal{S})$ containing $\mathcal{S}$ is called the *σ-algebra generated by* $\mathcal{S}$.

The σ-algebra $\mathcal{A}(\mathcal{S})$ is the intersection of the family of all σ-algebras containing $\mathcal{S}$. This family is nonempty since 2^X belongs to it. This way of defining $\mathcal{A}(\mathcal{S})$ is highly nonconstructive. A bottom-up construction of $\mathcal{A}(\mathcal{S})$ can be performed, but to guarantee closure under countable infinite unions, transfinite induction is required; see Schwartz [Schwartz (1993a)] (Chapter V, Section §2) or Folland [Folland (1999)] (Proposition 1.23).

Remark: Readers not familiar with ordinals should skip this remark. For a quick review of the notion of ordinal and their basic properties, see Chapter E. Recall that an ordinal $\alpha > 0$ is either a successor ordinal, which means that $\alpha = \beta + 1$ for some ordinal $\beta < \alpha$, or a limit ordinal, which means that $\alpha = \bigcup_{\beta<\alpha} \beta$. Given $\mathcal{S}$ we define the sequence $\mathcal{S}_\alpha$ by transfinite induction. In fact, it suffices to construct this sequence for countable ordinals. We set

$$\begin{aligned} \mathcal{S}_0 &= \mathcal{S} \\ \mathcal{S}_{\beta+1} &= \mathcal{S}_\beta \cup \left\{ \bigcup_{i=1}^{\infty} A_i \mid A_i \in \mathcal{S}_\beta \right\} \cup \{X - A \mid A \in \mathcal{S}_\beta\} \\ \mathcal{S}_\alpha &= \bigcup_{\beta<\alpha} \mathcal{S}_\beta \end{aligned}$$

where α is a limit ordinal. If Ω is the set of all countable ordinals, then we let

$$\mathcal{S}^\dagger = \bigcup_{\alpha\in\Omega} \mathcal{S}_\alpha.$$

Because every increasing sequence in Ω has a supremum, it can shown that $\mathcal{A}(\mathcal{S}) = \mathcal{S}^\dagger$; see Folland [Folland (1999)] (Proposition 1.23). The cardinal of the set $\mathbb{R}$ of real numbers is denoted by $\mathfrak{c}$ or $2^{\aleph_0}$. The proof also implies that if $\mathcal{S}$ is of cardinality $\aleph_0 \leq |\mathcal{S}| \leq \mathfrak{c}$, then $\mathcal{A}(\mathcal{S})$ has cardinality $\mathfrak{c}$.

An important example arises when X is a topological space $(X, \mathcal{O})$.

Definition 4.6. Let $(X, \mathcal{O})$ be a topological space. The σ-algebra $\mathcal{B}(X)$ generated by the family $\mathcal{O}$ of open sets is called the *Borel σ-algebra* of X. The subsets in $\mathcal{B}(X)$ are called *Borel sets*.

All open subsets and all closed sets are Borel sets. Countably infinite unions of closed sets and countable infinite intersections of open sets are Borel sets. But there are many more Borel sets.

Another way to construct σ-algebras is to use algebras and monotone classes. Although we are not going to use monotone classes in this book, there are a useful tool in constructing σ-algebras. They are used in the proof of Theorem 5.55 on the existence of measures on products of measure spaces. They are also useful in proving that certain functions defined on semi-algebras or algebras $\mathcal{B}$ having some of the properties of measures can be extended to measures on certain σ-algebras induced by $\mathcal{B}$.

Definition 4.7. Let X be any nonempty set. A nonempty family $\mathfrak{M}$ of subsets of X is a *monotone class* if for every countable family $(A_i)_{i\geq 1}$ of subsets of X, if $A_i \in \mathfrak{M}$ for all $i \geq 1$ then:

(1) If $A_i \subseteq A_{i+1}$ for all $i \geq 1$, then $\bigcup_{i=1}^{\infty} A_i \in \mathfrak{M}$. See Figure 4.6, Diagram (i).
(2) If $A_{i+1} \subseteq A_i$ for all $i \geq 1$, then $\bigcap_{i=1}^{\infty} A_i \in \mathfrak{M}$. See Figure 4.6, Diagram (ii).

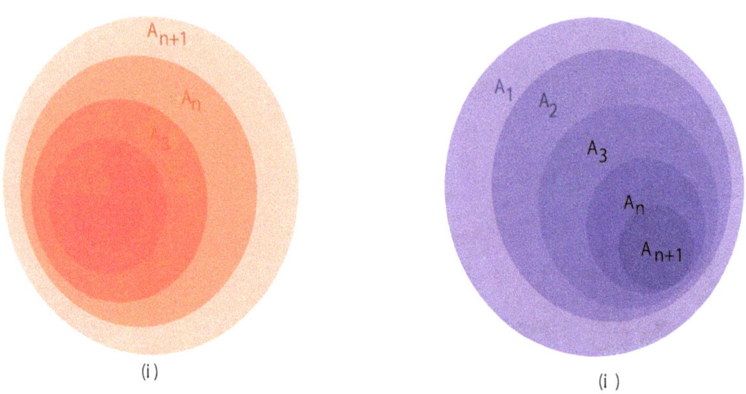

Fig. 4.6 The rose colored sets of Figure (i) satisfy the increasing nesting condition of $A_i \subseteq A_{i+1}$, while the periwinkle sets of Figure (ii) satisfy the decreasing nesting condition $A_{i+1} \subseteq A_i$.

Proposition 4.4. *Let X be any nonempty set. For any algebra $\mathcal{B}$, if $\mathcal{B}$ is a monotone class, then $\mathcal{B}$ is a σ-algebra.*

Proof. Let $(A_i)_{i\geq 1}$ be a countable family of subsets of X, such that $A_i \in \mathcal{B}$ for all $i \geq 1$. Since $\mathcal{B}$ is an algebra, it is closed under finite unions, so $B_n = \bigcup_{i=1}^{n} A_i \in \mathcal{B}$, and obviously $B_n \subseteq B_{n+1}$ for all $n \geq 1$, and $\bigcup_{i=1}^{\infty} A_i = \bigcup_{n=1}^{\infty} B_n$. Since $\mathcal{B}$ is a monotone class, $\bigcup_{i=1}^{\infty} A_i = \bigcup_{n=1}^{\infty} B_n \in \mathcal{B}$. □

Here is a version of Proposition 4.3 for monotone classes.

Proposition 4.5. *Let X be any nonempty set, and let $\mathcal{S}$ be any nonempty family of subsets of X. Then there is a monotone class $\mathfrak{M}(\mathcal{S})$ with the following properties:*

(a) $\mathcal{S} \subseteq \mathfrak{M}(\mathcal{S})$.
(b) If $\mathfrak{M}'$ is any monotone class such that $\mathcal{S} \subseteq \mathfrak{M}'$, then $\mathfrak{M}(\mathcal{S}) \subseteq \mathfrak{M}'$.

This means that $\mathfrak{M}(\mathcal{S})$ is the smallest monotone class containing $\mathcal{S}$.

Definition 4.8. Let X be any nonempty set, and let $\mathcal{S}$ be any nonempty family of subsets of X. The smallest monotone class $\mathfrak{M}(\mathcal{S})$ containing $\mathcal{S}$ is called the *monotone class generated by* $\mathcal{S}$.

The following theorem yields another way of generating a σ-algebra from an algebra.

Theorem 4.6. *Let X be any nonempty set. For any algebra $\mathcal{B}$, the σ-algebra $\mathcal{A}(\mathcal{B})$ generated by $\mathcal{B}$ and the monotone class $\mathfrak{M}(\mathcal{B})$ generated by $\mathcal{B}$ are identical; that is,*

$$\mathcal{A}(\mathcal{B}) = \mathfrak{M}(\mathcal{B}).$$

Theorem 4.6 is proven in Folland [Folland (1999)] (Lemma 2.35).

We now come to the very important notion of measure.

4.2 Measures

Definition 4.9. Let X be any nonempty set. A *measure* on X is a map μ satisfying the following properties:

(μ1) $\mu\colon \mathcal{A} \to [0, +\infty]$, where $\mathcal{A}$ is a σ-algebra of subsets of X.
(μ2) $\mu(\emptyset) = 0$.

($\mu 3$) For any countable sequence $(A_i)_{i\geq 1}$ of subsets A_i of $\mathcal{A}$ such that $A_i \cap A_j = \emptyset$ for all $i \neq j$,

$$\mu\left(\bigcup_{i=1}^{\infty} A_i\right) = \sum_{i=1}^{\infty} \mu(A_i).$$

This property is called *σ-additivity*.

A *measure space* is a triple $(X, \mathcal{A}, \mu)$, where $(X, \mathcal{A})$ is a measurable space and μ is a measure on X. A measure μ is also called a *positive measure*, to stress that its range is nonnegative.

Remarks:

(1) The degenerate situation where $\mu(A) = +\infty$ for all nonempty subsets in $\mathcal{A}$ is allowed. If μ is *nontrivial*, which means that $\mathcal{A}$ possesses some nonempty subset A such that $\mu(A)$ is finite, then by letting $A_1 = A$ and $A_i = \emptyset$ for all $i \geq 2$, by ($\mu 3$) we get $\mu(A) = \mu(A) + \sum_{i=2}^{\infty} \mu(\emptyset)$, which implies $\mu(\emptyset) = 0$. In this situation Axiom ($\mu 2$) is unnecessary. Rudin makes the assumption that a measure is nontrivial; see [Rudin (1987)].

(2) Axiom ($\mu 3$) raises a subtle point. If $(A_i)_{i\geq 1}$ is a countable family of pairwise disjoint subsets $A_i \in \mathcal{A}$, the subset $A = \bigcup_{i=1}^{\infty} A_i$ does not depend on the order of the A_i, so for any permutation σ of the positive integers we should have

$$\mu(A) = \sum_{i=1}^{\infty} \mu(A_{\sigma(i)}) = \sum_{i=1}^{\infty} \mu(A_i).$$

How do we know that this is the case?

But the numbers $\mu(A_i)$ are nonnegative, so the series $\sum_{i=1}^{\infty} \mu(A_i)$ converges absolutely, and thus *commutatively*. For example, see Schwartz [Schwartz (1991)] (Chapter II, Theorem 2.12.7 and Theorem 2.12.12, which says that in a normed vector space of finite dimension, a series is commutatively convergent iff its is absolutely convergent). Thus there is actually no problem with Axiom ($\mu 3$).

There are more general measures taking their values in $\mathbb{R}$ or $\mathbb{C}$, or even in a Banach space. For such measures, Condition ($\mu 3$) needs to be slightly strengthened.

(3) Some authors use the term *measured space* instead of *measure space*.

Definition 4.10. Let $(X, \mathcal{A}, \mu)$ be a measure space. The measure μ is *finite* if $\mu(X)$ is finite. If $\mu\colon \mathcal{A} \to [0,1]$ and if $\mu(X) = 1$, then $(X, \mathcal{A}, \mu)$ is called a *probability space*. The measure μ is a *σ-finite* if there exist a countable family $(A_i)_{i\geq 1}$ of subsets $A_i \in \mathcal{A}$ such that $X = \bigcup_{i=1}^{\infty} A_i$, and $\mu(A_i)$ is finite for all $i \geq 1$; see Figure 4.7. The measure μ is *complete* if for all $A \in \mathcal{A}$, if $\mu(A) = 0$, then $B \in \mathcal{A}$ for all $B \subseteq A$. A subset $A \in \mathcal{A}$ such that $\mu(A) = 0$ is called a *set of measure zero*.

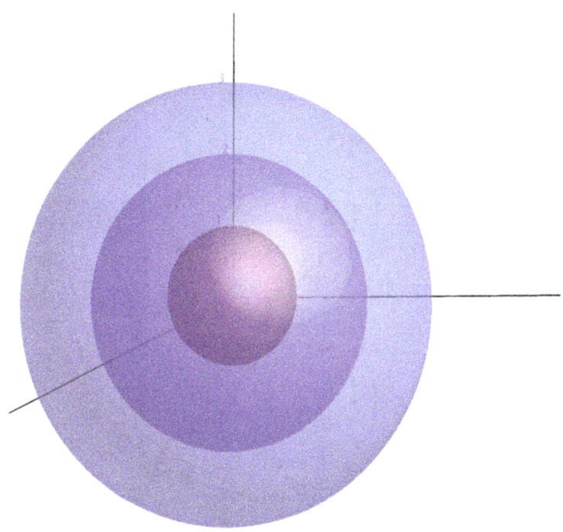

Fig. 4.7 Let $X = \mathbb{R}^3$ and let μ be the Lebesgue measure on $\mathbb{R}^3$. Then X is σ-finite since $X = \bigcup_{i=1}^{\infty} A_i$, where $A_i = \{x \in \mathbb{R} \mid \|x\| \leq i\}$. The illustration shows the solid spheres A_1, A_2, and A_3.

Example 4.3. Let X be any nonempty set, and consider the σ-algebra $\mathcal{A} = 2^X$. The map $\mu\colon 2^X \to [0, +\infty]$ given by

$$\mu(A) = \begin{cases} |A| & \text{if } A \text{ is finite} \\ +\infty & \text{if } A \text{ is infinite} \end{cases}$$

is a measure called the *counting measure* on X.

Here is a summary of useful properties of measures.

Proposition 4.7. *Let $(X, \mathcal{A}, \mu)$ be a measure space. The following properties hold:*

(1) *For any finite sequence* $(A_1, \ldots, A_n)$ *of subsets* $A_i \in \mathcal{A}$ *such that* $A_i \cap A_j = \emptyset$ *whenever* $i \neq j$*, we have*

$$\mu\left(\bigcup_{i=1}^{n} A_i\right) = \sum_{i=1}^{n} \mu(A_i).$$

(2) *For any two subsets* A, B *of* X*, if* $A, B \in \mathcal{A}$ *and if* $A \subseteq B$*, then* $\mu(A) \leq \mu(B)$.

(3) *For any countable sequence* $(A_i)_{i\geq 1}$ *of subsets* $A_i \in \mathcal{A}$*,*

$$\mu\left(\bigcup_{i=1}^{\infty} A_i\right) \leq \sum_{i=1}^{\infty} \mu(A_i).$$

(4) *For any countable sequence* $(A_i)_{i\geq 1}$ *of subsets* $A_i \in \mathcal{A}$*, if* $A_{i+1} \subseteq A_i$ *for all* $i \geq 1$ *and if* $\mu(A_1)$ *is finite, then*

$$\mu\left(\bigcap_{i=1}^{\infty} A_i\right) = \lim_{n\mapsto\infty} \mu(A_n).$$

(5) *For any countable sequence* $(A_i)_{i\geq 1}$ *of subsets* $A_i \in \mathcal{A}$*, if* $A_i \subseteq A_{i+1}$ *for all* $i \geq 1$*, then*

$$\mu\left(\bigcup_{i=1}^{\infty} A_i\right) = \lim_{n\mapsto\infty} \mu(A_n).$$

Proof. The proof of (1) and (2) is identical to the proof given just before Proposition 4.1, and (3) is Proposition 4.1. We prove (4), leaving the proof of (5) as an exercise.

We can write

$$A_n = \left(\bigcap_{i=1}^{\infty} A_i\right) \cup \left(\bigcup_{i=n}^{\infty} (A_i - A_{i+1})\right),$$

a union of pairwise disjoint subsets since $A_{i+1} \subseteq A_i$ for all $i \geq 1$. By ($\mu 3$), we have

$$\mu\left(\bigcap_{i=1}^{\infty} A_i\right) + \sum_{i=n}^{\infty} \mu(A_i - A_{i+1}) = \mu(A_n) \leq \mu(A_1) < +\infty,$$

since $A_{i+1} \subseteq A_i$ for all $i \geq 1$ and since $\mu(A_1)$ is assumed to be finite. See Figure 4.8.

Consequently, for $n = 1$ we deduce that the series $\sum_{i=1}^{\infty} \mu(A_i - A_{i+1})$ converges, which implies that

$$\lim_{n\mapsto\infty} \sum_{i=n}^{\infty} \mu(A_i - A_{i+1}) = 0.$$

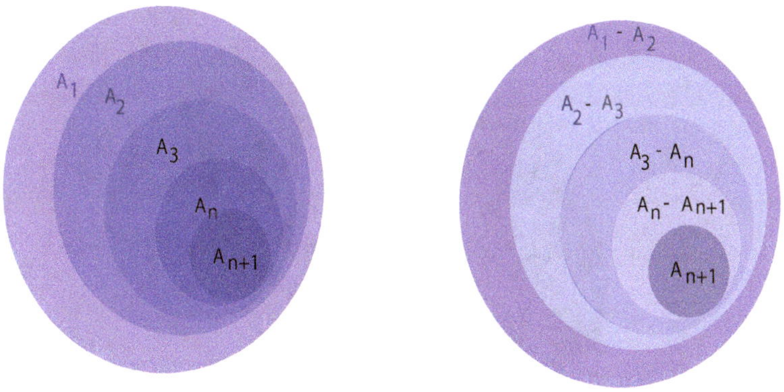

Fig. 4.8 Decomposing the decreasing nested sequences of periwinkle sets into disjoint rings. Note $A_1 = (A_1 - A_2) \cup (A_2 - A_3) \cup \cdots \cup (A_n - A_{n+1}) \cup A_{n+1}$.

Since

$$\mu\left(\bigcap_{i=1}^{\infty} A_i\right) + \sum_{i=n}^{\infty} \mu(A_i - A_{i+1}) = \mu(A_n),$$

we conclude that $\mu\left(\bigcap_{i=1}^{\infty} A_i\right) = \lim_{n \mapsto \infty} \mu(A_n)$. ☐

The following result shows that every measure can be completed; this is technically useful.

Proposition 4.8. *Let $(X, \mathcal{A}, \mu)$ be a measure space. A measure space $(X, \overline{\mathcal{A}}, \overline{\mu})$ with the following properties can be constructed:*

(a) $\mathcal{A} \subseteq \overline{\mathcal{A}}$.
(b) $\overline{\mu}$ *extends* μ*; that is,* $\overline{\mu}(A) = \mu(A)$ *for all* $A \in \mathcal{A}$.
(c) The measure $\overline{\mu}$ *is complete.*

We force the completeness property by defining $\overline{\mathcal{A}}$ as follows:

$$\overline{\mathcal{A}} = \{\overline{A} \subseteq X \mid (\exists A, A' \in \mathcal{A})(\exists B \subseteq A')(\overline{A} = A \cup B,\ \mu(A') = 0)\}.$$

The measure $\overline{\mu}$ is defined such that

$$\overline{\mu}(\overline{A}) = \overline{\mu}(A \cup B) = \mu(A).$$

See Figure 4.9.

Proposition 4.8 is proven in Rudin [Rudin (1987)] (Theorem 1.36). The verification that $\overline{\mathcal{A}}$ is a σ-algebra with the required properties and that $\overline{\mu}$ is a measure with the required properties is tedious (among other things, one needs to check that $\overline{\mu}(\overline{A})$ does not depend on the representation of $\overline{A}$).

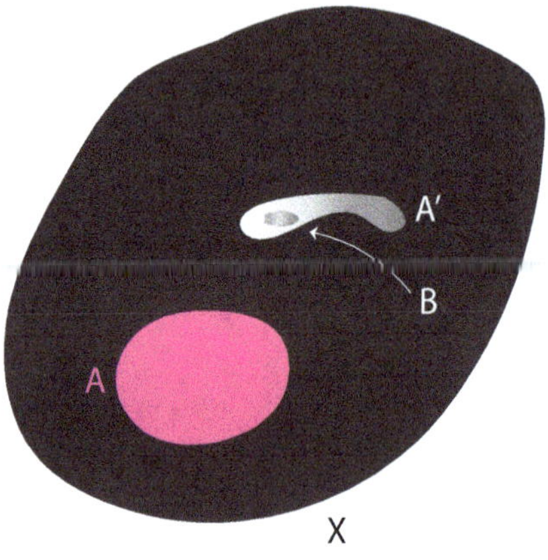

Fig. 4.9 A schematic illustration of a set in $\overline{\mathcal{A}}$. The magenta set A has positive measure, while the grayish set A', and all of its subsets, including B, have zero measure. Then $\overline{A} = A \cup B$.

Definition 4.11. The measure $\overline{\mu}$ given by Proposition 4.8 is called the *completed measure* of μ.

4.3 Null Subsets and Properties Holding Almost Everywhere

One of the secrets of measure theory is that subsets of measure zero should be ignored. Since a measure is not necessarily complete the correct technical definition is as follows.

Definition 4.12. Let $(X, \mathcal{A}, \mu)$ be a measure space. A subset $E \subseteq X$ is *null*[1] (or *negligible*) if there is some $A \in \mathcal{A}$ such that $E \subseteq A$ and $\mu(A) = 0$. A property P of the elements of X *holds almost everywhere*, abbreviated *holds a.e.*, if the subset where it fails is null; that is, the set $\{x \in X \mid P(x) = \textbf{false}\}$ is null.

To be very precise, we should say *μ-null* and *holds μ-a.e.*, since these notions depend on the measure μ. In most cases there is no risk of confusion, and we drop μ.

---

[1]Beware that in measure theory, the notion of null set has more than one meaning. Some authors mean something different from what we define here.

Observe that if the measure μ is complete, then a subset $E \subseteq X$ is *null* iff $\mu(E) = 0$, and a property P *holds a.e.* iff $\mu(\{x \in X \mid P(x) = \textbf{false}\}) = 0$. In general, a null set may either be measurable or nonmeasurable, and a nonmeasurable set has no reason to be null, but may be null.

Here are a few properties of null sets.

Proposition 4.9. *Let $(X, \mathcal{A}, \mu)$ be a measure space. Every subset of a null set is null. Every countable union of null sets is a null set.*

Proof. The first property follows immediately from the definition. Let $(A_i)_{i \geq 1}$ be a countable family of null sets. There are subsets $B_i \in \mathcal{A}$ such that $A_i \subseteq B_i$ and $\mu(B_i) = 0$ for all $i \geq 1$. We have

$$\bigcup_{i=1}^{\infty} A_i \subseteq \bigcup_{i=1}^{\infty} B_i \in \mathcal{A}$$

because $\mathcal{A}$ is a σ-algebra, so it remains to show that $\bigcup_{i=1}^{\infty} B_i$ has measure zero. For this, observe that

$$0 \leq \mu\left(\bigcup_{i=1}^{\infty} B_i\right) \leq \sum_{i=1}^{\infty} \mu(B_i) = 0,$$

so $\mu\left(\bigcup_{i=1}^{\infty} B_i\right) = 0$, as desired. □

Let P and P' be two properties of X. If P implies P' and if P holds a.e., then P' holds a.e.

Definition 4.13. Consider the set of functions $f \colon X \to \mathbb{R}$, where $(X, \mathcal{A}, \mu)$ is a measure space. We say that f *and* g *are equal a.e.* if the set $\{x \in X \mid f(x) \neq g(x)\}$ is null. Write $f = g$ (a.e.).

It is an easy exercise to show that equality a.e. is an equivalence relation.

It should be observed that the notion of equality a.e. is more subtle than one might think.

Example 4.4. For example, consider the function $\chi_{\mathbb{Q}} \colon \mathbb{R} \to \mathbb{R}$, given by

$$\chi_{\mathbb{Q}}(x) = \begin{cases} 1 & \text{if } x \in \mathbb{Q} \\ 0 & \text{if } x \notin \mathbb{Q}; \end{cases}$$

see Figure 4.10.

In other words, $\chi_{\mathbb{Q}}$ is the characteristic function of the rationals. It is easy to see that $\chi_{\mathbb{Q}}$ is discontinuous at every point $x \in \mathbb{R}$ (if x is irrational,

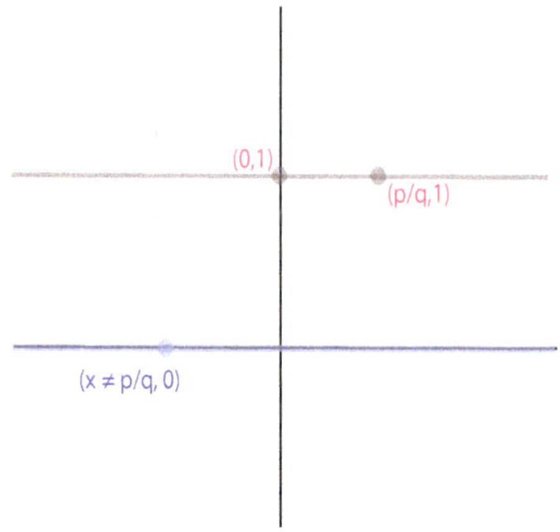

Fig. 4.10 The graph of $\chi_{\mathbb{Q}}$. The points on the light brown line have rational x-coordinates while the points on the light gray line have irrational x-coordinates.

then every small interval containing x contains some rational number; similarly, if x is rational, then every small interval containing x contains some irrational number, say of the form $x + \frac{\sqrt{2}}{2^n}$ for n large enough). Now, the Lebesgue measure μ_L discussed in Section 4.5 has the property that every countable set has measure zero, so in particular $\mathbb{Q}$ has Lebesgue measure zero. It follows that $\chi_{\mathbb{Q}}$ is equal to the zero function (on $\mathbb{Q}$) a.e., and the zero function is a "very nice" function; it is infinitely differentiable.

This is the beauty of equality a.e. Given a "very bad" function, we can ignore its bad behavior on a set of measure zero, as least from the point of view of integration.

An interesting variation of $\chi_{\mathbb{Q}}$ is the following function $D_{\mathbb{Q}}\colon \mathbb{R} \to \mathbb{R}$, given by

$$D_{\mathbb{Q}}(x) = \begin{cases} \frac{1}{q} & \text{if } x = p/q \in \mathbb{Q},\ q > 0,\ p \neq 0,\ \gcd(p, q) = 1, \\ 0 & \text{if } x \notin \mathbb{Q}, \\ 1 & \text{if } x = 0. \end{cases}$$

It is easy to show that $D_{\mathbb{Q}}$ is discontinuous at every rational point x, but is continuous at every irrational point x. In fact, $D_{\mathbb{Q}}$ is a regulated function. Again $D_{\mathbb{Q}}$ is equal to the zero function a.e. (with respect to the Lebesgue measure).

A property that will play an important role is *pointwise convergence a.e.*

Definition 4.14. Let $(X, \mathcal{A}, \mu)$ be a measure space, and let F be any topological space (in most cases a normed vector space). A sequence $(f_n)_{n\geq 1}$ of functions $f_n \colon X \to F$ *converges pointwise a.e.* to a function $f \colon X \to F$ if there is a null set $Z \subseteq X$ such that the sequence $(f_n(x))_{n\geq 1}$ converges to $f(x)$ for all $x \in X - Z$. See Figure 4.11.

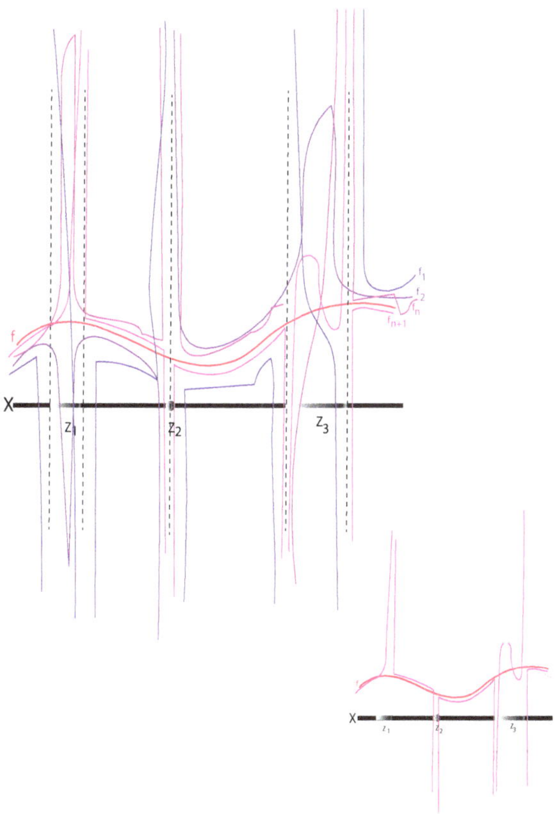

Fig. 4.11 A schematic illustration of pointwise convergence a.e. Let X be the solid black line, $F = \mathbb{R}$, and $Z = Z_1 \cup Z_2 \cup Z_3$, where each Z_i has measure zero. The sequence (f_n) converges pointwise to the graph f (in red) for all $x \in X - Z$. As shown in the bottom right corner, the magenta graph f_{n+1} is "close" to f outside of Z.

4.4 Construction of a Measure from an Outer Measure

It turns out that defining explicitly a function m satisfying Conditions (2) and (3) from the beginning of Section 4.1 on a σ-algebra is not easy, but defining a function μ^* on 2^X satisfying (1), (2), and (3'), is quite easy. Furthermore, given such a function μ^*, called an outer measure, there is a way of generating a σ-algebra and a measure on it.

If X is a locally compact topological space, then there is a way to construct a σ-algebra and a function m satisfying (2) and (3) on this σ-algebra using *Radon functionals*. This method will be explored in Chapter 7.

We now consider Option 1 from Section 4.1 and define outer measures.

Definition 4.15. Given a nonempty set X, an *outer measure* μ^* on X is a function satisfying the following properties:

(μ^*1) $\mu^*\colon 2^X \to [0, +\infty]$, that is, μ^* is defined for *all* subsets of X.
(μ^*2) $\mu^*(\emptyset) = 0$.
(μ^*3) For any countable sequence $(A_i)_{i\geq 1}$ of subsets A_i of X,

$$\mu^*\left(\bigcup_{i=1}^{\infty} A_i\right) \leq \sum_{i=1}^{\infty} \mu^*(A_i).$$

This property is called *σ-subadditivity*.
(μ^*4) If $A \subseteq B$, then $\mu^*(A) \leq \mu^*(B)$.

Example 4.5. (Outer measure of Dirac) Let X be any nonempty set, and let a be any point chosen in X. The map $\mu_a^*\colon 2^X \to [0, +\infty]$ given by

$$\mu_a^*(A) = \begin{cases} 1 & \text{if } a \in A \\ 0 & \text{if } a \notin A \end{cases}$$

is an outer measure called the *outer measure of Dirac*.

Here is a simple way to construct outer measures.

Proposition 4.10. *Let X be a nonempty set, and $\mathfrak{I} \subseteq 2^X$ be a family of subsets with the following properties:*

(a) $\emptyset \in \mathfrak{I}$.
(b) For every subset A of X, there is a countably infinite sequence $(I_n)_{n\geq 1}$ of subsets $I_n \in \mathfrak{I}$ such that $A \subseteq \bigcup_{n=1}^{\infty} I_n$.
(c) Moreover, let $\lambda\colon \mathfrak{I} \to [0, +\infty]$ be any function such that $\lambda(\emptyset) = 0$.

Then the map μ^* given by

$$\mu^*(A) = \inf\left\{\sum_{n=1}^{\infty} \lambda(I_n) \mid A \subseteq \bigcup_{n=1}^{\infty} I_n,\ I_n \in \mathfrak{I}\right\}$$

is an outer measure on X.

Proof. The verification of (μ^*1), (μ^*2), and (μ^*4) is immediate and left to the reader.

Let $(A_i)_{i\geq 1}$ be an arbitrary family of subsets A_i of X. We may assume that $\sum_{i=1}^{\infty} \mu^*(A_i) < +\infty$, since otherwise (μ^*3) holds trivially. Then we have $\mu^*(A_i) < +\infty$ for all $i \geq 1$. By definition of $\mu^*(A_i)$ as an infimum, for every $\epsilon > 0$, for every fixed $i \geq 1$, there is a countable family $(I_{i_n})_{n\geq 1}$ of subsets $I_{i_n} \in \mathfrak{I}$ such that

$$A_i \subseteq \bigcup_{n=1}^{\infty} I_{i_n} \text{ and } \mu^*(A_i) \leq \sum_{n=1}^{\infty} \lambda(I_{i_n}) \leq \mu^*(A_i) + \frac{\epsilon}{2^i}.$$

Since, as shown in Figure 4.12,

$$\bigcup_{i=1}^{\infty} A_i \subseteq \bigcup_{i=1}^{\infty} \bigcup_{n=1}^{\infty} I_{i_n},$$

by definition of $\mu^*\left(\bigcup_{i=1}^{\infty} A_i\right)$ as an infimum and since

$$\sum_{n=1}^{\infty} \lambda(I_{i_n}) \leq \mu^*(A_i) + \frac{\epsilon}{2^i},$$

we have

$$\begin{aligned} \mu^*\left(\bigcup_{i=1}^{\infty} A_i\right) &\leq \sum_{i=1}^{\infty}\sum_{n=1}^{\infty} \lambda(I_{i_n}), \\ &\leq \sum_{i=1}^{\infty} \mu^*(A_i) + \sum_{i=1}^{\infty} \frac{\epsilon}{2^i} \\ &= \sum_{i=1}^{\infty} \mu^*(A_i) + \epsilon, \end{aligned}$$

since

$$\sum_{i=1}^{\infty} \frac{1}{2^i} = \frac{1}{2}\left(\sum_{i=0}^{\infty} \frac{1}{2^i}\right) = \frac{1}{2} \times 2 = 1.$$

Since $\epsilon > 0$ is arbitrary, we conclude that

$$\mu^*\left(\bigcup_{i=1}^{\infty} A_i\right) \leq \sum_{i=1}^{\infty} \mu^*(A_i),$$

which is (μ^*3). $\square$

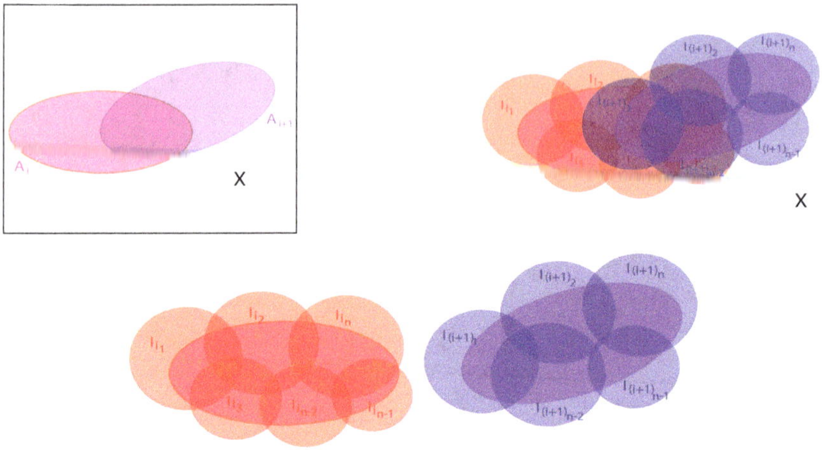

Fig. 4.12 A Venn diagram illustration of $\bigcup_{i=1}^{\infty} A_i \subseteq \bigcup_{i=1}^{\infty} \bigcup_{n=1}^{\infty} I_{i_n}$.

As an application of Proposition 4.10, we obtain the outer Lebesgue measure.

Example 4.6. Let $\mathfrak{I}$ consist of the set of all open intervals (a, b), where $a = -\infty$ or $b = +\infty$ is allowed. It is easy to see that Properties (a) and (b) of Proposition 4.10 are satisfied. Let $\lambda\colon \mathfrak{I} \to [0, +\infty]$ be given by $\lambda((a, b)) = b - a$. Obviously, Property (c) holds. The outer measure given by Proposition 4.10 is called the *outer Lebesgue measure* μ_L^* on $\mathbb{R}$.

A similar construction could be performed on $\mathbb{R}^n$ by using products of open intervals $(a_1, b_1) \times \cdots \times (a_n, b_n)$ and $\lambda((a_1, b_1) \times \cdots \times (a_n, b_n)) = \prod_{i=1}^{n}(b_i - a_i)$.

We now state a fundamental theorem due to C. Carathéodory which gives a method for constructing a measure space from an outer measure.

Theorem 4.11. *(Carathéodory) Let $\mu^*\colon 2^X \to [0, +\infty]$ be an outer measure. Define the family $\mathcal{A}$ of subsets of X as follows:*

$$\mathcal{A} = \{A \in 2^X \mid \mu^*(E) = \mu^*(E \cap A) + \mu^*(E \cap (X - A)), \textit{for all } E \subseteq X\}. \quad \text{(C)}$$

See Figure 4.13. Then the following properties hold:

(a) $\mathcal{A}$ is a σ-algebra which contains all subsets $A \subseteq X$ such that $\mu^*(A) = 0$.
(b) The restriction μ of μ^* to $\mathcal{A}$ is a measure. Furthermore, μ is a complete measure.

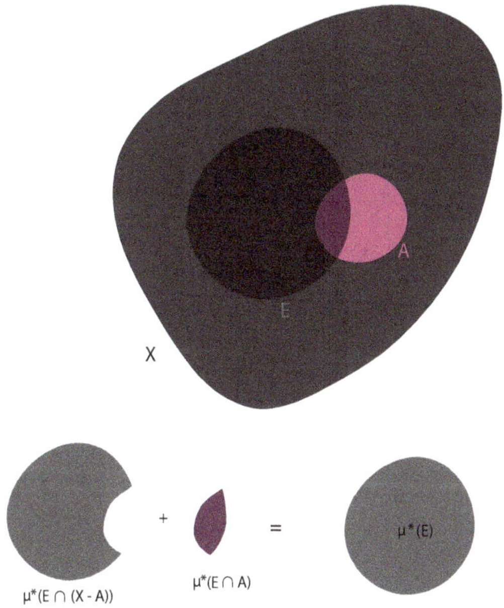

Fig. 4.13 A schematic illustration of the Carathéodory construction of $\mathcal{A}$. The σ-algebra $\mathcal{A}$ consists of those magenta sets A which "cut" (with respect to μ^*) arbitrary subsets E in a "nice" manner.

Proof. (a) To prove that $\mathcal{A}$ is a σ-algebra, we show that in order to prove the defining equation in (C) it suffices to prove the inequality (C') shown below. For this we prove

Claim 1. A subset A of X belongs to $\mathcal{A}$ if and only if, *for all* $E \subseteq X$ *such that* $\mu^*(E) < +\infty$,

$$\mu^*(E \cap A) + \mu^*(E \cap (X - A)) \leq \mu^*(E). \qquad \text{(C')}$$

Proof of Claim 1. We have $E = (E \cap A) \cup (E \cap (X - A))$ and by Condition (μ^*3), if $E = (E \cap A) \cup (E \cap (X - A))$, then

$$\mu^*(E) \leq \mu^*(E \cap A) + \mu^*(E \cap (X - A)),$$

so it suffices to prove the reverse inequality when $\mu^*(E) < +\infty$, because if $\mu^*(E) = +\infty$, then

$$\mu^*(E \cap A) + \mu^*(E \cap (X - A)) = +\infty = \mu^*(E),$$

since both sides are equal to $+\infty$. $\square$

Next the proof consists of several steps.

Step 1. Verification of (A1). By (μ^*2), for every $E \subseteq X$, we have

$$\mu^*(E \cap X) + \mu^*(E \cap (X - X)) = \mu^*(E) + \mu^*(\emptyset) = \mu^*(E) + 0 = \mu^*(E),$$

which shows that X satisfies Equation (C), and thus $X \in \mathcal{A}$.

Step 2. Verification of (A2). This follows from the fact that Equation (C) implies that $A \in \mathcal{A}$ iff $X - A \in \mathcal{A}$.

Step 3. Verification of (σ-A3). We begin by verifying (σ-A3) for finite unions. Since by *Step 2*, $\mathcal{A}$ is closed under complementation, this shows that $\mathcal{A}$ is an algebra.

Step 3a. The case of any finite union $\bigcup_{i=1}^{n} A_i$ reduces by induction to the case where $n = 2$, so it suffices to prove that for all $A_1, A_2 \subseteq X$, if $A_1, A_2 \in \mathcal{A}$, then $A_1 \cup A_2 \in \mathcal{A}$. In view of *Claim 1*, this is equivalent to checking that for all $E \subseteq X$ (with $\mu^*(E) < +\infty$),

$$\mu^*(E \cap (A_1 \cup A_2)) + \mu^*(E \cap (X - (A_1 \cup A_2))) \le \mu^*(E). \qquad (*_1)$$

We begin by rewriting the two terms $\mu^*(E \cap (A_1 \cup A_2))$ and $\mu^*(E \cap (X - (A_1 \cup A_2)))$. Since, (see Figure 4.14),

$$E \cap (A_1 \cup A_2) = (E \cap A_1) \cup (E \cap A_2) = (E \cap A_1) \cup (E \cap (X - A_1) \cap A_2),$$

by (μ^*3), we have

$$\mu^*(E \cap (A_1 \cup A_2)) \le \mu^*(E \cap A_1) + \mu^*(E \cap (X - A_1) \cap A_2). \qquad (*_2)$$

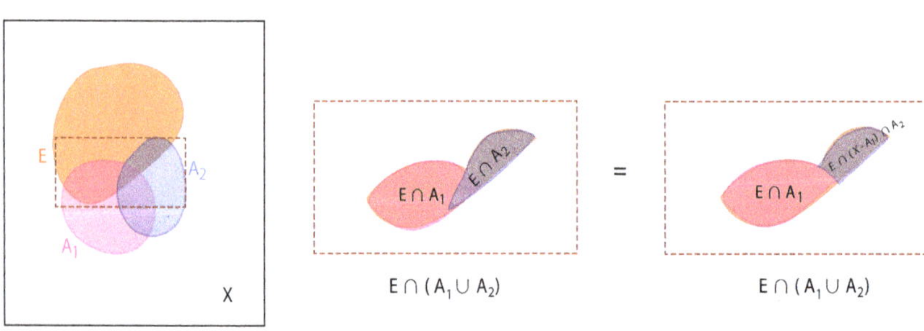

Fig. 4.14 A Venn diagram illustration of $E \cap (A_1 \cup A_2) = (E \cap A_1) \cup (E \cap A_2) = (E \cap A_1) \cup (E \cap (X - A_1) \cap A_2)$.

Since, (see Figure 4.15), we also have

$$E \cap (X - (A_1 \cup A_2)) = E \cap (X - A_1) \cap (X - A_2),$$

by $(*_2)$, we obtain

$$\begin{aligned}\mu^*(E \cap (A_1 \cup A_2)) + \mu^*(E \cap (X - (A_1 \cup A_2)))& \\ \leq \mu^*(E \cap A_1) + \mu^*(E \cap (X - A_1) \cap A_2)& \\ + \mu^*(E \cap (X - A_1) \cap (X - A_2)).& \quad (*_3)\end{aligned}$$

Since $A_1 \in \mathcal{A}$ and $A_2 \in \mathcal{A}$, for any $E \subseteq X$, by applying (C) to A_1 with E

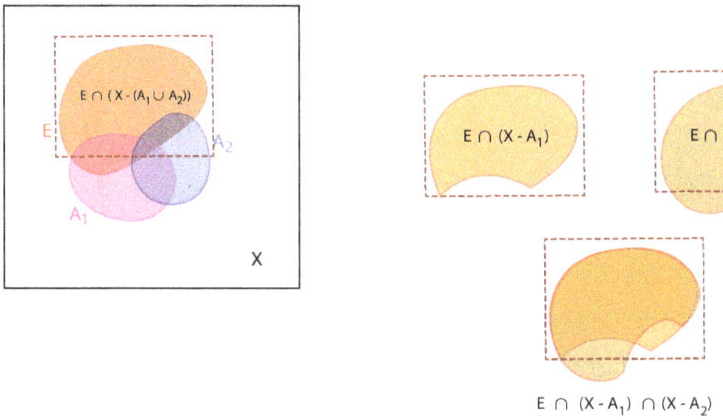

Fig. 4.15 A Venn diagram illustration of $E \cap (X - (A_1 \cup A_2)) = E \cap (X - A_1) \cap (X - A_2)$.

we have

$$\mu^*(E) = \mu^*(E \cap A_1) + \mu^*(E \cap (X - A_1)),$$

and by applying (C) to A_2 with $E \cap (X - A_1)$ we have

$$\begin{aligned}\mu^*(E \cap (X - A_1)) &= \mu^*((E \cap (X - A_1)) \cap A_2) \\ &\quad + \mu^*((E \cap (X - A_1)) \cap (X - A_2)) \\ &= \mu^*(E \cap (X - A_1) \cap A_2) \\ &\quad + \mu^*(E \cap (X - A_1) \cap (X - A_2)),\end{aligned}$$

so we obtain

$$\begin{aligned}\mu^*(E) = \mu^*(E \cap A_1) + \mu^*(E \cap (X - A_1) \cap A_2)& \\ + \mu^*(E \cap (X - A_1) \cap (X - A_2));& \quad (*_4)\end{aligned}$$

see Figure 4.16.

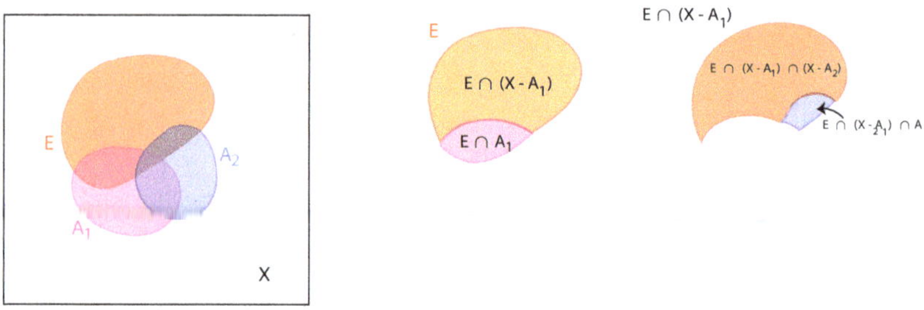

Fig. 4.16 Venn diagram illustrations associated with the identities $\mu^*(E) = \mu^*(E \cap A_1) + \mu^*(E \cap (X - A_1))$ and $\mu^*(E \cap (X - A_1)) = \mu^*((E \cap (X - A_1)) \cap A_2) + \mu^*((E \cap (X - A_1)) \cap (X - A_2))$.

Since the right-hand sides of $(*_3)$ and $(*_4)$ are identical, we obtain

$$\mu^*(E \cap (A_1 \cup A_2)) + \mu^*(E \cap (X - (A_1 \cup A_2))) \leq \mu^*(E),$$

as desired.

Step 3b. We prove that (σ-A3) holds for countably infinite unions $B = \bigcup_{i \geq 1} B_i$, with $B_i \in \mathcal{A}$ and $B_i \cap B_j = \emptyset$ for all $i \neq j$, which means that we have to show that for all $E \subseteq X$ such that $\mu^*(E) < +\infty$, we have

$$\mu^*(E \cap B) + \mu^*(E \cap (X - B)) \leq \mu^*(E).$$

We begin by analyzing the term $\mu^*(E \cap B)$. Since $E \cap B = \bigcup_{i=1}^{\infty}(E \cap B_i)$, by ($\mu^*3$) we have

$$\mu^*(E \cap B) = \mu^*\left(\bigcup_{i=1}^{\infty}(E \cap B_i)\right) \leq \sum_{i=1}^{\infty} \mu^*(E \cap B_i),$$

which implies the inequality

$$\mu^*(E \cap B) + \mu^*(E \cap (X - B)) \leq \sum_{i=1}^{\infty} \mu^*(E \cap B_i) + \mu^*(E \cap (X - B)). \quad (*_5)$$

Thus if we prove that

$$\sum_{i=1}^{\infty} \mu^*(E \cap B_i) + \mu^*(E \cap (X - B)) \leq \mu^*(E),$$

we are done. To deal with the infinite sum on the left-hand side we use *Step 3a*. By the result of *Step 3a*, we have $C_n = \bigcup_{i=1}^{n} B_i \in \mathcal{A}$ for all $n \geq 1$.

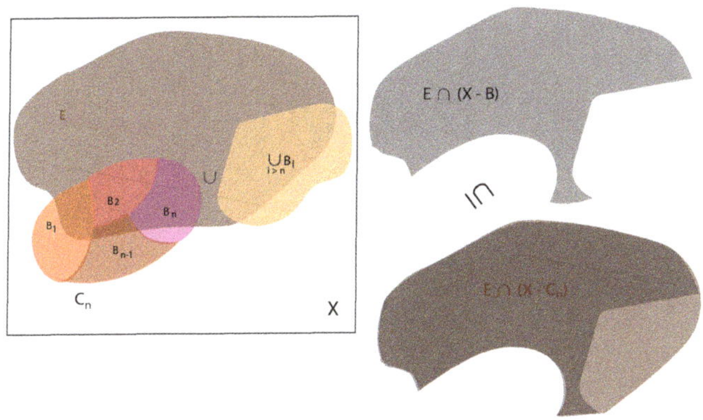

Fig. 4.17 Venn diagram illustration of $E \cap (X - B) \subseteq E \cap (X - C_n)$.

Since, as shown in Figure 4.17, $E \cap (X - B) \subseteq E \cap (X - C_n)$, by (μ^*4), we have

$$\mu^*(E \cap C_n) + \mu^*(E \cap (X - B)) \leq \mu^*(E \cap C_n) + \mu^*(E \cap (X - C_n)) = \mu^*(E). \quad (*_6)$$

On the other hand, since $B_i \in \mathcal{A}$, we can show by induction using the fact that $C_n = \bigcup_{i=1}^n B_i$ and the B_i are pairwise disjoint that

$$\begin{aligned} \mu^*(E \cap C_n) &= \mu^*(E \cap C_n \cap B_n) + \mu^*(E \cap C_n \cap (X - B_n)) \\ &= \mu^*(E \cap B_n) + \mu^*(E \cap C_{n-1}) \\ &= \sum_{i=1}^n \mu^*(E \cap B_i); \end{aligned}$$

see Figure 4.18.

Consequently, by $(*_6)$ and the above equation, we obtain

$$\mu^*(E \cap C_n) + \mu^*(E \cap (X - B)) = \sum_{i=1}^n \mu^*(E \cap B_i) + \mu^*(E \cap (X - B)) \leq \mu^*(E). \quad (*_7)$$

Since by hypothesis $\mu^*(E) < +\infty$ and by (μ^*4),

$$\mu^*(E \cap (X - B)) \leq \mu^*(E) < +\infty,$$

passing to the limit the inequality $(*_7)$ implies that

$$\sum_{i=1}^\infty \mu^*(E \cap B_i) < +\infty,$$

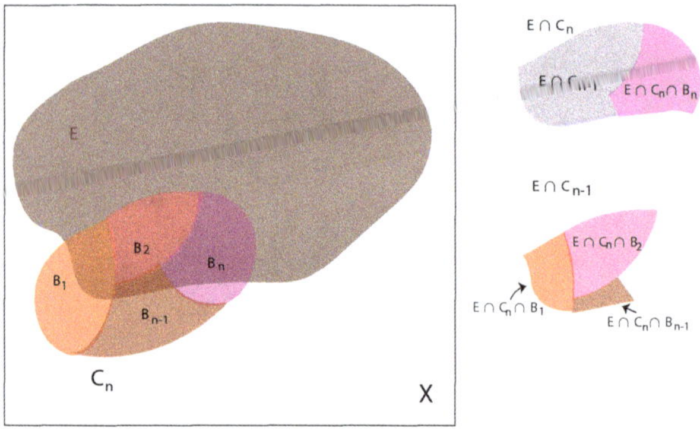

Fig. 4.18 Venn diagram illustration associated with $\mu^*(E \cap C_n) = \sum_{i=1}^n \mu^*(E \cap B_i)$.

and also that

$$\sum_{i=1}^{\infty} \mu^*(E \cap B_i) + \mu^*(E \cap (X - B)) \leq \mu^*(E), \qquad (\dagger)$$

as desired.

Step 3c. We prove that (σ-A3) holds for arbitrary countably infinite unions $A = \bigcup_{i \geq 1} A_i$, with $A_i \in \mathcal{A}$.

The trick (already used in the proof of Proposition 4.1) is to define the family $(B_i)_{i \geq 1}$ as follows:

$$B_1 = A_1$$

$$B_i = A_i - \left(\bigcup_{j=1}^{i-1} A_j \right) = X - \left(\left(\bigcup_{j=1}^{i-1} A_j \right) \cup (X - A_i) \right);$$

see Figure 4.3. Since $\mathcal{A}$ is an algebra, it is closed under finite unions and complementation, so $B_i \in \mathcal{A}$. Furthermore, by definition, $B_i \cap B_j = \emptyset$ for all $i \neq j$, and

$$\bigcup_{i=1}^{\infty} A_i = \bigcup_{i=1}^{\infty} B_i,$$

so by *Step 3b*, we get $\bigcup_{i=1}^{\infty} A_i = \bigcup_{i=1}^{\infty} B_i \in \mathcal{A}$.

Therefore, we proved that $\mathcal{A}$ is a σ-algebra.

Step 4. Proving (*a*). If $\mu^*(A) = 0$, since by (μ^*4) we have

$$\mu^*(E \cap A) \leq \mu^*(A) = 0$$

and

$$\mu^*(E \cap (X - A)) \leq \mu^*(E),$$

we obtain

$$\mu^*(E \cap A) + \mu^*(E \cap (X - A)) \leq \mu^*(E)$$

for all $E \subseteq X$ such that $\mu^*(E) < +\infty$, which by *Claim 1* means that $A \in \mathcal{A}$.

(b) We prove that the restriction μ of μ^* to $\mathcal{A}$ is a measure, which means that we need to check Condition ($\mu 1$), ($\mu 2$) and ($\mu 3$), which is achieved in three steps.

Step 5. Property ($\mu 1$) is obvious.

Step 6. Since $\mathcal{A}$ is a σ-algebra, $\emptyset \in \mathcal{A}$, so by ($\mu^* 2$),

$$\mu(\emptyset) = \mu^*(\emptyset) = 0,$$

which is ($\mu 2$).

Step 7. Let $B = \bigcup_{i=1}^{\infty} B_i$ be a countably infinite union of subsets $B_i \in \mathcal{A}$ which are pairwise disjoint. For all $E \subseteq X$ such that $\mu^*(E) < +\infty$, we proved in (†) that

$$\sum_{i=1}^{\infty} \mu^*(E \cap B_i) + \mu^*(E \cap (X - B)) \leq \mu^*(E).$$

If $\mu^*(B) < +\infty$, then we can let $E = B$ in the above inequality, and we get

$$\sum_{i=1}^{\infty} \mu^*(B_i) \leq \mu^*(B). \qquad (*_8)$$

By ($\mu^* 3$), since $B = \bigcup_{i=1}^{\infty} B_i$, we also have

$$\mu^*(B) \leq \sum_{i=1}^{\infty} \mu^*(B_i). \qquad (*_9)$$

Then ($*_8$) and ($*_9$) yield

$$\sum_{i=1}^{\infty} \mu^*(B_i) = \mu^*(B).$$

Since $B \in \mathcal{A}$ and $B_i \in \mathcal{A}$, $\mu(B) = \mu^*(B)$ and $\mu(B_i) = \mu^*(B_i)$, we get

$$\sum_{i=1}^{\infty} \mu(B_i) = \mu(B),$$

which is ($\mu 3$). If $\mu^*(B) = +\infty$, then ($*_8$) implies that trivially

$$\sum_{i=1}^{\infty} \mu(B_i) = \sum_{i=1}^{\infty} \mu^*(B_i) = \mu^*(B) = \mu(B) = +\infty.$$

Step 8. Finally it remains to show that μ is a complete measure. Let $A \in \mathcal{A}$ such that $\mu(A) = 0$ and consider any subset $B \subseteq A$. Since $A \in \mathcal{A}$, by definition of μ,

$$\mu^*(A) = \mu(A) = 0,$$

and by (μ^*4), $B \subseteq A$ implies that

$$\mu^*(B) \leq \mu^*(A) = 0,$$

so $\mu^*(B) = 0$, and we proved in *Step 4* that $B \in \mathcal{A}$. Therefore, μ is a complete measure. □

Example 4.7. If we apply Theorem 4.11 to the Dirac outer measure μ_a^* of Example 4.5, we find easily that $\mathcal{A} = 2^X$ and that $\mu = \mu_a^*$. The *Dirac measure* μ_a^* is usually denoted by δ_a.

If we apply Theorem 4.11 to the Lebesgue outer measure of Example 4.6, we obtain the *Lebesgue measure* on $\mathbb{R}$. It can be shown that the σ-algebra of *Lebesgue-measurable sets* obtained from the construction contains the σ-algebra of Borel sets of $\mathbb{R}$. This example is considered in slightly more details in the next section.

4.5 The Lebesgue Measure on $\mathbb{R}$

Recall that in Example 4.6 we defined the outer Lebesgue measure μ_L^* on $\mathbb{R}$. For this we considered the set $\mathfrak{I}$ consisting of all open intervals (a, b), where $a = -\infty$ or $b = +\infty$ is allowed. By Proposition 4.10 applied to the function $\lambda\colon \mathfrak{I} \to [0, +\infty]$ given by $\lambda((a, b)) = b - a$, we obtained the outer Lebesgue measure μ_L^* given by

$$\mu_L^*(A) = \inf \left\{ \sum_{i=1}^{\infty} \lambda(I_n) \mid A \subseteq \bigcup_{n=1}^{\infty} I_n,\ I_n \in \mathfrak{I} \right\}.$$

By applying Theorem 4.11 to the outer measure μ_L^*, we obtain the σ-algebra $\mathcal{L}(\mathbb{R})$ of *Lebesgue-measurable sets*, and the *Lebesgue measure* μ_L.

The construction used by Theorem 4.11 yields very little explicit information regarding what the Lebesgue-measurable sets look like, but it is possible to describe some of them. In particular, if $\mathcal{B}(\mathbb{R})$ denotes the Borel σ-algebra generated by the open sets of $\mathbb{R}$, it turns out that $\mathcal{B}(\mathbb{R}) \subset \mathcal{L}(\mathbb{R})$, a proper inclusion. Actually, every open subset of $\mathbb{R}$ can be expressed as a countable disjoint union of finite open intervals, so the Borel σ-algebra is generated by the open intervals (a, b). The following proposition gives convenient characterizations of the Borel sets.

Proposition 4.12. *The σ-algebra $\mathcal{B}(\mathbb{R})$ of Borel sets of $\mathbb{R}$ is generated by the following intervals:*

(1) $[a, b]$, with $a \leq b$ finite.
(2) $[a, b)$, with $a \leq b$ finite.
(3) (a, ∞) and $(-\infty, a)$, with a finite.

Proof. (1) We know that the open intervals (a, b) generate $\mathcal{B}(\mathbb{R})$. We have

$$(a, \infty) = \bigcup_{n=1}^{\infty} (a, a+n), \quad (-\infty, a) = \bigcup_{n=1}^{\infty} (a-n, a),$$

so (a, ∞) and $(-\infty, a)$ are elements of $\mathcal{B}(\mathbb{R})$; see Figure 4.19. Observe that

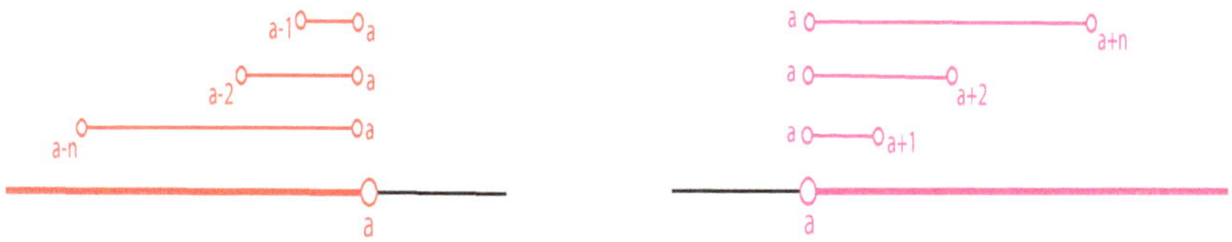

Fig. 4.19 The left figure illustrates $(-\infty, a) = \bigcup_{n=1}^{\infty}(a-n, a)$, while the right figure illustrates $(a, \infty) = \bigcup_{n=1}^{\infty}(a, a+n)$.

$$[a, b] = \overline{(-\infty, a)} \cap \overline{(b, \infty)} = [a, \infty) \cap (-\infty, b],$$

so $[a, b] \in \mathcal{B}(\mathbb{R})$. We also have

$$(a, b) = \bigcup_{n=1}^{\infty} \left[a + \frac{1}{n}, b - \frac{1}{n}\right],$$

so the closed intervals $[a, b]$ generate $\mathcal{B}(\mathbb{R})$; see Figure 4.20.

(2) We have

$$[a, b) = \bigcup_{n=1}^{\infty} \left[a, b - \frac{1}{n}\right],$$

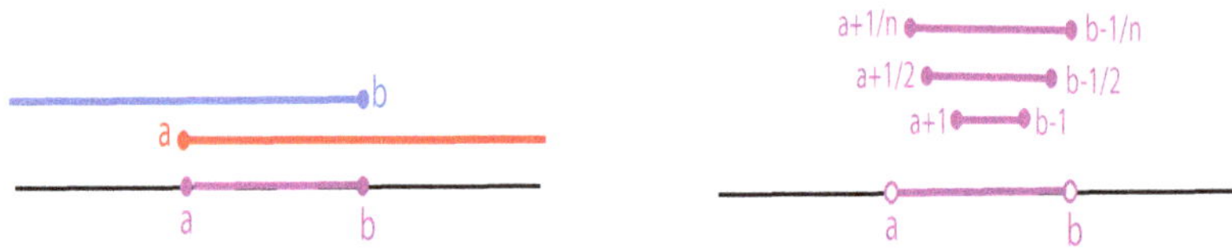

Fig. 4.20 The left figure illustrates $[a, b] = [a, \infty) \cap (-\infty, b]$, while the right figure illustrates $(a, b) = \bigcup_{n=1}^{\infty} \left[a + \frac{1}{n}, b - \frac{1}{n}\right]$.

so $[a, b) \in \mathcal{B}(\mathbb{R})$. Then

$$(a, b) = \bigcup_{n=1}^{\infty} \left[a + \frac{1}{n}, b\right),$$

so the intervals $[a, b)$ generate $\mathcal{B}(\mathbb{R})$. See Figure 4.21.

Fig. 4.21 The left figure illustrates $[a, b) = \bigcup_{n=1}^{\infty} \left[a, b - \frac{1}{n}\right]$, while the right figure illustrates $(a, b) = \bigcup_{n=1}^{\infty} \left[a + \frac{1}{n}, b\right)$.

(3) We already know from (1) that $(a, \infty) \in \mathcal{B}(\mathbb{R})$. This implies that

$$(-\infty, a] = \overline{(a, \infty)} \in \mathcal{B}(\mathbb{R}),$$

so

$$(-\infty, a) = \bigcup_{n=1}^{\infty} \left(-\infty, a - \frac{1}{n}\right] \in \mathcal{B}(\mathbb{R}),$$

and thus

$$(a, b) = (-\infty, b) \cap (a, \infty),$$

so the intervals (a, ∞) generate $\mathcal{B}(\mathbb{R})$. See Figure 4.22. □

Let's use the notation $\wr a, b \wr$ to denote any of the four types of intervals $(a, b), [a, b), (a, b]$, and $[a, b]$ (with $a = -\infty$ or $b = +\infty$ allowed, and $a = b$ allowed). The following result can be shown.

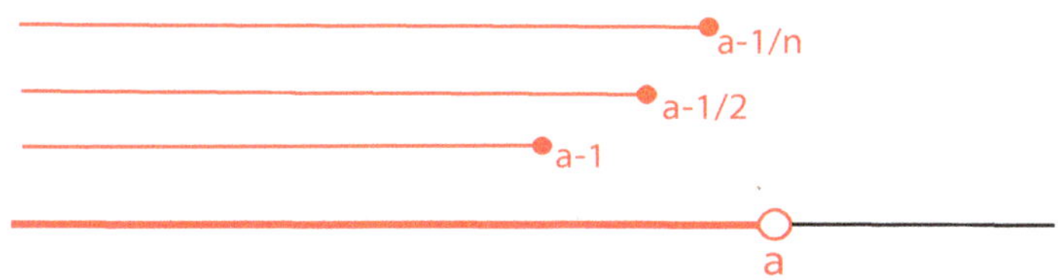

Fig. 4.22 An illustration of the identity $(-\infty, a) = \bigcup_{n=1}^{\infty} (-\infty, a - \frac{1}{n}]$.

Theorem 4.13. *Let $\mathcal{B}(\mathbb{R})$ be the Borel σ-algebra of open sets, $\mathcal{L}(\mathbb{R})$ be the σ-algebra of Lebesgue-measurable sets, and μ_L be the Lebesgue measure for $\mathbb{R}$.*

(1) *$\mathcal{L}(\mathbb{R}) \neq 2^{\mathbb{R}}$; that is, there exist non-measurable sets. The proof requires the axiom of choice.*
(2) *$\mathcal{B}(\mathbb{R}) \subset \mathcal{L}(\mathbb{R})$, the inclusion being strict. This is because $|\mathcal{B}(\mathbb{R})| = 2^{\aleph_0} = \mathfrak{c}$, but $|\mathcal{L}(\mathbb{R})| = 2^{\mathfrak{c}}$.*
(3) *The Borel σ-algebra contains all four types of intervals, and*

$$\mu_L(\wr a, b\wr) = \begin{cases} b - a & \text{if } a \neq -\infty \text{ and } b \neq +\infty \\ +\infty & \text{if } a = -\infty \text{ or } b = +\infty. \end{cases}$$

(4) *The restriction of the Lebesgue measure μ_L to the Borel σ-algebra is a measure μ_B. The completion of the measure space $(\mathbb{R}, \mathcal{B}(\mathbb{R}), \mu_B)$ given by Proposition 4.8 gives back the measure space $(\mathbb{R}, \mathcal{L}(\mathbb{R}), \mu_L)$ of Lebesgue-measurable sets.*

The proofs for most parts of Theorem 4.13 are given in Halmos [Halmos (1978)] (some of them as exercises). The fact that $|\mathcal{B}(\mathbb{R})| = 2^{\aleph_0}$ follows from the fact that $\mathcal{B}(\mathbb{R})$ is generated by the open intervals (a, b) and the remark just before Definition 4.6. It is surprising how much work it takes to prove Part (3) of Theorem 4.13. See also Folland [Folland (1999)] and Rudin [Rudin (1987)].

As a corollary, every one-point set $\{a\}$ has Lebesgue measure 0, and thus every countable subset has Lebesgue measure 0. There are also uncountable subsets of Lebesgue measure 0. The *Cantor set* is such an example; see Folland [Folland (1999)], Section 1.5.

The Lebesgue measure also has the following regularity properties which show that every Lebesgue-measurable set can be approximated either by an open set or by a closed set; see Folland [Folland (1999)] (Section 1.5).

Proposition 4.14. *For any subset A of $\mathbb{R}$, we have*

$$\mu_L^*(A) = \inf\{\mu_L(O) \mid A \subseteq O,\ O \textit{ is open}\}.$$

For every Lebesgue-measurable set $A \in \mathcal{L}(\mathbb{R})$, the following facts hold:

(a) *For every $\epsilon > 0$, there is some open subset O such that $A \subseteq O$ and $\mu_L(O - A) < \epsilon$.*
(b) *For every $\epsilon > 0$, there is some closed subset F such that $F \subseteq A$ and $\mu_L(A - F) < \epsilon$.*

As a corollary of Proposition 4.14 we have the following facts.

Proposition 4.15. *For every Lebesgue-measurable set $A \in \mathcal{L}(\mathbb{R})$:*

(a') $\mu_L(A) = \inf\{\mu_L(O) \mid A \subseteq O,\ O \textit{ is open}\}$.
(b') $\mu_L(A) = \sup\{\mu_L(F) \mid F \subseteq A,\ F \textit{ is closed}\}$;

see Figure 4.23.

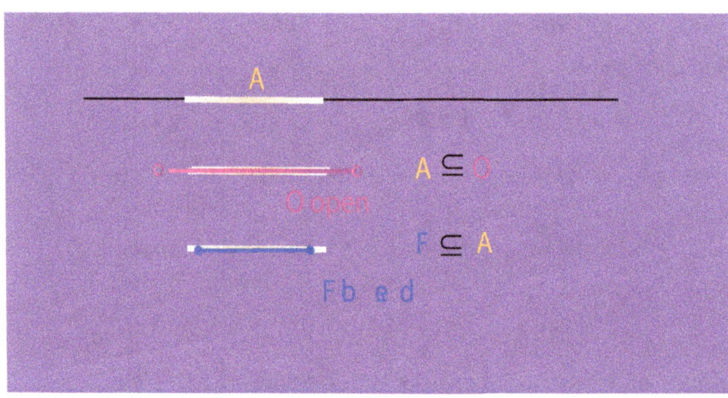

Fig. 4.23 A Lebesgue-measurable set A of $\mathbb{R}$ is approximated from the "outside" by an open set O; it is also approximated from the "inside" by a closed set F.

It should be noted that Properties (a') and (b') are weaker than Properties (a) and (b), because they imply Properties (a) and (b) only when $\mu(A)$ is finite.

It can also be shown that for every Lebesgue-measurable set $A \in \mathcal{L}(\mathbb{R})$, we have

$$\mu_L(A) = \sup\{\mu_L(K) \mid K \subseteq A,\ K \text{ is compact}\}.$$

Proposition 4.14 also holds for the Lebesgue-measurable subsets of $\mathbb{R}^n$.

Another important property of the Lebesgue measure is that it is translation-invariant.

Proposition 4.16. *For any Lebesgue measurable set $A \in \mathcal{L}(\mathbb{R})$, we have $\mu_L(x + A) = \mu_L(A)$ for all $x \in \mathbb{R}$, where $x + A = \{x + a \mid a \in \mathbb{R}\}$. This property is called translation-invariance.*

For a proof, see Section 8.5, Example 8.1.

Proposition 4.17. *There is no translation-invariant measure μ defined on all subsets of $\mathbb{R}$ such that $\mu([0, 1)) \neq 0$ and $\mu([0, 1)) \neq +\infty$. As a consequence, there is no translation-invariant measure defined on all subsets of $\mathbb{R}$ such that $\mu([a, b)] = b - a$.*

Proof. To prove the proposition, we consider the quotient set $\mathbb{R}/\mathbb{Q}$ of the reals modulo the equivalence relation $x \sim y$ iff $x - y \in \mathbb{Q}$. Using the axiom of choice, we can form a subset $E \subseteq [0, 1)$ which contains exactly one number from each equivalence class of $\mathbb{R}/\mathbb{Q}$. Let $R = \mathbb{Q} \cap [0, 1)$, and for each $r \in R$, let

$$E_r = \{x + r \mid x \in E \cap [0, 1 - r)\} \cup \{x + r - 1 \mid x \in E \cap [1 - r, 1)\};$$

see Figure 4.24. Clearly $E_r \subseteq [0, 1)$, and we claim that every $x \in [0, 1)$ belongs to some unique E_r.

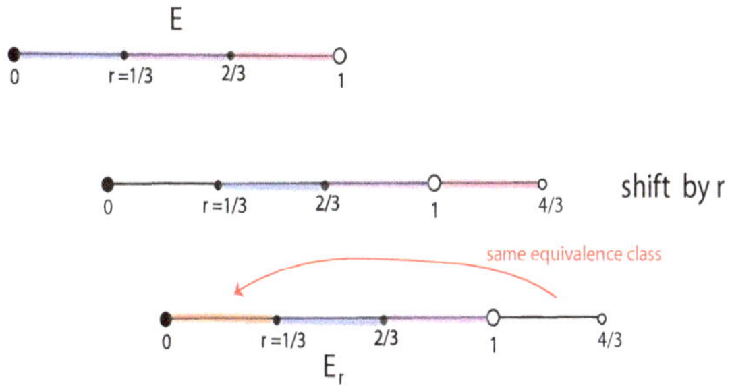

Fig. 4.24 The construction of E_r.

Indeed, if $y \in E$ belongs to the equivalence class of $x \in [0, 1)$, then $x \in E_r$ where $r = x - y$ if $x \geq y$ or $r = x - y + 1$ if $x < y$. Furthermore, if $x \in E_r \cap E_s$ with $r \neq s$, then $x - r$ (or $x - r + 1$) and $x - s$ (or $x - s + 1$) would be distinct elements of E belonging to the same equivalence class, which is impossible (since $r, s \in R \subset \mathbb{Q}$). It follows that $[0, 1)$ is the countable disjoint union of the E_r. If a translation-invariant measure μ exists, then for any $r \in R$ we have

$$\mu(E) = \mu(E \cap [0, 1 - r)) + \mu(E \cap [1 - r, 1)) = \mu(E_r).$$

Since $[0, 1)$ is the countable disjoint union of the E_r,

$$\mu([0, 1)) = \sum_{r \in E} \mu(E_r) = \sum_{r \in E} \mu(E).$$

Now by assumption $\mu([0, 1)) \neq 0$ and $\mu([0, 1)) \neq +\infty$, but the sum on the right-hand side is either 0 if $\mu(E) = 0$ or $+\infty$ otherwise, a contradiction. □

The above proof also implies that E is an uncountable subset of $[0, 1)$ which is not Lebesgue measurable (since the Lebesgue measure is translation-invariant).

We conclude by mentioning that if X is a topological space, given a function μ defined on the open subsets and the compact subsets of X, we can define the following maps for *every* subset A of X:

$$\mu^*(A) = \inf\{\mu(O) \mid A \subseteq O,\ A \text{ is open}\}$$
$$\mu_*(A) = \sup\{\mu(K) \mid K \subseteq A,\ K \text{ is compact}\}.$$

Then the measurable subsets are those subsets A of X such that

$$\mu^*(A) = \mu_*(A).$$

It can be shown that these subsets form a σ-algebra $\mathcal{A}$, and that the map μ with domain $\mathcal{A}$ given by $\mu(A) = \mu^*(A) = \mu_*(A)$ is a measure. This is the approach using Radon measures.

4.6 Problems

Problem 4.1. Let X and Y be two nonempty sets, and let $\mathcal{A}$ be an algebra on X and let $\mathcal{B}$ be an algebra on Y. Define the set $\mathcal{R}$ of *rectangles* in $X \times Y$ as $\mathcal{R}$

$$\mathcal{R} = \{A \times B \in X \times Y \mid A \in \mathcal{A},\ B \in \mathcal{B}\}.$$

Show that the set $\mathcal{B}(\mathcal{R})$ of finite unions of pairwise disjoint sets in $\mathcal{R}$ is the smallest algebra containing the semi-algebra $\mathcal{R}$.

Problem 4.2. Prove Proposition 4.2.

Problem 4.3. Prove Theorem 4.6. Hint: See Folland [Folland (1999)] (Lemma 2.35).

Problem 4.4. Prove Part (5) of Proposition 4.7.

Problem 4.5. Advanced Exercise: Prove Proposition 4.8. Hint: See Rudin [Rudin (1987)] (Theorem 1.36).

Problem 4.6. Let $X = (X, \mathcal{A}, \mu)$ be a measure space. Consider $S := \{f \mid f\colon X \to \mathbb{R}\}$. Show that equality *a.e.* is an equivalence relation on S.

Problem 4.7. Let $D_{\mathbb{Q}}\colon \mathbb{R} \to \mathbb{R}$ be defined as

$$D_{\mathbb{Q}}(x) = \begin{cases} \frac{1}{q} & \text{if } x = p/q \in \mathbb{Q},\ q > 0,\ p \neq 0,\ \gcd(p, q) = 1, \\ 0 & \text{if } x \notin \mathbb{Q}, \\ 1 & \text{if } x = 0. \end{cases}$$

Show that $D_{\mathbb{Q}}$ is discontinuous at every rational point x, but is continuous at every irrational point x. Also prove that $D_{\mathbb{Q}}$ is a regulated function.

Problem 4.8. Advanced Exercise: Prove Proposition 4.14. Hint: See Folland [Folland (1999)] (Section 1.5).

Problem 4.9. Prove Proposition 4.15.

Chapter 5

Integration

Given a measure space $(X, \mathcal{A}, \mu)$, we would like to define the integral of a real-valued function $f\colon X \to \mathbb{R}$, or more generally of a complex-valued function $f\colon X \to \mathbb{C}$, or even of a function $f\colon X \to F$, where F is a normed vector space. The key idea is that the integral of a very simple function f, such as a function taking only a finite number of nonzero values $y_1, \ldots, y_n$, should be "obvious." Namely, if $A_i = f^{-1}(y_i)$ is the subset of X over which f has the value y_i, then each A_i should be measurable (that is, $A_i \in \mathcal{A}$), and A_i should have finite measure, so that the expression

$$\sum_{i=1}^{n} y_i \mu(A_i) \in F$$

makes sense. Then we define the integral $\int f d\mu$ of our simple function f as

$$\int f d\mu = \sum_{i=1}^{n} \mu(A_i) y_i. \tag{$*$}$$

Observe that the function f can be written as

$$f = \sum_{i=1}^{n} y_i \chi_{A_i},$$

where χ_{A_i} is the characteristic function of the subset A_i. Such a function is called a *μ-step function.*

Observe that $(*)$ is a generalization of the notion of area under the curve. If the subsets A_i are closed adjacent intervals, then we are back to the notion of Riemann integral. However, in our new setting, the subsets A_i can be very complicated, but as long as they are measurable and have finite measure, the integral $(*)$ makes sense.

If we define $\|f\|$ as

$$\|f\| = \sum_{i=1}^{n} \|y_i\| \, \chi_{A_i},$$

(remember that our set F of values is a normed vector space), then the integral of $\|f\|$ is

$$\int \|f\| \, d\mu = \sum_{i=1}^{n} \mu(A_i) \, \|y_i\| \in \mathbb{R}_+.$$

If we define $N_1(f) = \int \|f\| \, d\mu$, then N_1 satisfies all the properties of a norm, except that $N_1(f) = 0$ does not necessarily imply that $f = 0$. However, $N_1(f) = 0$ iff $f = 0$ almost everywhere. The set $\mathcal{S}tep_\mu(X, \mathcal{A}, F)$ of μ-step functions is a vector space, and N_1 is almost a norm on it; it is a semi-norm. The integral given by $(*)$ is a linear continuous map on $\mathcal{S}tep_\mu(X, \mathcal{A}, F)$. However, the space $\mathcal{S}tep_\mu(X, \mathcal{A}, F)$ is not Cauchy-complete under the semi-norm N_1 (there are Cauchy sequences with respect to N_1 that do not have a limit). The problem then is to complete the space $\mathcal{S}tep_\mu(X, \mathcal{A}, \mu)$ and to extend the integral $(*)$ to this bigger set of functions.

There are several ways to proceed.

(1) If we let $\mathcal{SN}$ be subspace of $\mathcal{S}tep_\mu(X, \mathcal{A}, F)$ consisting of the μ-step functions equal to 0 a.e., then the quotient space $\mathrm{Step}_\mu(X, \mathcal{A}, F) = \mathcal{S}tep_\mu(X, \mathcal{A}, \mu)/\mathcal{SN}$ is a vector space and N_1 induces a (true) norm on it. Therefore we can apply the general completion theorem (Theorem A.72) to obtain a complete normed vector space $(\mathrm{L}_\mu(X, \mathcal{A}, F), \| \; \|_1)$. Since integration is a linear continuous map on $\mathcal{S}tep_\mu(X, \mathcal{A}, F)$, it extends uniquely to a linear continuous map on $\mathrm{L}_\mu(X, \mathcal{A}, F)$.

In theory we have achieved our goal of defining a complete normed vector space of functions containing the μ-step functions for which every function is integrable. However, the completion $(\mathrm{L}_\mu(X, \mathcal{A}, F), \| \; \|_1)$ is a very complicated object. It consists of equivalence classes of Cauchy sequences of functions in the quotient space $\mathrm{Step}_\mu(X, \mathcal{A}, F)$. It would be much more convenient if the objects in $\mathrm{L}_\mu(X, \mathcal{A}, F)$ could be described as functions, and this is indeed possible.

(2) The second approach is to first define a set $\mathcal{L}_\mu(X, \mathcal{A}, F)$ of *functions* using a limit process. Every function f in $\mathcal{L}_\mu(X, \mathcal{A}, F)$ is the limit pointwise a.e. of a N_1-Cauchy sequence $(f_n)_{n \geq 1}$ (called an approximation sequence) of functions f_n in $\mathcal{S}tep_\mu(X, \mathcal{A}, F)$. We also define

the space $\mathcal{M}_\mu(X, \mathcal{A}, F)$ of *μ-measurable functions*, and $\mathcal{L}_\mu(X, \mathcal{A}, F)$ is the subspace of $\mathcal{M}_\mu(X, \mathcal{A}, F)$ consisting of the functions for which the integral is well defined.

It turns out that $\mathcal{L}_\mu(X, \mathcal{A}, F)$ is complete with respect to an extension $\| \|_1$ of the semi-norm N_1, and the integral $\int f d\mu$ of any function $f \in \mathcal{L}_\mu(X, \mathcal{A}, F)$ can be defined by a limit process. There are technical complications when F is infinite-dimensional, and it also takes some work to show that the integral of a function $f \in \mathcal{L}_\mu(X, \mathcal{A}, F)$ does not depend on the approximation sequence used to define f, but all difficulties can be overcome. Finally, the subspace $\mathcal{N}$ of functions f such that $\|f\|_1 = 0$ is the set of functions equal to 0 a.e., and we obtain the complete space $(\mathrm{L}_\mu(X, \mathcal{A}, F), \| \|_1)$ of the first approach as the quotient space $\mathcal{L}_\mu(X, \mathcal{A}, F)/\mathcal{N}$. However, the construction of $\mathcal{L}_\mu(X, \mathcal{A}, F)$ is much more informative.

We also investigate convergence properties of $\mathcal{L}_\mu(X, \mathcal{A}, F)$, as well as other related spaces (the spaces $\mathcal{L}^p_\mu(X, \mathcal{A}, F)$, $p = 1, 2, \infty$). We conclude with the construction of the integral on a product space.

The vector valued-integral defined in this chapter (where the space F of values is a Banach space) was first discovered by Bochner in 1933. The version discussed here is due to Dunford (1935), and is presented in detail in Dunford and Schwartz [Dunford and Schwartz (1988)]. More recent expositions of this method are given in Lang [Lang (1996)] and Marle [Marle (1974)].

5.1 Measurable Maps

Measurable functions are functions between measurable spaces that are the analog of continuous functions between topological spaces, but as we will see, they are a lot more flexible, especially in terms of convergence properties. In this chapter our presentation follows Marle [Marle (1974)] and Lang [Lang (1996)] very closely.

Definition 5.1. Given any two measurable spaces $(X, \mathcal{A})$ and $(Y, \mathcal{B})$, a function $f \colon X \to Y$ is *measurable* if $f^{-1}(B) \in \mathcal{A}$ for every $B \in \mathcal{B}$. A measurable function is also called a *measurable map*.

If $(X, \mathcal{A})$ is a measurable space, then obviously the identity $\mathrm{id} \colon X \to X$ is measurable.

The composition of two measurable maps is also measurable.

Proposition 5.1. *Given three measurable spaces* $(X,\mathcal{A})$, $(Y,\mathcal{B})$, *and* $(Z,\mathcal{C})$, *if* $f\colon X\to Y$ *and* $g\colon Y\to Z$ *are measurable maps, then* $g\circ f\colon X\to Z$ *is a measurable map.*

Proof. Recall that one of the properties of inverse images is that $(g\circ f)^{-1}(C)=f^{-1}(g^{-1}(C))$, for any subset C of Z. But if $C\in\mathcal{C}$, since g is measurable, $g^{-1}(C)\in\mathcal{B}$, and since f is measurable, $f^{-1}(g^{-1}(C))\in\mathcal{A}$, which shows that $g\circ f$ is measurable. □

Remark: The above properties show that measurable spaces are the objects of a category whose morphisms are the measurable maps.

Proposition 5.2. *Let* X *and* Y *be any two nonempty sets, and let* $f\colon X\to Y$ *be a function between them.*

(1) If $\mathcal{A}$ *is a* σ*-algebra on* X*, then we can define* $\mathcal{A}_f$ *as the family of subsets of* Y *given by*

$$\mathcal{A}_f=\{B\in 2^Y\mid f^{-1}(B)\in\mathcal{A}\}.$$

Then $\mathcal{A}_f$ *is the largest* σ*-algebra on* Y *which makes* f *measurable.*

(2) If $\mathcal{B}$ *is a* σ*-algebra on* Y*, then let* $f^{-1}(\mathcal{B})$ *be the family of subsets of* X *given by*

$$f^{-1}(\mathcal{B})=\{f^{-1}(B)\in 2^X\mid B\in\mathcal{B}\}.$$

Then $f^{-1}(\mathcal{B})$ *is the smallest* σ*-algebra on* X *which makes* f *measurable.*

The proof of Proposition 5.2 is left as an exercise.

Using Proposition 5.2 we obtain the following proposition which gives simple criteria to check that a map is measurable.

Proposition 5.3. *Let* $(X,\mathcal{A})$ *and* $(Y,\mathcal{B})$ *be two measurable spaces.*

(1) If $\mathcal{S}$ *generates the* σ*-algebra* $\mathcal{B}$ *(which means that the smallest* σ*-algebra containing* $\mathcal{S}$ *is* $\mathcal{B}$*), then a function* $f\colon X\to Y$ *is measurable iff* $f^{-1}(S)\in\mathcal{A}$ *for all* $S\in\mathcal{S}$*.*

(2) If Y *is a topological space and if* $\mathcal{B}$ *is its Borel* σ*-algebra of open subsets, then a function* $f\colon X\to Y$ *is measurable iff* $f^{-1}(U)\in\mathcal{A}$ *for every open subset* U *of* Y *(or* $f^{-1}(U)\in\mathcal{A}$ *for every closed subset* U *of* Y*).*

(3) If X *and* Y *are both topological spaces and if* $\mathcal{A}$ *and* $\mathcal{B}$ *are their respective Borel* σ*-algebras, then every continuous map* $f\colon X\to Y$ *is measurable.*

Given any subset A of X, recall that the *characteristic function* χ_A of A is defined by

$$\chi_A(x) = \begin{cases} 1 & \text{if } x \in A \\ 0 & \text{if } x \notin A. \end{cases}$$

Then, as illustrated in Figure 5.1, it is easy to show that for any subset A of X, the function $\chi_A \colon X \to \mathbb{R}$ (where $\mathbb{R}$ is equipped with its σ-algebra of Borel sets) is measurable iff $A \in \mathcal{A}$, that is, A is measurable.

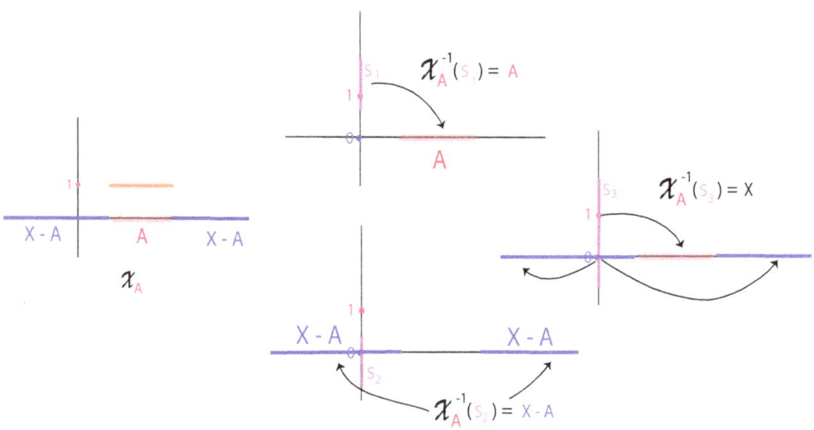

Fig. 5.1 The left figure illustrates $\chi_A \colon X \to \mathbb{R}$. If $S_1 \subset \mathbb{R}$ contains 1 but not 0, $\chi_A^{-1}(S_1) = A$. If $S_2 \subset \mathbb{R}$ contains 0 but not 1, $\chi_A^{-1}(S_1) = X - A$. Finally, if $S_3 \subset \mathbb{R}$ contains both 0 and 1, $\chi_A^{-1}(S_1) = A \cup (X - A) = X$.

In the theory of integration, all maps of interest will be measurable maps[1] $f \colon X \to F$ where $(X, \mathcal{A})$ is a measurable space, and $(F, \mathcal{B})$ is a measurable space such that either $F = \mathbb{R}$, or $F = \mathbb{C}$, or more generally F is a Banach space (a complete normed vector space over $\mathbb{R}$ or $\mathbb{C}$), and $\mathcal{B}$ is the Borel σ-algebra of open subsets of F. In this case various operations can be performed on functions $f \colon X \to F$.

Assume that F is a normed vector space over the field K, where $K = \mathbb{R}$ or $K = \mathbb{C}$, and that $f \colon X \to F$ is any function, not necessarily measurable.

[1] Actually, not quite in the most general case, but they will be equal to a measurable map a.e.

(1) Given any function $f\colon X \to F$, for any $\lambda \in K$, let $\lambda f\colon X \to F$ be the function given by

$$(\lambda f)(x) = \lambda f(x), \quad x \in X.$$

(2) Given any function $f\colon X \to F$, let $\|f\| : X \to \mathbb{R}_+$ be the function given by

$$\|f\|(x) = \|f(x)\|, \quad x \in X;$$

see Figure 5.2.

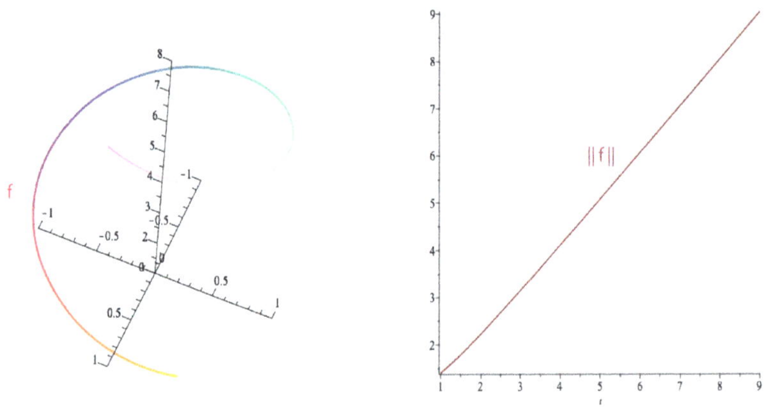

Fig. 5.2 Let $f\colon \mathbb{R} \to \mathbb{R}^3$ be $f(t) = (\sin t, \cos t, t)$, the graph of which is the space curve in the left figure. If we use the Euclidean norm on $\mathbb{R}^3$, $\|f\|(t) = \sqrt{t^2+1}$, the graph of which is shown in the right figure.

Beware that $\|f\|$ is *not* the norm of the function f, where $\|\ \|$ is the norm on some function space consisting of functions from X to F. Instead, $\|f\|$ is the *function* defined pointwise as $\|f(x)\|$ for every $x \in X$, where $\|f(x)\|$ is the norm of $f(x)$ in F. This notation is somewhat confusing but appears to be standard. Later on, we will equip our space of functions from X to F with a norm, but it will be denoted $\|\ \|_1$, or more generally $\|\ \|_p$, so there will be no risk of confusion.

(3) For any two functions $f\colon X \to \mathbb{R}$ and $g\colon X \to \mathbb{R}$, let $\sup(f,g)$ and $\inf(f,g)$ be the functions given by

$$\begin{aligned} \sup(f,g)(x) &= \max(f(x), g(x)), \quad x \in X, \\ \inf(f,g)(x) &= \min(f(x), g(x)), \quad x \in X; \end{aligned}$$

see Figure 5.3.

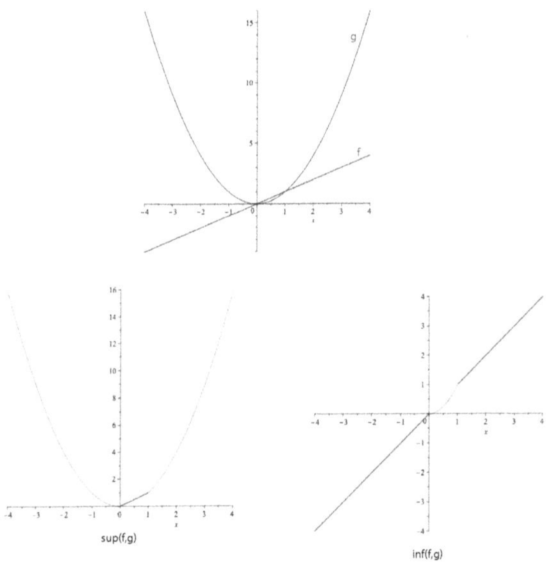

Fig. 5.3 Let $f\colon \mathbb{R} \to \mathbb{R}$ be $f(x) = x$, and let $g\colon \mathbb{R} \to \mathbb{R}$ be $g(x) = x^2$. The graph of $\sup(f, g)$ is the lower left figure, while the graph of $\inf(f, g)$ is the lower right figure.

(4) For any two functions $f\colon X \to \mathbb{R}$, let f^+ and f^- be the functions given by

$$f^+(x) = \begin{cases} 0 & \text{if } f(x) \leq 0 \\ f(x) & \text{if } f(x) > 0, \end{cases}$$

$$f^-(x) = \begin{cases} 0 & \text{if } f(x) \geq 0 \\ -f(x) & \text{if } f(x) < 0. \end{cases}$$

We also define $|f|$ as $|f| = f^+ + f^- = \sup(f, -f)$. Observe that $f = f^+ - f^-$. See Figures 5.4 through 5.6.

(5) For any two functions $f\colon X \to F$ and $g\colon X \to F$, let $f + g\colon X \to F$ be the function given by

$$(f+g)(x) = f(x) + g(x), \quad x \in X.$$

(6) For any two functions $f\colon X \to K$ and $g\colon X \to K$, where $K = \mathbb{R}$ of $K = \mathbb{C}$, let $fg\colon X \to K$ be the function given by

$$(fg)(x) = f(x)g(x), \quad x \in X.$$

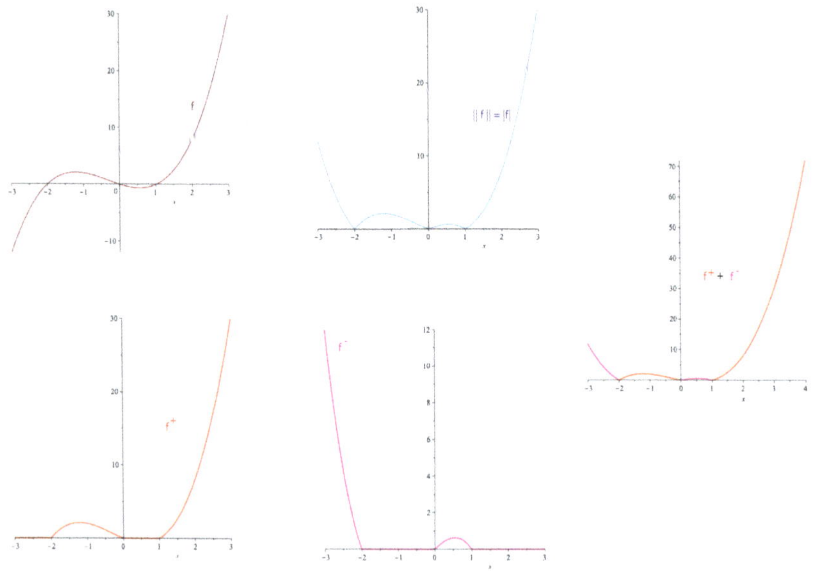

Fig. 5.4 Let $f\colon \mathbb{R} \to \mathbb{R}$ be $f(x) = x(x-1)(x+2)$. The lower figures illustrate f^+ and f^-, while the right figures illustrate the identity $|f| = f^+ + f^-$.

Definition 5.2. Let $(X, \mathcal{A})$ be a measurable space, and let $(F, \mathcal{B})$ be a measurable space such that $F = \mathbb{R}$, or $F = \mathbb{C}$, or more generally F is metric space (not necessarily complete), and $\mathcal{B}$ is the Borel σ-algebra of open subsets of F. The set of measurable maps $f\colon X \to F$ is denoted by $\mathcal{M}(X, \mathcal{A}, F)$.

The following technical result is needed; see Marle [Marle (1974)] (Proposition 2.1.10).

Proposition 5.4. *Let $(X, \mathcal{A})$ be any measurable space, and let $(F_1, \mathcal{B}_1)$, $(F_2, \mathcal{B}_2)$, and $(G, \mathcal{G})$ be three measurable spaces, where F_1, F_2, G are topological spaces, and $\mathcal{B}_1, \mathcal{B}_2, \mathcal{G}$ are their respective Borel σ-algebras. Let $h\colon F_1 \times F_2 \to G$ be a continuous map, and let $f_1\colon X \to F_1$ and $f_2\colon X \to F_2$ be two measurable maps. If the subspace topologies on $f_1(X) \subseteq F_1$ and $f_2(X) \subseteq F_2$ are second-countable (which means that they have a countable basis of open subsets), then $h \circ (f_1, f_2)\colon X \to G$ is measurable.*

Recall that a topological space E is separable if it contains a countable subset which is dense in E (see Definition A.42). If E is a metric space, then

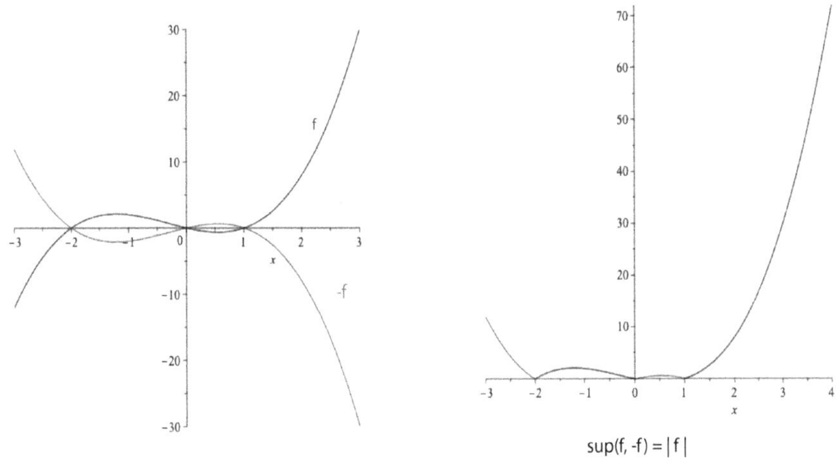

Fig. 5.5 Let $f\colon \mathbb{R} \to \mathbb{R}$ be $f(x) = x(x-1)(x+2)$. The right figure illustrates the identity $|f| = \sup(f, -f)$.

by Proposition A.46, the space E is separable if and only if it is second-countable. Using Proposition 5.4 we obtain the following important result stating various closure properties of $\mathcal{M}(X, \mathcal{A}, F)$; see Marle [Marle (1974)] (Corollary 2.1.11).

Proposition 5.5. *Let $(X, \mathcal{A})$ be any measurable space, and assume that F is a normed vector space over the field K, where $K = \mathbb{R}$ or $K = \mathbb{C}$. The following properties hold:*

(1) For any $f \in \mathcal{M}(X, \mathcal{A}, F)$ and any $\lambda \in K$, we have $\lambda f \in \mathcal{M}(X, \mathcal{A}, F)$.
(2) For any $f \in \mathcal{M}(X, \mathcal{A}, F)$, we have $\|f\| \in \mathcal{M}(X, \mathcal{A}, \mathbb{R})$.
(3) For any $f \in \mathcal{M}(X, \mathcal{A}, \mathbb{R})$ and any $g \in \mathcal{M}(X, \mathcal{A}, \mathbb{R})$, we have $\sup(f, g)$, $\inf(f, g), f^+, f^-, |f| \in \mathcal{M}(X, \mathcal{A}, \mathbb{R})$.
(4) For any $f \in \mathcal{M}(X, \mathcal{A}, F)$ and any $g \in \mathcal{M}(X, \mathcal{A}, F)$, if $f(X)$ and $g(X)$ are separable subsets of F, then $f + g \in \mathcal{M}(X, \mathcal{A}, F)$. In particular, if F is separable, then $\mathcal{M}(X, \mathcal{A}, F)$ is a vector space over K.
(5) For any $f \in \mathcal{M}(X, \mathcal{A}, K)$ and any $g \in \mathcal{M}(X, \mathcal{A}, K)$, we have $fg \in \mathcal{M}(X, \mathcal{A}, K)$. This implies that $\mathcal{M}(X, \mathcal{A}, K)$ is actually a K-algebra.

One will observe that in (4), if F is infinite-dimensional, the sum of two measurable maps may *not* be measurable. This is the first technical difficulty of the general theory of integration (with values in an infinite-dimensional vector space). As we will see, a second technical difficulty

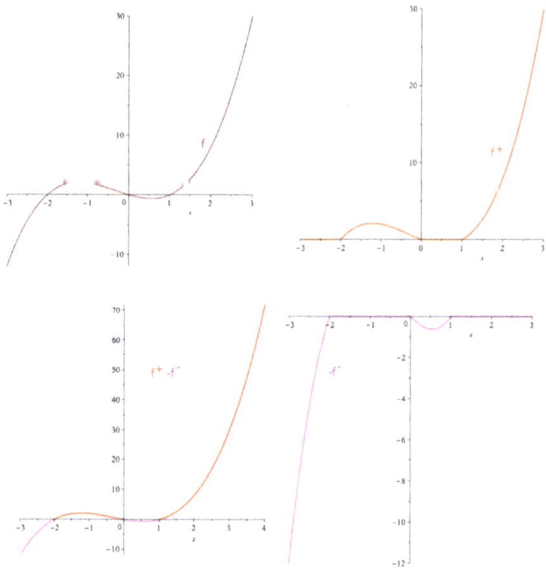

Fig. 5.6 Let $f\colon \mathbb{R} \to \mathbb{R}$ be $f(x) = x(x-1)(x+2)$. The lower left figure, when combined with the two right figures, illustrates the identity $f = f^+ - f^-$.

has to do with the approximation of a measurable map by step functions. Fortunately these technical difficulties can be overcome in a simple way.

The following important result shows that measurable maps behave better than continuous maps in terms of simple (pointwise) convergence.

Theorem 5.6. *Let $(X, \mathcal{A})$ and $(F, \mathcal{B})$ be two measurable spaces, where F is a metric space and $\mathcal{B}$ is the Borel σ-algebra on F. If $(f_n)_{n\geq 1}$ is a sequence of measurable maps $f_n \in \mathcal{M}(X, \mathcal{A}, F)$ which converges pointwise to a function $f\colon X \to F$, then $f \in \mathcal{M}(X, \mathcal{A}, F)$; that is, f is measurable.*

A proof of Theorem 5.6 can be found in Lang [Lang (1996)] (Chapter VI, Section 1, Property **M7**).

Our next goal is to generalize the notion of step function given in Definition 2.21 to the framework of measure spaces.

5.2 Step Maps on a Measurable Space

Let $(X, \mathcal{A})$ be a measurable space. The generalization of the notion of step map is obtained by replacing the intervals (a_i, a_{i+1}) by *arbitrary measurable sets*.

Definition 5.3. Let $(X, \mathcal{A})$ be a measurable space, and let F be any set. A function $f: X \to F$ is a *step map* (with respect to $\mathcal{A}$) if there is a finite partition $(A_1, \ldots, A_n)$ of X by pairwise disjoint nonempty subsets $A_i \in \mathcal{A}$ such that $X = \bigcup_{i=1}^n A_i$, and such that the restriction of f to each A_i is a constant function with some value $y_i \in F$. The partition $(A_1, \ldots, A_n)$ is said to be *adapted to* f; see Figure 5.7. The set of all step maps is denoted by $\mathcal{S}tep(X, \mathcal{A}, F)$.

Observe that every constant function is a step map, and that $f(X)$ is a finite subset of F. At this stage, no measure μ is involved, but for the theory of integration, we will have a measure space $(X, \mathcal{A}, \mu)$ and we will need to require each A_i for which $y_i \neq 0$ to have finite measure (this makes sense since in this case F is a vector space).

We gather some useful properties of step maps in the following proposition.

Proposition 5.7. *Let $(X, \mathcal{A})$ be a measurable space, and let F be any set.*

(1) For any σ-algebra $\mathcal{B}$ on F, every step map $\mathcal{S}tep(X, \mathcal{A}, F)$ is measurable.
(2) Let F_1, F_2, G be three sets, and let $h: F_1 \times F_2 \to G$ be any function. For any $f_1 \in \mathcal{S}tep(X, \mathcal{A}, F_1)$ and any $f_2 \in \mathcal{S}tep(X, \mathcal{A}, F_2)$, we have $h \circ (f_1, f_2) \in \mathcal{S}tep(X, \mathcal{A}, G)$.
(3) If $K = \mathbb{R}$ or $K = \mathbb{C}$, then $\mathcal{S}tep(X, \mathcal{A}, K)$ is a vector space and a ring under pointwise multiplication of functions. Thus, $\mathcal{S}tep(X, \mathcal{A}, K)$ is an algebra over K.
(4) If F is a vector space over K ($K = \mathbb{R}$ or $K = \mathbb{C}$), then $\mathcal{S}tep(X, \mathcal{A}, F)$ is a vector space over K and a module over $\mathcal{S}tep(X, \mathcal{A}, K)$, which means that if $f \in \mathcal{S}tep(X, \mathcal{A}, F)$ and $g \in \mathcal{S}tep(X, \mathcal{A}, K)$, then $gf \in \mathcal{S}tep(X, \mathcal{A}, F)$.
(5) If F is a normed vector space, and if $f \in \mathcal{S}tep(X, \mathcal{A}, F)$, then $\|f\| \in \mathcal{S}tep(X, \mathcal{A}, \mathbb{R})$.
(6) If $f \in \mathcal{S}tep(X, \mathcal{A}, \mathbb{R})$ and $g \in \mathcal{S}tep(X, \mathcal{A}, \mathbb{R})$, then we have $\sup(f, g)$, $\inf(f, g), f^+, f^-, |f| \in \mathcal{S}tep(X, \mathcal{A}, \mathbb{R})$.

Proposition 5.7 is proven in Marle [Marle (1974)] (Corollary 2.1.14).

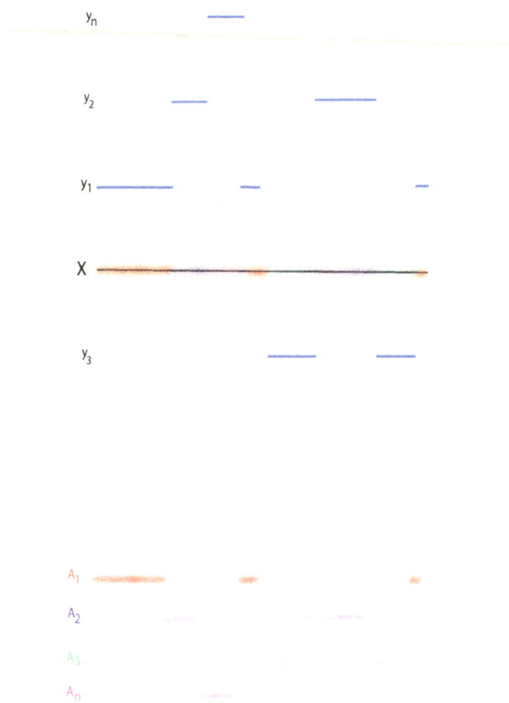

Fig. 5.7 Let $(X, \mathcal{A}) = (\mathbb{R}, \mathcal{B}(\mathbb{R}))$ and $F = \mathbb{R}$. A step map is shown in blue with values $\{y_i\}_{i=1}^n$. The partition $(A_1, \ldots, A_n)$ adapted to f is shown underneath the peach box.

Theorem 5.6 and Proposition 5.7 imply the following result.

Proposition 5.8. *Given a metric space F equipped with its σ-algebra of Borel sets, if a function $f\colon X \to F$ is the limit of a sequence $(f_n)_{n\geq 1}$ of step functions $f_n \in \mathcal{S}tep(X, \mathcal{A}, F)$ that converges pointwise, then the function $f\colon X \to F$ must be measurable.*

Unfortunately, in general, a measurable map $f\colon X \to F$ *may not* be the pointwise limit of a sequence of step maps if F has infinite dimension. For one thing, such a limit of steps maps has its image contained in the closure of a countable subset of F. This is the second technical difficulty of the general theory.

To overcome this second difficulty, we need to define a more refined notion of measurable map and of step map. We will do so shortly, but first we observe that if we only need to consider values in a finite-dimensional vector space, then there is no problem.

Proposition 5.9. *Let $(X, \mathcal{A})$ and $(F, \mathcal{B})$ be two measurable spaces, where F is a topological space and $\mathcal{B}$ is its Borel σ-algebra, and let $f: X \to F$ be a measurable map.*

(1) *If F is either a finite-dimensional vector space over $\mathbb{R}$ or $\mathbb{C}$, or $F = \overline{\mathbb{R}}_+$, then there is a sequence (f_n) of step maps $f_n \in \mathcal{S}tep(X, \mathcal{A}, F)$ that converges pointwise to f. If $F = \overline{\mathbb{R}}_+$, we may assume that the f_n take finite values.*
(2) *If $F = \mathbb{R}$ or $F = \overline{\mathbb{R}}_+$, and if $f \geq 0$, then we may assume that $f_n \geq 0$ and $f_n \leq f_{n+1}$ for all $n \geq 1$.*

A proof of Proposition 5.9 can be found in Lang [Lang (1996)] (Chapter VI, Section 1, Properties **M8** and **M9**).

5.3 μ-Step Maps

We explained in the previous sections that in general, the space $\mathcal{M}(X, \mathcal{A}, F)$ of measurable maps from X to F is not a vector space, and that a measurable map $f: X \to F$ may not be the pointwise limit of a sequence of step maps. This suggests modifying the notion of measurable map and the notion of step map to recover these properties. The second property is crucial in extending the notion of integral to more general functions.

So far, the space X was only a measurable space, but no measure was involved. The new ingredient is to define suitable notions of step maps and measurable maps relative to a *measure space* $(X, \mathcal{A}, \mu)$, where the measure μ plays a role.

The main trick is to relax the notion of pointwise convergence to pointwise convergence almost everywhere, and more generally, to consider that two functions are equivalent if they are equal almost everywhere (they differ on a null set). The plan is the following:

(1) Define the space $\mathcal{S}tep_\mu(X, \mathcal{A}, F)$ of *μ-step maps*.
(2) Define the space $\mathcal{M}_\mu(X, \mathcal{A}, F)$ of *μ-measurable maps*, where a μ-measurable map is the limit of a sequence (f_n) of μ-step maps $f_n \in \mathcal{S}tep_\mu(X, \mathcal{A}, F)$ converging pointwise almost everywhere.

(3) Prove that if F is a vector space, then $\mathcal{M}_\mu(X, \mathcal{A}, F)$ is a vector space.

Our presentation of the method that we just sketched follows Marle [Marle (1974)] and Lang [Lang (1996)] very closely. It is a generalization (with some simplifications) to functions with values in a Banach space of the approach followed by Halmos [Halmos (1978)]. The results that we state without proof are proved either in Marle [Marle (1974)] or in Lang [Lang (1996)].

Definition 5.4. Let $(X, \mathcal{A}, \mu)$ be a measure space, and let F be any vector space (over $\mathbb{R}$ or $\mathbb{C}$). A function $f : X \to F$ is a *μ-step map* if it is a step map, and if $\{x \in X \mid f(x) \neq 0\} \in \mathcal{A}$ and has finite measure; see Figure 5.8. The set of μ-step maps is denoted by $\mathcal{S}tep_\mu(X, \mathcal{A}, F)$.

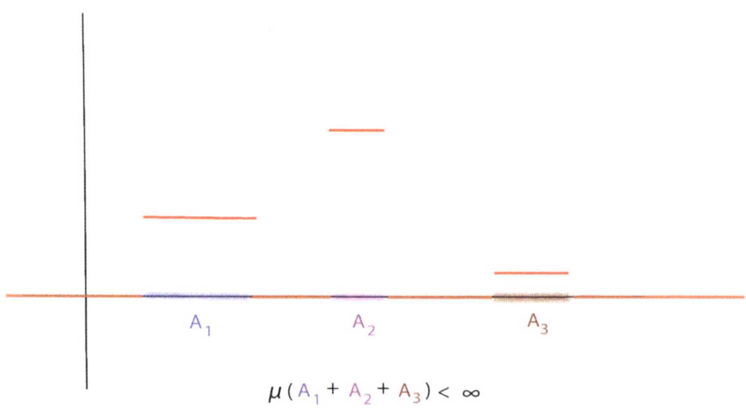

Fig. 5.8 Let $(X, \mathcal{A}) = (\mathbb{R}, \mathcal{B}(\mathbb{R}))$ and $F = \mathbb{R}$. A μ-step map is shown in red where $A_1 \cup A_2 \cup A_3 = \{x \in X \mid f(x) \neq 0\} \in \mathcal{A}$.

For technical reasons, it is useful to have the following equivalent characterization of a μ-step map.

Proposition 5.10. *Let $(X, \mathcal{A}, \mu)$ be a measure space, and let F be any vector space (over $\mathbb{R}$ or $\mathbb{C}$). A function $f : X \to F$ is a μ-step map iff there is a nonempty subset $A \in \mathcal{A}$ of finite measure such that f vanishes outside A, that is, $f(x) = 0$ for all $x \in X - A$, and if there is a finite partition $(A_1, \ldots, A_n)$ of A of subsets $A_i \in \mathcal{A}$ (nonempty pairwise disjoint subsets) such that the restriction of f to each A_i has a constant value y_i.*

Proof. Let f be a μ-step map, that is, a step map with respect to a partition $(A_1, \ldots, A_n)$ of X such that $\{x \in X \mid f(x) \neq 0\} \in \mathcal{A}$ and has finite measure. Then any A_i on which f has value $y_i \neq 0$ must have finite measure. If $f = 0$ on X, then pick A to be any A_i and the partition to be (A_i). Otherwise, let $J = \{j \in \{1, \ldots, n\} \mid f \neq 0 \text{ on } A_j\}$, and let $A = \bigcup_{j \in J} A_j$. Then, $(A_j)_{j \in J}$ is a partition of A with $A_j \in \mathcal{A}$, where A is a nonempty set of finite measure, and f vanishes on $X - A$; see Figure 5.9.

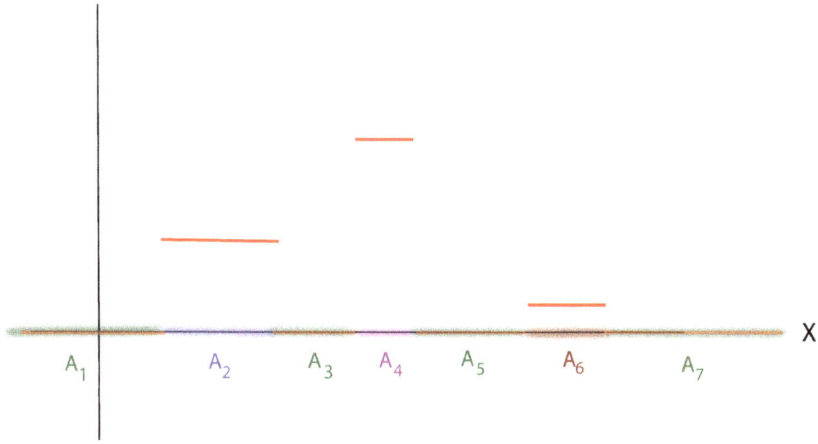

Fig. 5.9 Let $(X, \mathcal{A}) = (\mathbb{R}, \mathcal{B}(\mathbb{R}))$ and $F = \mathbb{R}$. A step map is shown in red with adapted partition $\mathbb{R} = \bigcup_{i=1}^{7} A_i$. To interpret this step map as a μ-step map, let $A = A_2 \cup A_4 \cup A_6$, where $\mu(A) < \infty$, $A = \{x \in X \mid f(x) \neq 0\} \in \mathcal{A}$. Then $f(x) = 0$ on $X - A$ where $X - A = A_1 \cup A_3 \cup A_5 \cup A_7$.

Conversely, since A has finite measure and since the A_i belongs to $\mathcal{A}$, each A_i has finite measure, so $\{x \in X \mid f(x) \neq 0\} \in \mathcal{A}$ is a set of finite measure. If $A = X$, then we already have a step map (as defined in Definition 5.3). Otherwise, $X - A \in \mathcal{A}$ and f vanishes on $X - A$, so $(A_1, \ldots, A_n, X - A)$ is partition of X, and f is a step map with respect to this partition; see Figure 5.10. $\square$

The condition that a μ-step map must vanish outside of a measurable set of finite measure is the measure-theoretic analog of the topological notion of compact support.

Proposition 5.10 suggests the following equivalent definition of a *μ-step map*.

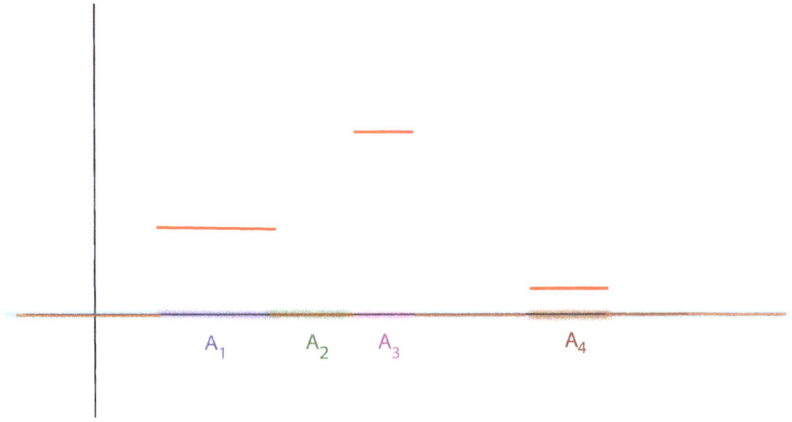

Fig. 5.10 Let $(X, \mathcal{A}) = (\mathbb{R}, \mathcal{B}(\mathbb{R}))$ and $F = \mathbb{R}$. A μ-step map is shown in red with $A = A_1 \cup A_2 \cup A_3 \cup A_4$. The turquoise set is $X - A$ and $(A_1, A_2, A_3, A_4, X - A)$ forms an adapted partition for the corresponding step map.

Definition 5.5. Let $(X, \mathcal{A}, \mu)$ be a measure space, and let F be any vector space (over $\mathbb{R}$ or $\mathbb{C}$). A function $f\colon X \to F$ is a *μ-step map* if there is a nonempty subset $A \in \mathcal{A}$ of finite measure such that f vanishes outside A, that is, $f(x) = 0$ for all $x \in X - A$, and if there is a finite partition $(A_1, \ldots, A_n)$ of A consisting of nonempty pairwise disjoint subsets in $\mathcal{A}$, such that the restriction of f to each A_i has a constant value y_i (possibly zero). The partition $(A_1, \ldots, A_n)$ of A is said to be *adapted to f*.

Technically, Definition 5.5 appears to be more convenient. Observe that a μ-step map can be expressed as a (necessarily finite) linear combination

$$f = \sum_{i=1}^{n} y_i \chi_{A_i},$$

for some $y_i \in F$ and for some nonempty pairwise disjoint measurable sets $A_i \in \mathcal{A}$ of finite measure, a concise and convenient representation.

Remark: The proof of Proposition 5.10 shows that if a μ-step function f is not identically zero, then we can find a subset A in $\mathcal{A}$ of finite measure, and a partition $(A_1, \ldots, A_n)$ of A of subsets in $\mathcal{A}$ such that the value of f on each A_i is nonzero, and f is zero outside of A. However, it turns out to be more convenient for certain proofs to allow f to be zero on some of the A_i, and this is why we allow this possibility in Definition 5.5.

Example 5.1. Consider the function $f: \mathbb{R} \to \mathbb{R}$, given by

$$f(x) = \begin{cases} 0 & \text{if } x < 0 \text{ or } x > 1 \\ 1 & \text{if } x \in [0, 1/2] - \mathbb{Q} \\ 0 & \text{if } x \in [0, 1/2] \cap \mathbb{Q} \\ 2 & \text{if } x \in [1/2, 1] - \mathbb{Q} \\ 0 & \text{if } x \in [1/2, 1] \cap \mathbb{Q}; \end{cases}$$

see Figure 5.11. If we let $A_1 = [0, 1/2] - \mathbb{Q}$, $A_2 = [0, 1/2] \cap \mathbb{Q}$, $A_3 = [1/2, 1] - \mathbb{Q}$, $A_4 = [1/2, 1] \cap \mathbb{Q}$, and $A = [0, 1]$, with the Lebesgue measure μ_L on $\mathbb{R}$, then A_1, A_2, A_3, A_4 are Lebesgue measurable, $\mu(A_1) = 1/2$, $\mu(A_2) = 0$, $\mu(A_3) = 1/2$, $\mu(A_3) = 0$, (A_1, A_2, A_3, A_4) is a partition of A, a set of measure 1. Thus f is a μ_L-step function.

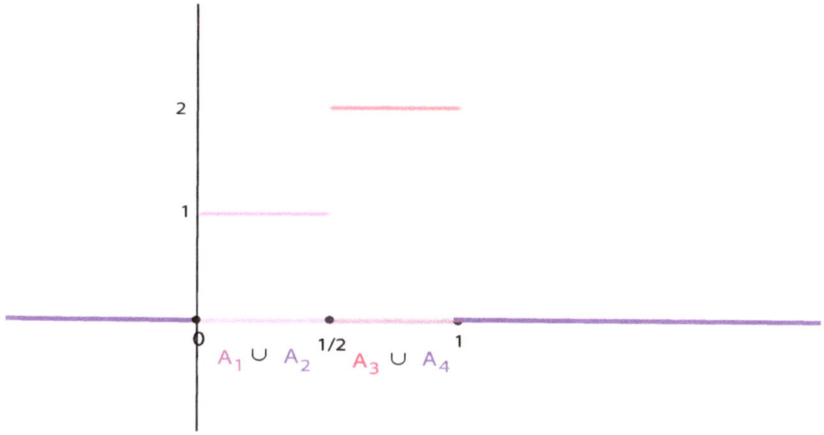

Fig. 5.11 The μ-step map of Example 5.1.

This example shows that a μ-step function can be very complicated, unlike the step functions of Definition 2.21.

Proposition 5.11. *Let $(X, \mathcal{A}, \mu)$ be a measure space, and let F be any vector space*

(1) *Let F_1, F_2, G be three Banach spaces over $\mathbb{R}$ or $\mathbb{C}$, and let $h: F_1 \times F_2 \to G$ be any function. If h satisfies $h(0, 0) = 0$, then for any $f_1 \in$*

$\mathcal{S}tep_\mu(X, \mathcal{A}, F_1)$ *and any* $f_2 \in \mathcal{S}tep_\mu(X, \mathcal{A}, F_2)$*, we have* $h \circ (f_1, f_2) \in \mathcal{S}tep_\mu(X, \mathcal{A}, G)$.

(2) *If* $K = \mathbb{R}$ *or* $\mathbb{C}$*, then* $\mathcal{S}tep_\mu(X, \mathcal{A}, K)$ *is a subspace of* $\mathcal{S}tep(X, \mathcal{A}, K)$*, and for any* $g \in \mathcal{S}tep(X, \mathcal{A}, K)$ *and any* $f \in \mathcal{S}tep_\mu(X, \mathcal{A}, K)$ *we have* $gf \in \mathcal{S}tep_\mu(X, \mathcal{A}, K)$*. Thus* $\mathcal{S}tep_\mu(X, \mathcal{A}, K)$ *is an ideal in* $\mathcal{S}tep(X, \mathcal{A}, K)$.

(3) *If* F *is a vector space over* K *(with* $K = \mathbb{R}$ *or* $K = \mathbb{C}$*), then* $\mathcal{S}tep_\mu(X, \mathcal{A}, F)$ *is a subspace of* $\mathcal{S}tep(X, \mathcal{A}, F)$ *and a module over* $\mathcal{S}tep(X, \mathcal{A}, K)$*, which means that if* $f \in \mathcal{S}tep_\mu(X, \mathcal{A}, F)$ *and* $g \in \mathcal{S}tep(X, \mathcal{A}, K)$*, then* $gf \in \mathcal{S}tep_\mu(X, \mathcal{A}, F)$.

(4) *If* F *is a normed vector space, and if* $f \in \mathcal{S}tep_\mu(X, \mathcal{A}, F)$*, then* $\|f\| \in \mathcal{S}tep_\mu(X, \mathcal{A}, \mathbb{R})$*. In fact, if* $f = \sum_{i=1}^n y_i \chi_{A_i}$*, then* $\|f\| = \sum_{i=1}^n \|y_i\| \chi_{A_i}$.

(5) *If* $f \in \mathcal{S}tep_\mu(X, \mathcal{A}, \mathbb{R})$ *and* $g \in \mathcal{S}tep_\mu(X, \mathcal{A}, \mathbb{R})$*, then* $\sup(f, g)$, $\inf(f, g), f^+, f^-, |f| \in \mathcal{S}tep_\mu(X, \mathcal{A}, \mathbb{R})$.

Proposition 5.11 is proven in Marle [Marle (1974)] (Proposition 2.2.3).

We now come to the crucial notion of μ-measurable map.

5.4 μ-Measurable Maps

Definition 5.6. Let $(X, \mathcal{A}, \mu)$ be a measure space, and let F be any vector space (over $\mathbb{R}$ or $\mathbb{C}$). A function $f \colon X \to F$ is a *μ-measurable* if there is a sequence $(f_n)_{n \geq 1}$ of μ-step maps $f_n \in \mathcal{S}tep_\mu(X, \mathcal{A}, F)$ which converges pointwise to f almost everywhere. See Figures 5.12 and 5.13. Recall that this means that there is a null set $Z \subseteq X$ such that for every $x \in X - Z$, the sequence $(f_n(x))$ converges to $f(x)$. The set of μ-measurable maps is denoted by $\mathcal{M}_\mu(X, \mathcal{A}, F)$.

Observe that a μ-measurable map is not necessarily measurable, so $\mathcal{M}_\mu(X, \mathcal{A}, F)$ is *not* a subspace of $\mathcal{M}(X, \mathcal{A}, F)$. However, we will see shortly that a μ-measurable map is equal to a measurable map almost everywhere, and this is good enough to construct the Lebesgue integral. The following proposition can be proved using Proposition 5.11 by passing to the limit (carefully).

Proposition 5.12. *Let* $(X, \mathcal{A}, \mu)$ *be a measure space, and let* F *be any vector space*

(1) *Let* F_1, F_2, G *be three Banach spaces over* $\mathbb{R}$ *or* $\mathbb{C}$*, and let* $h \colon F_1 \times F_2 \to G$ *be any function. If* h *satisfies* $h(0, 0) = 0$*, then for any*

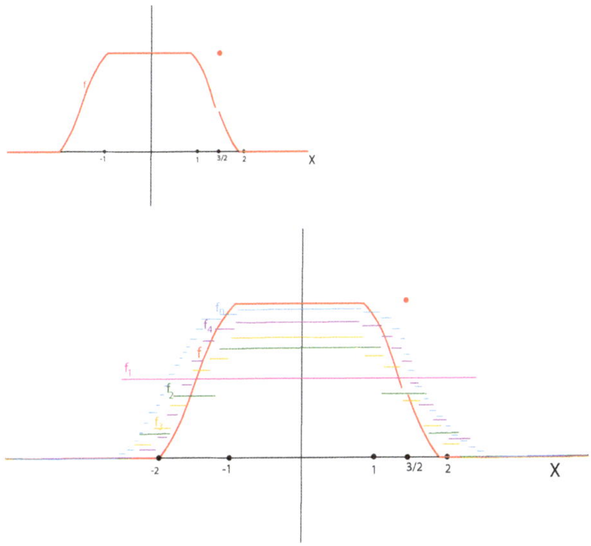

Fig. 5.12 Let $X = F = \mathbb{R}$. Assume μ is the Lebesgue measure on $\mathbb{R}$. The graph of the μ-measurable f is shown in the upper left corner. The middle figure illustrates the sequence $(f_n)_{n\geq 1}$ of μ-step maps $f_n \in \mathcal{S}tep_\mu(X, \mathcal{A}, F)$ which converges pointwise to f almost everywhere.

$f_1 \in \mathcal{M}_\mu(X, \mathcal{A}, F_1)$ and any $f_2 \in \mathcal{M}_\mu(X, \mathcal{A}, F_2)$, we have $h \circ (f_1, f_2) \in \mathcal{M}_\mu(X, \mathcal{A}, G)$.

(2) *If $K = \mathbb{R}$ or $\mathbb{C}$, then $\mathcal{M}_\mu(X, \mathcal{A}, K)$ is a vector space, and for all $f, g \in \mathcal{M}_\mu(X, \mathcal{A}, K)$ we have $fg \in \mathcal{M}_\mu(X, \mathcal{A}, K)$. Thus $\mathcal{M}_\mu(X, \mathcal{A}, K)$ is an algebra over K. For any $g \in \mathcal{M}(X, \mathcal{A}, K)$ and any $f \in \mathcal{M}_\mu(X, \mathcal{A}, K)$ we have $gf \in \mathcal{M}_\mu(X, \mathcal{A}, K)$.*

(3) *If F is a vector space over K ($K = \mathbb{R}$ or $\mathbb{C}$), then $\mathcal{M}_\mu(X, \mathcal{A}, F)$ is a vector space over K and a module over $\mathcal{M}(X, \mathcal{A}, K)$, which means that if $f \in \mathcal{M}_\mu(X, \mathcal{A}, F)$ and $g \in \mathcal{M}(X, \mathcal{A}, K)$, then $gf \in \mathcal{M}_\mu(X, \mathcal{A}, F)$. The space $\mathcal{M}_\mu(X, \mathcal{A}, F)$ is also a module over $\mathcal{M}_\mu(X, \mathcal{A}, K)$.*

(4) *If F is a normed vector space, and if $f \in \mathcal{M}_\mu(X, \mathcal{A}, F)$, then $\|f\| \in \mathcal{M}_\mu(X, \mathcal{A}, \mathbb{R})$.*

(5) *If $f \in \mathcal{M}_\mu(X, \mathcal{A}, \mathbb{R})$ and $g \in \mathcal{M}_\mu(X, \mathcal{A}, \mathbb{R})$, then we have $\sup(f, g)$, $\inf(f, g), f^+, f^-, |f| \in \mathcal{M}_\mu(X, \mathcal{A}, \mathbb{R})$.*

The following result gives a characterization of a μ-measurable map

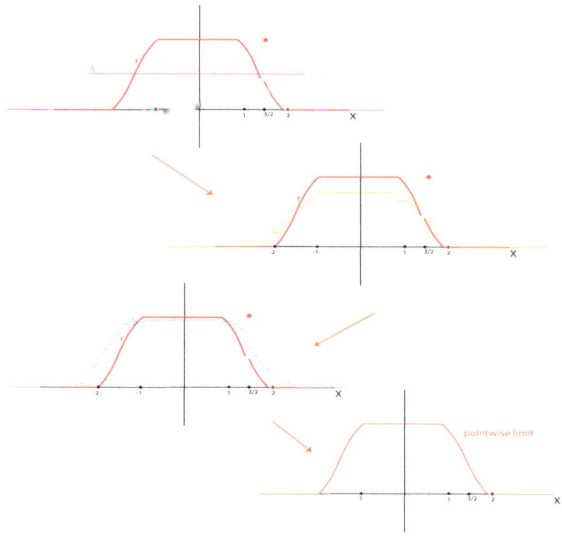

Fig. 5.13 A more detailed look at the pointwise convergence of converges the $(f_n)_{n\geq 1}$. The pointwise limit only differs from f on a set of measure zero, namely $\{3/2\}$.

which shows that a μ-measurable map is equal to a measurable map almost everywhere, and that there are strong countability restrictions on its domain and its range.

Proposition 5.13. *Let $(X, \mathcal{A}, \mu)$ be a measure space, and let F be any Banach space. A function $f\colon X \to F$ is μ-mesurable iff there is a null set Z such that the following three conditions hold:*

(1) There is a measurable map $g \in \mathcal{M}(X, \mathcal{A}, F)$ such that f and g are equal on $X - Z$.
(2) The function f vanishes outside of a measurable σ-finite subset of X (recall Definition 4.10).
(3) The image $f(X - Z)$ is separable in F, which means that $f(X - Z)$ contains a countable dense subset.

In particular, if μ is σ-finite and if F is separable, then $f\colon X \to F$ is μ-measurable iff f is measurable almost everywhere (there is a null set Z such that f agrees with a measurable map on $X - Z$).

A proof of Proposition 5.13 can be found in Lang [Lang (1996)] (Chapter VI, Section 1, Property **M11**). Again, Condition (2) is a measure-theoretic analog of the notion of compact support.

The version of Theorem 5.6 for μ-measurable maps is stated below.

Theorem 5.14. *Let $(X, \mathcal{A}, \mu)$ be a measure space and let $(F, \mathcal{B})$ be a measurable space, where F is a metric space and $\mathcal{B}$ is the Borel σ-algebra on F. If $(f_n)_{n\geq 1}$ is a sequence of μ-measurable maps $f_n \in \mathcal{M}_\mu(X, \mathcal{A}, F)$ which converges pointwise to a function $f\colon X \to F$, then $f \in \mathcal{M}_\mu(X, \mathcal{A}, F)$; that is, f is μ-measurable.*

A proof of Theorem 5.6 can be found in Lang [Lang (1996)] (Chapter VI, Section 1, Property **M12**).

We are now ready construct a very general version of the integral. The original construction was first proposed by Lebesgue, but the more general version presented here applying to functions with values in a Banach space is due to Bochner and Dunford.

5.5 The Integral of μ-Step Maps

Let $(X, \mathcal{A}, \mu)$ be a measure space and let $(F, \mathcal{B})$ be a measurable space consisting of a Banach space F and its Borel σ-algebra $\mathcal{B}$. There is an "obvious" definition of the integral of a μ-step map $f = \sum_{i=1}^n y_i \chi_{A_i}$ (where $y_i \in F$), namely

$$I(f) = \int f d\mu = \sum_{i=1}^n \mu(A_i) y_i.$$

Since by definition the A_i belong to $\mathcal{A}$ and have finite measure, the linear combination $\sum_{i=1}^n \mu(A_i) y_i$ is a well-defined vector in F. The only problem is that $I(f)$ seems to depend on the subset A (and its partition) chosen to express f, but it is easy to show that $I(f)$ is independent of the representation of f. Then it is easy to show that $I\colon \mathcal{S}tep_\mu(X, \mathcal{A}, F) \to F$ is a linear map. Furthermore, by Proposition 5.12, we have $\|f\| \in \mathcal{S}tep_\mu(X, \mathcal{A}, \mathbb{R})$, so we can define

$$N_1(f) = \int \|f\|\, d\mu,$$

and we have $\left\|\int f d\mu\right\| \leq \int \|f\|\, d\mu = N_1(f)$.

It turns out that N_1 satisfies all the axioms of a norm, except that $N_1(f) = 0$ does not necessarily imply that $f = 0$. We say that N_1 is a *semi-norm*, see Definition A.3. Fortunately, for any $f \in \mathcal{S}tep_\mu(X, \mathcal{A}, F)$, we have $N_1(f) = 0$ iff $f = 0$, except on a subset of measure zero.

We can define the notion of N_1-*Cauchy sequence* of a sequence (f_n) of functions $f_n \in \mathcal{S}tep_\mu(X, \mathcal{A}, F)$ as follows: for all $\epsilon > 0$, there is some $N > 0$, such that for all $m, n \geq N$, we have $N_1(f_m - f_n) < \epsilon$. We can also define the notion of N_1-*convergence* of a sequence (f_n) of functions $f_n \in \mathcal{S}tep_\mu(X, \mathcal{A}, F)$ to a limit $f \in \mathcal{S}tep_\mu(X, \mathcal{A}, F)$ as follows: for all $\epsilon > 0$, there is some $N > 0$, such that for all $n \geq N$, we have $N_1(f - f_n) < \epsilon$. A convergent N_1-sequence does not necessarily have a unique limit, but we will see that any two limits are equal a.e.

The problem is that an N_1-Cauchy sequence *may not* have a limit in $\mathcal{S}tep_\mu(X, \mathcal{A}, F)$. Thus we are led to completing $\mathcal{S}tep_\mu(X, \mathcal{A}, F)$ with respect to the semi-norm N_1. This can be done and we obtain a vector space $\mathcal{L}_\mu(X, \mathcal{A}, F)$ which is a subspace of $\mathcal{M}_\mu(X, \mathcal{A}, F)$. The integral map I and the semi-norm N_1 can be extended to $\mathcal{L}_\mu(X, \mathcal{A}, F)$ as a semi-norm denoted $\| \, \|_1$, the space $\mathcal{L}_\mu(X, \mathcal{A}, F)$ is Cauchy-complete with respect to the semi-norm $\| \, \|_1$, and $\mathcal{S}tep_\mu(X, \mathcal{A}, F)$ is dense in $\mathcal{L}_\mu(X, \mathcal{A}, F)$ with respect to the semi-norm $\| \, \|_1$. This situation is schematically illustrated in Figure 5.14.

It also turns out that the subspace $\mathcal{SN}$ of $\mathcal{S}tep_\mu(X, \mathcal{A}, F)$ consisting of all functions f such that $N_1(f) = 0$ is the set of functions in $\mathcal{S}tep_\mu(X, \mathcal{A}, F)$ that are equal to 0 a.e. Similarly, the subspace $\mathcal{N}$ of $\mathcal{L}_\mu(X, \mathcal{A}, F)$ consisting of all functions f such that $\|f\|_1 = 0$ is the set of functions in $\mathcal{L}_\mu(X, \mathcal{A}, F)$ that are equal to 0 a.e. Thus, we can form the quotients spaces $\mathrm{Step}_\mu(X, \mathcal{A}, F) = \mathcal{S}tep_\mu(X, \mathcal{A}, F)/\mathcal{SN}$ and $\mathrm{L}_\mu(X, \mathcal{A}, F) = \mathcal{L}_\mu(X, \mathcal{A}, F)/\mathcal{N}$. In $\mathrm{Step}_\mu(X, \mathcal{A}, F)$ and in $\mathrm{L}_\mu(X, \mathcal{A}, F)$ the semi-norm $\| \, \|_1$ is really a norm, and $\mathrm{L}_\mu(X, \mathcal{A}, F)$ is the completion of $\mathrm{Step}_\mu(X, \mathcal{A}, F)$.

Theoretically, we could define $\mathrm{L}_\mu(X, \mathcal{A}, F)$ directly as the Cauchy completion (see Theorem A.62 and Theorem A.72) of $\mathrm{Step}_\mu(X, \mathcal{A}, F)$, but we obtain equivalence classes of Cauchy sequences of equivalence classes of functions in $\mathcal{S}tep_\mu(X, \mathcal{A}, F)$, which are not easily interpretable as functions. The same space $\mathrm{L}_\mu(X, \mathcal{A}, F)$ is obtained, see the diagram below.

$$
\begin{array}{ccc}
\mathcal{S}tep_\mu(X, \mathcal{A}, F) & \xrightarrow{\textit{completion}} & \mathcal{L}_\mu(X, \mathcal{A}, F) \\
{\scriptstyle \textit{quotient}} \big\downarrow & & \big\downarrow {\scriptstyle \textit{quotient}} \\
\mathrm{Step}_\mu(X, \mathcal{A}, F) & \xrightarrow[\textit{completion}]{} & \mathrm{L}_\mu(X, \mathcal{A}, F).
\end{array}
$$

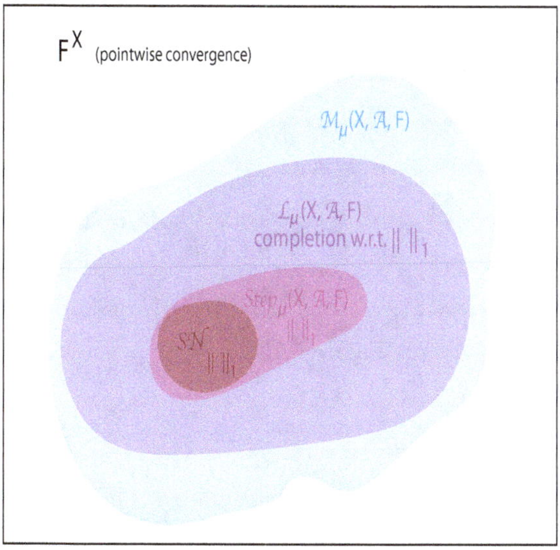

Fig. 5.14 Let $(X, \mathcal{A}, \mu)$ be a measure space and $(F, \mathcal{B})$ be a Banach space with the Borel σ-algbera. The completion of $\mathcal{S}tep_\mu(X, \mathcal{A}, F)$ with respect to the semi-norm $\| \ \|_1$ is $\mathcal{L}_\mu(X, \mathcal{A}, F)$.

The construction that we alluded to, although involving some extra work, yields a very clear description of these equivalence classes in terms of functions (in $\mathcal{L}_\mu(X, \mathcal{A}, F)$). The completeness of $\mathrm{L}_\mu(X, \mathcal{A}, F)$ (under the $\| \ \|_1$-norm) is also immediately obtained.

As in the previous section the results that we state without proof are proved either in Marle [Marle (1974)] or in Lang [Lang (1996)].

We now return to the definition of the integral of a μ-step maps.

Proposition 5.15. *Let $(X, \mathcal{A}, \mu)$ be a measure space and let $(F, \mathcal{B})$ be a measurable space, with F a Banach space and $\mathcal{B}$ its Borel σ-algebra. For any μ-step map $f \in \mathcal{S}tep_\mu(X, \mathcal{A}, F)$, for any two partitions $(A_1, \ldots, A_m)$ and $(B_1, \ldots, B_n)$ adapted to f, so that $f = \sum_{i=1}^m y_i \chi_{A_i} = \sum_{j=1}^n z_j \chi_{B_j}$, we have*

$$\sum_{i=1}^m \mu(A_i) y_i = \sum_{j=1}^n \mu(B_j) z_j.$$

Proposition 5.15 justifies the following definition.

Definition 5.7. Let $(X, \mathcal{A}, \mu)$ be a measure space and let $(F, \mathcal{B})$ be a measurable space, with F a Banach space and $\mathcal{B}$ its Borel σ-algebra. For any μ-step map $f \in \mathcal{S}tep_\mu(X, \mathcal{A}, F)$, the common value

$$\int f d\mu$$

of the expression

$$I(f) = \sum_{i=1}^{n} \mu(A_i) y_i$$

for any partition $(A_1, \ldots, A_n)$ adapted to f is called the *integral of f (relative to the measure μ)*; see Figure 5.15.[2]

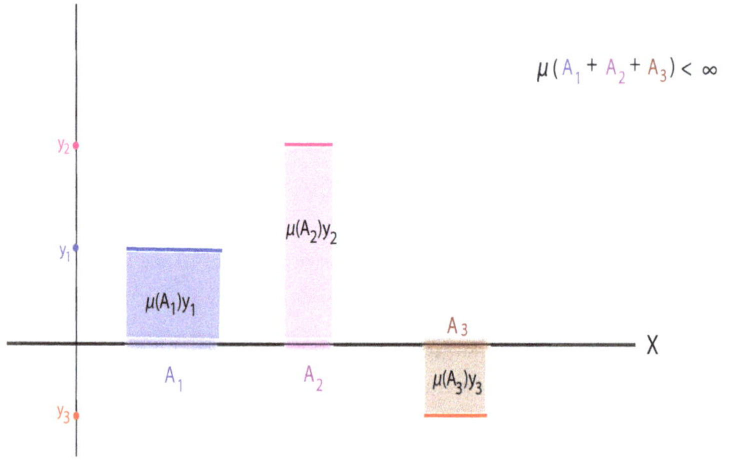

Fig. 5.15 Let $(X, \mathcal{A}) = (\mathbb{R}, \mathcal{B}(\mathbb{R}))$ and $F = \mathbb{R}$. The integral of the μ-step map f is the signed area of the pastel "boxes".

Recall that if the μ-step map f is expressed as $f = \sum_{i=1}^{n} y_i \chi_{A_i}$, then the μ-step map $\|f\|$ is expressed as $\|f\| = \sum_{i=1}^{n} \|y_i\| \chi_{A_i}$.

Definition 5.8. We define the *semi-norm* $N_1(f)$ of the μ-step map $f = \sum_{i=1}^{n} y_i \chi_{A_i}$ as

$$N_1(f) = \int \|f\| \, d\mu = \sum_{i=1}^{n} \mu(A_i) \|y_i\| .$$

[2]This integral is usually called the *Lebesgue integral* or *Bochner integral*. A more appropriate name might be the *Bochner–Dunford integral*.

For any measurable subset $E \in \mathcal{A}$, since $\chi_E f \in \mathcal{S}tep_\mu(X, \mathcal{A}, F)$, we let

$$\int_E f d\mu = \int \chi_E f d\mu.$$

For simplicity of notation, we often write $\int_E f$ instead of $\int_E f d\mu$, and if $E = X$, we write $\int f$ instead of $\int f d\mu$.

We stress that the integral $\int f d\mu$ or $\int_E f d\mu$ is *always finite*; that is, an element of F, but not ∞. This is in contrast with the approach where the integral of a step function may have the value $+\infty$, as in Rudin [Rudin (1987)] (Chapter 1). At some later stage, in defining the space $\mathrm{L}^1(X, \mathcal{A}, F)$, it is necessary to require the integral to be finite anyway. We find the approach where the integral is finite in the first place less confusing. It also yields a more explicit definition of $\mathrm{L}^1(X, \mathcal{A}, F)$.

Example 5.2. The special case in which X is a countable set, $\mathcal{A} = 2^X$, μ is the counting measure defined in Example 4.3, and $F = \mathbb{C}$ is of particular interest. Say $X = \mathbb{N}$. A μ-step function is of the form

$$f = \sum_{i=1}^{n} y_i \chi_{A_i},$$

where A_i must be a finite subset of $\mathbb{N}$, and $y_i \in \mathbb{C}$. By definition of μ, we have $\mu(A_i) = |A_i|$, so

$$\int f d\mu = \sum_{i=1}^{n} y_i |A_i|.$$

But f is the function with finite support $A = \bigcup_{i=1}^{n} A_i$, such that $f(j) = y_i$ for all $j \in A_i$, and $f(j) = 0$ for all $j \notin A$, so

$$\int f d\mu = \sum_{j \in A} f(j) = \sum_{j \in \mathbb{N}} f(j),$$

the sum of the (finite) sequence $(f(j))_{j \in \mathbb{N}}$. Similarly, if $X = \mathbb{Z}$, then for any sequence $(f_j)_{j \in \mathbb{Z}}$ with only finitely many nonzero entries,

$$\int f d\mu = \sum_{j \in \mathbb{Z}} f(j).$$

Example 5.3. Recall from Example 4.7 that for any $a \in X$, the Dirac measure δ_a is defined such that for any $A \subseteq X$,

$$\delta_a(A) = \begin{cases} 1 & \text{if } a \in A \\ 0 & \text{if } a \notin A. \end{cases}$$

Here the σ-algebra is $\mathcal{A} = 2^X$. Then it is easy to check that for any μ-step function

$$f = \sum_{i=1}^{n} y_i \chi_{A_i},$$

we have

$$\int f\, d\delta_a = f(a).$$

So $f(a) = y_i$ iff $a \in A_i$, and $f(a) = 0$ otherwise.

Here are some of the main properties of the integral.

Proposition 5.16. *Let $(X, \mathcal{A}, \mu)$ be a measure space and let $(F, \mathcal{B})$ be a measurable space, with F a Banach space and $\mathcal{B}$ its Borel σ-algebra. The following properties hold:*

(1) The integral map $\int \colon \mathcal{S}tep_\mu(X, \mathcal{A}, F) \to F$ is a linear map.

(2) If A and B are any two disjoint measurable subsets, then

$$\int_{A \cup B} f d\mu = \int_A f d\mu + \int_B f d\mu.$$

(3) For any map $f \in \mathcal{S}tep_\mu(X, \mathcal{A}, F)$, we have $\|f\| \in \mathcal{S}tep_\mu(X, \mathcal{A}, \mathbb{R})$, and

$$\left\| \int f d\mu \right\| \leq \int \|f\|\, d\mu = N_1(f).$$

We also have

$$\int \|f\|\, d\mu \leq \mu(\{x \in X \mid f(x) \neq 0\}) \|f\|_\infty .$$

(4) For any two maps $f, g \in \mathcal{S}tep_\mu(X, \mathcal{A}, F)$, if $f = g$ a.e., then $\int f d\mu = \int g d\mu$.

(5) For any two maps $f, g \in \mathcal{S}tep_\mu(X, \mathcal{A}, \mathbb{R})$, if $f \leq g$ a.e., then $\int f d\mu \leq \int g d\mu$. In particular, if $f \geq 0$ a.e., then $\int f d\mu \geq 0$.

(6) N_1 is a semi-norm on $\mathcal{S}tep_\mu(X, \mathcal{A}, F)$. Furthermore, for any $f \in \mathcal{S}tep_\mu(X, \mathcal{A}, F)$, we have $N_1(f) = 0$ iff $f = 0$, except on a subset of measure zero.

(7) If F_1 and F_2 are two Banach spaces over $\mathbb{R}$ or $\mathbb{C}$, and if $h \colon F_1 \to F_2$ is a continuous linear map, then for any $f \in \mathcal{S}tep_\mu(X, \mathcal{A}, F_1)$, we have $h \circ f \in \mathcal{S}tep_\mu(X, \mathcal{A}, F_2)$, and

$$\int (h \circ f) d\mu = h\left(\int f d\mu\right).$$

If $F = \mathbb{C}$, the above property holds for any semi-linear map.

Proof. We prove (3) and (6), leaving the other properties as exercises.

(3) If $f = \sum_{i=1}^n y_i \chi_{A_i}$ then $\int f d\mu = \sum_{i=1}^n \mu(A_i) y_i$. We also have $\|f\| = \sum_{i=1}^n \|y_i\| \chi_{A_i}$ and $\int \|f\| \, d\mu = \sum_{i=1}^n \mu(A_i) \|y_i\|$. It follows that

$$\left\| \int f d\mu \right\| = \left\| \sum_{i=1}^n \mu(A_i) y_i \right\| \leq \sum_{i=1}^n \|\mu(A_i) y_i\| = \sum_{i=1}^n \mu(A_i) \|y_i\| = \int \|f\| \, d\mu.$$

We also have

$$\begin{aligned} \int \|f\| \, d\mu &= \sum_{i=1}^n \mu(A_i) \|y_i\| \leq \mu(\{x \in X \mid f(x) \neq 0\}) \max_{1 \leq i \leq n} \|y_i\| \\ &= \mu(\{x \in X \mid f(x) \neq 0\}) \|f\|_\infty . \end{aligned}$$

(6) Since by (1) the integral is linear, we have

$$N_1(\lambda f) = \int \|\lambda f\| \, d\mu = \int |\lambda| \, \|f\| \, d\mu = |\lambda| \int \|f\| \, d\mu = |\lambda| N_1(f).$$

Since $\|(f+g)(x)\| \leq \|f(x)\| + \|g(x)\|$ for all $x \in X$, by (5) we have

$$N_1(f+g) = \int \|f+g\| \, d\mu \leq \int \|f\| \, d\mu + \int \|g\| \, d\mu = N_1(f) + N_1(g).$$

Assume that $N_1(f) = 0$, which means that $\int \|f\| \, d\mu = 0$. Since f is a μ-step function, we can write

$$f = \sum_{i=1}^n y_i \chi_{A_i},$$

for a finite sequence $(A_1, \ldots, A_n)$ of nonempty pairwise disjoint subsets $A_i \in \mathcal{A}$ of finite measure. Since

$$\|f\| = \sum_{i=1}^n \|y_i\| \chi_{A_i},$$

so

$$N_1(f) = \int \|f\| \, d\mu = \sum_{i=1}^n \|y_i\| \mu(A_i) = 0.$$

Since $\|y_i\| \geq 0$ and $\mu(A_i) \geq 0$, the following must hold:

$$\text{if } \mu(A_i) \neq 0, \text{ then } \|y_i\| = 0, \text{ that is } y_i = 0.$$
$$\text{if } y_i \neq 0, \text{ that is } \|y_i\| \neq 0, \text{ then } \mu(A_i) = 0.$$

Consequently

$$\{x \in X \mid f(x) \neq 0\} = \bigcup_{i \in I} A_i, \quad \text{with} \quad I = \{i \mid 1 \leq i \leq n \mid y_i \neq 0\},$$

where $\bigcup_{i \in I} A_i \in \mathcal{A}$ is a set of measure 0, since $i \in I$ implies that $\mu(A_i) = 0$. ∎

By Proposition 5.16(6), the set

$$\begin{aligned}\mathcal{SN} &= \{f \in \mathcal{S}tep_\mu(X, \mathcal{A}, F) \mid N_1(f) = 0\}\\ &= \{f \in \mathcal{S}tep_\mu(X, \mathcal{A}, F) \mid f = 0 \text{ a.e.}\}\end{aligned}$$

is a subspace of $\mathcal{S}tep_\mu(X, \mathcal{A}, F)$.

Definition 5.9. Let $\mathrm{Step}_\mu(X, \mathcal{A}, F)$ be the quotient space $\mathcal{S}tep_\mu(X, \mathcal{A}, F)/\mathcal{SN}$.

For every equivalence class $\mathbf{f} \in \mathrm{Step}_\mu(X, \mathcal{A}, F)$, we can define

$$\int \mathbf{f} d\mu = \int f d\mu$$

for any function $f \in \mathcal{S}tep_\mu(X, \mathcal{A}, F)$ in the equivalence class of $\mathbf{f}$, because if $f = g$ a.e., then $\int f d\mu = \int g d\mu$, so $\int \mathbf{f} d\mu$ does not depend on the representative chosen in the equivalence class $\mathbf{f}$. Similarly, we define $N_1(\mathbf{f})$ by

$$N_1(\mathbf{f}) = N_1(f) = \int \|f\| \, d\mu,$$

for any function $f \in \mathcal{S}tep_\mu(X, \mathcal{A}, F)$ in the equivalence class of $\mathbf{f}$. Again if $f = g$ a.e., then $\|f\| = \|g\|$ a.e., so $N_1(f) = N_1(g)$, which means that $N_1(\mathbf{f})$ is well defined. It is immediately verified that N_1 is a semi-norm, and in fact a norm, since $N_1(\mathbf{f}) = 0$ iff $N_1(f) = 0$ for any representative $f \in \mathcal{S}tep_\mu(X, \mathcal{A}, F)$ in the equivalence class $\mathbf{f}$ iff $f = 0$ a.e., which means that $\mathbf{f} = 0$. Therefore, $(\mathrm{Step}_\mu(X, \mathcal{A}, F), N_1)$ is a normed vector space. It is easy to see that the inequality

$$\left\| \int \mathbf{f} d\mu \right\| \leq \int \|\mathbf{f}\| \, d\mu = N_1(\mathbf{f})$$

holds, which shows that the map $\int : \mathrm{Step}_\mu(X, \mathcal{A}, F) \to F$ is continuous (in fact, uniformly continuous). The space $(\mathrm{Step}_\mu(X, \mathcal{A}, F), N_1)$ is not complete, so we can apply Theorem A.72) to form its completion $\mathrm{L}_\mu(X, \mathcal{A}, F)$ and extend the map $\int$ to it. Theoretically we have achieved our goal of defining a notion of integral on a normed vector space $\mathrm{L}_\mu(X, \mathcal{A}, F)$ which is complete and in which $\mathrm{Step}_\mu(X, \mathcal{A}, F)$ is dense, but the elements in this abstract completion are equivalence classes of Cauchy sequences, and are not easily identifiable with functions.

We will follow a different path, still very much inspired by the completion method involving Cauchy sequences, the twist being that we consider Cauchy sequences whose limit is known ahead of time, but where we use pointwise convergence *almost everywhere*, instead of pointwise convergence.

5.6 Integrable Functions; the Spaces $\mathcal{L}_\mu(X, \mathcal{A}, F)$ and $\mathrm{L}_\mu(X, \mathcal{A}, F)$

In this section we construct the completion $\mathcal{L}_\mu(X, \mathcal{A}, F)$ of the vector space $\mathcal{S}tep_\mu(X, \mathcal{A}, F)$ equipped with the semi-norm N_1, and construct the integral of a function in $\mathcal{L}_\mu(X, \mathcal{A}, F)$. The semi-norm N_1 is extended to $\mathcal{L}_\mu(X, \mathcal{A}, F)$ as a semi-norm $\| \|_1$ called the L^1-semi-norm, and we find that the space of functions such that $\|f\|_1 = 0$ is the set $\mathcal{N}$ of functions in $\mathcal{L}_\mu(X, \mathcal{A}, F)$ that are zero a.e. Then we define the quotient space $\mathrm{L}_\mu(X, \mathcal{A}, F) = \mathcal{L}_\mu(X, \mathcal{A}, F)/\mathcal{N}$. The space $\mathrm{L}_\mu(X, \mathcal{A}, F)$ is the completion of $\mathrm{Step}_\mu(X, \mathcal{A}, F)$; this is one of the most important results of this section (the Fischer–Riesz theorem).

As in the previous section the results that we state without proof are proved either in Marle [Marle (1974)] or in Lang [Lang (1996)].

Recall the following definitions.

Definition 5.10. A sequence (f_n) of functions $f_n \in \mathcal{S}tep_\mu(X, \mathcal{A}, F)$ is a *N_1-Cauchy sequence* if for every $\epsilon > 0$, there is some $N > 0$, such that for all $m, n \geq N$, we have $N_1(f_m - f_n) < \epsilon$, where $N_1(f_m - f_n) = \int \|f_m - f_n\| \, d\mu$. A sequence (f_n) of maps $f_n \in \mathcal{S}tep_\mu(X, \mathcal{A}, F)$ *converges pointwise almost everywhere* to a limit $f \colon X \to F$ if there is a null set Z such that for every $x \in X - Z$, for every $\epsilon > 0$, there is some $N > 0$, such that $\|f(x) - f_n(x)\| < \epsilon$ for all $n \geq N$.

We define the space $\mathcal{L}_\mu(X, \mathcal{A}, F)$ as follows.

Definition 5.11. Let $(X, \mathcal{A}, \mu)$ be a measure space and let $(F, \mathcal{B})$ be a measurable space, with F a Banach space and $\mathcal{B}$ its Borel σ-algebra. The set $\mathcal{L}_\mu(X, \mathcal{A}, F)$ of *μ-integrable functions* consists of all functions $f \colon X \to F$ such that there is some N_1-Cauchy sequence $(f_n)_{n \geq 1}$ of μ-step maps $f_n \in \mathcal{S}tep_\mu(X, \mathcal{A}, F)$ which converges pointwise almost everywhere to f. A sequence $(f_n)_{n \geq 1}$ of μ-step maps as above is called an *approximation sequence* for f.

Observe that not only do we require that the sequence $(f_n)_{n \geq 1}$ converges pointwise to f a.e., which makes f a μ-measurable map, but also that this sequence is N_1-Cauchy. This is the key to defining the notion of integral of the function f, as shown technically in Proposition 5.17.

We will see that $\mathcal{L}_\mu(X, \mathcal{A}, F)$ is a vector space containing $\mathcal{S}tep_\mu(X, \mathcal{A}, F)$, and a subspace of $\mathcal{M}_\mu(X, \mathcal{A}, F)$. Also, and this is the point of the construction, $\mathcal{L}_\mu(X, \mathcal{A}, F)$ is complete with respect to the extension $\| \|_1$ of the semi-norm N_1 to $\mathcal{L}_\mu(X, \mathcal{A}, F)$, a fact that is not obvious at all from the definition.

The *crucial point* is that Definition 5.11 is designed so that the following fact holds

Proposition 5.17. *For any N_1-Cauchy sequence $(f_n)_{n\geq 1}$ of μ-step maps, the sequence of integrals $\left(\int f_n d\mu\right)_{n\geq 1}$ is a Cauchy sequence in F.*

Proof. Indeed, by Proposition 5.16(3), we have

$$\left\| \int f_n d\mu - \int f_m d\mu \right\| = \left\| \int (f_n - f_m) d\mu \right\| \leq \int \|f_n - f_m\| \, d\mu$$
$$= N_1(f_n - f_m),$$

and since by hypothesis (f_n) is an N_1-Cauchy sequence, the sequence $\left(\int f_n d\mu\right)_{n\geq 1}$ is a Cauchy sequence in F. Indeed, for every $\epsilon > 0$, since the sequence (f_n) is N_1-Cauchy, there is some $N > 0$ such that $N_1(f_n - f_m) < \epsilon$ for all $m, n \geq N$, which implies that $\left\| \int f_n d\mu - \int f_m d\mu \right\| < \epsilon$ for all $m, n \geq N$. ☐

Then, since F is complete, the sequence $\left(\int f_n d\mu\right)_{n\geq 1}$ converges to an element of F, and if $(f_n)_{n\geq 1}$ is an approximation sequence for $f \in \mathcal{L}_\mu(X, \mathcal{A}, F)$, it is natural to define the integral of f as

$$\int f d\mu = \lim_{n \mapsto \infty} \int f_n d\mu.$$

The problem is that the definition of $\int f d\mu$ depends on the approximation sequence $(f_n)_{n\geq 1}$ chosen for f.

Actually, the definition of $\int f d\mu$ does *not depend* on the approximation sequence $(f_n)_{n\geq 1}$ chosen for f, but proving this is nontrivial. The proof relies on a remarkable fact called the *fundamental lemma of integration* by Serge Lang; see [Lang (1996)], Chapter VI, §3.

Proposition 5.18. *Let $(f_n)_{n\geq 1}$ be any N_1-Cauchy sequence of maps $f_n \in \mathcal{S}tep_\mu(X, \mathcal{A}, F)$. There exists a subsequence (g_k) which converges pointwise almost everywhere to a limit $f \in \mathcal{L}_\mu(X, \mathcal{A}, F)$. Furthermore, for any $\epsilon > 0$, there is a measurable subset $Z_\epsilon \in \mathcal{A}$ such that $\mu(Z_\epsilon) \leq \epsilon$, and the subsequence (g_k) converges uniformly to f on $X - Z_\epsilon$ (recall Definition 2.6).*

Proof. We follow Lang; see Lang [Lang (1996)] (Chapter VI, §3, Lemma 3.1). Since $(f_n)_{n\geq 1}$ is an N_1-Cauchy sequence, for every $k \geq 1$, there is some M_k such that if $m, n \geq M_k$, then

$$N_1(f_m - f_n) < \frac{1}{2^{2k}}.$$

By induction we can define the sequence (M_k) such that $M_k < M_{k+1}$ for all $k \geq 1$. We define the subsequence (g_k) such that $g_k = f_{M_k}$. By construction, we have

$$N_1(g_m - g_n) < \frac{1}{2^{2n}} \quad \text{if } m \geq n. \tag{$*_1$}$$

In particular, the sequence (g_k) is N_1-Cauchy.

Our next goal is to prove that the series

$$g_1(x) + \sum_{k=1}^{\infty} (g_{k+1}(x) - g_k(x)) \tag{$*_2$}$$

converges absolutely (thus pointwise) to a function $f \in \mathcal{L}_\mu(X, \mathcal{A}, F)$ outside a subset Z of measure 0, and in fact the convergence is uniform except on a set of arbitrary small measure. Observe that the partial sums of the series $(*_2)$ are $g_n(x)$, so this establishes the statement of the proposition.

Let Y_n be the set of all $x \in X$ such that

$$\|g_{n+1}(x) - g_n(x)\| \geq \frac{1}{2^n}.$$

Since the g_k are μ-step maps and since by Proposition 5.11(3) and 5.11(4) the function $\|g_{n+1} - g_n\|$ is measurable, the set Y_n has finite measure. Since

$$\|g_{n+1}(x) - g_n(x)\| \geq \frac{1}{2^n}$$

on Y_n, using $(*_1)$ we have

$$\frac{1}{2^n}\mu(Y_n) = \int_{Y_n} \frac{1}{2^n} d\mu \leq \int_X \|g_{n+1}(x) - g_n(x)\| \, d\mu = N_1(g_{n+1} - g_n) < \frac{1}{2^{2n}}.$$

The above implies that

$$\mu(Y_n) < \frac{1}{2^n}. \tag{$*_3$}$$

If we let

$$Z_n = \bigcup_{k\geq n} Y_n,$$

then we have

$$\mu(Z_n) \leq \sum_{k\geq n} \mu(Y_n) \leq \sum_{k\geq n} \frac{1}{2^k} = \frac{1}{2^n}\left(\sum_{k=0}^{\infty} \frac{1}{2^k}\right) < \frac{1}{2^{n-1}}.$$

If $x \notin Z_n$, then for all $k \geq n$ we have

$$\|g_{k+1}(x) - g_k(x)\| < \frac{1}{2^k},$$

and this implies that we have

$$\sum_{k=n}^{\infty} \|g_{k+1}(x) - g_k(x)\| < \sum_{k=n}^{\infty} \frac{1}{2^k} < \frac{1}{2^{n-1}},$$

so the series

$$\|g_1(x)\| + \sum_{k=1}^{\infty} \|g_{k+1}(x) - g_k(x)\|$$

converges, and since F is complete, the series in $(*_2)$ is uniformly convergent outside Z_n to a limit $f(x)$. For every $\epsilon > 0$ there is an n such that $\frac{1}{2^{n-1}} < \epsilon$, so the statement about uniform convergence holds with $Z = Z_n$. If we define Z by

$$Z = \bigcap_{n\geq 1} Z_n,$$

then $\mu(Z) = 0$ and since $x \in X - Z$ iff there is some n such that $x \in X - Z_n$, so the series $(*_2)$ converges to $f(x)$ and thus $g_n(x)$ converges to $f(x)$, which means that the sequence (g_n) converges pointwise to f outside the subset Z of measure zero. ☐

It should be mentioned that in general, the original sequence (f_n) may not converge pointwise, even a.e. An example of such a sequence (f_n) which is N_1-Cauchy, yet $(f_n(x))$ diverges for every $x \in X$, is given in Schwartz [Schwartz (1993a)] (Chapter 5, §6).

Using Proposition 5.18, the following result is obtained. This result implies that the integral $\int f d\mu$ is well defined.

Proposition 5.19. *Let $(f_n)_{n\geq 1}$ and $(g_n)_{n\geq 1}$ be two N_1-Cauchy sequences of μ-step maps $f_n, g_n \in \mathcal{S}tep_\mu(X, \mathcal{A}, F)$ which approximate the same function f. The sequences $\left(\int f_n d\mu\right)_{n\geq 1}$ and $\left(\int g_n d\mu\right)_{n\geq 1}$ converge to the same limit, and*

$$\lim_{n\mapsto\infty} \int \|f_n - g_n\|\, d\mu = 0,$$

that is, $\lim_{n\mapsto\infty} N_1(f_n - g_n) = 0$.

Proof. We follow Lang's proof; see Lang [Lang (1996)] (Chapter VI, §3, Lemma 3.2). The convergence of the sequences $\left(\int f_n d\mu\right)_{n\geq 1}$ and $\left(\int g_n d\mu\right)_{n\geq 1}$ follows from Proposition 5.17. Note that

$$\left\|\int f_n d\mu - \int f_m d\mu\right\| \leq N_1(f_n - f_m). \tag{$*$}$$

Next let $h_n = f_n - g_n$. Since the maps f_n and g_n approximate the same function f, the fact that

$$\begin{aligned}\int \|h_n - h_m\|\, d\mu &= \int \|f_n - g_n - (f_m - g_m)\|\, d\mu \\ &\leq \int \|f_n - f_m\|\, d\mu + \int \|g_n - g_m\|\, d\mu\end{aligned}$$

implies that the sequence (h_n) is N_1-Cauchy and converges almost everywhere to the zero function. We will prove that $N_1(h_n) = \int \|h_n\|\, d\mu$ converges to 0, and since

$$\left\|\int h_n d\mu\right\| \leq \int \|h_n\|\, d\mu,$$

the integral $\int h_n d\mu$ also converges to 0.

Since (h_n) is N_1-Cauchy, for every $\epsilon > 0$ there is some $N > 0$ such that for all $m, n \geq N$ we have

$$N_1(h_n - h_m) < \epsilon. \tag{$*_1$}$$

Since f_n and g_n are μ-step functions for all n there is some subset A of finite measure such that h_N vanishes outside A. Then for all $n \geq N$ we have

$$\begin{aligned}\int_{X-A} \|h_n\|\, d\mu &= \int_{X-A} \|h_n - h_N\|\, d\mu \\ &\leq \int_X \|h_n - h_N\|\, d\mu = N_1(h_n - h_N) < \epsilon,\end{aligned}$$

so

$$\int_{X-A} \|h_n\|\, d\mu < \epsilon, \quad n \geq N. \tag{$*_2$}$$

By Proposition 5.18, there is a subset Z of A such that

$$\mu(Z) < \frac{\epsilon}{1 + \|h_N\|_\infty}, \tag{$*_3$}$$

and a subsequence (h_m) that tends to 0 uniformly on $A - Z$. The reason for using $1 + \|h_N\|_\infty$ is to avoid division by zero. The point is that in all cases we have $\mu(Z)\left\|h_N\right\|_\infty < \epsilon$. Then for $m \geq N$ large enough we conclude that

$$\int_{A-Z} \|h_m\|\, d\mu < \epsilon. \tag{$*_4$}$$

Finally for m large enough we have

$$\begin{aligned}\int_Z \|h_m\|\, d\mu &\leq \int_Z \|h_n - h_N\|\, d\mu + \int_Z \|h_N\|\, d\mu \\ &\leq N_1(h_n - h_N) + \mu(Z)\left\|h_N\right\|_\infty < 2\epsilon,\end{aligned}$$

so

$$\int_Z \|h_m\|\, d\mu < 2\epsilon. \tag{$*_5$}$$

Using $(*_2)$, $(*_2)$ and $(*_5)$, we obtain

$$\begin{aligned}N_1(h_m) = \int_X \|h_m\|\, d\mu = \int_{X-A} &\|h_m\|\, d\mu + \int_Z \|h_m\|\, d\mu \\ &+ \int_{A-Z} \|h_m\|\, d\mu < \epsilon + \epsilon + 2\epsilon = 4\epsilon,\end{aligned}$$

proving our result. □

Proposition 5.19 justifies the following definition.

Definition 5.12. Let $(X, \mathcal{A}, \mu)$ be a measure space and let $(F, \mathcal{B})$ be a measurable space, with F a Banach space and $\mathcal{B}$ its Borel σ-algebra. For any function $f \in \mathcal{L}_\mu(X, \mathcal{A}, F)$, we define the *integral*[3] of f (*with respect to* μ) by

$$\int f d\mu = \lim_{n \mapsto \infty} \int f_n d\mu,$$

where $(f_n)_{n\geq 1}$ is any approximation sequence of f by μ-step maps.

Proposition 5.20. *For any function $f \in \mathcal{L}_\mu(X, \mathcal{A}, F)$ and any approximation sequence (f_n) of f with $f_n \in \mathcal{S}tep_\mu(X, \mathcal{A}, F)$, we have $\|f\| \in \mathcal{L}_\mu(X, \mathcal{A}, \mathbb{R})$, and the sequence $(\|f_n\|)$ is an approximation sequence of $\|f\|$ with $\|f_n\| \in \mathcal{S}tep_\mu(X, \mathcal{A}, \mathbb{R})$. Furthermore,*

$$\int \|f\|\, d\mu = \lim_{n \mapsto \infty} \int \|f_n\|\, d\mu = \lim_{n \mapsto \infty} N_1(f_n).$$

[3]This integral is usually called the *Lebesgue integral* or *Bochner integal*.

Proof. Since the sequence (f_n) converges pointwise to f a.e., we verify immediately that the sequence $(\|f_n\|)$ converges pointwise to $\|f\|$ a.e. Since

$$|\,\|f_n\| - \|f_m\|\,| \leq \|f_n - f_m\|$$

(see just after Definition A.3), by Proposition 5.16(5) we have

$$N_1(\|f_n\| - \|f_m\|) = \int |\,\|f_n\| - \|f_m\|\,| d\mu \leq \int \|f_n - f_m\|\, d\mu = N_1(f_n - f_m),$$

and since (f_n) is an N_1-Cauchy sequence, the sequence $(\|f_n\|)$ is an N_1-Cauchy sequence. Therefore $\|f\| \in \mathcal{L}_\mu(X, \mathcal{A}, \mathbb{R})$, and $(\|f_n\|)$ is an approximation sequence of $\|f\|$. By definition of the integral,

$$\int \|f\|\, d\mu = \lim_{n \mapsto \infty} \int \|f_n\|\, d\mu = \lim_{n \mapsto \infty} N_1(f_n),$$

as claimed. ☐

Definition 5.13. For any function $f \in \mathcal{L}_\mu(X, \mathcal{A}, F)$, we define the L^1-*semi-norm* $\|f\|_1$ of f as

$$\|f\|_1 = \int \|f\|\, d\mu.$$

Observe that if $f \in \mathcal{S}tep_\mu(X, \mathcal{A}, F)$, then $\|f\|_1 = N_1(f)$. The following proposition is easily shown by passing to the limit.

Proposition 5.21. *The set $\mathcal{L}_\mu(X, \mathcal{A}, F)$ is a vector space, and $\|\;\|_1$ is a semi-norm on $\mathcal{L}_\mu(X, \mathcal{A}, F)$. The space $\mathcal{S}tep_\mu(X, \mathcal{A}, F)$ is a subspace of $\mathcal{L}_\mu(X, \mathcal{A}, F)$, which is a subspace of $\mathcal{M}_\mu(X, \mathcal{A}, F)$.*

We are almost ready to prove that $\mathcal{L}_\mu(X, \mathcal{A}, F)$ is complete with respect to the L^1-semi-norm, but first we need the following result.

Proposition 5.22. *The subspace $\mathcal{S}tep_\mu(X, \mathcal{A}, F)$ is dense in $\mathcal{L}_\mu(X, \mathcal{A}, F)$ with respect to the L^1-semi-norm $\|\;\|_1$. Furthermore, any approximation sequence $(f_n)_{n \geq 1}$ of f by μ-step maps converges to f according to the semi-norm $\|\;\|_1$.*

Proof. Pick any $f \in \mathcal{L}_\mu(X, \mathcal{A}, \mathbb{R})$ and let (f_n) be any approximation sequence for f. This means that the sequence (f_n) is a N_1-Cauchy sequence of μ-step maps which converges pointwise to f a.e. We will prove that

$$\lim_{n \mapsto \infty} \|f - f_n\|_1 = 0,$$

which shows that the sequence (f_n) converges to f in the L^1-semi-norm.

First we claim that for any fixed $n \geq 1$, the sequence $(\|f_p - f_n\|)_{p\geq 1}$ is an N_1-Cauchy sequence which converges to $\|f - f_n\|$ a.e. Indeed, we have

$$\begin{aligned}\int \big|\,\|f_p - f_n\| - \|f_q - f_n\|\,\big| d\mu &\leq \int \|f_p - f_n - (f_q - f_n)\|\, d\mu \\ &= \int \|f_p - f_q\|\, d\mu = N_1(f_p - f_q),\end{aligned}$$

and since (f_n) is a N_1-Cauchy sequence, for every $\epsilon > 0$, there is some $N > 0$ such that $N_1(f_p - f_q) < \epsilon$ for all $p, q \geq N$, which shows that $(\|f_p - f_n\|)_{p\geq 1}$ is a N_1-Cauchy sequence (in $\mathbb{R}$). The fact that $(f_p)_{p\geq 1}$ converges pointwise a.e. to f immediately implies that $\|f_p - f_n\|$ converges to $\|f - f_n\|$ a.e. By definition of $\|\ \|_1$ and of the integral

$$\|f - f_n\|_1 = \int \|f - f_n\|\, d\mu = \lim_{p\mapsto\infty} \int \|f_p - f_n\|\, d\mu = \lim_{p\mapsto\infty} N_1(f_p - f_n).$$

Thus for every $\epsilon > 0$, there is some $M_1 > 0$ such that

$$\big|\,\|f - f_n\|_1 - N_1(f_p - f_n)\big| < \frac{\epsilon}{2} \quad \text{for all } p \geq M_1,$$

and since (f_n) is an N_1-Cauchy sequence, there is some $M_2 > 0$ such that

$$N_1(f_p - f_n) < \frac{\epsilon}{2} \quad \text{for all } n, p \geq M_2,$$

so for all $n, p \geq \max(M_1, M_2)$ we have

$$\|f - f_n\|_1 \leq \big|\,\|f - f_n\|_1 - N_1(f_p - f_n)\big| + N_1(f_p - f_n) < \frac{\epsilon}{2} + \frac{\epsilon}{2} = \epsilon,$$

which proves that $\lim_{n\mapsto\infty} \|f - f_n\|_1 = 0$; that is, the sequence (f_n) converges to f in the L^1-semi-norm. $\square$

Remark: It appears that Lang [Lang (1996)] skipped this step, which is used in the proof his Theorem 3.4, and the proof of the next theorem.

Now we can prove one of our main theorems.

5.7 The Fischer–Riesz Theorem

Theorem 5.23. *(Fischer–Riesz) The space $\mathcal{L}_\mu(X, \mathcal{A}, F)$ is complete with respect to the L^1-semi-norm. This means that for every sequence $(f_n)_{n\geq 1}$ of functions $f_n \in \mathcal{L}_\mu(X, \mathcal{A}, F)$, if (f_n) is $\|\ \|_1$-Cauchy, then there is some function $f \in \mathcal{L}_\mu(X, \mathcal{A}, F)$ such that for every $\epsilon > 0$, there is some $N > 0$ such that $\|f - f_n\|_1 < \epsilon$ for all $n \geq N$.*

Proof. Let $(f_n)_{n\geq 1}$ be an $\| \|_1$-Cauchy sequence of functions $f_n \in \mathcal{L}_\mu(X, \mathcal{A}, F)$. By Proposition 5.22, for every n there is an approximation sequence $(g_{n,m})_{m\geq 1}$ of μ-step maps that converges to f_n pointwise a.e. and in the $\| \|_1$-semi-norm. Thus, for every $n \geq 1$, there is some $m(n)$ such that

$$\left\|f_n - g_{n,m(n)}\right\|_1 \leq \frac{1}{n}. \tag{$*_6$}$$

Each sequence $(g_{n,m(n)})_{n\geq 1}$ is N_1-Cauchy, because

$$\begin{aligned} N_1(g_{p,m(p)} - g_{q,m(q)}) &= \left\|g_{p,m(p)} - g_{q,m(q)}\right\|_1 \\ &\leq \left\|g_{p,m(p)} - f_p\right\|_1 + \left\|f_p - f_q\right\|_1 + \left\|f_q - g_{q,m(q)}\right\|_1 \\ &\leq \frac{1}{p} + \frac{1}{q} + \left\|f_p - f_q\right\|_1, \end{aligned}$$

and the right-hand side tends to 0 when p and q tend to $+\infty$, since the sequence (f_n) is $\| \|_1$-Cauchy. By Proposition 5.18, for each sequence $(g_{n,m(n)})_{m\geq 1}$, we can extract a subsequence $(g_{n_k,m(n_k)})_{k\geq 1}$ that converges pointwise a.e. to some function $f \in \mathcal{L}_\mu(X, \mathcal{A}, F)$, and is also N_1-Cauchy. By the second part of Proposition 5.22, the subsequence $(g_{n_k,m(n_k)})_{k\geq 1}$ converges to f for the semi-norm $\| \|_1$. Since $(g_{n,m(n)})_{n\geq 1}$ is N_1-Cauchy and has a subsequence $(g_{n_k,m(n_k)})_{k\geq 1}$ $\| \|_1$-convergent to f, it also $\| \|_1$-converges to the function f. Using ($*_6$) and the inequality

$$\begin{aligned} \left\|f - f_n\right\|_1 &\leq \left\|f - g_{n,m(n)}\right\|_1 + \left\|g_{n,m(n)} - f_n\right\|_1 \\ &\leq \left\|f - g_{n,m(n)}\right\|_1 + \frac{1}{n}, \end{aligned}$$

and since the sequence $(g_{n,m(n)})_{n\geq 1}$ $\| \|_1$ converges to the function f, we deduce that the sequence $(f_n)_{n\geq 1}$ converges to f for the semi-norm $\| \|_1$.

In the following diagram, the original sequence $(f_n)_{n\geq 1}$ is shown as the top horizontal row. Below each f_n, we have the approximation sequence $(g_{n,m})_{m\geq 1}$ shown as an ascending column. The sequence of $g_{n,m(n)}$ chosen for each n is shown in boldface, and its subsequence in red as illustrated below.

$$\begin{array}{ccccccccc}
f_1 & f_2 & f_3 & f_4 & f_5 & f_6 & \cdots & f_n & \cdots \\
\vdots & \vdots & \vdots & \vdots & \vdots & \vdots & & \vdots & \\
g_{1,m(n)} & g_{2,m(n)} & g_{3,m(n)} & g_{4,m(n)} & g_{5,m(n)} & g_{6,m(n)} & \cdots & \mathbf{g_{n,m(n)}} & \cdots \\
\vdots & \vdots & \vdots & \vdots & \vdots & \vdots & & \vdots & \\
g_{1,6} & g_{2,6} & \mathbf{g_{3,6}} & g_{4,6} & \mathbf{g_{5,6}} & g_{6,6} & \cdots & g_{n,6} & \cdots \\
g_{1,5} & \mathbf{g_{2,5}} & g_{3,5} & g_{4,5} & g_{5,5} & g_{6,5} & \cdots & g_{n,5} & \cdots \\
g_{1,4} & g_{2,4} & g_{3,4} & \mathbf{g_{4,4}} & g_{5,4} & g_{6,4} & \cdots & g_{n,4} & \cdots \\
\mathbf{g_{1,3}} & g_{2,3} & g_{3,3} & g_{4,3} & g_{5,3} & g_{6,3} & \cdots & g_{n,3} & \cdots \\
g_{1,2} & g_{2,2} & g_{3,2} & g_{4,2} & g_{5,2} & \mathbf{g_{6,2}} & \cdots & g_{n,2} & \cdots \\
g_{1,1} & g_{2,1} & g_{3,1} & g_{4,1} & g_{5,1} & g_{6,1} & \cdots & g_{n,1} & \cdots
\end{array}$$

This concludes the proof. □

The following properties of the integral are easily obtained by passing to the limit.

Proposition 5.24. *Let $(X, \mathcal{A}, \mu)$ be a measure space and let $(F, \mathcal{B})$ be a measurable space, with F a Banach space and $\mathcal{B}$ its Borel σ-algebra. The following properties hold:*

(1) For any $f \in \mathcal{L}_\mu(X, \mathcal{A}, F)$, if $f = 0$ a.e., then $\int f d\mu = 0$. More generally, if $f, g \in \mathcal{L}_\mu(X, \mathcal{A}, F)$ and if $f = g$ a.e., then $\int f d\mu = \int g d\mu$.

(2) For any $f \in \mathcal{L}_\mu(X, \mathcal{A}, F)$, and for any measurable subset $A \in \mathcal{A}$, the integral $\int_A f d\mu = \int f \chi_A d\mu$ exists, and

$$\left\| \int_A f d\mu \right\| \leq \int_A \|f\| \, d\mu \leq \|f\|_\infty \, \mu(A).$$

Furthermore, if $A, B \in \mathcal{A}$ are disjoint, then

$$\int_{A \cup B} f d\mu = \int_A f d\mu + \int_B f d\mu.$$

(3) The integral $\int : \mathcal{L}_\mu(X, \mathcal{A}, F) \to F$ is linear.

(4) For any $f \in \mathcal{L}_\mu(X, \mathcal{A}, F)$, we have $\|f\| \in \mathcal{L}_\mu(X, \mathcal{A}, \mathbb{R})$, and

$$\left\| \int f d\mu \right\| \leq \int \|f\| \, d\mu = \|f\|_1 .$$

(5) If $f, g \in \mathcal{L}_\mu(X, \mathcal{A}, \mathbb{R})$, then $\sup(f, g), \inf(f, g), f^+, f^-, |f| \in \mathcal{L}_\mu(X, \mathcal{A}, \mathbb{R})$. Since $f^+ = (|f| + f)/2$ and $f^- = (|f| - f)/2$, we have $f \in \mathcal{L}_\mu(X, \mathcal{A}, \mathbb{R})$ iff $f^+ \in \mathcal{L}_\mu(X, \mathcal{A}, \mathbb{R})$ and $f^- \in \mathcal{L}_\mu(X, \mathcal{A}, \mathbb{R})$.

(6) If $f, g \in \mathcal{L}_\mu(X, \mathcal{A}, \mathbb{R})$ and $f \leq g$ a.e., then $\int f d\mu \leq \int g d\mu$. In particular, if $f \geq 0$ a.e., then $\int f d\mu \geq 0$.

(7) *Let F_1 and F_2 be two Banach spaces, and let $h\colon F_1 \to F_2$ be a linear map (or semi-linear map when the field is $\mathbb{C}$). If $f \in \mathcal{L}_\mu(X, \mathcal{A}, F_1)$, then $h \circ f \in \mathcal{L}_\mu(X, \mathcal{A}, F_2)$, and*

$$\int (h \circ f) d\mu = h\left(\int f d\mu\right).$$

(8) *Let F_1 and F_2 be two Banach spaces, and let $F_1 \times F_2$ be the product space (under any of the product norms defined just before Definition A.13). Then there is an isomorphism between $\mathcal{L}_\mu(X, \mathcal{A}, F_1 \times F_2)$ and $\mathcal{L}_\mu(X, \mathcal{A}, F_1) \times \mathcal{L}_\mu(X, \mathcal{A}, F_2)$, and if $f = (f_1, f_2)$, then*

$$\int f d\mu = \left(\int f_1 d\mu, \int f_2 d\mu\right).$$

In particular, since $\mathbb{C}$ is isomorphic to $\mathbb{R} \times \mathbb{R}$, a function $f \in \mathcal{L}_\mu(X, \mathcal{A}, \mathbb{C})$ corresponds uniquely to a function $f = u + iv$ with $u, v \in \mathcal{L}_\mu(X, \mathcal{A}, \mathbb{R})$, and we have

$$\int f d\mu = \int u d\mu + i \int v d\mu.$$

Remark: Observe that in our approach, if f is a real-valued function or a complex-valued function, the integral $\int f d\mu$ is defined directly. There is another approach in which the integral is first defined for real-valued *positive* functions. Then the integral of a real-valued function f is defined in terms of the integrals of f^+ and f^-, and the integral of a complex valued function $f = u + iv$ is defined in terms of the integrals of u^+, u^-, v^+, v^-. See Rudin [Rudin (1987)], Definition 1.31.

5.8 Characterizing Which Functions Satisfy $\|f\|_1 = 0$

The next step is to identify the functions f in $\mathcal{L}_\mu(X, \mathcal{A}, \mathbb{R})$ such that $\|f\|_1 = 0$. For this, we need two propositions.

Proposition 5.25. *For any function $f \in \mathcal{L}_\mu(X, \mathcal{A}, F)$, and for any real $a > 0$, the subset $E_a = \{x \in X \mid \|f(x)\| \geq a\}$ can be written as $E_a = (B - Z) \cup N$, with B a measurable subset of finite measure, and Z and N two null subsets. The function f vanishes outside of a σ-finite measurable set.*

Proof. We begin by showing that E_a is a measurable set with finite measure. Since $f \in \mathcal{L}_\mu(X, \mathcal{A}, F)$, by Proposition 5.5(2), the function $\|f\|$ is measurable, so E_a is measurable. By Proposition 5.18, there is an N_1-Cauchy sequence (f_n) of μ-step maps which converges pointwise to f a.e., and for every $\epsilon > 0$, there is a measurable subset Z_1 of measure $\mu(Z_1) < \epsilon$ such that f_n converges uniformly to f on $X - Z_1$. Pick $\epsilon = a/2$. The uniform convergence implies that there is some $M > 0$ such that for all $n \geq M$ and all $x \in X - Z_1$,

$$\|f(x) - f_n(x)\| \leq a/2,$$

and since $\|f(x)\| \leq \|f(x) - f_n(x)\| + \|f_n(x)\|$, we have

$$\|f(x)\| \leq \|f_n(x)\| + a/2,$$

and thus $\|f_n(x)\| \geq \|f(x)\| - a/2$, so $\|f(x)\| \geq a$ implies $\|f_n(x)\| \geq a/2$, which implies that

$$E_a \subseteq \{x \in Z_1 \mid \|f(x)\| \geq a\} \cup \{x \in X - Z_1 \mid \|f_n(x)\| \geq a/2\},$$

where both sets on the right-hand side have finite measure (the second one because f_n is a μ-step function, and so is $\|f_n\|$, and a μ-step function vanishes outside of a set of finite measure). See Figure 5.16. Since both sets are measurable and the set on the right-hand side has finite measure, by Proposition 4.7(2) we deduce that E_a has finite measure. Since the function $\|f\|$ is μ-measurable, by Proposition 5.13(1), it is equal a.e. to a measurable function g, so there is a null set such Z that $\|f\|(x) = g(x)$ for all $x \in X - Z$. Then we have

$$\begin{aligned}
E_a &= \{x \in X \mid \|f(x)\| \geq a\} \\
&= \{x \in X - Z \mid \|f(x)\| \geq a\} \cup \{x \in Z \mid \|f(x)\| \geq a\} \\
&= \{x \in X - Z \mid g(x) \geq a\} \cup \{x \in Z \mid \|f(x)\| \geq a\} \\
&= (\{x \in X \mid g(x) \geq a\} - Z) \cup \{x \in Z \mid \|f(x)\| \geq a\} \\
&= (B - Z) \cup N,
\end{aligned}$$

with $B = \{x \in X \mid g(x) \geq a\}$ and $N = \{x \in Z \mid \|f(x)\| \geq a\}$. Since g is measurable and $[a, \infty)$ is closed, B is measurable, and N as a subset of a null set is a null set; see Figure 5.17.

What we showed above with $\|f\|$ replaced by g measurable implies that B has finite measure. The second statement of the proposition follows from Proposition 5.13(2). ☐

Proposition 5.18 can be promoted to $\mathcal{L}_\mu(X, \mathcal{A}, F)$ as follows.

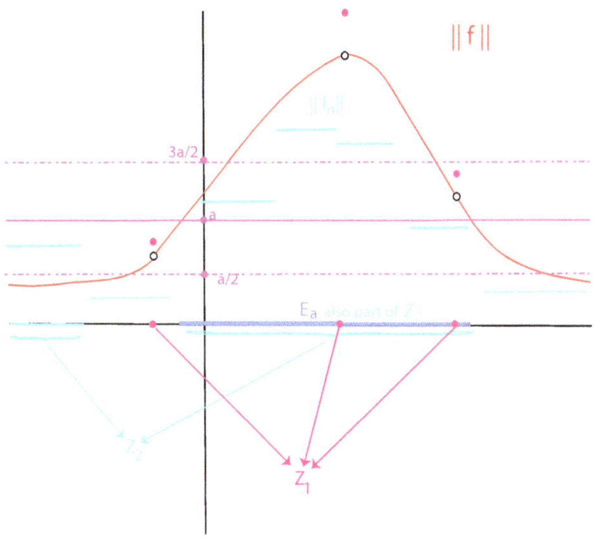

Fig. 5.16 Let $X = \mathbb{R}$. The red curve is the graph of $\|f\|$, while the aqua graph is the μ-step function $\|f_n\|$. The set Z_1 corresponds to the three magenta dots on the x-axis. The purple horizontal line segment is E_a, the two horizontal aqua line segments are Z_2, and $E_a \subseteq \{x \in Z_1 \mid \|f(x)\| \geq a\} \cup Z_2$, where $Z_2 = \{x \in X - Z_1 \mid \|f_n(x)\| \geq a/2\}$.

Proposition 5.26. *Let $(f_n)_{n\geq 1}$ be any $\| \, \|_1$-Cauchy sequence of maps $f_n \in \mathcal{L}_\mu(X, \mathcal{A}, F)$ that converges to some function $f \in \mathcal{L}_\mu(X, \mathcal{A}, F)$ in the semi-norm $\| \, \|_1$. There exists a subsequence (g_k) which converges pointwise almost everywhere to f. Furthermore, for any $\epsilon > 0$, there is a measurable subset $Z_\epsilon \in \mathcal{A}$ such that $\mu(Z_\epsilon) \leq \epsilon$, and the subsequence $(g_k$ converges uniformly to f on $X - Z_\epsilon$ (recall Definition 2.6).*

Proposition 5.26 is proven in Lang [Lang (1996)] (Chapter VI, Theorem 5.2). The proof is very similar to the proof of Proposition 5.18. However, the f_n are no longer μ-step functions so we need Proposition 5.25 to justify the fact that the sets Y_n have finite measure.

Here are some corollaries of Proposition 5.26.

Proposition 5.27. *For any function $f \in \mathcal{L}_\mu(X, \mathcal{A}, F)$, we have $\|f\|_1 = 0$ iff $f = 0$ a.e.*

Proof. If $f = 0$ a.e., then $\|f\| = 0$ a.e., and by Proposition 5.24(1), we have $\|f\|_1 = 0$. Conversely, the sequence (f_n) where f_n is the zero function

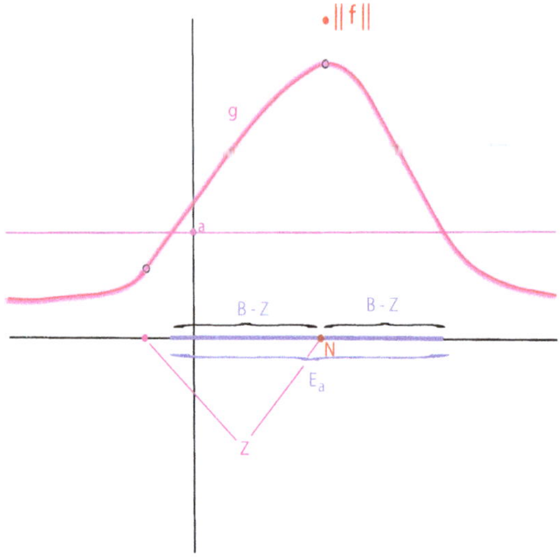

Fig. 5.17 A continuation of Figure 5.16 where $\|f\|$ is replaced by the magenta bell curve g. Note that Z is the union of the two reddish dots while N is the darker red dot contained within E_a. In this particular illustration $B = E_a$.

is $\| \; \|_1$-Cauchy and converges to f in the $\| \; \|_1$-norm. By Proposition 5.26 there is a subsequence that converges pointwise a.e. to f. But since f_n is the zero function for all n, this subsequence also converges pointwise a.e. to the zero function, so $f = 0$ a.e. $\square$

Proposition 5.27 is the second main important result of this section because it provides a very natural characterization of the functions f such that $\|f\|_1 = 0$.

Proposition 5.28. *Let (f_n) be a sequence of functions $f_n \in \mathcal{L}_\mu(X, \mathcal{A}, F)$. If (f_n) is an $\| \; \|_1$-Cauchy sequence which converges pointwise a.e. to a function $f \colon X \to F$, then $f \in \mathcal{L}_\mu(X, \mathcal{A}, F)$, and (f_n) converges to f in the semi-norm $\| \; \|_1$.*

Proof. Since the sequence (f_n) is an $\| \; \|_1$-Cauchy sequence, by the Fischer–Riesz theorem (Theorem 5.23) it converges to some function $g \in \mathcal{L}_\mu(X, \mathcal{A}, F)$ in the $\| \; \|_1$-semi-norm. By Proposition 5.26, some subsequence $(f_{n_k})_{k \geq 1}$ of (f_n) converges pointwise a.e. to g. Since (f_n) converges pointwise a.e. to f, the subsequence $(f_{n_k})_{k \geq 1}$ also converges pointwise a.e. to

f, so $f = g$ a.e., and since $g \in \mathcal{L}_\mu(X, \mathcal{A}, F)$ and (f_n) converges to g in the semi-norm $\| \ \|_1$, we also have $f \in \mathcal{L}_\mu(X, \mathcal{A}, F)$, and (f_n) converges to f in the semi-norm $\| \ \|_1$. □

The main disadvantage of the space $\mathcal{L}_\mu(X, \mathcal{A}, F)$ is that it is not a normed vector space under the semi-norm $\| \ \|_1$. Thus it is natural to consider the quotient of $\mathcal{L}_\mu(X, \mathcal{A}, F)$ by the subspace $\mathcal{N}$ consisting of the functions such that $\|f\|_1 = 0$.

Definition 5.14. Let $\mathcal{N}$ be the subspace of $\mathcal{L}_\mu(X, \mathcal{A}, F)$ given by

$$\mathcal{N} = \{f \in \mathcal{L}_\mu(X, \mathcal{A}, F) \mid \|f\|_1 = 0\},$$

which is just the subspace of function equal to 0 a.e. Then we define $\mathrm{L}_\mu(X, \mathcal{A}, F)$ as the quotient space

$$\mathrm{L}_\mu(X, \mathcal{A}, F) = \mathcal{L}_\mu(X, \mathcal{A}, F)/\mathcal{N}.$$

For any equivalence class $\mathbf{f} \in \mathrm{L}_\mu(X, \mathcal{A}, F)$, since for any two representatives $f, g \in \mathcal{L}_\mu(X, \mathcal{A}, F)$ in the equivalence class $\mathbf{f}$, we have $f = g$ a.e., by Proposition 5.24(1),

$$\int f d\mu = \int g d\mu,$$

so we can define $\int \mathbf{f} d\mu$ as

$$\int \mathbf{f} d\mu = \int f d\mu.$$

Similarly, $\|\mathbf{f}\|_1$ is defined as

$$\|\mathbf{f}\|_1 = \|f\|_1,$$

for any $f \in \mathcal{L}_\mu(X, \mathcal{A}, F)$ in the equivalence class $\mathbf{f}$.

The following theorem is immediately obtained from Theorem 5.23 by passing to the quotient.

Theorem 5.29. *(Fischer–Riesz) The semi-norm* $\| \ \|_1$ *on* $\mathrm{L}_\mu(X, \mathcal{A}, F)$ *induced by the semi-norm* $\| \ \|_1$ *on* $\mathcal{L}_\mu(X, \mathcal{A}, F)$ *by passing to the quotient is a norm on* $\mathrm{L}_\mu(X, \mathcal{A}, F)$ *called the* L^1*-norm. With this norm, the space* $\mathrm{L}_\mu(X, \mathcal{A}, F)$ *is complete (it is a Banach space). The subspace* $\mathrm{Step}_\mu(X, \mathcal{A}, F)$ *is dense in* $\mathrm{L}_\mu(X, \mathcal{A}, F)$.

Finally, the following proposition confirms one of our earlier claims.

Proposition 5.30. *The space* $\mathrm{L}_\mu(X, \mathcal{A}, F) = \mathcal{L}_\mu(X, \mathcal{A}, F)/\mathcal{N}$ *is isomophic to the Cauchy completion of the space* $\mathcal{S}tep_\mu(X, \mathcal{A}, F)/\mathcal{SN}$*; see the diagram*

$$\begin{array}{ccc}
\mathcal{S}tep_\mu(X, \mathcal{A}, F) & \xrightarrow{\textit{completion}} & \mathcal{L}_\mu(X, \mathcal{A}, F) \\
\Big\downarrow \textit{quotient} & & \Big\downarrow \textit{quotient} \\
\mathrm{Step}_\mu(X, \mathcal{A}, F) & \xrightarrow[\textit{completion}]{} & \mathrm{L}_\mu(X, \mathcal{A}, F).
\end{array}$$

In the next section we consider some fundamental convergence theorems. A very useful corollary of these theorems is that a function f belongs to $\mathcal{L}_\mu(X, \mathcal{A}, F)$ iff it belongs to $\mathcal{M}_\mu(X, \mathcal{A}, F)$ (it is μ-measurable), and if $\int \|f\|\, d\mu$ exists. By Proposition 5.21, the space $\mathcal{L}_\mu(X, \mathcal{A}, F)$ is a subspace of $\mathcal{M}_\mu(X, \mathcal{A}, F)$, and we already know from Proposition 5.24(4) that if $f \in \mathcal{L}_\mu(X, \mathcal{A}, F)$ then $\|f\| \in \mathcal{L}_\mu(X, \mathcal{A}, \mathbb{R})$. The converse is not trivial, but it will be shown as a corollary of the dominated convergence theorem discussed in Section 5.9.

5.9 Fundamental Convergence Theorems

Besides the fact that the Lebesgue–Bochner integral is defined for a much bigger class of functions than the regulated functions (or the Riemann-integrable functions), one of its main advantages is that it leads to simple and flexible criteria to tell whether the limit of a sequence of integrable functions is integrable. Most of these results allow interchanging a limit and an integral. We begin with criteria applying to real-valued functions. These results actually apply to extended functions with values in $\mathbb{R} \cup \{+\infty\}$, but for simplicity we stick to functions $f\colon X \to \mathbb{R}$. As in the previous section the results that we state without proof are proven either in Marle [Marle (1974)] or in Lang [Lang (1996)].

Theorem 5.31. *(Monotone Convergence Theorem) Let* $(f_n)_{n\geq 1}$ *be a sequence of functions* $f_n \in \mathcal{L}_\mu(X, \mathcal{A}, \mathbb{R})$ *such that* $f_n \leq f_{n+1}$ *for all* $n \geq 1$*, and assume that there is some* $M > 0$ *such that*

$$\left| \int f_n d\mu \right| \leq M \quad \textit{for all } n \geq 1.$$

Then the sequence $(f_n)_{n\geq 1}$ *converges pointwise a.e., and also in the* $\| \, \|_1$*-norm, to a function* $f \in \mathcal{L}_\mu(X, \mathcal{A}, \mathbb{R})$*. We also have* $\lim_{n\mapsto\infty} \|f_n\|_1 = \|f\|_1$

and

$$\lim_{n\mapsto\infty}\int f_n d\mu = \int f d\mu.$$

The same result applies to a nonincreasing sequence (f_n) *(with* $f_n \geq f_{n+1}$ *for all* $n \geq 1$*).*

Proof. We follow Lang [Lang (1996)] (Chapter VI, §5, Theorem 5.5). Let

$$\alpha = \sup_k \int f_k d\mu,$$

which is well defined since

$$\left|\int f_n d\mu\right| \leq M \quad \text{for all } n \geq 1.$$

For $n \geq m$, since $f_n \leq f_{n+1}$ for all $n \geq 1$, we have

$$\begin{aligned}\|f_n - f_m\|_1 &= \int (f_n - f_m) d\mu \\ &= \int f_n d\mu - \int f_m d\mu \\ &\leq \alpha - \int f_m d\mu,\end{aligned}$$

which implies that (f_n) is a $\| \|_1$-Cauchy sequence. By the Fischer–Riesz theorem (Theorem 5.23), the sequence converges to some limit $f \in \mathcal{L}_\mu(X, \mathcal{A}, \mathbb{R})$ in the $\| \|_1$-norm. By Proposition 5.26, there is a subsequence $(f_{n_k})_{k\geq 1}$ of (f_n) that converges a.e. to $f \in \mathcal{L}_\mu(X, \mathcal{A}, \mathbb{R})$, and since the sequence (f_n) is increasing, by a standard ϵ-argument, it also converges a.e. to f. Since

$$| \|f_n\|_1 - \|f\|_1 | \leq \|f_n - f\|_1$$

and (f_n) converges to f in the $\| \|_1$-norm, we deduce that $\lim_{n\mapsto\infty} \|f_n\|_1 = \|f\|_1$. We have

$$\begin{aligned}\left|\int f_n d\mu - \int f d\mu\right| &= \left|\int (f_n - f) d\mu\right| \\ &\leq \int |f_n - f| d\mu \\ &= \|f_n - f\|_1,\end{aligned}$$

and since (f_n) converges to f in the $\| \|_1$-norm, this implies that

$$\lim_{n\mapsto\infty}\int f_n d\mu = \int f d\mu,$$

as claimed. □

The following theorem has a different flavor. It asserts the existence of the sup of a sequence of functions.

Theorem 5.32. *(Beppo–Levi) Let (f_n) be a sequence of functions $f_n \in \mathcal{L}_\mu(X, \mathcal{A}, \mathbb{R})$. If there is a function $g \in \mathcal{L}_\mu(X, \mathcal{A}, \mathbb{R})$ such that $g \geq 0$ and $|f_n| \leq g$ for all $n \geq 1$, then $\sup_{n\geq 1} f_n$ and $\inf_{n\geq 1} f_n$ belong to $\mathcal{L}_\mu(X, \mathcal{A}, \mathbb{R})$, and we have*

$$\sup_{n\geq 1} \int f_n d\mu \leq \int (\sup_{n\geq 1} f_n) d\mu \quad \textit{and} \quad \int (\inf_{n\geq 1} f_n) d\mu \leq \inf_{n\geq 1} \int f_n d\mu.$$

Proof. We follow Lang [Lang (1996)] (Chapter VI, §5, Corollary 5.6). By Proposition 5.24(5), the functions

$$g_n = \sup\{f_1, \ldots, f_n\}$$

belong to $\mathcal{L}_\mu(X, \mathcal{A}, \mathbb{R})$ and form an increasing sequence bounded by g. Since

$$\int g_n d\mu \leq \int g_{n+1} d\mu \quad \text{and} \quad \int g_n d\mu \leq \int g d\mu,$$

there is some $M > 0$ such that $|\int g_n d\mu| \leq M$ for all $n \geq 1$. Therefore, by the monotone convergence theorem (Theorem 5.31), the sequence (g_n) converges pointwise a.e. to some function in $\mathcal{L}_\mu(X, \mathcal{A}, \mathbb{R})$, but (g_n) converges pointwise to $\sup_{n\geq 1} f_n$, so the sequence (g_n) converges pointwise a.e. to $\sup_{n\geq 1} f_n$. Since $f_n \leq \sup_{n\geq 1} f_n$, we have

$$\int f_n d\mu \leq \int (\sup_{n\geq 1} f_n) d\mu,$$

which implies

$$\sup_{n\geq 1} \int f_n d\mu \leq \int (\sup_{n\geq 1} f_n) d\mu,$$

as claimed. □

Given a sequence $(f_n)_{n\geq 1}$ of functions $f_n \colon X \to \mathbb{R}$ such that $f_n \geq 0$, recall that

$$\liminf f_n = \lim_{k\mapsto\infty} \inf_{n\geq k} f_n.$$

Theorem 5.33. *(Fatou's Lemma) Let $(f_n)_{n\geq 1}$ be a sequence of functions $f_n \in \mathcal{L}_\mu(X, \mathcal{A}, \mathbb{R})$ such that $f_n \geq 0$. If $\liminf \|f_n\|_1 = \liminf \int f_n d\mu$ exists, then there is a function $f \in \mathcal{L}_\mu(X, \mathcal{A}, \mathbb{R})$ such that $\liminf f_n$ converges pointwise to f a.e., and*

$$\int f d\mu \leq \liminf \int f_n d\mu.$$

A proof of Theorem 5.33 is given in Lang [Lang (1996)] (Chapter VI, §5).

The next theorem applies to functions with values in any Banach space F and is the most important convergence theorem.

Theorem 5.34. *(Lebesgue Dominated Convergence Theorem) Let $(f_n)_{n\geq 1}$ be a sequence of functions $f_n \in \mathcal{L}_\mu(X, \mathcal{A}, F)$. If (f_n) converges pointwise a.e. to a function $f\colon X \to F$, and if there is some function $g \in \mathcal{L}_\mu(X, \mathcal{A}, \mathbb{R})$ such that $g \geq 0$ and $\|f_n\| \leq g$ for all $n \geq 1$, then $f \in \mathcal{L}_\mu(X, \mathcal{A}, F)$ and $(f_n)_{n\geq 1}$ converges to f in the $\| \|_1$-norm. Consequently*

$$\lim_{n\mapsto\infty} \int f_n d\mu = \int f d\mu.$$

Proof. We follow Marle [Marle (1974)] (Chapter 2, Section 4, Theorem 2.4.7). For each $n, p \geq 1$, let

$$g_{n,p} = \sup_{\substack{n\leq m\leq n+p \\ n\leq r\leq n+p}} \|f_m - f_r\|$$

$$g_n = \sup_{m,r\geq n} \|f_m - f_r\| = \sup_{p\geq 1} g_{n,p}.$$

By Proposition 5.24(4,5), for all $n, p \geq 1$, we have $g_{n,p} \in \mathcal{L}_\mu(X, \mathcal{A}, \mathbb{R})$, $g_{n,p} \leq g_{n,p+1}$, and

$$0 \leq g_{n,p} \leq 2g.$$

We get

$$\left| \int g_{n,p}\, d\mu \right| \leq 2 \int g\, d\mu,$$

so by the monotone convergence theorem (Theorem 5.31), the sequence $(g_{n,p})_{p\geq 1}$ converges pointwise a.e. to a limit in $\mathcal{L}_\mu(X, \mathcal{A}, \mathbb{R})$. However, by construction this limit is g_n. Thus $g_n \in \mathcal{L}_\mu(X, \mathcal{A}, \mathbb{R})$, and we also have

$$\int g_n d\mu \leq 2 \int g\, d\mu.$$

The sequence $(g_n)_{n\geq 1}$ is nonincreasing and since by hypothesis (f_n) converges pointwise a.e. to f, the sequence (g_n) converges pointwise a.e. to 0. By the monotone convergence theorem (Theorem 5.31),

$$\lim_{n\mapsto\infty} \int g_n\, d\mu = 0.$$

Hence, by definition of g_n, the sequence (f_n) is actually an $\| \|_1$-Cauchy sequence, and by Proposition 5.28, we have $f \in \mathcal{L}_\mu(X, \mathcal{A}, F)$ and (f_n) converges to f in the $\| \|_1$-norm. We have

$$\begin{aligned} \left\| \int f_n d\mu - \int f d\mu \right\| &= \left\| \int (f_n - f) d\mu \right\| \\ &\leq \int \|f_n - f\| \, d\mu \\ &= \|f_n - f\|_1 , \end{aligned}$$

and since (f_n) converges to f in the $\| \|_1$-norm, this implies that

$$\lim_{n \mapsto \infty} \int f_n d\mu = \int f d\mu,$$

as claimed. ☐

The first important application of Theorem 5.34 is to provide a characterization of the integrability of a function $f \in \mathcal{M}_\mu(X, \mathcal{A}, F)$ in terms of $\int \|f\| \, d\mu$.

Theorem 5.35. *A function $f\colon X \to F$ is integrable, that is, $f \in \mathcal{L}_\mu(X, \mathcal{A}, F)$, iff $f \in \mathcal{M}_\mu(X, \mathcal{A}, F)$ and $\|f\| \in \mathcal{L}_\mu(X, \mathcal{A}, \mathbb{R})$. More generally, if $f \in \mathcal{M}_\mu(X, \mathcal{A}, F)$ and if there is a function $g \in \mathcal{L}_\mu(X, \mathcal{A}, \mathbb{R})$ such that $g \geq 0$ and $\|f\| \leq g$, then $f \in \mathcal{L}_\mu(X, \mathcal{A}, F)$.*

Proof. By Proposition 5.21, the space $\mathcal{L}_\mu(X, \mathcal{A}, F)$ is a subspace of $\mathcal{M}_\mu(X, \mathcal{A}, F)$, and we already know from Proposition 5.24(4) that if $f \in \mathcal{L}_\mu(X, \mathcal{A}, F)$ then $\|f\| \in \mathcal{L}_\mu(X, \mathcal{A}, \mathbb{R})$.

For the converse, we may assume that f and g are measurable, since a μ-measurable function is equal a.e. to a measurable function. There is a sequence $(h_n)_{n \geq 1}$ of μ-step maps that converges pointwise a.e. to f. For every $x \in X$ and every $n \geq 1$, let

$$h'_n(x) = \begin{cases} h_n(x) & \text{if } \|h_n(x)\| \leq 2g(x) \\ 0 & \text{if } \|h_n(x)\| > 2g(x). \end{cases}$$

For every $n \geq 1$, the function h'_n is a μ-step function and $\|h'_n\| \leq 2g \in \mathcal{L}_\mu(X, \mathcal{A}, \mathbb{R})$. We claim that for every $x \in X$ such that $(h_n(x))$ converges to $f(x)$, the sequence $(h'_n(x))$ also converges to $f(x)$. If $g(x) = 0$, then $\|f(x)\| = 0$, so $f(x) = 0$, and then $h'_n(x) = 0$ for all $n \geq 1$. If $g(x) \neq 0$, since the sequence $(h_n(x))$ converges to $f(x)$ and since $\|f(x)\| < 2g(x)$, there is some $M > 0$ such that $\|h_n(x)\| \leq 2g(x)$ for all $n \geq M$, which implies that $h'_n(x) = h_n(x)$. It follows that the sequence $(h'_n)_{n\geq 1}$ converges pointwise a.e. to f. By Theorem 5.34, since $\|h'_n\| \leq 2g$, we conclude that $f \in \mathcal{L}_\mu(X, \mathcal{A}, F)$. □

A useful corollary of Theorem 5.35 is the following result.

Proposition 5.36. *The following facts hold:*

(1) *If $f \in \mathcal{L}_\mu(X, \mathcal{A}, F)$, $g \in \mathcal{M}_\mu(X, \mathcal{A}, K)$ with $K = \mathbb{R}$ or $K = \mathbb{C}$, and $\|g\|$ is bounded, then $fg \in \mathcal{L}_\mu(X, \mathcal{A}, F)$.*
(2) *Let $h\colon E \times F \to G$ be a continuous bilinear map, where E, F, G are Banach spaces. If $f \in \mathcal{L}_\mu(X, \mathcal{A}, E)$ and $g \in \mathcal{M}_\mu(X, \mathcal{A}, F)$ with $\|g\|$ bounded, then $h(f, g) \in \mathcal{L}_\mu(X, \mathcal{A}, G)$.*
(3) *Let $f \in \mathcal{L}_\mu(X, \mathcal{A}, \mathbb{R})$, with $f \geq 0$, and let $g \in \mathcal{M}_\mu(X, \mathcal{A}, \mathbb{R})$ with values in an interval $[m, M]$. Then $fg \in \mathcal{L}_\mu(X, \mathcal{A}, \mathbb{R})$, and we have*
$$m \int f d\mu \leq \int g f d\mu \leq M \int f d\mu.$$

Another corollary involves series of functions in $\mathcal{L}_\mu(X, \mathcal{A}, F)$.

Proposition 5.37. *Let $(f_n)_{n\geq}$ be a sequence of functions $f_n \in \mathcal{L}_\mu(X, \mathcal{A}, F)$. If the series*
$$\sum_{n=1}^{\infty} \int \|f_n\| \, d\mu$$
converges, then the series
$$f(x) = \sum_{n=1}^{\infty} f_n(x)$$
converges a.e., $f \in \mathcal{L}_\mu(X, \mathcal{A}, F)$, and
$$\int f d\mu = \sum_{n=1}^{\infty} \int f_n \, d\mu$$

The following proposition is needed for the proof of several results stated in Chapter 7.

Proposition 5.38. *(Averaging Theorem) Let $f \in \mathcal{L}_\mu(X, \mathcal{A}, F)$ be any function and let S be any closed subset of F. If for any measurable subset A of finite measure $\mu(A) > 0$ we have*
$$\frac{1}{\mu(A)} \int_A f d\mu \in S,$$
and if $0 \in S$ or if X is σ-finite, then $f(x) \in S$ for all almost all $x \in X$.

Proposition 5.38 is proven in Lang [Lang (1996)] (Chapter VI, Theorem 5.15). By applying Proposition 5.38 to the set $S = \{0\}$, we obtain the following useful corollary.

Proposition 5.39. *For any function $f \in \mathcal{L}_\mu(X, \mathcal{A}, F)$, if*

$$\int_A f d\mu = 0$$

for every measurable subset A of finite measure, then $f = 0$ almost everywhere.

We conclude this section with two results about the continuity and the differentiability of a function defined by an integral.

Proposition 5.40. *Let $(X, \mathcal{A}, \mu)$ be a measure space, let U be metric space, let F be a Banach space (over $\mathbb{R}$ or $\mathbb{C}$), and let $f\colon U \times X \to F$ be a function.*

(1) (Continuity of the integral) Assume that f has the following properties:

(a) For every $u \in U$, the map $f_{u,-}\colon X \to F$ given by

$$f_{u,-}(x) = f(u, x) \quad x \in X,$$

belongs to $\mathcal{L}_\mu(X, \mathcal{A}, F)$,

(b) For every $x \in X$, the map $f_{-,x}\colon U \to F$ given by

$$f_{-,x}(u) = f(u, x) \quad u \in U,$$

is continuous.

(c) There is some $g \in \mathcal{L}_\mu(X, \mathcal{A}, \mathbb{R})$, $g \geq 0$, such that

$$\|f(u, x)\| \leq g(x) \quad \text{for all } u \in U, \text{ and all } x \in X.$$

Then the map $h\colon U \to F$ given by

$$h(u) = \int f_{u,-}\, d\mu$$

is continuous.

(2) (Taking a derivative under the integral sign) Suppose U is an open subset of a Banach space G, and let $\mathcal{L}(G; F)$ be the space of linear continuous maps from G to F with the operator norm (see Definition A.50). Assume that f has the following properties:

(d) For every $u \in U$, the map $f_{u,-}\colon X \to F$ given by

$$f_{u,-}(x) = f(u, x) \quad x \in X,$$

belongs to $\mathcal{L}_\mu(X, \mathcal{A}, F)$,

(e) *For every $x \in X$, the map $f_{-,x}: U \to F$ is differentiable, and let $Df_{-,x}$ be this derivative (a map from U to $\mathcal{L}(G; F)$).*

(f) *For every $u \in U$, the map from X to $\mathcal{L}(G; F)$ given by*

$$x \mapsto Df_{-,x}(u)$$

belongs to $\mathcal{L}_\mu(X, \mathcal{A}, \mathcal{L}(G; F))$, and there is some $g \in \mathcal{L}_\mu(X, \mathcal{A}, \mathbb{R})$, $g \geq 0$, such

$$\|Df_{-,x}(u)\| \leq g(x) \quad \text{for all } u \in U, \text{ and all } x \in X.$$

Then the map $h: U \to F$ given by

$$h(u) = \int f_{u,-}\, d\mu$$

is differentiable in U, and its derivative at $u \in U$ is given by

$$Dh_u = \int Df_{-,x}(u)\, d\mu.$$

Proposition 5.40 is proven in Marle [Marle (1974)] (Proposition 2.4.10).

More could be said about the applications of the convergence theorems, but we have everything we need.

Remark: There is another approach to the definition of the integral that applies only to real and complex-valued functions, presented in various texts such as Rudin [Rudin (1987)]. In this approach, positive functions play a central role. This approach relies on the fact that for any measurable function $f: X \to [0, +\infty]$ there is a monotonic sequence (f_n) of positive step functions that converges pointwise to f; this method is presented in Rudin [Rudin (1987)] (Chapter 1, Theorem 1.17). The integral of a step function is defined in the usual way. Then given any *measurable* function $f: X \to [0, +\infty]$, the integral of f is defined as

$$\int f d\mu = \sup_{0 \leq s \leq f} \int s\, d\mu,$$

where s is a step function.

A main difference with the approach we followed is that this definition of the integral allows it to take the value $+\infty$. Of course, later on, in order to define what it means for a measurable complex-valued function $f: X \to \mathbb{C}$ to be integrable, the condition

$$\int |f| d\mu < +\infty$$

is required. Thus in this approach, the space $\mathcal{L}_\mu(X, \mathcal{A}, \mathbb{C})$ is defined as the space of measurable functions such that the positive function $|f|$ has a finite integral.

In the approach that we followed, due to Bochner and Dunford, the space $\mathcal{L}_\mu(X, \mathcal{A}, \mathbb{C})$ is defined in terms of various Cauchy sequences, and the fact that if a function $f\colon X \to \mathbb{C}$ is measurable and if $|f|$ has a finite integral, then $f \in \mathcal{L}_\mu(X, \mathcal{A}, \mathbb{C})$, is a *theorem* (Theorem 5.35). Ultimately, it is proved that $\mathcal{L}_\mu(X, \mathcal{A}, \mathbb{C})$ is complete (see Rudin [Rudin (1987)] Chapter 3, Theorem 3.11), and it is observed that as a corollary, from a $\| \|_1$-Cauchy sequence, one can extract a subsequence that converges pointwise a.e. (Rudin [Rudin (1987)] Chapter 3, Theorem 3.12). It is also shown that the μ-step functions are dense in $\mathcal{L}_\mu(X, \mathcal{A}, \mathbb{C})$ (Rudin [Rudin (1987)] Chapter 3, Theorem 3.13).

The circle is closed. What we took as a definition of $\mathcal{L}_\mu(X, \mathcal{A}, \mathbb{C})$ is obtained as a corollaries in the other approach, and the two approaches yield the same notion of integrability (the same space $\mathrm{L}_\mu(X, \mathcal{A}, \mathbb{C})$).

One might argue that the approach relying on the integral of positive functions is simpler, or at least takes less efforts. For one thing, it does not need the refined notion of μ-step maps and μ-measurable maps. However, our feeling is that the approach we followed provides a better understanding of the structure of $\mathcal{L}_\mu(X, \mathcal{A}, \mathbb{C})$. Also, it can't be avoided if one wants to integrate functions with values in an infinite-dimensional vector space.

5.10 The Spaces $\mathcal{L}^p_\mu(X, \mathcal{A}, F)$ and $\mathrm{L}^p_\mu(X, \mathcal{A}, F)$; $p = 1, 2$

Theorem 5.35 suggests the definition of other families of integrable functions.

Definition 5.15. Let $(X, \mathcal{A}, \mu)$ be a measure space and let $(F, \mathcal{B})$ be a measurable space, with F a Banach space and $\mathcal{B}$ its Borel σ-algebra. For any $p \geq 1$, the set of functions $\mathcal{L}^p_\mu(X, \mathcal{A}, F)$ is the set of functions $f \in \mathcal{M}_\mu(X, \mathcal{A}, F)$ such that $\|f\|^p \in \mathcal{L}_\mu(X, \mathcal{A}, \mathbb{R})$, or equivalently

$$\int \|f\|^p \, d\mu < +\infty.$$

By Theorem 5.35, we have $\mathcal{L}^1_\mu(X, \mathcal{A}, F) = \mathcal{L}_\mu(X, \mathcal{A}, F)$, and we know that $\mathcal{L}^1_\mu(X, \mathcal{A}, F)$ is a vector space. Although it is possible to develop a theory of $\mathcal{L}^p$ spaces for any $p \geq 1$, for our applications to harmonic analysis we only need the cases $p = 1, 2$. The case where $p = \infty$ arises when we consider duality, but we postpone the definition of $\mathcal{L}^\infty_\mu(X, \mathcal{A}, F)$.

The space $\mathcal{L}^2_\mu(X, \mathcal{A}, F)$ is particularly interesting because if F is a Hilbert space, then it can be given a Hilbert space structure which uses the inner product on F (not quite, because the Hermitian form $\langle -, - \rangle_\mu$ that we obtain is not positive definite, which means that $\langle f, f \rangle_\mu = 0$ does necessarily imply that $f = 0$).

Let us start with the simple case where $F = \mathbb{C}$. If $f \colon X \to \mathbb{C}$ is a complex-valued function, then by $|f|^2$ we mean the function defined such that

$$|f|^2(x) = f(x)\overline{f(x)} \quad \text{for all } x \in X.$$

For any two functions $f, g \colon X \to \mathbb{C}$, by $\langle f, g \rangle$ we mean the function defined such that

$$\langle f, g \rangle(x) = f(x)\overline{g(x)} \quad \text{for all } x \in X.$$

For the more general case where F is a Hilbert space with Hermitian inner product $\langle -, - \rangle_F$, for any two functions $f, g \colon X \to F$, then by $\langle f, g \rangle$ we mean the function defined such that

$$\langle f, g \rangle(x) = \langle f(x), g(x) \rangle_F \quad \text{for all } x \in X.$$

In particular, since $\langle f, f \rangle$ is the function given by $\langle f, f \rangle(x) = \langle f(x), g(x) \rangle_F$ and $\|f\|^2$ is the function given by

$$\|f\|^2 (x) = \|f(x)\|^2 = \langle f(x), f(x) \rangle_F,$$

we have

$$\|f\|^2 = \langle f, f \rangle.$$

To simplify notation we will drop the subscript F when referring to the inner product on F.

From now on, *when dealing with $\mathcal{L}^2_\mu(X, \mathcal{A}, F)$, we assume that F is a Hilbert space* (over $\mathbb{C}$). If the reader feels more comfortable, he/she may assume that $F = \mathbb{C}$, but significant simplifications do not arise.

Proposition 5.41. *The set $\mathcal{L}^2_\mu(X, \mathcal{A}, F)$ is a vector space. For any two maps $f, g \in \mathcal{L}^2_\mu(X, \mathcal{A}, F)$, we have $\langle f, g \rangle \in \mathcal{L}^1_\mu(X, \mathcal{A}, \mathbb{C})$, and the map*

$$(f, g) \mapsto \int \langle f, g \rangle d\mu$$

is a Hermitian positive map (not necessarily positive definite).

Proof. It is easy to see that $\langle f, g\rangle$ is a limit of step maps a.e., so $\langle f, g\rangle \in \mathcal{M}_\mu(X, \mathcal{A}, \mathbb{C})$, and by the Cauchy–Schwarz inequality, we have the standard inequality

$$2|\langle f, g\rangle| \leq \|f\|^2 + \|g\|^2,$$

with $\|f\|^2 + \|g\|^2 \in \mathcal{L}^1_\mu(X, \mathcal{A}, \mathbb{R})$ since $f, g \in \mathcal{L}^2_\mu(X, \mathcal{A}, F)$. By Theorem 5.35, we have $\langle f, g\rangle \in \mathcal{L}^1_\mu(X, \mathcal{A}, \mathbb{C})$, so $\int \langle f, g\rangle d\mu$ is well defined. If $f \in \mathcal{L}^2_\mu(X, \mathcal{A}, F)$ and $g \in \mathcal{L}^2_\mu(X, \mathcal{A}, F)$, then since

$$\|f + g\|^2 \leq \|f\|^2 + 2|\langle f, g\rangle| + \|g\|^2,$$

as all the functions on the right-hand side are in $\mathcal{L}^1_\mu(X, \mathcal{A}, \mathbb{R})$, we have $f + g \in \mathcal{L}^2_\mu(X, \mathcal{A}, F)$. For any $\lambda \in \mathbb{C}$, we have $|\lambda|^2 \|f\|^2 \in \mathcal{L}^1_\mu(X, \mathcal{A}, \mathbb{R})$, so $\lambda f \in \mathcal{L}^2_\mu(X, \mathcal{A}, F)$. Thus, $\mathcal{L}^2_\mu(X, \mathcal{A}, F)$ is a vector space. Using the linearity of the integral, it is easy to check that the map

$$(f, g) \mapsto \int \langle f, g\rangle d\mu$$

is a Hermitian positive map. □

Definition 5.16. For any two functions $f, g \in \mathcal{L}^2_\mu(X, \mathcal{A}, F)$, the Hermitian map $\langle f, g\rangle_\mu$ is defined by

$$\langle f, g\rangle_\mu = \int \langle f, g\rangle d\mu.$$

The L^2-semi-norm $\|f\|_2$ is given by

$$\|f\|_2 = \sqrt{\langle f, f\rangle_\mu} = \left(\int \langle f, f\rangle d\mu\right)^{1/2} = \left(\int \|f\|^2 d\mu\right)^{1/2}.$$

It is a standard result of linear algebra that the *Cauchy–Schwarz* inequality holds:

$$|\langle f, g\rangle_\mu| \leq \|f\|_2 \|g\|_2 .$$

As a consequence $\| \; \|_2$ is a semi-norm.

Proposition 5.42. *For any $f \in \mathcal{L}^2_\mu(X, \mathcal{A}, F)$, we have $\|f\|_2 = 0$ iff $f = 0$ a.e.*

Proof. If $f = 0$ a.e., then $\langle f, f\rangle = 0$ a.e., so $\|f\|_2^2 = \int \langle f, f\rangle d\mu = 0$. Conversely, if $\|f\|_2 = 0$, then this means that $\int \langle f, f\rangle d\mu = 0$, but $\langle f, f\rangle \in \mathcal{L}^1_\mu(X, \mathcal{A}, \mathbb{R})$ is a positive function, so we know from Proposition 5.27 that $\langle f, f\rangle = 0$ a.e., that is, $f = 0$ a.e. □

If X has finite measure, then $\mathcal{L}^2_\mu(X, \mathcal{A}, F)$ is contained in $\mathcal{L}^1_\mu(X, \mathcal{A}, F)$.

Proposition 5.43. *If X has finite measure, then for any $f \in \mathcal{L}^2_\mu(X, \mathcal{A}, F)$, we have $\|f\|_1 \leq \|f\|_2 \|1_X\|_2$, and $\mathcal{L}^2_\mu(X, \mathcal{A}, F)$ is contained in $\mathcal{L}^1_\mu(X, \mathcal{A}, F)$.*

Proof. The function $\|f\|$ (namely $x \mapsto \|f(x)\|$) is complex-valued so we can apply the Cauchy–Schwarz inequality to $\|f\|$ and to the constant function 1_X equal to 1 on X. To be more specific, since

$$\langle \|f\|, 1_X \rangle = \|f\| \overline{1_X} = \|f\|,$$

we have

$$\langle \|f\|, 1_X \rangle_\mu = \int \langle \|f\|, 1_X \rangle d\mu = \int \|f\| \, d\mu = \|f\|_1,$$

and

$$\|\|f\|\|_2^2 = \langle \|f\|, \|f\| \rangle_\mu = \int \langle \|f\|, \|f\| \rangle d\mu = \int \|f\|^2 \, d\mu = \|f\|_2^2,$$

and so the Cauchy–Schwarz inequality (for functions in $\mathcal{L}^2_\mu(X, \mathcal{A}, \mathbb{C})$)

$$\langle \|f\|, 1_X \rangle_\mu \leq \|f\|_2 \|1_X\|_2$$

implies that $\|f\|_1 \leq \|f\|_2 \|1_X\|_2$. Obviously, this inequality shows that $\mathcal{L}^2_\mu(X, \mathcal{A}, F)$ is contained in $\mathcal{L}^1_\mu(X, \mathcal{A}, F)$. $\square$

It should be noted that if X has finite measure then the inclusion can be strict, and if X has infinite measure, then in general there are no inclusion properties.

Example 5.4.

(1) If $X = (0, 1)$, with the Lebesgue measure, then $\frac{1}{\sqrt{x}} \in \mathcal{L}^1((0, 1), \mu_L)$ but $\frac{1}{\sqrt{x}} \notin \mathcal{L}^2((0, 1), \mu_L)$; see Figure 5.18.
(2) If $X = (1, \infty)$ with the Lebesgue measure, then $\frac{1}{x} \in \mathcal{L}^2((1, \infty), \mu_L)$ but $\frac{1}{x} \notin \mathcal{L}^1((1, \infty), \mu_L)$; see Figure 5.19.
(3) If $X = (0, \infty)$ with the Lebesgue measure, then $\frac{1}{(x+1)\sqrt{x}} \in \mathcal{L}^1((0, \infty), \mu_L)$ but $\frac{1}{(x+1)\sqrt{x}} \notin \mathcal{L}^2((0, \infty), \mu_L)$; see Figure 5.20.

One of the main properties of $\mathcal{L}^2_\mu(X, \mathcal{A}, F)$ is that it is complete for the semi-norm $\| \; \|_2$. By taking the quotient of $\mathcal{L}^2_\mu(X, \mathcal{A}, F)$ by the space of function equal to 0 a.e., we obtain a Hilbert space.

Theorem 5.44. *Let $(f_n)_{n \geq 1}$ be an $\| \; \|_2$-Cauchy sequence of functions $f_n \in \mathcal{L}^2_\mu(X, \mathcal{A}, F)$. Then there is a function $f \in \mathcal{L}^2_\mu(X, \mathcal{A}, F)$ with the following properties:*

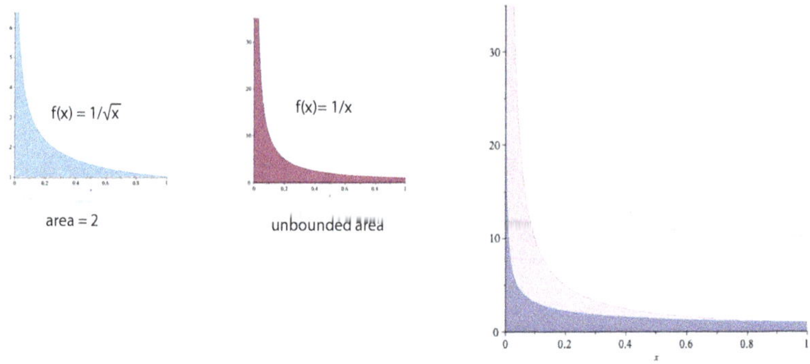

Fig. 5.18 The first figure shows that $\frac{1}{\sqrt{x}} \in \mathcal{L}^1((0,1),\mu_L)$ since the area between f and the x-axis has finite value, while the second figure shows that $\frac{1}{\sqrt{x}} \notin \mathcal{L}^2((0,1),\mu_L)$. The third figure shows a direct comparison between the areas under the respective graphs.

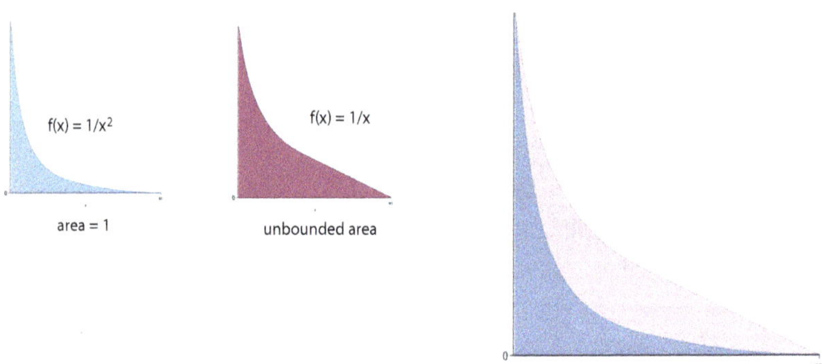

Fig. 5.19 The first figure shows that $\frac{1}{x} \in \mathcal{L}^2((1,\infty),\mu_L)$ since the area between f^2 and the x-axis has finite value, while the second figure shows that $\frac{1}{x} \notin \mathcal{L}^1((1,\infty),\mu_L)$. The third figure shows a direct comparison between the areas under the respective graphs.

(1) *The sequence* $(f_n)_{n\geq 1}$ *converges to* f *in the* $\|\ \|_2$*-semi-norm. Thus* $\mathcal{L}^2_\mu(X,\mathcal{A},F)$ *is complete.*

There is a subsequence $(f_{n_k})_{k\geq 1}$ *of* $(f_n)_{n\geq 1}$ *with the following properties:*

(2) *The subsequence* $(f_{n_k})_{k\geq 1}$ *converges pointwise a.e. to* f.

(3) *For every* $\epsilon > 0$, *there is a subset* Z *such that* $\mu(Z) < \epsilon$ *and the subsequence* $(f_{n_k})_{k\geq 1}$ *converges uniformly to* f *on* $X - Z$.

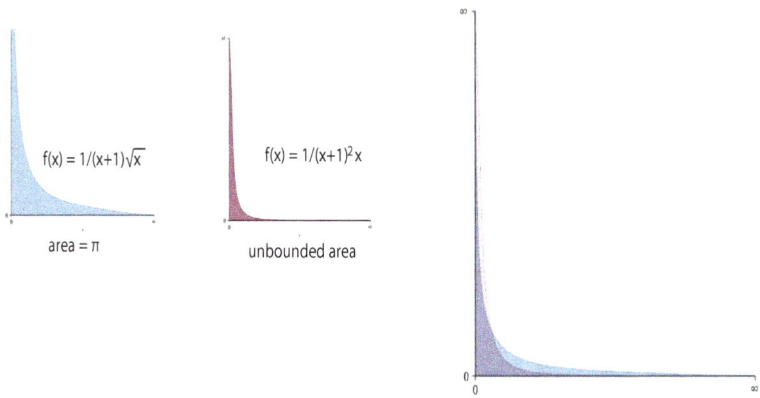

Fig. 5.20 The first figure shows that $\frac{1}{(x+1)\sqrt{x}} \in \mathcal{L}^1((0,\infty), \mu_L)$s since the area between f and the x-axis has finite value, while the second figure shows that $\frac{1}{(x+1)\sqrt{x}} \notin \mathcal{L}^2((0,\infty), \mu_L)$. The third figure shows a direct comparison between the areas under the respective graphs.

A proof of Theorem 5.44 is given in Lang [Lang (1996)] (Chapter VII, §1).

In view of Proposition 5.42, we make the following definition.

Definition 5.17. Let $\mathrm{L}^2_\mu(X, \mathcal{A}, F)$ be the quotient of the vector space $\mathcal{L}^2_\mu(X, \mathcal{A}, F)$ by the subspace of functions equal to 0 a.e. (which is the set of functions f such that $\|f\|_2 = 0$). The norm induced by the semi-norm $\| \, \|_2$ on $\mathrm{L}^2_\mu(X, \mathcal{A}, F)$ is called the *L^2-norm*.

Obviously the positive Hermitian form $\langle f, g\rangle_\mu$ induces a positive definite Hermitian form on $\mathrm{L}^2_\mu(X, \mathcal{A}, F)$. Theorem 5.44 immediately implies the following result.

Theorem 5.45. *(Fischer–Riesz) The space $\mathrm{L}^2_\mu(X, \mathcal{A}, F)$ is a Hilbert space under the positive definite Hermitian form induced by $\langle -, -\rangle_\mu$.*

Definition 5.18. The norm $\| \, \|_2$ associated with the inner product $\langle -, -\rangle_\mu$ on $\mathrm{L}^2_\mu(X, \mathcal{A}, F)$ is called the *L^2-norm*.

Example 5.5. In the special case where $X = \mathbb{N}$ (or $X = \mathbb{Z}$), $\mathcal{A} = 2^X$, μ is the counting measure, and $F = \mathbb{C}$, as in Example 5.2, we see that for

$p = 1, 2$, we have

$$\mathrm{L}^p_\mu(X, \mathcal{A}, \mathbb{C}) = \{(x_n)_{n\in\mathbb{N}} \mid x_n \in \mathbb{C}, \sum_{n\in\mathbb{N}} |x_n|^p < \infty\}.$$

It is customary to denote this space by $\ell^p(\mathbb{N})$. We define $\ell^p(\mathbb{Z})$ similarly by replacing $\mathbb{N}$ by $\mathbb{Z}$.

We will show shortly that the space of μ-step functions is dense in $\mathrm{L}^2_\mu(X, \mathcal{A}, F)$ (for the L^2-norm). First here is a corollary of Theorem 5.45.

Proposition 5.46. *If $(f_n)_{n\geq 1}$ is a $\| \, \|_2$-Cauchy sequence of functions $f_n \in \mathcal{L}^2_\mu(X, \mathcal{A}, F)$, and if $(f_n)_{n\geq 1}$ converges pointwise a.e. to a function $f \colon X \to F$, then $f \in \mathcal{L}^2_\mu(X, \mathcal{A}, F)$, and $(f_n)_{n\geq 1}$ converges to f in the $\| \, \|_2$-semi-norm.*

The Lebesgue dominated convergence theorem also holds for $\mathcal{L}^2_\mu(X, \mathcal{A}, F)$.

Theorem 5.47. *(Lebesgue Dominated Convergence Theorem for $\mathcal{L}^2_\mu$) Let $(f_n)_{n\geq 1}$ be a sequence of functions $f_n \in \mathcal{L}^2_\mu(X, \mathcal{A}, F)$. If (f_n) converges pointwise a.e. to a function $f \colon X \to F$, and if there is some function $g \in \mathcal{L}^2_\mu(X, \mathcal{A}, \mathbb{R})$ such that $g \geq 0$ and $\|f_n\| \leq g$ for all $n \geq 1$, then $f \in \mathcal{L}^2_\mu(X, \mathcal{A}, F)$ and $(f_n)_{n\geq 1}$ converges to f in the $\| \, \|_2$-norm. Consequently*

$$\lim_{n\mapsto\infty} \int f_n d\mu = \int f d\mu.$$

A proof of Theorem 5.47 is given in Lang [Lang (1996)] (Chapter VII, §1).

The following version of Theorem 5.35 also holds for $\mathcal{L}^2_\mu(X, \mathcal{A}, F)$.

Theorem 5.48. *A function $f \colon X \to F$ is $\mathcal{L}^2$-integrable, that is, $f \in \mathcal{L}^2_\mu(X, \mathcal{A}, F)$, iff $f \in \mathcal{M}_\mu(X, \mathcal{A}, F)$ and $\|f\|_2 \in \mathcal{L}^2_\mu(X, \mathcal{A}, \mathbb{R})$.*

As a corollary of Theorem 5.47 we can show that the μ-step functions are dense in $\mathcal{L}^2_\mu(X, \mathcal{A}, F)$.

Proposition 5.49. *The subspace $\mathcal{S}tep_\mu(X, \mathcal{A}, F)$ is dense in $\mathcal{L}^2_\mu(X, \mathcal{A}, F)$ with respect to the L^2-semi-norm.*

Proof. Let $f \in \mathcal{L}^2_\mu(X, \mathcal{A}, F)$. Since f is μ-measurable, there is a sequence $(f_n)_{n\geq 1}$ of μ-step functions f_n that converges pointwise a.e. to f. For every

$n \geq 1$ and every $x \in X$, define g_n by

$$g_n(x) = \begin{cases} f_n(x) & \text{if } \|f_n(x)\| \leq 2\|f(x)\| \\ 0 & \text{if } \|f_n(x)\| > 2\|f(x)\|. \end{cases}$$

We may assume that f is measurable since it differs from a measurable function on a set of measure zero. Then the functions g_n are μ-step functions, they satisfy the inequality $\|g_n\| \leq 2\|f\|$ with $2\|f\| \in \mathcal{L}^2_\mu(X, \mathcal{A}, \mathbb{R})$, and the sequence (g_n) converges a.e. to f. By Theorem 5.47, the sequence (g_n) converges to f in the $\| \; \|_2$-norm, which proves that $\mathcal{S}tep_\mu(X, \mathcal{A}, F)$ is dense in $\mathcal{L}^2_\mu(X, \mathcal{A}, F)$ with respect to the L^2-semi-norm $\square$

We now would like to understand the duals of $\mathrm{L}^1_\mu(X, \mathcal{A}, F)$ and $\mathrm{L}^2_\mu(X, \mathcal{A}, F)$, that is, the spaces of continuous linear forms on $\mathrm{L}^1_\mu(X, \mathcal{A}, F)$ and $\mathrm{L}^2_\mu(X, \mathcal{A}, F)$ (with values in $\mathbb{C}$). In the case of $\mathrm{L}^2_\mu(X, \mathcal{A}, F)$, it is a classical theorem (the *Riesz respresentation theorem*) that the dual of a Hilbert space is isomorphic to itself, so the dual of $\mathrm{L}^2_\mu(X, \mathcal{A}, F)$ is isomorphic to $\mathrm{L}^2_\mu(X, \mathcal{A}, F)$. In the case of $\mathrm{L}^1_\mu(X, \mathcal{A}, F)$, it turns out that its dual is isomorphic to a space denoted $\mathrm{L}^\infty_\mu(X, \mathcal{A}, F)$. Here we assumed that F is a Hilbert space.

5.11 The Spaces $\mathcal{L}^\infty_\mu(X, \mathcal{A}, F)$ and $\mathrm{L}^\infty_\mu(X, \mathcal{A}, F)$

To define $\mathcal{L}^\infty_\mu(X, \mathcal{A}, F)$, we only need the fact that F is a Banach space. The space $\mathcal{L}^\infty_\mu(X, \mathcal{A}, F)$ consists of all functions $f \colon X \to F$ that are equal to a bounded μ-measurable function a.e. We can define a semi-norm on $\mathcal{L}^\infty_\mu(X, \mathcal{A}, F)$ as follows.

Definition 5.19. For any function $f \in \mathcal{M}_\mu(X, \mathcal{A}, F)$, define the *essential sup* or *semi-norm* $N_\infty(f)$ of f by

$$N_\infty(f) = \inf\{\alpha \in \mathbb{R}_+ \mid \mu(\{x \in X \mid \|f(x)\| \geq \alpha\}) = 0\};$$

see Figure 5.21. The space $\mathcal{L}^\infty_\mu(X, \mathcal{A}, F)$ is the set of functions $f \in \mathcal{M}_\mu(X, \mathcal{A}, F)$ such that $N_\infty(f) < +\infty$.

Remark: We decided to use the notation $N_\infty(f)$ for the essential sup semi-norm to avoid the confusion with the sup norm, $\|f\|_\infty$, since these norms differ in general. In the case of the semi-norms $\|f\|_1$ and $\|f\|_2$ there is little risk of confusion. A number of authors prefer the notation $N_p(f)$, but the notation $\| \; \|_p$ seems more prevalent (if $1 \leq p < \infty$). Another way to avoid confusion is to use the notation $\| \; \|_{\mathrm{L}^p}$ (even if $p = \infty$).

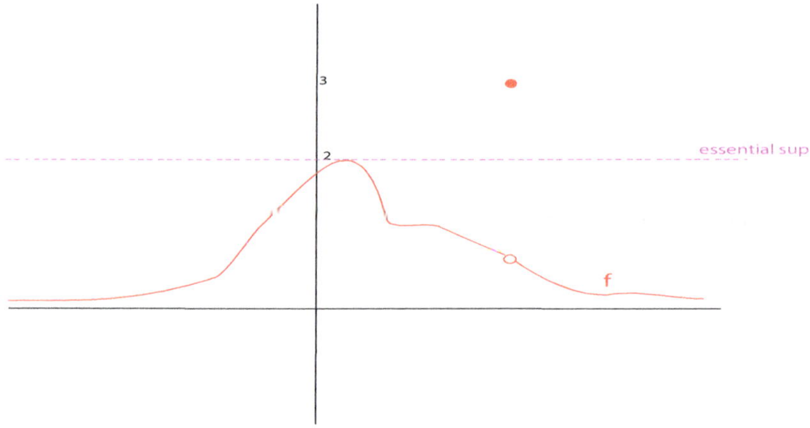

Fig. 5.21 Let $X = F = \mathbb{R}$ with absolute value as norm and $\mathcal{A}$ as the Borel σ-algebra. The graph of $f \in \mathcal{M}_\mu(X, \mathcal{A}, F)$ is in red and has essential sup $N_\infty(f) = 2$. Note this is not the same as the sup norm for $f \in (F^X)_b$, which in this particular case is $\|f\|_\infty = 3$.

The definition of $N_\infty(f)$ makes is clear that $N_\infty(f) = 0$ iff $f = 0$ a.e. Observe that $N_\infty(f)$ is the greastest lower bound of the numbers $\alpha \geq 0$ such that a μ-measurable function f has the property that $\|f(x)\| \geq \alpha$ on a set of measure zero, in other words, such a μ-measurable function is bounded a.e.

The space $\mathcal{L}^\infty_\mu(X, \mathcal{A}, F)$ is a vector space. We also have the following result showing that $\mathcal{L}^\infty_\mu(X, \mathcal{A}, F)$ is complete in the semi-norm N_∞, but unless X has finite measure, the μ-step maps are *not* dense in $\mathcal{L}^\infty_\mu(X, \mathcal{A}, F)$.

Theorem 5.50. *The following properties hold.*

(1) The space $\mathcal{L}^\infty_\mu(X, \mathcal{A}, F)$ is complete in the semi-norm N_∞. Furthermore, if $(f_n)_{n\geq 1}$ is an N_∞-Cauchy sequence, then there is a set Z of measure zero such that $(f_n)_{n\geq 1}$ converges uniformly to f on $X - Z$.

(2) If F is finite-dimensional, then the step maps (not the μ-step maps) are dense in $\mathcal{L}^\infty_\mu(X, \mathcal{A}, F)$.

(3) If X has finite measure, then for every $\epsilon > 0$ and every $f \in \mathcal{L}^\infty_\mu(X, \mathcal{A}, F)$, there is a μ-step map s and a subset Z with $\mu(Z) < \epsilon$ such that

$$\|f - s\| < \epsilon \quad on \quad X - Z.$$

Theorem 5.50 is proven in Lang [Lang (1996)] (Chapter VII, Theorem 2.1).

Note that the constant with value 1 belongs to $\mathcal{L}_\mu^\infty(X, \mathcal{A}, \mathbb{C})$, so if X has infinite measure, there is no way that it is a uniform limit of μ-step maps, since a μ-step map vanishes outside of a set of finite measure.

Remark: If X has finite measure, then we have the inclusion $\mathcal{L}_\mu^\infty(X, \mathcal{A}, F) \subseteq \mathcal{L}_\mu^2(X, \mathcal{A}, F)$. In fact, $\mathcal{L}_\mu^\infty(X, \mathcal{A}, F) \subseteq \mathcal{L}_\mu^p(X, \mathcal{A}, F) \subseteq \mathcal{L}_\mu^q(X, \mathcal{A}, F)$ for all $p, q \geq 1$ with $p > q$; see Marle [Marle (1974)] (Chapter 4, Proposition 4.5.7).

Definition 5.20. Let $\mathrm{L}_\mu^\infty(X, \mathcal{A}, F)$ be the quotient of the vector space $\mathcal{L}_\mu^\infty(X, \mathcal{A}, F)$ by the subspace of functions equal to 0 a.e. (which is the set of functions f such that $N_\infty(f) = 0$). The norm induced by the semi-norm N_∞ on $\mathrm{L}_\mu^\infty(X, \mathcal{A}, F)$ is called the *L^∞-norm*.

It should be noted that both the monotone convergence theorem and the dominated convergence theorem *fail* for $\mathcal{L}_\mu^\infty(X, \mathcal{A}, F)$. Convergence in the N_∞-semi-norm fails; see Marle [Marle (1974)] (Chapter 4, Section 2).

We now consider the duality between the spaces $\mathrm{L}_\mu^1(X, \mathcal{A}, F)$ and $\mathrm{L}_\mu^\infty(X, \mathcal{A}, F)$. The field $\mathbb{C}$ is a Hilbert space, but for the general case we need to assume that F is a Hilbert space. The key point is that by Proposition 5.36(2), for any $f \in \mathrm{L}_\mu^1(X, \mathcal{A}, F)$ and any $g \in \mathrm{L}_\mu^\infty(X, \mathcal{A}, F)$, then $\langle f, g\rangle \in \mathrm{L}_\mu^1(X, \mathcal{A}, \mathbb{C})$.

Definition 5.21. For any functions $f \in \mathrm{L}_\mu^1(X, \mathcal{A}, F)$ and $g \in \mathrm{L}_\mu^\infty(X, \mathcal{A}, F)$, define $[f, g]_\mu$ by

$$[f, g]_\mu = \int \langle f, g\rangle d\mu.$$

We obtain a map

$$[-, -]_\mu \colon \mathrm{L}_\mu^1(X, \mathcal{A}, F) \times \mathrm{L}_\mu^\infty(X, \mathcal{A}, F) \to \mathbb{C}$$

which is a sesquilinear pairing.

For simplicity, let us consider the special case where $F = \mathbb{C}$. In this case, we can define a bilinear (as opposed to sesquilinear) pairing $[-, -]_\mu \colon \mathrm{L}_\mu^1(X, \mathcal{A}, \mathbb{C}) \times \mathrm{L}_\mu^\infty(X, \mathcal{A}, \mathbb{C}) \to \mathbb{C}$ given by

$$[f, g]_\mu = \int fg\, d\mu.$$

Observe that we intentionally used fg instead of $f\overline{g}$, because we simply want a *bilinear* pairing.

Whenever we have a bilinear pairing $\varphi\colon E \times F \to \mathbb{C}$, recall that we define the linear maps $l_\varphi\colon E \to F^*$ and $r_\varphi\colon F \to E^*$ such that, for every $u \in E$,

$$l_\varphi(u)(y) = \varphi(u, y) \quad \text{for all } y \in F,$$

and for every $v \in F$,

$$r_\varphi(v)(x) = \varphi(x, v) \quad \text{for all } x \in E.$$

Definition 5.22. A bilinear pairing φ is *nondegenerate* if for every $u \in E$, if $\varphi(u, v) = 0$ for all $v \in F$, then $u = 0$, and for every $v \in F$, if $\varphi(u, v) = 0$ for all $u \in E$, then $v = 0$.

Then if φ is nondegenerate, then the maps l_φ and r_φ are injective. They are not surjective in general.

If E is a normed vector space, then its dual E' is the space of all continuous linear maps from E to $\mathbb{C}$. We have $E' \subseteq E^*$, and the inclusion is strict if E is infinite-dimensional.

The following result holds. For simplicity of notation, we drop φ when writing l_φ and r_φ.

Theorem 5.51. *Assume $(X, \mathcal{A}, \mu)$ is a measure space and that μ is σ-finite. Then the bilinear pairing*

$$[-,-]_\mu\colon \mathrm{L}^1_\mu(X, \mathcal{A}, \mathbb{C}) \times \mathrm{L}^\infty_\mu(X, \mathcal{A}, \mathbb{C}) \to \mathbb{C}$$

is nondegenerate. It satisfies the inequality

$$|[f, g]_\mu| \leq \|fg\|_1 \leq \|f\|_1 \|g\|_\infty .$$

The map l is a norm-preserving injective linear map between $\mathrm{L}^1_\mu(X, \mathcal{A}, \mathbb{C})$ and the dual $\mathrm{L}^\infty_\mu(X, \mathcal{A}, \mathbb{C})'$ of $\mathrm{L}^\infty_\mu(X, \mathcal{A}, \mathbb{C})$, and the map r is a norm-preserving injective linear map between $\mathrm{L}^\infty_\mu(X, \mathcal{A}, \mathbb{C})$ and the dual $\mathrm{L}^1_\mu(X, \mathcal{A}, \mathbb{C})'$ of $\mathrm{L}^1_\mu(X, \mathcal{A}, \mathbb{C})$. Furthermore, the map

$$r\colon \mathrm{L}^\infty_\mu(X, \mathcal{A}, \mathbb{C}) \to \mathrm{L}^1_\mu(X, \mathcal{A}, \mathbb{C})'$$

is an isomorphism.

A proof of Theorem 5.51 is given in Lang [Lang (1996)] (Chapter VII, §2). Theorem 5.51 can be generalized to a Hilbert space F, one just has to exercise caution in defining l and r to deal with sequilinearity.

The map $l\colon \mathrm{L}^1_\mu(X, \mathcal{A}, \mathbb{C}) \to \mathrm{L}^\infty_\mu(X, \mathcal{A}, \mathbb{C})'$ is not surjective, and understanding which linear forms in $\mathrm{L}^\infty_\mu(X, \mathcal{A}, \mathbb{C})'$ can be represented by functions in $\mathrm{L}^1_\mu(X, \mathcal{A}, \mathbb{C})$ is a natural question. A partial answer to this question is the *Radon–Nikodym theorem*, but will this would lead us too far. The interested reader is referred to Lang [Lang (1996)] or Rudin [Rudin (1987)].

5.12 Products of Measure Spaces and Fubini's Theorem

The purpose of this section is to define, given two measure spaces $(X, \mathcal{A}, \mu)$ and $(Y, \mathcal{B}, \nu)$, the notion of product measure space and product measure. Then we will state *Fubini's theorem* (also known as *the theorem of Lebesgue–Fubini*), which allows us to compute the integral on a product space as two successive integrals. The technical details are surprisingly involved.

We begin by recalling what we did in Example 4.1. We defined the set $\mathcal{R}$ of *rectangles* in $X \times Y$ as follows:

$$\mathcal{R} = \{A \times B \in X \times Y \mid A \in \mathcal{A},\ B \in \mathcal{B}\}.$$

The set $\mathcal{R}$ is a semi-algebra, and it can be shown that the set $\mathcal{B}(\mathcal{R})$ of finite unions of pairwise disjoint sets in $\mathcal{R}$ is the smallest algebra containing the semi-algebra $\mathcal{R}$.

Definition 5.23. Let $\mathcal{A} \otimes \mathcal{B}$ be the smallest σ-algebra generated by $\mathcal{R}$ (and thus by $\mathcal{B}(\mathcal{R})$); see Proposition 4.3.[4]

The hard part is now to define a product measure λ on $\mathcal{A} \otimes \mathcal{B}$ which satisfies the natural identity

$$\lambda(A \times B) = \mu(A)\nu(B)$$

for all rectangles $A \times B$. Here as in Section 4.1 we use extended multiplication on $\overline{\mathbb{R}}_+$, where

$$a \cdot (+\infty) = (+\infty) \cdot a = +\infty$$

if $0 < a \leq +\infty$, and

$$0 \cdot (+\infty) = (+\infty) \cdot 0 = 0.$$

We need a few definitions.

Definition 5.24. Given any subset $E \subseteq X \times Y$, for any $x \in X$, we define the *section of E (determined by x)* as the subset E_x given by

$$E_x = \{y \in Y \mid (x, y) \in E\} \subseteq Y.$$

Similarly, for any $y \in Y$, we define the *section of E (determined by y)* as the subset E_y given by

$$E_y = \{x \in X \mid (x, y) \in E\} \subseteq X.$$

See Figure 5.22.

[4]The meaning of the tensor sign $\otimes$ in the notation $\mathcal{A} \otimes \mathcal{B}$ is a completely different from its meaning in a tensor product of vector spaces. Hopefully, the two notions will never appear together!

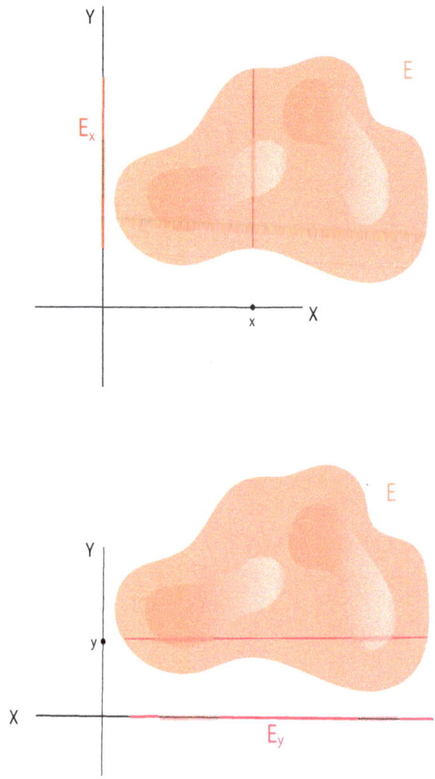

Fig. 5.22 Let $X = Y = \mathbb{R}$. The top figure illustrates an x-section of the peach set E, while the bottom figure illustrates a y-section.

Proposition 5.52. *The sections of any subset $E \in \mathcal{A} \otimes \mathcal{B}$ are measurable.*

Proof idea. Let $\mathcal{E}$ be the family of subsets of $X \times Y$ defined as follows:

$$\mathcal{E} = \{F \subseteq X \times Y \mid F_x \in \mathcal{B} \quad \text{for all } x \in X, \quad \text{and} \quad F_y \in \mathcal{A} \quad \text{for all } y \in Y\}.$$

These are the subsets of $X \times Y$ whose sections are measurable. Then prove that $\mathcal{E}$ is a σ-algebra containing $\mathcal{R}$, which implies that $\mathcal{E} = \mathcal{A} \otimes \mathcal{B}$. $\square$

Definition 5.25. Given any function $f\colon X \times Y \to F$ (where F is any set), for any $x \in X$, we define the *section of f (determined by x)* as the function $f_x\colon Y \to F$ given by

$$f_x(y) = f(x, y) \quad \text{for all } y \in Y.$$

Similarly, for any $y \in Y$, we define the *section of f (determined by y)* as the function $f_y\colon X \to F$ given by

$$f_y(x) = f(x, y) \quad \text{for all } x \in X.$$

See Figure 5.23.

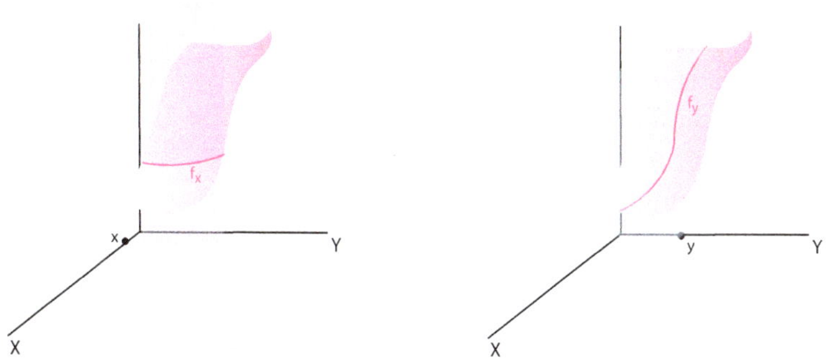

Fig. 5.23 Let $X = Y = F = \mathbb{R}$. The graph of f is the pink surface. The left figure illustrates a section of f determined by x, while the right figure illustrates a section of f determined by y

Proposition 5.53. *If $f\colon X \times Y \to \mathbb{R}$ is a measurable function (on $(X \times Y, \mathcal{A} \otimes \mathcal{B})$, then every section of f is measurable.*

Proof. By Proposition 4.12, it suffices to show that the inverse image of every open subset of the form $(-\infty, \alpha)$ is measurable.

For any $x \in X$, for any $\alpha \in \mathbb{R}$, we have

$$\begin{aligned}\{y \in Y \mid f_x(y) < \alpha\} &= \{y \in Y \mid f(x, y) < \alpha\} \\ &= \{(x, y) \in X \times Y \mid f(x, y) < \alpha\}_x,\end{aligned}$$

and this last subset is measurable by Proposition 5.52. The proof for f_y is similar. ☐

Definition 5.26. Given an algebra $\mathfrak{A}$ of sets, a *measure* on $\mathfrak{A}$ satisfies the same axioms as a measure on a σ-algebra; see Definition 4.9.

The next two results take a lot more work.

Proposition 5.54. *Let $(X, \mathcal{A}, \mu)$ and $(Y, \mathcal{B}, \nu)$ be two measure spaces, and assume that μ and ν are σ-finite. Then the map $\lambda\colon \mathcal{R} \to [0, +\infty]$ given by*

$$\lambda(A \times B) = \mu(A)\nu(B)$$

has a unique extension to a σ-finite measure on the algebra $\mathcal{B}(\mathcal{R})$.

A proof of Proposition 5.54 can be found in course notes given by Philippe G. Ciarlet in 1970–1971 at ENPC (Paris, France). Interestingly, the proof uses the monotone convergence theorem. A related treatment is given in Halmos [Halmos (1978)] (Chapter VII); see also Lang [Lang (1996)] (Chapter VI, §8) and Marle [Marle (1974)] (Chapter 5, Section 2).

Theorem 5.55. *Let $(X, \mathcal{A}, \mu)$ and $(Y, \mathcal{B}, \nu)$ be two measure spaces, and assume that μ and ν are σ-finite. Then the map $\lambda\colon \mathcal{R} \to [0, +\infty]$ given by*

$$\lambda(A \times B) = \mu(A)\nu(B)$$

has a unique extension to a measure $\lambda = \mu \otimes \nu$ is on the σ-algebra $\mathcal{A} \otimes \mathcal{B}$. The measure $\mu \otimes \nu$ is σ-finite.

The following properties hold for any measurable subset $E \in \mathcal{A} \times \mathcal{B}$:

(1) We have

$$(\mu \otimes \nu)(E) = \inf\left\{\sum_{i=1}^{\infty} \mu(A_i)\nu(B_i) \mid E \subseteq \bigcup_{i=1}^{\infty}(A_i \times B_i),\ A_i \in \mathcal{A},\ B_i \in \mathcal{B}\right\}. \qquad (*)$$

See Figure 5.24.

(2) The map ν_E from X to $\mathbb{R}_+$ given by $x \mapsto \nu(E_x)$ is measurable (w.r.t. $\mathcal{A}$), and the map μ_E from Y to $\mathbb{R}_+$ given by $y \mapsto \mu(E_y)$ is measurable (w.r.t. $\mathcal{B}$). One of these maps is integrable iff the other is integrable.

(3) We have

$$(\mu \otimes \nu)(E) = \begin{cases} \int \nu_E \, d\mu = \int \mu_E \, d\nu & \text{if both } \nu_E \text{ and } \mu_E \text{ are integrable} \\ +\infty & \text{otherwise.} \end{cases} \qquad (**)$$

See Figure 5.25.

A proof of Theorem 5.55 can be found in course notes given by Philippe G. Ciarlet in 1970–1971 at ENPC (Paris, France). Again, the proof uses

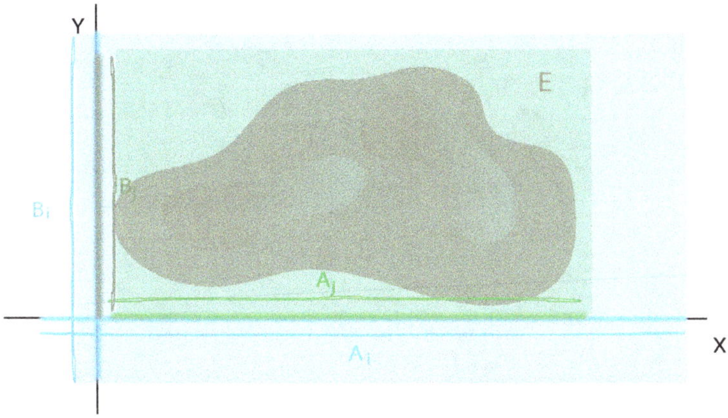

Fig. 5.24 A schematic illustration of Equation (∗) in Theorem 5.55, where the measure of the peach set E is calculated by the "area" of the rectangles.

the monotone convergence theorem. A related treatment is given in Halmos [Halmos (1978)] (Chapter VII); see also Lang [Lang (1996)] (Chapter VI, §8) and Marle [Marle (1974)] (Chapter 5, Section 2, Proposition 5.2.3).

If $(X, \mathcal{A}, \mu)$ and $(Y, \mathcal{B}, \nu)$ are two measure spaces with μ and ν both σ-finite, then for any Banach space F, we have the space of integrable functions $\mathcal{L}_{\mu\otimes\nu}(X \times Y, \mathcal{A} \otimes \mathcal{B}, F)$. The problem is to find a way to compute an integral $\iint f\, d(\mu \otimes \nu)$, also written $\iint f\, d\mu \otimes d\nu$, as two successive integrals. The answer is given by a theorem known as Fubini's theorem. The first version of this theorem was proved by Lebesgue and then was generalized by Fubini. For this reason some authors refer to this theorem as the Lebesgue–Fubini theorem, but it seems more common to call it simply Fubini's theorem.

Theorem 5.56. *(Fubini's Theorem, Part 1) Let $(X, \mathcal{A}, \mu)$ and $(Y, \mathcal{B}, \nu)$ be two measure spaces with μ and ν both σ-finite. Consider a function $f\colon X \times Y \to F$, where F is a Banach space. If $f \in \mathcal{L}_{\mu\otimes\nu}(X \times Y, \mathcal{A} \otimes \mathcal{B}, F)$ then:*

(1) The section $f_x\colon Y \to F$ is ν-integrable for almost all $x \in X$, the section $f_y\colon X \to F$ is μ-integrable for almost all $y \in Y$.
(2) The map from X to F defined a.e. by

$$x \mapsto \int f_x\, d\nu$$

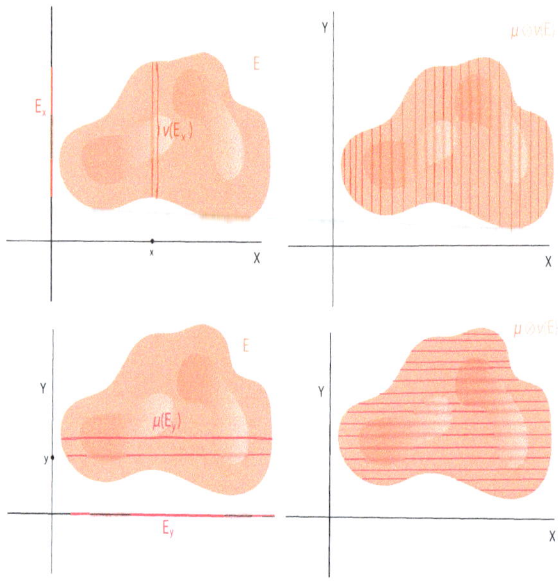

Fig. 5.25 Two ways of calculating $(\mu \otimes \nu)(E)$. The top row of figures illustrates $(\mu \otimes \nu)(E) = \int \nu_E \, d\mu$, where the vertical slices represent $\nu(E_x)$. The bottom row of figures illustrates $(\mu \otimes \nu)(E) = \int \mu_E \, d\nu$, where the horizontal slices represent $\mu(E_y)$.

is μ-integrable, and the map from Y to F defined a.e. by

$$y \mapsto \int f_y \, d\mu$$

is ν-integrable.

Then

$$\iint f d\mu \otimes d\nu = \int_X \left(\int_Y f_x \, d\nu \right) d\mu = \int_Y \left(\int_X f_y \, d\mu \right) d\nu;$$

see Figure 5.26.

Theorem 5.56 is proved in Marle [Marle (1974)] (Chapter 5, Section 2, Theorem 5.2.10), and Lang [Lang (1996)] (Chapter VI, §8, Theorem 8.4).

Theorem 5.56 assumes that f is *integrable*. It is possible to weaken this assumption at the price of strengthening the other conditions. However, this is worth it in practice.

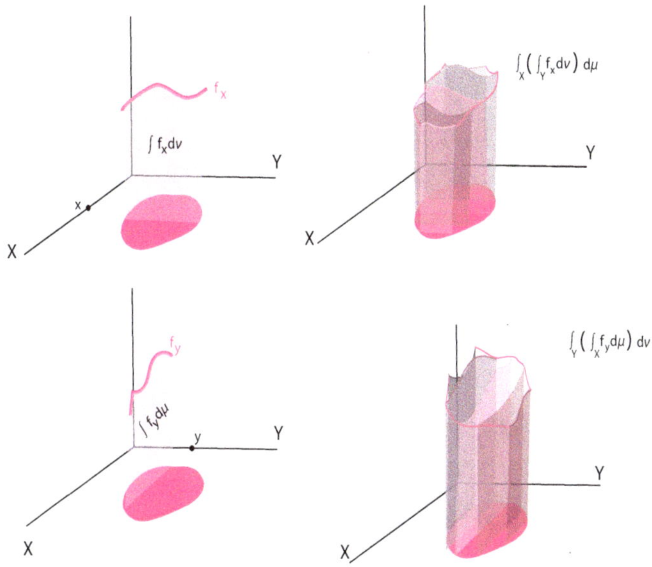

Fig. 5.26 Two ways of calculating $\iint f d\mu \otimes d\nu$ when $F = \mathbb{R}$. The top row of figures illustrates $\iint f d\mu \otimes d\nu = \int_X \left(\int_Y f_x \, d\nu\right) d\mu$ as the "volume" under the graph calculated by the stacked "areas" of $\int_Y f_x \, d\nu$ "sheets". The bottom row of figures illustrates $\iint f d\mu \otimes d\nu = \int_Y \left(\int_X f_y \, d\mu\right) d\nu$ as the "volume" under the graph calculated by the stacked "areas" of $\int_X f_y \, d\mu$ "sheets".

Theorem 5.57. *(Fubini's Theorem, Part 2) Let $(X, \mathcal{A}, \mu)$ and $(Y, \mathcal{B}, \nu)$ be two measure spaces with μ and ν both σ-finite. Consider a function $f: X \times Y \to F$, where F is a Banach space. If $f \in \mathcal{M}_{\mu\otimes\nu}(X \times Y, \mathcal{A} \otimes \mathcal{B}, F)$ and if the following conditions hold:*

(1) The section $f_x: Y \to F$ is ν-integrable for almost all $x \in X$, the section $f_y: X \to F$ is μ-integrable for almost all $y \in Y$.

(2) The map from X to $\mathbb{R}$ defined a.e. by

$$x \mapsto \int \|f_x\| \, d\nu$$

is μ-integrable, and the map from Y to $\mathbb{R}$ defined a.e. by

$$y \mapsto \int \|f_y\| \, d\mu$$

is ν-integrable.

Then $f \in \mathcal{L}_{\mu\otimes\nu}(X \times Y, \mathcal{A} \otimes \mathcal{B}, F)$ *and*

$$\iint f d\mu \otimes d\nu = \int_X \left(\int_Y f_x \, d\nu \right) d\mu = \int_Y \left(\int_X f_y \, d\mu \right) d\nu.$$

Theorem 5.57 is proved in Lang [Lang (1996)] (Chapter VI, §8, Theorem 8,7); see also Marle [Marle (1974)] (Chapter 5, Section 2).

In practice, it is customary to use a less formal notation to express Fubini's theorem, namely

$$\begin{aligned}\iint f(x,y)d\mu(x) \otimes d\nu(y) &= \int_X \left(\int_Y f(x,y) \, d\nu(y) \right) d\mu(x) \\ &= \int_Y \left(\int_X f(x,y) \, d\mu(x) \right) d\nu(y),\end{aligned}$$

and the measure $d\mu(x) \otimes d\nu(y)$ is often denoted simply by $d\mu(x)d\nu(y)$.

As an application of the product measure, we define the Lebesgue measure in $\mathbb{R}^n$.

5.13 The Lebesgue Measure in $\mathbb{R}^n$

As an application of Theorem 5.55, since the Lebesgue measure μ_L on $\mathbb{R}$ is σ-finite, we see that the product measure $\mu_{L,n}$ of n copies of μ_L is a measure on $\mathbb{R}^n$. The completed σ-algebra (see Proposition 4.8) obtained from the product algebra $\underbrace{\mathcal{L}(\mathbb{R}) \otimes \cdots \otimes \mathcal{L}(\mathbb{R})}_{n}$ is called the σ-algebra of *Lebesgue measurable subsets of* $\mathbb{R}^n$; it is denoted $\mathcal{L}(\mathbb{R}^n)$. To simplify notation, we may write μ_n instead of $\mu_{L,n}$, and $\mathcal{L}^1(\mu_n)$ instead of $\mathcal{L}^1_{\mu_n}(\mathbb{R}^n, \mathcal{L}(\mathbb{R}^n), \mathbb{C})$.

A crucial property of the Lebesgue integral is that the space $\mathcal{K}^\infty_{\mathbb{C}}(\mathbb{R}^n)$ of *smooth* functions with compact support is dense in $\mathcal{L}^1(\mu_n)$. To prove this, one needs to show the existence of smooth "bump functions" in order to approximate the characteristic function χ_A of a rectangle $A = [a_1, b_1] \times \cdots \times [a_n, b_n]$. The following results are shown in Lang [Lang (1996)] (Chapter VI, Section 9).

Proposition 5.58. *For any function* $f \in \mathcal{L}^1(\mu_n)$, *if*

$$\int f\varphi \, d\mu_n = 0 \quad \textit{for all } \varphi \in \mathcal{K}^\infty_{\mathbb{C}}(\mathbb{R}^n),$$

then $f = 0$ *a.e.*

Proposition 5.59. *For every rectangle $A = [a_1, b_1] \times \cdots \times [a_n, b_n]$, for every $\epsilon > 0$, there exits some functions $\varphi, \psi \in \mathcal{K}_{\mathbb{C}}^{\infty}(\mathbb{R}^n)$ such that*

(1) $0 \leq \varphi \leq \chi_A \leq \psi \leq 1$.
(2) $\int (\psi - \varphi) d\mu_n < \epsilon$.

Furthermore, ψ vanishes outside the rectangle $[a_1 - \epsilon, b_1 + \epsilon] \times \cdots \times [a_n - \epsilon, b_n + \epsilon]$, and $\varphi \equiv 1$ on the rectangle $[a_1 + \epsilon, b_1 - \epsilon] \times \cdots \times [a_n + \epsilon, b_n - \epsilon]$.

Using the above results, we obtain the following theorem.

Theorem 5.60. *The space $\mathcal{K}_{\mathbb{C}}^{\infty}(\mathbb{R}^n)$ is dense in $\mathcal{L}^1(\mu_n)$ (for the L^1-semi-norm).*

Theorem 5.60 is proven in Lang [Lang (1996)] (Chapter VI, Section 9).

The Lebesgue measure μ_n on $\mathbb{R}^n$ has the same regularity properties as the Lebesgue measure on $\mathbb{R}$, and we have the following version of Proposition 4.14.

Proposition 5.61. *For every Lebesgue-measurable set $A \in \mathcal{L}(\mathbb{R}^n)$, the following facts hold:*

(a)

$$\mu_n(A) = \inf\{\mu_n(O) \mid A \subseteq O,\ O \text{ is open}\}$$
$$\mu_n(A) = \sup\{\mu_n(K) \mid K \subseteq A,\ K \text{ is compact}\}.$$

(b) For every $\epsilon > 0$, if $\mu_n(A)$ has finite measure then there is some open subset O such that $A \subseteq O$ and $\mu_n(O - A) < \epsilon$, and there is some compact subset F such that $F \subseteq A$ and $\mu_n(A - F) < \epsilon$.

Proposition 5.61 is proven in Lang [Lang (1996)] (Chapter VI, Section 9).

The Lebesgue measure on $\mathbb{R}^n$ is translation-invariant, which means that $\mu_n(x + A) = \mu_n(A)$ for all $x \in \mathbb{R}^n$ and all $A \in \mathcal{L}(\mathbb{R}^n)$, where $x + A = \{x + a \mid a \in A\}$. This will be proved in Section 8.9.

We conclude this section with the change of variables formula. Without this formula, it would be basically impossible to compute the integrals of familiar functions. The proof is not really difficult but quite long and tedious. The interested reader is referred to Lang [Lang (1996)] (Chapter XXI, Section 2, Theorem 2.6).

Given an injective C^1 function $f\colon U \to \mathbb{R}^n$ where U is some open subset of $\mathbb{R}^n$, which means that the derivative $df\colon U \to \mathcal{L}(\mathbb{R}^n, \mathbb{R}^n)$ is defined and continuous on U, we denote the Jacobian matrix of df_x at $x \in U$ (in the canonical basis of $\mathbb{R}^n$) by $J_f(x)$ (where $\mathcal{L}(\mathbb{R}^n, \mathbb{R}^n)$ denotes the vector space of linear maps from $\mathbb{R}^n$ to itself).

Theorem 5.62. *(Change of variables formula, I) Let U be an open subset of $\mathbb{R}^n$, and let $f\colon U \to \mathbb{R}^n$ be an injective C^1 function. For every function $g \in \mathcal{L}^1(f(U), \mu_n)$, we have $(g \circ f)|\det(J_f)| \in \mathcal{L}^1(U, \mu_n)$, and*

$$\int_{f(U)} g(x) d\mu_n(x) = \int_U (g \circ f)(x) |\det(J_f(x))| \, d\mu_n(x).$$

In some cases, for example using polar coordinates, we deal with a C^1 function $f\colon U \to \mathbb{R}^n$ which is only injective on the interior of a measurable subset A of U whose boundary has measure zero. In this case, the following theorem can be used. For a proof, see Lang [Lang (1996)] (Chapter XXI, Section 2, Corollary 2.67).

Theorem 5.63. *(Change of variables formula, II) Let U be an open subset of $\mathbb{R}^n$, and let $f\colon U \to \mathbb{R}^n$ be an injective C^1 function. Let A be a measurable subset of U whose boundary has measure zero, and such that f is injective on the interior of A. For every function $g \in \mathcal{L}^1(f(A), \mu_n)$, we have $(g \circ f)|\det(J_f)| \in \mathcal{L}^1(A, \mu_n)$, and*

$$\int_{f(A)} g(x) d\mu_n(x) = \int_A (g \circ f)(x) |\det(J_f(x))| \, d\mu_n(x).$$

5.14 Problems

Problem 5.1. Prove Proposition 5.2.

Problem 5.2. Prove Proposition 5.3.

Problem 5.3. Prove Proposition 5.5. Hint: Use Proposition 5.4. Alternatively, see Marle [Marle (1974)] (Corollary 2.1.11).

Problem 5.4. Prove Theorem 5.6. Hint: See Lang [Lang (1996)] (Chapter VI, Section 1, Property **M7**).

Problem 5.5. Prove Proposition 5.7. Hint: See Marle [Marle (1974)] (Corollary 2.1.14).

Problem 5.6. Prove Proposition 5.11. Hint: See Marle [Marle (1974)] (Proposition 2.2.3).

Problem 5.7. Prove Proposition 5.13. Hint: See Lang [Lang (1996)] (Chapter VI, Section 1, Property **M11**).

Problem 5.8. Prove Properties (1), (2), (4), (5), and (7) of Proposition 5.16.

Problem 5.9. Prove Proposition 5.26. Hint: Use Proposition 5.25 to adjust the proof of Proposition 5.18. Alternatively, see [Lang (1996)] (Chapter VI, Theorem 5.2).

Problem 5.10. Prove Fatou's Lemma, Theorem 5.33. Hint: See Lang [Lang (1996)] (Chapter VI, §5).

Problem 5.11. Prove Propositions 5.36 and 5.37.

Problem 5.12. Advanced Exercise: Prove Proposition 5.40. Hint: See Marle [Marle (1974)] (Proposition 2.4.10).

Problem 5.13. Advanced Exercise: Prove Theorem 5.44. Hint: See Lang [Lang (1996)] (Chapter VII, §1).

Problem 5.14. Prove Proposition 5.46.

Problem 5.15. Prove the Lebesgue Dominated Convergence Theorem of $\mathcal{L}^2_\mu$, Theorem 5.47. Hint: See Lang [Lang (1996)] (Chapter VII, §1).

Problem 5.16. Advanced Exercise: Prove Theorem 5.50. Hint: See Lang [Lang (1996)] (Chapter VII, Theorem 2.1).

Problem 5.17. Advanced Exercise: Prove Theorem 5.51. Hint: See Lang [Lang (1996)] (Chapter VII, §2).

Problem 5.18. Complete the proof of Proposition 5.52.

Problem 5.19. Advanced Exercise: Prove Fubini's Theorem, Part 1, Theorem 5.56. Hint: See Marle [Marle (1974)] (Chapter 5, Section 2, Theorem 5.2.10) or Lang [Lang (1996)] (Chapter VI, §8, Theorem 8.4).

Problem 5.20. Advanced Exercise: Prove Fubini's Theorem, Part 2, Theorem 5.57. Hint: See Lang [Lang (1996)] (Chapter VI, §8, Theorem 8.7) or Marle [Marle (1974)] (Chapter 5, Section 2).

Problem 5.21. Prove Proposition 5.61. Hint: See Lang [Lang (1996)] (Chapter VI, Section 9).

Problem 5.22. Prove Proposition 5.62. Hint: See Lang [Lang (1996)] (Chapter XXI, Section 2, Theorem 2.6).

Chapter 6

The Fourier Transform and the Fourier Cotransform on $\mathbb{T}^n$, $\mathbb{Z}^n$, $\mathbb{R}^n$

Historically, trigonometric series were first used by D'Alembert (1747) to solve the equation of a vibrating string, elaborated by Euler a year later, and then solved in a different way essentially using Fourier series by D. Bernoulli (1753). However it was Fourier who introduced and developed Fourier series in order to solve the heat equation, in a sequence of works on heat diffusion, starting in 1807, and culminating with his famous book, *Théorie analytique de la chaleur*, published in 1822.

Originally, the theory of Fourier series is meant to deal with $\mathbb{T} = \mathbf{U}(1) = \{z \in \mathbb{C} \mid |z| = 1\} \cong \mathbb{R}/(2\pi\mathbb{Z})$, say functions with period 2π. Remarkably (but we must apologize for the oversimplification), the theory of Fourier series is captured by the following two equations:

$$f(\theta) = \sum_{m \in \mathbb{Z}} c_m e^{im\theta}, \tag{1}$$

$$c_m = \int_{-\pi}^{\pi} f(\theta) e^{-im\theta} \frac{d\theta}{2\pi}. \tag{2}$$

Equation (1) involves a series, and Equation (2) involves an integral. There are two ways of interpreting these equations.

The first way consists of starting with a convergent series as given by the right-hand side of (1) (of course $c_n \in \mathbb{C}$), and to ask what kind of function is obtained. A second question is the following: are the coefficients in (1) computable in terms of the formulae given by (2)?

Such questions were considered by Riemann and then Cantor and Lebesgue. Since they deal with the notion of integral, it is not surprising that they motivated the invention of the Riemann integral and then the Lebesgue integral.

The second way is to start with a periodic function f, apply Equation (2) to obtain the c_m, called *Fourier coefficients*, and then to consider Equation (1). Does the series $\sum_{m\in\mathbb{Z}} c_m e^{im\theta}$ (called *Fourier series*) converge at all? Does it converge to f?

Observe that the expression $f(\theta) = \sum_{m\in\mathbb{Z}} c_m e^{im\theta}$ may be interpreted as a countably infinite superposition of elementary periodic functions, intuitively representing simple wave functions, the functions $\theta \mapsto e^{im\theta}$. We can think of m as the frequency of this wave function.

The above questions were first considered by Fourier. Fourier boldly claimed that *every* function can be represented by a Fourier series. Of course this is false, and for several reasons. First, one needs to define what is an integrable function, and there are plenty of nonintegrable functions. Second, it depends on the kind of convergence that are we dealing with. The nth partial sum $S_{n,f}$ of the Fourier series $\sum_{m\in\mathbb{Z}} c_m e^{im\theta}$ for f (where the c_m are given by Equation (2)) is given by

$$S_{n,f}(\theta) = \sum_{k=-n}^{n} c_k e^{ik\theta}.$$

The most common type of convergence is *pointwise convergence*, which means that for every θ, we have $\lim_{n\mapsto\infty} |f(\theta) - S_{n,f}(\theta)| = 0$. Even if f is a continuous function, there are examples of Fourier series that do not converge pointwise for $\theta = 0$ (du Bois-Reymond). There is even a function in $L^1(\mathbb{T})$ whose Fourier series diverges for all θ (Kolmogoroff). The convergence of Fourier series is a subtle matter.

But Fourier was almost right. If we consider a function f in $L^2(\mathbb{T})$, a famous and deep theorem of Carleson states that its Fourier series converges to f pointwise almost everywhere. Other ways to ensure the convergence of the Fourier series of a function is to either restrict the class of functions being considered (Dirichlet, Jordan), or to use different kinds of summation (Abel, Cesàro). Abel summation leads to the Poisson kernel, and Cesàro summation leads to the Féjer kernel; see Example 8.10, Section 6.1, and Stein and Shakarchi [Stein and Shakarchi (2003)] (Chapter 2).

In Section 6.1, as a motivation for Fourier analysis on $\mathbb{T}$, we solve the wave equation for a vibrating string. We are led immediately to the problem of Fourier inversion.

Given a periodic function f, the problem of determining when f can be reconstructed as the Fourier series (Equation (1)) given by its Fourier coefficients c_m (Equation (2)) is called the problem of *Fourier inversion.* To discuss this problem, it is useful to adopt a more general point of view of the correspondence between functions and Fourier coefficients, and Fourier coefficients and Fourier series.

Given a function $f \in \mathrm{L}^1(\mathbb{T})$, Equation (2) yields the $\mathbb{Z}$-indexed sequence $(c_m)_{m\in\mathbb{Z}}$ of Fourier coefficients of f, with

$$c_m = \int_{-\pi}^{\pi} f(\theta)e^{-im\theta}\,\frac{d\theta}{2\pi},$$

which we call the *Fourier transform* of f and denote by $\widehat{f}$, or $\mathcal{F}(f)$. We can view the Fourier transform $\mathcal{F}(f)$ of f as a function $\mathcal{F}(f)\colon \mathbb{Z} \to \mathbb{C}$ with domain $\mathbb{Z}$.

On the other hand, given a $\mathbb{Z}$-indexed sequence $c = (c_m)_{m\in\mathbb{Z}}$ of complex numbers c_m, we can define the Fourier series $\overline{\mathcal{F}}(c)$ associated with c, or *Fourier cotransform* of c, given by

$$\overline{\mathcal{F}}(c)(\theta) = \sum_{m\in\mathbb{Z}} c_m e^{im\theta}.$$

This time $\overline{\mathcal{F}}(c)$ is a function $\overline{\mathcal{F}}(c)\colon \mathbb{T} \to \mathbb{C}$ with domain $\mathbb{T}$. Of course there is an issue of convergence. If $c = (c_m) \in \ell^1(\mathbb{Z})$, then the series $\overline{\mathcal{F}}(c)$ converges uniformly. In general, if $c = (c_m) \notin \ell^1(\mathbb{Z})$, then $\overline{\mathcal{F}}(c)(\theta)$ may be undefined. If $(\overline{\mathcal{F}} \circ \mathcal{F})(f))(\theta)$ is defined, Fourier inversion can be stated as the equation

$$f(\theta) = ((\overline{\mathcal{F}} \circ \mathcal{F})(f))(\theta).$$

In general, even if $f \in \mathrm{L}^1(\mathbb{T})$, the above equation fails.

There are special cases for which Fourier inversion holds. One case is if $\mathcal{F}(f) \in \ell^1(\mathbb{Z})$, which means that the sum $\sum_{m\in\mathbb{Z}} |c_m|$ is finite. Another case is if $f \in \mathrm{L}^2(\mathbb{T})$. In fact, Plancherel's theorem asserts that the map $f \mapsto \widehat{f}$ is an isometric isomorphism between $\mathrm{L}^2(\mathbb{T})$ and $\ell^2(\mathbb{Z})$.

In Section 6.3 we return to the issue of pointwise convergence of Fourier series on $\mathbb{T}$. We give examples of functions for which the Fourier series does not converge pointwise, or worse. We show that for the class of functions of bounded variation there is a pointwise convergence theorem due to Dirichlet and Jordan.

In Section 6.4 we generalize the results of Section 6.1 to $\mathbb{T}^n$ and $\mathbb{Z}^n$. In addition to the definition of the Fourier transform on $\mathbb{T}^n$, we define the Fourier cotransform on $\mathbb{T}^n$, and in addition to the definition of the cotransform on $\mathbb{Z}^n$, we define the Fourier transform on $\mathbb{Z}^n$. We also generalize the Poisson kernel to $\mathbb{T}^n$ and prove generalizations of the results of Section 6.1 on spectral synthesis and Abel summation. Plancherel's theorem asserts that the map $f \mapsto \widehat{f}$ is an isometric isomorphism between $\mathrm{L}^2(\mathbb{T}^n)$ and $\ell^2(\mathbb{Z}^n)$.

In Section 6.5 we discuss the Fourier transform of functions defined on the entire real line $\mathbb{R}$ that are not necessarily periodic. Because $\mathbb{R}$ is not compact, $\mathrm{L}^1(\mathbb{R})$ and $\mathrm{L}^2(\mathbb{R})$ are incomparable (with respect to inclusion), and the theory of the Fourier transform on $\mathbb{R}$ is more delicate than the Fourier theory on $\mathbb{T}$.

In Section 6.6 we consider a classical problem in signal processing, which is to reconstruct a function $f\colon \mathbb{R} \to \mathbb{C}$ which is *band-limited*, which means that its Fourier transform $\widehat{f}$ vanishes outside some interval $[-\Omega, \Omega]$. Then f can be completely reconstructed by sampling at the points $t_n = n\pi/\Omega$, for $n \in \mathbb{N}$. We obtain the sampling theorem (Theorem 6.25).

The results of Section 6.5 are generalized to $\mathbb{R}^n$ in Section 6.7.

In Section 6.8 we define a class $\mathcal{S}(\mathbb{R}^n)$ of smooth functions that decay quickly when $\|x\|$ goes to infinity called the *Schwartz space*. The space $\mathcal{S}(\mathbb{R}^n)$ is not a normed vector space, but its topology can be defined by a countable family of semi-norms. It is a metrizable space that is complete, called a Fréchet space. The Schwartz space is closed under the Fourier transform and cotransform and is generally well-behaved. Fourier inversion holds and taking the Fourier transform of a derivative is just multiplication of the Fourier transform by a variable.

This last property can be exploited to solve certain partial differential equations by converting them to ordinary differential equations *via* the Fourier transform. We illustrate this method by solving the steady-state heat equation in the upper half-plane.

In Section 6.9 we discuss the Poisson summation formula, which is a way of finding the Fourier coefficients of the periodic function obtained from a nonperiodic function by applying the process of periodization.

In Section 6.10 we show that roughly, a function f and its Fourier transform $\widehat{f}$ can't be both highly localized. This can be stated precisely in terms of the *dispersion* of f about the point a given by

$$\Delta_a f = \int (x-a)^2 |f(x)|^2\, dx \Big/ \int |f(x)|^2\, dx.$$

The Heisenberg inequality states that if f is a function in $L^2(\mathbb{R})$, then for all $a, b \in \mathbb{R}$, we have

$$(\Delta_a f)(\Delta_b \widehat{f}) \geq \frac{1}{4}.$$

We briefly discuss the interpretation of this inequality in quantum mechanics, called the *Heisenberg uncertainty principle*.

In the last section, Section 6.11, we give a brief summary of Fourier's captivating life.

6.1 Fourier Analysis on $\mathbb{T}$

We begin this chapter with a preview of Fourier analysis on one of the simplest locally compact abelian groups, namely $\mathbb{T}$.

Definition 6.1. The *circle group* $\mathbb{T} = \mathbf{U}(1)$ is the group $\{z \in \mathbb{C} \mid |z| = 1\}$ of complex numbers of unit length under multiplication. We give $\mathbb{T}$ the subspace topology induced by $\mathbb{C}$.

The circle group $\mathbb{T}$ is abelian (commutative). As a set,

$$\mathbb{T} = \{e^{i\theta} \mid \theta \in [-\pi, \pi)\}.$$

Geometrically, this is the unit circle $S = S^1$.

The map $\sigma\colon \mathbb{R} \to \mathbb{T}$ given by

$$\sigma(\theta) = e^{i\theta}$$

is clearly a surjective group homomorphism (with $\mathbb{R}$ under addition, and $\mathbb{T}$ under multiplication); see Figure 6.1. Since $e^{i\theta} = 1$ iff $\theta = k2\pi$ with $k \in \mathbb{Z}$, we see that the kernel of σ is $2\pi\mathbb{Z}$, so by the first isomorphism theorem the *additive group* $\mathbb{R}/(2\pi\mathbb{Z})$ *is isomorphic to the multiplicative group* $\mathbb{T}$. This isomorphism allows to view a complex number of unit length as $e^{i\theta}$, with θ defined modulo 2π, which is often more convenient than picking a representative of the equivalence class of $\theta \pmod{2\pi}$ in $[-\pi, \pi)$.

Functions on the unit circle $\mathbb{T}$ are equivalent to periodic functions on $\mathbb{R}$ as defined next.

Definition 6.2. A function $f\colon \mathbb{R} \to \mathbb{C}$ is *periodic with period T* (for some $T \in \mathbb{R}$, with $T > 0$), if $f(x + T) = f(x)$ for all $x \in \mathbb{R}$.

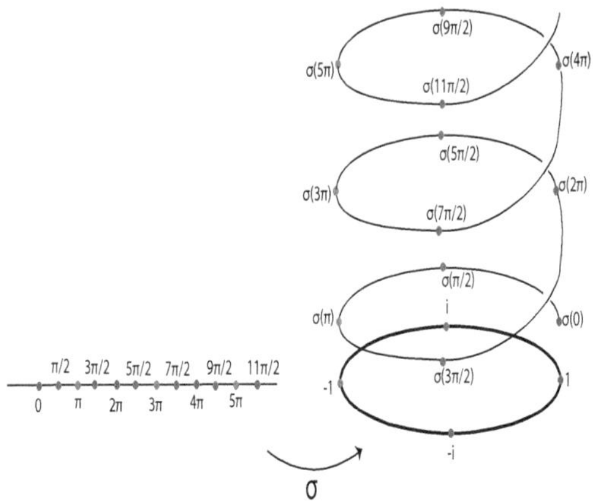

Fig. 6.1 The map $\sigma\colon \mathbb{R} \to \mathbb{T}$ which "wraps" the line around the unit circle.

Obviously, a periodic function is completely defined by its restriction to the interval $[-T/2, T/2)$. In most cases, the periods $T = 1$ or $T = 2\pi$ are considered, and which is picked is a matter of taste. We pick $T = 2\pi$. Then we have the following two transformations.

Given a periodic function $f\colon \mathbb{R} \to \mathbb{C}$ (with period 2π), let $f_{\mathbb{T}}\colon \mathbb{T} \to \mathbb{C}$ be the function given by

$$f_{\mathbb{T}}(e^{i\theta}) = f(\theta), \quad -\pi \le \theta < \pi.$$

Given a function $g\colon \mathbb{T} \to \mathbb{C}$, let $g_{\mathbb{R}}\colon \mathbb{R} \to \mathbb{C}$ be the periodic function (with period 2π) given by

$$g_{\mathbb{R}}(\theta) = g(e^{i\theta}), \quad \theta \in \mathbb{R}.$$

Observe that because the map $\theta \mapsto e^{i\theta}$ is a bijection between $[-\pi, \pi)$ and $\mathbb{T}$, we have

$$(f_{\mathbb{T}})_{\mathbb{R}} = f, \qquad (g_{\mathbb{R}})_{\mathbb{T}} = g,$$

which shows that there is a bijection between the space of periodic function $f\colon \mathbb{R} \to \mathbb{C}$ (with period 2π), and the space of functions $g\colon \mathbb{T} \to \mathbb{C}$. This bijection restricts to the space of periodic L^p functions that are integrable over $[-\pi, \pi]$, and the space $\mathrm{L}^p(\mathbb{T})$, for $p = 1, 2, \infty$.

The identification between $\mathbb{R}/(2\pi\mathbb{Z})$ and $\mathbb{T}$, and between the space of functions defined on $\mathbb{T}$ and the space of periodic function on $\mathbb{R}$ is often implicit, and in what follows, we take the view that functions on $\mathbb{T}$ are periodic (with period 2π). The reader should be cautioned that other authors use the period 1, so the factor $1/(2\pi)$ showing up in our formulae is missing in the other version (assuming period 1).

To be completely rigorous, we need to equip the abelian group $\mathbb{T}$ with an invariant measure called a Haar measure. This will be done very thoroughly in Chapter 8. For the time being, it suffices to know that in Example 8.10, we show that a normalized Haar measure on $\mathbb{T}$ is given by $dx/2\pi$, where dx is the Lebesgue on $\mathbb{R}$ (so that $\mathbb{T}$ has measure 1). Readers not familiar with the Lebesgue theory of integration should not be concerned, and they should replace this fancy notion with the notion of integral that they are familiar with. After reading this motivating chapter, they should return to the chapters presenting measure theory and integration.

The solution of the wave equation for a vibrating string provides an excellent motivation for using Fourier series on $\mathbb{T}$.

Consider a homogeneous string in the (x, y)-plane, stretched along the x-axis between $x = 0$ and $x = \pi$. The constant π is chosen for mathematical convenience; we could use any constant $L > 0$, but by a change of units, we may assume that it is equal to π. If the string is set to vibrate, its displacement $u(x, t)$ is then a function of x and t. We assume that its endpoints are fixed, so that we have the initial conditions

$$u(0, t) = u(\pi, t) = 0 \qquad \text{for all } t.$$

We also assume that the initial position and velocity of the string are given by two functions f and g defined on $[0, \pi]$ (with $f(0) = f(\pi) = 0$), so that

$$u(x, 0) = f(x), \qquad \frac{\partial u}{\partial t}(x, 0) = g(x).$$

We extend the functions f and g to $[-\pi, \pi]$ by making them odd, namely, we set $f(-x) = -f(x)$ and $g(-x) = -g(x)$ for $x \in [0, \pi]$, and then we extend f and g to $\mathbb{R}$ by making them periodic of period 2π (so, $f(x + 2\pi k) = f(x)$ and $g(x + 2\pi k) = g(x)$, for all $k \in \mathbb{Z}$ and $x \in [-\pi, \pi]$).

Using some physics, it can be shown that u is a solution of the *one-dimensional wave-equation*,

$$\frac{\partial^2 u}{\partial x^2} = \frac{1}{c^2} \frac{\partial^2 u}{\partial t^2},$$

for some constant c. Again, by a change of units, we may assume that $c = 1$, so the wave equation becomes

$$\frac{\partial^2 u}{\partial x^2} = \frac{\partial^2 u}{\partial t^2}. \tag{$*$}$$

Equation $(*)$ can be solved by two methods:

(1) Using *traveling waves*.
(2) Using *standing waves*.

The method of traveling waves was used by d'Alembert, and the method of standing waves by D. Bernoulli; see Stein and Shakarchi [Stein and Shakarchi (2003)] (Chapter 1).

The method of standing waves leads immediately to Fourier series. In this method we use the technique of *separation of variables*, which means that we express the solution $u(x,t)$ as the product $u(x,t) = \varphi(x)\psi(t)$, where $\varphi(x)$ and $\psi(t)$ are functions of the two independent variables x and t. Equation $(*)$ yields the equation

$$\varphi(x)\psi''(t) = \varphi''(x)\psi(t),$$

which can be written as

$$\frac{\psi''(t)}{\psi(t)} = \frac{\varphi''(x)}{\varphi(x)}.$$

Since the left-hand side depends only on t, and the right-hand side depends only on x, the above equation can hold only if both sides are equal to the same constant, say λ, so we deduce that

$$\begin{aligned}\varphi''(x) - \lambda\varphi(x) &= 0\\ \psi''(t) - \lambda\psi(t) &= 0.\end{aligned}$$

These equations have well-known solutions. If $\lambda > 0$, then

$$\varphi(x) = Ae^{\sqrt{\lambda}x} + Be^{-\sqrt{\lambda}x},$$

and we obtain a solution which is not physically possible since the displacement of the string is unbounded, so we must have $\lambda \leq 0$, say $\lambda = -m^2$. The solution is given by

$$\varphi(x) = \alpha e^{imx} + \beta e^{-imx}$$

with $\alpha, \beta \in \mathbb{C}$, or equivalently

$$\begin{aligned}\varphi(x) &= \frac{\alpha}{2}(\cos mx + i\sin mx) + \frac{\beta}{2}(\cos mx - i\sin mx)\\ &= \frac{(\alpha+\beta)}{2}\cos mx + i\frac{(\alpha-\beta)}{2}\sin mx.\end{aligned}$$

Since we are seeking real functions as solutions, the solutions are given by

$$\begin{aligned}\varphi(x) &= C\cos mx + D\sin mx\\ \psi(t) &= A\cos mt + B\sin mt,\end{aligned}$$

with $A, B, C, D \in \mathbb{R}$. Since $\varphi(0) = \varphi(\pi) = 0$, we get $C = 0$, and if $D \neq 0$, then m must be an integer in order to have $\sin m\pi = 0$. If $m = 0$, then $\varphi(x) = 0$ for all x, and if $m \leq -1$, we can rename the constants and reduce to the case $m \geq 1$ (since cos is even and sin is odd). Finally, we arrive at the solution

$$u_m(x,t) = (A_m \cos mt + B_m \sin mt) \sin mx, \quad m \geq 1. \tag{$**$}$$

Since the wave equation is linear, any linear combination of the functions in $(**)$ is also a solution, so we are led to the fact that a general solution $u(x,t)$ of the wave equation is a superposition of the solutions u_m, that is,

$$u(x,t) = \sum_{m=1}^{\infty} (A_m \cos mt + B_m \sin mt) \sin mx.$$

There is obviously an issue of convergence, but we will not worry about this yet. The last step is to impose the boundary conditions

$$u(x,0) = f(x), \qquad \frac{\partial u}{\partial t}(x,0) = g(x),$$

which yield the equations

$$f(x) = \sum_{m=1}^{\infty} A_m \sin mx$$
$$g(x) = \sum_{m=1}^{\infty} mB_m \sin mx.$$

Thus we arrived at the following question: given a "reasonable" periodic function $f\colon \mathbb{T} \to \mathbb{C}$ (say $f \in \mathbb{L}^1(\mathbb{T})$), can we find some coefficients $c_m \in \mathbb{C}$ such that

$$f(\theta) = \sum_{m\in\mathbb{Z}} c_m e^{im\theta},$$

where the series on the right-hand side is the Fourier series associated with c_m?

This is the basic problem that motivated Fourier in his quest for solving the heat equation on various domains.

The integer $m \geq 1$ is the *frequency* of the wave component $(A_m \cos mt + B_m \sin mt) \sin mx$, which is called a *harmonic* or *tone*. The general solution is thus a superpositions of harmonics. The case $m = 1$ corresponds to the *first harmonic* or *fundamental tone*. If the vibrating string is the string of a violin, then the first harmonic is the sound of lowest pitch.

If $f \in \mathbb{L}^1(\mathbb{T})$, then we can compute the Fourier coefficients c_m by the formula

$$c_m = \int_{-\pi}^{\pi} f(t)e^{-imt}\,\frac{dx(t)}{2\pi},$$

and then the question is whether the Fourier series $\sum_{m\in\mathbb{Z}} c_m e^{im\theta}$ converges to f (and in what sense).

Recall that the nth partial sum $S_{n,f}$ of the Fourier series $\sum_{m\in\mathbb{Z}} c_m e^{im\theta}$ for f is given by

$$S_{n,f}(\theta) = \sum_{k=-n}^{n} c_k e^{ik\theta},$$

and the average $A_{n,f}$ of these partial sums is given by

$$A_{n,f} = \frac{1}{n}(S_{0,f} + \cdots + S_{n-1,f}).$$

It would be desirable that the partial sums $S_{n,f}$ converge pointwise to f, but in general, this is not the case, even for continuous functions. We will see that if $f \in \mathrm{L}^2(\mathbb{T})$, then $S_{n,f}$ converge to f in the L^2-sense (Proposition 6.2), but the convergence may fail to be pointwise.

In Example 8.10 we discuss Cesàro sums and Féjer's theorem. We show that the average sums $A_{n,f}$ converge uniformly to f if f is continuous. We now discuss *Abel's sums* and *Poisson kernels*, which yield another kind of convergence.

The *Poisson kernel* on the unit disk is the family of functions $P_r(\theta)$, parametrized by $r \in [0,1)$, and given by

$$P_r(\theta) = \sum_{n=-\infty}^{n=\infty} r^{|n|}e^{in\theta} = \frac{1-r^2}{1-2r\cos\theta + r^2};$$

see Example 8.11 for the derivation of this formula. Also see Figures 6.2 and 6.3.

A key concept in Fourier analysis is the notion of convolution. To discuss convolution rigorously requires some work so in this chapter we content ourselves with a definition leaving justifications to Section 8.12.

Definition 6.3. The *convolution* $f * g$ of two functions $f, g \in \mathrm{L}^1(\mathbb{T})$ is given by

$$(f*g)(\theta) = \int_{\mathbb{T}} f(\theta-\varphi)g(\varphi)\,\frac{dx(\varphi)}{2\pi} = \int_{\mathbb{T}} f(\varphi)g(\theta-\varphi)\,\frac{dx(\varphi)}{2\pi},$$

where dx is the Lebesgue measure on $\mathbb{R}$.

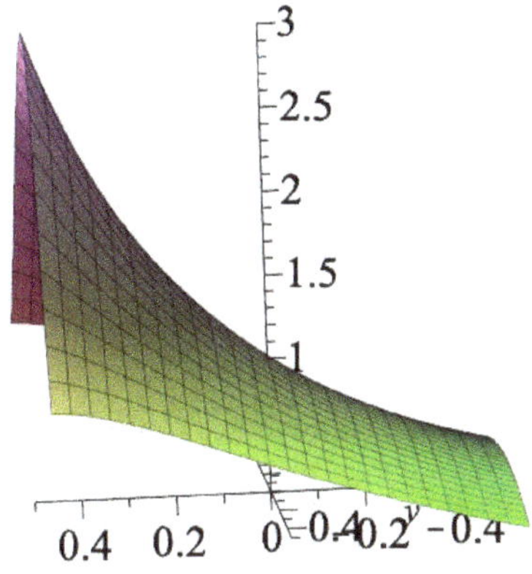

Fig. 6.2 The graph $P_r(\theta) = \frac{1-r^2}{1-2r\cos\theta+r^2} = \frac{1-x^2-y^2}{1-2x+x^2+y^2}$ over the region $-1/4 \leq x \leq 1/4$ and $-1/4 \leq y \leq 1/4$. When $r = 0$, the z-coordinate is 1.

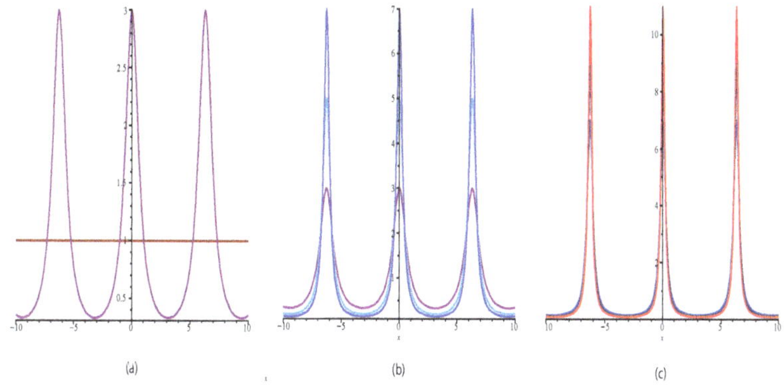

Fig. 6.3 Another graphical interpretation of $P_r(\theta) = \frac{1-r^2}{1-2r\cos\theta+r^2}$ when r is fixed. Figure (a) shows the graphs of $P_0(\theta) = 1$ and $P_{1/2}(\theta)$. Figure (b) shows the graphs of $P_{1/2}(\theta)$, $P_{2/3}(\theta)$, and $P_{3/4}(\theta)$, while Figure (c) shows the graphs of $P_{3/4}(\theta)$, $P_{4/5}(\theta)$, and $P_{5/6}(\theta)$. As $r \to 1$, the sinusoid curves have "narrower" peaks centered at $\theta = 2\pi k$, $k \in \mathbb{Z}$, and outside of those peaks, the function limits to the constant value of zero.

By Proposition 8.48, we have $f * g \in \mathrm{L}^1(\mathbb{T})$.

We have the following result using the Poisson kernel which gives a preview of Fourier analysis.

Proposition 6.1. *For any $r \in [0,1)$, if $f \in \mathrm{L}^1(\mathbb{T})$ and if P_r is the Poisson kernel, then for all $\theta \in \mathbb{R}/2\pi\mathbb{Z}$, we have*

$$(P_r * f)(\theta) = \sum_{m=-\infty}^{m=\infty} c_m r^{|m|} e^{im\theta},$$

where c_m is the mth Fourier coefficient of f,

$$c_m = \int_{-\pi}^{\pi} f(t) e^{-imt} \frac{dx(t)}{2\pi}.$$

Proof. For $0 \leq r < 1$, the series defining P_r is absolutely convergent, so

$$\begin{aligned}
(P_r * f)(\theta) &= \int_{-\pi}^{\pi} P_r(\theta - \varphi) f(\varphi) \frac{dx(\varphi)}{2\pi} \\
&= \int_{-\pi}^{\pi} \sum_{m=-\infty}^{m=\infty} r^{|m|} e^{im(\theta-\varphi)} f(\varphi) \frac{dx(\varphi)}{2\pi} \\
&= \sum_{m=-\infty}^{m=\infty} \int_{-\pi}^{\pi} r^{|m|} e^{im(\theta-\varphi)} f(\varphi) \frac{dx(\varphi)}{2\pi} \\
&= \sum_{m=-\infty}^{m=\infty} r^{|m|} e^{im\theta} \int_{-\pi}^{\pi} f(\varphi) e^{-im\varphi} \frac{dx(\varphi)}{2\pi} \\
&= \sum_{m=-\infty}^{m=\infty} c_m r^{|m|} e^{im\theta},
\end{aligned}$$

as claimed. □

The functions D_n and K_n are defined as

$$D_n(x) = \sum_{k=-n}^{n} e^{ikx}$$

$$K_n(x) = \frac{1}{n} \sum_{m=0}^{n-1} \sum_{k=-m}^{m} e^{ikx} = \frac{1}{n}(D_0(x) + \cdots + D_{n-1}(x)).$$

It can be shown that

$$D_n(x) = \frac{\sin((2n+1)x/2)}{\sin(x/2)}$$

$$K_n(x) = \frac{1}{n}\left(\frac{\sin(nx/2)}{\sin(x/2)}\right)^2.$$

The functions D_n are known as *Dirichlet kernels*, and the functions K_n are *Fejér kernels*. See Figures 6.4 and 6.5. Also see Section 8.15 for applications.

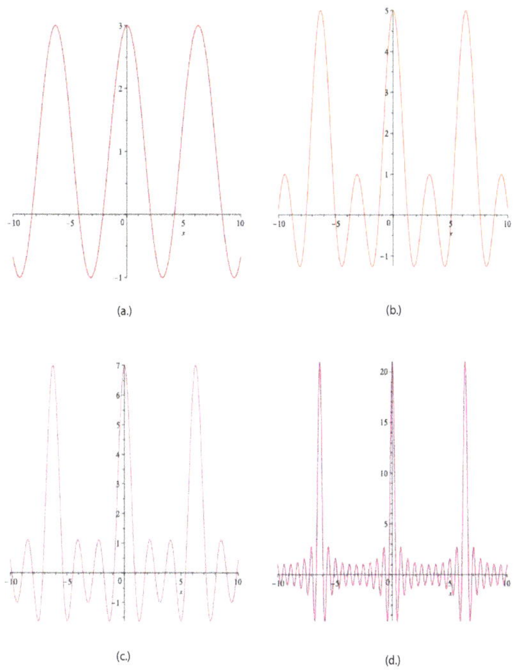

Fig. 6.4 Figure (a) is the graph of $D_1(x)$, Figure (b) is the graph of $D_2(x)$, Figure (c) is the graph of $D_3(x)$, while Figure (d) is the graph of $D_{10}(x)$. In all cases the "spike" at $x = 0$ has y-value $2n + 1$.

Observe that for $r = 1$, the partial sum $\sum_{m=-n}^{m=n} c_m r^{|m|} e^{im\theta}$ is the partial sum $S_{n,f}$ of the Fourier series for f, and the partial sum $\sum_{m=-n}^{m=n} r^{|m|} e^{im\theta}$ of the Poisson kernel is the Dirichlet kernel D_n. A slight modification of the proof of Proposition 6.1 shows that

$$D_n * f = S_{n,f},$$

and this immediately implies that

$$K_n * f = A_{n,f}.$$

Recall that for any $p \geq 1$, the space $\ell^p(\mathbb{Z})$ is the set of sequences $x = (x_n)_{n \in \mathbb{Z}}$ with $x_n \in \mathbb{C}$ such that $\sum_{n \in \mathbb{Z}} |x_n|^p < \infty$. Also, if $1 \leq p < q$, then $\ell^p(\mathbb{Z}) \subseteq \ell^q(\mathbb{Z})$; see Figure 6.6.

Indeed, since the sequence $|x_m|^p$ converges, for some $M > 0$ we have $|x_m| < 1$ for all $|m| \geq M$, and since if $q > p$ we have $|x_m|^q \leq |x_m|^p$

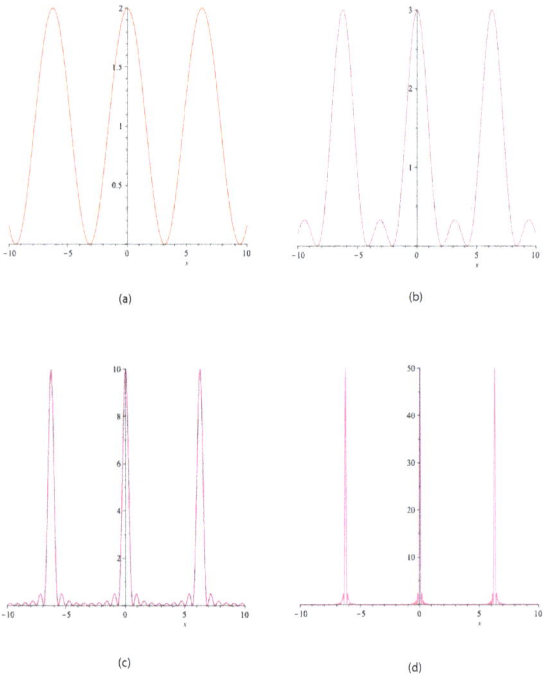

Fig. 6.5 Figure (a) is the graph of $K_2(x)$, Figure (b) is the graph of $K_3(x)$, Figure (c) is the graph of $K_{10}(x)$, while Figure (d) is the graph of $K_{50}(x)$. In all cases the "spike" at $x = 0$ has y-value n.

(because $|x_m|^p - |x_m|^q = |x_m|^p(1 - |x_m|^{q-p}) \geq 0$ since $|x_m| < 1$), thus $\sum_{|m|\geq M} |x_m|^q \leq \sum_{|m|\geq M} |x_m|^p$, and since $\sum_{n\in\mathbb{Z}} |x_n|^p < \infty$, we also have $\sum_{n\in\mathbb{Z}} |x_n|^q < \infty$.

Each space $\ell^p(\mathbb{Z})$ $(p \geq 1)$ is a normed vector space with the norm

$$\|(x_m)_{m\in\mathbb{Z}}\| = \left(\sum_{m\in\mathbb{Z}} |x_m|^p\right)^{1/p}.$$

The space $\ell^p(\mathbb{Z})$ $(p \geq 1)$ is a Banach space (it is complete). This is proven by a simple modification of the proof of Proposition D.14.

In general, given a function $f \in \mathrm{L}^1(\mathbb{T})$, the Fourier series $\sum_{m\in\mathbb{Z}} c_m e^{im\theta}$ *does not* converge pointwise. However, if $0 \leq r < 1$, then

$$f_r(\theta) = (P_r * f)(\theta) = \sum_{m\in\mathbb{Z}} c_m r^{|m|} e^{im\theta},$$

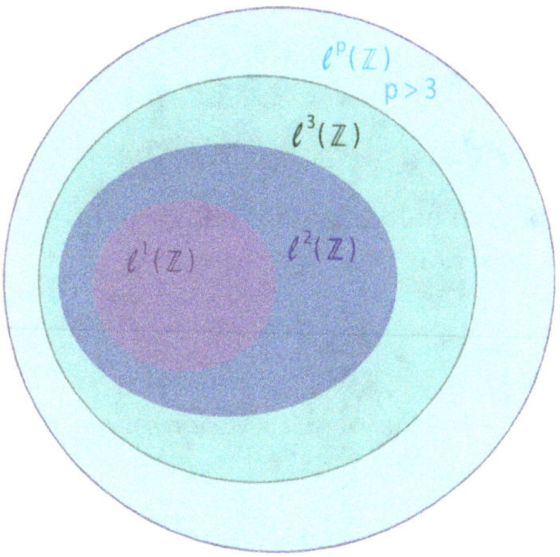

Fig. 6.6 A Venn diagram of the containments $\ell^1(\mathbb{Z}) \subseteq \ell^2(\mathbb{Z}) \subseteq \ell^3(\mathbb{Z}) \subseteq \ell^p(\mathbb{Z})$, where $p > 3$.

so the series on the right-hand side converges pointwise. The following results shows that if r tends to 1, then f_r is an approximation of f that tends to f (in a technical sense). Since $\mathbb{T}$ is compact, we have $\|f\|_{\mathrm{L}^1(\mathbb{T})} \leq \|f\|_{\mathrm{L}^2(\mathbb{T})}$ and so $\mathrm{L}^2(\mathbb{T}) \subseteq \mathrm{L}^1(\mathbb{T})$. Then if $f \in \mathrm{L}^2(\mathbb{T})$, the partial sums of the Fourier series of f converge to f in the L^2-norm.

Theorem 6.2. *(Spectral Synthesis)*

(1) If $f \in \mathrm{L}^p(\mathbb{T})$ for $p = 1, 2$, and if $r \in [0, 1)$, for all $\theta \in \mathbb{R}/2\pi\mathbb{Z}$, write

$$f_r(\theta) = (P_r * f)(\theta) = \sum_{m=-\infty}^{m=\infty} c_m r^{|m|} e^{im\theta},$$

with

$$c_m = \int_{-\pi}^{\pi} f(t) e^{-imt} \frac{dx(t)}{2\pi}.$$

Then $\lim_{r\mapsto 1} \|f - f_r\|_p = 0$.

(2) If $f \in \mathcal{C}(\mathbb{T})$, then $\lim_{r\mapsto 1} \|f - f_r\|_\infty = 0$.

(3) If $f \in \mathrm{L}^2(\mathbb{T})$, then

$$\lim_{n\mapsto\infty} \left\| f - \sum_{m=-n}^{m=n} c_m e^{im\theta} \right\|_2 = 0.$$

Furthermore, we have the Parseval theorem:

$$\|f\|_2^2 = \sum_{m=-\infty}^{m=\infty} |c_m|^2.$$

The above implies that $c = (c_m)_{m\in\mathbb{Z}} \in \ell^2(\mathbb{Z})$.

Theorem 6.2 is proven in Malliavin [Malliavian *et al.* (1995)] (Chapter 3, Section 2.2.5). The function

$$f_r(\theta) = (P_r * f)(\theta) = \sum_{m=-\infty}^{m=\infty} c_m r^{|m|} e^{im\theta}$$

is known as the *rth Abel mean* of the Fourier series

$$\sum_{m=-\infty}^{m=\infty} c_m e^{im\theta}$$

of f. The Fourier series does not always converge pointwise, but the rth Abel mean f_r converges uniformly for all $r < 1$ $(r \geq 0)$.

The results of Theorem 6.2 are examples of *spectral synthesis*, namely, the reconstruction of a function from its Fourier coefficients. Facts (1) and (2) are not very practical because they require first summing the series f_r. Fact (2) for continuous functions is better because it shows uniform convergence. Fact (3) is very satisfactory since it shows convergence of the partial sums of the Fourier series in the L^2-sense, but convergence pointwise generally fails. If $(c_m) \in \ell^2(\mathbb{Z})$, for some $\theta \in \mathbb{R}/(2\pi\mathbb{Z})$, the sums $\sum_{m=-n}^{m=n} c_m e^{im\theta}$ may not converge. For more about this phenomenon, see Section 6.3.

Remark: Lennart Carleson showed in 1966 that for any function $f \in \mathrm{L}^2(\mathbb{T})$, the partial sums of the Fourier series of f converge pointwise almost everywhere to f, putting a close to a problem that had been open for fifty years.

6.2 Fourier Inversion on $\mathbb{T}$

Recall that for any $p \geq 1$, the space $\ell^p(\mathbb{Z})$ is the set of sequences $x = (x_n)_{n\in\mathbb{Z}}$ with $x_n \in \mathbb{C}$ such that $\sum_{n\in\mathbb{Z}} |x_n|^p < \infty$. For $p = 2$, the space $\ell^2(\mathbb{Z})$ is a Hilbert space with the inner product

$$\langle (x_m)_{m\in\mathbb{Z}}, (y_m)_{m\in\mathbb{Z}} \rangle = \sum_{m\in\mathbb{Z}} x_m \overline{y_m}$$

and norm

$$\|(x_m)_{m\in\mathbb{Z}}\| = \left(\sum_{m\in\mathbb{Z}} |x_m|^2\right)^{1/2};$$

see Proposition D.14.

The following result shows that if the sequence $c = (c_m)_{m\in\mathbb{Z}}$ of Fourier coefficients of f is well-behaved, then f can be reconstructed from c.

Theorem 6.3. *(Fourier inversion formula) Let $f \in \mathrm{L}^1(\mathbb{T})$. If $c = (c_m)_{m\in\mathbb{Z}} \in \ell^1(\mathbb{Z})$, that is, if the series $\sum_{m=-\infty}^{m=\infty} |c_m|$ converges, where c_m is the Fourier coefficient*

$$c_m = \int_{-\pi}^{\pi} f(t)e^{-imt}\frac{dx(t)}{2\pi},$$

then

$$f(\theta) = \sum_{m=-\infty}^{m=\infty} c_m e^{im\theta}$$

for all almost all $\theta \in \mathbb{R}/2\pi\mathbb{Z}$. Furthermore, if f is continuous, then equality holds everywhere.

Proof. Write

$$\varphi(\theta) = \sum_{m=-\infty}^{m=\infty} c_m e^{im\theta}$$

and recall that

$$f_r(\theta) = \sum_{m=-\infty}^{m=\infty} c_m r^{|m|} e^{im\theta}.$$

Since the series $\sum_{m=-\infty}^{m=\infty} |c_m|$ converges, the series defining φ converges absolutely, so φ is continuous. We claim that

$$\lim_{r\mapsto 1} \|\varphi - f_r\|_\infty = 0.$$

We have

$$\|\varphi - f_r\|_\infty \le \sum_{m=-\infty}^{m=\infty} |c_m|(1 - r^{|m|}).$$

Given $\epsilon > 0$, we can find p so that $\sum_{|m|>p} |c_m| \le \epsilon/2$. Then $\sum_{|m|\le p} |c_m|$ $(1 - r^{|m|})$ is the sum of $2p + 1$ terms that tend to 0 as r tends to 1, so for r close enough to 1 so that $\sum_{|m|>p} |c_m|(1 - r^{|m|}) < \epsilon/2$, we have

$$\sum_{|m|>p} |c_m|(1 - r^{|m|}) + \sum_{|m|\le p} |c_m|(1 - r^{|m|}) < \epsilon/2 + \epsilon/2 = \epsilon,$$

which shows that $\lim_{r\mapsto 1} \|\varphi - f_r\|_\infty = 0$.

Since by Proposition 5.24(2), $\|\varphi - f_r\|_1 \leq 2\pi \|\varphi - f_r\|_\infty$, we also have

$$\lim_{r \mapsto 1} \|\varphi - f_r\|_1 = 0.$$

Since $f \in \mathrm{L}^1(\mathbb{T})$, by Theorem 6.2(1),

$$\lim_{r \mapsto 1} \|f - f_r\|_1 = 0,$$

and since

$$\|f - \varphi\|_1 \leq \|f - f_r\|_1 + \|f_r - \varphi\|_1,$$

we deduce that

$$\|f - \varphi\|_1 = 0,$$

which means that $f = \varphi$ almost everywhere. If f is continuous, since φ is also continuous, $f - \varphi = h$ is continuous. But if $h \neq 0$, then h is nonzero on some interval, which contradicts the fact that $f = \varphi$ almost everywhere. □

Definition 6.4. Given any function $f \in \mathrm{L}^1(\mathbb{T})$, the function $\mathcal{F}(f) \colon \mathbb{Z} \to \mathbb{C}$ given by $\mathcal{F}(f)(m) = c_m$, where c_m is the *Fourier coefficient*

$$c_m = \int_{-\pi}^{\pi} f(t) e^{-imt} \frac{dx(t)}{2\pi},$$

is called the *Fourier transform* of f. We identify the sequence $\mathcal{F}(f)$ with the sequence $(c_m)_{m \in \mathbb{Z}}$, which is also denoted by $\widehat{f}$.

Theorem 6.2(3) (Parseval's theorem) implies that if $f \in \mathrm{L}^2(\mathbb{T})$, then $\widehat{f} \in \ell^2(\mathbb{Z})$. However, if $f \in \mathrm{L}^2(\mathbb{T})$, then it *may not* be the case that $\widehat{f} \in \ell^1(\mathbb{Z})$.

Theorem 6.3 says that if $f \in \mathrm{L}^1(\mathbb{T})$ and if $\widehat{f} \in \ell^1(\mathbb{Z})$, then f can be reconstructed by its Fourier series $\sum_{m=-\infty}^{m=\infty} c_m e^{im\theta}$. Theorem 6.2(3) says that if $f \in \mathrm{L}^2(\mathbb{T})$ then the partial sums $S_{n,f}$ (with $S_{n,f}(\theta) = \sum_{m=-n}^{m=n} c_m e^{im\theta}$) of the Fourier series of f converge to f in the L^2-norm. In fact, we have a stronger result.

Theorem 6.4. *(Plancherel) The map $\mathcal{F} \colon f \mapsto \widehat{f}$ is an isometric isomorphism of the Hilbert spaces $\mathrm{L}^2(\mathbb{T})$ and $\ell^2(\mathbb{Z})$.*

Proof sketch. Theorem 6.4 is classical theorem of Hilbert theory. Its proof can be found in Rudin [Rudin (1987)] (Chapter 4) or Malliavin [Malliavian *et al.* (1995)] (Chapter 3, Section 2.2.5). Consider the map $\mathcal{F} \colon \mathrm{L}^2(\mathbb{T}) \to \ell^2(\mathbb{Z})$ given by $\mathcal{F}f = \widehat{f}$. The fact that the linear map $\mathcal{F}$ is an isometry is an immediate consequence of Parseval's theorem. This fact implies that $\mathcal{F}$

is injective. We prove the surjectivity of the map $\mathcal{F}$ by a density argument. Since $\mathcal{F}$ is an isometry, its image $\mathcal{F}(\mathrm{L}^2(\mathbb{T}))$ is complete in $\ell^2(\mathbb{Z})$, and thus closed. Consider the subset W of $\ell^2(\mathbb{Z})$ given by

$$W = \{(c_m) \in \ell^2(\mathbb{Z}) \mid c_m = 0 \text{ for all but finitely many } m \in \mathbb{Z}\}.$$

The subset W is dense in $\ell^2(\mathbb{Z})$, and obviously $W \subseteq \ell^1(\mathbb{Z})$. For any $c = (c_m) \in W$, the series $\varphi(\theta) = \sum_{m=-\infty}^{m=\infty} c_m e^{im\theta}$ only has finitely many nonzero terms $c_m e^{im\theta}$, so φ is continuous, and thus in $\mathrm{L}^2(\mathbb{T})$. It is also immediately verified that $\mathcal{F}(\varphi) = c$. It follows that $W \subseteq \mathcal{F}(\mathrm{L}^2(\mathbb{T}))$, and since W is dense in $\ell^2(\mathbb{Z})$ and $\mathcal{F}(\mathrm{L}^2(\mathbb{T}))$ is closed in $\ell^2(\mathbb{Z})$, we have $\mathcal{F}(\mathrm{L}^2(\mathbb{T})) = \ell^2(\mathbb{Z})$. □

Definition 6.5. Given a sequence $c = (c_m)_{m\in\mathbb{Z}} \in \ell^1(\mathbb{Z})$, we define the *Fourier cotransform* $\overline{\mathcal{F}}(c)$ of c as the function $\overline{\mathcal{F}}(c)\colon \mathbb{T} \to \mathbb{C}$ defined on $\mathbb{T}$ given by

$$\overline{\mathcal{F}}(c)(\theta) = \sum_{m=-\infty}^{m=\infty} c_m e^{im\theta} = \sum_{m=-\infty}^{m=\infty} c_m (e^{i\theta})^m,$$

the Fourier series associated with c (with $\theta \in \mathbb{R}/2\pi\mathbb{Z}$). Given a function $f \in \mathrm{L}^1(\mathbb{T})$, if $\widehat{f}$ is the Fourier transform of f, then the Fourier cotransform $\overline{\mathcal{F}}(\widehat{f}) = \sum_{m=-\infty}^{m=\infty} \widehat{f}_m e^{im\theta}$ of $\widehat{f}$ is called the *Fourier series* of f.

Note that $e^{im\theta}$ is used instead of the term $e^{-im\theta}$ occurring in the Fourier transform.

If $c \in \ell^1(\mathbb{Z}) \subseteq \ell^2(\mathbb{Z})$, then the series $\overline{\mathcal{F}}(c)(\theta)$ converges uniformly. On the other hand, if $c \in \ell^2(\mathbb{Z}) - \ell^1(\mathbb{Z})$, the series $\overline{\mathcal{F}}(c)$ may not converge pointwise or uniformly, although it converges to a function in $\mathrm{L}^2(\mathbb{T})$ in the L^2-norm. In general, for an arbitrary function $f \in \mathrm{L}^1(\mathbb{T}) - \mathrm{L}^2(\mathbb{T})$, we must have $\widehat{f} \notin \ell^2(\mathbb{Z})$, so the Fourier series $\overline{\mathcal{F}}(\widehat{f}) = \sum_{m=-\infty}^{m=\infty} \widehat{f}_m e^{im\theta}$ may not converge to f pointwise or in the L^1 sense.

Remark: The maps $e^{i\theta} \mapsto e^{im\theta} = (e^{i\theta})^m$, for $m \in \mathbb{Z}$ and $\theta \in \mathbb{R}/(2\pi\mathbb{Z})$, are continuous homomorphisms of the group $\mathbb{T} = \mathbf{U}(1)$ into itself. In fact, it can be shown that they are the only ones of this kind. They are called the *characters* of $\mathbb{T}$; see Section 10.1 and more generally Chapter 10 for a detailed treatment. Obviously the set of characters of $\mathbb{T}$ is in bijection with $\mathbb{Z}$. Thus the Fourier transform $\mathcal{F}(f)$ of a function $f \subset \mathrm{L}^2(\mathbb{T})$, a sequence of complex numbers indexed by $\mathbb{Z}$, can be viewed as a function of the characters of $\mathbb{T}$.

The characters of $\mathbb{Z}$ are the group homomorphisms $\varphi\colon \mathbb{Z} \to \mathbb{T}$. Since $\mathbb{Z}$ is generated by 1, a homomorphism satisfies the equation

$$\varphi(m) = (\varphi(1))^m, \qquad m \in \mathbb{Z},$$

so it is uniquely determined by picking $\varphi(1) = e^{i\theta} \in \mathbb{T}$ (with $\theta \in \mathbb{R}/(2\pi\mathbb{Z})$), and is of the form $\varphi(m) = (e^{i\theta})^m = e^{im\theta}$ for all $m \in \mathbb{Z}$. Thus the set of characters of $\mathbb{Z}$ is in bijection with $\mathbb{T}$. Then the Fourier cotransform $\overline{\mathcal{F}}(c)$ of a "function" $c \in \ell^2(\mathbb{Z})$ ($\overline{\mathcal{F}}(c)$ is the Fourier series associated with c) can also be viewed as a function on the characters of $\mathbb{Z}$, namely a function on $\mathbb{T}$. This fact generalizes to an arbitrary abelian locally compact group and is the key to the definition of the Fourier transform on such a group; see Chapter 10.

Sometimes it is more convenient to express the Fourier series $\overline{\mathcal{F}}(c)(\theta) = \sum_{m=-\infty}^{m=\infty} c_m e^{im\theta}$ in terms of $\cos m\theta$ and $\sin m\theta$ instead of the complex exponentials $e^{im\theta}$. Here we are assuming that $c = (c_m)_{m\in\mathbb{Z}} \in \ell^1(\mathbb{Z})$, so the series $\sum_{m=-\infty}^{m=\infty} c_m e^{im\theta}$ is absolutely convergent and it is permissible to permute terms. Since $e^{im\theta} = \cos m\theta + i \sin m\theta$, the Fourier series $\overline{\mathcal{F}}(c)(\theta)$ can be expressed as

$$\begin{aligned}
\overline{\mathcal{F}}(c)(\theta) &= \sum_{m=-\infty}^{m=\infty} c_m e^{im\theta} = \sum_{m=-\infty}^{m=\infty} c_m(\cos m\theta + i\sin m\theta)\\
&= \sum_{m=-\infty}^{-1} c_m(\cos m\theta + i\sin m\theta) + \sum_{m=0}^{\infty} c_m(\cos m\theta + i\sin m\theta)\\
&= c_0 + \sum_{m=1}^{\infty} c_{-m}(\cos m\theta - i\sin m\theta) + \sum_{m=1}^{\infty} c_m(\cos m\theta + i\sin m\theta)\\
&= c_0 + \sum_{m=1}^{\infty} ((c_m + c_{-m})\cos m\theta + i(c_m - c_{-m})\sin m\theta).
\end{aligned}$$

Therefore, if we let

$$a_0 = 2c_0, \qquad a_m = c_m + c_{-m}, \qquad b_m = i(c_m - c_{-m}), \qquad m \geq 1,$$

then we have

$$\overline{\mathcal{F}}(c)(\theta) = \frac{1}{2}a_0 + \sum_{m=1}^{\infty} (a_m \cos m\theta + b_m \sin m\theta). \tag{$\dagger$}$$

Equation (†) makes it very clear that the function $\overline{\mathcal{F}}(c)(\theta)$ can be viewed as the countably infinite superposition of the basic periodic functions $\cos m\theta$ and $\sin m\theta$, often called *harmonics*. The number $m \in \mathbb{N} - \{0\}$ is called a *frequency*.

Conversely, if

$$\overline{\mathcal{F}}(c)(\theta) = \frac{1}{2}a_0 + \sum_{m=1}^{\infty}(a_m \cos m\theta + b_m \sin m\theta),$$

then if we let

$$c_0 = \frac{1}{2}a_0, \qquad c_m = \frac{1}{2}(a_m - ib_m), \qquad c_{-m} = \frac{1}{2}(a_m + ib_m), \qquad m \geq 1,$$

then

$$\overline{\mathcal{F}}(c)(\theta) = \sum_{m=-\infty}^{m=\infty} c_m e^{im\theta}.$$

From

$$c_m = \int_{-\pi}^{\pi} f(t)e^{-imt}\,\frac{dx(t)}{2\pi},$$

and

$$a_0 = 2c_0, \qquad a_m = c_m + c_{-m}, \qquad b_m = i(c_m - c_{-m}), \qquad m \geq 1,$$

for $m \geq 1$, we get

$$a_m = c_m + c_{-m} = \int_{-\pi}^{\pi} f(t)(e^{-imt} + e^{imt})\,\frac{dx(t)}{2\pi} = 2\int_{-\pi}^{\pi} f(t)\cos mt\,\frac{dx(t)}{2\pi}$$

and

$$b_m = i(c_m - c_{-m}) = \int_{-\pi}^{\pi} f(t)i(e^{-imt} - e^{imt})\,\frac{dx(t)}{2\pi} = 2\int_{-\pi}^{\pi} f(t)\sin mt\,\frac{dx(t)}{2\pi},$$

that is, for $m \geq 1$, we have

$$a_m = \frac{1}{\pi}\int_{-\pi}^{\pi} f(t)\cos mt\,dt$$
$$b_m = \frac{1}{\pi}\int_{-\pi}^{\pi} f(t)\sin mt\,dt.$$

We also have

$$a_0 = 2c_0 = 2\int_{-\pi}^{\pi} f(t)\,\frac{dx(t)}{2\pi} = \frac{1}{\pi}\int_{-\pi}^{\pi} f(t)\,dt.$$

Therefore we can combine the above equations and we obtain

$$a_m = \frac{1}{\pi}\int_{-\pi}^{\pi} f(t)\cos mt\,dt \qquad (m \geq 0)$$
$$b_m = \frac{1}{\pi}\int_{-\pi}^{\pi} f(t)\sin mt\,dt \qquad (m \geq 1).$$

Note that the equation for a_m also holds for $m = 0$. This is the reason for the term $(1/2)a_0$ in equation (†). The numbers a_m and b_m are also called the *Fourier coefficients* of f. If the function f is real-valued, then the coefficients a_m and b_m are real.

Observe that

$$c_0 = \frac{1}{2}a_0 = \frac{1}{2\pi}\int_{-\pi}^{\pi} f(t)\, dt$$

is the mean value of f over the interval $[-\pi, \pi]$.

Here are a few examples of Fourier transforms. Many more examples can be found in Folland [Folland (1992)].

Example 6.1. Let $f\colon [-\pi, \pi] \to \mathbb{R}$ be the periodic function given by

$$f(\theta) = |\theta|, \qquad -\pi \leq \theta \leq \pi.$$

The graph of $f(\theta)$ is shown in Figure 6.7.

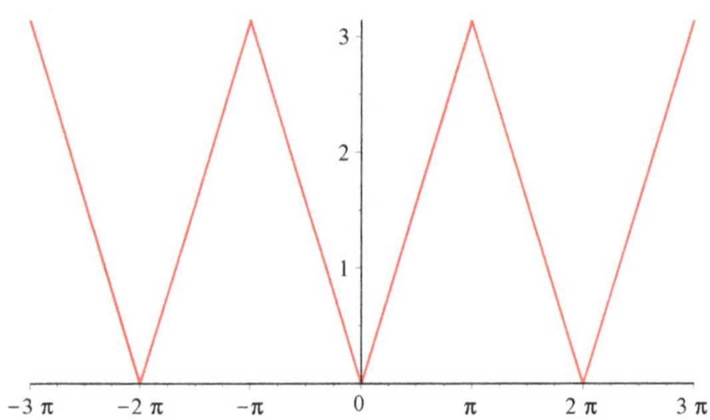

Fig. 6.7 The graph of the periodic function $f(\theta) = |\theta|$, where $-\pi \leq \theta \leq \pi$.

Let us compute the coefficients a_m and b_m. Since the function f is even, we have $b_m = 0$ for all $m \geq 1$, and

$$a_m = \frac{1}{\pi}\int_{-\pi}^{\pi} f(\theta)\cos m\theta\, d\theta = \frac{2}{\pi}\int_{0}^{\pi} f(\theta)\cos m\theta\, d\theta.$$

Thus for $m = 0$ we have

$$a_0 = \frac{2}{\pi}\int_0^{\pi} \theta\, d\theta = \frac{1}{\pi}[\theta^2]_0^{\pi} = \pi,$$

and for $m \geq 1$, integrating by parts we have

$$a_m = \frac{2}{\pi}\left[\frac{\theta \sin m\theta}{m}\right]_0^\pi - \frac{2}{\pi}\int_0^\pi \frac{\sin m\theta}{m}\, d\theta = \frac{2}{\pi}\left[\frac{\cos m\theta}{m^2}\right]_0^\pi = \frac{2}{\pi}\frac{(-1)^m - 1}{m^2}$$

since $\sin m\pi = 0$ and $\cos m\pi = (-1)^m$. Now $(-1)^m - 1 = -2$ when m is odd and 0 when m is even, so we find that the Fourier series for f is given by

$$\frac{\pi}{2} - \frac{4}{\pi}\sum_{m \text{ odd}} \frac{1}{m^2}\cos m\theta = \frac{\pi}{2} - \frac{4}{\pi}\sum_{k=1}^{\infty} \frac{\cos(2k-1)\theta}{(2k-1)^2}.$$

If we plot the graphs of the partial sums for a few terms (say five terms), we see that they provide a very good approximation to f. See Figures 6.8 and 6.9. The series converges uniformly to f due to the presence of the term $1/(2k-1)^2$.

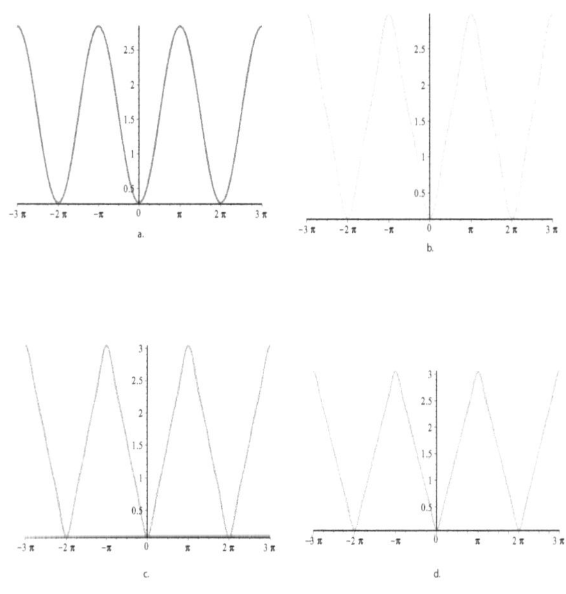

Fig. 6.8 Let $S_M = \frac{\pi}{2} - \frac{4}{\pi}\sum_{k=1}^{M} \frac{\cos(2k-1)\theta}{(2k-1)^2}$. Figure (a) is the graph of S_1; Figure (b) is the graph of S_2; Figure (c) is the graph of S_3, and Figure (d) is the graph of S_4.

Example 6.2. Let $f\colon [-\pi, \pi] \to \mathbb{R}$ be the periodic function given by

$$f(\theta) = \theta, \qquad -\pi < \theta \leq \pi.$$

The graph of $f(\theta)$ is shown in Figure 6.10.

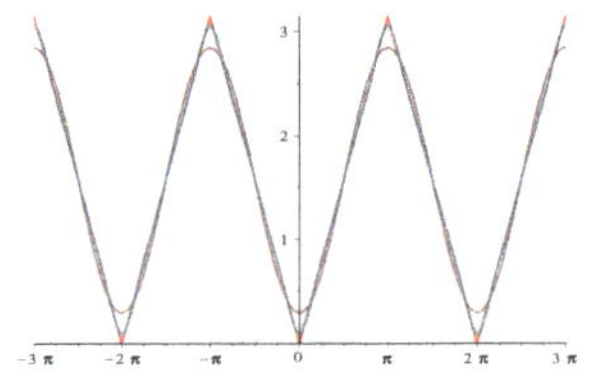

Fig. 6.9 The partial sums S_1 through S_4 approximating $f(\theta)$ of Example 6.1.

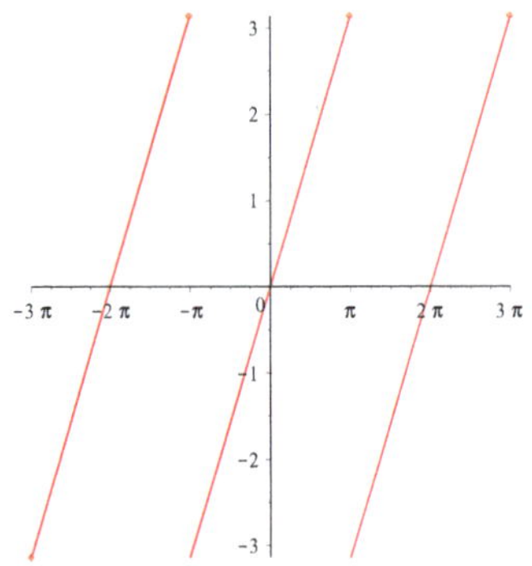

Fig. 6.10 The graph of the periodic function $f(\theta) = \theta$, where $-\pi < \theta \leq \pi$.

This time let us compute the coefficients c_m. We have

$$c_0 = \frac{1}{2\pi} \int_{-\pi}^{\pi} \theta \, d\theta = 0,$$

and for $m \neq 0$, by integrating by parts, we have

$$\begin{aligned} c_m &= \frac{1}{2\pi} \int_{-\pi}^{\pi} \theta e^{-im\theta} \, d\theta \\ &= \frac{1}{2\pi} \left[\frac{\theta e^{-im\theta}}{-im} \right]_{-\pi}^{\pi} - \frac{1}{2\pi} \int_{-\pi}^{\pi} \frac{e^{-im\theta}}{-im} \, d\theta \\ &= \frac{1}{2\pi} \left[e^{-im\theta} \left(\frac{\theta}{-im} + \frac{1}{m^2} \right) \right]_{-\pi}^{\pi} = \frac{(-1)^{m+1}}{im}, \end{aligned}$$

since $e^{-im\pi} = e^{im\pi} = (-1)^m$. Hence the Fourier series for f is

$$\sum_{m \neq 0} \frac{(-1)^{m+1}}{im} e^{im\theta}.$$

Since $(-1)^m = (-1)^{-m}$, the mth and the $(-m)$th term can be combined to give

$$(-1)^{m+1}\left(\frac{e^{im\theta}}{im} + \frac{e^{-im\theta}}{-im}\right) = \frac{2(-1)^{m+1}}{m}\sin m\theta,$$

and we obtain the Fourier series

$$2\sum_{m=1}^{\infty}\frac{(-1)^{m+1}}{m}\sin m\theta. \qquad (*)$$

Here to be rigorous we should consider the partial sums

$$S_{m,f}(\theta) = \sum_{k=-m}^{m} c_k e^{ik\theta},$$

in which the terms corresponding to the indices $-m$ and m can be combined. The details are left as an exercise. Note that $f \in \mathrm{L}^2(\mathbb{T})$ and $c_m \notin \ell^1(\mathbb{Z})$. The series belongs to $\mathrm{L}^2(\mathbb{T})$ but it does not converge to f pointwise or uniformly.

This time if we plot the graphs of the partial sums, we see that they approximate the function f, but the quality of the approximation is inferior to that of Example 6.1. See Figures 6.11 and 6.12.

This is due to the fact that the function of Example 6.1 is continuous, but the function of Example 6.2 has jump discontinuities. The other reason why the quality of approximation is not as good as in Example 6.1 is that the terms of the series in Example 6.1 tend to zero faster than the terms of the series in Example 6.2. Thus, in Example 6.2, the influence of the higher order terms is much more significant in Example 6.2. The point is that the rougher the function is, the more difficult it is to approximate it by smooth functions such as $\cos m\theta$ and $\sin m\theta$. In fact, it is not obvious that the series

$$2\sum_{m=1}^{\infty}\frac{(-1)^{m+1}}{m}\sin m\theta$$

converges pointwise. It does, with

$$2\sum_{m=1}^{\infty}\frac{(-1)^{m+1}}{m}\sin m\theta = \begin{cases}\theta & \text{if } -\pi < \theta < \pi\\ 0 & \text{if } \theta = \pm\pi.\end{cases}$$

This series converges pointwise to the function f of Example 6.2, except for $\theta = (2k+1)\pi$ where $f((2k+1)\pi-) = \pi$ and $f((2k+1)\pi+) = -\pi$, according to a theorem of Dirichlet (see Section 6.3).

A phenomenon that shows up in Example 6.2 is the *Gibbs phenomenon*. Even for a partial sum of 40 terms, we observe some spikes near the discontinuities. These spikes tend to zero in width, but not in height; see Folland [Folland (1992)] (Chapter 2, Section 2.6).

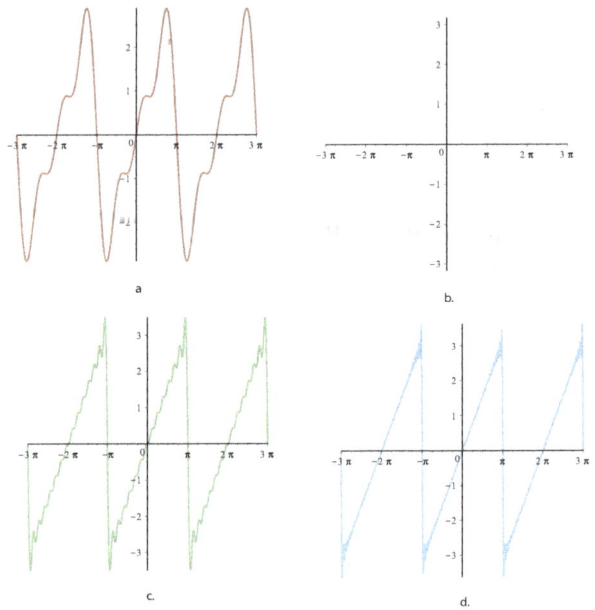

Fig. 6.11 Let $S_M = 2\sum_{m=1}^{M} \frac{(-1)^{m+1}}{m} \sin m\theta$. Figure (a) is the graph of S_3; Figure (b) is the graph of S_5; Figure (c) is the graph of S_{14}, and Figure (d) is the graph of S_{40}.

6.3 Pointwise Convergence of Fourier Series on $\mathbb{T}$

By Theorem 6.2, if $f \in \mathrm{L}^2(\mathbb{T})$, then

$$\lim_{n \mapsto \infty} \left\| f - \sum_{m=-n}^{m=n} c_m e^{im\theta} \right\|_2 = 0,$$

where c_m is the mth Fourier coefficient of f,

$$c_m = \int_{-\pi}^{\pi} f(t) e^{-imt} \frac{dx(t)}{2\pi}.$$

Thus the partial sums

$$S_{m,f}(\theta) = \sum_{k=-m}^{k=m} c_k e^{ik\theta}$$

converge to f in the L^2-norm. However, even if f is continuous, the partial sums $S_{m,f}$ may not converge to f pointwise.

The first example of a function whose Fourier series diverges at 0 was given by du Bois–Reymond in 1873. This is a fairly complicated example

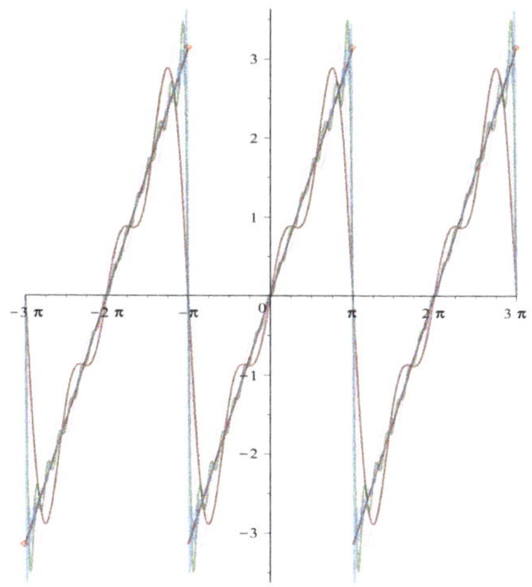

Fig. 6.12 The partial sums S_3,S_5,S_{14},S_{40} approximating $f(\theta)$ of Example 6.2.

involving a piecewise monotone function that oscillates indefinitely near 0. Simpler examples were given later by Fejér and Lebesgue.

Fejér's example makes use of the functions

$$F_n(x) = \sin x + \frac{1}{2}\sin 2x + \cdots + \frac{1}{n}\sin nx,$$

which are uniformly bounded; see Figures 6.13 and 6.14.

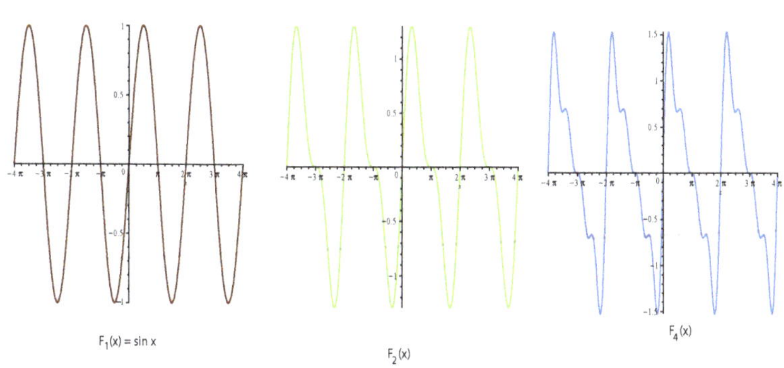

Fig. 6.13 The graphs of $F_1(x)$, $F_2(x)$, and $F_4(x)$.

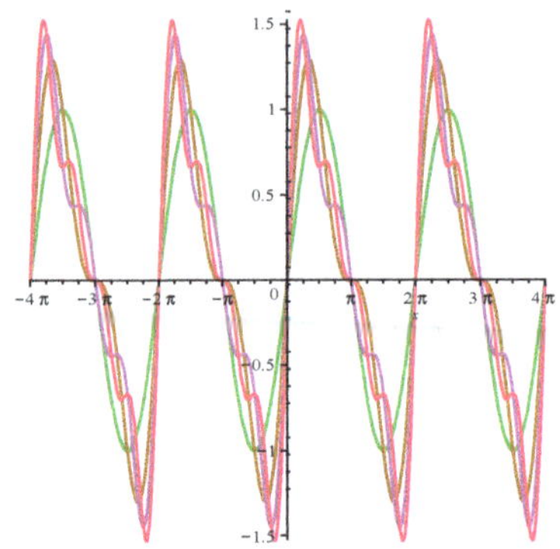

Fig. 6.14 The graphs of $F_1(x)$ through $F_4(x)$ superimposed on each other.

The series

$$\sum_{n=1}^{\infty} \frac{1}{n^2} \sin(3^{n^2} x) F_{2^{n^2}}(x)$$

defines a continuous function f, but it can be shown that its Fourier series diverges for $x = 0$. See also the example in Stein and Shakarchi [Stein and Shakarchi (2003)] (Chapter 3, Section 2.2).

In 1926 Kolmogoroff gave an example of a function $f \in \mathrm{L}^1(\mathbb{T})$ whose Fourier series diverges for all x.

Later it was found that a systematic method for producing functions with a "bad" Fourier series was to use the Banach–Steinhaus theorem.

Definition 6.6. For any fixed $\theta \in \mathbb{R}/2\pi\mathbb{Z}$ and for any continuous function $f \in \mathcal{C}(\mathbb{T}; \mathbb{C})$, let

$$S^*(f, \theta) = \sup_{m \in \mathbb{N}} |S_{m,f}(\theta)|.$$

Also recall the following definition about Borel sets.

Definition 6.7. Let X be a topological space. Countable unions of closed subsets of X are called *F_σ-sets*, and countable intersections of open sets of X are called *G_δ-sets*.

The following result is proven in Rudin [Rudin (1987)] (Chapter 5, Page 102). The proof uses the Banach–Steinhaus theorem.

Proposition 6.5. *For every $\theta \in \mathbb{R}/2\pi\mathbb{Z}$, there is a subset $E_\theta \subseteq \mathcal{C}(\mathbb{T};\mathbb{C})$ of continuous functions which is a dense G_δ set in $\mathcal{C}(\mathbb{T};\mathbb{C})$ such that $S^*(f,\theta) = \infty$ for all $f \in E_\theta$. Consequently, the Fourier series of every $f \in E_\theta$ diverges at θ.*

Using Baire's theorem a stronger result can be obtained, as shown in Rudin [Rudin (1987)] (Chapter 5, Theorem 5.12).

Proposition 6.6. *There is a set $E \subseteq \mathcal{C}(\mathbb{T};\mathbb{C})$ of continuous functions which is a dense G_δ set in $\mathcal{C}(\mathbb{T};\mathbb{C})$ and which has the following property: For every $f \in E$, the set*

$$Q_f = \{\theta \in \mathbb{R}/2\pi\mathbb{Z} \mid S^*(f,\theta) = \infty\}$$

is a dense G_δ set in $\mathbb{R}/2\pi\mathbb{Z}$.

As a consequence, the Fourier series of every continuous function $f \in E$ diverges for infinitely many points. In fact, E and Q_f are uncountable; see Rudin [Rudin (1987)] (Chapter 5, Theorem 5.13).

We just saw that in general, the partial sums $S_{m,f}$ do not behave well, so if we want to approximate a continuous function on $\mathbb{T}$, we should not count on the partial sums to do the job. We will see in Example 8.10 that the Cesàro means,

$$A_{n,f} = \frac{1}{n}(S_{0,f} + \cdots + S_{n-1,f}),$$

have a much better behavior, since they converge uniformly to f.

We are led to the conclusion that in order to obtain positive results for pointwise convergence of the partial sums $S_{m,f}$, we *must restrict* the class of functions that we are considering. Dirichlet was the first to obtain a significant result. In 1829 he proved that the partial sums $S_{m,f}$ converge pointwise to $(f(x+) + f(x-))/2$, for every piecewise continuous and piecewise monotone function f. Here $f(x+)$ is the limit when y tends to x from above, and $f(x-)$ is the limit when y tends to x from below (see Definition 2.19). His paper is not only significant because of its results but because it raised the standards of rigor in mathematical exposition to a new level. Dirichlet's full paper is reproduced in Kahane and Lemarié–Rieusset [Kahane and Lemaré-Rieusset (1998)]. As Kahane comments, "Dirichlet's style is superb and incredibly modern."

Later on it was realized that what is really needed for this pointwise convergence result to hold is that the functions have bounded variation.

Camille Jordan, whose mathematical interests were group theory, algebra, and its relations to geometry, introduced the notion of a function of bounded variation in 1881 and published one paper generalizing Dirichlet's paper to this class of functions.

Let $f\colon [a,b] \to \mathbb{R}$ be a function. The intuition is that the variation of f over $[a,b]$ is the total distance travelled from time a to time b. If f' exits and is continuous, then the variation is $\int_a^b |f'(t)|\, dt$. Otherwise, we approximate the curve by a piecewise affine function. For this we subdivide the interval $[a,b]$ into smaller intervals, $[t_{j-1}, t_j]$ and approximate f on this subinterval by the line segment from $(t_{j-1}, f(t_{j-1}))$ to $(t_j, f(t_j))$; see Figure 6.15.

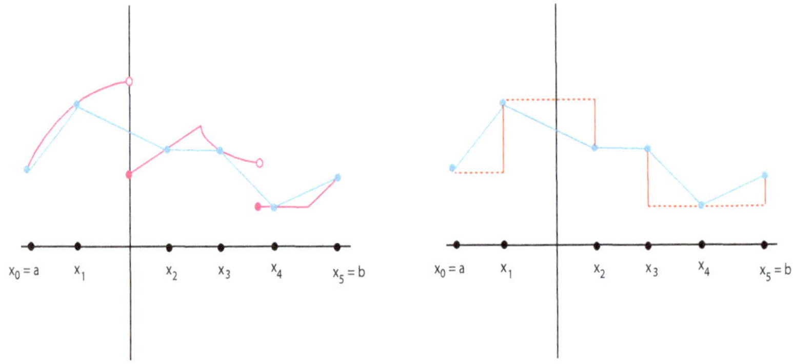

Fig. 6.15 The graph of f is represented in pink. The linear approximation from $(t_{j-1}, f(t_{j-1}))$ to $(t_j, f(t_j))$, where $0 \leq j \leq 5$, is represented in blue. A variation of f over $[a,b]$ is the sum of lengths of the solid orange vertical lines found in the right figure.

More precisely, consider a function $f\colon \mathbb{R} \to \mathbb{C}$. For any $x \in \mathbb{R}$, we consider subdivisions of the interval $(-\infty, x]$ using finite sequences $x_0 < x_1 < \cdots < x_n = x$.

Definition 6.8. Let $f\colon \mathbb{R} \to \mathbb{C}$ be a function. The *total variation function* T_f of f is the function given by

$$T_f(x) = \sup \left\{ \sum_{j=1}^{n} |f(x_j) - f(x_{j-1})| \;\middle|\; -\infty < x_0 < x_1 < \cdots < x_n = x \right\},$$

with $n \in \mathbb{N} - \{0\}$, where the supremum is taken over all finite subdivisions $x_0 < x_1 < \cdots < x_n = x$. If $[a,b]$ is a finite interval ($a \leq b$), then the *total variation of* f *on* $[a,b]$ is the quantity

$$V(f,a,b) = \sup\left\{\sum_{j=1}^{n} |f(x_j) - f(x_{j-1})| \mid a = x_0 < x_1 < \cdots < x_n = b\right\},$$

with $n \in \mathbb{N} - \{0\}$. The set BV of functions of *bounded variation* is the set of functions $f\colon \mathbb{R} \to \mathbb{C}$ such that $\lim_{x \mapsto +\infty} T_f(x) < \infty$. The set $BV([a,b])$ of functions of *bounded variation over* $[a,b]$ is the set of functions $f\colon [a,b] \to \mathbb{C}$ such that $V(f,a,b)$ is finite.

If $f\colon \mathbb{R} \to \mathbb{C}$ is a function in BV, since a can be chosen as a subdivision point we see immediately that

$$V(f,a,b) = T_f(b) - T_f(a),$$

so $f \in BV([a,b])$. We also see that $T_f(x)$ is an increasing function. The limits $\lim_{x \mapsto -\infty} f(x)$ and $\lim_{x \mapsto +\infty} f(x)$ also exist, as a corollary of Proposition 6.9.

Example 6.3.

(1) If $f\colon \mathbb{R} \to \mathbb{R}$ is bounded and increasing, then $f \in BV$. In fact, $T_f(x) = f(x) - f(-\infty)$; see Figure 6.16.

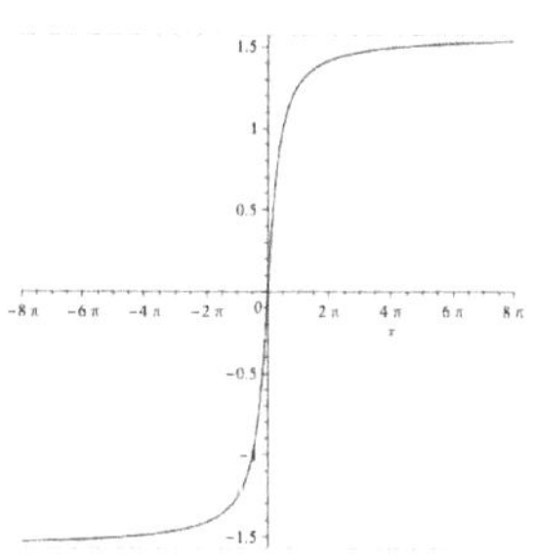

Fig. 6.16 Let $f(x) = \tan^{-1}(x)$. Then $f(x)$ is a bounded, increasing function whose total variation $T_f(x) = f(x) - f(-\infty) = f(x) + \frac{\pi}{2}$.

(2) The space BV is a complex vector space.
(3) If f is differentiable on $\mathbb{R}$ and if f' is bounded, then $f \in BV([a,b])$ for every finite interval $[a,b]$ (by the mean value theorem); see Figure 6.17.

(4) If $f(x) = \sin x$, then $f \in BV([a,b])$ for every finite interval $[a,b]$, but $f \notin BV$ (it oscillates forever); see Figure 6.18.

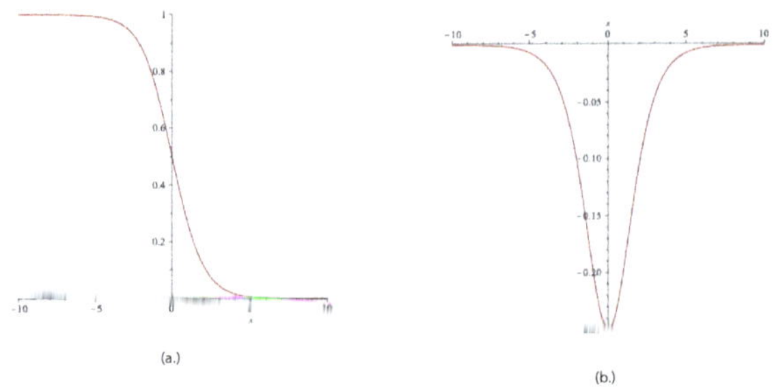

Fig. 6.17 Figure (a) is the graph of $f(x) = \frac{1}{e^x+1}$, a bounded decreasing function. Its derivative $f'(x) = -\frac{e^x}{(e^x+1)^2}$, whose graph is shown in Figure (b), is also bounded. Hence $f \in BV([a,b])$ for every finite interval $[a,b]$.

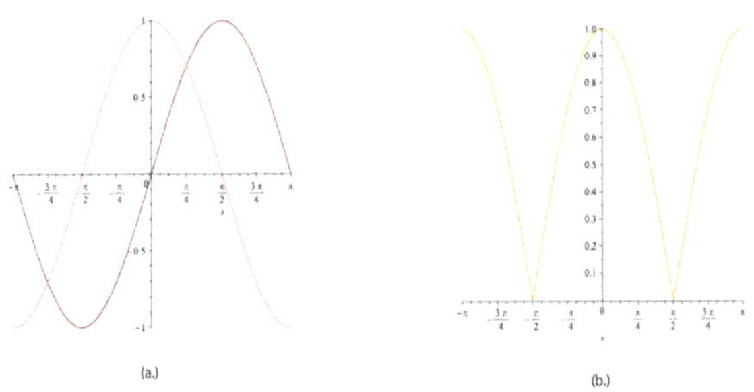

Fig. 6.18 In Figure (a), the graph of $f(x) = \sin x$ is the dark red curve, while $f'(x) = \cos x$ is the lighter red curve. The total variation of $f(x)$ on $[\pi,\pi]$ is given by $\int_{-\pi}^{\pi} |f'(x)|\, dx = \int_{-\pi}^{\pi} |\cos(x)|\, dx = 4$ and is visualized as the area under the graph of $y = |\cos(x)|$, as illustrated by Figure (b).

(5) If $f(x) = x\sin(x^{-1})$ for $x \neq 0$ and $f(0) = 0$, then $f \notin BV([a,b])$ for $a \leq 0 < b$ or $a < 0 \leq b$; see Figure 6.19.

Here are some of the main properties of functions of bounded variation; proofs can be found in Folland [Folland (1999)] (Chapter 3, Section 3.5).

Proposition 6.7. *Let $f\colon \mathbb{R} \to \mathbb{C}$ be any function. If $f \in BV$, then* $\lim_{x \mapsto -\infty} T_f(x) = 0$.

Proof. By definition of $T_f(x)$ as a least upper bound, for every $\epsilon > 0$, there

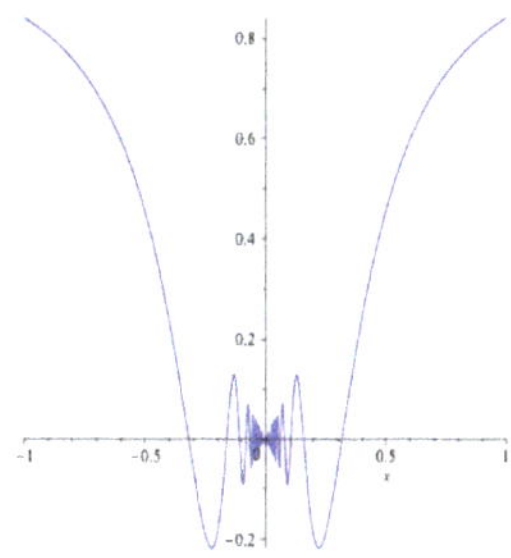

Fig. 6.19 The graph of $f(x) = x\sin(x^{-1})$ for $x \neq 0$ and $f(0) = 0$. This function has too much oscillation around 0 for it to be of bounded variation.

is some subdivision $x_0 < x_1 < \cdots < x_n = x$ such that

$$\sum_{j=1}^{n} |f(x_j) - f(x_{j-1})| \geq T_f(x) - \epsilon.$$

By definition of $V(f, x_0, x)$, this implies that

$$T_f(x) - T_f(x_0) \geq T_f(x) - \epsilon,$$

so $T_f(x_0) \leq \epsilon$. Since T_f is increasing, we also have $T_f(y) \leq \epsilon$ for all $y \leq x_0$. Since ϵ is arbitrary, we must have $\lim_{x \mapsto -\infty} f(x) = 0$. $\square$

Proposition 6.8. *Let $f\colon \mathbb{R} \to \mathbb{R}$ be a function. If $f \in BV$, then $T_f + f$ and $T_f - f$ are increasing functions.*

Proposition 6.9. *Let $f\colon \mathbb{R} \to \mathbb{C}$ be a function.*

(1) We have $f \in BV$ iff both the real part and the imaginary part of f belong to BV.
(2) If $f\colon \mathbb{R} \to \mathbb{R}$, then $f \in BV$ iff f can be written as the difference of two bounded increasing functions; these can be chosen as $(T_f + f)/2$ and $(T_f - f)/2$. This is called a **Jordan decomposition**.
(3) If $f\colon \mathbb{R} \to \mathbb{C}$ and if $f \in BV$, then the left limit $f(x-)$ and the right limit $f(x+)$ exist for all $x \in \mathbb{R}$, including $x = -\infty$ and $x = +\infty$.
(4) If $f\colon \mathbb{R} \to \mathbb{C}$ and if $f \in BV$, then f has at most countably many discontinuities.
(5) If $f \in BV$ and if we let $g(x) = f(x+)$, then f' and g' exist almost everywhere and are equal almost everywhere.

Remark: The space NBV consists of the functions $f\colon \mathbb{R} \to \mathbb{C}$ in BV which are right continuous and such that $f(-\infty) = 0$. There is a relationship between the space NBV and the complex Borel measures on $\mathbb{R}$. If μ is a complex Borel measure, then the function $F(x) = \mu((-\infty, x])$ is in NBV, and conversely, given any function $f \in NBV$, there is a unique complex measure μ_f such that $f(x) = \mu_f((-\infty, x])$; see Folland [Folland (1999)] (Chapter 3, Section 3.5, Theorem 3.29).

We now return to Fourier series on $\mathbb{T}$ and state the following theorem essentially due to Jordan which generalizes an historically famous result of Dirichlet.

Theorem 6.10. *For any $f \in \mathrm{L}^1(\mathbb{T})$, if $f \in BV([-\pi, \pi])$, then*

$$\lim_{m \mapsto \infty} S_{m,f}(x) = \frac{f(x+) + f(x-)}{2}$$

for all $x \in [-\pi, \pi]$. In particular, $\lim_{m \mapsto \infty} S_{m,f}(x) = f(x)$ whenever f is continuous at x.

Theorem 6.10 is proven in Folland [Folland (1999)] (Chapter 8, Section 8.5, Theorem 8.43).

Other convergence theorems (some about pointwise convergence) are discussed in Folland [Folland (1992)] and Stein and Shakarchi [Stein and Shakarchi (2003)].

The behavior of $S_{m,f}$ at a jump discontinuity $x = m$, with $m \in \mathbb{Z}$, (a point x where $f(x) \neq f(x-)$ or $f(x) \neq f(x+)$) is a little strange. It turns out that near an integer value of x, the function $S_{m,f}$ contains spikes that overshoot or undershoot the function f, and when m tends to infinity, the width of the spikes tends to zero but the height does not. This behavior is known as *Gibbs phenomenon*. For example, the function

$$\varphi(x) = 2\pi \left(\frac{1}{2} - (x - \lfloor x \rfloor) \right)$$

(where $\lfloor x \rfloor$ is the greatest integer $\leq x$) is periodic (of period 2π) and exhibits the Gibbs phenomenon. One easily computes the Fourier coefficients, which are

$$c_0 = 0, \quad c_m = \frac{1}{im}, \quad m \neq 0.$$

Then we get

$$S_{m,\varphi}(x) = \sum_{k=1}^{m} \frac{2 \sin kx}{k}.$$

See Figures 6.20 and 6.21.

For more details see Folland [Folland (1999)] (Chapter 8, Section 8.5).

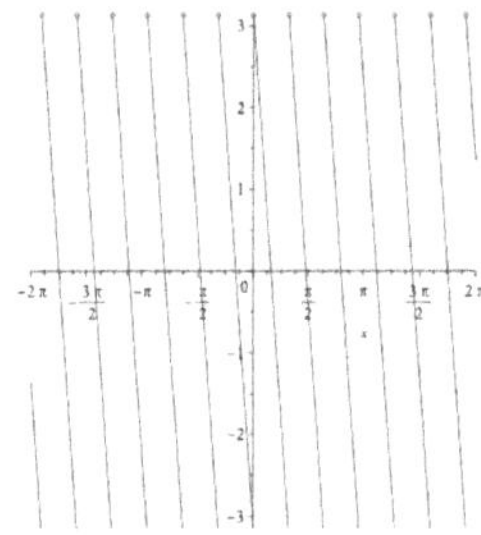

Fig. 6.20 The graph of $\varphi(x) = 2\pi\left(\frac{1}{2} - (x - \lfloor x \rfloor)\right)$.

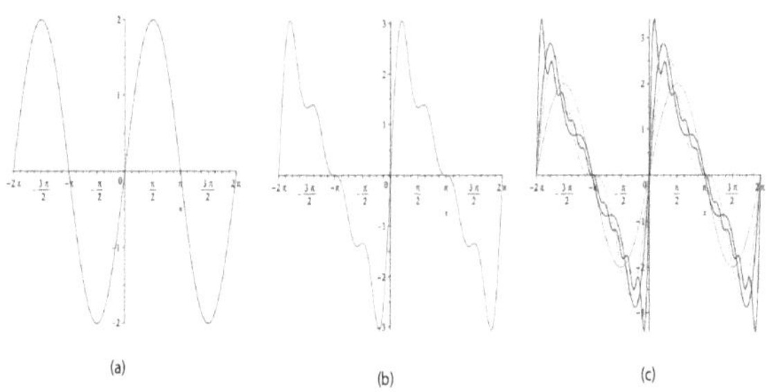

Fig. 6.21 Figure (a.) is the graph of $S_{1,\varphi}(x)$, Figure (b.) is the graph of $S_{4,\varphi}(x)$, while Figure (c.) shows the supersition of the graphs of $S_{1,\varphi}(x)$, $S_{2,\varphi}(x)$, $S_{3,\varphi}(x)$, $S_{4,\varphi}(x)$, and $S_{10,\varphi}(x)$.

6.4 The Fourier Transform and the Fourier Cotransform on $\mathbb{T}^n$ and $\mathbb{Z}^n$

In Section 6.1 we introduced the Fourier transform on $\mathbb{T}$ and the Fourier cotransform on $\mathbb{Z}$. In this section we briefly present the generalization to $\mathbb{T}^n = \underbrace{\mathbb{T} \times \cdots \times \mathbb{T}}_{n}$, called the *n-torus*, and to $\mathbb{Z}^n$. As in Section 6.1, a normalized Haar measure on $\mathbb{T}^n$ is $dx_n/(2\pi)^n$, where dx_n the Lebesgue measure on $\mathbb{R}^n$ (so that $\mathbb{T}^n$ has measure 1).

Recall that given any function $f \in \mathrm{L}^1(\mathbb{T})$, the function $\mathcal{F}(f)\colon \mathbb{Z} \to \mathbb{C}$ given by $\mathcal{F}(f)(m) = c_m$, where c_m is the *Fourier coefficient*

$$c_m = \int_{-\pi}^{\pi} f(t)e^{-imt}\,\frac{dx(t)}{2\pi},$$

is called the *Fourier transform* of f. We identify the sequence $\mathcal{F}(f)$ with the sequence $(c_m)_{m\in\mathbb{Z}}$, which is also denoted by $\widehat{f}$.

Given a sequence $c = (c_m)_{m\in\mathbb{Z}} \in \ell^1(\mathbb{Z})$, we define the *Fourier cotransform* $\overline{\mathcal{F}}(c)$ of c as the function $\overline{\mathcal{F}}(c)\colon \mathbb{T} \to \mathbb{C}$ defined on $\mathbb{T}$ given by

$$\overline{\mathcal{F}}(c)(\theta) = \sum_{m=-\infty}^{m=\infty} c_m e^{im\theta},$$

the Fourier series associated with c (with $\theta \in \mathbb{R}/2\pi\mathbb{Z}$). Note that $e^{im\theta}$ is used instead of the term $e^{-im\theta}$ occurring in the Fourier transform.

For symmetry reasons, it seems natural to define a Fourier cotransform on $\mathbb{T}$ and a Fourier transform on $\mathbb{Z}$.

Definition 6.9. The *Fourier cotransform* $\overline{\mathcal{F}}(f)$ of a function $f \in \mathrm{L}^1(\mathbb{T})$ is the $\mathbb{Z}$-indexed sequence $\overline{\mathcal{F}}(f)\colon \mathbb{Z} \to \mathbb{C}$ given by

$$\overline{\mathcal{F}}(f)(m) = \int_{-\pi}^{\pi} f(t)e^{imt}\,\frac{dx(t)}{2\pi},$$

and the *Fourier transform* $\mathcal{F}(c)$ of a sequence $c = (c_m)_{m\in\mathbb{Z}} \in \ell^1(\mathbb{Z})$ is the function $\mathcal{F}(c)\colon \mathbb{T} \to \mathbb{C}$ given by

$$\mathcal{F}(c)(\theta) = \sum_{m=-\infty}^{m=\infty} c_m e^{-im\theta},$$

with $\theta \in \mathbb{R}/2\pi\mathbb{Z}$.

Observe that if $f \in \mathrm{L}^1(\mathbb{T})$, then

$$\overline{\mathcal{F}}(f)(m) = \mathcal{F}(f)(-m) = \overline{\mathcal{F}(\overline{f})(m)},$$

with $m \in \mathbb{Z}$, and if $c \in \ell^1(\mathbb{Z})$, then

$$\overline{\mathcal{F}}(c)(\theta) = \mathcal{F}(c)(-\theta) = \overline{\mathcal{F}(\overline{c})(\theta)},$$

where $\theta \in \mathbb{R}/2\pi\mathbb{Z}$. Thus only one of the two transforms is really needed, but it is convenient to use both (especially in stating Fourier inversion).

Remark: Note a certain asymmetry in the measure chosen on $\mathbb{T}$ and $\mathbb{Z}$. The measure on $\mathbb{T}$ is $dx/2\pi$, so that $\mathbb{T}$ has measure 1, and the measure on $\mathbb{Z}$ is the counting measure.

The main results are:

(1) The spectral synthesis, Theorem 6.2.

(2) The Fourier inversion formula, Theorem 6.3. This result can be expressed as follows. If $f \in L^1(\mathbb{T})$ and if $\widehat{f} = \mathcal{F}(f) \in \ell^1(\mathbb{Z})$, then

$$f(\theta) = (\overline{\mathcal{F}} \circ \mathcal{F})(f)(\theta) = \sum_{m \in \mathbb{Z}} \widehat{f}_m e^{im\theta}.$$

(3) Plancherel's theorem, Theorem 6.4. This theorem asserts that $\mathcal{F}$ is an isometric isomorphisms between the Hilbert spaces $L^2(\mathbb{T})$ and $\ell^2(\mathbb{Z})$.

All three results stated above generalize to $\mathbb{T}^n$ and $\mathbb{Z}^n$. First we need a bit of notation.

Definition 6.10. A *multi-index* is a sequence $\alpha = (\alpha_1, \ldots, \alpha_n)$ of natural numbers $\alpha_i \in \mathbb{N}$. Define $|\alpha|, \alpha!, \partial^\alpha$, and x^α by

$$|\alpha| = \sum_{i=1}^{n} \alpha_i, \quad \alpha! = \alpha_1! \times \cdots \times \alpha_n!, \quad \partial^\alpha = \left(\frac{\partial}{\partial x_1}\right)^{\alpha_1} \cdots \left(\frac{\partial}{\partial x_n}\right)^{\alpha_n},$$

and

$$x^\alpha = \prod_{i=1}^{n} x_i^{\alpha_i}.$$

Example 6.4. For a specific example of Definition 6.10, let $n = 3$ and $\alpha = (\alpha_1, \alpha_2, \alpha_3) = (1, 3, 4)$. Then $|\alpha| = \alpha_1 + \alpha_2 + \alpha_3 = 1 + 3 + 4 = 8$, $\alpha! = \alpha_1!\alpha_2!\alpha_3! = 1!3!4! = 144$, $\partial^\alpha = \frac{\partial}{\partial x_1}\left(\frac{\partial}{\partial x_2}\right)^3\left(\frac{\partial}{\partial x_3}\right)^4$, and $x^\alpha = x_1^{\alpha_1}x_2^{\alpha_2}x_3^{\alpha_3} = x_1x_2^3x_3^4$.

We now generalize the Poisson kernel and the Fourier transform (and cotransform) to $\mathbb{T}^n$ and $\mathbb{Z}^n$.

Observe that a function $z\colon \mathbb{Z}^n \to \mathbb{C}$ can be viewed as a $\mathbb{Z}^n$-indexed sequence $z = (z_m)_{m \in \mathbb{Z}^n}$, with $z_m \in \mathbb{C}$.

Example 6.5. To gain some insight into a $\mathbb{Z}^n$-indexed sequence, set $n = 2$ and $z\colon \mathbb{Z}^2 \to \mathbb{C}$. The indices of z are the integer-indexed lattice points of $\mathbb{R}^2$. In particular, if we assume that the nonzero elements of z are entries whose lattice points lie in the closed unit square centered at the origin, z is the finite sequence

$$z = \left(z_{(-1,-1)}, z_{(-1,0)}, z_{(-1,1)}, z_{(0,-1)}, z_{(0,0)}, z_{(0,1)}, z_{(1,-1)}, z_{(1,0)}, z_{(1,1)}\right),$$

where we implicitly made use of the following total ordering for $\mathbb{Z}^2$: given $(i,j), (p,q) \in \mathbb{Z}^2$, $(i,j) < (p,q)$ if either $i < p$ or $i = p$ and $j < q$; see Figure 6.22.

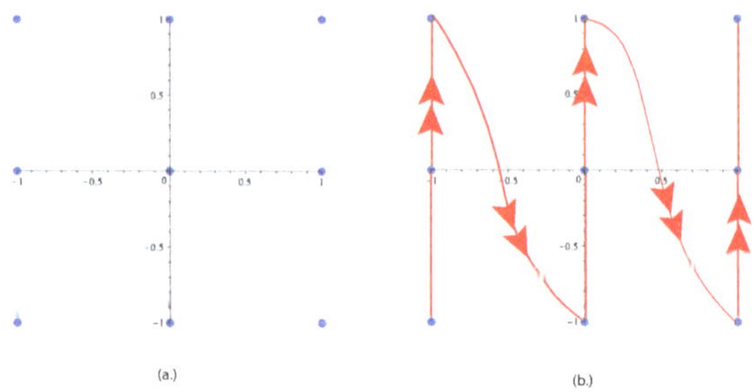

Fig. 6.22 Figure (a) illustrates the lattice points in $\mathbb{R}^2$ associated with the $\mathbb{Z}^2$-indexed sequence of Example 6.5. The directed red curve of Figure (b) illustrates the total ordering of $\mathbb{Z}^2$ used in Example 6.5.

Let $\ell^p(\mathbb{Z}^n)$ $(p \geq 1)$ be the space

$$\ell^p(\mathbb{Z}^n) = \left\{ z = (z_m)_{m\in\mathbb{Z}^n},\ z_m \in \mathbb{C} \ \middle|\ \sum_{m\in\mathbb{Z}^n} |z_m|^p < \infty \right\}.$$

As in the case $n = 1$, if $1 \leq p < q$, then $\ell^p(\mathbb{Z}^n) \subseteq \ell^q(\mathbb{Z}^n)$.

We denote the product measure on $\mathbb{T}^n$ by

$$dx_n/(2\pi)^n = (1/(2\pi)^n)\underbrace{dx \otimes \cdots \otimes dx}_{n},$$

where dx is the Lebesgue measure on $\mathbb{R}$. With this measure, $\mathbb{T}^n$ has measure 1.

Definition 6.11. The *Poisson kernel* $P_r(\theta)$ on $\mathbb{T}^n$ (with $\theta = (\theta_1, \ldots, \theta_n) \in \mathbb{R}^n/2\pi\mathbb{Z}^n$) is the family of functions $P_r(\theta)$, parametrized by $r \in [0, 1)$, given by

$$P_r(\theta) = \prod_{k=1}^{n} P_r(\theta_k),$$

with

$$P_r(\theta_k) = \sum_{n=-\infty}^{n=\infty} r^{|n|}\, e^{in\theta_k}.$$

Definition 6.12. For any function $f \in \mathrm{L}^1(\mathbb{T}^n)$, the *Fourier transform* $\widehat{f} = \mathcal{F}(f)$ of f is the $\mathbb{Z}^n$-indexed sequence $\mathcal{F}(f)\colon \mathbb{Z}^n \to \mathbb{C}$ given by

$$\widehat{f}(m) = \mathcal{F}(f)(m) = \int_{\mathbb{T}^n} f(\theta)e^{-im\cdot\theta}\,\frac{dx_n(\theta)}{(2\pi)^n},$$

and the *Fourier cotransform* $\overline{\mathcal{F}}(f)$ of f is the $\mathbb{Z}^n$-indexed sequence $\overline{\mathcal{F}}(f)\colon \mathbb{Z}^n \to \mathbb{C}$ given by

$$\overline{\mathcal{F}}f(m) = \int_{\mathbb{T}^n} f(\theta)e^{im\cdot\theta}\,\frac{dx_n(\theta)}{(2\pi)^n},$$

with $m \in \mathbb{Z}^n$, $\theta \in \mathbb{R}^n/2\pi\mathbb{Z}^n$, and with $m\cdot\theta = \sum_{k=1}^n m_k\theta_k$, the inner product of the vectors $m = (m_1, \dots, m_n)$ and $\theta = (\theta_1, \dots, \theta_n)$.

For any $c \in \ell^1(\mathbb{Z}^n)$, the *Fourier transform* $\mathcal{F}(c)$ of c is the function $\mathcal{F}(c)\colon \mathbb{T}^n \to \mathbb{C}$ defined on $\mathbb{T}^n$ given by

$$\mathcal{F}(c)(\theta) = \sum_{m\in\mathbb{Z}^n} c_m e^{-im\cdot\theta},$$

and the *Fourier cotransform* $\overline{\mathcal{F}}(c)$ of c is the function $\overline{\mathcal{F}}(c)\colon \mathbb{T}^n \to \mathbb{C}$ defined on $\mathbb{T}^n$ given by

$$\overline{\mathcal{F}}(c)(\theta) = \sum_{m\in\mathbb{Z}^n} c_m e^{im\cdot\theta},$$

with $\theta \in \mathbb{R}^n/2\pi\mathbb{Z}^n$.

Remark: The Fourier cotransform is also called the *inverse Fourier transform* by some authors, including Hewitt and Ross.

It can be shown that $|\widehat{f}(m)|$ tends to zero when $|m|$ tends to infinity. This is a special case of Proposition 10.18.

Definition 6.13. The *convolution* $f * g$ of two functions $f, g \in \mathrm{L}^1(\mathbb{T}^n)$ is given by

$$(f*g)(\theta) = \int_{\mathbb{T}^n} f(\theta-\varphi)g(\varphi)\,\frac{dx_n(\varphi)}{(2\pi)^n} = \int_{\mathbb{T}^n} f(\varphi)g(\theta-\varphi)\,\frac{dx_n(\varphi)}{(2\pi)^n},$$

where dx_n is the Lebesgue measure on $\mathbb{R}^n$.

By Proposition 8.48, we have $f * g \in \mathrm{L}^1(\mathbb{T}^n)$.

One of the main reasons why the Fourier transform is useful is that it converts a convolution into a product.

Proposition 6.11. *For any two functions $f, g \in \mathrm{L}^1(\mathbb{T}^n)$, we have*

$$\mathcal{F}(f*g) = \mathcal{F}(f)\mathcal{F}(g), \qquad \overline{\mathcal{F}}(f*g) = \overline{\mathcal{F}}(f)\overline{\mathcal{F}}(g).$$

The equation $\mathcal{F}(f*g) = \mathcal{F}(f)\mathcal{F}(g)$ can also be written as $\widehat{f*g} = \widehat{f}\,\widehat{g}$.

Proposition 6.11 actually holds in the more general framework of locally compact abelian groups, and a proof is given in Proposition 10.5 (see also Proposition 10.18).

It is not hard to adapt the proof of Proposition 6.1 to prove that for any $f \in \mathrm{L}^1(\mathbb{T}^n)$, for all $\theta \in \mathbb{R}^n/2\pi\mathbb{Z}^n$, we have

$$(f * P_r)(\theta) = \sum_{m\in\mathbb{Z}^n} \widehat{f}(m) r^{\|m\|_1} e^{im\cdot\theta},$$

where $\|m\|_1 = |m_1| + \cdots + |m_n|$. As a consequence we have the following result.

Theorem 6.12. *(Spectral Synthesis)*

(1) If $f \in \mathrm{L}^p(\mathbb{T}^n)$ for $p = 1, 2$ and if $r \in [0, 1)$, for any $\theta \in \mathbb{R}^n/2\pi\mathbb{Z}^n$, write

$$f_r(\theta) = (P_r * f)(\theta) = \sum_{m\in\mathbb{Z}^n} \widehat{f}(m) r^{\|m\|_1} e^{im\cdot\theta}.$$

Then $\lim_{r\mapsto 1} \|f - f_r\|_p = 0$.

(2) If $f \in \mathcal{C}(\mathbb{T}^n)$, then $\lim_{r\mapsto 1} \|f - f_r\|_\infty = 0$.

For any $p \in \mathbb{N}$, let

$$S_p = \{m \in \mathbb{Z}^n \mid |m_k| \leq p,\ k = 1, \ldots, n\}.$$

Note that the sequence z of Example 6.5 is the case of S_1 (when $n = 2$).

Recall that the inner product of two functions $f, g \in \mathrm{L}^2(\mathbb{T}^n)$ is given by

$$\langle f, g\rangle = \int_{\mathbb{T}^n} f(\theta)\overline{g(\theta)}\, \frac{dx_n(\theta)}{(2\pi)^n},$$

and the inner product of two sequences $x, y \in \ell^2(\mathbb{Z}^n)$ is given by

$$\langle x, y\rangle = \sum_{m\in\mathbb{Z}^n} x_m \overline{y_m}.$$

Theorem 6.13. *Let $f \in \mathrm{L}^2(\mathbb{T}^n)$. Then we have the equality (Parseval)*

$$\|f\|_2^2 = \sum_{m\in\mathbb{Z}^n} |\widehat{f}(m)|^2.$$

Define $s_p(\theta)$ by

$$s_p(\theta) = \sum_{m\in S_p} \widehat{f}(m) e^{im\cdot\theta}.$$

Then we have

$$\lim_{p\mapsto\infty} \|f - s_p\|_2 = 0.$$

Plancherel's theorem holds.

Theorem 6.14. *(Plancherel) The map $f \mapsto \widehat{f}$ is an isometric isomorphism of the Hilbert spaces* $\mathrm{L}^2(\mathbb{T}^n)$ *and* $\ell^2(\mathbb{Z}^n)$.

The Fourier inversion formula is generalized as follows.

Theorem 6.15. *(Fourier inversion formula) Let $f \in \mathrm{L}^1(\mathbb{T}^n)$. If $\widehat{f} \in \ell^1(\mathbb{Z}^n)$ then*

$$f(\theta) = \sum_{m \in \mathbb{Z}^n} \widehat{f}_m e^{im\cdot\theta} = (\overline{\mathcal{F}}(\widehat{f}))(\theta),$$

for all almost all $\theta \in \mathbb{R}^n/2\pi\mathbb{Z}^n$. Furthermore, if f is continuous, then equality holds everywhere.

Theorem 6.14 and Theorem 6.15 are proven in Malliavin [Malliavian *et al.* (1995)]. They allow the extension of the Fourier cotransform $\overline{\mathcal{F}}$ on $\ell^1(\mathbb{Z}^n)$ to $\ell^2(\mathbb{Z}^n)$ in such a way that $\mathcal{F}$ and $\overline{\mathcal{F}}$ are mutual inverses.

We now turn to the Fourier transform on $\mathbb{R}$.

6.5 The Fourier Transform and the Fourier Cotransform on $\mathbb{R}$

In this section we discuss the Fourier transform of functions defined on the entire real line $\mathbb{R}$ that are not necessarily periodic. Because $\mathbb{R}$ is not compact, $\mathrm{L}^1(\mathbb{R})$ and $\mathrm{L}^2(\mathbb{R})$ are incomparable (with respect to inclusion), and the theory of the Fourier transform on $\mathbb{R}$ is more delicate than the Fourier theory on $\mathbb{T}$. In particular, although Plancherel's theorem holds (Theorem 6.14), its proof is more complicated.

Definition 6.14. For any function $f \in \mathrm{L}^1(\mathbb{R})$, the *Fourier transform* $\widehat{f} = \mathcal{F}(f)$ of f is the function $\mathcal{F}(f)\colon \mathbb{R} \to \mathbb{C}$ defined on $\mathbb{R}$ given by

$$\widehat{f}(x) = \mathcal{F}(f)(x) = \int_{\mathbb{R}} f(y)e^{-iyx}\,\frac{dx(y)}{\sqrt{2\pi}},$$

and the *Fourier cotransform* $\overline{\mathcal{F}}(f)$ of f is the function $\overline{\mathcal{F}}(f)\colon \mathbb{R} \to \mathbb{C}$ defined on $\mathbb{R}$ given by

$$\overline{\mathcal{F}}f(x) - \int_{\mathbb{R}} f(y)e^{iyx}\,\frac{dx(y)}{\sqrt{2\pi}},$$

where dx is the Lebesgue measure on $\mathbb{R}$.

Remark: The Fourier cotransform is also called the *inverse Fourier transform* by some authors, including Hewitt and Ross.

The formula for $\mathcal{F}(f)$ (and $\overline{\mathcal{F}}(f)$) is reminiscent of the formula

$$\mathcal{F}(c)(x) = \sum_{m \in \mathbb{Z}} c_m e^{-imx},$$

where $(c_m)_{m \in \mathbb{Z}}$ is a sequence, except that the infinite sum is replaced by an integral. The integer m is replaced by the real number y, the coefficient c_m is replaced by the value $f(y)$ of the function f at y, and the exponential $e^{-im\theta}$ is replaced by e^{-iyx}. Thus we can view $\mathcal{F}(f)(x)$ as a continuous superposition of the basic periodic functions $y \mapsto e^{-iyx}$. However, this time, $\mathcal{F}(f)(x)$ is not necessarily periodic. We can still think of y as a frequency. In fact, in signal analysis, the domain of the Fourier transform is called the frequency domain.

The reader might be puzzled by the presence of the scale factor $1/\sqrt{2\pi}$. The reason why it is included is that it makes certain formulae more symmetric, for example, the Fourier inversion formula and the Plancherel isomorphism. The deep reason for its need has to do with the fact that the domain of a Fourier transform $\widehat{f}$ is not actually $\mathbb{R}$, but an isomorphic copy $\widehat{\mathbb{R}}$ of $\mathbb{R}$, with a certain measure which is not necessarily identical to the measure on $\mathbb{R}$.

In order for certain results to hold, such as Fourier inversion, if $\mathbb{R}$ is given the Lebesgue measure dx, then $\widehat{\mathbb{R}}$ should be given the measure $dx/2\pi$. Some authors use this normalization. Following Rudin [Rudin (1987, 1991)], a more symmetric normalization is to use the same scale factor $1/\sqrt{2\pi}$ for both $\mathbb{R}$ and $\widehat{\mathbb{R}}$. Another approach is to incorporate the factor 2π in the exponential; that is, to use $e^{-2\pi iyx}$ instead of e^{-iyx}. In this case, the Lebesgue measure can be used for both $\mathbb{R}$ and $\widehat{\mathbb{R}}$; see Folland [Folland (1999, 1995)]. All of this will be elucidated in Chapter 10.

A consequence of using the measure $dx/\sqrt{2\pi}$ is that the *convolution* of two functions $f, g \in \mathrm{L}^1(\mathbb{R})$ is

$$(f * g)(x) = \int_{\mathbb{R}} f(x-y)g(y)\,\frac{dx(y)}{\sqrt{2\pi}} = \int_{\mathbb{R}} f(y)g(x-y)\,\frac{dx(y)}{\sqrt{2\pi}},$$

and the inner product of two functions $f, g \in \mathrm{L}^2(\mathbb{R})$ is given by

$$\langle f, g\rangle = \int_{\mathbb{R}} f(x)\overline{g(x)}\,\frac{dx}{\sqrt{2\pi}}.$$

By Proposition 8.48, we have $f * g \in \mathrm{L}^1(\mathbb{R})$.

It is immediately verified that $\overline{\mathcal{F}}(f)(x) = \mathcal{F}(f)(-x) = \overline{\mathcal{F}(\overline{f})(x)}$.

We will now state the most important results about the Fourier theory for $\mathbb{R}$ without proof. Proofs of these results can be found in Folland [Folland (1999)], Rudin [Rudin (1987, 1991)], Stein and Shakarchi [Stein and Shakarchi (2003)], and Malliavin [Malliavian *et al.* (1995)]. The most important results will be proven in Chapter 10.

First, following Stein and Shakarchi [Stein and Shakarchi (2003)] (Chapter 5 Section 1), observe that there is a nice class $\mathrm{Mod}(\mathbb{R})$ of continuous functions f such that $\mathrm{Mod}(\mathbb{R}) \subseteq \mathrm{L}^1(\mathbb{R})$, and such that the Fourier transform $\widehat{f} = \mathcal{F}(f)$ of f is well-defined.

Definition 6.15. A continuous function $f\colon \mathbb{R} \to \mathbb{C}$ is of *moderate decrease* is there is some $A > 0$ such that

$$|f(x)| \leq \frac{A}{1+x^2} \qquad \text{for all} \in \mathbb{R};$$

see Figure 6.23. The set of functions of moderate decrease is denoted by $\mathrm{Mod}(\mathbb{R})$.

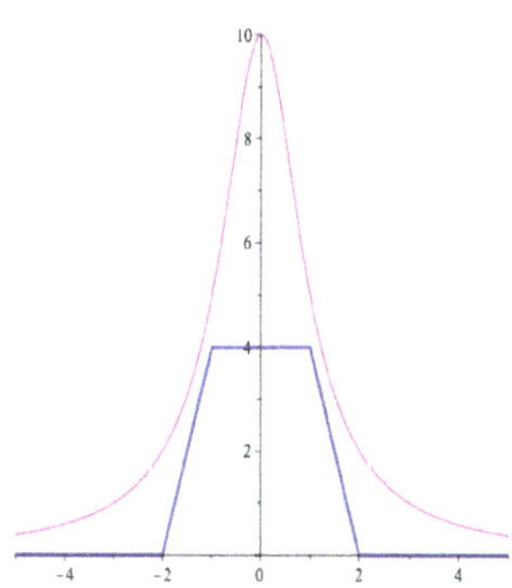

Fig. 6.23 The blue real-valued "bump" function is of moderate decrease since it is under magenta graph of $g(x) = \frac{10}{1+x^2}$.

It is shown in Stein and Shakarchi [Stein and Shakarchi (2003)] (Chapter 5 Section 1) that $\mathrm{Mod}(\mathbb{R})$ is a vector space contained in $\mathrm{L}^1(\mathbb{R})$, and that the Fourier transform $\widehat{f} = \mathcal{F}(f)$ of every function $f \in \mathrm{Mod}(\mathbb{R})$ is well-defined. However, the Fourier transform $\widehat{f}$ *may not* be of moderate decrease.

Let us give a few examples of Fourier transforms.

Example 6.6.

(1) Let f be the characteristic function $\chi_{[-a,a]}$ of the interval $[-a,a]$, with $a > 0$. Then we have

$$\mathcal{F}(f)(x) = \int_{\mathbb{R}} \chi_{[-a,a]}(y) e^{-iyx} \frac{dy}{\sqrt{2\pi}} = \int_{-a}^{a} e^{-iyx} \frac{dy}{\sqrt{2\pi}}$$
$$= \frac{e^{-iax} - e^{iax}}{-ix\sqrt{2\pi}} = \frac{2}{\sqrt{2\pi}} \frac{\sin ax}{x}.$$

Therefore,

$$\mathcal{F}(f)(x) = \frac{2}{\sqrt{2\pi}} \frac{\sin ax}{x}.$$

We also have

$$\overline{\mathcal{F}}(f)(x) = \frac{2}{\sqrt{2\pi}} \frac{\sin ax}{x},$$

because

$$\overline{\mathcal{F}}(f)(x) = \int_{\mathbb{R}} \chi_{[-a,a]}(y) e^{iyx} \frac{dy}{\sqrt{2\pi}} = \int_{-a}^{a} e^{iyx} \frac{dy}{\sqrt{2\pi}}$$
$$= \frac{e^{iax} - e^{-iax}}{ix\sqrt{2\pi}} = \frac{2}{\sqrt{2\pi}} \frac{\sin ax}{x};$$

see Figure 6.24.

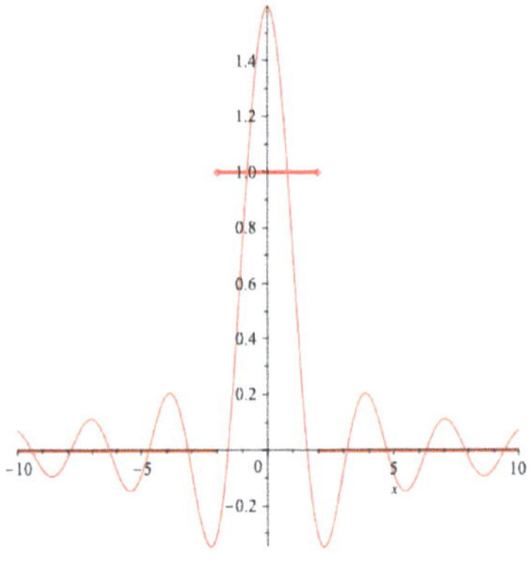

Fig. 6.24 The red graph is the plot of $\chi_{[-2,2]}$, while the wavy salmon graph is the plot of $\mathcal{F}(f)(x) = \frac{2}{\sqrt{2\pi}} \frac{\sin 2x}{x} = \overline{\mathcal{F}}(f)(x)$.

Remark: The function sinc is defined by

$$\operatorname{sinc}(x) = \begin{cases} \frac{\sin \pi x}{\pi x} & \text{if } x \neq 0 \\ 1 & \text{if } x = 0. \end{cases}$$

Since clearly $\chi_{[-a,a]} \in L^1(\mathbb{R}) \cap L^2(\mathbb{R})$, by Plancherel theorem (Theorem 6.22), $\widehat{\chi_{[-a,a]}} = \frac{2}{\sqrt{2\pi}} \frac{\sin ax}{x} \in L^2(\mathbb{R})$, so by setting $a = \pi$ we see that sinc $\in L^2(\mathbb{R})$. This can also be shown directly by showing that $(\sin \pi x/(\pi x))^2$ is continuous and bounded near zero. However, the function sinc is not in $L^1(\mathbb{R})$, because

$$\int_{-\infty}^{\infty} \left| \frac{\sin \pi x}{\pi x} \right| dx = \infty;$$

see Figure 6.25. As a consequence, its Fourier transform is undefined.

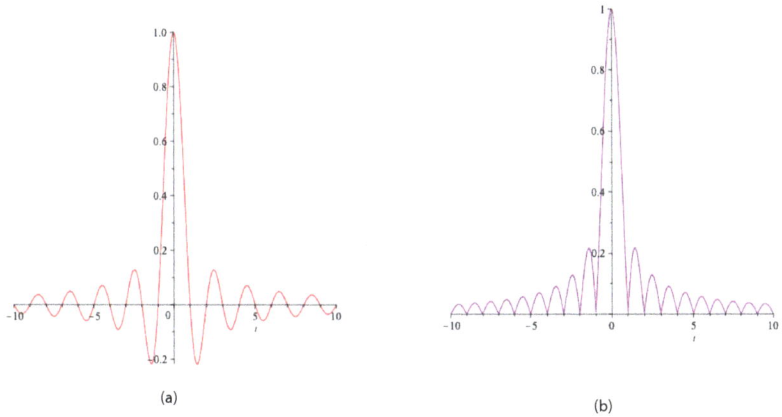

Fig. 6.25 Figure (a) is the graph of sinc(x), while Figure (b) is the graph of $|\text{sinc}(x)|$. Since sinc $\notin L^1(\mathbb{R})$, the area under the graph of the purple curve is unbounded.

However, by Plancherel's theorem (Theorem 6.22) the Fourier transform has a unique extension to $L^2(\mathbb{R})$, and the Fourier inversion formula holds. This implies that

$$f(x) = (\overline{\mathcal{F}}(\widehat{f}))(x) = (\mathcal{F}(\mathcal{F}(f))(-x),$$

so

$$(\mathcal{F}(\mathcal{F}(f))(x) = f(-x).$$

Consequently, since

$$\widehat{\chi_{[-\pi,\pi]}} = \sqrt{2\pi}\, \frac{\sin \pi x}{\pi x},$$

the (extended) Fourier transform of sinc is $(1/\sqrt{2\pi})\chi_{[-\pi,\pi]}$.

The function sinc plays a crucial role in the *sampling theorem*, which gives a nice expression for a function $f \in L^2(\mathbb{R})$ which is band-limited, which means that $\widehat{f}(x) = 0$ for all x such that $|x| > a$.

(2) Let f be the function given by

$$f(x) = \frac{y}{x^2 + y^2},$$

with $y > 0$ fixed, and let g be the function given by

$$g(x) = \frac{\pi}{\sqrt{2\pi}} e^{-y|x|};$$

see Figure 6.26. Then we have

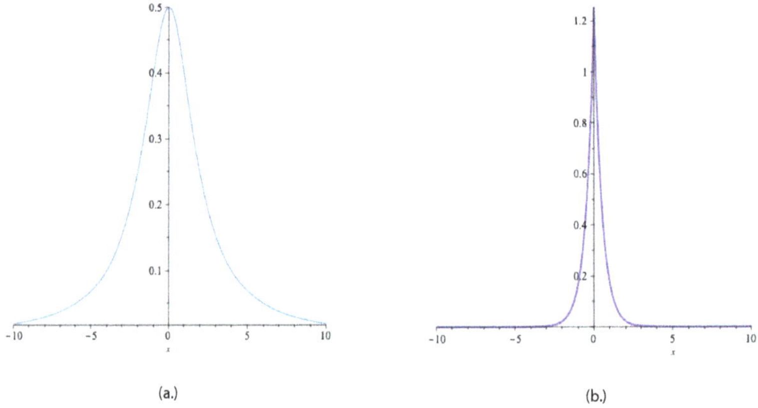

Fig. 6.26 Let $y = 2$. Figure (a) is the graph of $f(x) = \frac{2}{x^2+4}$, while Figure (b) is the graph of $g(x) = \frac{\pi}{\sqrt{2\pi}} e^{-2|x|}$.

$$\mathcal{F}(f)(x) = g(x)$$
$$\overline{\mathcal{F}}(g)(x) = f(x).$$

The second formula is proven as follows. Using the fact that $y > 0$, we have

$$\begin{aligned}
\overline{\mathcal{F}}(g)(x) &= \int_{\mathbb{R}} g(t) e^{itx} \frac{dt}{\sqrt{2\pi}} \\
&= \int_{-\infty}^{0} \frac{\pi}{\sqrt{2\pi}} e^{yt} e^{itx} \frac{dt}{\sqrt{2\pi}} + \int_{0}^{\infty} \frac{\pi}{\sqrt{2\pi}} e^{-yt} e^{itx} \frac{dt}{\sqrt{2\pi}} \\
&= \int_{-\infty}^{0} \frac{\pi}{\sqrt{2\pi}} e^{i(x-iy)t} \frac{dt}{\sqrt{2\pi}} + \int_{0}^{\infty} \frac{\pi}{\sqrt{2\pi}} e^{i(x+iy)t} \frac{dt}{\sqrt{2\pi}}
\end{aligned}$$

$$
\begin{aligned}
&= \left[\frac{\pi}{2\pi}\frac{e^{i(x-iy)t}}{i(x-iy)}\right]_{-\infty}^{0} + \left[\frac{\pi}{2\pi}\frac{e^{i(x+iy)t}}{i(x+iy)}\right]_{0}^{\infty} \\
&= \frac{1}{2i(x-iy)} - \frac{1}{2i(x+iy)} \\
&= \frac{y}{x^2+y^2}.
\end{aligned}
$$

The first formula is harder to prove directly, but it follows from Fourier inversion (see Theorem 6.20).

We now return to the general case of functions in $\mathrm{L}^1(\mathbb{R})$.

Proposition 6.16. *(Riemann–Lebesgue) For any function $f \in \mathrm{L}^1(\mathbb{R})$, the Fourier transform $\widehat{f}$ (and the Fourier cotransform $\overline{\mathcal{F}}(f)$) is continuous and tends to zero at infinity; that is, $\widehat{f} \in \mathcal{C}_0(\mathbb{R}; \mathbb{C})$. Furthermore*

$$\|\widehat{f}\|_\infty \leq \|f\|_1 .$$

Proposition 6.16 is proven in Malliavin [Malliavian *et al.* (1995)].

As for the Fourier transform on $\mathbb{T}$, the Fourier transform converts a convolution into a product. The following proposition is a special case of Proposition 10.18 and Proposition 10.19, Parts (3) and (4). First we need some notation. For any function $f\colon \mathbb{R}^n \to \mathbb{R}$, for any $y \in \mathbb{R}^n$, the function $\lambda_y(f)\colon \mathbb{R}^n \to \mathbb{R}$ is given by

$$(\lambda_y(f))(x) = f(x-y) \quad \text{for all } x \in \mathbb{R}^n.$$

The above is a special case of Definition 8.7 for the abelian group $\mathbb{R}^n$. The operator λ_y is often called a *translation operator*.

Proposition 6.17. *For any two functions $f, g \in \mathrm{L}^1(\mathbb{R})$, the following properties hold:*

(1) $\widehat{f * g} = \widehat{f}\,\widehat{g}$.
(2) $(\lambda_y(f))\widehat{\ }(x) = e^{-iyx}\widehat{f}(x)$.
(3) $(e^{iyx}f)\widehat{\ }(x) = \lambda_y(\widehat{f})(x)$.
(4) If $\alpha > 0$ and $h(x) = f(x/\alpha)$, then $\widehat{h}(x) = \alpha\widehat{f}(\alpha x)$.

For spectral synthesis in $\mathrm{L}^1(\mathbb{R})$, the Poisson kernel is replaced by a function G_μ defined using the following result.

Proposition 6.18. *For any $\mu > 0$, we have*

$$e^{-\frac{\mu x^2}{2}} = \frac{1}{\sqrt{\mu}} \int e^{-\frac{y^2}{2\mu}}\, e^{ixy}\, \frac{dx(y)}{\sqrt{2\pi}}.$$

Let φ be the function given by

$$\varphi(x) = e^{-\frac{x^2}{2}}.$$

Then $\widehat{\varphi} = \varphi$, and $\varphi(0) = \int \varphi(x)\, dx$.

For a proof of Proposition 6.18, see Rudin [Rudin (1991)] (Chapter 7, Lemma 7.6) or Folland [Folland (1999)] (Chapter 8, Section 8.3, Proposition 8.24).

Definition 6.16. The function φ given by

$$\varphi(x) = e^{-\frac{x^2}{2}}$$

is called a *Gauss kernel* or *Weierstrass kernel.*

For any $\mu > 0$, let G_μ be the following function:

$$G_\mu(x) = \frac{1}{\sqrt{\mu}} e^{-\frac{x^2}{2\mu}}.$$

In view of Proposition 6.18 (replacing μ by $1/\mu$) we have

$$G_\mu(x) = \frac{1}{\sqrt{\mu}} e^{-\frac{x^2}{2\mu}} = \int e^{-\frac{\mu y^2}{2}} e^{ixy} \frac{dx(y)}{\sqrt{2\pi}}.$$

Here is our first result about spectral synthesis analogous to Theorem 6.2(1)–(2). First, we leave it as an exercise to prove that

$$(f * G_\mu)(x) = \int e^{iyx} \widehat{f}(y) e^{-\frac{\mu y^2}{2}} \frac{dx(y)}{\sqrt{2\pi}}.$$

Proposition 6.19. *(Spectral Synthesis) Let $f \in \mathrm{L}^1(\mathbb{R})$, let $\widehat{f}$ be its Fourier transform, and for any $\mu > 0$, let*

$$g_\mu(x) = (f * G_\mu)(x) = \int e^{iyx} \widehat{f}(y) e^{-\frac{\mu y^2}{2}} \frac{dx(y)}{\sqrt{2\pi}}.$$

If $f \in \mathrm{L}^1(\mathbb{R})$, then

$$\lim_{\mu \mapsto 0} \|f - g_\mu\|_1 = 0,$$

and if $f \in \mathrm{L}^1(\mathbb{R}) \cap \mathrm{L}^2(\mathbb{R})$, then

$$\lim_{\mu \mapsto 0} \|f - g_\mu\|_2 = 0.$$

Proposition 6.19 is proven in Malliavin [Malliavian *et al.* (1995)] (Chapter 3, Section 2.4, Theorem 2.4.5). The proof uses Fubini's theorem and some technical properties about g_μ that are proven in Malliavin [Malliavian *et al.* (1995)].

In general, given a function $f \in \mathrm{L}^1(\mathbb{R})$, the integral

$$\int e^{iyx}\widehat{f}(y)\,\frac{dx(y)}{\sqrt{2\pi}} = (\overline{\mathcal{F}}(\widehat{f}))(x)$$

does not converge. However, for $\mu > 0$, the function $g_\mu(x) = (f * G_\mu)(x)$ is well-defined and when μ tends to 0, the function g_μ is an approximation of f that tends to f.

Now comes our first Fourier inversion theorem analogous to Theorem 6.3.

Theorem 6.20. *(Fourier inversion formula) Let $f \in \mathrm{L}^1(\mathbb{R})$. If $\widehat{f} \in \mathrm{L}^1(\mathbb{R})$, then*

$$f(x) = \int e^{iyx}\widehat{f}(y)\,\frac{dx(y)}{\sqrt{2\pi}} = (\overline{\mathcal{F}}(\widehat{f}))(x),$$

almost everywhere. If f is continuous, the equation holds for all $x \in \mathbb{R}$.

Theorem 6.20 is proven in Rudin [Rudin (1987)] (Chapter 9, Theorem 9.11) Folland [Folland (1999)] (Chapter 8, Section 8.3, Theorem 8.26) and Malliavin [Malliavian *et al.* (1995)] (Chapter 3, Section 2.4).

Proposition 6.21. *If $f \in \mathrm{L}^1(\mathbb{R}) \cap \mathrm{L}^2(\mathbb{R})$, then*

$$\|f\|_2 = \|\widehat{f}\|_2.$$

Proposition 6.21 is proven in Malliavin [Malliavian *et al.* (1995)] (Chapter 3, Section 4.2).

Here is the version of Plancherel's theorem for $\mathrm{L}^2(\mathbb{R})$.

Theorem 6.22. *(Plancherel) If $f \in \mathrm{L}^1(\mathbb{R}) \cap \mathrm{L}^2(\mathbb{R})$, then $\widehat{f} \in \mathrm{L}^2(\mathbb{R})$. The Fourier transform defined on $\mathrm{L}^1(\mathbb{R}) \cap \mathrm{L}^2(\mathbb{R})$ has a unique extension $\mathcal{F}$ to $\mathrm{L}^2(\mathbb{R})$ which is an isometric isomorphism of the Hilbert space $\mathrm{L}^2(\mathbb{R})$ whose inverse is the (extension of) Fourier cotransform $\overline{\mathcal{F}}$.*

Theorem 6.22 proven in Rudin [Rudin (1987)] (Chapter 9, Theorem 9.13) Folland [Folland (1999)] (Chapter 8, Section 8.3, Theorem 8.29) and Malliavin [Malliavian *et al.* (1995)] (Chapter 3, Section 2.4)

Although Theorem 6.22 says that the Fourier transform $\mathcal{F}$ extends to an isometric isomorphism of the Hilbert space $\mathrm{L}^2(\mathbb{R})$, this result is useless in practice because for a function $f \in \mathrm{L}^2(\mathbb{R})$ not in $\mathrm{L}^1(\mathbb{R}) \cap \mathrm{L}^2(\mathbb{R})$, the extension of $\mathcal{F}$ to f is given by a limit.

The Fourier inversion formula also holds in the following situation.

Proposition 6.23. *(Fourier inversion formula, II) Let $f \in \mathrm{L}^2(\mathbb{R})$. If $\widehat{f} \in \mathrm{L}^1(\mathbb{R})$, then*

$$f(x) = \int e^{iyx} \widehat{f}(y) \, \frac{dx(y)}{\sqrt{2\pi}} = (\overline{\mathcal{F}}(\widehat{f}))(x),$$

almost everywhere. If f is continuous, the equation holds for all $x \in \mathbb{R}$.

Proposition 6.23 is proven in Rudin [Rudin (1987)] (Chapter 9, Theorem 9.14).

Definition 6.17. Let $B(\mathbb{R}) = \{f \in \mathrm{L}^1(\mathbb{R}) \mid \widehat{f} \in \mathrm{L}^1(\mathbb{R})\}$.

Proposition 6.24. *The space $B(\mathbb{R})$ is dense in $\mathrm{L}^1(\mathbb{R})$, $\mathrm{L}^2(\mathbb{R})$, and $\mathcal{C}_0(\mathbb{R};\mathbb{C})$.*

Proposition 6.24 is proven in Malliavin [Malliavian *et al.* (1995)] (Chapter 3, Section 2.4).

6.6 The Sampling Theorem

In signal analysis a function $f\colon \mathbb{R} \to \mathbb{C}$ represents a physical signal, and a common problem is to try to reconstruct this signal by sampling it, which means to compute its values at some sequence $t_1 < t_2 < \ldots$ of times. A basic issue is to determine how much information can be gained this way.

It turns out that if the signal f is *band-limited*, which means that its Fourier transform $\widehat{f}$ vanishes outside some interval $[-\Omega, \Omega]$, then f can be completely reconstructed by sampling at the points $t_n = n\pi/\Omega$, for $n \in \mathbb{N}$.

Theorem 6.25. *(Sampling theorem) Suppose that $f \in \mathrm{L}^2(\mathbb{R})$ and that there is some $\Omega > 0$ such that $\widehat{f}(x) = 0$ for all $|x| \geq \Omega$. Then f is completely determined by its values at the points $t_n = n\pi/\Omega$, $n \in \mathbb{N}$. In fact, we have*

$$f(t) = \sum_{n=-\infty}^{\infty} f\left(\frac{n\pi}{\Omega}\right) \frac{\sin(\Omega t - n\pi)}{\Omega t - n\pi}.$$

Proof. We follow Folland [Folland (1992)] (Chapter 7, Section 7.3). We can extend $\widehat{f}$ to a periodic function of period 2Ω, and expand it as a Fourier series over the interval $[-\Omega, \Omega]$. For reason of later convenience, we use the index $-n$ instead of n, so we write

$$\widehat{f}(t) = \sum_{n\in\mathbb{Z}} c_{-n} e^{-in\pi t/\Omega}, \qquad (|t| \leq \Omega).$$

By Plancherel's theorem (extended to $\mathrm{L}^2(\mathbb{R})$), $\widehat{f} \in \mathrm{L}^2(\mathbb{R})$, and since $\widehat{f}(t) = 0$ for $|t| \geq \Omega$, by Proposition 5.43, we have $\mathrm{L}^2(\mathbb{R}) \subseteq \mathrm{L}^1(\mathbb{R})$ and so $\widehat{f} \in \mathrm{L}^1(\mathbb{R})$. By adjusting the computation below, we can show that the Fourier coefficients c_{-n} are given by

$$c_{-n} = \frac{1}{2\Omega}\int_{-\Omega}^{\Omega} \widehat{f}(t)e^{in\pi t/\Omega}\,dt = \frac{1}{2\Omega}\int_{-\infty}^{\infty} \widehat{f}(t)e^{in\pi t/\Omega}\,dt = \frac{\sqrt{2\pi}}{2\Omega} f\left(\frac{n\pi}{\Omega}\right),$$

where we used Fourier inversion (Theorem 6.23) and the fact that $\widehat{f}(t) = 0$ for $|t| \geq \Omega$. Again, using these two facts we have

$$\begin{aligned} f(t) = \int_{-\Omega}^{\Omega} \widehat{f}(\omega)e^{i\omega t}\frac{d\omega}{\sqrt{2\pi}} &= \int_{-\Omega}^{\Omega} \sum_{n=-\infty}^{\infty} \frac{\sqrt{2\pi}}{2\Omega} f\left(\frac{n\pi}{\Omega}\right) e^{-in\pi\omega/\Omega}\, e^{i\omega t}\frac{d\omega}{\sqrt{2\pi}} \\ &= \frac{1}{2\Omega}\int_{-\Omega}^{\Omega} \sum_{n=-\infty}^{\infty} f\left(\frac{n\pi}{\Omega}\right) e^{i(\Omega t - n\pi)\omega/\Omega)}\,d\omega \\ &= \frac{1}{2\Omega}\sum_{n=-\infty}^{\infty} f\left(\frac{n\pi}{\Omega}\right)\left[\frac{e^{i(\Omega t - n\pi)\omega/\Omega)}}{i(\Omega t - n\pi)/\Omega}\right]_{\omega=-\Omega}^{\omega=\Omega} \\ &= \sum_{n=-\infty}^{\infty} f\left(\frac{n\pi}{\Omega}\right)\frac{\sin(\Omega t - n\pi)}{\Omega t - n\pi}. \end{aligned}$$

Since $\widehat{f} \in \mathrm{L}^2(\mathbb{R})$ the above manipulations are legitimate. □

Observe that variants of the function $\mathrm{sinc}(x) = \frac{\sin \pi x}{\pi x}$ show up.

Theorem 6.25 is due independently to E.T. Whittaker and Shannon (a similar result was published by Kotelnikov).

It worth noting that the functions

$$s_n(t) = \frac{\sin(\Omega t - n\pi)}{\Omega t - n\pi}$$

form an orthonormal Hilbert basis for the Hilbert space of functions f in $\mathrm{L}^2(\mathbb{R})$ such that $\widehat{f} = 0$ a.e. outside $(-\Omega, \Omega)$; see Figure 6.27. This is because the computations in the proof of the sampling theorem show that s_n is the Fourier cotransform (inverse Fourier transform) of the function

$$\widehat{s}_n(t) = \begin{cases} \frac{\pi}{\Omega} e^{-in\pi t/\Omega} & \text{if } |t| < \Omega \\ 0 & \text{otherwise.} \end{cases}$$

By Plancherel theorem and the fact that the functions $t \mapsto e^{-in\pi t/\Omega}$ constitute an orthonormal Hilbert basis for $L^2(-\Omega, \Omega)$, we deduce that the s_n form a Hilbert basis.

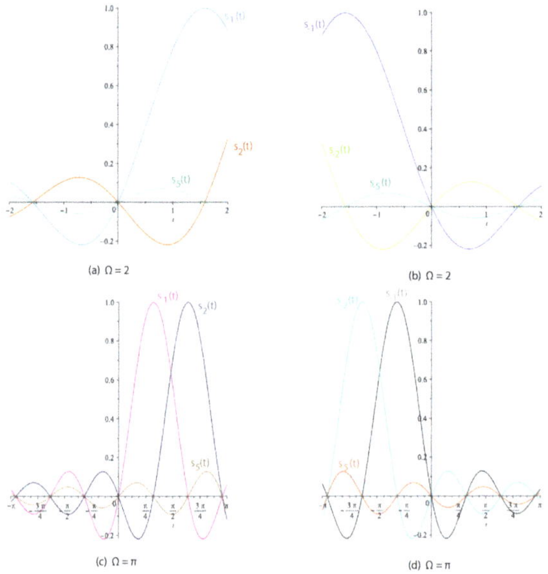

Fig. 6.27 The graphs of various $s_n(t)$. Note that $n \to -n$ results in a reflection over the y-axis.

From a practical point of view, the expansion of f given by the sampling theorem has the disadvantage that it generally does not converge very rapidly. By oversampling, that is, evaluating f at a more closely spaced sequence of points $n\pi/\lambda\Omega$, with $\lambda > 1$, we can replace the functions s_n by functions $g_\lambda(t - n\pi/\lambda\Omega)$ that vanish like $1/t^2$ when t goes to infinity. The function g_λ is given by

$$g_\lambda(t) = \frac{\cos \Omega t - \cos \lambda\Omega t}{\pi(\lambda - 1)\Omega t^2};$$

see Folland [Folland (1992)] (Chapter 7, Section 7.3, Exercise 8) and Stein and Shakarchi [Stein and Shakarchi (2003)] (Chapter 5, Exercise 20). Also see Figure 6.28.

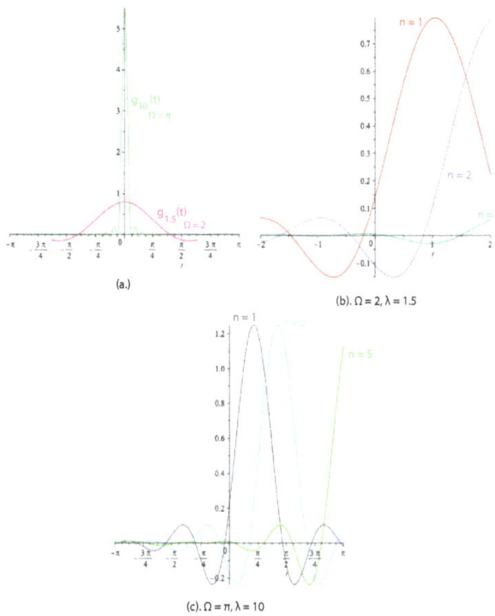

Fig. 6.28 Figure (a) shows the graphs of various $g_\lambda(t)$. Figure (b) shows graphs of $g_{1.5}(t)$ when $t \to t - n\pi/\lambda\Omega$ and $\Omega = 2$. Figure (c) shows graphs of $g_{10}(t)$ when $t \to t - n\pi/\lambda\Omega$ and $\Omega = \pi$.

There is also a dual version of the sampling theorem for functions $f \in L^2(\mathbb{R})$ that vanish outside an interval $[-L, L]$. Then the Fourier transform $\widehat{f}$ of f is determined by sampling at the points $\omega = n\pi/L$, and $\widehat{f}$ is given by the formula

$$\widehat{f}(t) = \sum_{n=-\infty}^{\infty} \widehat{f}\left(\frac{n\pi}{\Omega}\right) \frac{\sin(\Omega t - n\pi)}{\Omega t - n\pi},$$

which is obtained from the formula of Theorem 6.25 by replacing f with $\widehat{f}$.

6.7 The Fourier Transform and the Fourier Cotransform on $\mathbb{R}^n$

The generalization of the results of Section 6.5 to $\mathbb{R}^n$ is straightforward.

Definition 6.18. For any function $f \in L^1(\mathbb{R}^n)$, the *Fourier transform* $\widehat{f} = \mathcal{F}(f)$ of f is the function $\mathcal{F}(f)\colon \mathbb{R}^n \to \mathbb{C}$ defined on $\mathbb{R}^n$ given by

$$\widehat{f}(x) = \mathcal{F}(f)(x) = \int_{\mathbb{R}^n} f(y)e^{-iy\cdot x}\,\frac{dx_n(y)}{(2\pi)^{n/2}},$$

and the *Fourier cotransform* $\overline{\mathcal{F}}(f)$ of f is the function $\overline{\mathcal{F}}(f)\colon \mathbb{R}^n \to \mathbb{C}$ defined on $\mathbb{R}^n$ given by

$$\overline{\mathcal{F}}f(x) = \int_{\mathbb{R}^n} f(y)e^{iy\cdot x}\,\frac{dx_n(y)}{(2\pi)^{n/2}},$$

where dx_n is the Lebesgue measure on $\mathbb{R}^n$, and $x\cdot y = \sum_{k=1}^n x_k y_k$ is the inner product of $x, y \in \mathbb{R}^n$.

Remark: The Fourier cotransform is also called the *inverse Fourier transform* by some authors, including Hewitt and Ross.

Again, we are using Rudin's normalization scale factor $1/(2\pi)^{n/2}$, so we are really using the measure $dx_n/(2\pi)^{n/2}$. In particular, the *convolution* of two functions $f, g \in L^1(\mathbb{R}^n)$ is

$$(f * g)(x) = \int_{\mathbb{R}} f(x-y)g(y)\,\frac{dx_n(y)}{(2\pi)^{n/2}} = \int_{\mathbb{R}} f(y)g(x-y)\,\frac{dx_n(y)}{(2\pi)^{n/2}},$$

and the inner product of two functions $f, g \in L^2(\mathbb{R}^n)$ is given by

$$\langle f, g\rangle = \int_{\mathbb{R}^n} f(x)\overline{g(x)}\,\frac{dx_n}{(2\pi)^{n/2}}.$$

By Proposition 8.48, we have $f * g \in L^1(\mathbb{R}^n)$.

It is immediately verified that $\overline{\mathcal{F}}(f)(x) = \mathcal{F}(f)(-x) = \overline{\mathcal{F}(\overline{f})(x)}$.

Proposition 6.26. *(Riemann–Lebesgue) For any function $f \in L^1(\mathbb{R}^n)$, the Fourier transform $\widehat{f}$ (and the Fourier cotransform $\overline{\mathcal{F}}(f)$) is continuous and tends to zero at infinity; that is, $\widehat{f} \in \mathcal{C}_0(\mathbb{R}^n;\mathbb{C})$. Furthermore*

$$\|\widehat{f}\|_\infty \le \|f\|_1 .$$

As for the Fourier transform on $\mathbb{T}^n$, the Fourier transform converts a convolution into a product.

Proposition 6.27. *For any two functions $f, g \in L^1(\mathbb{R}^n)$, the following properties hold for all $x, y \in \mathbb{R}^n$:*

(1) $\widehat{f * g} = \widehat{f}\,\widehat{g}$.
(2) $(\lambda_y(f))\widehat{\ }(x) = e^{-iy\cdot x}\widehat{f}(x)$.

(3) $(e^{iy\cdot x}f)\widehat{\ }(x) = \lambda_y(\widehat{f})(x)$.
(4) If $\alpha > 0$ *and* $h(x) = f(x/\alpha)$, *then* $\widehat{h}(x) = \alpha^n \widehat{f}(\alpha x)$.

Proposition 6.27 is proven in Rudin [Rudin (1991)] (Chapter 7, Theorem 7.2).

Another useful property of convolution is that under certain conditions it allows differentiation under the integral sign. This property is another regularization feature of convolution. By convolving a function with a "nice" function, we obtain a "nice" function.

Proposition 6.28. *If* $f \in L^1(\mathbb{R}^n)$, $g \in C^k(\mathbb{R}^n)$, *and* $\partial^\alpha g$ *is bounded for all* α *such that* $|\alpha| \leq k$, *then* $f * g \in C^k(\mathbb{R}^n)$ *and*

$$\partial^\alpha(f * g) = f * (\partial^\alpha g), \quad |\alpha| \leq k.$$

See Folland [Folland (1999)] (Proposition 8.10).

For any $\mu > 0$, an for any $x \in \mathbb{R}^n$, let G_μ be the following function:

$$G_\mu(x) = \frac{1}{\mu^{n/2}} e^{-\frac{\|x\|^2}{2\mu}},$$

where $\|x\|^2 = x_1^2 + \cdots + x_n^2$; see Figure 6.29. Using Proposition 6.18, it is easy to see that

$$G_\mu(x) = \int e^{-\frac{\mu\|y\|^2}{2}} e^{ix\cdot y} \frac{dx_n(y)}{(2\pi)^{n/2}}.$$

We also easily verify that

$$(f * G_\mu)(x) = \int e^{iy\cdot x}\widehat{f}(y)e^{-\frac{\mu\|y\|^2}{2}} \frac{dx_n(y)}{(2\pi)^{n/2}}.$$

Proposition 6.29. *(Spectral Synthesis) Let* $f \in L^1(\mathbb{R}^n)$, *let* $\widehat{f}$ *be its Fourier transform, and for any* $\mu > 0$, *let*

$$g_\mu(x) = (f * G_\mu)(x) = \int e^{iy\cdot x}\widehat{f}(y)e^{-\frac{\mu\|y\|^2}{2}} \frac{dx_n(y)}{(2\pi)^{n/2}}.$$

If $f \in L^1(\mathbb{R}^n)$, *then*

$$\lim_{\mu\mapsto 0} \|f - g_\mu\|_1 = 0,$$

and if $f \in L^1(\mathbb{R}^n) \cap L^2(\mathbb{R}^n)$, *then*

$$\lim_{\mu\mapsto 0} \|f - g_\mu\|_2 = 0.$$

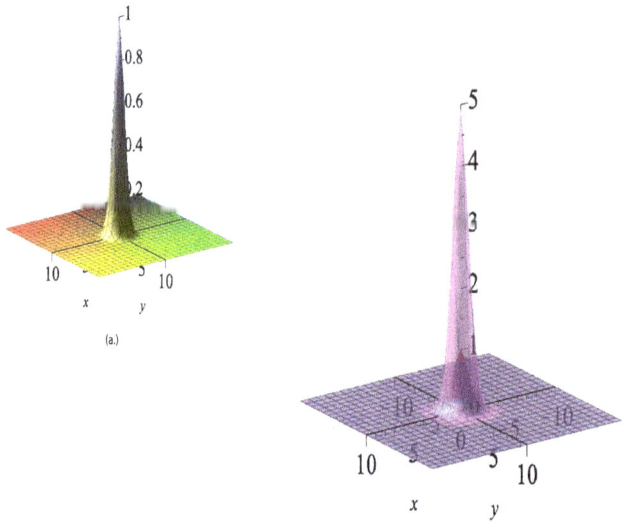

Fig. 6.29 Let $n = 2$, Figure (a) is the graph of $G_1(x)$, while Figure (b) shows graph of $G_5(x)$ nested inside the graph of $G_1(x)$, which itself is nested inside $G_{\frac{1}{5}}(x)$. A smaller μ leads to a larger peak above the origin.

Proposition 6.19 is proven in Malliavin [Malliavian *et al.* (1995)] (Chapter 3, Section 2.4).

Theorem 6.30. *(Fourier inversion formula) Let $f \in \mathrm{L}^1(\mathbb{R}^n)$. If $\widehat{f} \in \mathrm{L}^1(\mathbb{R}^n)$, then*

$$f(x) = \int e^{iy \cdot x} \widehat{f}(y) \, \frac{dx_n(y)}{(2\pi)^{n/2}} = (\overline{\mathcal{F}}(\widehat{f}))(x),$$

almost everywhere. If f is continuous, the equation holds for all $x \in \mathbb{R}^n$.

Theorem 6.20 is proven in Rudin [Rudin (1991)] (Chapter 7, Theorem 7.7) Folland [Folland (1999)] (Chapter 8, Section 8.3, Theorem 8.26) and Malliavin [Malliavian *et al.* (1995)] (Chapter 3, Section 2.4).

Definition 6.19. Let $B(\mathbb{R}^n) = \{f \in \mathrm{L}^1(\mathbb{R}^n) \mid \widehat{f} \in \mathrm{L}^1(\mathbb{R}^n)\}$.

Proposition 6.31. *The space $B(\mathbb{R}^n)$ is dense in $\mathrm{L}^1(\mathbb{R}^n)$, $\mathrm{L}^2(\mathbb{R}^n)$, and $\mathcal{C}_0(\mathbb{R}^n; \mathbb{C})$.*

Proposition 6.24 is proven in Malliavin [Malliavian *et al.* (1995)] (Chapter 3, Section 2.4)

Here is the version of Plancherel's theorem for $\mathrm{L}^2(\mathbb{R}^n)$.

Theorem 6.32. *(Plancherel) If $f \in \mathrm{L}^1(\mathbb{R}^n) \cap \mathrm{L}^2(\mathbb{R}^n)$, then $\widehat{f} \in \mathrm{L}^2(\mathbb{R}^n)$. The Fourier transform defined on $\mathrm{L}^1(\mathbb{R}^n) \cap \mathrm{L}^2(\mathbb{R}^n)$ has a unique extension $\mathcal{F}$ to $\mathrm{L}^2(\mathbb{R}^n)$ which is an isometric isomorphism of the Hilbert space $\mathrm{L}^2(\mathbb{R}^n)$ whose inverse is the (extension of) Fourier cotransform $\overline{\mathcal{F}}$.*

Theorem 6.22 proven in Rudin [Rudin (1991)] (Chapter 7, Theorem 7.9), Folland [Folland (1999)] (Chapter 8, Section 8.3, Theorem 8.29) and Malliavin [Malliavian *et al.* (1995)] (Chapter 3, Section 2.4)

6.8 The Schwartz Space

It turns out that $\mathrm{L}^1(\mathbb{R}^n)$ contains an important subspace $\mathcal{S}(\mathbb{R}^n)$ of rapidly decreasing smooth functions and that the Fourier transform is an isomorphism of this space, whose inverse is the Fourier cotransform. Functions in the space $\mathcal{S}(\mathbb{R}^n)$ and all their derivatives vanish at infinity faster than any power of $\|x\|$, where $\|x\|$ is the Euclidean norm on $\mathbb{R}^n$. Technically, we introduce the following family of norms.

Definition 6.20. A continuous function $f \in \mathcal{C}(\mathbb{R}^n, \mathbb{C})$ is *rapidly decreasing* if for every integer $m \geq 0$, there is some $C > 0$ such that $(1 + \|x\|^2)^m |f(x)|$ remains bounded for all x such that $\|x\| \geq C$. Let $\mathcal{C}_{0,0}(\mathbb{R}^n)$ be the set of rapidly decreasing functions. For every $m \in \mathbb{N}$, define the norm $\|f\|_{m,0}$ by

$$\|f\|_{m,0} = \sup_{x \in \mathbb{R}^n} (1 + \|x\|^2)^m |f(x)|.$$

Observe that Definition 6.20 immediately implies that $\mathcal{C}_{0,0}(\mathbb{R}^n)$ is a subspace of $\mathcal{C}_0(\mathbb{R}^n; \mathbb{C})$. Also, since $(1+\|x\|^2)^m |f(x)| \leq (1+\|x\|^2)^{m+1} |f(x)|/(1+\|x\|^2)$, if $(1 + \|x\|^2)^{m+1} |f(x)|$ is bounded for all x such that $\|x\|$ is large enough, we see that

$$\lim_{\|x\| \mapsto \infty} (1 + \|x\|^2)^m |f(x)| = 0, \qquad \text{for all } m \in \mathbb{N}. \tag{$*$}$$

Conversely, Condition (*) implies that $(1 + \|x\|^2)^{m+1} |f(x)|$ is bounded for all x such that $\|x\|$ is large enough. Therefore, $(*)$ is equivalent to the condition used in Definition 6.20. In view of all this, we have

$$\mathcal{C}_{0,0}(\mathbb{R}^n) = \{f \in \mathcal{C}_0(\mathbb{R}^n; \mathbb{C}) \mid \|f\|_{m,0} < \infty, \text{ for all } m \geq 0\}.$$

Definition 6.21. The *Schwartz space* $\mathcal{S}(\mathbb{R}^n)$ consists of all smooth functions (that is, differentiable at all orders) given by

$$\mathcal{S}(\mathbb{R}^n) = \{f \in \mathcal{C}^\infty(\mathbb{R}^n) \cap \mathcal{C}_{0,0}(\mathbb{R}^n) \mid \partial^\alpha f \in \mathcal{C}_{0,0}(\mathbb{R}^n) \quad \text{for every } \alpha\}.$$

For all $m, p \in \mathbb{N}$, define the norm $\|f\|_{m,p}$ by

$$\|f\|_{m,p} = \sup_{|\alpha| \le p} \|\partial^\alpha f\|_{m,0} = \sup_{x \in \mathbb{R}^n, |\alpha| \le p} (1 + \|x\|^2)^m |\partial^\alpha f(x)|.$$

Definition 6.21 is due to Laurent Schwartz. Observe that when $p = 0$, the norm $\|f\|_{m,p}$ is just the norm $\|f\|_{m,0}$ introduced in Definition 6.20, and that by definition,

$$\mathcal{S}(\mathbb{R}^n) = \{f \in \mathcal{C}_0(\mathbb{R}^n; \mathbb{C}) \mid \|f\|_{m,p} < \infty, \quad \text{for all } m, p \in \mathbb{N}\}.$$

Functions such as $x^k e^{-x^2}$ where $k \in \mathbb{N}$ belong to $\mathcal{S}(\mathbb{R})$; see Figure 6.30. The functions $e^{-c\|x\|^{2m}}$ where m is a positive integer and $e^{-c(1+\|x\|^2)^\alpha}$ with $c > 0$ and $\alpha > 0$ belong to $\mathcal{S}(\mathbb{R}^n)$; see Figures 6.31 and 6.32.

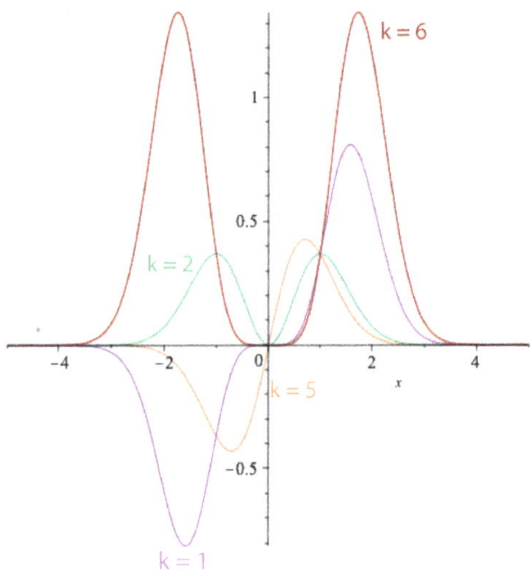

Fig. 6.30 Various graphs of $x^k e^{-x^2}$, where k is a nonnegative integer.

Remark: Although it decreases very fast at infinity, the function $x \mapsto e^{-y|x|}$ (with $y > 0$ fixed) does not belong to $\mathcal{S}(\mathbb{R})$, because it is not differentiable at $x = 0$; see the sharp peak at $x = 0$ in Figure 6.26(b).

The space $\mathcal{D}(\mathbb{R}^n)$ of smooth functions with compact support is obviously a subspace of $\mathcal{S}(\mathbb{R}^n)$. In Section 5.13 the space $\mathcal{D}(\mathbb{R}^n)$ was denoted $\mathcal{K}^\infty_{\mathbb{C}}(\mathbb{R}^n)$, but the notation $\mathcal{D}(\mathbb{R}^n)$ is more common.

Since the Schwartz space is a subspace of $\mathcal{C}_0(\mathbb{R}^n; \mathbb{C})$, we can make it a normed vector space by giving it the norm $\| \|_\infty$. Unfortunately, with this

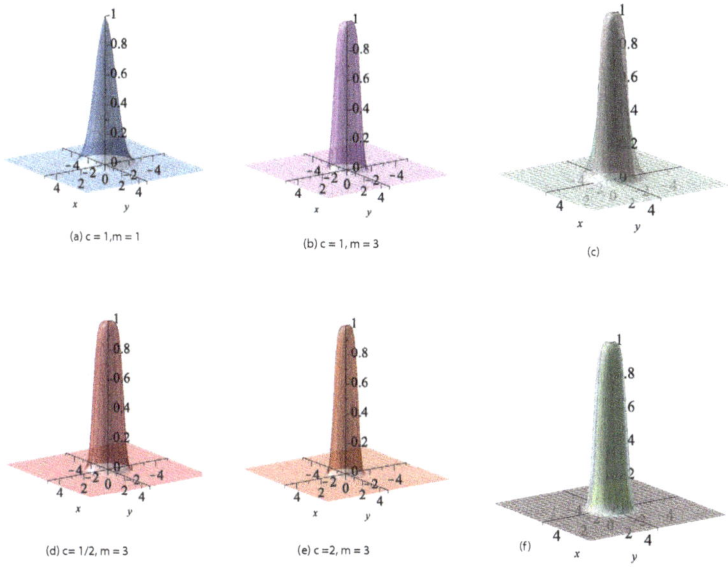

Fig. 6.31 Let $n = 2$. Figure (a) is the graph of $e^{-\|x\|^2}$, while Figure (b) is the graph of $e^{-\|x\|^6}$. Figure (c) is the juxtaposition of these two graphs and shows for fixed c, as m increases, the peak becomes wider. Figure (d) is $e^{-\frac{1}{2}\|x\|^6}$, while Figure (e) is $e^{-2\|x\|^6}$. Figure (f) is the juxtaposition of Figures (b), (e), and (d), and shows that for fixed m, as c increases, the peak becomes thinner.

norm it is *not* complete. We can give it a topology induced by the family of norms $\| \|_{m,p}$, according to the standard process for defining a topology in terms of a family of semi-norms described in Section 2.7. Moreover, because this topology is Hausdorff and the family of norms is countable, $\mathcal{S}(\mathbb{R}^n)$ is actually a metric space; better, a complete metric space.

Let us now use the family of semi-norms $\| \|_{m,p}$ to define a topology on the Schwartz space $\mathcal{S}(\mathbb{R}^n)$. The semi-norms $\| \|_{m,p}$ are actually norms, so by Proposition 2.18, the space $\mathcal{S}(\mathbb{R}^n)$ is Hausdorff.

Definition 6.22. The vector space $\mathcal{S}(\mathbb{R}^n)$ endowed with the topology induced by the countable family of norms $\| \|_{m,p}$ is a Hausdorff space called the *topological Schwartz space*.

We usually omit the word topological in topological Schwartz space. The value of the topology defined above is that $\mathcal{S}(\mathbb{R}^n)$ is complete.

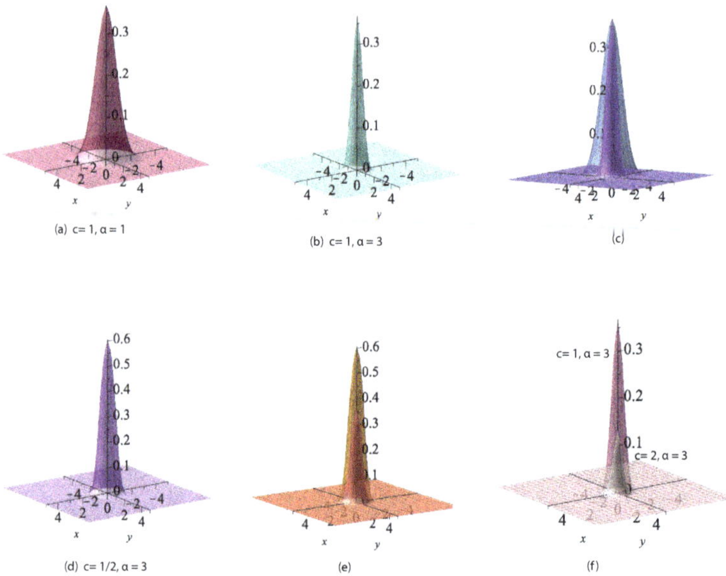

Fig. 6.32 Let $n = 2$. Figure (a) is the graph of $e^{-(1+\|x\|^2)}$, while Figure (b) is the graph of $e^{-(1+\|x\|^2)^3}$. Figure (c) is the juxtaposition of these two graphs and shows for fixed c, as α increases, the peak becomes narrower. Figure (d) is $e^{-\frac{1}{2}(1+\|x\|^2)^3}$, while Figure (e) is the juxtaposition of Figures (b) and (d), and shows that for fixed α, as c increases, the peak becomes shorter and narrower. This phenomenon is also see in Figure (f), which is the juxtaposition of the graphs of $e^{-(1+\|x\|^2)^3}$ and $e^{-2(1+\|x\|^2)^3}$.

Theorem 6.33. *The topological Schwartz space $\mathcal{S}(\mathbb{R}^n)$ is a Fréchet space; that is, it is complete for the metric given by*

$$d(x, y) = \sum_{m=0,p=0}^{\infty} \frac{1}{2^{m+p}} \frac{\|y - x\|_{m,p}}{1 + \|y - x\|_{m,p}}.$$

The space $\mathcal{D}(\mathbb{R}^n)$ of smooth functions with compact support is dense in $\mathcal{S}(\mathbb{R}^n)$.

Note that the above metric is the metric used in Proposition 2.20. Theorem 6.33 is proven in Rudin [Rudin (1991)] (Chapter 7, Theorem 7.4 and Theorem 7.10).

The following result is proven in Malliavin [Malliavian *et al.* (1995)] using the technique of regularization by some suitable convolution (Chapter 3, Section 3.2, Proposition 3.2.4).

Proposition 6.34. *The space $\mathcal{D}(\mathbb{R}^n)$ of smooth functions with compact support is dense in $\mathrm{L}^1(\mathbb{R}^n)$ and $\mathrm{L}^2(\mathbb{R}^n)$ (with the Lebesgue measure). As a corollary, the Schwartz space $\mathcal{S}(\mathbb{R}^n)$ is dense in $\mathrm{L}^1(\mathbb{R}^n)$ and $\mathrm{L}^2(\mathbb{R}^n)$.*

The Fourier theory of $\mathcal{S}(\mathbb{R}^n)$ is particularly nice because the Fourier transform is a map from $\mathcal{S}(\mathbb{R}^n)$ to itself. The following results can be shown.

Theorem 6.35. *If f is any function in $\mathcal{S}(\mathbb{R}^n)$, then the following properties hold:*

(1) *We have $\widehat{f} = \mathcal{F}(f) \in \mathrm{L}^1(\mathbb{R}^n)$ and Fourier inversion holds:*

$$f(x) = \int \widehat{f}(y)e^{iy\cdot x} \frac{dx_n(y)}{(2\pi)^{n/2}} = (\overline{\mathcal{F}}(\widehat{f}))(x).$$

(2) *Actually, $\widehat{f} \in \mathcal{S}(\mathbb{R}^n)$ and there exist constants $c_{r,s}$ such that*

$$\|\widehat{f}\|_{r,s} \leq c_{r,s} \, \|f\|_{m+s,r}\,, \qquad m > n.$$

(3) *The map $f \mapsto \widehat{f} = \mathcal{F}(f)$ is an algebra isomorphism and a homeomorphism from $\mathcal{S}(\mathbb{R}^n)$ to itself whose inverse is $\overline{\mathcal{F}}$, under both algebra structures given by pointwise multiplication and convolution.*

(4) $\widehat{x_k f(x)} = i\dfrac{\partial}{\partial x_k}\widehat{f}(x)$.

(5) $\widehat{\left(\dfrac{\partial}{\partial x_k} f\right)}(x) = ix_k\widehat{f}(x)$.

(6) *If $f, g \in \mathcal{S}(\mathbb{R}^n)$, then $fg \in \mathcal{S}(\mathbb{R}^n)$ and $\widehat{fg} = \widehat{f} * \widehat{g}$.*

(7) *If $f, g \in \mathcal{S}(\mathbb{R}^n)$, then $f * g \in \mathcal{S}(\mathbb{R}^n)$ and $\widehat{f * g} = \widehat{f}\,\widehat{g}$.*

Theorem 6.35 is proven Malliavin [Malliavian *et al.* (1995)] (Chapter 3, Section 4, Theorem 4.2) and Rudin [Rudin (1991)] (Chapter 7, Sections 7.3 pages 184–189). Parts of it are also proven in Folland [Folland (1999)] (Chapter 8, Section 8.3).

Equation (5) is a small miracle since it says that the Fourier transform of a derivative acts as multiplication of the Fourier transform by ix_k, and it can be used to solve certain partial differential equations. Several examples of this technique are presented in Folland [Folland (1992)] and Stein and Shakarchi [Stein and Shakarchi (2003)]. We give an example involving the heat equation.

Consider a region of the plane. Given an initial heat distribution, we are interested in finding the temperature $u(x, y, t)$ of the point (x, y) at time t.

Using Newton's law of cooling, it can be shown that u satisfies the *partial differential equation* called the *time-dependent heat equation*

$$\frac{\partial^2 u}{\partial x^2} + \frac{\partial^2 u}{\partial y^2} = \frac{\sigma}{\kappa}\frac{\partial u}{\partial t};$$

see Stein and Shakarchi [Stein and Shakarchi (2003)] (Chapter 1) or Folland [Folland (1999)] (Section 8.7). After a long period of time, there is no more heat exchange, so that the system reaches a thermal equilibrium, and then $\frac{\partial u}{\partial t} = 0$. In this case, u depends only on x and y, and the time-dependent equation reduces to the *steady-state heat equation*

$$\frac{\partial^2 u}{\partial x^2} + \frac{\partial^2 u}{\partial y^2} = 0. \tag{1}$$

The expression Δu on the left-hand side of (1) is the *Laplacian* of u. Suppose our domain is the upper half plane

$$\mathbb{R}^2_+ = \{(x, y) \in \mathbb{R}^2 \mid y > 0\}.$$

We would like to find a (the?) solution $u(x, y)$ of the above equation, given that

$$u(x, 0) = f(x) \tag{2}$$

on the boundary, where f is some given function (in $\mathcal{S}(\mathbb{R})$).

The method for finding the solution u proceeds in two steps.

Step 1. The first trick is to apply the Fourier transform with respect to x to both (1) and (2). We assume that $u \in \mathcal{S}(\mathbb{R}^2)$ even though the solution is only defined on the closure of the upper half plane and may not be extendable to a function in $\mathcal{S}(\mathbb{R}^2)$. The goal of this step is to show that u must be given in terms of a convolution defined on the upper-half plane. After guessing a solution using this step and the next, it is still necessary to prove that it works.

In view of Equation (5) of Theorem 6.35, we get the two equations

$$\begin{aligned} -x^2\widehat{u}(x, y) + \frac{\partial^2 \widehat{u}}{\partial y^2}(x, y) &= 0 \\ \widehat{u}(x, 0) &= \widehat{f}(x). \end{aligned}$$

Observe that we now have a much simpler problem, namely an *ordinary differential equation* with respect to y in the unknown $\widehat{u}(x, y)$. The solution of the first equation is well-known:

$$\widehat{u}(x, y) = C_1(x)e^{|x|y} + C_2(x)e^{-|x|y},$$

with

$$C_1(x) + C_2(x) = \widehat{f}(x).$$

Since the first term has exponential increase, it has to be discarded (because we are seeking solutions in $\mathcal{S}(\mathbb{R})$), so we must have $C_1 = 0$, and we get

$$\widehat{u}(x, y) = \widehat{f}(x)e^{-|x|y}. \tag{†}$$

Step 2. The second trick is that if we can find the Fourier cotransform (inverse Fourier transform) $x \mapsto \mathcal{P}_y(x)$ of $x \mapsto e^{-|x|y}$, since $\mathcal{F}(\mathcal{P}_y)(x) = e^{-|x|y}$, we have (by Proposition 6.17(1)),

$$\widehat{u}(x, y) = \widehat{f}(x)e^{-|x|y} = \mathcal{F}(f)(x)\mathcal{F}(\mathcal{P}_y)(x) = \mathcal{F}(f * \mathcal{P}_y)(x), \quad y > 0.$$

Note that for y fixed, $\mathcal{P}_y \notin \mathcal{S}(\mathbb{R})$, but $\mathcal{P}_y \in \mathrm{L}^1(\mathbb{R})$, so $f * \mathcal{P}_y \in \mathrm{L}^1(\mathbb{R})$ for y fixed. Since $\widehat{u}(x, y) \in \mathcal{S}(\mathbb{R}) \subseteq \mathrm{L}^1(\mathbb{R})$ for $y > 0$ fixed, by Fourier inversion (Theorem 6.20) we deduce that

$$u(x, y) = (f * \mathcal{P}_y)(x) \quad \text{for all } x \in \mathbb{R} \text{ and all } y > 0.$$

But we showed in Example 6.6(2) that the Fourier cotransform (inverse Fourier transform) of

$$g(x) = \frac{\pi}{\sqrt{2\pi}} e^{-y|x|}$$

(with $y > 0$) is

$$f(x) = \frac{y}{x^2 + y^2},$$

so the Fourier cotransform (inverse Fourier transform) of $x \mapsto e^{-|x|y}$ is

$$\mathcal{P}_y(x) = \frac{\sqrt{2\pi}}{\pi} \frac{y}{x^2 + y^2}.$$

Therefore we obtain the solution

$$u(x, y) = (f * \mathcal{P}_y)(x).$$

Explicitly, we have

$$u(x, y) = \int_{\mathbb{R}} \frac{yf(x - t)}{\pi(x^2 + y^2)}\, dt,$$

which is called the *Poisson integral formula*, and the function

$$\mathcal{P}_y(x) = \frac{\sqrt{2\pi}}{\pi} \frac{y}{x^2 + y^2}$$

is called the *Poisson kernel* for the upper half plane (there are variants of $\mathcal{P}_y(x)$ with different constants).

To be honest, we still need to check carefully that $u(x, y) = (f * \mathcal{P}_y)(x)$ is indeed a solution of the problem. For this we use Proposition 6.28. It can be shown that $\Delta u = 0$ on $\mathbb{R}^2_+$, but $u(x, 0)$ may not be equal to $f(x)$ on the boundary. What we can claim is that $u(x, y)$ tends to $f(x)$ uniformly as y tends to 0. For details see Folland [Folland (1999)] (Theorem 8.53) and Stein and Shakarchi [Stein and Shakarchi (2003)] (Chapter 5, Theorem 2.6).

Various other problems involving the wave equation or the heat equation can be solved using the above method; see Stein and Shakarchi [Stein and Shakarchi (2003)] and Folland [Folland (1992)].

It turns out that if we use certain kinds of generalized functions, called *distributions*, then we can apply a more general version of Theorem 6.35 and obtain more general solutions for various partial differential equations.

6.9 The Poisson Summation Formula

Given a function $f \in \mathcal{S}(\mathbb{R})$ it is sometimes desirable to make a periodic function from f. One way to do this is to define the function F_1 as follows.

Definition 6.23. Given a function $f \in \mathcal{S}(\mathbb{R})$, the function $F_1 : \mathbb{R} \to \mathbb{C}$ is given by

$$F_1(x) = \sum_{n \in \mathbb{Z}} f(x + 2\pi n).$$

Since $f \in \mathcal{S}(\mathbb{R})$, the series converges absolutely and uniformly on every compact subset of $\mathbb{R}$, so F_1 is continuous. It is also clear that

$$F_1(x) = F_1(x + 2\pi n), \quad n \in \mathbb{Z},$$

so F_1 is indeed periodic. We call F_1 the *periodization* of f.

There is another way to make f periodic, which is to use the sequence of numbers $(\widehat{f}(n))_{n \in \mathbb{Z}}$ and to make the Fourier series from it,

$$F_2(x) = \sum_{n \in \mathbb{Z}} \widehat{f}(n) e^{inx}.$$

Again, since $f \in \mathcal{S}(\mathbb{R})$, the sum converges absolutely and uniformly since $\widehat{f} \in \mathcal{S}(\mathbb{R})$, so F_2 is continuous. The remarkable fact is that $F_1 = F_2$.

Theorem 6.36. *(Poisson summation formula) For any function $f \in \mathcal{S}(\mathbb{R})$, we have*

$$\sum_{n\in\mathbb{Z}} f(x+2\pi n) = \sum_{n\in\mathbb{Z}} \widehat{f}(n)e^{inx}.$$

In other words, the Fourier coefficients of $F_1(x) = \sum_{n\in\mathbb{Z}} f(x+2\pi n)$ are the numbers $\widehat{f}(n)$. In particular,

$$\sum_{n\in\mathbb{Z}} f(2\pi n) = \sum_{n\in\mathbb{Z}} \widehat{f}(n).$$

The proof of Theorem 6.36 can be found in Stein and Shakarchi [Stein and Shakarchi (2003)] (Chapter 5, Theorem 3.1). It consists in computing the Fourier coefficients of F_1.

Theorem 6.36 also holds when both f and $\widehat{f}$ are of moderate decrease (recall Definition 6.15).

Remark: There is a relationship between the Poisson kernel on the unit disk,

$$P_r(\theta) = \sum_{n=-\infty}^{n=\infty} r^{|n|}e^{in\theta} = \frac{1-r^2}{1-2r\cos\theta + r^2},$$

and the Poisson kernel on the upper-half plane,

$$\mathcal{P}_y(x) = \frac{\sqrt{2\pi}}{\pi}\frac{y}{x^2+y^2},$$

obtained by applying the Poisson summation formula to $f(x) = \mathcal{P}_y(x)$ and $\widehat{f}(\theta) = e^{-|\theta|y}$. We get

$$\sum_{n\in\mathbb{Z}} \mathcal{P}_y(x+2\pi n) = \sum_{n\in\mathbb{Z}} e^{-|n|y}e^{inx} = \sum_{n\in\mathbb{Z}} (e^{-y})^{|n|}e^{inx} = P_{e^{-y}}(x).$$

In summary, (for $y > 0$), we have

$$P_{e^{-y}}(x) = \sum_{n\in\mathbb{Z}} \mathcal{P}_y(x+2\pi n).$$

6.10 The Heisenberg Uncertainly Principle

A fundamental fact about Fourier series is that it is *impossible* for a nonzero function $f \in \mathrm{L}^2(\mathbb{R})$ that both f and its Fourier transform $\widehat{f}$ vanish outside of some finite interval. This can be shown easily using some elementary complex analysis; see Folland [Folland (1992)] (Chapter 7).

There is an even stronger limitation. Roughly, f and $\widehat{f}$ can't be both highly localized. A precise way to state this fact is to define the notion of dispersion.

Definition 6.24. For any function $f \in \mathrm{L}^2(\mathbb{R})$, the *dispersion* of f about the point a is given by

$$\Delta_a f = \int (x-a)^2 |f(x)|^2 \, dx \Big/ \int |f(x)|^2 \, dx.$$

Then we have the following theorem.

Theorem 6.37. *(Heisenberg inequality) Let f be a function in $\mathrm{L}^2(\mathbb{R})$. Then for all $a, b \in \mathbb{R}$, we have*

$$(\Delta_a f)(\Delta_b \widehat{f}) \geq \frac{1}{4}.$$

Theorem 6.37 is proven in Stein and Shakarchi [Stein and Shakarchi (2003)] in the case where $f \in \mathcal{S}(\mathbb{R})$ (Chapter 5, Section 4), and in Folland [Folland (1992)] (Chapter 7) in a more general situation.

Theorem 6.37 has an interpretation in quantum mechanics (we apologize to those who are familiar with quantum mechanics for the vagueness of our comments). In quantum mechanics, among other things, one studies the motion of particles. For this, we need to know the position and the momentum of the particle, but these are not known exactly, but instead described in terms of probabilities. For simplicity, assume that we are dealing with an electron that travels along the real line. There is a function ψ, called a state function (or wave function), which we assume to be in $\mathcal{S}(\mathbb{R})$, normalized so that

$$\int |\psi(t)|^2 \, dt = 1,$$

such that the probability that the electron is located in the interval $[a, b]$ is

$$\int_a^b |\psi(t)|^2 \, dt.$$

The *expectation* of where the particle might be is the best guess of the position of the particle, and it is given by

$$\overline{x} = \int_{\mathbb{R}} t |\psi(t)|^2 \, dt.$$

The *uncertainty* attached to our expectation, or *variance*, is given by the quantity

$$\Delta_{\overline{x}} \psi = \int_{\mathbb{R}} (t - \overline{x})^2 |\psi(t)|^2 \, dt.$$

By differentiating under the integral sign with respect to a, we can show that the expectation $\overline{x}$ is the choice of a that minimizes the variance $\int_{\mathbb{R}}(t-a)^2|\psi(t)|^2\,dt$.

Now in quantum mechanics the momentum ξ of the particle is determined by the Fourier transform $\widehat{\psi}$ of ψ, in the sense that the probability that the electron has momentum ξ in the interval $[a,b]$ is

$$\int_a^b |\widehat{\psi}(t)|^2\,dt.$$

As above, we also have the expectation

$$\overline{\xi} = \int_{\mathbb{R}} t|\widehat{\psi}(t)|^2\,dt,$$

and the variance

$$\Delta_{\overline{\xi}}\widehat{\psi} = \int_{\mathbb{R}} (t-\overline{\xi})^2|\widehat{\psi}(t)|^2\,dt.$$

Theorem 6.37 states that

$$(\Delta_{\overline{x}}\psi)(\Delta_{\overline{\xi}}\widehat{\psi}) \geq \frac{1}{4},$$

which is the *Heisenberg uncertainty principle*. Intuitively, it says that the more certain we are about the position of the particle, the less certain we are about its momentum, and vice versa. Actually, we have ignored units of measurements, and in fact Planck's constant $\hbar$ should be inserted, so the physically correct statement of Heisenberg uncertainty principle is that

$$(\Delta_{\overline{x}}\psi)(\Delta_{\overline{\xi}}\widehat{\psi}) \geq \frac{\hbar}{4}.$$

For more details, see Stein and Shakarchi [Stein and Shakarchi (2003)] (Chapter 5, Section 4), and Folland [Folland (1992)] (Chapter 7), and for even more, any text on quantum mechanics.

6.11 Fourier's Life; a Brief Summary

Joseph Fourier was born on March 21, 1768, in Auxerre, a town in northern Burgundy, France, and died in 1830. Because he was 21 during the French revolution (1789), he had a particularly exciting life. In this section we give a very condensed summary of his life, based on the wonderful account in Chapter 1 of Kahane [Kahane and Lemaré-Rieusset (1998)].

Fourier's family was poor. At age 10 Fourier had already lost his mother and his father. The organist of the cathedral had noticed that Fourier was exceptionally gifted, so he arranged to have him attend the military college in Auxerre. Teaching was provided by Benedictine monks. Fourier fell in love with mathematics through the writings of Bézout and Clairaut. He worked very hard and completed his studies early at age 14. It was arranged that he stayed in the college, in preparation for starting teaching there at age 16. He already sent some papers on locating the roots of algebraic equations to the Institut, which were noticed by Legendre. Legendre requested that Fourier join the army in the artillery (the most scientific branch), but his request was denied because Fourier was not a "noble" (he was of humble extraction). So in 1787 he entered the benedictine abbey of Saint-Benoit in Fleury in preparation for becoming a monk.

Fourier stayed in Fleury until 1789, where he taught mathematics. He was going to become a monk on November 5, 1798, but the revolution had taken place and put a hold on new religious positions on November 2. Fourier never became a monk!

Between 1789 and 1793 Fourier continues working on mathematics, but also gets involved in the revolution. He is involved in the supply of food and weapons to Orléans, and being a good politician, does a very good job at that.

In 1794 he is sent to Paris where a new school called "École Normale" has been created. There he meets other mathematicians such as Laplace and Monge. The "École Polytechnique" is created in 1794, and Fourier teaches there between 1795 and 1798.

Apparently, Fourier is noticed by Napoleon, and he follows Napoleon for the expedition to Egypt. There, Napoleon creates a replica of the "Institut de France," headed by Monge, and with Fourier as "perpetual secretary." So Fourier becomes an archeologist.

Napoleon goes back to France where he proclaims himself emperor. Still in Egypt, Fourier negotiates the retreat of the French defeated by the British. Fourier returns to France in 1801. At his return Napoleon charges Fourier with the important administrative position of "préfet" (sort of superintendent) of the department of Isère. France was divided in 90 departments (districts), and the main city in Isère is Grenoble. One might think that this would signal the end of Fourier's mathematical life, but not at all. Fourier was also an astute politician, and a good administrator, so he excelled at everything he did. He started working on his theory of heat propagation.

In 1807 he submitted a paper on this subject to the Institut. Lagrange, Laplace, Lacroix, and Monge were the referees. Lagrange felt that the paper was not rigorous enough, and the paper was rejected. The topic of heat propagation was then proposed for a competition. Fourier reworked his paper which was submitted in 1811, and this time the same referees awarded him the price. However, the commentaries, although they praised the originality of the work, especially the heat equation, pointed out some lack of rigor.

Fourier continued to work on a major manuscript on the analytic theory of heat, but this manuscript was not published until 1822.

In the meantime, Napoleon abdicated in 1815. Life is hectic. Fourier is opposed to Napoleon III. Although he is promoted as préfet of the Rhone, he resigned from this position and returns to Paris. There, with the help of a former student, he finds the position of director of the bureau of statistics! He is elected at the Academy of Sciences in 1817.

Fourier continues working on his book on the analytic theory of heat, but also does some work in statistics. In 1822 he finally publishes his book, *Théorie anaytique de la chaleur*, which includes all of his work on the subject, starting with his work between 1807 and 1811, and then 1816, 1821.

Laplace, Monge, Liouville, Dirichlet, Navier, Sturm had great respect for Fourier. However, Poisson and Cauchy, who were his rivals, were not his friends. The obituaries by Arago and Cousin did not do justice to Fourier's work. It is sad that the collected works of Fourier were never gathered and published. Darboux collected some of Fourier's papers, but ignored all his work on what is now called linear programming, saying that Fourier attributed an exaggerated importance to this type of work. After all, Fourier adapted Gaussian elimination to linear inequalities (Fourier–Motzkin elimination).

However, Fourier's work had a tremendous influence in mathematics, physics, and engineering, so even if he did not get the recognition that he deserved from his peers, the public voted with their feet.

We must conclude with a famous note of Jacobi to Legendre, sent on July 2, 1831, after Fourier' death.

Fourier deeply believed that the main goal of mathematics was to provide a clear explanation of natural phenomena. In his book he writes:

"L'étude approfondie de la nature est la source la plus féconde des découvertes mathématiques."

Jacobi (1804–1851) complains to Legendre that Poisson included in a report that Fourier made the reproach to Abel and Jacobi that they did not work enough on the theory of heat, but instead on number theory. Jacobi says:

"... mais un philosophe comme lui aurait dû savoir que le but unique de la science, c'est l'honneur de l'esprit humain, et que, sous ce titre, une question de nombres vaut autant qu'une question du système du monde."

Roughly translated: But such a philosopher should have known that the unique goal of science is the honor of the human spirit, and that, as such, a question about numbers is as worthy as a question about the system of the world.

A very complete account of the mathematical history of Fourier series and its influence on mathematics can be found in the captivating book by Kahane and Lemarié–Rieusset [Kahane and Lemaré-Rieusset (1998)].

6.12 Problems

Problem 6.1. Recall that D_n and K_n are defined as

$$D_n(x) = \sum_{k=-n}^{n} e^{ikx}$$

$$K_n(x) = \frac{1}{n}\sum_{m=0}^{n-1}\sum_{k=-m}^{m} e^{ikx} = \frac{1}{n}(D_0(x) + \cdots + D_{n-1}(x)).$$

Show that

$$D_n(x) = \frac{\sin((2n+1)x/2)}{\sin(x/2)}, \qquad K_n(x) = \frac{1}{n}\left(\frac{\sin(nx/2)}{\sin(x/2)}\right)^2.$$

Also show that $D_n * f = S_{n,f}$.

Problem 6.2. Recall that for any $p \geq 1$, the space $\ell^p(\mathbb{Z})$ is the set of sequences $x = (x_n)_{n\in\mathbb{Z}}$ with $x_n \in \mathbb{C}$ such that $\sum_{n\in\mathbb{Z}} |x_n|^p < \infty$. Verify that $\ell^p(\mathbb{Z})$ $(p \geq 1)$ is a normed vector space with the norm

$$\|(x_m)_{m\in\mathbb{Z}}\| = \left(\sum_{m\in\mathbb{Z}} |x_m|^p\right)^{1/p}.$$

Prove that the space $\ell^p(\mathbb{Z})$ $(p \geq 1)$ is a Banach space. Hint: Adapt the proof of Proposition D.14.

Problem 6.3. Prove Theorem 6.2. Hint: See Malliavin [Malliavian *et al.* (1995)] (Chapter 3, Section 2.2.5).

Problem 6.4. Consider the periodic function (over $(-\pi, \pi)$) given by

$$f(\theta) = \begin{cases} 0 & \text{if } -\pi < \theta < 0 \\ \theta & \text{if } 0 \leq \theta < \pi. \end{cases}$$

Compute the real Fourier coefficients a_n and b_n of f as defined in Section 6.2 and prove that the corresponding Fourier series defined by (†) is given by

$$\frac{\pi}{4} - \frac{2}{\pi} \sum_{n=1}^{\infty} \frac{\cos(2n-1)\theta}{(2n-1)^2} + \sum_{n=1}^{\infty} \frac{(-1)^{n+1}}{n} \sin n\theta.$$

Problem 6.5. Consider the periodic function (over $(-\pi, \pi)$) given by

$$f(\theta) = \sin^2 \theta.$$

Compute the real Fourier coefficients a_n and b_n of f as defined in Section 6.2 and prove that the corresponding Fourier series defined is given by

$$\frac{1}{2} - \frac{1}{2} \cos 2\theta.$$

Problem 6.6. Consider the periodic function (over $(-\pi, \pi)$) given by

$$f(\theta) = \begin{cases} 0 & \text{if } -\pi < \theta < 0 \\ 1 & \text{if } 0 < \theta < \pi. \end{cases}$$

Compute the real Fourier coefficients a_n and b_n of f as defined in Section 6.2 and prove that the corresponding Fourier series is given by

$$\frac{1}{2} + \frac{2}{\pi} \sum_{n=1}^{\infty} \frac{\sin(2n-1)\theta}{2n-1}.$$

Problem 6.7. Consider the periodic function (over $(-\pi, \pi)$) given by

$$f(\theta) = \theta^2.$$

Compute the real Fourier coefficients a_n and b_n of f as defined in Section 6.2 and prove that the corresponding Fourier series is given by

$$\frac{\pi^2}{3} + 4 \sum_{n=1}^{\infty} \frac{(-1)^n}{n^2} \cos n\theta.$$

Problem 6.8. Prove Proposition 6.6. Hint: See Rudin [Rudin (1987)] (Chapter 5, Page 102).

Problem 6.9. Let $f\colon \mathbb{R} \to \mathbb{C}$ be a function. Recall that the total variation function T_f of f is given by

$$T_f(x) = \sup\left\{\sum_{j=1}^{n} |f(x_j) - f(x_{j-1})| \mid -\infty < x_0 < x_1 < \cdots < x_n = x\right\},$$

with $n \in \mathbb{N} - \{0\}$, where the supremum is taken over all finite subdivisions $x_0 < x_1 < \cdots < x_n = x$. Show that $T_f(x)$ is an increasing function.

Problem 6.10. Recall that BV is the set of functions $f\colon \mathbb{R} \to \mathbb{C}$ such that $\lim_{x\mapsto+\infty} T_f(x) < \infty$. Prove that BV is a complex vector space.

Problem 6.11. Prove that if f is differentiable on $\mathbb{R}$ and if f' is bounded, then $f \in BV([a,b])$ for every finite interval $[a,b]$. Hint: Use the mean value theorem.

Problem 6.12. Show that $\sin x \notin BV$.

Problem 6.13. Prove Proposition 6.8.

Problem 6.14. Prove Proposition 6.9. Hint: See Folland [Folland (1999)] (Chapter 3, Section 3.5).

Problem 6.15. Prove Theorem 6.10. Hint: See Folland [Folland (1999)] (Chapter 8, Section 8.5, Theorem 8.43).

Problem 6.16. Adapt the proof of Proposition 6.1 to prove that for any $f \in \mathrm{L}^1(\mathbb{T}^n)$, for all $\theta \in \mathbb{R}^n/2\pi\mathbb{Z}^n$, we have

$$(f * P_r)(\theta) = \sum_{m\in\mathbb{Z}^n} \widehat{f}(m) r^{\|m\|_1} e^{im\cdot\theta},$$

where $\|m\|_1 = |m_1| + \cdots + |m_n|$.

Problem 6.17. Prove Theorem 6.12.

Problem 6.18. Prove Theorem 6.13.

Problem 6.19. Prove Theorem 6.15.

Problem 6.20. Let f be the function given by

$$f(x) = \frac{y}{x^2 + y^2},$$

with $y > 0$ fixed, and let g be the function given by

$$g(x) = \frac{\pi}{\sqrt{2\pi}} e^{-y|x|}.$$

Show that $\mathcal{F}(f)(x) = g(x)$.

Problem 6.21. Prove Proposition 6.16.

Problem 6.22. Prove Proposition 6.17.

Problem 6.23. Prove Proposition 6.18. Hint: See Rudin [Rudin (1991)] (Chapter 7, Lemma 7.6) or Folland [Folland (1999)] (Chapter 8, Section 8.3, Proposition 8.24).

Problem 6.24. Prove Proposition 6.19. Hint: See Malliavin [Malliavian *et al.* (1995)] (Chapter 3, Section 2.4, Theorem 2.4.5).

Problem 6.25. Prove Theorem 6.20. Hint: See Rudin [Rudin (1987)] (Chapter 9, Theorem 9.11) or Folland [Folland (1999)] (Chapter 8, Section 8.3, Theorem 8.26) or Malliavin [Malliavian *et al.* (1995)] (Chapter 3, Section 2.4).

Problem 6.26. Prove Proposition 6.21. Hint: See Malliavin [Malliavian *et al.* (1995)] (Chapter 3, Section 4.2).

Problem 6.27. Prove Proposition 6.23. Hint: See Rudin [Rudin (1987)] (Chapter 9, Theorem 9.14).

Problem 6.28. State a dual version of the sampling theorem for functions $f \in \mathrm{L}^2(\mathbb{R})$ that vanish outside an interval $[-L, L]$. In this case the Fourier transform $\widehat{f}$ of f is determined by sampling at the points $\omega = n\pi/L$, and $\widehat{f}$ is given by the formula

$$\widehat{f}(t) = \sum_{n=-\infty}^{\infty} \widehat{f}\left(\frac{n\pi}{\Omega}\right) \frac{\sin(\Omega t - n\pi)}{\Omega t - n\pi}.$$

Problem 6.29. Prove Proposition 6.26.

Problem 6.30. Prove Proposition 6.27. Hint: See Rudin [Rudin (1991)] (Chapter 7, Theorem 7.2).

Problem 6.31. Prove Proposition 6.29. Hint: See Malliavin [Malliavian *et al.* (1995)] (Chapter 3, Section 2.4).

Problem 6.32. Prove Theorem 6.30. Hint: See Rudin [Rudin (1991)] (Chapter 7, Theorem 7.7) or Folland [Folland (1999)] (Chapter 8, Section 8.3, Theorem 8.26) or Malliavin [Malliavian *et al.* (1995)] (Chapter 3, Section 2.4).

Problem 6.33. Prove Theorem 6.35: Hint: See Malliavin [Malliavian *et al.* (1995)] (Chapter 3, Section 4, Theorem 4.2) or Rudin [Rudin (1991)] (Chapter 7, Sections 7.3 pages 184–189).

Problem 6.34. Prove Theorem 6.36. Hint: See Stein and Shakarchi [Stein and Shakarchi (2003)] (Chapter 5, Theorem 3.1).

Problem 6.35. Prove Theorem 6.37 for $f \in \mathcal{S}(\mathbb{R})$. Hint: See Stein and Shakarchi [Stein and Shakarchi (2003)] (Chapter 5, Section 4).

Problem 6.36. Consider the periodic function f given by

$$f(\theta) = \begin{cases} \dfrac{1}{\sqrt{|\theta|}} & \text{if } -\pi \leq \theta < 0,\ 0 < \theta \leq \pi \\ 0 & \text{if } \theta = 0. \end{cases}$$

(1) Prove that $f \in \mathrm{L}^1(\mathbb{T}) - \mathrm{L}^2(\mathbb{T})$.
(2) Prove that the Fourier coefficients c_m are given by

$$c_0 = \frac{2}{\sqrt{\pi}}, \quad c_m = 2\int_0^{\pi} \frac{\cos m\theta}{\sqrt{\theta}} \frac{d\theta}{2\pi}, \ m \neq 0,$$

so that $c_{-m} = c_m$ if $m \neq 0$.
(3) By making a suitable change of variable twice, prove that for $m > 0$ we have

$$c_m = \frac{2}{\pi\sqrt{m}} \int_0^{\sqrt{m\pi}} \cos(\theta^2)\, d\theta$$

and that the corresponding Fourier series is

$$\frac{2}{\sqrt{\pi}} + \sum_{m=1}^{\infty} \frac{4}{\pi\sqrt{m}} \left(\int_0^{\sqrt{m\pi}} \cos(\theta^2)\, d\theta \right) \cos m\varphi.$$

Remark: The integral $C(\sqrt{m\pi}) = \int_0^{\sqrt{m\pi}} \cos(\theta^2)\, d\theta$ is a *Fresnel integral*. It can be shown that it is bounded by 1 and that we have $\lim_{m \mapsto \infty} C(\sqrt{m\pi}) = \sqrt{\frac{\pi}{8}}$.
(4) Using the above fact prove that (c_m) does not belong to $\ell^2(\mathbb{Z})$.

Chapter 7

Radon Functionals and Radon Measures on Locally Compact Spaces

After having considered a very general theory of integration of functions defined on an arbitrary measure space and taking their values in any Banach space, we turn to the special case of complex-valued or real-valued functions defined on a locally compact space X. This corresponds to measure spaces $(X, \mathcal{B}, \mu)$, where X is a locally compact space, $\mathcal{B}$ is the σ-algebra of Borel sets (which is the smallest σ-algebra containing the open subsets of X), and μ is any (positive) measure on $\mathcal{B}$, which we call a *Borel measure*.

The theme of this chapter is that *a Borel measure μ can be used to define linear forms on various function spaces*. For example, pick the space $\mathcal{K}_{\mathbb{C}}(X)$ of continuous functions on X with compact support. For every function $f \in \mathcal{K}_{\mathbb{C}}(X)$ we can compute the integral

$$\varphi_\mu(f) = \int f d\mu.$$

We have to check that functions in $f \in \mathcal{K}_{\mathbb{C}}(X)$ are integrable, which is indeed true if $\mu(K)$ is finite for every compact subset. We obtain a map $\varphi_\mu \colon \mathcal{K}_{\mathbb{C}}(X) \to \mathbb{C}$, and since the integral is a linear operator, the map φ_μ is linear. In general it is not continuous, but it satisfies some weaker continuity properties. It is also a positive map, which means that $\varphi_\mu(f) \geq 0$ for every positive function $f \geq 0$.

What F. Riesz and J. Radon discovered is that, in some sense to be made precise, a special class of Borel measures is in one-to-one correspondence with the positive linear forms on the space $\mathcal{K}_{\mathbb{C}}(X)$. This means that for every positive linear form Φ on $\mathcal{K}_{\mathbb{C}}(X)$, there is a (unique) Borel measure m_Φ with some special properties such that Φ is represented by m_Φ, in the sense that

$$\Phi(f) = \int f dm_\Phi \quad \text{for all } f \in \mathcal{K}_{\mathbb{C}}(X).$$

There are two versions of this correspondence theorem known as the Radon–Riesz theorem, depending on the conditions imposed on the Borel measures.

These results are similar in flavor to the fact known from linear algebra that, in a finite-dimensional vector space E with an inner product $\langle -, - \rangle$, every linear form $\varphi \in E^*$ is represented by a unique vector $u \in E$, in the sense that

$$\varphi(v) = \langle v, u \rangle \quad \text{for all } v \in E.$$

If $(E, \langle -, - \rangle)$ is an infinite-dimensional vector space which is a Hilbert space (it is complete for the norm $u \mapsto \sqrt{\langle u, u \rangle}$), then by the *Riesz representation theorem*, every *continuous* linear form $\varphi \in E'$ is represented by a unique vector $u \in E$, in the sense that

$$\varphi(v) = \langle v, u \rangle \quad \text{for all } v \in E.$$

The Radon–Riesz theorems show that certain kinds of (possibly discontinuous) linear forms on $\mathcal{K}_{\mathbb{C}}(X)$ can be represented *using integration instead of an inner product.*

The main limitation of this approach is that the linear forms Φ induced by a positive measure are positive, which means that $\Phi(f) \geq 0$ if $f \geq 0$. In particular, it is impossible to represent an arbitrary continuous linear form on $\mathcal{K}_{\mathbb{C}}(X)$ using integration. The solution to overcome this limitation is to generalize the notion of measure so that a measure can take negative, or even complex values! We will show how to do this. We will also see that, in the end, complex measures can be expressed in terms of four positive measures, but these positive measures only take finite values in $\mathbb{R}_+$. Then we will obtain a third Radon–Riesz correspondence between the continuous linear forms on $\mathcal{K}_{\mathbb{C}}(X)$ and certain kinds of complex Borel measures. This correspondence plays a crucial role in defining the notion of convolution on a locally compact group.

In this chapter every topological space X is assumed to be locally compact (and Hausdorff).

7.1 Positive Radon Functionals Induced by Borel Measures

For the record a Borel measure is defined as follows.

Definition 7.1. A *Borel measure* is any (positive) measure on a measurable space $(X, \mathcal{B})$ where X is a locally compact space and $\mathcal{B}$ is the σ-algebra of

Borel sets (which is the smallest σ-algebra containing the open subsets of X).

One direction of the correspondence (Borel measures $\Longrightarrow$ linear forms) is easy to describe. It is the observation that the linear forms induced by Borel measures are positive.

Definition 7.2. For any function $f\colon X \to \mathbb{C}$, we write $f \geq 0$ if $f(X) \subseteq [0, \infty)$. If $f, g\colon X \to \mathbb{R}$, we write $f \leq g$ iff $g - f \geq 0$. A linear form $\Phi\colon \mathcal{K}_{\mathbb{C}}(X) \to \mathbb{C}$ is *positive* if for every $f \in \mathcal{K}_{\mathbb{C}}(X)$, if $f \geq 0$, then $\Phi(f) \in \mathbb{R}$ and $\Phi(f) \geq 0$.

A positive linear form has the following properties.

Proposition 7.1. *If $\Phi\colon \mathcal{K}_{\mathbb{C}}(X) \to \mathbb{C}$ is a positive linear form, then the following properties hold:*

(1) For any real-valued function $f \in \mathcal{K}_{\mathbb{R}}(X)$ we must have $\Phi(f) \in \mathbb{R}$.
(2) For any two real-valued functions $f, g \in \mathcal{K}_{\mathbb{R}}(X)$, if $f \leq g$, then $\Phi(f) \leq \Phi(g)$.

Proof. Indeed, a real-valued function f can be written uniquely as $f = f^+ - f^-$, with $f^+, f^- \in \mathcal{K}_{\mathbb{R}}(X)$, $f^+ \geq 0$ and $f^- \geq 0$. Since Φ is linear,

$$\Phi(f) = \Phi(f^+) - \Phi(f^-) \in \mathbb{R},$$

since $\Phi(f^+) \geq 0$ and $\Phi(f^-) \geq 0$ as Φ is positive.

We have $f \leq g$ iff $g - f \geq 0$, and since Φ is positive, $\Phi(g - f) \geq 0$, but since Φ is linear and positive, $\Phi(g) - \Phi(f) \geq 0$ with $\Phi(f), \Phi(g) \in \mathbb{R}$, that is, $\Phi(f) \leq \Phi(g)$. $\square$

The following proposition yields the mapping from Borel measures to positive linear forms.

Proposition 7.2. *Assume that the Borel measure μ has the property that $\mu(K)$ is finite for every compact subset of X (since X is Hausdorff, a compact set is closed, and thus a Borel set). Every function $f \in \mathcal{K}_{\mathbb{C}}(X)$ is integrable. Furthermore, the map $\varphi_\mu\colon \mathcal{K}_{\mathbb{C}}(X) \to \mathbb{C}$ given by*

$$\varphi_\mu(f) = \int f d\mu$$

is a positive linear form.

Proof. Since f has compact support, say K, and since it is continuous, it is bounded, say $|f| \leq M\chi_K$. Since f is continuous, it is measurable, and the function $M\chi_K$ is a step function which is integrable since $\mu(K)$ is finite. By Theorem 5.35, the function f is integrable. By Proposition 5.24, the map φ_μ is linear and positive. $\square$

Remark: As a point of terminology, the map $\varphi_\mu\colon \mathcal{K}_{\mathbb{C}}(X) \to \mathbb{C}$ just is just a linear form, but since its domain is a function space ($\mathcal{K}_{\mathbb{C}}(X)$), it is customary to call it a *linear functional.*

The remarkable fact is that any positive linear functional $\Phi\colon \mathcal{K}_{\mathbb{C}}(X) \to \mathbb{C}$ determines a Borel measure m_Φ (with some special properties) such that

$$\Phi(f) = \int f dm_\Phi \quad \text{for all } f \in \mathcal{K}_{\mathbb{C}}(X).$$

Knowing how to integrate functions in $\mathcal{K}_{\mathbb{C}}(X)$ is sufficient to determine the measure m_Φ completely. In some sense, continuous functions with compact support play the role of μ-step functions.

Recall that for any compact subset K of X, we denote by $\mathcal{K}(K;\mathbb{C})$ the set of complex-valued continuous functions whose support is contained in K (and similarly $\mathcal{K}(K;\mathbb{R})$ for real-valued functions). Interestingly, every positive linear functional on $\mathcal{K}_{\mathbb{C}}(X)$ is continuous on $\mathcal{K}(K;\mathbb{C})$ for every compact subset K of X.

Proposition 7.3. *If $\Phi\colon \mathcal{K}_{\mathbb{C}}(X) \to \mathbb{C}$ is a positive linear functional on $\mathcal{K}_{\mathbb{C}}(X)$, then for every compact subset K of X, there is some real number $c_K \geq 0$ such that $|\Phi(f)| \leq c_K \|f\|_\infty$ for all $f \in \mathcal{K}(K;\mathbb{C})$.*

Proof. Every function f in $\mathcal{K}(K;\mathbb{C})$ can be written uniquely as $f = f_1 + if_2$ with $f_1, f_2 \in \mathcal{K}_{\mathbb{R}}(X)$. Since Φ is a positive linear functional, we have $\Phi(f_1) \in \mathbb{R}$, $\Phi(f_2) \in \mathbb{R}$ and $\Phi(f) = \Phi(f_1 + if_2) = \Phi(f_1) + i\Phi(f_2)$, so

$$|\Phi(f)| = \sqrt{\Phi(f_1)^2 + \Phi(f_2)^2}.$$

Since

$$\begin{aligned} \|f\|_\infty &= \sup_{x\in K} |f(x)| = \sup_{x\in K} \sqrt{(f_1(x))^2 + (f_2(x))^2} \\ &= \sqrt{\sup_{x\in K}(f_1(x))^2 + \sup_{x\in K}(f_2(x))^2}, \end{aligned}$$

we obtain the inequalities

$$\|f_1\|_\infty = \sup_{x\in K} |f_1(x)| \leq \|f\|_\infty,$$

and

$$\|f_2\|_\infty = \sup_{x\in K} |f_2(x)| \leq \|f\|_\infty .$$

Using the above inequalities, if we can show that $|\Phi(f_1)| \leq c_1 \|f_1\|_\infty$ and $|\Phi(f_2)| \leq c_2 \|f_2\|_\infty$, then we get

$$|\Phi(f)| \leq \sqrt{c_1^2 \|f_1\|_\infty^2 + c_2^2 \|f_2\|_\infty^2} \leq \sqrt{c_1^2 + c_2^2}\, \|f\|_\infty .$$

Therefore we may assume that $f \in \mathcal{K}(K;\mathbb{R})$. By Proposition A.39, there is a continuous function with compact support $g \in \mathcal{K}_{\mathbb{C}}(X)$ (a bump function) such that $g(x) = 1$ for all $x \in K$. For any $f \in \mathcal{K}_{\mathbb{R}}(X)$, we have

$$-g \|f\|_\infty \leq f \leq g \|f\|_\infty$$

and since Φ is a positive linear functional, by Proposition 7.1(2), we get

$$-\Phi(g) \|f\|_\infty \leq \Phi(f) \leq \Phi(g) \|f\|_\infty$$

that is

$$|\Phi(f)| \leq \Phi(g) \|f\|_\infty ,$$

as desired. ☐

Proposition 7.3 suggests that the linear functionals $\Phi\colon \mathcal{K}_{\mathbb{C}}(X) \to \mathbb{C}$ satisfying the conclusion of the proposition are of particular interest, and they are. In fact the measure theory and the integration theory for complex-valued functions on a locally compact space can be developed entirely in terms of these functionals. This approach is presented in Dieudonné [Dieudonné (2005)], Bourbaki [Bourbaki (1965a, 1967a, 1968b)], and Schwartz [Schwartz (1993a)]. Dieudonné and Bourbaki even go as far as calling such functionals *measures*, which we feel is unfortunate because this term already has a well established meaning. Unlike these two previous sources, Schwartz actually develops in parallel both the theory of integration using measure theory, and the theory of integration using certain linear functionals that he calls *Radon measures*. Again, we find this terminology unfortunate because these are functionals and not measures in the traditional sense. We propose to use the term *Radon functional*.

Definition 7.3. A linear functional $\Phi\colon \mathcal{K}_{\mathbb{C}}(X) \to \mathbb{C}$ is a *Radon functional* if for every compact subset K of X, there is some real number $c_K \geq 0$ such that $|\Phi(f)| < c_K \|f\|_\infty$ for all $f \in \mathcal{K}(K;\mathbb{C})$. The set of Radon functionals is denoted $\mathrm{M}_{\mathbb{C}}(X)$, or simply, $\mathrm{M}(X)$. The set of *positive Radon functionals* is denoted $\mathrm{M}^+(X)$, and the set of *continuous (or bounded) Radon functionals* is denoted $\mathrm{M}^1(X)$. See Figure 7.1.

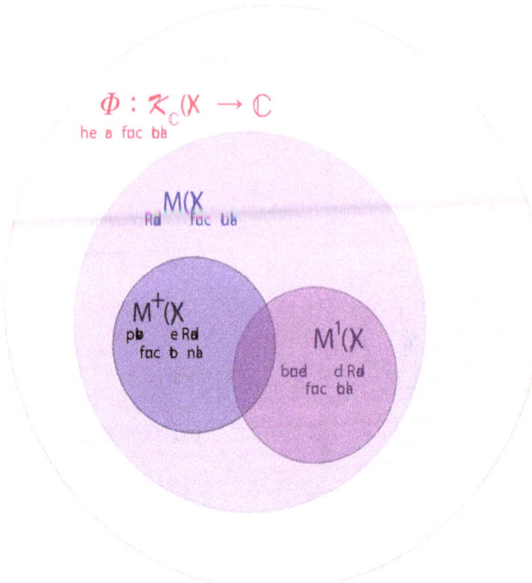

Fig. 7.1 A Venn diagram classification of Radon functionals.

Equivalently, a linear functional is a Radon functional if it is continuous when restricted to $\mathcal{K}(K; \mathbb{C})$, for every compact subset K of X.

In general, a Radon functional is *not* continuous on $\mathcal{K}_{\mathbb{C}}(X)$ for the sup norm $\| \, \|_\infty$. For a continuous Radon functional, there is a *uniform constant* $c \geq 0$ such that

$$|\Phi(f)| \leq c \|f\|_\infty \quad \text{for all } f \in \mathcal{K}_{\mathbb{C}}(X).$$

Continuous Radon functionals are often called *bounded Radon functionals.*

Proposition 7.3 immediately implies the following result.

Proposition 7.4. *Any positive linear functional $\Phi \colon \mathcal{K}_{\mathbb{C}}(X) \to \mathbb{C}$ is a positive Radon functional.*

Observe that a Radon functional $\Phi \colon \mathcal{K}_{\mathbb{C}}(X) \to \mathbb{C}$ is completely determined by its restriction $\Phi_{\mathbb{R}} \colon \mathcal{K}_{\mathbb{R}}(X) \to \mathbb{C}$ to the space of real-valued functions in $\mathcal{K}_{\mathbb{R}}(X)$. Indeed, every function $f \in \mathcal{K}_{\mathbb{C}}(X)$ can be written uniquely as $f = f_1 + if_2$ with $f_1, f_2 \in \mathcal{K}_{\mathbb{R}}(X)$, and by $\mathbb{C}$-linearity,

$$\Phi(f) = \Phi(f_1 + if_2) = \Phi_{\mathbb{R}}(f_1) + i\Phi_{\mathbb{R}}(f_2).$$

Furthermore, if Φ is a positive Radon functional, then by Proposition 7.1 we have $\Phi(f) \in \mathbb{R}$ for all $f \in \mathcal{K}_{\mathbb{R}}(X)$, so $\Phi_{\mathbb{R}} \colon \mathcal{K}_{\mathbb{R}}(X) \to \mathbb{R}$. Therefore, there is a bijection between the space $\mathrm{M}^+(X)$ of positive linear functionals $\Phi \colon \mathcal{K}_{\mathbb{C}}(X) \to \mathbb{C}$ and the space $\mathrm{M}^+_{\mathbb{R}}(X)$ of positive linear functionals $\Psi \colon \mathcal{K}_{\mathbb{R}}(X) \to \mathbb{R}$ as illustrated by Figure 7.2

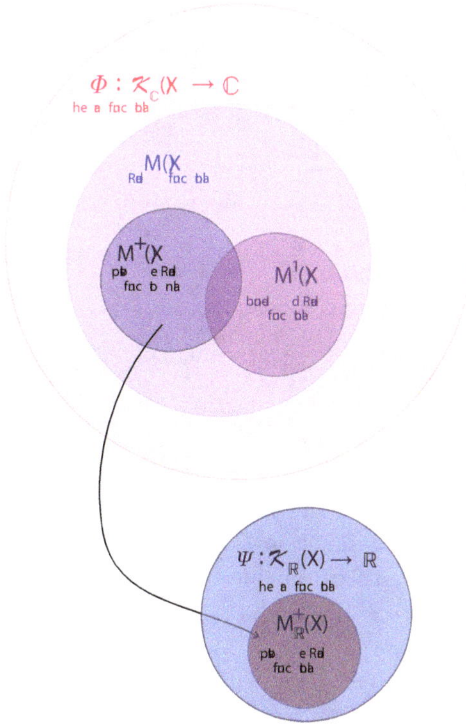

Fig. 7.2 The correspondence between $\mathrm{M}^+(X)$ and $\mathrm{M}^+_{\mathbb{R}}(X)$.

Also observe that $\mathrm{M}(X)$ and $\mathrm{M}^1(X)$ are vector spaces. The operator norm $\| \ \|$ is well defined on the vector space $\mathrm{M}^1(X)$. For any bounded linear functional Φ, by definition

$$\|\Phi\| = \sup\{|\Phi(f)| \mid f \in \mathcal{K}_{\mathbb{C}}(X),\ \|f\|_\infty = 1\}.$$

Using Proposition 2.17 it is easy to show that $\mathrm{M}^1(X)$ is isomorphic to the dual $\mathcal{C}_0(X;\mathbb{C})'$ of the space $\mathcal{C}_0(X;\mathbb{C})$, that is, the space of all continuous linear forms on $\mathcal{C}_0(X;\mathbb{C})$. Recall that $\mathcal{C}_0(X;\mathbb{C})$ is the space of continuous functions which tend to 0 at infinity; see Definition 2.16.

Proposition 7.5. *Let X be a locally compact space. The space* $\mathrm{M}^1(X)$ *of bounded Radon functionals is isomorphic to the dual* $\mathcal{C}_0(X;\mathbb{C})'$ *of* $\mathcal{C}_0(X;\mathbb{C})$*, that is, the space of all continuous linear forms on* $\mathcal{C}_0(X;\mathbb{C})$*. Consequently* $\mathrm{M}^1(X)$ *is a Banach space (w.r.t. the sup norm).*

Proof. By Proposition 2.17, the space $\mathcal{C}_0(X;\mathbb{C})$ is the closure of $\mathcal{K}_\mathbb{C}(X)$. By definition, $\mathrm{M}^1(X)$ is the space of continuous linear forms on $\mathcal{K}_\mathbb{C}(X)$. By Theorem A.73, every continuous linear form has a unique continuous extension to $\mathcal{C}_0(X;\mathbb{C})$. Therefore $\mathrm{M}^1(X)$ is isomorphic to the dual of $\mathcal{C}_0(X;\mathbb{C})$. Since $\mathbb{C}$ is complete, it is known that the set of continuous linear maps from any vector space into $\mathbb{C}$ is complete. ☐

Here are some example of Radon functionals.

Example 7.1.

(1) Pick any $a \in X$. The map δ_a given by

$$\delta_a(f) = f(a)$$

for all $f \in \mathcal{K}_\mathbb{C}(X)$ is a Radon functional called (with an abuse of terminlogy) the *Dirac measure*. Since $|f(a)| \leq \|f\|_\infty$, it is a bounded Radon functional.

(2) Consider the space $\mathcal{K}_\mathbb{C}(\mathbb{R})$ of continuous functions $f\colon \mathbb{R} \to \mathbb{C}$ with compact support. For each function $f \in \mathcal{K}_\mathbb{C}(\mathbb{R})$, there is a compact interval $[a, b]$ such that f vanishes outside of $[a, b]$, and from Section 3.1, the Riemann integral 5

$$I(f) = \int_a^b f(t)dt$$

is defined. We obtain a map $I\colon \mathcal{K}_\mathbb{C}(\mathbb{R}) \to \mathbb{C}$ which is obviously linear. Since

$$\left|\int_a^b f(t)dt\right| \leq (b-a)\,\|f\|_\infty\,,$$

this map is a Radon functional. Actually, this functional is positive. We will see later that this Radon functional corresponds to the Lebesgue measure.

(3) Let Φ be any Radon functional and pick any continuous function $g \in \mathcal{C}(X;\mathbb{C})$. It is clear that if $f \in \mathcal{K}_\mathbb{C}(X)$, then $gf \in \mathcal{K}_\mathbb{C}(X)$, and we have a map Ψ given by

$$\Psi(f) = \Phi(gf) \quad \text{for all } f \in \mathcal{K}_\mathbb{C}(X).$$

Clearly, this is a linear functional. For any compact subset K of X, if $f \in \mathcal{K}_{\mathbb{C}}(X)$, then we have

$$\|gf\|_\infty \leq \|f\|_\infty \sup_{x \in K} |g(x)|.$$

Since Φ is a Radon functional, there is some real $c_K \geq 0$ such that

$$|\Phi(gf)| \leq c_K \|gf\|_\infty ,$$

so we obtain

$$|\Phi(gf)| \leq c_K \sup_{x \in K} |g(x)| \, \|f\|_\infty ,$$

which shows that Ψ is a Radon functional. The Radon functional Ψ is called the *Radon functional with density g relative to Φ*, and it is denoted $g \cdot \Phi$. Such Radon functionals play an important role in the definition of the notion of convolution in the theory of integration based on Radon functionals developed in Dieudonné [Dieudonné (2005)] and Bourbaki [Bourbaki (1965a, 1967a, 1968b, 1965b)].

In the next section we state the most important theorem of the theory of Radon functionals, which is that every positive Radon functional arises from a unique Borel measure with some regularity properties.

7.2 The Radon–Riesz Theorem and Positive Radon Functionals

In this section we deal with the direction of the correspondence positive Radon functionals $\Longrightarrow$ Borel measures. Our first goal is to show that for every positive Radon functional Φ, there is a σ-algebra $\mathfrak{M}$ and a unique positive measure m_Φ on $\mathfrak{M}$ (with certain properties) representing Φ as an integral, which means that

$$\Phi(f) = \int f dm_\Phi \quad \text{for all } f \in \mathcal{K}_{\mathbb{C}}(X).$$

For instance, the positive Radon functional of Example 7.1(2) yields the Lebesgue measure. In a second stage, by imposing some reasonable conditions on the measure, we obtain a bijective correspondence.

Complete proofs of these results are quite long and intricate. Such proofs can be found in Rudin [Rudin (1987)] (Chapter 2), Lang [Lang (1996)] (Chapter IX), Folland [Folland (1999)] (Chapter 7, Theorem 7.2), and Schwartz [Schwartz (1993a)] (Chapters 5 and 7). Going back and forth between Rudin, Folland, and Lang is a possible strategy to understanding the proof.

Theorem 7.6 is often referred to as the *Riesz representation theorem*. A version of this theorem for $X = [0, 1]$ was first proven by Frigyes Riesz[1] in 1909. In 1913, Radon extended Riesz' result to a compact subset of $\mathbb{R}^n$ in terms of regular measures rather than a Stieltjes integral. Following Malliavin [Malliavin *et al.* (1995)], it seems appropriate to call it the Radon–Riesz theorem, but it should be noted that other versions of this theorem were obtained by Banach, Saks, Markov, and Kakutani, which gives the most general version stated in Theorem 7.30; see Dunford and Schwartz [Dunford and Schwartz (1988)].

Theorem 7.6. *(Radon–Riesz) Let X be a locally compact (Hausdorff) space. For every positive linear functional $\Phi\colon \mathcal{K}_{\mathbb{C}}(X) \to \mathbb{C}$, there is a σ-algebra $\mathfrak{M}$ containing the Borel σ-algebra, and there is a unique positive measure m_Φ on $\mathfrak{M}$ with the following properties:*

(1) The linear functional Φ is represented by m_Φ, that is,

$$\Phi(f) = \int f dm_\Phi \quad \text{for all } f \in \mathcal{K}_{\mathbb{C}}(X).$$

(2) The measure $m_\Phi(K)$ is finite for every compact subset K of X.

(3) We have

$$m_\Phi(E) = \inf\{m_\Phi(V) \mid E \subseteq V,\ V \text{ open}\}$$

for every $E \in \mathfrak{M}$.

(4) We have

$$m_\Phi(E) = \sup\{m_\Phi(K) \mid K \subseteq E,\ K \text{ compact}\}$$

for every open subset E, and for every $E \in \mathfrak{M}$ with $m_\Phi(E) < +\infty$.

(5) For any $E \in \mathfrak{M}$ and any $A \subseteq E$, if $m_\Phi(E) = 0$, then $m_\Phi(A) = 0$, in other words, m_Φ is a complete measure.

Let us make a few comments about the proof. The uniqueness of m_Φ is not so bad. Observe that by (3) and (4), the measure m_Φ is determined by its values on compact subsets. Hence it suffices to prove that if two measures μ_1 and μ_2 satisfy the theorem, then they agree on all compact subsets.

[1]Not to be confused with his younger brother Marcel Riesz.

Pick any compact K and any $\epsilon > 0$. By (3) and (4), there is some open subset V such that $K \subseteq V$ and $\mu_2(V) < \mu_2(K) + \epsilon$. By Proposition A.39, there is a continuous function $f\colon X \to [0,1]$ such that $f(x) = 1$ for all $x \in K$, and such that $\mathrm{supp}(f)$ is compact and $\mathrm{supp}(f) \subseteq V$; this implies that

$$\begin{aligned}\mu_1(K) &= \int \chi_K \, d\mu_1 \\ &\leq \int f d\mu_1 = \Phi(f) = \int f d\mu_2 \\ &\leq \int \chi_V \, d\mu_2 = \mu_2(V) \\ &< \mu_2(K) + \epsilon.\end{aligned}$$

Therefore, $\mu_1(K) \leq \mu_2(K)$. By swapping μ_1 and μ_2, we obtain $\mu_2(K) \leq \mu_1(K)$, and thus $\mu_1(K) = \mu_2(K)$. Observe that the above derivation also shows that $\mu_1(K)$ is finite for every compact subset K.

To construct m_Φ we proceed as follows; for simplicity of notation, write μ instead of m_Φ.

(a) For every open set V in X, for every continuous function $g\colon X \to \mathbb{R}$, write $g \prec V$ if $g\colon X \to [0,1]$, $\mathrm{supp}(g)$ is compact, and $\mathrm{supp}(g) \subseteq V$. Let

$$\mu(V) = \sup\{\Phi(g) \mid g \prec V\}.$$

This will force Condition (4).

(b) Next, to force Condition (3), we extend μ to arbitrary subsets. For every $E \subseteq X$, let

$$\mu(E) = \inf\{\mu(V) \mid E \subseteq V,\ V \text{ open}\}.$$

It can be checked that μ is an outer measure.

(c) In order to obtain a σ-algebra and a measure, we need to cut down the family of subsets, still forcing Conditions (3) and (4). Let $\mathcal{A}$ be the family of all subsets A of X such that $\mu(A) < +\infty$ and

$$\mu(A) = \sup\{\mu(K) \mid K \subseteq A,\ K \text{ compact}\}.$$

Then $\mathcal{A}$ is an algebra containing all compact sets and all open sets of finite measure. The map μ is a measure on $\mathcal{A}$, and if $\mu(A) < +\infty$, then $A \in \mathcal{A}$.

(d) Let $\mathfrak{M}$ be the family of all subsets Y of X such that $Y \cap K$ lies in $\mathcal{A}$ for all compact subsets K. Then $\mathfrak{M}$ is the desired σ-algebra containing the Borel sets, and μ is a positive measure on $\mathfrak{M}$. The algebra $\mathcal{A}$ consists of the sets of finite measure in $\mathfrak{M}$.

Having done all this, one still needs to check that Conditions (1), (3), and (4) hold. Proposition A.40 (existence of finite partitions of unity) is used for some of these checks.

Theorem 7.6 shows that the measure that arises from a positive linear functional has special regularity properties that we already encountered when we met the Lebesgue measure in Section 4.5.

7.3 σ-Regular Borel Measures

Definition 7.4. A Borel measure μ on the Borel σ-algebra $\mathcal{B}$ of a locally compact space X is *σ-regular* if the following two conditions hold:

For every $E \in \mathcal{B}$,

$$\mu(E) = \inf\{\mu(V) \mid E \subseteq V,\ V \text{ open}\}. \tag{$*$}$$

For every open subset E, and for every $E \in \mathcal{B}$ with $\mu(E) < +\infty$,

$$\mu(E) = \sup\{\mu(K) \mid K \subseteq E,\ K \text{ compact}\}. \tag{$**_\sigma$}$$

Condition $(*)$ is called *outer regularity*, and Condition $(**_\sigma)$ is called *σ-inner regularity*.

We say that μ is *locally finite* if $\mu(K)$ is finite for every compact subset K.

The following proposition justifies the terminology σ-inner regularity.

Proposition 7.7. *Let X be a locally compact (Hausdorff) space. If a Borel measure μ is σ-inner regular, then*

$$\mu(E) = \sup\{\mu(K) \mid K \subseteq E,\ K \text{ compact}\} \tag{$**_\sigma$}$$

holds for every σ-finite subset $E \in \mathcal{B}$.

Proof. Say $E = \bigcup_{i=1}^{\infty} E_i$ with $E_i \in \mathcal{B}$ and $\mu(E_i) < +\infty$. We may assume that $\mu(E) = +\infty$, since if $\mu(E) < +\infty$ then we already have σ-inner regularity by definition. For every $M > 0$, there is some $n \geq 1$ such that

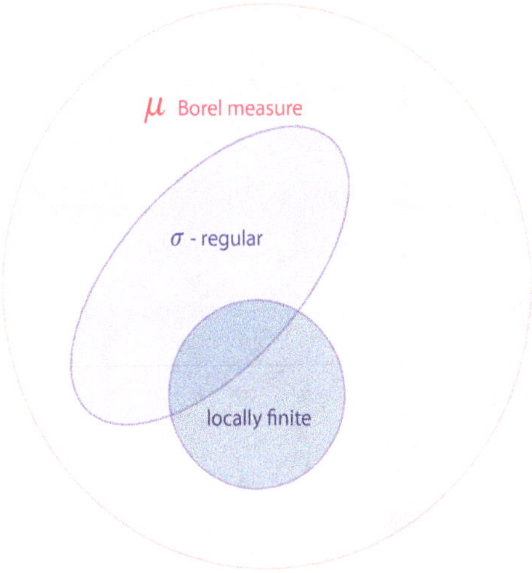

Fig. 7.3 A Venn diagram classification of Borel measures.

$\mu\left(\bigcup_{i=1}^{n} E_i\right) > M$. Since $\bigcup_{i=1}^{n} E_i$ has finite measure, σ-inner regularity applies, so there is some compact subset K such that $\mu(K) > M$. This shows that

$$\sup\{\mu(K) \mid K \subseteq E,\ K \text{ compact}\} = +\infty = \mu(E),$$

which shows σ-inner regularity for E. □

Definition 7.5. Let X be a locally compact (Hausdorff) space. A Borel measure μ is called a *(positive) σ-Radon measure* if it is σ-regular and locally finite. The space of σ-Radon measures is denoted by $\mathcal{M}_\sigma^+(X)$. See Figure 7.3.

Theorem 7.6 immediately implies the following correspondence which we illustrate in Figure 7.4.

Theorem 7.8. *(Radon–Riesz Correspondence, I) Let X be a locally compact (Hausdorff) space. The maps $m\colon \mathrm{M}^+(X) \to \mathcal{M}_\sigma^+(X)$ and $\varphi\colon \mathcal{M}_\sigma^+(X) \to \mathrm{M}^+(X)$ given by*

$$m(\Phi) = m_\Phi \quad \textit{for all } \Phi \in \mathrm{M}^+(X)$$
$$\varphi(\mu) = \varphi_\mu \quad \textit{for all } \mu \in \mathcal{M}_\sigma^+(X)$$

are mutual inverses that define a bijection between the space $\mathrm{M}^+(X)$ *of positive Radon functionals and the space* $\mathcal{M}_\sigma^+(X)$ *of (positive)* σ*-Radon measures (recall from Proposition 7.2 that*

$$\varphi_\mu(f) = \int f d\mu$$

for any $f \in \mathcal{K}_{\mathbb{C}}(X)$*.)*

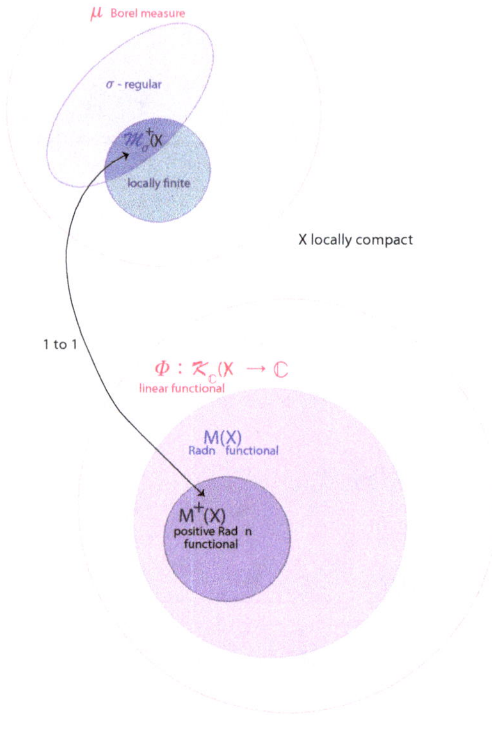

Fig. 7.4 Radon–Riesz Correspondence, Version I.

Measurable functions on a locally compact space with a σ-regular, locally finite, Borel measure are very close to being continuous as stated in the following theorem of Lusin.

Theorem 7.9. *(Lusin's Theorem) Let* X *be a locally compact space equipped with a* σ*-regular, locally finite, Borel measure* μ*, and let* f *be any* *measurable function on X. If f vanishes outside of a set A of finite measure, for any $\epsilon > 0$, there is some function $g \in \mathcal{K}_{\mathbb{C}}(X)$ and a measurable set Z with $\mu(Z) < \epsilon$, such that $f(x) = g(x)$ for all $x \in X - Z$, and $\|g\|_\infty \leq \|f\|_\infty$.*

Theorem 7.9 is proven in Rudin [Rudin (1987)] (Chapter 2, Theorem 2.24) and Lang [Lang (1996)] (Chapter IX, Theorem 3.3).

The Vitali–Carathéodory theorem states that every function in $\mathrm{L}^1_\mu(X, \mathcal{B}, \mathbb{C})$ can be approximated from below and from above by certain kinds of functions called upper semicontinuous and lower semicontinuous, see Rudin [Rudin (1987)] (Chapter 2, Theorem 2.25).

We have the following density result which uses Lusin's theorem (Theorem 7.9).

Theorem 7.10. *Let X be a locally compact space equipped with a σ-regular, locally finite, Borel measure μ. The space $\mathcal{K}_{\mathbb{C}}(X)$ is dense in $\mathrm{L}^p_\mu(X, \mathcal{B}, \mathbb{C})$ for $p = 1, 2$.*[2]

Theorem 7.10 is proven in Rudin [Rudin (1987)] (Chapter 3, Theorem 3.14) and Lang [Lang (1996)] (Chapter IX, Theorem 3.1).

The following corollary of Theorem 7.10 is used in Vol II, Chapter 3.

Theorem 7.11. *Let X be a locally compact, metrizable, separable space equipped with a σ-regular, locally finite, Borel measure μ. Then $\mathrm{L}^p_\mu(X, \mathcal{B}, \mathbb{C})$ is separable for $p = 1, 2$.*

Theorem 7.11 follows immediately from Theorem 7.10 and Theorem 2.16.

The following proposition is needed for proving the uniqueness of the Haar measure up to a constant.

Proposition 7.12. *Let X be a locally compact space equipped with a σ-regular, locally finite, Borel measure μ. For any function $f \in \mathcal{L}^1_\mu(X, \mathcal{B}, \mathbb{C})$, if*

$$\int fg\, d\mu = 0 \quad \text{for all } g \in \mathcal{K}_{\mathbb{C}}(X),$$

then $f = 0$ almost everywhere.

[2] Even for all p with $1 \leq p < +\infty$.

Proof. We use Proposition 5.39, recalling the fact that $\int_A f d\mu = \int f\chi_A\, d\mu$. Let A be any subset of finite measure. By Theorem 7.10, χ_A is the L^1-limit of a sequence (g_n) of functions $g_n \in \mathcal{K}_{\mathbb{C}}(X)$ with $g_n(X) \subseteq [0,1]$. By Proposition 5.26, there is a subsequence $(g_{n_k})_{k\geq 1}$ that converges pointwise to χ_A a.e., and thus (fg_{n_k}) converges pointwise to $f\chi_A$ a.e. By Proposition 7.2, the functions g_{n_k} are integrable, so the functions (fg_{n_k}) are also integrable, and since $g_{n_k}(X) \subseteq [0,1]$, by the dominated convergence theorem, we conclude that $\int_A f d\mu = \int f\chi_A\, d\mu = 0$ for all subsets A of finite measure, and by Proposition 5.39, we have $f = 0$ a.e. ☐

In the next section we show that by requiring the locally compact space X to be also σ-compact, then we obtain Borel measures that are not only σ-regular, but regular as well, which means that inner regularity holds for all $E \in \mathfrak{B}$.

7.4 Regular Borel Measures

In Theorem 7.6 outer regularity holds, but σ-inner regularity holds only for open subsets and measurable sets of finite measure. It is often desirable for inner regularity to hold for *arbitrary* subsets $E \in \mathcal{B}$, possibly not σ-finite. It turns out that making some mild restrictions on X, we obtain a bijection between positive linear functionals and these regular measures. On this subject, Rudin's exposition seems clearer than Lang's exposition.

Definition 7.6. A Borel measure μ on the Borel σ-algebra $\mathcal{B}$ of a locally compact space X is *regular* if the following two conditions hold for every $E \in \mathcal{B}$:

$$\mu(E) = \inf\{\mu(V) \mid E \subseteq V,\ V \text{ open}\} \tag{$*$}$$

and

$$\mu(E) = \sup\{\mu(K) \mid K \subseteq E,\ K \text{ compact}\}. \tag{$**$}$$

Condition $(*)$ is called *outer regularity*, and Condition $(**)$ is called *inner regularity*. See Figure 7.5.

Observe that if a Borel measure μ is σ-finite (on X) and if it is σ-regular, then it is actually regular. Another sufficient condition is given in the next proposition.

Proposition 7.13. *Let X be a locally compact (Hausdorff) space in which every open subset is σ-compact. If μ is a locally finite Borel measure, then μ is a regular measure.*

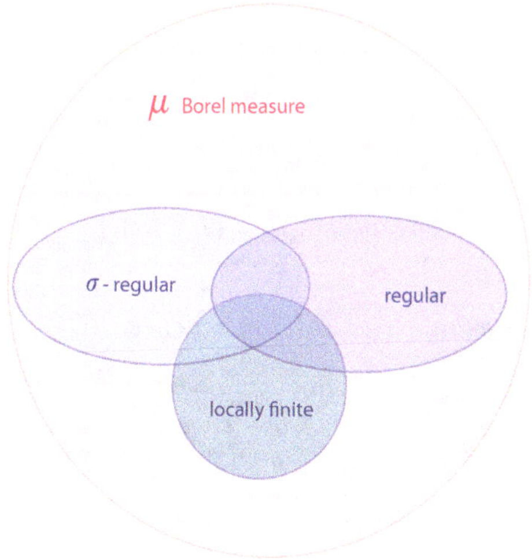

Fig. 7.5 Another Venn diagram classification of Borel measures.

Proposition 7.13 is proven in Rudin [Rudin (1987)] (Chapter 2, Theorem 2.18).

Observe that $X = \mathbb{R}^n$ satisfies the condition of Proposition 7.13. Thus a locally finite Borel measure on $\mathbb{R}^n$ is a regular measure.

A way to obtain the Radon–Riesz correspondence between positive Radon functionals and regular locally finite Borel measures is to require X to be σ-compact, which means that X is the countable union of compact subsets (see Definition A.43).

Theorem 7.14. *Let X be a locally compact (Hausdorff), σ-compact space. For every positive linear functional $\Phi\colon \mathcal{K}_{\mathbb{C}}(X) \to \mathbb{C}$, if $\mathfrak{M}$ and m_Φ are the σ-algebra and the measure obtained in Theorem 7.6, then the following properties holds:*

(1) For any $E \in \mathfrak{M}$ and any $\epsilon > 0$, there is a closed set F and an open set O such that $F \subseteq E \subseteq O$ and $\mu(O - F) < \epsilon$.
(2) The measure m_Φ is a regular, locally finite Borel measure on the Borel σ-algebra $\mathcal{B}$.

Theorem 7.14 is proven in Rudin [Rudin (1987)] (Chapter 2, Theorem 2.17). The following theorem allows us to get a bijective correspondence between positive linear functional and regular locally finite Borel measures, and to state this theorem it is convenient to introduce the following definition.

Definition 7.7. Let X be a locally compact (Hausdorff) space. A Borel measure μ is called a *(positive) Radon measure* if it is regular and locally finite. The space of Radon measures is denoted by $\mathcal{M}^+_{\rm rad}(X)$, or simply $\mathcal{M}^+(X)$. See Figure 7.5.

Theorem 7.15. *(Radon–Riesz Correspondence, II) Let X be a locally compact (Hausdorff), σ-compact space. The maps $m\colon \mathrm{M}^+(X) \to \mathcal{M}^+(X)$ and $\varphi\colon \mathcal{M}^+(X) \to \mathrm{M}^+(X)$ given by*

$$m(\Phi) = m_\Phi \quad \textit{for all } \Phi \in \mathrm{M}^+(X)$$
$$\varphi(\mu) = \varphi_\mu \quad \textit{for all } \mu \in \mathcal{M}^+(X)$$

are mutual inverses that define a bijection between the space $\mathrm{M}^+(X)$ of positive Radon functionals and the space $\mathcal{M}^+(X)$ of (positive) Radon measures. See Figure 7.6.

An interesting application of Theorem 7.15 is obtained by choosing $X = \mathbb{R}$ and Φ to be the Radon functional I induced by the Riemann integral defined in Example 7.1(2). The Radon measure m_I given by Theorem 7.15 turns out to be the Lebesgue measure μ_L. For details, see Rudin [Rudin (1987)] (Chapter 2).

7.5 Complex and Real Measures

By Proposition 7.2, the functionals induced by Borel measures are positive, but there are Radon functionals that are not positive, so it is natural to ask if such functionals arise from some generalized measures allowed to take negative values, or even complex values. The answer is yes. It is even possible to define measures with values in any Banach space. Such measures are discussed in Lang [Lang (1996)], Schwartz [Schwartz (1993a)] and Marle [Marle (1974)], but for simplicity we will only consider real and complex measures. In this section we take a small detour to define complex measures. Then we will show how they relate to functionals on $\mathcal{K}_{\mathbb{C}}(X)$ that are not necessarily positive, but continuous.

Going back to Definition 4.9, a (positive) measure on a measurable set $(X, \mathcal{A})$ is a map μ satisfying the following properties:

(μ1) $\mu\colon \mathcal{A} \to [0, +\infty]$, where $\mathcal{A}$ is a σ-algebra of subsets of X.

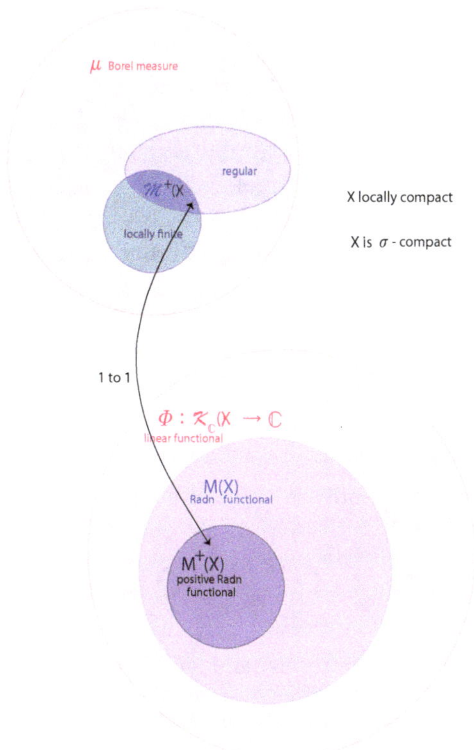

Fig. 7.6 Radon–Riesz Correspondence, Version 2.

($\mu 2$) $\mu(\emptyset) = 0$.

($\mu 3$) For any countable sequence $(A_i)_{i \geq 1}$ of subsets A_i of $\mathcal{A}$ such that $A_i \cap A_j = \emptyset$ for all $i \neq j$,

$$\mu\left(\bigcup_{i=1}^{\infty} A_i\right) = \sum_{i=1}^{\infty} \mu(A_i).$$

Such a function may have the value $+\infty$, but in ($\mu 3$), if $A = \bigcup_{i=1}^{\infty} A_i$ and if $\mu(A)$ is finite, then the series $\sum_{i=1}^{\infty} \mu(A_i)$ converges, and since it consists of nonnegative numbers, it converges absolutely, and thus *commutatively*, which means that for any permutation σ of $\mathbb{N}_+$, we have

$$\mu(A) = \mu\left(\bigcup_{i=1}^{\infty} A_\sigma(i)\right) = \sum_{i=1}^{\infty} \mu(A_{\sigma(i)}).$$

If we replace $[0, +\infty]$ by $\mathbb{R}$ or $\mathbb{C}$, then a new problem arises, namely that the convergence of the sum $\sum_{i=1}^{\infty} \mu(A_i)$ generally depends on the order of the A_i. The solution is to *require* commutative convergence of the series arising in (μ3). It is known from analysis that for $\mathbb{R}$ or $\mathbb{C}$, a series is commutatively convergent iff it is absolutely convergent, so we require the latter. We also require $\mu(A)$ be an element of $\mathbb{R}$ or $\mathbb{C}$, that is, $\mu(A)$ must be "finite." There is a way to define measures with values in $\mathbb{R} \cup \{+\infty\}$, and even in $\mathbb{R} \cup \{-\infty, +\infty\}$, but we have no need for such generality (see Schwartz [Schwartz (1993a)], Chapter V, §9).

Definition 7.8. Let $(X, \mathcal{A})$ be a measurable space. A *complex measure* on $(X, \mathcal{A})$ is a map μ satisfying the following properties:

(μ1) $\mu\colon \mathcal{A} \to \mathbb{C}$.
(μ2) $\mu(\emptyset) = 0$.
(μ3) For any countable family $(A_i)_{i \geq 1}$ of subsets A_i of $\mathcal{A}$ such that $A_i \cap A_j = \emptyset$ for all $i \neq j$,

$$\mu\left(\bigcup_{i=1}^{\infty} A_i\right) = \sum_{i=1}^{\infty} \mu(A_i),$$

where the series on the right-hand side is absolutely convergent.

A *real measure* (or *signed measure*) is a complex measure such that $\mu(\mathcal{A}) \subseteq \mathbb{R}$.

Observe that a real measure which is also positive is a positive measure according to Definition 4.9, but since a positive measure may take the value $+\infty$, there are positive measures that are not real measures in the sense of Definition 7.8. When we use the term *positive real measure*, we mean that this measure only takes finite values. By *positive measure*, we mean a measure that may take the value $+\infty$.

One might wonder if interesting real or complex measures exist. Indeed, for any *arbitrary measure space* $(X, \mathcal{A}, \mu)$, every function $f \in \mathcal{L}_{\mu}(X, \mathcal{A}, \mathbb{C})$ gives rise to such a measure.

Proposition 7.16. *Let $(X, \mathcal{A}, \mu)$ be a measure space (here, μ is a positive measure). For every integrable map $f \in \mathcal{L}_{\mu}(X, \mathcal{A}, \mathbb{C})$, the function $\mu_f\colon \mathcal{A} \to \mathbb{C}$ given by*

$$\mu_f(A) = \int_A f d\mu = \int f \chi_A d\mu \quad \textit{for all } A \in \mathcal{A}$$

is a complex measure.

What is not obvious is that (μ3) holds. This follows from Proposition 5.37 (a consequence of the Lebesgue dominated convergence theorem). A detailed proof is given in Marle [Marle (1974)] (Chapter 2, Proposition 2.5.2).

The new twist here is that given a measure μ, rather than defining a functional by *varying the function being integrated*, we fix a function but we integrate by *varying the subset over which we integrate*.

It is trivial to check that the complex measures (and the real measures) form a vector space.

Remarkably, every complex measure μ arises as a measure of the form $|\mu|_h$ for some suitable positive measure $|\mu|$ and some well chosen function $h \in \mathcal{L}_{|\mu|}(X, \mathcal{A}, \mathbb{C})$; see Theorem 7.21. The measure $|\mu|$ is defined as follows.

Definition 7.9. Let $(X, \mathcal{A})$ be a measurable space, and let μ be a complex measure on $(X, \mathcal{A})$. Define the map $|\mu|\colon \mathcal{A} \to [0, +\infty]$ by

$$|\mu|(A) = \sup \sum_{i=1}^{\infty} |\mu(A_i)|,$$

for all $A \in \mathcal{A}$ and for all countable partitions $(A_i)_{i \geq 1}$ of A with $A_i \in \mathcal{A}$. The map $|\mu|$ is called the *total variation measure* (for short *total variation*) of μ.

Obviously, if μ is a real positive measure, then $|\mu| = \mu$. It is easy to see that by definition,

$$|\mu(A)| \leq |\mu|(A) \quad \text{for all } A \in \mathcal{A}.$$

In fact, it is minimal with this property. We have the following remarkable theorems.

Theorem 7.17. *Let $(X, \mathcal{A})$ be a measurable space, and let μ be a complex measure on $(X, \mathcal{A})$. The map $|\mu|\colon \mathcal{A} \to [0, +\infty]$ is a positive measure. The positive measure $|\mu|$ is the minimal measure such that*

$$|\mu(A)| \leq |\mu|(A) \quad \textit{for all } A \in \mathcal{A},$$

in the sense that if λ is any positive measure such that

$$|\mu(A)| \leq \lambda(A) \quad \textit{for all } A \in \mathcal{A},$$

then $|\mu| \leq \lambda$ (which means that $|\mu|(A) \leq \lambda(A)$ for all $A \in \mathcal{A}$).

A proof of Theorem 7.17 is given in Rudin [Rudin (1987)] (Chapter 6, Theorem 6.2) and Lang [Lang (1996)] (Chapter VII, Theorem 3.1).

The next theorem is even more surprising.

Theorem 7.18. *Let $(X, \mathcal{A})$ be a measurable space, and let μ be a complex measure on $(X, \mathcal{A})$. The map $|\mu|\colon \mathcal{A} \to [0, +\infty]$ is a finite positive measure; that is, $|\mu|(X) < +\infty$.*

A proof of Theorem 7.18 is given in Rudin [Rudin (1987)] (Chapter 6, Theorem 6.4) and Lang [Lang (1996)] (Chapter VII, Theorem 3.2). Theorem 7.18 implies that $\mu(X)$ is bounded: it is contained in a closed disk of finite radius. This fact shows that the convergence requirement of Condition ($\mu 3$) is quite strong.

Theorem 7.18 allows us to make the space of complex measures into a normed vector space.

Definition 7.10. Let $(X, \mathcal{A})$ be measurable space. For any complex measure μ, define $\|\mu\|$ as $\|\mu\| = |\mu|(X)$. The vector space of complex measures equipped with the norm defined above is denoted $\mathbb{C}\mathcal{M}^1(X, \mathcal{A})$.

It is not hard to show that $\mathbb{C}\mathcal{M}^1(X, \mathcal{A})$ is a Banach space.

Proposition 7.19. *Let $(X, \mathcal{A})$ be a measurable space. The normed vector space $\mathbb{C}\mathcal{M}^1(X, \mathcal{A})$ is a Banach space (it is complete).*

Another interesting fact is that if μ is a positive measure (possibly taking the value $+\infty$) then $\mathcal{L}^1_\mu(X, \mathcal{A}, \mathbb{C})$ can be embedded in $\mathbb{C}\mathcal{M}^1(X, \mathcal{A})$.

Proposition 7.20. *Let $(X, \mathcal{A}, \mu)$ be a measure space. The map $f \mapsto \mu_f$ is a linear embedding of $\mathcal{L}^1_\mu(X, \mathcal{A}, \mathbb{C})$ into $\mathbb{C}\mathcal{M}^1(X, \mathcal{A})$, and*

$$\|\mu_f\| = \|f\|_1 \quad \text{for all } f \in \mathcal{L}_\mu(X, \mathcal{A}, \mathbb{C}).$$

Proposition 7.20 is proven in Lang [Lang (1996)] (Chapter VII, §3, Theorem 3.3). The proof uses Proposition 5.18.

The next theorem shows an important fact that we mentioned earlier, namely that every complex measure μ arises as a measure of the form $|\mu|_h$ for some well chosen function $h \in \mathcal{L}_{|\mu|}(X, \mathcal{A}, \mathbb{C})$. This result is a special case of the Radon–Nikodym theorem, but for now we prefer not discussing this theorem.

Theorem 7.21. *For every complex measure μ on a measurable space $(X, \mathcal{A})$, there is a function $h \in \mathcal{L}^1_{|\mu|}(X, \mathcal{A}, \mathbb{C})$ such that $|h| = 1$ and*

$$\mu(A) = \int_A h\, d|\mu| \quad \text{for all } A \in \mathcal{A}.$$

In other words, $\mu = |\mu|_h$ (recall that $|\mu|$ is a positive measure). Furthermore, any two functions $h_1, h_2 \in \mathcal{L}^1_{|\mu|}(X, \mathcal{A}, \mathbb{C})$ satisfying the conditions of the theorem are equal $|\mu|$-a.e.

For a proof of Theorem 7.21, see Rudin [Rudin (1987)] (Chapter 6, Theorem 6.12) and Lang [Lang (1996)] (Chapter VII, §2 and §4).

Let us now turn our attention to real measures. We will see that any real measure can be expressed in terms of two positive real measures. This implies that any complex measure can be expressed in terms of four positive real measures. This will allow us to explain how to integrate with respect to a complex measure.

7.6 Real Measures and the Hahn–Jordan Decomposition

We begin by showing that a real measure can be expressed as the difference of two finite positive measures. If μ is a real measure, since $|\mu|$ is a finite measure, we can define two finite positive measures μ^+ and μ^- such that $\mu = \mu^+ - \mu^-$.

Definition 7.11. If μ is a real measure, the real measures μ^+ and μ^- are defined by

$$\mu^+ = \frac{1}{2}(|\mu| + \mu), \quad \mu^- = \frac{1}{2}(|\mu| - \mu).$$

It is immediately checked that μ^+ and μ^- are *finite positive* measures, and we have

$$\mu = \mu^+ - \mu^-, \quad |\mu| = \mu^+ + \mu^-.$$

Definition 7.12. Given a real measure μ, the positive real measures μ^+ and μ^- are called the *positive variation* and *negative variation* of μ. The expression of μ as $\mu = \mu^+ - \mu^-$ is called the *Jordan decomposition* of μ.

The Jordan decomposition has certain minimality properties that we are going to describe.

Definition 7.13. Let $(X, \mathcal{A})$ be a measurable space. A complex measure μ is *concentrated* on (or *carried by*) a measurable subset A if $\mu(E) = 0$ for all $E \in \mathcal{A}$ such that $E \cap A = \emptyset$. Two complex measures μ_1 and μ_2 are *mutually singular* if there exist two disjoint measurable subsets A_1 and A_2 such that μ_1 is concentrated on A_1 and μ_2 is concentrated on A_2. We sometimes write $\mu_1 \perp \mu_2$.

Every real measure has a Hahn–Jordan decomposition as described by the following theorem.

Theorem 7.22. *(Hahn–Jordan Decomposition) Let $(X, \mathcal{A})$ be a measurable space. For any real measure μ, there is a partition (X^+, X^-) of X into two disjoint subsets of X such that if*

$$\mu = \mu^+ - \mu^-$$

is the Jordan decomposition of μ, then μ^+ is concentrated on X^+, and μ^- is concentrated on X^-. Furthermore, for any $E \in \mathcal{A}$, we have

$$\mu^+(E) = \sup\{\mu(A) \mid A \subseteq E,\ A \in \mathcal{A}\},$$
$$\mu^-(E) = \sup\{-\mu(A) \mid A \subseteq E,\ A \in \mathcal{A}\}.$$

For any other partition (Y^+, Y^-) of X such that μ^+ is concentrated on Y^+ and μ^- is concentrated on Y^-,

$$\mu^+(E \cap X^+) = \mu^+(E \cap Y^+), \quad \mu^-(E \cap X^-) = \mu^-(E \cap Y^-),$$

for all $E \in \mathcal{A}$.

Let us now consider a complex measure $\mu\colon \mathcal{A} \to \mathbb{C}$.

Definition 7.14. Given a complex measure $\mu\colon \mathcal{A} \to \mathbb{C}$, the function $\overline{\mu}\colon \mathcal{A} \to \mathbb{C}$ called *the conjugate* of μ is defined by $\overline{\mu}(A) = \overline{\mu(A)}$ for all $A \in \mathcal{A}$. We also define $\mu_1\colon \mathcal{A} \to \mathbb{R}$ and $\mu_2\colon \mathcal{A} \to \mathbb{R}$ by

$$\mu_1(A) = \frac{1}{2}(\mu(A) + \overline{\mu(A)}), \quad \mu_2(A) = \frac{1}{2i}(\mu(A) - \overline{\mu(A)})$$

for all $A \in \mathcal{A}$. We call μ_1 the *real part* of μ and μ_2 the *imaginary part* of μ.

It is immediately checked that $\overline{\mu}$ is a complex measure, and that μ_1 and μ_2 are *real* measures such that

$$\mu = \mu_1 + i\mu_2$$
$$\overline{\mu} = \mu_1 - i\mu_2.$$

Using the Hahn–Jordan decomposition of μ_1 an μ_2, we see that we can write μ uniquely in terms of four positive real measure $\mu_1^+, \mu_1^-, \mu_2^+, \mu_2^-$, as

$$\mu = \mu_1^+ - \mu_1^- + i(\mu_2^+ - \mu_2^-).$$

Definition 7.15. For any complex measure $\mu\colon \mathcal{A} \to \mathbb{C}$, the expression

$$\mu = \mu_1^+ - \mu_1^- + i(\mu_2^+ - \mu_2^-)$$

is called the *Jordan decomposition* of μ.

Proposition 7.23. *For any complex measure $\mu\colon \mathcal{A} \to \mathbb{C}$, we have $|\mu_1| \leq |\mu|$, $|\mu_2| \leq |\mu|$, and that $|\mu| \leq |\mu_1| + |\mu_2|$. A function f is $|\mu|$-integrable iff it is integrable for all four positive real measures $\mu_1^+, \mu_1^-, \mu_2^+$, and μ_2^-.*

Proof. It is easy to check that $|\mu_1| \leq |\mu|$, $|\mu_2| \leq |\mu|$, and that $|\mu| \leq |\mu_1| + |\mu_2|$. It follows easily that f is $|\mu|$-integrable if f is $|\mu_1|$-integrable and $|\mu_2|$-integrable. Since $|\mu_1| = \mu_1^+ + \mu_1^-$ and $|\mu_2| = \mu_2^+ + \mu_2^-$, it is also easy to see that f is $|\mu_1|$-integrable iff f is μ_1^+-integrable and μ_1^--integrable, and similarly f is $|\mu_2|$-integrable iff f is μ_2^+-integrable and μ_2^--integrable. Therefore, f is $|\mu|$-integrable iff it is integrable for all four positive measures $\mu_1^+, \mu_1^-, \mu_2^+$, and μ_2^-. ☐

The Jordan decomposition of the complex measure μ suggests defining the integral $\int f d\mu$ for any function $f \in \mathcal{L}^1_{|\mu|}(X, \mathcal{A}, \mathbb{C})$; see Dieudonné [Dieudonné (2005)] (Chapter XIII, Section 16, no. 13.16.2), or Folland [Folland (1999)] (end of Section 3.1 and Section 3.3).

Definition 7.16. Given any complex measure $\mu\colon \mathcal{A} \to \mathbb{C}$, for any function $f \in \mathcal{L}^1_{|\mu|}(X, \mathcal{A}, \mathbb{C})$, we define the integral $\int f d\mu$ as

$$\int f d\mu = \int f d\mu_1 + i \int f d\mu_2 = \int f d\mu_1^+ - \int f d\mu_1^- + i \int f d\mu_2^+ - i \int f d\mu_2^-.$$

By Proposition 7.23, the above expression is well defined since f is $|\mu|$-integrable iff it is integrable for all four positive real measures $\mu_1^+, \mu_1^-, \mu_2^+$, and μ_2^-.

Remark: Alternatively, if μ is a complex measure, $\int f d\mu$ can be defined using Theorem 7.21 as $\int f h \, d|\mu|$, as in Rudin [Rudin (1987)] (Chapter 6, Section 6.18).

The following fact will be needed later.

Proposition 7.24. *Given a complex measure μ, if $\overline{\mu}$ is the conjugate measure of μ, for any function $f \in \mathcal{L}^1_{|\mu|}(X, \mathcal{A}, \mathbb{C})$ we have*

$$\int f d\overline{\mu} = \overline{\int \overline{f} \, d\mu}, \quad \text{or equivalently} \quad \int \overline{f} \, d\mu = \overline{\int f d\overline{\mu}}.$$

As a consequence, $\overline{\mu}$ is the unique complex measure such that

$$\int f d\overline{\mu} = \overline{\int \overline{f} \, d\mu}, \qquad \text{for all } f \in \mathcal{C}_0(X; \mathbb{C}).$$

Proof. Write $\mu = \mu_1 + i\mu_2$ as above, where μ_1 and μ_2 are real measures. We have $\overline{\mu} = \mu_1 - i\mu_2$, and the measures μ_1 and μ_2 are written as $\mu_1 = \mu_1^+ - \mu_1^-$ and $\mu_2 = \mu_2^+ - \mu_2^-$, where $\mu_1^+, \mu_1^-, \mu_2^+$, and μ_2^-, are real positive measures. Now for any function f integrable for all four positive measures above it is obvious that

$$\int \overline{f} d\mu_i^+ = \overline{\int f d\mu_i^+}, \qquad \int \overline{f} d\mu_i^- = \overline{\int f d\mu_i^-},$$

so

$$\int \overline{f} d\mu_1 = \overline{\int f d\mu_1}, \qquad \int \overline{f} d\mu_2 = \overline{\int f d\mu_2},$$

thus

$$\begin{aligned} \int \overline{f} d\mu &= \int \overline{f} d\mu_1 + i \int \overline{f} d\mu_2 \\ &= \overline{\int f d\mu_1} + i \overline{\int f d\mu_2} \\ &= \overline{\int f d\mu_1 - i \int f d\mu_2} \\ &= \overline{\int f d\overline{\mu}}, \end{aligned}$$

as claimed. Since $\mathcal{C}_0(X; \mathbb{C})$ is obviously contained in $\mathcal{L}^1_{|\mu|}(X, \mathcal{A}, \mathbb{C})$, the last statement follows from Theorem 7.30 (Radon–Riesz III), which will be proven in Section 7.8. ☐

Since the measures $\mu_1^+, \mu_1^-, \mu_2^+$, and μ_2^- are positive *real* measures, they are finite. This immediately implies that the Radon functional φ_μ induced by a complex measure μ *is bounded*. Therefore, complex measures represent only bounded Radon functionals. Actually they represent all of them, which is the object of Section 7.8.

To show the above fact, we need to decompose a bounded Radon functional in terms of (four) positive bounded Radon functionals, and for this we introduce the notion of total variation of a Radon functional.

7.7 Total Variation of a Radon Functional

The notion of total variation of a Radon functional allows the decomposition of a bounded Radon functional into four positive bounded functionals in a way that is similar to the Jordan decomposition of a complex measure. This fact is the key to the representation of a bounded Radon functional by a complex measure.

Recall that for any function $g\colon X \to \mathbb{C}$, we denote by $|g|$ the function $|g|\colon X \to \mathbb{R}$ given by $|g|(x) = |g(x)|$ for all $x \in X$.

The following result is shown in Dieudonné [Dieudonné (2005)] (Chapter XIII, Section 3).

Theorem 7.25. *For any Radon functional $\Phi\colon \mathcal{K}_{\mathbb{C}}(X) \to \mathbb{C}$ on a locally compact space X, there is a smallest positive Radon functional $|\Phi|\colon \mathcal{K}_{\mathbb{C}}(X) \to \mathbb{C}$ such that*

$$|\Phi(f)| \leq |\Phi|(|f|) \quad \text{for all } f \in \mathcal{K}_{\mathbb{C}}(X).$$

The functional $|\Phi|$ is completely defined by its restriction to positive functions $f \geq 0$ in $\mathcal{K}_{\mathbb{R}}(X)$ by

$$|\Phi|(f) = \sup\{|\Phi(g)| \mid g \in \mathcal{K}_{\mathbb{C}}(X),\ |g| \leq f\}.$$

Proof sketch. We know from the remark just after Proposition 7.4 that a Radon functional $\Phi\colon \mathcal{K}_{\mathbb{C}}(X) \to \mathbb{C}$ is completely determined by its restriction $\Phi_{\mathbb{R}}\colon \mathcal{K}_{\mathbb{R}}(X) \to \mathbb{C}$ to the space of real-valued functions in $\mathcal{K}_{\mathbb{R}}(X)$. The first step in the proof of Theorem 7.25 is to show that the formula

$$|\Phi|(f) = \sup\{|\Phi(g)| \mid g \in \mathcal{K}_{\mathbb{C}}(X),\ |g| \leq f\}$$

defined on positive functions $f \geq 0$ in $\mathcal{K}_{\mathbb{R}}(X)$ is yields a finite number. Let K be the support of f, which is compact. Since $|g| \leq f$, the support of g is contained in K, so

$$|\Phi(g)| \leq c_K \left\|g\right\|_\infty \leq c_K \left\|f\right\|_\infty,$$

which shows that $|\Phi|(f)$ is finite. Next we show that $|\Phi|$ is additive, which is left as an exercise.

The second step is to extend $|\Phi|$ to arbitrary functions $f \in \mathcal{K}_{\mathbb{R}}(X)$ by writing $f = f' - f''$, where $f', f'' \in \mathcal{K}_{\mathbb{R}}(X)$ and $f', f'' \geq 0$, by setting

$$|\Phi|(f) = |\Phi|(f') - |\Phi|(f'').$$

This expression does not depend on the decomposition of f because if $f = f_1' - f_1'' = f_2' - f_2''$, then $f_1' + f_2'' = f_1'' + f_2'$, so $|\Phi|(f_1') + |\Phi|(f_2'') = |\Phi|(f_1'') + |\Phi|(f_2')$, which implies $|\Phi|(f_1') - |\Phi|(f_2') = |\Phi|(f_1'') - |\Phi|(f_2'')$.

The last step is to prove that $|\Phi|(\lambda f) = \lambda|\Phi|(f)$, which is clear $\lambda \geq 0$. For $\lambda < 0$, we write $f = f' - f''$ with $f', f'' \geq 0$, and then

$$\begin{aligned}
|\Phi|(\lambda f) &= |\Phi|(\lambda f' - \lambda f'') \\
&= |\Phi|(\lambda f') + |\Phi|(-\lambda f'') \\
&= -|\Phi|(-\lambda f') - \lambda|\Phi|(f'') \\
&= -(-\lambda)|\Phi|(f') - \lambda|\Phi|(f'') \\
&= \lambda(|\Phi|(f') - |\Phi|(f'')) \\
&= \lambda|\Phi|(f).
\end{aligned}$$

In summary, $|\Phi|$ is a positive linear functional. By Proposition 7.4, the functional $|\Phi|$ is a positive Radon functional. □

Definition 7.17. Given any Radon functional $\Phi\colon \mathcal{K}_{\mathbb{C}}(X) \to \mathbb{C}$, the positive Radon functional $|\Phi|$ is called *total variation* (or *absolute value*) of Φ.

If Φ is a positive Radon functional, then

$$|\Phi| = \Phi.$$

Definition 7.18. Given a Radon functional $\Phi\colon \mathcal{K}_{\mathbb{C}}(X) \to \mathbb{C}$, we define the *conjugate* $\overline{\Phi}$ of Φ by

$$\overline{\Phi}(f) = \overline{\Phi(\overline{f})}, \quad f \in \mathcal{K}_{\mathbb{C}}(X).$$

If we write $f = f_1 + if_2$ with $f_1, f_2 \in \mathcal{K}_{\mathbb{R}}(X)$, then we have

$$\begin{aligned}
\overline{\Phi}(f) = \overline{\Phi}(f_1 + if_2) &= \overline{\Phi((\overline{f_1 + if_2}))} = \overline{\Phi(f_1 - if_2)} \\
&= \overline{(\Phi(f_1) - i\Phi(f_2))} = \overline{\Phi(f_1)} + i\overline{\Phi(f_2)}.
\end{aligned}$$

Definition 7.19. A Radon functional $\Phi\colon \mathcal{K}_{\mathbb{C}}(X) \to \mathbb{C}$ is *real* if $\overline{\Phi} = \Phi$.

Proposition 7.26. *A Radon functional $\Phi\colon \mathcal{K}_{\mathbb{C}}(X) \to \mathbb{C}$ is real iff its restriction $\Phi_{\mathbb{R}}$ to $\mathcal{K}_{\mathbb{R}}(X)$ is a real-valued function $\Phi_{\mathbb{R}}\colon \mathcal{K}_{\mathbb{R}}(X) \to \mathbb{R}$.*

Proof. In view of the above computation, a Radon functional Φ is real iff

$$\Phi(f_1) + i\Phi(f_2) = \overline{\Phi(f_1)} + i\overline{\Phi(f_2)}$$

for all $f_1, f_2 \in \mathcal{K}_{\mathbb{R}}(X)$, which by setting $f_2 = 0$ or $f_1 = 0$ means that $\Phi(f_i) \in \mathbb{R}$ for all $f_i \in \mathcal{K}_{\mathbb{R}}(X)$, for $i = 1, 2$. Equivalently, a Radon functional $\Phi\colon \mathcal{K}_{\mathbb{C}}(X) \to \mathbb{C}$ is real iff its restriction $\Phi_{\mathbb{R}}$ to $\mathcal{K}_{\mathbb{R}}(X)$ is a real-valued function $\Phi_{\mathbb{R}}\colon \mathcal{K}_{\mathbb{R}}(X) \to \mathbb{R}$. □

Since a Radon functional is Φ is completely determined by its restriction $\Phi_{\mathbb{R}}$ to $\mathcal{K}_{\mathbb{R}}(X)$, we often think of a real Radon functional as a linear map $\Phi\colon \mathcal{K}_{\mathbb{R}}(X) \to \mathbb{R}$.

Definition 7.20. Given a Radon functional $\Phi\colon \mathcal{K}_{\mathbb{C}}(X) \to \mathbb{C}$, we define Φ_r and Φ_i by

$$\Phi_r = \frac{1}{2}(\Phi + \overline{\Phi}), \quad \Phi_i = \frac{1}{2i}(\Phi - \overline{\Phi}).$$

It is immediately verified that Φ_r and Φ_i are *real* Radon functionals such that

$$\Phi = \Phi_r + i\Phi_i, \quad \overline{\Phi} = \Phi_r - i\Phi_i.$$

We also have

$$|\Phi_r| \leq |\Phi|, \quad |\Phi_i| \leq |\Phi|, \quad |\Phi| \leq |\Phi_r| + |\Phi_i|.$$

Definition 7.21. If $\Phi\colon \mathcal{K}_{\mathbb{C}}(X) \to \mathbb{R}$ is a real Radon functional, then as in the case of real measures we can define Φ^+ and Φ^- by

$$\Phi^+ = \frac{1}{2}(|\Phi| + \Phi), \quad \Phi^- = \frac{1}{2}(|\Phi| - \Phi).$$

It is immediately checked that Φ^+ and Φ^- are *positive* Radon functionals, and we have

$$\Phi = \Phi^+ - \Phi^-, \quad |\Phi| = \Phi^+ + \Phi^-.$$

In the end, we have the following decomposition result analogous to the Jordan decomposition for complex measures.

Proposition 7.27. *Every Radon functional $\Phi\colon \mathcal{K}_{\mathbb{C}}(X) \to \mathbb{C}$ can be expressed in terms of four positive Radon functionals:*

$$\Phi = \Phi_r^+ - \Phi_r^- + i(\Phi_i^+ - \Phi_i^-).$$

By the Radon–Riesz I theorem (Theorem 7.8), there exist four positive σ-Radon measures m_1, m_2, m_3, m_4 such that

$$\Phi(f) = \int f dm_1 - \int f dm_2 + i\left(\int f dm_3 - \int f dm_4\right) \quad \text{for all } f \in \mathcal{K}_{\mathbb{C}}(X).$$

It is tempting to define the complex measure m by

$$m = m_1 - m_2 + i(m_3 - m_4),$$

but there is a problem, which is that the positive measures m_i may take the value $+\infty$, so expressions of the form $+\infty - (+\infty)$ may arise, but they do not make any sense!

We are not aware of a way around this problem in general. If X is compact, then the Radon–Riesz II theorem yields positive Radon measures m_i such that $m_i(X)$ is finite for $i = 1, \ldots, 4$, in which case the expression m is indeed a measure. It is even possible to define a bijective correspondence by adding disjointness conditions on the subsets over which the m_i are concentrated. Such results are given in Malliavin [Malliavian *et al.* (1995)] (Chapter II, Section 5).

Another situation where m is a complex measure is the case where the Radon functional Φ is bounded (continuous). This is the object of the next section.

7.8 The Radon–Riesz Theorem and Bounded Radon Functionals

Let $\Phi\colon \mathcal{K}_{\mathbb{C}}(X) \to \mathbb{C}$ be a bounded Radon functional. In this case the operator norm $\|\Phi\|$ is finite. Recall that

$$\begin{aligned}\|\Phi\| &= \sup\{|\Phi(f)| \mid f \in \mathcal{K}_{\mathbb{C}}(X),\ \|f\|_\infty \leq 1\}\\ &= \sup\{|\Phi(f)| \mid f \in \mathcal{K}_{\mathbb{C}}(X),\ \|f\|_\infty = 1\}.\end{aligned}$$

The following result is shown in Dieudonné [Dieudonné (2005)] (Chapter VII, Section 20).

Proposition 7.28. *Given a Radon functional $\Phi\colon \mathcal{K}_{\mathbb{C}}(X) \to \mathbb{C}$, the norm $\|\Phi\|$ is finite, that is, Φ is bounded, iff $|\Phi|$ is bounded. In this case, $\|\Phi\| = \||\Phi|\|$.*

We deduce that $\Phi = \Phi_r + i\Phi_i$ is bounded iff Φ_r and Φ_i are bounded (see Definition 7.20). But we also see that a real bounded Radon functional $\Psi = \Psi^+ - \Psi^-$ is bounded iff the positive Radon functionals Ψ^+ and Ψ^- are bounded (see Definition 7.21).

Proposition 7.29. *A Radon functional Φ is bounded iff the positive Radon functional $\Phi_r^+, \Phi_r^-, \Phi_i^+, \Phi_i^-$ are bounded.*

If m_1, m_2, m_3, m_4 are the positive σ-Radon measures representing $\Phi_r^+, \Phi_r^-, \Phi_i^+, \Phi_i^-$ given by the Radon–Riesz I theorem (Theorem 7.8), it turns out that they are all finite measures, so $m = m_1 - m_2 + i(m_3 - m_4)$ is a complex measure, and it represents Φ on functions in $\mathcal{C}_0(X)$. In order to state a suitable version of the Radon–Riesz correspondence, we need the following definition.

Definition 7.22. Let X be a locally compact (Hausdorff) space. A complex measure μ on the σ-algebra $\mathcal{B}$ of Borel sets of X is a *regular complex Borel measure* if the positive measure $|\mu|$ is a finite Radon measure, that is, a positive Borel measure that is regular and finite ($|\mu|(X)$ is finite). We denote the vector space of regular complex Borel measures $\mathcal{M}^1_{\mathrm{reg},\mathbb{C}}(X)$. See Figure 7.7.

Since $|\mu|(X)$ is finite, the measure $|\mu|(K)$ of every compact subset K of X is also finite (since X is Hausdorff, every compact subset K of X is closed and thus measurable, and since $K \subseteq X$, we have $|\mu|(K) \leq |\mu|(X)$). Thus the positive Borel measure $|\mu|$ is locally finite.

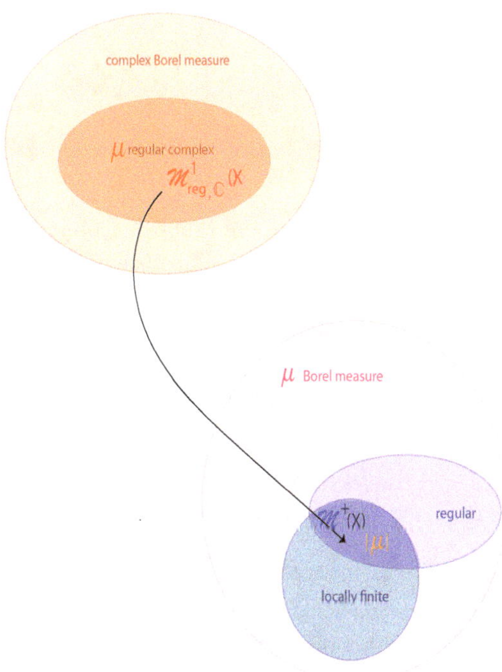

Fig. 7.7 A Venn diagram representation of $\mathcal{M}^1_{\mathrm{reg},\mathbb{C}}(X)$.

We have the following beautiful theorem. Theorem 7.30 is also often referred to as the *Riesz representation theorem*, which is somewhat confusing.

Theorem 7.30. *(Radon–Riesz Correspondence, III) Let X be a locally compact (Hausdorff) space. There are bijections $m\colon \mathrm{M}^1(X) \to \mathcal{M}^1_{\mathrm{reg},\mathbb{C}}(X)$ and $\varphi\colon \mathcal{M}^1_{\mathrm{reg},\mathbb{C}}(X) \to \mathrm{M}^1(X)$ between the Banach space $\mathrm{M}^1(X) = \mathcal{C}_0(X,\mathbb{C})'$ of bounded Radon functionals, the dual of the space $\mathcal{C}_0(X,\mathbb{C})$ of continuous functions that tend to zero at infinity, and the Banach space $\mathcal{M}^1_{\mathrm{reg},\mathbb{C}}(X)$ of regular complex Borel measures. For every regular complex Borel measure $m \in \mathcal{M}^1_{\mathrm{reg},\mathbb{C}}(X)$, the bounded Radon functionals $\varphi(m) = \varphi_m$ is given by*

$$\varphi_m(f) = \int f dm, \quad \text{for all } f \in \mathcal{C}_0(X,\mathbb{C}).$$

For every bounded Radon functional $\Phi \in \mathrm{M}^1(X) = \mathcal{C}_0(X,\mathbb{C})'$, the regular complex Borel measure m_Φ represents Φ in the sense that

$$\begin{aligned}\Phi(f) &= \int f dm_\Phi \\ &= \int f d(m_\Phi)^+_r - \int f d(m_\Phi)^-_r + i\left(\int f d(m_\Phi)^+_i - \int f d(m_\Phi)^-_i\right)\end{aligned}$$

or all $f \in \mathcal{C}_0(X,\mathbb{C})$. Furthermore, these bijections are norm preserving, that is, $\|\Phi\| = \|m_\Phi\| = |m_\Phi|(X)$. See Figure 7.8.

Theorem 7.30 is proven in Lang [Lang (1996)] (Chapter IX, §4, Theorem 4.2), Rudin [Rudin (1987)] (Chapter 6 Theorem 6.19), Folland [Folland (1999)] (Chapter 7, Theorem 7.17), and Marle [Marle (1974)] (Chapter 9, Section 7, Proposition 9.7.3). The proof is quite involved. Among other things it uses Lusin's theorem (Theorem 7.9). It also uses the corollary of the Radon–Nikodym theorem (Theorem 7.21) and the fact that $\mathcal{K}_{\mathbb{C}}(X)$ is dense in $\mathcal{L}_{|\mu|}(X,\mathcal{B},\mathbb{C})$ to prove injectivity.

To prove surjectivity, by Proposition 7.27 we express the bounded Radon functional $\Phi\colon \mathcal{K}_{\mathbb{C}}(X) \to \mathbb{C}$ in terms of four positive Radon functionals:

$$\Phi = \Phi^+_r - \Phi^-_r + i(\Phi^+_i - \Phi^-_i).$$

By Proposition 7.29, these positive Radon functionals are bounded. By the Radon–Riesz theorem I (Theorem 7.8), there exist four positive σ-Radon measures m_1, m_2, m_3, m_4 such that

$$\Phi(f) = \int f dm_1 - \int f dm_2 + i\left(\int f dm_3 - \int f dm_4\right) \quad \text{for all } f \in \mathcal{K}_{\mathbb{C}}(X).$$

The reason why the σ-Radon measure m corresponding to a positive bounded Radon functional Φ is finite is that this measure is inner regular, that is,

$$m_\Phi(E) = \sup\{\mu(K) \mid K \subseteq E,\ K \text{ compact}\}$$

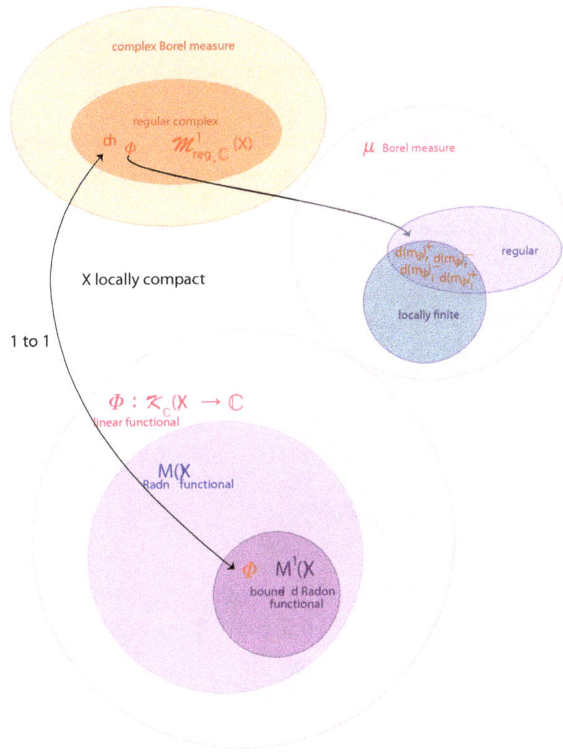

Fig. 7.8 Radon-Riesz Correspondence, Version 3.

for every $E \in \mathcal{B}$. We use this to compute $m_\Phi(X)$. For every compact subset K, by Proposition A.39, there is a continuous function $f\colon X \to [0,1]$ of compact support such that $f(x) = 1$ for all $x \in K$. Then since Φ is bounded we have

$$m_\Phi(K) \leq \int f dm_\Phi = \Phi(f) \leq \|\Phi\| \, \|f\|_\infty = \|\Phi\|$$

since f has maximum value 1. Therefore,

$$m_\Phi(X) = \sup\{\mu(K) \mid K \subseteq X, \, K \text{ compact}\} \leq \|\Phi\|$$

is indeed finite. Since $m_\Phi(X)$ is finite, every measurable subset has finite measure and so the σ-regular measure m_Φ is actually regular.

We also need to check that $\varphi_m(f) = \int f dm$ is finite for every function $f \in \mathcal{C}_0(X;\mathbb{C})$ and every positive finite Borel measure m. Since $\mathcal{C}_0(X;\mathbb{C})$

is the closure of $\mathcal{K}_{\mathbb{C}}(X)$, there is a sequence (f_n) of functions $f_n \in \mathcal{K}_{\mathbb{C}}(X)$ that converges to f according to the sup norm, and thus converges pointwise to f. Also f is a bounded function, so there is some $M > 0$ such that $|f_n| \leq M$ for all $n \geq 1$. Since $m(X)$ is finite, the constant function M is integrable, and the continuous functions f_n are integrable. By the dominated convergence theorem (Theorem 5.34), f is integrable.

Theorem 7.30 plays a crucial role in defining the notion of convolutions of two measures in $\mathcal{M}^1_{\mathrm{reg},\mathbb{C}}(X)$. We will need the following simple fact.

Proposition 7.31. *Let X be any locally compact space, and let μ be any positive Borel measure on $\mathcal{B}$. For any function $f \in \mathcal{L}^1_\mu(X, \mathcal{B}, \mathbb{C})$, the functional $\Phi_{f,\mu} \colon \mathcal{C}_0(X; \mathbb{C}) \to \mathbb{C}$ given by*

$$\Phi_{f,\mu}(g) = \int fg\, d\mu \quad \text{for all } g \in \mathcal{C}_0(X; \mathbb{C})$$

is a bounded Radon functional.

Proof. Since $f \in \mathcal{L}^1_\mu(X, \mathcal{B}, \mathbb{C})$ and g is continuous, g is measurable, and $|g|$ is bounded by some $M > 0$, so by Proposition 5.36(1) $fg \in \mathcal{L}^1_\mu(X, \mathcal{B}, \mathbb{C})$. We have

$$|\Phi_{f,\mu}(g)| = \left| \int fg\, d\mu \right| \leq \int |fg|\, d\mu = \int |f||g|\, d\mu \leq \|g\|_\infty \int |f| d\mu,$$

which shows that $\Phi_{f,\mu}$ is bounded. ∎

By Theorem 7.30, the bounded Radon functional $\Phi_{f,\mu}$ corresponds to a unique regular complex Borel measure m such that

$$\int fg\, d\mu = \int g\, dm \quad \text{for all } g \in \mathcal{C}_0(X; \mathbb{C}).$$

The measure m is usually denoted by $f d\mu$. Proposition 7.31 gives us an embedding of $\mathcal{L}^1_\mu(X, \mathcal{B}, \mathbb{C})$ into $\mathcal{M}^1_{\mathrm{reg},\mathbb{C}}(X)$ as stated in the next proposition.

Proposition 7.32. *Let X be a locally compact space. For every positive Borel measure μ on $\mathcal{B}$, the map $f \mapsto f d\mu$ is a norm-preserving embedding of $\mathcal{L}^1_\mu(X, \mathcal{B}, \mathbb{C})$ into the space $\mathcal{M}^1_{\mathrm{reg},\mathbb{C}}(X)$ of regular complex Borel measures on X, with the property that*

$$\int fg\, d\mu = \int g\, f d\mu \quad \text{for all } g \in \mathcal{C}_0(X; \mathbb{C}).$$

The reason why the embedding is norm-preserving is quite subtle. By Theorem 7.30, $\|f d\mu\| = \|\Phi_{f,\mu}\|$, where $\Phi_{f,\mu}\colon \mathcal{C}_0(X;\mathbb{C}) \to \mathbb{C}$ is the bounded Radon functional given by

$$\Phi_{f,\mu}(g) = \int fg\, d\mu \quad \text{for all } g \in \mathcal{C}_0(X;\mathbb{C}).$$

By an exercise in Folland [Folland (1999)] (Chapter 7, Section 7.2, Exercise 9), the measure $f d\mu$ associated with the functional $\Phi_{f,\mu}$ is equal to the measure μ_f of Proposition 7.16, with

$$\mu_f(A) = \int_A f\, d\mu, \quad A \in \mathcal{B},$$

when f is a positive continuous function. In this case, $\mu_f = f d\mu$ is a positive Radon measure. By Proposition 7.20,

$$\|\mu_f\| = \|f\|_1,$$

so

$$\|\Phi_{f,\mu}\| = \|f d\mu\| = \|\mu_f\| = \|f\|_1.$$

This fact is extended to continuous functions $f\colon X \to \mathbb{C}$ by writing $f = f_1 - f_2 + i(f_3 - f_4)$, where f_1, f_2, f_3, f_4 are four positive continuous functions. Finally, since $\mathcal{K}_{\mathbb{C}}(X)$ is dense in $\mathcal{L}^1(X, \mathcal{B}, \mathbb{C})$, the fact that $\|f d\mu\| = \|\Phi_{f,\mu}\| = \|f\|_1$ is extended to functions in $\mathcal{L}^1(X, \mathcal{B}, \mathbb{C})$.

This embedding is technically important because if X is a locally compact group and if μ is a Haar measure, convolution can be defined on both $\mathcal{L}^1_\mu(X, \mathcal{B}, \mathbb{C})$ and $\mathcal{M}^1_{\mathrm{reg},\mathbb{C}}(X)$, but there is no identity element for convolution on $\mathcal{L}^1_\mu(X, \mathcal{B}, \mathbb{C})$ while there is one for convolution on $\mathcal{M}^1_{\mathrm{reg},\mathbb{C}}(X)$. Technically $\mathcal{M}^1_{\mathrm{reg},\mathbb{C}}(X)$ is a unital normed Banach algebra but $\mathcal{L}^1_\mu(X, \mathcal{B}, \mathbb{C})$ is a nonunital normed Banach algebra. This point will be significant in Chapter 9 and in Chapter 10.

7.9 Problems

Problem 7.1. Verify that $\mathrm{M}(X)$, the set of Radon linear functionals, is a vector space. Verify that $\mathrm{M}^1(X)$, the set of continuous Radon functionals, is also a vector space. Explain why $\mathrm{M}^+(X)$, the set of positive linear functionals, is *not* a vector space.

Problem 7.2. Refer to either Rudin [Rudin (1987)] (Chapter 2), Lang [Lang (1996)] (Chapter IX), Folland [Folland (1999)] (Chapter 7, Theorem 7.2), or Schwartz [Schwartz (1993a)] (Chapters 5 and 7) to complete the proof sketch of the Riesz representation theorem, Theorem 7.6.

Problem 7.3. Prove Theorem 7.9, Lusin's theorem. Hint: See Rudin [Rudin (1987)] (Chapter 2, Theorem 2.24) or Lang [Lang (1996)] (Chapter IX, Theorem 3.3).

Problem 7.4. Prove Theorem 7.10. Hint: Theorem 7.10 is proven in Rudin [Rudin (1987)] (Chapter 3, Theorem 3.14) or Lang [Lang (1996)] (Chapter IX, Theorem 3.1).

Problem 7.5. Prove Proposition 7.13. Hint: See Rudin [Rudin (1987)] (Chapter 2, Theorem 2.18).

Problem 7.6. Prove Theorem 7.14. Hint: See Rudin [Rudin (1987)] (Chapter 2, Theorem 2.17).

Problem 7.7. Prove Proposition 7.16. Hint: See Marle [Marle (1974)] (Chapter 2, Proposition 2.5.2).

Problem 7.8. Verify that the set of complex measures is a vector space. Verify that the set of real measures is also a vector space.

Problem 7.9. Prove Theorem 7.17. Hint: See Rudin [Rudin (1987)] (Chapter 6, Theorem 6.2) or Lang [Lang (1996)] (Chapter VII, Theorem 3.1).

Problem 7.10. Prove Theorem 7.18. Hint: See Rudin [Rudin (1987)] (Chapter 6, Theorem 6.4) or Lang [Lang (1996)] (Chapter VII, Theorem 3.2).

Problem 7.11. Prove that $\mathbb{C}\mathcal{M}^1(X, \mathcal{A})$ is a Banach space.

Problem 7.12. Prove Proposition 7.20. Hint: See Lang [Lang (1996)] (Chapter VII, §3, Theorem 3.3).

Problem 7.13. Prove Theorem 7.21. Hint: See Rudin [Rudin (1987)] (Chapter 6, Theorem 6.12) or Lang [Lang (1996)] (Chapter VII, §2 and §4).

Problem 7.14. Advanced Exercise: Prove Theorem 7.22, the Hahn-Jordan decomposition.

Problem 7.15. Complete the details of the proof sketch of Theorem 7.25.

Problem 7.16. Prove Proposition 7.28. Hint: See Dieudonné [Dieudonné (2005)] (Chapter VII, Section 20).

Problem 7.17. Advanced Exercise: Complete the details of Radon–Riesz Correspondence, Theorem 7.30. Hint: See Lang [Lang (1996)] (Chapter IX, §4, Theorem 4.2), Rudin [Rudin (1987)] (Chapter 6 Theorem 6.19), Folland [Folland (1999)] (Chapter 7, Theorem 7.17), or Marle [Marle (1974)] (Chapter 9, Section 7, Proposition 9.7.3).

Chapter 8

The Haar Measure and Convolution

Let G be a locally compact group. Haar proved (1933) the remarkable fact that there is a positive σ-regular locally finite Borel measure μ on G such that $\mu(U) > 0$ for every nonempty open subset U, and such that μ is left-invariant, which means that

$$\mu(A) = \mu(sA) \quad \text{for all } s \in G \text{ and all } A \in \mathcal{B},$$

where $\mathcal{B}$ is the σ-algebra of Borel sets on G. Furthermore, such a left-invariant measure is unique up to a positive scalar.

Actually, Haar proved the existence of a left-invariant measure in a special case. This result was established in full generality later by André Weil [Weil (1979)]. All proofs we are aware of (Weil [Weil (1979)], Halmos [Halmos (1978)], Bourbaki [Bourbaki (1965b)], Dieudonné [Dieudonné (2005)], Lang [Lang (1996)], Folland [Folland (1995)]) make use of Haar's original clever idea (1933). Except for Halmos who constructs directly a measure (as Haar did), all the other proofs are essentially André Weil's proof (which constructs a Haar functional) from his famous little book [Weil (1979)] first published in 1940.

In this chapter we sketch the existence of the (left) Haar measure, providing most details, and we also prove its uniqueness up to a scalar; see Sections 8.2, 8.3, 8.4. Some Examples are given in Section 8.5.

For any $s \in G$ and any measure μ on G, let $\rho_s(\mu)$ be the measure given by

$$(\rho_s(\mu))(A) = \mu(As) \quad \text{for all } A \in \mathcal{B}.$$

If μ is a left Haar measure, then it is easy to see that $\rho_s(\mu)$ is a left Haar measure, so by uniqueness up to a scalar, there is a unique positive number $\Delta(s)$ such that

$$\rho_s(\mu) = \Delta(s)\mu \qquad (*)$$

The function $\Delta\colon G \to \mathbb{R}_+^*$ (given by $\Delta(s)$ for every $s \in G$) is called the *modular function* of G. We investigate properties of the modular function in Section 8.6. We say that the group G is *unimodular* if $\Delta(s) = 1$ for all $s \in G$, equivalently, if and only if a left Haar measure is also a right Haar measure. If G is abelian, compact, or a connected semisimple Lie group, then G is unimodular. More examples of Haar measures are given in Section 8.7.

Let G be a locally compact group, and let $u\colon G \to G$ be an automorphism of G. For every left Haar measure μ, define the measure $u^{-1}(\mu)$ by

$$(u^{-1}(\mu))(A) = \mu(u(A)), \quad \text{for all } A \in \mathcal{B}.$$

It can be shown that there is a unique positive number $\text{mod}(u)$ such that

$$u^{-1}(\mu) = \text{mod}(u)\mu$$

for all left Haar measures μ. The number $\text{mod}(u)$ is called the *modulus of the automorphism* u. Properties of the modulus of an automorphism are discussed Section 8.8. As an application, we obtain formulae for the measure (volume) of a parallelotope and of a simplex.

Some applications of the Haar measure are discussed in Section 8.9. In particular, we prove Theorem 8.36, a basic tool in representation theory.

Let G be a locally compact group, let X be a locally compact space, and let $\cdot\colon G \times X \to X$ be a continuous left action of G on X. A Borel measure μ on X is *G-invariant* if

$$\mu(s^{-1} \cdot A) = \mu(A) \quad \text{for all } s \in G \text{ and all } A \in \mathcal{B}.$$

Our goal is to find sufficient conditions to ensure that X has some G-invariant measure. We will consider the case where $X = G/H$, with the left action of G on G/H given by

$$a \cdot (bH) = abH, \qquad a, b \in G.$$

In this case, by Proposition 8.6, the space X is also locally compact (and Hausdorff).

A G-invariant measure on G/H does not always exist. It turns out that there is a necessary and sufficient condition for a G-invariant σ-Radon measure to exist on G/H in terms of Δ_G and Δ_H: Δ_H must the equal to the restriction of Δ_G on H. This topic is discussed in Section 8.10.

One of the main applications of the Haar measure is the definition of the notion of convolution on a locally compact group. Recall that $\mathcal{M}^1_{\mathrm{reg},\mathbb{C}}(G)$ denotes the Banach space of complex regular Borel measures on G (see Definition 7.22), and that $\mathrm{L}^1_\lambda(G, \mathcal{B}, \mathbb{C})$ denotes the space of integrable functions on the measure space $(G, \mathcal{B}, \lambda)$, where $\mathcal{B}$ is the σ-algebra of Borel sets of G. To simplify notation, we write $\mathcal{M}^1(G)$ for $\mathcal{M}^1_{\mathrm{reg},\mathbb{C}}(G)$, and $\mathrm{L}^1(G)$ for $\mathrm{L}^1_\lambda(G, \mathcal{B}, \mathbb{C})$. The vector space $\mathcal{M}^1(G)$ is a Banach space with the norm $\|\mu\| = |\mu|(G)$, and $\mathrm{L}^1(G)$ is a Banach space with the L^1-norm. There are three flavors of convolutions but we will use mostly the first two of them:

(1) Convolutions $\mu * \nu$ of two measures $\mu, \nu \in \mathcal{M}^1(G)$. This makes $\mathcal{M}^1(G)$ into a Banach algebra with identity and with an involution.
(2) Convolution $f * g$ of two functions $f, g \in \mathrm{L}^1(G)$, which makes $\mathrm{L}^1(G)$ into a Banach algebra with involution, but without a multiplicative unit element, unless G is discrete. A closely related concept heavily used in signal processing and computer vision is the notion of *cross-correlation* $k \star f$ of two functions k and f. The idea is that f is a signal, say an image, and k is a pattern. Then $k \star f$ is a measure of how much the pattern k, when moved around by all transformations $g \in G$, occurs in f. The cross-correlation $k \star f$ is equal to the convolution $f * \overline{\check{k}}$ (where $\overline{\check{k}}$ is the reflected kernel k).
(3) There is also a notion of convolution $\mu * f$ of a measure $\mu \in \mathcal{M}^1(G)$ and of a function $f \in \mathrm{L}^1(G)$, and convolution $f * \mu$ of a function $f \in \mathrm{L}^1(G)$ and a measure $\mu \in \mathcal{M}^1(G)$.

These notions of convolution (and cross-correlation) are discussed in Sections 8.11, 8.12, 8.13.

Convolution applied to functions and measures can be used as a regularization (or filtering) process; see Section 8.14.

8.1 Topological Groups

Since locally compact groups (and Lie groups) are topological groups, it is useful to gather a few basic facts about topological groups.

Definition 8.1. A set G is a *topological group* iff

(a) G is a Hausdorff topological space;
(b) G is a group (with identity 1);

(c) Multiplication $\cdot\colon G \times G \to G$, and the inverse operation $G \longrightarrow G\colon g \mapsto g^{-1}$, are continuous, where $G \times G$ has the product topology.

It is easy to see that the two requirements of Condition (c) are equivalent to

(c′) The map $G \times G \longrightarrow G\colon (g, h) \mapsto gh^{-1}$ is continuous.

Proposition 8.1. *If G is a topological group and H is any subgroup of G, then the closure $\overline{H}$ of H is a subgroup of G. If H is a normal subgroup of G, then $\overline{H}$ is also a normal subgroup of G.*

Proof. We use the fact that if $f\colon X \to Y$ is a continuous map between two topological spaces X and Y, then $f(\overline{A}) \subseteq \overline{f(A)}$ for any subset A of X. For any $a \in \overline{A}$, we need to show that for any open subset $W \subseteq Y$ containing $f(a)$, we have $W \cap f(A) \neq \emptyset$. Since f is continuous, $V = f^{-1}(W)$ is an open subset containing a, and since $a \in \overline{A}$, we have $f^{-1}(W) \cap A \neq \emptyset$, so there is some $x \in f^{-1}(W) \cap A$, which implies that $f(x) \in W \cap f(A)$, so $W \cap f(A) \neq \emptyset$, as desired. The map $f\colon G \times G \to G$ given by $f(x, y) = xy^{-1}$ is continuous, and since H is a subgroup of G, $f(H \times H) \subseteq H$. By the above property, if $a \in \overline{H}$ and if $b \in \overline{H}$, that is, $(a, b) \in \overline{H \times H}$, then $f(a, b) = ab^{-1} \in \overline{H}$, which shows that $\overline{H}$ is a subgroup of G.

For every $g \in G$, the map $C_g\colon G \to G$ given by $C_g(x) = gxg^{-1}$ for all $x \in G$ is continuous, and if H is a normal subgroup of G, then $C_g(H) \subseteq H$. It follows that $C_g(\overline{H}) \subseteq \overline{H}$ for all $g \in G$, which means that $\overline{H}$ is a normal subgroup of G. ☐

Given a topological group G, for every $a \in G$ we define the *left translation* L_a as the map $L_a\colon G \to G$ such that $L_a(b) = ab$, for all $b \in G$, and the *right translation* R_a as the map $R_a\colon G \to G$ such that $R_a(b) = ba$, for all $b \in G$. Observe that $L_{a^{-1}}$ is the inverse of L_a and similarly, $R_{a^{-1}}$ is the inverse of R_a. As multiplication is continuous, we see that L_a and R_a are continuous. Moreover, since they have a continuous inverse, they are homeomorphisms. As a consequence, if U is an open subset of G, then so is $gU = L_g(U)$ (resp. $Ug = R_gU$), for all $g \in G$. Therefore, the topology of a topological group is *determined* by the knowledge of the open subsets containing the identity 1.

Given any subset $S \subseteq G$, let $S^{-1} = \{s^{-1} \mid s \in S\}$; let $S^0 = \{1\}$, and $S^{n+1} = S^n S$, for all $n \geq 0$. Property (c) of Definition 8.1 has the following useful consequences, which shows there exists an open set containing 1 which has a special symmetrical structure.

Proposition 8.2. *If G is a topological group and U is any open subset containing 1, then there is some open subset $V \subseteq U$, with $1 \in V$, so that $V = V^{-1}$ and $V^2 \subseteq U$. Furthermore, $\overline{V} \subseteq U$.*

Proof. Since multiplication $G \times G \longrightarrow G$ is continuous and $G \times G$ is given the product topology, there are open subsets U_1 and U_2, with $1 \in U_1$ and $1 \in U_2$, so that $U_1 U_2 \subseteq U$. Let $W = U_1 \cap U_2$ and $V = W \cap W^{-1}$. Then V is an open set containing 1, and clearly $V = V^{-1}$ and $V^2 \subseteq U_1 U_2 \subseteq U$. If $g \in \overline{V}$, then gV is an open set containing g (since $1 \in V$) and thus, $gV \cap V \neq \emptyset$. This means that there are some $h_1, h_2 \in V$ so that $gh_1 = h_2$, but then, $g = h_2 h_1^{-1} \in VV^{-1} = VV \subseteq U$. □

Definition 8.2. A subset U containing 1 and such that $U = U^{-1}$ is called *symmetric*.

Proposition 8.2 is used in the proofs of many the propositions and theorems on the structure of topological groups. For example, it is key in verifying the following proposition regarding discrete topological subgroups.

Definition 8.3. A subgroup H of a topological group G is *discrete* iff the induced topology on H is discrete; that is, for every $h \in H$, there is some open subset U of G so that $U \cap H = \{h\}$.

Proposition 8.3. *If G is a topological group and H is a discrete subgroup of G, then H is closed.*

Proof. As H is discrete, there is an open subset U of G so that $U \cap H = \{1\}$, and by Proposition 8.2, we may assume that $U = U^{-1}$. Our goal is to show $H = \overline{H}$. Clearly $H \subset \overline{H}$. Thus it remains to show $\overline{H} \subset H$. If $g \in \overline{H}$, as gU is an open set containing g, we have $gU \cap H \neq \emptyset$. Consequently, there is some $y \in gU \cap H = gU^{-1} \cap H$, so $g \in yU$ with $y \in H$. We claim that $yU \cap H = \{y\}$. Note that $x \in yU \cap H$ means $x = yu_1$ with $yu_1 \in H$ and $u_1 \in U$. Since H is a subgroup of G and $y \in H$, $y^{-1}yu_1 = u_1 \in H$. Thus $u_1 \in U \cap H$, which implies $u_1 = 1$ and $x = yu_1 = y$, and we have

$$g \in yU \cap \overline{H} \subseteq \overline{yU \cap H} = \overline{\{y\}} = \{y\}$$

since G is Hausdorff. Therefore, $g = y \in H$. □

Using Proposition 8.2, we can give a very convenient characterization of the Hausdorff separation property in a topological group.

Proposition 8.4. *If G is a topological group, then the following properties are equivalent:*

(1) G is Hausdorff;
(2) The set $\{1\}$ is closed;
(3) The set $\{g\}$ is closed, for every $g \in G$.

Proof. The implication (1) $\longrightarrow$ (2) is true in any Hausdorff topological space. We just have to prove that $G - \{1\}$ is open, which goes as follows: For any $g \neq 1$, since G is Hausdorff, there exists disjoint open subsets U_g and V_g, with $g \in U_g$ and $1 \in V_g$. Thus, $\bigcup U_g = G - \{1\}$, showing that $G - \{1\}$ is open. Since L_g is a homeomorphism, (2) and (3) are equivalent. Let us prove that (3) $\longrightarrow$ (1). Let $g_1, g_2 \in G$ with $g_1 \neq g_2$. Then, $g_1^{-1}g_2 \neq 1$ and if U and V are disjoint open subsets such that $1 \in U$ and $g_1^{-1}g_2 \in V$, then $g_1 \in g_1U$ and $g_2 \in g_1V$, where g_1U and g_1V are still open and disjoint. Thus, it is enough to separate 1 and $g \neq 1$. Pick any $g \neq 1$. If every open subset containing 1 also contained g, then 1 would be in the closure of $\{g\}$, which is absurd since $\{g\}$ is closed and $g \neq 1$. Therefore, there is some open subset U such that $1 \in U$ and $g \notin U$. By Proposition 8.2, we can find an open subset V containing 1, so that $VV \subseteq U$ and $V = V^{-1}$. We claim that V and gV are disjoint open sets with $1 \in V$ and $g \in gV$.

Since $1 \in V$, it is clear that $g \in gV$. If we had $V \cap gV \neq \emptyset$, then by the last sentence in the proof of Proposition 8.2 we would have $g \in VV^{-1} = VV \subseteq U$, a contradiction. $\square$

If H is a subgroup of G (not necessarily normal), we can form the set of left cosets G/H, and we have the projection $p\colon G \to G/H$, where $p(g) = gH = \overline{g}$. If G is a topological group, then G/H can be given the *quotient topology*, where a subset $U \subseteq G/H$ is open iff $p^{-1}(U)$ is open in G. With this topology, p is continuous. The trouble is that G/H is not necessarily Hausdorff. However, we can neatly characterize when this happens.

Proposition 8.5. *If G is a topological group and H is a subgroup of G, then the following properties hold:*

(1) The map $p\colon G \to G/H$ is an open map, which means that $p(V)$ is open in G/H whenever V is open in G.
(2) The space G/H is Hausdorff iff H is closed in G.
(3) If H is open, then H is closed and G/H has the discrete topology (every subset is open).

(4) The subgroup H is open iff $1 \in \overset{\circ}{H}$ (i.e., there is some open subset U so that $1 \in U \subseteq H$).

Proof. (1) Observe that if V is open in G, then $VH = \bigcup_{h \in H} Vh$ is open, since each Vh is open (as right translation is a homeomorphism). However, it is clear that

$$p^{-1}(p(V)) = VH,$$

i.e., $p^{-1}(p(V))$ is open which, by definition of the quotient topology, means that $p(V)$ is open.

(2) If G/H is Hausdorff, then by Proposition 8.4, every point of G/H is closed, i.e., each coset gH is closed, so H is closed. Conversely, assume H is closed. Let $\overline{x}$ and $\overline{y}$ be two distinct point in G/H and let $x, y \in G$ be some elements with $p(x) = \overline{x}$ and $p(y) = \overline{y}$. As $\overline{x} \neq \overline{y}$, the elements x and y are not in the same coset, so $x \notin yH$. As H is closed, so is yH, and since $x \notin yH$, there is some open containing x which is disjoint from yH, and we may assume (by translation) that it is of the form Ux, where U is an open containing 1. By Proposition 8.2, there is some open V containing 1 so that $VV \subseteq U$ and $V = V^{-1}$. Thus, we have

$$V^2 x \cap yH = \emptyset$$

and in fact,

$$V^2 xH \cap yH = \emptyset,$$

since H is a group; if $z \in V^2xH \cap yH$, then $z = v_1 v_2 x h_1 = y h_2$ for some $v_1, v_2 \in V$, and some $h_1, h_2 \in H$, but then $v_1 v_2 x = y h_2 h_1^{-1}$ so that $V^2 x \cap yH \neq \emptyset$, a contradiction. Since $V = V^{-1}$, we get

$$VxH \cap VyH = \emptyset,$$

and then, since V is open, both VxH and VyH are disjoint, open, so $p(VxH)$ and $p(VyH)$ are open sets (by (1)) containing $\overline{x}$ and $\overline{y}$ respectively and $p(VxH)$ and $p(VyH)$ are disjoint (because $p^{-1}(p(VxH)) = VxHH = VxH$, $p^{-1}(p(VyH)) = VyHH = VyH$, and $VxH \cap VyH = \emptyset$). See Figure 8.1.

(3) If H is open, then every coset gH is open, so every point of G/H is open and G/H is discrete. Also, $\bigcup_{g \notin H} gH$ is open, i.e., H is closed.

(4) Say U is an open subset such that $1 \in U \subseteq H$. Then for every $h \in H$, the set hU is an open subset of H with $h \in hU$, which shows that H is open. The converse is trivial. ☐

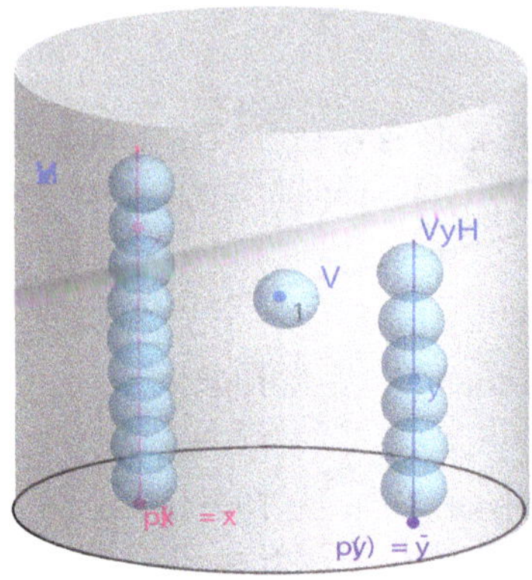

Fig. 8.1 A schematic illustration of $VxH \cap VyH = \emptyset$, where G is the pink cylinder, H is the vertical edge, and G/H is the circular base. Note xH and yH are vertical fibres.

Recall that a topological space X is *locally compact* iff for every point $p \in X$, there is a compact neighborhood C of p; that is, there is a compact C and an open U, with $p \in U \subseteq C$. For example, manifolds are locally compact.

The next two propositions will be needed.

Proposition 8.6. *Let G be a topological group and let H be a closed subgroup of G. The following properties hold.*

(1) If G is locally compact, then so is G/H.
(2) If H is a normal subgroup of G, then G/H is a topological group.

Proof. (1) Since H is closed, we already know from Proposition 8.5(2) that G/H is Hausdorff. Let K be a compact neighborhood of 1 in G, so that there is an open subset U such that $1 \in U \subseteq K$ with K compact. By Proposition 8.5(1) the quotient map $p\colon G \to G/H$ is an open map, and it is continuous, so for any $g \in G$, we have $g \in gU \subseteq gK$ with gU open and gK compact, so $p(g) \in p(gU) \subseteq p(gK)$ with $p(gU)$ open and $p(gK)$ compact, which shows that G/H is locally compact.

(2) If H is a closed normal subgroup, then G/H is a group, and we already know from Proposition 8.5(2) that G/H is Hausdorff. We have to show that multiplication and inversion in G/H are continuous. For any two cosets g_1H and g_2H in G/H, if W is an open subset in G/H containing

$p(g_1g_2) = p(g_1)p(g_2) = (g_1H)(g_2H) = g_1g_2H$, then because the projection map p is continuous, there are open subsets U_1 and U_2 of G with $g_1 \in U_1$ and $g_2 \in U_2$, such that $p(U_1U_2) \subseteq W$. Since p is an open map, $p(U_1)$ is an open subset containing $p(g_1) = g_1H$ and $p(U_2)$ is an open subset containing $p(g_2) = g_2H$, and we have $p(U_1)p(U_2) \subseteq W$, so multiplication in G/H is continuous. A similar proof shows that inversion is continuous in G/H. □

Proposition 8.7. *Let G be a locally compact topological group and let H be a closed subgroup of G. For any compact subset K' in G/H, there is compact subset K of G such that $p(K) = K'$.*

Proof. Since G is locally compact, there is an open subset U and a compact subset V such that $1 \in U \subseteq V$. Since p is an open map, the subsets of the form $p(gU)$ for $g \in G$ form an open cover of K', and since K' is compact, there is a finite subcover $\{p(g_1U), \ldots, p(g_nU)\}$ of K'. Since p is continuous and K' is compact and thus closed (since G/H is Hausdorff), $p^{-1}(K')$ is closed and $g_1V \cup \cdots \cup g_nV$ is compact; then $K = p^{-1}(K') \cap (g_1V \cup \cdots \cup g_nV)$ is compact in G, and we have $p(K) = K'$. □

We next provide a criterion relating the connectivity of G with that of G/H.

Proposition 8.8. *Let G be a topological group and H be any subgroup of G. If H and G/H are connected, then G is connected.*

Proof. It is a standard fact of topology that a space G is connected iff every continuous function f from G to the discrete space $\{0, 1\}$ is constant; see Proposition A.17. Pick any continuous function f from G to $\{0, 1\}$. As H is connected and left translations are homeomorphisms, all cosets gH are connected. Thus, f is constant on every coset gH. It follows that the function $f\colon G \to \{0, 1\}$ induces a continuous function $\overline{f}\colon G/H \to \{0, 1\}$ such that $f = \overline{f} \circ p$ (where $p\colon G \to G/H$; the continuity of $\overline{f}$ follows immediately from the definition of the quotient topology on G/H). As G/H is connected, $\overline{f}$ is constant, and so $f = \overline{f} \circ p$ is constant. □

The next three propositions describe how to generate a topological group from its symmetric neighborhoods of 1.

Proposition 8.9. *If G is a connected topological group, then G is generated by any symmetric neighborhood V of 1. In fact,*

$$G = \bigcup_{n \geq 1} V^n.$$

Proof. Since $V = V^{-1}$, it is immediately checked that $H = \bigcup_{n\geq 1} V^n$ is the group generated by V. As V is a neighborhood of 1, there is some open subset $U \subseteq V$, with $1 \in U$, and so $1 \in \overset{\circ}{H}$. From Proposition 8.5(3), the subgroup H is open and closed, and since G is connected, $H = G$. $\square$

Proposition 8.10. *Let G be a topological group and let V be any connected symmetric open subset containing 1. Then if G_0 is the connected component of the identity, we have*

$$G_0 = \bigcup_{n\geq 1} V^n,$$

and G_0 is a normal subgroup of G. Moreover, the group G/G_0 is discrete.

Proof. First, as V is open, every V^n is open, so the group $\bigcup_{n\geq 1} V^n$ is open, and thus closed, by Proposition 8.5(3). For every $n \geq 1$, we have the continuous map

$$\underbrace{V \times \cdots \times V}_{n} \longrightarrow V^n : (g_1, \ldots, g_n) \mapsto g_1 \cdots g_n.$$

As V is connected, $V \times \cdots \times V$ is connected, and so V^n is connected; this follows from Proposition A.18 because a finite product of connected spaces is connected. Since $1 \in V^n$ for all $n \geq 1$ and every V^n is connected, we use Lemma A.19 to conclude that $\bigcup_{n\geq 1} V^n$ is connected. Now, $\bigcup_{n\geq 1} V^n$ is connected, open and closed, so it is the connected component of 1. Finally, for every $g \in G$, the group gG_0g^{-1} is connected and contains 1, so it is contained in G_0, which proves that G_0 is normal. Since G_0 is open, Proposition 8.5(3) implies that the group G/G_0 is discrete. $\square$

Proposition 8.11. *Let G be a topological group and assume that G is connected and locally compact. Then G is countable at infinity, which means that G is the union of a countable family of compact subsets. In fact, if V is any symmetric compact neighborhood of 1, then*

$$G = \bigcup_{n\geq 1} V^n.$$

Proof. Since G is locally compact, there is some compact neighborhood K of 1. Then, $V = K \cap K^{-1}$ is also compact and a symmetric neighborhood of 1. By Proposition 8.9, we have

$$G = \bigcup_{n\geq 1} V^n.$$

An argument similar to the one used in the proof of Proposition 8.10 to show that V^n is connected if V is connected proves that each V^n compact if V is compact. $\square$

If G is a locally compact group but G is not connected, and if G_0 is the connected component of the identity, then G is the disjoint union of the cosets gG_0, and each coset gG_0 is homeomorphic to G_0, connected, and countable at infinity (σ-compact). This observation plays a crucial role in the proof of the uniqueness of the Haar measure (Theorem 8.21), because it guarantees that the use of Fubini's theorem is legitimate.

The notion of uniform continuity can be generalized to functions defined on a group.

Definition 8.4. Given a topological group G and a subset S of G, for any normed vector space F, a function $f\colon G \to F$ is *left uniformly continuous on* S if for any $\epsilon > 0$, there is an open subset U of G containing 1 such that

$$\|f(y) - f(x)\| < \epsilon \quad \text{for all } x, y \in S \text{ such that } xy^{-1} \in U.$$

The function $f\colon G \to F$ is *right uniformly continuous on* S if for any $\epsilon > 0$, there is an open subset U of G containing 1 such that

$$\|f(y) - f(x)\| < \epsilon \quad \text{for all } x, y \in S \text{ such that } x^{-1}y \in U.$$

Observe that if $xy^{-1} \in U$, then we can write $xy^{-1} = z$ for some $z \in U$, so $y = z^{-1}x$, and $\|f(y) - f(x)\| = \left\|f(z^{-1}x) - f(x)\right\| < \epsilon$.

It is customary to introduce a left action λ of G on functions $f\colon G \to F$ defined on G by

$$(\lambda_s(f))(x) = f(s^{-1}x) \quad \text{for all } x, s \in G.$$

Observe that

$$\lambda_{st}(f)(x) = f((st)^{-1}x) = f(t^{-1}s^{-1}x) = \lambda_t(f)(s^{-1}x) = \lambda_s(\lambda_t(f))(x),$$

so

$$\lambda_{st} = \lambda_s \circ \lambda_t,$$

which is the reason why we used s^{-1} instead of s in the definition of λ_s.

Then $\left\|f(z^{-1}x) - f(x)\right\| = \|\lambda_z(f)(x) - f(x)\|$, so the condition of the definition is equivalent to

$$\|\lambda_z(f)(x) - f(x)\| < \epsilon \quad \text{for all } x \in S \text{ and all } z \in U.$$

Informally, the above condition can be written as

$$\lim_{z\mapsto 1}\sup_{x\in S}\|\lambda_z(f)(x) - f(x)\| = 0.$$

It is also customary to introduce a right action ρ of G on functions $f\colon G \to F$ defined on G by

$$(\rho_s(f))(x) = f(xs) \quad \text{for all } x, s \in G.$$

Observe that

$$\rho_{st}(f)(x) = f(xst) = \rho_t(f)(xs) = \rho_s(\rho_t(f))(x),$$

so

$$\rho_{st} = \rho_s \circ \rho_t.$$

Observe that if $x^{-1}y \in U$, then we can write $x^{-1}y = z$ for some $z \in U$, so $y = xz$, and $\|f(y) - f(x)\| = \|f(xz) - f(x)\| = \|\rho_z(f)(x) - f(x)\| < \epsilon$.

Thus the condition of the definition is equivalent to

$$\|\rho_z(f)(x) - f(x)\| < \epsilon \quad \text{for all } x \in S \text{ and all } z \in U.$$

Informally, the above condition can be written as

$$\lim_{z\mapsto 1}\sup_{x\in S}\|\rho_z(f)(x) - f(x)\| = 0.$$

Proposition 8.12. *Let G be a topological group and let S be a subset of G. For any function $f\colon S \to F$, where F is any normed vector space, if f is continuous with compact support K, then f is left (resp. right) uniformly continuous on K.*

Proof. We prove that f is left uniformly continuous, the proof that f is right uniformly continuous being similar and left as an exercise. Since f is continuous, for every $y \in K$, there is some open subset U_y with $1 \in U_y$ such that

$$\|f(y) - f(x)\| < \frac{\epsilon}{2} \quad \text{for all } x \in U_y y.$$

We can find an open subset V_y containing 1 such that $V_y V_y \subseteq U_y$. The open subsets of the form $V_y y$ for $y \in K$ form an open cover of K, and since K is compact, there is a finite subcover $\{V_{y_1}y_1 \ldots, V_{y_n}y_n\}$ of K with $y_1, \ldots, y_n \in K$. Let

$$V = V_{y_1} \cap \cdots \cap V_{y_n}.$$

Consider $x, y \in K$ such that $xy^{-1} \in V$, that is, $x \in Vy$. Then $y \in V_{y_i}y_i \subseteq U_{y_i}y_i$ for some i, and so

$$x \in Vy \in VV_{y_i}y_i \subseteq V_{y_i}V_{y_i}y_i \subseteq U_{y_i}y_i,$$

which implies that

$$\|f(y) - f(x)\| \leq \|f(y) - f(y_i)\| + \|f(y_i) - f(x)\| < \frac{\epsilon}{2} + \frac{\epsilon}{2} = \epsilon,$$

as desired. □

We end this section by combining the various properties of a topological group G to characterize when G/G_x is homeomorphic to X. The reader should review the notion of group action and the related concepts of stabilizer and orbit; see Appendix C, Sections C.2 and C.3.

First we need two definitions.

Definition 8.5. Let G be a topological group and let X be a topological space. An action $\varphi\colon G \times X \to X$ is *continuous* (and G *acts continuously on* X) if the map φ is continuous.

If an action $\varphi\colon G \times X \to X$ is continuous, then each map $\varphi_g\colon X \to X$ is a homeomorphism of X (recall that $\varphi_g(x) = g \cdot x$, for all $x \in X$). Indeed, the map $x \mapsto g \cdot x$ is a continuous bijection whose inverse $x \mapsto g^{-1} \cdot x$ is also continuous.

Under some mild assumptions on G and X, the quotient space G/G_x is homeomorphic to X. For example, this happens if X is a Baire space.

Definition 8.6. A *Baire space* X is a topological space with the property that if $\{F\}_{i\geq 1}$ is any countable family of closed sets F_i such that each F_i has empty interior, then $\bigcup_{i\geq 1} F_i$ also has empty interior. By complementation, this is equivalent to the fact that for every countable family of open sets U_i such that each U_i is dense in X (i.e., $\overline{U_i} = X$), then $\bigcap_{i\geq 1} U_i$ is also dense in X.

Remark: A subset $A \subseteq X$ is *rare* if its closure $\overline{A}$ has empty interior. A subset $Y \subseteq X$ is *meager* if it is a countable union of rare sets. Then it is immediately verified that a space X is a Baire space iff every nonempty open subset of X is not meager.

The following theorem shows that there are plenty of Baire spaces.

Theorem 8.13. *(Baire) (1) Every locally compact topological space is a Baire space.*

(2) Every complete metric space is a Baire space.

A proof of Theorem 8.13 can be found in Bourbaki [Bourbaki (1990b)], Chapter IX, Section 5, Theorem 1

Theorem 8.14. *Let G be a topological group which is locally compact and countable at infinity, X a Hausdorff topological space which is a Baire space, and assume that G acts transitively and continuously on X. Then for any $x \in X$, the map $\varphi\colon G/G_x \to X$ is a homeomorphism.*

Proof. We follow the proof given in Bourbaki [Bourbaki (1990b)], Chapter IX, Section 5, Proposition 6 (essentially the same proof can be found in Mneimné and Testard [Mneimné and Testard (1997)], Chapter 2). First observe that if a topological group acts continuously and transitively on a Hausdorff topological space, then for every $x \in X$, the stabilizer G_x is a closed subgroup of G. This is because, as the action is continuous, the projection $\pi_x\colon G \longrightarrow X\colon g \mapsto g \cdot x$ is continuous, and $G_x = \pi^{-1}(\{x\})$, with $\{x\}$ closed. Therefore, by Proposition 8.5, the quotient space G/G_x is Hausdorff. As the map $\pi_x\colon G \longrightarrow X$ is continuous, the induced map $\varphi_x\colon G/G_x \to X$ is continuous, and by Proposition C.14, it is a bijection. Therefore, to prove that φ_x is a homeomorphism, it is enough to prove that φ_x is an open map. For this, it suffices to show that π_x is an open map. Given any open U in G, we will prove that for any $g \in U$, the element $\pi_x(g) = g \cdot x$ is contained in the interior of $U \cdot x$. However, observe that this is equivalent to proving that x belongs to the interior of $(g^{-1} \cdot U) \cdot x$. Therefore, we are reduced to the following case: if U is any open subset of G containing 1, then x belongs to the interior of $U \cdot x$.

Since G is locally compact, using Proposition 8.2, we can find a compact neighborhood of the form $W = \overline{V}$, such that $1 \in W$, $W = W^{-1}$ and $W^2 \subseteq U$, where V is open with $1 \in V \subseteq U$. As G is countable at infinity, $G = \bigcup_{i \geq 1} K_i$, where each K_i is compact. Since V is open, all the cosets gV are open, and as each K_i is covered by the gV's, by compactness of K_i, finitely many cosets gV cover each K_i, and so

$$G = \bigcup_{i \geq 1} g_i V = \bigcup_{i \geq 1} g_i W,$$

for countably many $g_i \in G$, where each $g_i W$ is compact. As our action is transitive, we deduce that

$$X = \bigcup_{i \geq 1} g_i W \cdot x,$$

where each $g_i W \cdot x$ is compact, since our action is continuous and the $g_i W$ are compact. As X is Hausdorff, each $g_i W \cdot x$ is closed, and as X is a Baire space expressed as a union of closed sets, one of the $g_i W \cdot x$ must have nonempty interior; that is, there is some $w \in W$, with $g_i w \cdot x$ in the interior of $g_i W \cdot x$, for some i. But then, as the map $y \mapsto g \cdot y$ is a homeomorphism for any given $g \in G$ (where $y \in X$), we see that x is in the interior of

$$w^{-1} g_i^{-1} \cdot (g_i W \cdot x) = w^{-1} W \cdot x \subseteq W^{-1} W \cdot x = W^2 \cdot x \subseteq U \cdot x,$$

as desired. □

By Theorem 8.13, we get the following important corollary:

Theorem 8.15. *Let G be a topological group which is locally compact and countable at infinity, X a Hausdorff locally compact topological space, and assume that G acts transitively and continuously on X. Then for any $x \in X$, the map $\varphi_x \colon G/G_x \to X$ is a homeomorphism.*

Readers who wish to learn more about topological groups may consult Sagle and Walde [Sagle and Walde (1973)] and Chevalley [Chevalley (1946)] for an introductory account, and Bourbaki [Bourbaki (1990a)], Weil [Weil (1979)] and Pontryagin [Pontryagin (1939, 1960)], for a more comprehensive account (especially the last two references).

8.2 Existence of the Haar Measure; Preliminaries

Let G be a locally compact group. We are going to show there is a positive σ-regular locally finite Borel measure μ on G such that $\mu(U) > 0$ for every nonempty open subset U, and such that μ is left-invariant, which means that

$$\mu(A) = \mu(sA) \quad \text{for all } s \in G \text{ and all } A \in \mathcal{B},$$

where $\mathcal{B}$ is the σ-algebra of Borel sets on G.

Recall that for any $x \in G$, the maps $L_x \colon G \to G$ (left translation) and $R_x \colon G \to G$ (right translation) are defined by

$$L_x(z) = xz, \qquad R_x(z) = zx, \quad \text{for all } x, z \in G.$$

It is obvious that

$$L_{xy} = L_x \circ L_y \quad \text{and} \quad R_{xy} = R_y \circ R_x,$$

and that L_x and R_y commute for all $x, y \in G$.

It is customary to introduce a left action λ of G and a right action ρ of G on functions $f\colon G \to F$. We did this in the previous section, but for the sake of completeness, we repeat these definitions.

Definition 8.7. Let G be a group, and let F be any set. The *left action* λ of G on a function $f\colon G \to F$ is the function $\lambda_s(f)$ is given by

$$(\lambda_s(f))(x) = f(s^{-1}x) \quad \text{for all } x, s \in G,$$

and the *right action* ρ of G on a function $f\colon G \to F$ is the function $\rho_s(f)$ given by

$$(\rho_s(f))(x) = f(xs) \quad \text{for all } x, s \in G.$$

It might help the reader to remember that λ_s is a *left* action and that ρ_s is a *right action* by noticing that $\lambda =$ *l*ambda begins with an "l" as in *l*eft and that $\rho =$ *r*ho begin with an "r" as in *r*ight.

Observe that

$$\lambda_{st}(f)(x) = f((st)^{-1}x) = f(t^{-1}s^{-1}x) = \lambda_t(f)(s^{-1}x) = \lambda_s(\lambda_t(f))(x),$$

so

$$\lambda_{st} = \lambda_s \circ \lambda_t,$$

which is the reason why we used s^{-1} instead of s in the definition of λ_s. Observe that

$$\rho_{st}(f)(x) = f(xst) = \rho_t(f)(xs) = \rho_s(\rho_t(f))(x),$$

so

$$\rho_{st} = \rho_s \circ \rho_t.$$

Given a subset A of G , we usually write sA for $L_s(A)$ and As for $R_s(A)$.

We define a left action of λ_s and a right action ρ_s on measures and Radon functionals as follows.

Definition 8.8. Let G be a locally compact topological group. The *left action* λ of G on a measure μ on $(G, \mathcal{B})$ is the measure $\lambda_s(\mu)$ given by

$$(\lambda_s(\mu))(A) = \mu(s^{-1}A) \quad \text{for all } s \in G \text{ and all } A \in \mathcal{B},$$

and the *right action* ρ of G on a measure μ on $(G, \mathcal{B})$ is the measure $\rho_s(\mu)$ given by

$$(\rho_s(\mu))(A) = \mu(As) \quad \text{for all } s \in G \text{ and all } A \in \mathcal{B}.$$

The *left action* λ of G on a Radon functional $\Phi\colon \mathcal{K}_{\mathbb{C}}(G) \to \mathbb{C}$ is the Radon functional $\lambda_s(\Phi)$ given by

$$(\lambda_s(\Phi))(f) = \Phi(\lambda_{s^{-1}}(f)) \quad \text{for all } s \in G \text{ and all } f \in \mathcal{K}_{\mathbb{C}}(G),$$

and the *right action* ρ of G on a Radon functional $\Phi\colon \mathcal{K}_{\mathbb{C}}(G) \to \mathbb{C}$ is the Radon functional $\rho_s(\Phi)$ given by

$$(\rho_s(\Phi))(f) = \Phi(\rho_{s^{-1}}(f)) \quad \text{for all } s \in G \text{ and all } f \in \mathcal{K}_{\mathbb{C}}(G).$$

If m_Φ is the Borel measure corresponding to a positive Radon functional Φ, $m_{\lambda_s(\Phi)}$ is the Borel measure corresponding to $\lambda_s(\Phi)$, and $m_{\rho_s(\Phi)}$ is the Borel measure corresponding to $\rho_s(\Phi)$, given by Theorem 7.8, then we have

$$\begin{aligned}\int f(sx)dm_\Phi(x) &= \int (\lambda_{s^{-1}}(f))(x)dm_\Phi(x) = \Phi(\lambda_{s^{-1}}(f)) \\ &= (\lambda_s(\Phi))(f) = \int f(x)d(m_{\lambda_s(\Phi)})(x)\end{aligned}$$

and

$$\begin{aligned}\int f(xs^{-1})dm_\Phi(x) &= \int (\rho_{s^{-1}}(f))(x)dm_\Phi(x) = \Phi(\rho_{s^{-1}}(f)) \\ &= (\rho_s(\Phi))(f) = \int f(x)d(m_{\rho_s(\Phi)})(x).\end{aligned}$$

Therefore, we have the change of variable formulae

$$\int f(x)d(m_{\lambda_s(\Phi)})(x) = \int f(sx)dm_\Phi(x),$$

and

$$\int f(x)d(m_{\rho_s(\Phi)})(x) = \int f(xs^{-1})dm_\Phi(x)$$

for all $f \in \mathcal{K}_{\mathbb{C}}(G)$ and all $s \in G$. See Figures 8.2 and 8.3.

Definition 8.8 has been designed so that for every measure μ on G we have

$$\lambda_{st}(\mu) = \lambda_s(\lambda_t(\mu)), \qquad \rho_{st}(\mu) = \rho_s(\rho_t(\mu)) \quad \text{for all } s, t \in G.$$

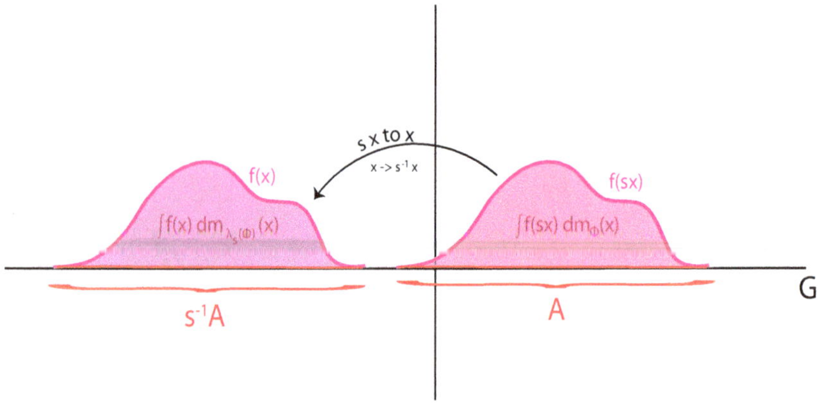

Fig. 8.2 A schematic illustration of the change of variable $x \to s^{-1}x$ associated with $\int f(x)d(m_{\lambda_s(\Phi)})(x) = \int f(sx)dm_{\Phi}(x)$.

For every Radon functional Φ and for every function $f \in \mathcal{K}_{\mathbb{C}}(G)$, we have

$$\begin{aligned}(\lambda_{st}(\Phi))(f) &= \Phi(\lambda_{(st)^{-1}}(f)) = \Phi(\lambda_{t^{-1}s^{-1}}(f)) \\ &= \Phi(\lambda_{t^{-1}}(\lambda_{s^{-1}}(f))) = (\lambda_t(\Phi))(\lambda_{s^{-1}}(f)) = (\lambda_s(\lambda_t(\Phi)))(f),\end{aligned}$$

and a similar computation shows that

$$(\rho_{st}(\Phi))(f) = (\rho_s(\rho_t(\Phi)))(f).$$

Therefore, for every Radon functional Φ, we have

$$\lambda_{st}(\Phi) = \lambda_s(\lambda_t(\Phi)), \qquad \rho_{st}(\Phi) = \rho_s(\rho_t(\Phi)) \quad \text{for all } s, t \in G.$$

The left actions λ_s and the right actions ρ_s are summarized in the following table.

Left action	Right action
On functions	
$(\lambda_s(f))(x) = f(s^{-1}x)$	$(\rho_s(f))(x) = f(xs)$
On measures	
$(\lambda_s(\mu))(A) = \mu(s^{-1}A)$	$(\rho_s(\mu))(A) = \mu(As)$
On functionals	
$(\lambda_s(\Phi))(f) = \Phi(\lambda_{s^{-1}}(f))$	$(\rho_s(\Phi))(f) = \Phi(\rho_{s^{-1}}(f))$

Definition 8.9. Let G be a locally compact group. A *left Haar measure* μ is a σ-regular, locally finite, Borel measure on the σ-algebra $\mathcal{B}$ of Borel

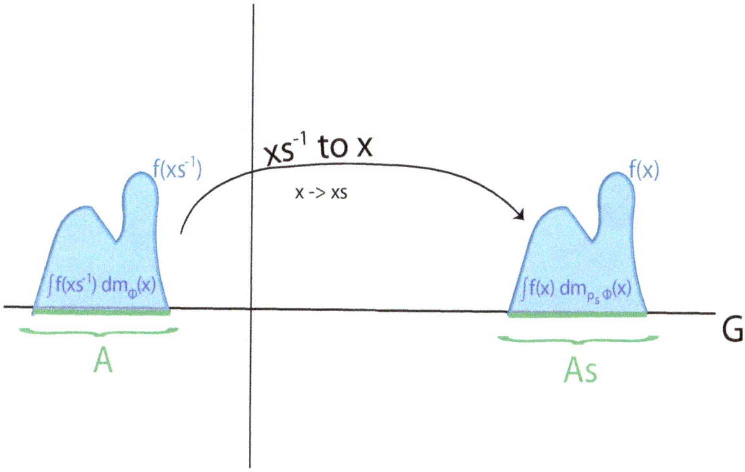

Fig. 8.3 A schematic illustration of the change of variable $x \to xs$ associated with $\int f(x)d(m_{\rho_s(\Phi)})(x) = \int f(xs^{-1})dm_{\Phi}(x)$.

sets of G, such that $\mu(U) > 0$ for all nonempty open subsets $U \in \mathcal{B}$ and μ is *left-invariant*, which means that

$$\lambda_s(\mu) = \mu \quad \text{for all } s \in G.$$

The above condition means that

$$\mu(s^{-1}A) = \mu(A) \quad \text{for all } A \in \mathcal{B} \text{ and all } s \in G,$$

or equivalently,

$$\mu(sA) = \mu(A) \quad \text{for all } A \in \mathcal{B} \text{ and all } s \in G.$$

A *right Haar measure* μ is a Borel measure satisfying the same conditions as a left Haar measure, except that it is *right-invariant*, which means that

$$\rho_s(\mu) = \mu \quad \text{for all } A \in \mathcal{B} \text{ and all } s \in G.$$

The above condition means that

$$\mu(As) = \mu(A) \quad \text{for all } A \in \mathcal{B} \text{ and all } s \in G.$$

Note that according to Definition 7.5, a left (resp. right) Haar measure is a σ-Radon measure which is left-invariant (resp. right-invariant), and such that $\mu(U) > 0$ for all nonempty open subsets $U \in \mathcal{B}$.

In order to prove that a left (resp. right) Haar measure exists, we will use Theorem 7.8, which motivates the following definition.

Definition 8.10. Let G be a locally compact group. A *left Haar functional* Φ is a positive non-zero Radon functional $\Phi\colon \mathcal{K}_{\mathbb{C}}(G) \to \mathbb{C}$ which is *left-invariant*, which means that

$$\lambda_s(\Phi) = \Phi \quad \text{for all } s \in G.$$

A *right Haar functional* Φ is a positive non-zero Radon functional $\Phi\colon \mathcal{K}_{\mathbb{C}}(G) \to \mathbb{C}$ which is *right-invariant*, which means that

$$\rho_s(\Phi) = \Phi \quad \text{for all } s \in G.$$

If m_Φ is the Borel measure associated with Φ, since

$$\int f(sx)dm_\Phi(x) = \int (\lambda_{s^{-1}}(f))(x)dm_\Phi(x) = \Phi(\lambda_{s^{-1}}(f)) = (\lambda_s(\Phi))(f),$$

then the left-invariance of Φ means that

$$(\lambda_s(\Phi))(f) = \Phi(f) = \int f(x)dm_\Phi(x),$$

so we have the change of variable formula

$$\int f(x)dm_\Phi(x) = \int f(sx)dm_\Phi(x)$$

for all $f \in \mathcal{K}_{\mathbb{C}}(G)$ and all $s \in G$. Similarly, since

$$\int f(xs^{-1})dm_\Phi(x) = \int (\rho_{s^{-1}}(f))(x)dm_\Phi(x) = \Phi(\rho_{s^{-1}}(f)) = (\rho_s(\Phi))(f),$$

the right-invariance of Φ means that

$$(\rho_s(\Phi))(f) = \Phi(f) = \int f(x)dm_\Phi(x),$$

so we have the change of variable formula

$$\int f(x)dm_\Phi(x) = \int f(xs^{-1})dm_\Phi(x)$$

for all $f \in \mathcal{K}_{\mathbb{C}}(G)$ and all $s \in G$.

The following operation will allow us to convert a left-invariant measure (resp. functional) to a right-invariant measure (resp. functional).

Definition 8.11. Let G be any locally compact group and F be any set. For any function $f\colon G \to F$, define the function $\check{f}\colon G \to F$ by

$$\check{f}(s) = f(s^{-1}) \quad \text{for all } s \in G.$$

For any Borel measure μ on $(G, \mathcal{B})$, define the Borel measure $\check{\mu}$ by

$$\check{\mu}(A) = \mu(A^{-1}) \quad \text{for all } A \in \mathcal{B}.$$

For any Radon functional $\Phi\colon \mathcal{K}_{\mathbb{C}}(G) \to \mathbb{C}$, define the Radon functional $\check{\Phi}\colon \mathcal{K}_{\mathbb{C}}(G) \to \mathbb{C}$ by

$$\check{\Phi}(f) = \Phi(\check{f}) \quad \text{for all } f \in \mathcal{K}_{\mathbb{C}}(G).$$

Observe that

$$\begin{aligned}(\lambda_s(\mu))\check{}(A) &= (\lambda_s(\mu))(A^{-1}) = \mu(s^{-1}A^{-1}) \\ &= \mu((As)^{-1}) = \check{\mu}(As) = (\rho_s(\check{\mu}))(A),\end{aligned}$$

so

$$(\lambda_s(\mu))\check{} = \rho_s(\check{\mu}).$$

Similarly,

$$\begin{aligned}(\rho_s(\mu))\check{}(A) &= (\rho_s(\mu))(A^{-1}) = \mu(A^{-1}s) = \mu((s^{-1}A)^{-1}) \\ &= \check{\mu}(s^{-1}A) = (\lambda_s(\check{\mu}))(A),\end{aligned}$$

so

$$(\rho_s(\mu))\check{} = \lambda_s(\check{\mu}).$$

For any function $f\colon G \to F$, we have

$$\begin{aligned}(\lambda_s(f))\check{}(x) &= (\lambda_s(f))(x^{-1}) = f(s^{-1}x^{-1}) = f((xs)^{-1}) \\ &= \check{f}(xs) = (\rho_s(\check{f}))(x),\end{aligned}$$

and similarly

$$\begin{aligned}(\rho_s(f))\check{}(x) &= (\rho_s(f))(x^{-1}) = f(x^{-1}s) = f((s^{-1}x)^{-1}) \\ &= \check{f}(s^{-1}x) = (\lambda_s(\check{f}))(x).\end{aligned}$$

Therefore,

$$(\lambda_s(f))\check{} = \rho_s(\check{f}), \qquad (\rho_s(f))\check{} = \lambda_s(\check{f}) \quad \text{for all } s \in G \text{ and all } f\colon G \to F.$$

Using the above equations, for every Radon functional $\Phi\colon \mathcal{K}_{\mathbb{C}}(G) \to \mathbb{C}$, we have

$$\begin{aligned}(\lambda_s(\Phi))\check{}(f) &= (\lambda_s(\Phi))(\check{f}) = \Phi(\lambda_{s^{-1}}(\check{f})) = \Phi((\rho_{s^{-1}}(f))\check{}) \\ &= \check{\Phi}(\rho_{s^{-1}}(f)) = (\rho_s(\check{\Phi}))(f).\end{aligned}$$

Similarly,

$$\begin{aligned}(\rho_s(\Phi))\check{}(f) &= (\rho_s(\Phi))(\check{f}) = \Phi(\rho_{s^{-1}}(\check{f})) = \Phi((\lambda_{s^{-1}}(f))\check{}) \\ &= \check{\Phi}(\lambda_{s^{-1}}(f)) = (\lambda_s(\check{\Phi}))(f).\end{aligned}$$

Therefore, we have

$$(\lambda_s(\Phi))\check{} = \rho_s(\check{\Phi}), \qquad (\rho_s(\Phi))\check{} = \lambda_s(\check{\Phi}) \quad \text{for all } s \in G.$$

The definition of the cech operation $(\check{})$ is summarized in the following table.

On functions $\check{f}(s) = f(s^{-1})$
On measures $\check{\mu}(A) = \mu(A^{-1})$
On functionals $\check{\Phi}(f) = \Phi(\check{f})$

Proposition 8.16. *Let G be a locally compact group, and let μ be a σ-regular, locally finite, Borel measure on G (a σ-Radon measure). The following properties hold:*

(1) *We have*

$$(\lambda_s(\mu))\check{} = \rho_s(\check{\mu}), \qquad (\rho_s(\mu))\check{} = \lambda_s(\check{\mu}) \quad \text{for all } s \in G.$$

Consequently, μ is a left-invariant measure iff $\check{\mu}$ is a right-invariant measure. For any Radon functional $\Phi\colon \mathcal{K}_{\mathbb{C}}(G) \to \mathbb{C}$, we have

$$(\lambda_s(\Phi))\check{} = \rho_s(\check{\Phi}), \qquad (\rho_s(\Phi))\check{} = \lambda_s(\check{\Phi}) \quad \text{for all } s \in G.$$

Consequently, Φ is left-invariant iff $\check{\Phi}$ is right-invariant.

(2) *If the Haar measure μ is left-invariant then*

$$\int \lambda_{s^{-1}}(f)\, d\mu = \int f d\mu$$

for all $f \in \mathcal{L}^1_\mu(G, \mathcal{B}, \mathbb{C})$ and all $s \in G$. If the Haar measure μ is right-invariant then

$$\int \rho_{s^{-1}}(f)\, d\mu = \int f d\mu$$

for all $f \in \mathcal{L}^1_\mu(G, \mathcal{B}, \mathbb{C})$ and all $s \in G$.

(3) *We have*

$$\int \lambda_{s^{-1}}(f)\, d\mu = \int f d\lambda_s(\mu) \quad \text{for all } f \in \mathcal{L}^1_\mu(G, \mathcal{B}, \mathbb{C}) \text{ and all } s \in G.$$

If

$$\int \lambda_{s^{-1}}(f)\, d\mu = \int f d\mu$$

for all $f \in \mathcal{K}_{\mathbb{C}}(G)$ and all $s \in G$, then μ is left-invariant. We have

$$\int \rho_{s^{-1}}(f)\, d\mu = \int f d\rho_s(\mu) \quad \text{for all } f \in \mathcal{L}^1_\mu(G, \mathcal{B}, \mathbb{C}) \text{ and all } s \in G.$$

If

$$\int \rho_{s^{-1}}(f)\, d\mu = \int f d\mu$$

for all $f \in \mathcal{K}_{\mathbb{C}}(G)$ and all $s \in G$, then μ is right-invariant.

Proof. We already proved (1).

(2) Let f be any μ-step function

$$f = \sum_{k=1}^{n} y_k \chi_{A_k},$$

where the A_k are measurable Borel sets of finite measure. For all $s, x \in G$, we see that $(\lambda_{s^{-1}} f)(x) = f(sx) = y_k$ iff $sx \in A_k$ iff $x \in s^{-1}A_k$, which means that

$$\lambda_{s^{-1}} f = \sum_{k=1}^{n} y_k \chi_{s^{-1}A_k},$$

so

$$\int (\lambda_{s^{-1}} f) d\mu = \sum_{k=1}^{n} y_k \mu(s^{-1}A_k) = \sum_{k=1}^{n} y_k (\lambda_s(\mu))(A_k) = \int f d\lambda_s(\mu).$$

See Figure 8.4. If μ is left-invariant, then $\lambda_s(\mu) = \mu$, so

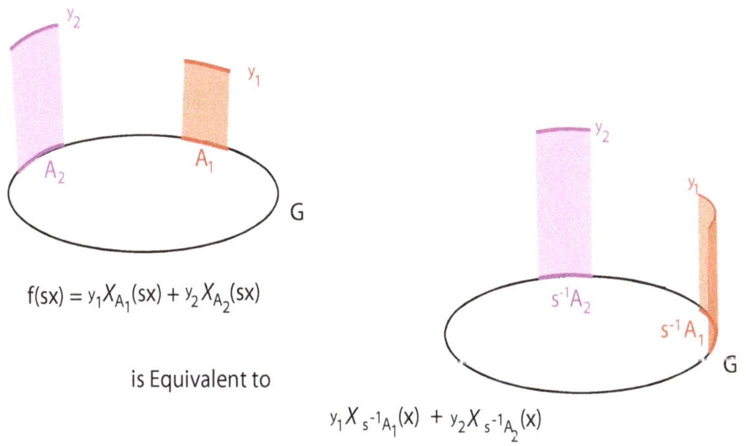

Fig. 8.4 Let $G = S^1$. A step function on S^1 is represented by the top arcs of the colored vertical "rectangular" sheets. The step function $f(sx) = \sum_{k=1}^{2} y_k \chi_{A_k}(sx)$ is equivalent to $\lambda_{s^{-1}} f(x) = \sum_{k=1}^{2} y_k \chi_{s^{-1}A_k}(x)$.

$$\sum_{k=1}^{n} y_k (\lambda_s(\mu))(A_k) = \sum_{k=1}^{n} y_k \mu(A_k) = \int f d\mu,$$

and we deduce that

$$\int (\lambda_{s^{-1}} f) d\mu = \int f d\mu.$$

Every function $f \in \mathcal{L}^1_\mu(G, \mathcal{B}, \mathbb{C})$ has some approximation sequence (f_n) by μ-step functions that converges to f a.e. and in the L^1-norm. It follows that the sequence $(\lambda_{s^{-1}} f_n)$ converges a.e. to $\lambda_{s^{-1}}(f)$. We check immediately that it is a Cauchy sequence because

$$\int (\lambda_{s^{-1}} f_n) d\mu = \int f_n \, d\mu$$

for all n, and it follows that

$$\int \lambda_{s^{-1}}(f) \, d\mu = \int f d\mu.$$

If f is any μ-step function

$$f = \sum_{k=1}^{n} y_k \chi_{A_k},$$

we have $(\rho_{s^{-1}} f)(x) = f(xs^{-1}) = y_k$ iff $xs^{-1} \in A_k$ iff $x \in A_k s$, which means that

$$\rho_{s^{-1}} f = \sum_{k=1}^{m} y_k \chi_{A_k s},$$

so

$$\int (\rho_{s^{-1}} f) d\mu = \sum_{k=1}^{n} y_k \mu(A_k s) = \sum_{k=1}^{n} y_k (\rho_s(\mu))(A_k) = \int f d\rho_s(\mu).$$

See Figure 8.5.

We finish the argument as in the previous case.

(3) The proof in (2) actually shows that

$$\int \lambda_{s^{-1}}(f) \, d\mu = \int f d\lambda_s(\mu)$$

and

$$\int \rho_{s^{-1}}(f) \, d\mu = \int f d\rho_s(\mu)$$

for all $f \in \mathcal{L}^1_\mu(G, \mathcal{B}, \mathbb{C})$. If

$$\int \lambda_{s^{-1}}(f) \, d\mu = \int f d\mu$$

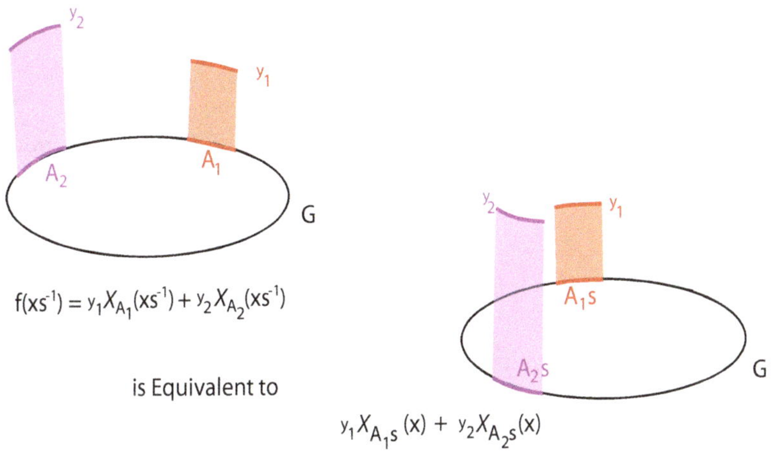

Fig. 8.5 Let $G = S^1$. A step function on S^1 is represented by the top arcs of the colored vertical "rectangular" sheets. The step function $f(xs^{-1}) = \sum_{k=1}^{2} y_k \chi_{A_k}(xs^{-1})$ is equivalent to $\rho_{s^{-1}} f(x) = \sum_{k=1}^{2} y_k \chi_{A_k s}(x)$.

then

$$\int f d\lambda_s(\mu) = \int f d\mu$$

for all $f \in \mathcal{K}_{\mathbb{C}}(G)$ and all $s \in G$, and by the uniqueness of the Borel measure corresponding to the Radon functional $f \mapsto \int f d\mu$ from Theorem 7.8, we see that $\lambda_s(\mu) = \mu$ for all $s \in G$, which means that μ is left-invariant. The right-invariant case is similar. ☐

The condition

$$\int \lambda_{s^{-1}}(f)\, d\mu = \int f d\mu \quad \text{for all } s \in G$$

is also written as

$$\int f(sx) d\mu(x) = \int f(x) d\mu(x) \quad \text{for all } s \in G.$$

The condition

$$\int \rho_{s^{-1}}(f)\, d\mu = \int f d\mu \quad \text{for all } s \in G$$

is also written as

$$\int f(xs^{-1}) d\mu(x) = \int f(x) d\mu(x) \quad \text{for all } s \in G.$$

Since G is a group and s is any arbitrary element of G, the above condition is also equivalent to

$$\int f(xs) d\mu(x) = \int f(x) d\mu(x) \quad \text{for all } s \in G.$$

8.3 Existence of the Haar Measure

We are now going to sketch the proof that a left-invariant Haar measure exists on any locally compact group. All proofs we are aware of (Weil [Weil (1979)], Halmos [Halmos (1978)], Bourbaki [Bourbaki (2005b)], Dieudonné [Dieudonné (2005)], Lang [Lang (1996)], Folland [Folland (1995)]) make use of Haar's original clever idea (1933). Except for Halmos who constructs directly a measure (as Haar did), all the other proofs are essentially André Weil's proof (which constructs a Haar functional) from his famous little book [Weil (1979)] first published in 1940.

As we noted just after Proposition 7.4, there is a bijection between the space $\mathrm{M}^+(X)$ of positive linear functionals $\Phi\colon \mathcal{K}_{\mathbb{C}}(X) \to \mathbb{C}$ and the space of positive linear functionals $\Psi\colon \mathcal{K}_{\mathbb{R}}(X) \to \mathbb{R}$, so it is enough to construct a left (or right) real Haar functional on $\mathcal{K}_{\mathbb{R}}(G)$.

Theorem 8.17. *(Haar) Every locally compact group G possesses a left-invariant Haar measure.*

Proof sketch. Folland [Folland (1995)] (Chapter 2, Section 2.2) is kind enough to provide the intuition behind the construction. In this method a measure is not constructed directly. Instead, a left Haar functional is constructed. Then Theorem 7.8 is used to obtain a left-invariant Borel measure which is a left Haar measure.

Suppose we have positive function $\varphi \in \mathcal{K}_{\mathbb{R}}(G)$ bounded by 1, equal to 1 on a small open set U, and whose support is a compact subset slightly larger than U. If $f \in \mathcal{K}_{\mathbb{R}}(G)$ is any other function slowly varying so that it is essentially constant on the left translates of U, then f can be approximated by a linear combination $f \approx \sum c_j \lambda_{s_j}(\varphi)$. If μ were a left Haar measure, then we would have

$$\int f d\mu \approx \left(\sum_j c_j\right) \int \varphi \, d\mu.$$

See Figure 8.6.

This approximation gets better and better as the support of φ shrinks to a point, and if we introduce a normalization to cancel out the factor $\int \varphi \, d\mu$, then we obtain $\int f d\mu$ as the limit of the sums $\sum_j c_j$. If $G = \mathbb{R}$, and if φ is the characteristic function of a small interval, this is reminiscent of the approximation of $\int f d\mu$ by Riemann sums. The issue is to make this idea precise and formal.

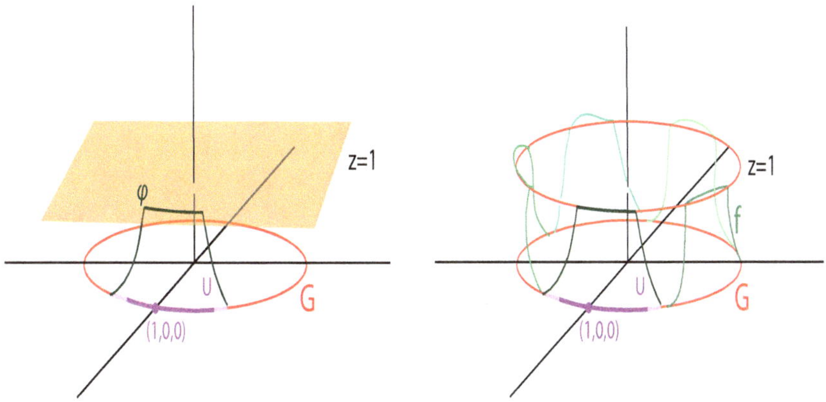

Fig. 8.6 Let G be the unit circle $\mathbb{T}$ in the xy-plane. The left figure shows the "bump" function φ, while the right figure illustrates f as five translates of φ, namely $f \approx \sum_{j=1}^{5} c_j \lambda_{s_j}(\varphi)$.

The first step is to pick some positive nonzero function $\varphi \in \mathcal{K}_{\mathbb{R}}(G)$. This function remains fixed until Proposition 8.18. Then we claim that for every function $f \in \mathcal{K}_{\mathbb{R}}(G)$, there exists a finite set $\{s_1, \ldots, s_n\}$ of elements of G and a finite sequence $(c_1, \ldots, c_n)$ of reals $c_j \in \mathbb{R}$, such that

$$f \leq \sum_{j=1}^{n} c_j \lambda_{s_j}(\varphi).$$

This is because f has compact support, so its support can be covered by finitely many translates of the open subset U given by

$$U = \left\{ s \in G \mid \varphi(s) > \frac{1}{2} \|\varphi\|_\infty \right\},$$

and if we pick $c_j = \|f\|_\infty / a$ where $a = (n/2) \|\varphi\|_\infty$, then on each translate $s_j U$ we have

$$\lambda_{s_j}(\varphi)(x) > \frac{1}{2} \|\varphi\|_\infty, \quad x \in s_j U,$$

which implies

$$\sum_{j=1}^{n} c_j \lambda_{s_j}(\varphi) = \sum_{j=1}^{n} \frac{2 \|f\|_\infty}{n \|\varphi\|_\infty} \lambda_{s_j}(\varphi) > \|f\|_\infty,$$

so

$$f \leq \sum_{j=1}^{n} c_j \lambda_{s_j}(\varphi).$$

Define $(f : \varphi)$ as the greatest lower bound of the sums $\sum_{j=1}^{n} c_j$, over all sets $\{s_1, \ldots, s_n\}$ of elements of G and all finite sequences $(c_1, \ldots, c_n)$ of reals $c_j \in \mathbb{R}$ such that

$$f \leq \sum_{j=1}^{n} c_j \lambda_{s_j}(\varphi).$$

Then it is not hard to show that the quantity $(f : \varphi)$ has the following properties:

$$\begin{aligned}
(f : \varphi) &= (\lambda_s(f) : \varphi) && \text{for all } s \in G \\
(f_1 + f_2 : \varphi) &\leq (f_1 : \varphi) + (f_2 : \varphi) && \text{for all } f_1, f_2 \in \mathcal{K}_{\mathbb{R}}(G) \\
(cf : \varphi) &= c(f : \varphi) && \text{for all } c \geq 0 \\
(f_1 : \varphi) &\leq (f_2 : \varphi) && \text{whenever } f_1 \leq f_2 \\
(f : \varphi) &\geq \|f\|_\infty / \|\varphi\|_\infty && \\
(f : \varphi) &\leq (f : \psi)(\psi : \varphi) && \text{for all positive } \psi \in \mathcal{K}_{\mathbb{R}}(G) \\
0 < \frac{1}{(f_0 : f)} &\leq \frac{(f : \varphi)}{(f_0 : \varphi)} \leq (f : f_0) && \text{for all positive } f, f_0 \in \mathcal{K}_{\mathbb{R}}(G).
\end{aligned}$$

We now make a normalization by fixing some positive nonzero $f_0 \in \mathcal{K}_{\mathbb{R}}(G)$, and defining

$$I_\varphi(f) = \frac{(f : \varphi)}{(f_0 : \varphi)}$$

for every *positive* function $f \in \mathcal{K}_{\mathbb{R}}(G)$. The above properties show that I_φ is a functional which is left-invariant, subadditive, homogeneous of degree 1, and monotone. It also satisfies the following property:

$$\frac{1}{(f_0 : f)} \leq I_\varphi(f) \leq (f : f_0). \tag{$*$}$$

If I_φ were additive rather than subadditive, it would be the restriction to the positive functions in $\mathcal{K}_{\mathbb{R}}(G)$ of a positive linear functional on $\mathcal{K}_{\mathbb{R}}(G)$, and we would be done. To make a linear functional, we need to shrink to the domain of φ, and this is the part of the argument which is the most subtle. Let $\mathcal{K}_{\mathbb{R}}^{+}(G)$ denote the set of positive functions in $\mathcal{K}_{\mathbb{R}}(G)$.

The following technical proposition whose proof is given in Folland [Folland (1995)] (Chapter 2, Lemma 2.18) is needed.

Proposition 8.18. *If f_1 and f_2 are any two positive functions in $\mathcal{K}_{\mathbb{R}}^{+}(G)$ and if $\epsilon > 0$, then there is an open subset V containing 1 such that $I_\varphi(f_1) + I_\varphi(f_2) \leq I_\varphi(f_1 + f_2) + \epsilon$, whenever $\mathrm{supp}(\varphi) \subseteq V$.*

To shrink the domain of φ we use a compactness argument. For every positive function $f \in \mathcal{K}^+_{\mathbb{R}}(G)$, let X_f be the interval

$$X_f = \left[\frac{1}{(f_0 : f)}, (f : f_0)\right].$$

Let $X = \prod_f X_f$. More precisely, X is the set of all functions from $\mathcal{K}^+_{\mathbb{R}}(G)$ to $(0, +\infty)$ mapping f into X_f. We put the topology of Definition 2.2 on X. Since each X_f is compact, by Tychonoff's theorem X is also compact. By $(*)$, we have $I_\varphi(f) \in X$ for all positive nonzero $\varphi \in \mathcal{K}_{\mathbb{R}}(G)$. For every compact neighborhood V containing 1, let $K(V)$ be the closure in X of the subset $\{I_\varphi \mid \mathrm{supp}(\varphi) \subseteq V\}$. The family of subsets $K(V)$ has the finite intersection property since $K(\bigcap_{j=1}^n V_j) \subseteq \bigcap_{j=1}^n K(V_j)$. Since X is compact, there is some $I \in X$ which lies in every $K(V)$. This means that every neighborhood of I in X contains some I_φ with $\mathrm{supp}(\varphi)$ arbitrarily small. In other words, for any open subset V containing 1, any $\epsilon > 0$, and for any positive functions $f_1, \ldots, f_n \in \mathcal{K}_{\mathbb{R}}(G)$, there exist some positive nonzero $\varphi \in \mathcal{K}_{\mathbb{R}}(G)$ with $\mathrm{supp}(\varphi) \subseteq V$ such that $|I(f_j) - I_\varphi(f_j)| < \epsilon$ for all j. By the properties of I_φ listed above and by Proposition 8.18, we conclude that I commutes with left translation, addition, and multiplication by positive scalars.

We can extend I to arbitrary functions $f \in \mathcal{K}_{\mathbb{R}}(G)$ as follows. We can write $f = f_1 - f_2$ with f_1, f_2 positive functions in $\mathcal{K}_{\mathbb{R}}(G)$, and we let $I(f) = I(f_1) - I(f_2)$. If we also have $f = f'_1 - f'_2$ with f'_1, f'_2 positive functions in $\mathcal{K}_{\mathbb{R}}(G)$, then

$$f_1 + f'_2 = f_2 + f'_1,$$

so by linearity of I on positive functions we get

$$I(f_1) + (f'_2) = I(f_2) + I(f'_1),$$

thus

$$I(f_1) - I(f_2) = I(f'_1) - I(f'_2),$$

which means that $I(f)$ is well defined. The functional I is a left Haar functional, and we are done. By Proposition 8.16(3), since the Haar functional I is left-invariant, the corresponding σ-Radon measure is also left-invariant. □

Remark: The proof in Bourbaki [Bourbaki (1965b)] uses an argument involving an ultrafilter instead of Tychonoff's theorem, but otherwise it is identical. Dieudonné [Dieudonné (2005)] assumes that the locally compact group G is separable and metrizable. This allows him to avoid using Tychonoff's theorem, but does not make the proof simpler.

Let μ be the left Haar measure associated with the left Haar functional I given by Theorem 8.17. Here is an immediate consequence of Theorem 8.17.

Proposition 8.19. *If μ is a left Haar measure on G, then for every nonempty open subset U, we have $\mu(U) > 0$. For every positive nonzero function $f \in \mathcal{K}_{\mathbb{R}}(G)$, we have $\int f d\mu > 0$.*

Proof. Assume U is a nonempty open set with $\mu(U) = 0$. Then since μ is left-invariant $\mu(gU) = 0$ for all $g \in G$, and since any compact subset K can be covered by finitely many translates of U, we have $\mu(K) = 0$. Since μ is a σ-regular Borel measure, it is σ-inner regular, that is,

$$\mu(G) = \sup\{\mu(K) \mid K \subseteq G,\ K \text{ compact}\}$$

so $\mu(G) = 0$, contradicting the fact that μ is not the zero measure because it arises from a non-zero left Haar functional by Radon–Riesz I.

For any positive nonzero function $f \in \mathcal{K}_{\mathbb{R}}(G)$, let $U = \{g \in G \mid f(g) > \frac{1}{2} \|f\|_\infty\}$. Then $\int f d\mu > \frac{1}{2} \|f\|_\infty \mu(U) > 0$. ☐

Remark: If G is a Lie group, there are much simpler methods for obtaining a left Haar measure on G. Suppose G has dimension n. Pick an n-differential form ω_0 on $\mathfrak{g}$, and transport it on all tangent spaces by left translation, obtaining a left-invariant volume form ω. Then $f \mapsto \int f\omega$ is a left-invariant Haar functional that induces a left Haar measure.

We now turn to the uniqueness of the Haar measure.

8.4 Uniqueness of the Haar Measure

Any two left Haar measures on a locally compact group are proportional up to a positive factor. All the proofs we are aware of use tricks involving a double integration and Fubini's theorem. These proofs are attributed to von Neumann. In our opinion, the proof using the least devious trick is that of Dieudonné [Dieudonné (2005)] (Chapter XIV, Section 1), also used in Bourbaki [Bourbaki (1965b)] in a slightly more concise form. Since Dieudonné uses a theory of integration based on Radon functionals rather than on measure theory, some minor adaptations need to be made; specifically, Proposition 7.12 is needed instead of Proposition 13.15.3 in Dieudonné [Dieudonné (2005)]. The first step is the following crucial result.

Proposition 8.20. *Given a left Haar functional Φ and a right Haar functional Ψ on a locally compact group G, if ν is the corresponding right Haar measure, for any function $f \in \mathcal{K}_{\mathbb{R}}(G)$, if $\Phi(f) \neq 0$, then the function D_f given by*

$$D_f(s) = \Phi(f)^{-1} \int f(t^{-1}s)\, d\nu(t)$$

for all $s \in G$ is continuous.

Proof. It suffices to show that the function

$$s \mapsto \int f(t^{-1}s)\, d\nu(t)$$

is continuous. Let K be the compact subset of G which is the support of f. Pick any $s_0 \in G$, and let V_0 by any compact neighborhood of s_0. For every $\epsilon > 0$, we have to find an open subset V containing s_0 such that $V \subseteq V_0$ and

$$\left| \int (f(t^{-1}s) - f(t^{-1}s_0)) d\nu(t) \right| < \epsilon$$

for all $s \in V$. In order to have $t^{-1}s, t^{-1}s_0 \in K$, since $s_0, s \in V_0$, it suffices that $t \in V_0K^{-1}$. If we let $L = V_0K^{-1}$, then for $s_0 \in V_0$ and $s \in V \subseteq V_0$,

$$\int (f(t^{-1}s) - f(t^{-1}s_0)) d\nu(t) = \int_L (f(t^{-1}s) - f(t^{-1}s_0)) d\nu(t).$$

By Proposition 8.12, the function f is right uniformly continuous, so there is some open subset W containing 1 such that

$$|f(t^{-1}s) - f(t^{-1}s_0)| < \frac{\epsilon}{\nu(L)}$$

for all $(t^{-1}s_0)^{-1}t^{-1}s \in W$, that is, $s_0^{-1}s \in W$, namely $s \in Ws_0$, and all $t \in G$. If we take $V = V_0 \cap Ws_0$ (see Figure 8.7), then

$$\left| \int_L (f(t^{-1}s) - f(t^{-1}s_0)) d\nu(t) \right| < \int_L |f(t^{-1}s) - f(t^{-1}s_0)| d\nu(t) < \epsilon,$$

as desired. ∎

We are now ready to prove our uniqueness result.

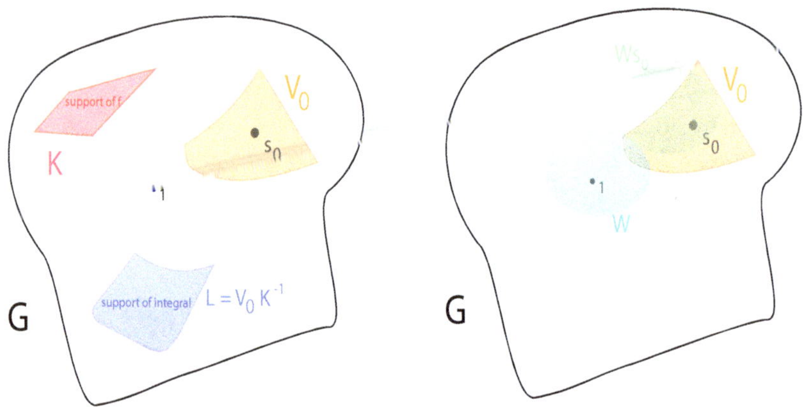

Fig. 8.7 A schematic representation of the sets used in the proof of Proposition 8.20. Observe that $V = V_0 \cap W s_0$ is the intersection of the light green ellipse and peach triangle.

Theorem 8.21. *(Haar) If μ and ν are any two left-invariant Haar measures on a locally compact group G, then there is some $c > 0$ such that $\mu = c\nu$.*

Proof. Since a Haar functional Ψ is right-invariant iff $\check{\Psi}$ is left-invariant, it suffices to prove that if Φ is a left-invariant Haar functional and if Ψ is a right-invariant Haar functional, then there is some $c > 0$ such that $\Phi = c\check{\Psi}$. Let μ be the left Haar measure associated with Φ and let ν be right Haar measure associated with Ψ.

Let $f \in \mathcal{K}_{\mathbb{R}}(G)$ be any function such that $\Phi(f) \neq 0$ and let $g \in \mathcal{K}_{\mathbb{R}}(G)$ be any other function. The function from $G \times G$ to $\mathbb{R}$ given by $(s,t) \mapsto f(s)g(ts)$ is continuous and has compact support. Recall that D_f is given by

$$D_f(s) = \Phi(f)^{-1} \int f(t^{-1}s)\, d\nu(t).$$

For $s = 1$, we have

$$D_f(1) = \Phi(f)^{-1}\check{\Psi}(f).$$

Therefore, if we can show that D_f is independent of f, we are done. We evaluate $\Phi(f)\Psi(g)$ using Fubini's theorem.

$$\begin{aligned}
\Phi(f)\Psi(g) &= \left(\int f(s)d\mu(s)\right)\left(\int g(t)\,d\nu(t)\right) \\
&= \int f(s)\left(\int g(t)\,d\nu(t)\right)d\mu(s) \\
&= \int\left(\int f(s)g(t)\,d\nu(t)\right)d\mu(s) \\
&= \int\left(\int f(s)g(ts)\,d\nu(t)\right)d\mu(s) && \text{by right-invariance of } \nu \\
&= \int\left(\int f(s)g(ts)\,d\mu(s)\right)d\nu(t) && \text{by Fubini} \\
&= \int\left(\int f(t^{-1}s)g(s)\,d\mu(s)\right)d\nu(t) && \text{by left-invariance of } \mu \\
&= \int\left(\int f(t^{-1}s)g(s)\,d\nu(t)\right)d\mu(s) && \text{by Fubini} \\
&= \int g(s)\left(\int f(t^{-1}s)\,d\nu(t)\right)d\mu(s) \\
&= \int g(s)\Phi(f)D_f(s)\,d\mu(s) && \text{by definition of } D_f \\
&= \Phi(f)\Phi(D_f\cdot g),
\end{aligned}$$

where $D_f \cdot g$ is the function given by $(D_f \cdot g)(s) = D_f(s)g(s)$ for all $s \in G$. Since $\Phi(f) \neq 0$, we deduce that

$$\Psi(g) = \Phi(D_f \cdot g).$$

The above equation shows that D_f is independent of f because if f' is another function $f' \in \mathcal{K}_{\mathbb{R}}(G)$ such that $\Phi(f') \neq 0$, then

$$\Phi(D_f \cdot g) = \int D_f(s)g(s)\,d\mu(s) = \int D_{f'}(s)g(s)\,d\mu(s) = \Phi(D_{f'} \cdot g)$$

for all $g \in \mathcal{K}_{\mathbb{R}}(G)$. By Proposition 7.12, we deduce that D_f and $D_{f'}$ are equal a.e. (The version of Proposition 7.12 is stated for complex-valued functions, but it also holds for real-valued functions). However, D_f and $D_{f'}$ are continuous and the subset N where they differ is open and a null set, thus empty by Proposition 8.19. Therefore $D_f = D_{f'} = D$, and by definition of D we have

$$\Phi(f) = D(1)^{-1}\check{\Psi}(f) \quad \text{for all } f \in \mathcal{K}_{\mathbb{R}}(G) \text{ with } \Phi(f) \neq 0.$$

Now Φ and $\check{\Psi}$ are two linear functionals that agree in the complement of the hyperplane H in $\mathcal{K}_{\mathbb{R}}(G)$ of equation $\Phi(f) = 0$, so they agree everywhere.

To see this, pick a basis of $\mathcal{K}_{\mathbb{R}}(G)$ consisting of a basis $(h_j)_{j\in J}$ of H and a function v not in H. We claim that the family consisting of $(h_j+v)_{j\in J}$ and v is a basis of $\mathcal{K}_{\mathbb{R}}(G)$. This family obviously spans $\mathcal{K}_{\mathbb{R}}(G)$ (since every h_j is obtained as h_j+v-v), and it is linearly independent because if we have a finite linear combination

$$\sum_{i\subset I}\lambda_i(h_i+v)+\mu v=0.$$

for any finite subset I of J, then

$$\sum_{i\in I}\lambda_i h_i+\left(\mu+\sum_{i\in I}\lambda_i\right)v=0,$$

and by linear independence, $\lambda_i=0$ for all $i\in I$ and $\mu+\sum_{i\in I}\lambda_i=0$, which implies $\mu=0$, and since this holds for any finite subset I of J, the family consisting of $(h_j+v)_{j\in J}$ and v is linearly independent. Since Φ and Ψ are linear and they agree on a basis, they must be identical.

Since $\Psi\neq 0$, we must have $D(1)\neq 0$, thus, $\Phi=D(1)^{-1}\check{\Psi}$. Since Φ and Ψ are positive functionals, we must have $D(1)>0$.

As we observed earlier, since a locally compact group is the disjoint union of σ-compact cosets, it is legitimate to use Fubini's theorem. □

8.5 Examples of Haar Measures

Here are some examples of Haar measures on various locally compact groups. In most cases, a Haar measure μ on a locally compact group G is defined indirectly by a Haar functional $f\mapsto\int f d\mu$, for all $f\in\mathcal{K}_{\mathbb{C}}(G)$. This Haar functional is denoted by $d\mu$.

Example 8.1. The additive group $\mathbb{R}$ is a locally compact group, and the Lebesgue measure μ_L is a left (and right) Haar measure on it. For a proof see Lang [Lang (1996)] (Chapter VI, Theorem 9.7). An alternative is to use Proposition 8.16(3). By the simple change of variable $x=t+s$, for any function $f\in\mathcal{K}_{\mathbb{C}}(\mathbb{R})$, we have

$$\int_{-\infty}^{\infty}f(t)dt=\int_{-\infty}^{\infty}f(t+s)dt.$$

Example 8.2. The additive group $\mathbb{R}^n$ is a locally compact group, and the product Lebesgue measure μ_L on it (see Section 5.13) is a left (and right) Haar measure on it. This will be shown as an application of Proposition 8.37.

Example 8.3. The multiplicative group $\mathbb{R}_+^*$ is a locally compact group. We claim that $d\mu = dx/x$ is a left Haar measure, where dx is the restriction of the Lebesgue measure to $\mathbb{R}_+^*$. Indeed, using the change of variable $t \mapsto st$, for any function $f \in \mathcal{K}_{\mathbb{C}}(\mathbb{R}_+^*)$, we have

$$\int_0^\infty \frac{f(t)dt}{t} = \int_0^\infty \frac{f(st)sd(t)}{st} = \int_0^\infty \frac{f(st)d(t)}{t},$$

establishing left-invariance. One might wonder what is the measure $\mu([a,b])$ of a closed interval, with $0 < a < b$. We have

$$\mu([a,b]) = \int \chi_{[a,b]} d\mu = \int_{[a,b]} d\mu = \int_a^b \frac{dt}{t} = [\log t]_a^b = \log \frac{b}{a}.$$

For any $s > 0$, we have $s \cdot [a,b] = [sa, sb]$, and

$$\mu(s \cdot [a,b]) = \mu([sa, sb]) = \log \frac{sb}{sa} = \log \frac{b}{a}.$$

This measure is indeed left invariant.

Example 8.4. Let $\mathbb{T} = \mathbf{U}(1) = \{z \in \mathbb{C} \mid |z| = 1\}$, the circle group, that is, the group of complex numbers of unit length. Let $\sigma\colon \mathbb{T} \to \mathbb{R}$ be the injection given by

$$\sigma(e^{i\theta}) = \theta, \quad -\pi \leq \theta < \pi.$$

Define the measure ν_1 on $\mathbb{T}$ by

$$\nu_1(A) = \mu_L(\sigma(A)),$$

on the σ-algebra $\sigma^{-1}(\mathcal{B}(\mathbb{R}))$ defined in Proposition 5.2(2) (where μ_L is the Lebesgue measure on $\mathbb{R}$). For any $f \in \mathcal{L}_{\nu_1}(\mathbb{T})$, we have

$$\int_{\mathbb{T}} f d\nu_1 = \int_{-\pi}^{\pi} f(e^{i\theta})\, d\mu_L(\theta),$$

also written as $\int_{-\pi}^{\pi} f(e^{i\theta})\, d\theta$. Observe that $\int_{\mathbb{T}} d\nu_1 = 2\pi$. It is easy to check that ν_1 is left-invariant. Indeed, for $\theta_0 \in [-\pi, \pi)$, if we let $\varphi = \theta + \theta_0$, we have

$$\int_{-\pi}^{\pi} f(e^{i(\theta+\theta_0)})\, d\theta = \int_{-\pi+\theta_0}^{\pi} f(e^{i\varphi})\, d\varphi + \int_{\pi}^{\pi+\theta_0} f(e^{i\varphi})\, d\varphi.$$

Using the change of variable $\varphi = u + 2\pi$ in the second integral, we get

$$\int_{\pi}^{\pi+\theta_0} f(e^{i\varphi})\, d\varphi = \int_{-\pi}^{-\pi+\theta_0} f(e^{iu}e^{i2\pi})\, du = \int_{-\pi}^{-\pi+\theta_0} f(e^{iu})\, du,$$

and so

$$\int_{-\pi}^{\pi} f(e^{i(\theta+\theta_0)})\, d\theta = \int_{-\pi+\theta_0}^{\pi} f(e^{i\varphi})\, d\varphi + \int_{-\pi}^{-\pi+\theta_0} f(e^{iu})\, du = \int_{-\pi}^{\pi} f(e^{i\theta})\, d\theta.$$

Example 8.5. Let $G = \mathbf{GL}(n, \mathbb{R})$, the group of invertible $n \times n$ real matrices. It can be shown that a left (and right) Haar measure on $\mathbf{GL}(n, \mathbb{R})$ is given by

$$d\mu = \frac{dA}{|\det(A)|^n} = |\det(A)|^{-n} \bigotimes_{i,j} da_{ij}$$

with $A = (a_{ij})$, where da_{ij} is the Lebesgue measure on $\mathbb{R}$, and dA is the Lebesgue measure on $\mathbb{R}^{n^2}$.

In the next section, we explore the relationship between a left Haar measure μ and the left Haar measure $\rho_s(\mu)$.

8.6 The Modular Function

Let μ be a left Haar measure on the locally compact group G. For all $s, t \in G$, since λ_t and ρ_s commute (on functions, measures, and Radon functionals), since μ is a left Haar measure, $\lambda_t(\mu) = \mu$, so we have

$$\lambda_t(\rho_s(\mu)) = \rho_s(\lambda_t(\mu)) = \rho_s(\mu),$$

which means that $\rho_s(\mu)$ is also a left Haar measure. By the uniqueness result of Theorem 8.21, there is a constant $a > 0$ such that

$$\rho_s(\mu) = a\mu.$$

If ν is another left Haar measure, again by theorem Theorem 8.21, we have $\nu = c\mu$ for some $c > 0$, but $\rho_s(\nu)(A) = \nu(As) = c\mu(As) = c\rho_s(\mu)(A)$ for all $A \in \mathcal{B}$, that is, $\rho_s(\nu) = c\rho_s(\mu)$, so

$$\rho_s(\nu) = c\rho_s(\mu) = ca\mu = ac\mu = a\nu.$$

Therefore, the number a such that $\rho_s(\mu) = a\mu$ is independent of μ. It is customary to denote this number by $\Delta(s)$.

Definition 8.12. Let G be a locally compact group. For every $s \in G$, there is a unique positive number $\Delta(s)$ such that

$$\rho_s(\mu) = \Delta(s)\mu \qquad (*)$$

for all left Haar measures μ. The function $\Delta\colon G \to \mathbb{R}_+^*$ (given by $\Delta(s)$ for every $s \in G$) is called the *modular function* of G (if necessary, we denote it by Δ_G to avoid ambiguities).

Observe that $(*)$ can be expressed as

$$\mu(As) = \Delta(s)\mu(A) \quad \text{for all } A \in \mathcal{B} \text{ and all } s \in G.$$

Proposition 8.22. *Let G be a locally compact group and let μ be a left Haar measure on G. For any $f \in \mathcal{L}^1_\mu(G, \mathcal{B}, \mathbb{C})$, we have*

$$\int \rho_s(f)\, d\mu = \Delta(s^{-1}) \int f d\mu.$$

The function $\Delta\colon G \to \mathbb{R}_+^*$ is a continuous homomorphism.

Proof sketch. Let $f = \sum_{i=1}^n y_i \chi_{A_i}$ be a μ-step function. For all $x \in G$, we have $\rho_s(f)(x) = f(xs) = y_i$ iff $xs \in A_i$ iff $x \in A_i s^{-1}$, which shows that

$$\rho_s(f) = \sum_{i=1}^n y_i \chi_{A_i s^{-1}}; \quad \text{see Figure 8.8.}$$

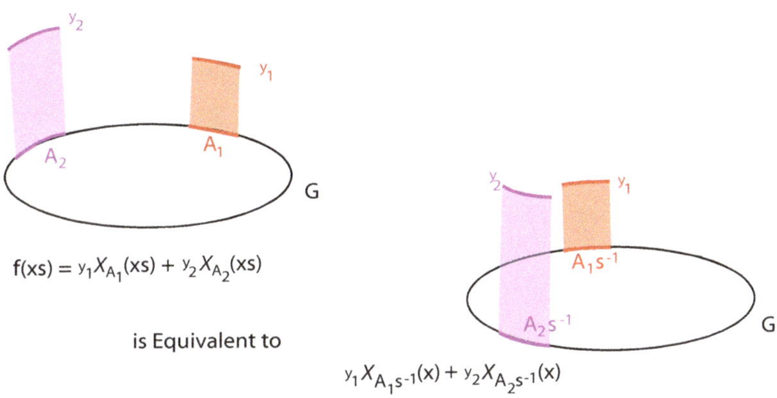

Fig. 8.8 Let $G = S^1$. A step function on S^1 is represented by the top arcs of the colored vertical "rectangular" sheets. The step function $f(xs) = \sum_{k=1}^2 y_k \chi_{A_k}(xs)$ is equivalent to $\rho_s f(x) = \sum_{k=1}^2 y_k \chi_{A_k s^{-1}}(x)$.

Consequently,

$$\int \rho_s(f)\, d\mu = \sum_{i=1}^n y_i \mu(A_i s^{-1}) = \Delta(s^{-1}) \sum_{i=1}^n y_i \mu(A_i) = \Delta(s^{-1}) \int f d\mu,$$

by $(*)$. As in the proof of Proposition 8.16, every function $f \in \mathcal{L}^1_\mu(G, \mathcal{B}, \mathbb{C})$ has some approximation sequence (f_n) by μ-step functions that converges to f a.e. and in the L^1-norm, which allows us to conclude that

$$\int \rho_s(f)\, d\mu = \Delta(s^{-1}) \int f d\mu$$

for every $f \in \mathcal{L}^1_\mu(G, \mathcal{B}, \mathbb{C})$.

By $(*)$ we have

$$\Delta(st)\mu(A) = \mu(Ast) = \Delta(t)\mu(As) = \Delta(t)\Delta(s)\mu(A)$$

for all $A \in \mathcal{B}$, and since $\mu(A) > 0$ if A is open and nonempty (and $\mathbb{R}_+$ is commutative under multiplication!), we deduce that

$$\Delta(st) = \Delta(s)\Delta(t).$$

Thus Δ is a homomorphism from G to the multiplicative group $\mathbb{R}_+^*$. Proposition 8.12 implies that the map $s \mapsto \rho_s(f)$ is uniformly continuous, and so it can be shown that the map $s \mapsto \int \rho_s(f)\, d\mu$ is continuous, and since

$$\int \rho_s(f)\, d\mu = \Delta(s^{-1}) \int f d\mu,$$

we deduce that Δ is continuous. □

The equation

$$\int \rho_s(f)\, d\mu = \Delta(s^{-1}) \int f d\mu$$

is also written as

$$\int f(xs) d\mu(x) = \Delta(s^{-1}) \int f(x) d\mu(x),$$

or equivalently as

$$\int f(xs^{-1}) d\mu(x) = \Delta(s) \int f(x) d\mu(x).$$

Since $\Delta \colon G \to \mathbb{R}_+^*$ is a group homomorphism, we have $\Delta(s^{-1}) = (\Delta(s))^{-1}$.

Definition 8.13. We write Δ^{-1} for the function given by $\Delta^{-1}(s) = \Delta(s^{-1})$ for all $s \in G$.

Let G and G' be two locally compact groups. An *isomorphism* is map $\varphi \colon G \to G'$ which is a group isomorphism and a homeomorphism. Then it is easy to check that φ maps Borel sets of G to Borel sets of G'.

Definition 8.14. Let G and G' be two locally compact groups, and $\varphi \colon G \to G'$ be an isomorphism. Given a measure μ' on G', we define the map $\varphi^{-1}(\mu')$ with domain $\mathcal{B}(G)$ by

$$(\varphi^{-1}(\mu'))(A) = \mu'(\varphi(A)) \quad \text{for all } A \in \mathcal{B}(G);$$

see Figure 8.9.

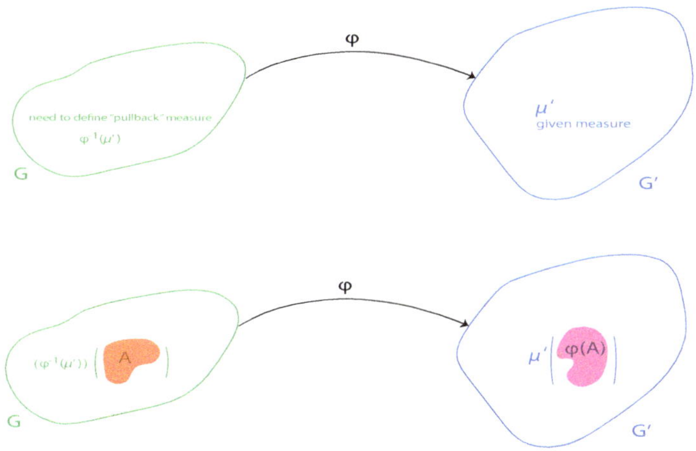

Fig. 8.9 A schematic representation of Definition 8.14.

Proposition 8.23. *Let G and G' be two locally compact groups, and let $\varphi\colon G \to G'$ be an isomorphism. For any left Haar measure μ' on G', the map $\varphi^{-1}(\mu')$ is a left Haar measure on G. We have*

$$\Delta_G = \Delta_{G'} \circ \varphi.$$

Proof sketch. The fact that $\varphi^{-1}(\mu')$ is a measure follows from the fact that φ maps Borel sets to Borel sets and is a bijection, so it preserves union and disjointness. The details are left as an exercise. Since μ' is a left Haar measure and φ is a homomorphism,

$$\varphi^{-1}(\mu')(sA) = \mu'(\varphi(sA)) = \mu'(\varphi(s)\varphi(A)) = \mu'(\varphi(A)) = \varphi^{-1}(\mu')(A),$$

so $\varphi^{-1}(\mu')$ is a left Haar measure.

We have

$$\begin{aligned}(\varphi^{-1}(\mu'))(As) &= \Delta_G(s)(\varphi^{-1}(\mu'))(A)\\ &= \Delta_G(s)\mu'(\varphi(A)),\end{aligned}$$

and

$$\begin{aligned}(\varphi^{-1}(\mu'))(As) &= \mu'(\varphi(As))\\ &= \mu'(\varphi(A)\varphi(s))\\ &= \Delta_{G'}(\varphi(s))\mu'(\varphi(A)),\end{aligned}$$

which implies

$$\Delta_{G'}(\varphi(s)) = \Delta_G(s) \quad \text{for all } s \in G$$

since we can pick a nonempty open subset A of G, and $\varphi(A)$ is a nonempty open subset of G'. □

Corollary 8.24. *If $G' = G$, that is, $\varphi\colon G \to G$ is an automorphism, then $\Delta \circ \varphi = \Delta$.*

Definition 8.15. Let G be a locally compact group. We say that G is *unimodular* if $\Delta(s) = 1$ for all $s \in G$, equivalently, if and only if a left Haar measure is also a right Haar measure.

Luckily, many familiar groups are unimodular (but unfortunately, not the affine groups of rigid motions). Obviously, abelian locally compact groups are unimodular.

Given a group G, recall that its *commutator subgroup* $[G, G]$ is the subgroup generated by all elements $[s, t] = sts^{-1}t^{-1}$. The group $[G, G]$ is a normal subgroup of G.

Proposition 8.25. *Let G be a locally compact group.*

(1) If there is a compact neighborhood V of 1 such that $s^{-1}Vs = V$ for all $s \in G$, then G is unimodular. Consequently, if G is compact, discrete, or commutative, then G is unimodular.
(2) If K is any compact subgroup of G, then $\Delta \mid K \equiv 1$.
(3) If $G/[G, G]$ is compact, then G is unimodular. As a consequence, every connected semisimple Lie group is unimodular. Recall that semisimple Lie group is a Lie group G such that the Killing form on its Lie algebra $\mathfrak{g}$ is nondegenerate.

Proof. (1) Let μ be any left Haar measure. Since μ is left-invariant

$$\mu(V) = \mu(s^{-1}Vs) = \mu(Vs) = \Delta(s)\mu(V),$$

but $\mu(V) > 0$ because V contains a nonempty open subset, so $\Delta(s) = 1$ for all $s \in G$. The corollaries are left as an easy exercise.

(2) Since Δ is continuous, $\Delta(K)$ is a compact subgroup of $\mathbb{R}_+^*$, which implies $\Delta(K) = \{1\}$.

(3) Since $\mathbb{R}_+^*$ is abelian, we have

$$\begin{aligned}\Delta([s, t]) &= \Delta(sts^{-1}t^{-1}) = \Delta(s)\Delta(t)\Delta(s)^{-1}\Delta(t)^{-1} \\ &= \Delta(s)\Delta(s)^{-1}\Delta(t)\Delta(t)^{-1} = 1,\end{aligned}$$

so Δ vanishes on $[G, G]$. It follows that Δ factors through $G/[G, G]$ as $\Delta = \pi \circ \theta$ where $\theta\colon G/[G, G] \to \mathbb{R}_+^*$ is a continuous homomorphism. Since $G/[G, G]$ is compact, we have $\theta(G/[G, G]) = \{1\}$, so $\Delta(G) = \{1\}$.

If G is a connected semisimple Lie group, it is known that $G = [G, G]$, so G is unimodular. $\square$

In order to discuss the behavior of the operator $\mu \mapsto \check{\mu}$ we need the following proposition.

Proposition 8.26. *Let μ and ν be two Radon measures on a locally compact topological space X. If there is a continuous function $g\colon X \to \mathbb{R}_+^*$ such that*

$$\int f d\nu = \int f g \, d\mu \quad \text{for all } f \in \mathcal{K}_{\mathbb{C}}(X),$$

and if $\widetilde{\nu}$ is the Radon measure given by

$$\widetilde{\nu}(E) = \int_E g \, d\mu \quad \text{for all } E \in \mathcal{B}(X),$$

then $\nu = \widetilde{\nu}$.

A proof of Proposition 8.26 is given in Folland [Folland (1995)] (Chapter 2, Proposition 2.23).

We propose to denote the Radon measure $\widetilde{\nu}$ by $g \cdot \mu$, by analogy with the definition of the Radon functional $g \cdot \Phi$ in Example 7.1(3). The notation $g \, d\mu$ is also used.

The following proposition shows the behavior of the operator $\mu \mapsto \check{\mu}$.

Proposition 8.27. *Let G be a locally compact group. For every left Haar measure μ, the measure $\check{\mu}$ is a right Haar measure, and we have*

$$\int \check{f} \, d\mu = \int f d\check{\mu}, \quad \check{\mu} = \Delta^{-1} \cdot \mu,$$

equivalently

$$\int f(t^{-1}) \, d\mu(t) = \int f(t)\Delta(t^{-1}) \, d\mu(t),$$

and

$$\int f(s) d\check{\mu}(s) = \int f(s)\Delta(s^{-1}) d\mu(s), \quad \int f(s^{-1})\Delta(s^{-1}) d\mu(s) = \int f(s) d\mu(s)$$

for all $f \in \mathcal{L}_{\mu}^1(G, \mathcal{B}, \mathbb{C})$.

Proof sketch. For every μ-step function

$$f = \sum_{i=1}^{n} y_i \chi_{A_i},$$

since $\check{f}(s) = f(s^{-1})$, we immediately obtain

$$\check{f} = \sum_{i=1}^{n} y_i \chi_{A_i^{-1}},$$

(see Figure 8.10) and since $\check{\mu}(A_i) = \mu(A_i^{-1})$, we get

$$\int \check{f}\, d\mu = \int f d\check{\mu}.$$

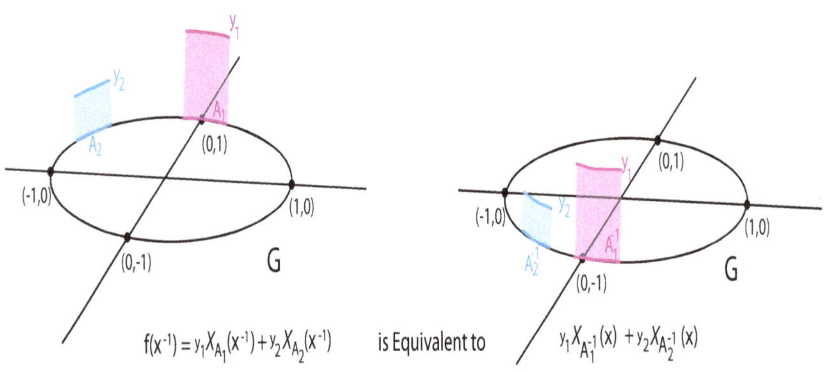

Fig. 8.10 Let $G = S^1$. A step function on S^1 is represented by the top arcs of the colored vertical "rectangular" sheets. The step function $f(x^{-1}) = \sum_{k=1}^{2} y_k \chi_{A_k}(x^{-1})$ is equivalent to $\check{f}(x) = \sum_{k=1}^{2} y_k \chi_{A_k-1}(x)$.

Then by a familiar argument using approximations sequences, we deduce that

$$\int \check{f}\, d\mu = \int f d\check{\mu} \qquad (*_{cech})$$

for all $f \in \mathcal{L}^1_\mu(G, \mathcal{B}, \mathbb{C})$.

Using the fact that Δ is a group homomorphism, for any $f \in \mathcal{K}_{\mathbb{C}}(G)$, we have

$$\begin{aligned}
\int (\rho_{t^{-1}}(f))(s)\Delta(s^{-1})d\mu(s) &= \Delta(t^{-1}) \int f(st^{-1})\Delta(t)\Delta(s^{-1})d\mu(s) \\
&\qquad \text{by Definition of } \rho_{t^{-1}}(f) \\
&= \Delta(t^{-1}) \int f(st^{-1})\Delta^{-1}(st^{-1})d\mu(s) \\
&= \Delta(t^{-1})\Delta(t) \int f(s)\Delta^{-1}(s)d\mu(s) \\
&\qquad \text{by Proposition 8.22 applied to } f\Delta^{-1} \\
&= \int f(s)\Delta(s^{-1})d\mu(s),
\end{aligned}$$

which shows that the Radon functional $f \mapsto \int f(s)\Delta(s^{-1})d\mu(s)$ is right-invariant. The corresponding Haar measure ν is a right Haar measure, and since $\check{\mu}$ is a right Haar measure, there is some $a > 0$ such that $a\check{\mu} = \nu$. Then we have

$$a \int f(s)d\check{\mu}(s) = \int f(s)d\nu = \int f(s)\Delta(s^{-1})d\mu(s),$$

for all $f \in \mathcal{K}_{\mathbb{C}}(G)$, and since Δ^{-1} is a positive continuous function, by Proposition 8.26, $\nu = \Delta^{-1} \cdot \mu$, so

$$a\check{\mu} = \Delta^{-1} \cdot \mu.$$

It remains to show that $a = 1$.

Assume that $a \neq 1$. Since Δ is continuous, there is a symmetric neighborhod U of 1 such that $|\Delta(s^{-1}) - 1| \leq 1/2|a - 1|$ on U. Since U is symmetric, $\mu(U) = \check{\mu}(U)$, and we have

$$|a-1|\mu(U) = |a\check{\mu}(U) - \mu(U)| = \left| \int_U (\Delta(s^{-1}) - 1)d\mu(s) \right| \leq \frac{1}{2}|a-1|\mu(U),$$

a contradiction.

Therefore,

$$\int f(s)d\check{\mu}(s) = \int f(s)\Delta(s^{-1})d\mu(s),$$

for all $f \in \mathcal{L}^1_{\mu}(G, \mathcal{B}, \mathbb{C})$, so

$$\int f(s)\Delta(s^{-1})d\mu(s) = \int \check{f}(s)\, d\mu(s),$$

and by changing f to $\check{f}$, we obtain the desired equation. $\square$

As a corollary, if G is unimodular, then we have

$$\int f(sx)d\mu(x) = \int f(xs)d\mu(x) = \int f(x^{-1})d\mu(x) = \int f(x)d\mu(x)$$

for all $f \in \mathcal{L}^1_\mu(G, \mathcal{B}, \mathbb{C})$, and

$$\mu(A) = \mu(As) = \mu(sA) = \mu(A^{-1}),$$

for all $A \in \mathcal{B}$.

Remark: If G is a Lie group, then it can be shown that the modular function Δ is given by

$$\Delta(s) = |\det \mathrm{Ad}(s^{-1})|;$$

see Gallier and Quaintance [Gallier and Quaintance (2020b)] (Chapter 6, Proposition 6.25).

8.7 More Examples of Haar Measures

In the examples of Section 8.5, the groups under consideration were unimodular. The groups of the next examples are not unimodular.

Example 8.6. Let $G = \mathbf{GA}(n, \mathbb{R})$, the affine group of $\mathbb{R}^n$, which consists of pairs (A, u) with $A \in \mathbf{GL}(n, \mathbb{R})$ and $u \in \mathbb{R}^n$, acting on $\mathbb{R}^n$ by $(A, u)(X) = Ax + u$. It can be shown that a left Haar measure on $\mathbf{GA}(n, \mathbb{R})$ is given by

$$d\mu_L = |\det(A)|^{-n-1} \bigotimes_{i,j} da_{ij} \otimes \bigotimes_i du_i$$

with $A = (a_{ij})$, and $u = (u_i)$, where da_{ij} and du_i is the Lebesgue measure on $\mathbb{R}$. A right Haar measure is given by

$$d\mu_R = |\det(A)|^{-n} \bigotimes_{i,j} da_{ij} \otimes \bigotimes_i du_i,$$

and the modular function is given by

$$\Delta((A, u)) = |\det(A)|^{-1}.$$

A proof of these facts can be found in Bourbaki [Bourbaki (1965b)] (Chapter VII, Section 2, no. 10, Proposition 14, and Section 3, no. 3, Example 2). In particular, if $n = 1$, then an affine bijection is a map $x \mapsto ax + b$ with $a \neq 0$, and we have $d\mu_L = da\, db/a^2$, $d\mu_R = da\, db/|a|$, and $\Delta((a, b)) = |a|^{-1}$.

Remark: In view of Proposition 8.27, the value of the modular function is not unexpected.

Example 8.7. Let $G = \mathbf{T}(n,\mathbb{R})$, the group of invertible upper triangular matrices. It can be shown that a left Haar measure on $\mathbf{T}(n,\mathbb{R})$ is given by

$$d\mu_L = \prod_{i=1}^{n} |a_{ii}|^{i-n-1} \bigotimes_{i\leq j} da_{ij}$$

with $A = (a_{ij})$, and da_{ij} is the Lebesgue measure on $\mathbb{R}$. A right Haar measure is given by

$$d\mu_R = \prod_{i=1}^{n} |a_{ii}|^{-i} \bigotimes_{i\leq j} da_{ij},$$

and the modular function is given by

$$\Delta(A) = \prod_{i=1}^{n} |a_{ii}|^{2i-n-1}.$$

A proof of these facts can be found in Bourbaki [Bourbaki (1965b)] (Chapter VII, Section 2, no. 10, Proposition 14, and Section 3, no. 3, Example 4).

Remark: In view of Proposition 8.27, the value of the modular function is not unexpected.

More examples can be found in Bourbaki [Bourbaki (1965b)] (Chapter VII, Section 3, no. 3). The group $\mathbf{SL}(n,\mathbb{R})$ is unimodular, but finding a Haar measure for it is nontrivial.

8.8 The Modulus of an Automorphism

We now consider the effect of an automorphism $u\colon G \to G$ on a Haar measure. Recall that u is a group isomorphism and a homeomorphism.

Definition 8.16. Let G be a locally compact group and let $u\colon G \to G$ be an automorphism of G. For every function $f\colon G \to \mathbb{C}$, define the function $u(f)$ by

$$(u(f))(s) = f(u^{-1}(s)), \quad \text{for all } s \in G,$$

(see Figure 8.11), and for every left Haar measure μ, define the measure $u^{-1}(\mu)$ by

$$(u^{-1}(\mu))(A) = \mu(u(A)), \quad \text{for all } A \in \mathcal{B};$$

see Figure 8.12.

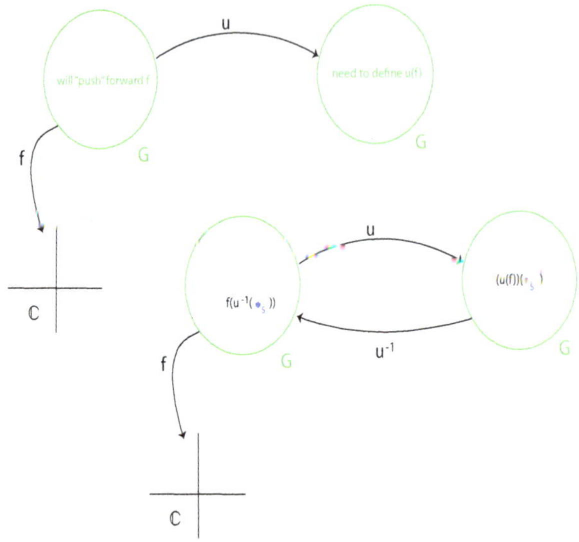

Fig. 8.11 A schematic illustration of the "push forward" function $u(f)$.

It is immediately verified that if u and v are two automorphisms of G, then

$$(u \circ v)(f) = (u(v(f)), \quad (u \circ v)(\mu) = (u(v(\mu)).$$

Also observe that since μ is left-invariant and u is an automorphism,

$$(u^{-1}(\mu))(sA) = \mu(u(sA)) = \mu(u(s)u(A)) = \mu(u(A)) = u^{-1}(\mu)(A),$$

so $u^{-1}(\mu)$ is left-invariant. By the uniqueness of a left Haar measure up to a constant, there is a real $a > 0$ such that $u^{-1}(\mu) = a\mu$. For any other left Haar measure $\nu = c\mu$, we have

$$\begin{aligned}(u^{-1}(\nu))(A) = \nu(u(A)) &= c\mu(u(A)) = cu^{-1}(\mu)(A)\\ &= ca\mu(A) = ac\mu(A) = a\nu(A).\end{aligned}$$

Therefore, the constant a is independent of the left Haar measure μ.

Definition 8.17. Let G be a locally compact group. For every automorphism $u\colon G \to G$, there is a unique positive number $\mathrm{mod}(u)$ such that

$$u^{-1}(\mu) = \mathrm{mod}(u)\mu$$

for all left Haar measures μ. The number $\mathrm{mod}(u)$ is called the *modulus of the automorphism u*.

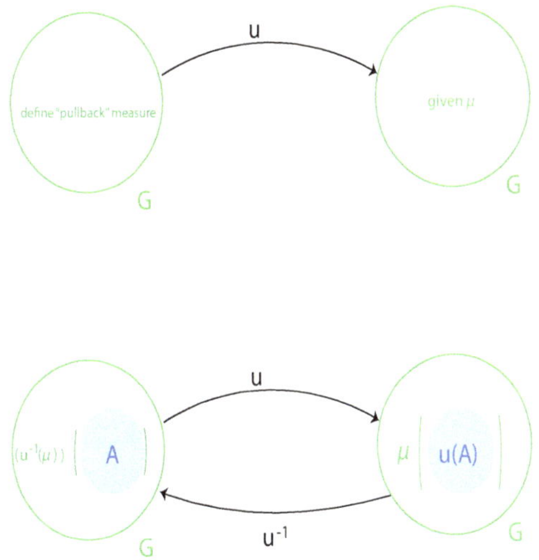

Fig. 8.12 A schematic illustration of the "pullback" measure $u^{-1}(\mu)$.

Note that the condition of Definition 8.17 can also be expressed as

$$\mu(u(A)) = \text{mod}(u)\mu(A) \quad \text{for all } A \in \mathcal{B}. \tag{$**$}$$

Proposition 8.28. *Let G be a locally compact group and let μ be any left Haar measure on G. For every automorphism $u\colon G \to G$, we have*

$$\int u(f)d\mu = \int f du^{-1}(\mu) = \text{mod}(u) \int f d\mu$$

for all $f \in \mathcal{L}^1_\mu(G, \mathcal{B}, \mathbb{C})$.

Proof sketch. For every μ-step function

$$f = \sum_{i=1}^n y_i \chi_{A_i},$$

since $(u(f))(s) = f(u^{-1}(s))$, we have $f(u^{-1}(s)) = y_i$ iff $u^{-1}(s) \in A_i$ iff $s \in u(A_i)$, which means that

$$u(f) = \sum_{i=1}^n y_i \chi_{u(A_i)};$$

see Figure 8.13.

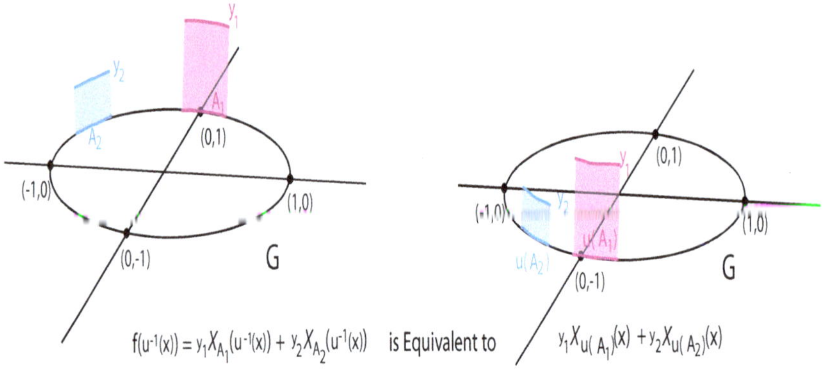

Fig. 8.13 Let $G = S^1$. A step function on S^1 is represented by the top arcs of the colored vertical "rectangular" sheets. The step function $f(u^{-1}(x)) = \sum_{k=1}^{2} y_k \chi_{A_k}(u^{-1}(x))$ is equivalent to $(u(f))(x) = \sum_{k=1}^{2} y_k \chi_{u(A_i)}(x)$.

Thus

$$\int u(f)\, d\mu = \sum_{i=1}^{n} y_i \mu(u(A_i)) = \sum_{i=1}^{n} y_i u^{-1}(\mu)(A_i) = \int f du^{-1}(\mu).$$

Then by a familiar argument using approximations sequences, we deduce that

$$\int u(f) d\mu = \int f du^{-1}(\mu)$$

for all $f \in \mathcal{L}^1_\mu(G, \mathcal{B}, \mathbb{C})$. Since $u^{-1}(\mu) = \text{mod}(u)\mu$, we get the second equation. $\square$

Proposition 8.28 can also be stated as

$$\int f(u^{-1}(s)) d\mu(s) = \text{mod}(u) \int f(s) d\mu(s).$$

Suppose that μ is a right Haar measure. As in the case of a left Haar measure, we define the measure $u^{-1}(\mu)$ by

$$(u^{-1}(\mu))(A) = \mu(u(A)) \quad \text{for all } A \in \mathcal{B}(G).$$

The measure $u^{-1}(\mu)$ is a right Haar measure because

$$(u^{-1}(\mu))(As) = \mu(u(As)) = \mu(u(A)u(s)) = \mu(u(A)) = (u^{-1}(\mu))(A).$$

As in the left-invariant case, for every automorphism $u\colon G \to G$, there a constant $c > 0$ such that $u^{-1}(\mu) = c\mu$. Interestingly, $c = \text{mod}(u)$, so there is *no difference* between the left modulus and the right modulus of an automorphism.

Proposition 8.29. *Let G be a locally compact group and let $u\colon G \to G$ be an automorphism. Then for all left Haar measures and all right Haar measures μ on G, we have*

$$u^{-1}(\mu) = \mathrm{mod}(u)\mu,$$

where $\mathrm{mod}(u)$ *is the modulus of u defined for left Haar measures (see Definition 8.17).*

Proof. We use Corollary 8.24 which implies that $\Delta \circ u^{-1} = \Delta$, since u^{-1} is also an automorphism when u is an automorphism. As a consequence,

$$\Delta(s^{-1}) = \Delta(u^{-1}(s^{-1})) = \Delta((u^{-1}(s))^{-1}) = \Delta^{-1}(u^{-1}(s)),$$

that is,

$$\Delta(s^{-1}) = \Delta^{-1}(u^{-1}(s)). \tag{$\dagger$}$$

Recall that if μ is a left Haar measure, then $\check{\mu}$ is a right Haar measure, and by Proposition 8.27 we have $\check{\mu} = \Delta^{-1} \cdot \mu$. Then for every $f \in \mathcal{K}_{\mathbb{C}}(G)$, since $(u(f))(s) = f(u^{-1}(s))$, we have

$$\begin{aligned}
\int f(s)\, du^{-1}(\Delta^{-1} \cdot \mu)(s) &= \int (u(f))(s) d(\Delta^{-1} \cdot \mu)(s) && \text{by Prop 8.28} \\
&= \int f(u^{-1}(s))\Delta(s^{-1}) d\mu(s) \\
&= \int f(u^{-1}(s))\Delta^{-1}(u^{-1}(s)) d\mu(s) && \text{by } (\dagger) \\
&= \mathrm{mod}(u) \int f(s)\Delta^{-1}(s) d\mu(s) && \text{by Prop 8.28} \\
&= \mathrm{mod}(u) \int f(s)\Delta(s^{-1}) d\mu(s) \\
&= \mathrm{mod}(u) \int f(s)\, d(\Delta^{-1} \cdot \mu)(s).
\end{aligned}$$

By the uniqueness of the Radon measure associated with a Radon functional, this proves that

$$u^{-1}(\Delta^{-1} \cdot \mu) = \mathrm{mod}(u)\, \Delta^{-1} \cdot \mu,$$

and by Proposition 8.27, we obtain, $u^{-1}(\check{\mu}) = \mathrm{mod}(u)\check{\mu}$. Since every right Haar measure is of the form $\check{\mu}$ for some left Haar measure μ, we proved our result. $\square$

For every $s \in G$, if C_s is the automorphism conjugation by s, namely $C_s(t) = sts^{-1}$, then we have the following result.

Proposition 8.30. *Let G be a locally compact group. For every $s \in G$, we have*

$$\operatorname{mod}(C_s) = \Delta(s^{-1}).$$

Proof. We prove that $\operatorname{mod}(C_{s^{-1}}) = \Delta(s)$, which is equivalent to the equation of the Proposition. By Definition 8.16, for any left Haar measure μ,

$$(C_{s^{-1}}^{-1}(\mu))(A) = \mu(C_{s^{-1}}(A)) = \mu(s^{-1}As) = (\rho_s(\lambda_s(\mu)))(A).$$

Since μ is left-invariant, $\lambda_s(\mu) = \mu$, and by definition of the modulus $\rho_s(\mu) = \Delta(s)\mu$, so

$$C_{s^{-1}}^{-1}(\mu) = \rho_s(\lambda_s(\mu)) = \rho_s(\mu) = \Delta(s)\mu,$$

which by Definition 8.17 shows that

$$\operatorname{mod}(C_{s^{-1}}) = \Delta(s),$$

as claimed. ☐

Proposition 8.31. *Let G be a locally compact group.*

(1) If G is compact or discrete, then $\operatorname{mod}(u) = 1$ for any automorphism $u\colon G \to G$.

(2) For any two automorphisms $u\colon G \to G$ and $v\colon G \to G$, we have

$$\operatorname{mod}(u \circ v) = \operatorname{mod}(u)\operatorname{mod}(v).$$

Proof. (1) Since u is an automorphism $u(G) = G$ and $u(\{1\}) = \{1\}$. If G is compact, let $A = G$ in Equation (∗∗) to obtain

$$\mu(G) = \mu(u(G)) = \operatorname{mod}(u)\mu(G),$$

and if G is discrete, let $A = \{1\}$, where μ is any left Haar measure.

(2) Using Equation (∗∗), we have

$$\mu((u \circ v)(A)) = \mu(u(v(A))) = \operatorname{mod}(u)\mu(v(A)) = \operatorname{mod}(u)\operatorname{mod}(v)\mu(A)),$$

and we choose A to be any open nonempty subset. ☐

If $G = \mathbb{R}^n$ (which is a locally compact group under addition with the topology induced by any norm), and u a linear automorphism of $\mathbb{R}^n$, that is, an invertible linear map of $\mathbb{R}^n$, then we have the following interesting characterization of $\operatorname{mod}(u)$.

Proposition 8.32. *Let $u\colon \mathbb{R}^n \to \mathbb{R}^n$ be an invertible linear map, with $\mathbb{R}^n$ as an additive group with the Lebesgue measure. Then*

$$\operatorname{mod}(u) = |\det(u)|.$$

Sketch of proof. This result makes use of the following fact from linear algebra which is stated in Gallier and Quaintance [Gallier and Quaintance (2020c)] Chapter 7, Proposition 7.18, which can be restated as stating that every real $n \times n$ invertible matrix can be expressed as the product of elementary matrices $E_{i,j;\beta} = I_n + \beta E_{ij}$ and $I_n + (\alpha - 1)E_{nn}$. Then one must check the formula of the proposition,

$$\int u(f)(x_1, \ldots, x_n)d\mu(x_1, \ldots, x_n) = \mathrm{mod}(u) \int f(x_1, \ldots, x_n)d\mu(x_1, \ldots, x_n),$$

by integrating the functions of the form

$$f(x_1, \ldots, x_{n-1}, \alpha x_n)$$

and

$$f(x_1, \ldots, x_j + \beta x_i, \ldots, x_n),$$

with $f \in \mathcal{K}_{\mathbb{R}}(\mathbb{R}^n)$, using a change of variables. Details can be found in Dieudonné [Dieudonné (2005)] (Chapter XIV, Proposition 14.3.9.1). ☐

As an application of Proposition 8.32, we obtain formulae for the measure (volume) of a parallelotope and of a simplex in $\mathbb{R}^n$.

Let $(v_1, \ldots, v_n)$ be n linearly independent vectors in $\mathbb{R}^n$. Then the set

$$P = \{\lambda_1 v_1 + \cdots + \lambda_n v_n \mid (\lambda_1, \ldots, \lambda_n) \in \mathbb{R}^n,\ 0 \leq \lambda_i \leq 1\}$$

is called a *parallelotope*; see Figure 8.14. The set

$$S = \{\lambda_1 v_1 + \cdots + \lambda_n v_n \mid (\lambda_1, \ldots, \lambda_n) \in \mathbb{R}^n,\ \lambda_i \geq 0,\ \lambda_1 + \cdots + \lambda_n \leq 1\}$$

is called a *simplex*; see Figure 8.15.

Proposition 8.33. *Let $(v_1, \ldots, v_n)$ be n linearly independent vectors in $\mathbb{R}^n$, and let P be the parallelotope and S be the simplex determined by $(v_1, \ldots, v_n)$. If μ is the Lebesgue measure on $\mathbb{R}^n$, then*

$$\mu(P) = |\det(v_1, \ldots, v_n)|, \quad \mu(S) = \frac{1}{n!}|\det(v_1, \ldots, v_n)|.$$

Proof sketch. Since $(v_1, \ldots, v_n)$ are linearly independent, there is a unique linear map u such that $u(e_i) = v_i$, for $i = 1, \ldots, n$, where e_i is the canonical basis vector of $\mathbb{R}^n$. Then $P = u(K)$, where K is the n-cube determined by $(e_1 \ldots, e_n)$, and

$$\mu(P) = \int \chi_P d\mu = \int u(\chi_K) d\mu = \mathrm{mod}(u) \int \chi_K d\mu = |\det(u)|\mu(K).$$

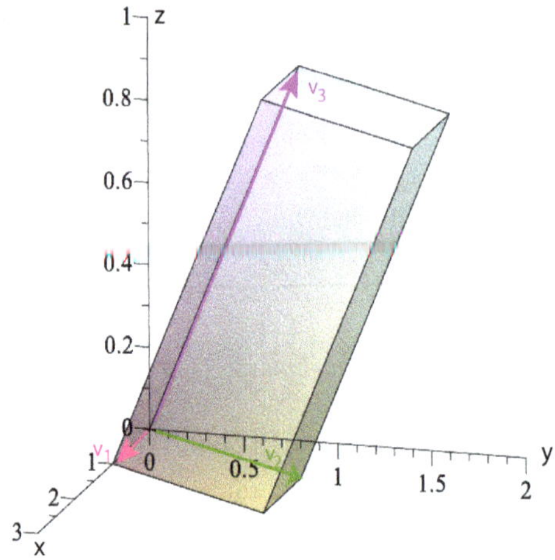

Fig. 8.14 The parallelotope in $\mathbb{R}^3$ spanned by the vectors $v_1 = (1, 0, 0)$, $v_2 = (1, 1, 0)$, and $v_3 = (1, 1, 1)$.

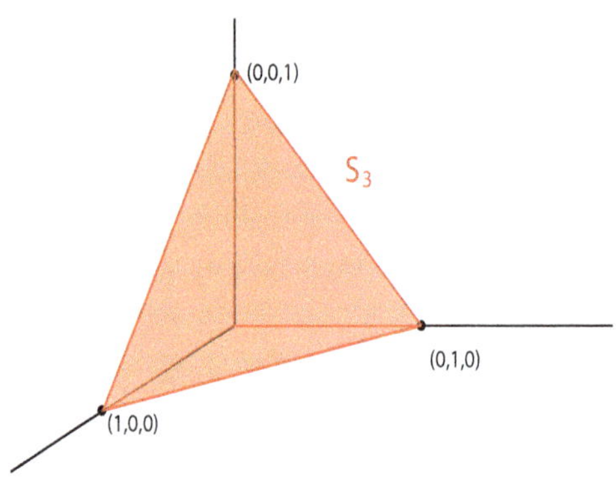

Fig. 8.15 The standard simplex S_3 is the solid tetrahedron spanned by the basis vectors e_1, e_2, and e_3.

But under the Lebesgue measure, $\mu(K) = 1$, and we get

$$\mu(P) = |\det(u)| = |\det(v_1, \ldots, v_n)|,$$

as claimed.

For the simplex, write S_m for simplex determined by the canonical basis vectors $e_1, \ldots, e_m$, and write μ_m for the Lebesgue measure in $\mathbb{R}^m$. Then

$S = u(S_n)$, so by a similar reasoning

$$\mu(S) = |\det(u)|\mu(S_n),$$

and we are reduced to computing $\mu(S_n)$. We view $\mathbb{R}^n$ as $\mathbb{R}^{n-1} \times \mathbb{R}$ and we consider the section $S_n(\lambda)$ of S_n consisting of the set of points $(x_1, \ldots, x_{n-1}) \in \mathbb{R}^{n-1}$ such that

$$x_1, \ldots, x_{n-1} \geq 0, \quad x_1 + \cdots + x_{n-1} \leq 1 - \lambda,$$

where $0 \leq \lambda \leq 1$; see Figure 8.16.

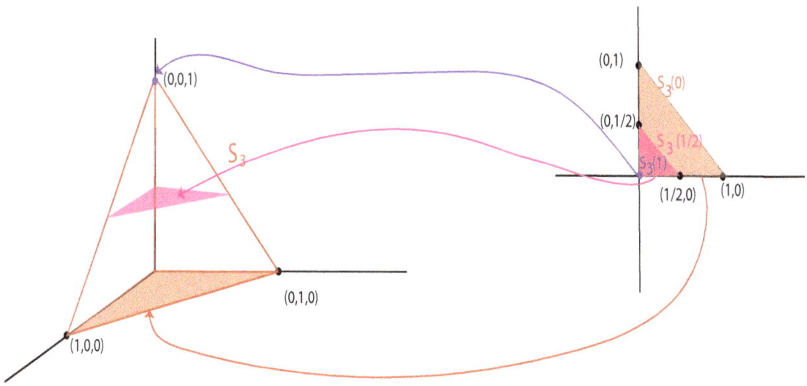

Fig. 8.16 The standard simplex S_3 with its embedded cross sections $S_n(\lambda)$, where $0 \leq \lambda \leq 1$.

This section is the image of $S_n(0)$ under the scaling by $1 - \lambda$, so we get

$$\mu_{n-1}(S_n(\lambda)) = (1-\lambda)^{n-1}\mu_{n-1}(S_{n-1}).$$

Then by Fubini we get

$$\mu_n(S_n) = \int_0^1 (1-\lambda)^{n-1}\mu_{n-1}(S_{n-1})d\lambda = \frac{1}{n}\mu_{n-1}(S_{n-1}).$$

By induction

$$\mu_{n-1}(S_{n-1}) = \frac{1}{(n-1)!},$$

so we get

$$\mu_n(S_n) = \frac{1}{n!}$$

as claimed. $\square$

As another application of Proposition 8.32, the computation of the measure (volume) of a closed ball in $\mathbb{R}^n$ can be found in Dieudonné [Dieudonné (2005)] (Chapter XIV, Proposition 14.3.11).

More on the modular function and the modulus of an automorphism can be found in Bourbaki [Bourbaki (1965b)], Chapter VII, Section 1.

8.9 Some Properties and Applications of the Haar Measure

Since the Haar measure is σ-regular and locally finite, Theorem 7.11 implies the following result which will be needed in Vol II, Chapter 3.

Theorem 8.34. *Let G be a locally compact, metrizable, separable group equipped with a left Haar measure. Then $\mathrm{L}^p_\mu(G, \mathbb{C})$ is separable for $p = 1, 2$.*

The following result is proven in Dieudonné [Dieudonné (2005)] (Chapter XIV, Proposition 14.2.3).

Proposition 8.35. *Let G be a locally compact group, and let μ be a left Haar measure on G. Then G is discrete if and only if $\mu(\{1\}) > 0$, and G is compact if and only if $\mu(G) < +\infty$.*

An interesting and important application of the Haar measure is the construction of a Hermitian inner product invariant under the representation of a compact group. The idea of such a construction originates with Hurwitz and was generalized by H. Weyl.

Let K be a topological group, and let H be a (complex) finite-dimensional Hermitian space (with inner product $\langle -, - \rangle$ and corresponding norm $\| \ \|$). A *representation* of K in H is a group homomorphism $U \colon K \to \mathbf{GL}(H)$, where $\mathbf{GL}(H)$ is the group of invertible linear maps on H.

Theorem 8.36. *Let K be a compact group, let H be a finite-dimensional Hermitian space, and let $U \colon K \to \mathbf{GL}(H)$ be a representation of K which is continuous when $\mathbf{GL}(H)$ is equipped with the operator norm. Then there is a Hermitian inner product φ on H such that*

$$\varphi(U_s(x), U_s(y)) = \varphi(x, y) \quad \textit{for all } s \in K \textit{ and all } x, y \in H.$$

In other words, the linear maps U_s are unitary transformations with respect to φ. Furthermore, the norms $\| \ \|$ and $x \mapsto \sqrt{\varphi(x, x)}$ on H are equivalent.

Proof. Let μ be a right Haar measure on K (which is also a left Haar measure since K is compact). By hypothesis the map

$$s \mapsto \langle U_s(x), U_s(y)\rangle$$

is continuous for all $x, y \in H$. Define φ by

$$\varphi(x, y) = \int \langle U_s(x), U_s(y)\rangle \, d\mu(s).$$

It is immediately verified that φ is a sesquilinear form on H. Since H is finite-dimensional, the sphere $S = \{x \in H \mid \|x\| = 1\}$ is compact. Since U is continuous, the map $\theta\colon K \times S \to \mathbb{R}$ given by

$$\theta(s, x) = \|U_s(x)\|$$

is continuous, and since K and S are compact, $K \times S$ is compact so θ achieves a minimum $m > 0$ and a maximum $M > 0$ (every map U_s is invertible and for $x \in S$, $U_s(x) \neq 0$ since $x \neq 0$). We deduce that for every $x \neq 0$,

$$m \|x\| \leq \|x\| \left\| U_s\left(\frac{x}{\|x\|}\right)\right\| \leq M \|x\|,$$

that is,

$$m \|x\| \leq \|U_s(x)\| \leq M \|x\|.$$

As a consequence, we get

$$m^2\mu(K) \|x\|^2 \leq \varphi(x, x) \leq M^2\mu(K) \|x\|^2,$$

which shows that φ is indeed positive definite, and that $\| \ \|$ is equivalent to the norm induced by φ. Finally, for every $t \in K$, since μ is right-invariant we have

$$\begin{aligned}
\varphi(U_t(x), U_t(y)) &= \int \langle U_s(U_t(x)), U_s(U_t(y))\rangle \, d\mu(s) \\
&= \int \langle U_{st}(x)), U_{st}(y)\rangle \, d\mu(s) \\
&= \int \langle U_s(x)), U_s(y)\rangle \, d\mu(s) \\
&= \varphi(x, y),
\end{aligned}$$

as claimed. ∎

Theorem 8.36 is a basic tool in representation theory. For example, if G is a Lie group and if V is a finite-dimensional vector space, for any representation $\rho\colon G \to \mathbf{GL}(V)$, there is a G-invariant inner product on V iff $\overline{\rho(G)}$ is compact; see Gallier and Quaintance [Gallier and Quaintance (2020a)] (Chapter 21, Theorem 21.5).

Theorem 8.36 will also be used in Vol II, Section 3.2 to show that every linear representation of a compact group is the sum of irreducible representations.

Regarding the product of Haar measures, we have the following result.

Proposition 8.37. *Let G_1 and G_2 be two locally compact groups, and let μ_1 be a left Haar measure on G_1 and μ_2 be a left Haar measure on G_2. Then the linear functional $\Phi_1 \otimes \Phi_2\colon \mathcal{K}_{\mathbb{C}}(G_1 \times G_2) \to \mathbb{C}$ given by*

$$(\Phi_1 \otimes \Phi_2)(f) = \int f(x_1, x_2) d\mu_1(x_1) \otimes d\mu_2(x_2)$$

is a left-invariant positive Radon functional. If G_1 and G_2 are σ-compact, then the Radon measure $\mu_{\Phi_1 \otimes \Phi_2}$ on $G_1 \times G_2$ associated with $\Phi_1 \otimes \Phi_2$ given by Theorem 7.8 is a left Haar measure extending the product measure $\mu_1 \otimes \mu_2$. Furthermore, if G_1 and G_2 are also second-countable, then $\mu_{\Phi_1 \otimes \Phi_2} = \mu_1 \otimes \mu_2$.

Proof sketch. By Fubini's Theorem (which applies since f vanishes outside of a compact subset), we have

$$\begin{aligned}
(\lambda_{(s_1,s_2)}(\Phi_1 \otimes \Phi_2))(f) &= (\Phi_1 \otimes \Phi_2)(\lambda_{(s_1^{-1}, s_2^{-1})}(f)) \\
&= \int (\lambda_{(s_1^{-1}, s_2^{-1})} f)(x_1, x_2) d\mu_1(x_1) \otimes d\mu_2(x_2) \\
&= \int f(s_1 x_1, s_2 x_2) d\mu_1(x_1) \otimes d\mu_2(x_2) \\
&= \int \left(\int f(s_1 x_1, s_2 x_2) d\mu_1(x_1) \right) d\mu_2(x_2) \\
&= \int \left(\int f(x_1, s_2 x_2) d\mu_1(x_1) \right) d\mu_2(x_2) \\
&= \int \left(\int f(x_1, s_2 x_2) d\mu_2(x_2) \right) d\mu_1(x_1) \\
&= \int \left(\int f(x_1, x_2) d\mu_2(x_2) \right) d\mu_1(x_1)
\end{aligned}$$

$$= \int \left(\int f(x_1, x_2) d\mu_1(x_1) \right) d\mu_2(x_2)$$

$$= \int f(x_1, x_2) d\mu_1(x_1) \otimes d\mu_2(x_2) = (\Phi_1 \otimes \Phi_2)(f).$$

Therefore, $\Phi_1 \otimes \Phi_2$ is left-invariant. The other two statements are explained in Folland [Folland (1995)] (Chapter 2, Section 2.2). □

Remark: Since $\Phi_1 \otimes \Phi_2$ is a positive linear functional, by Theorem 7.8, the corresponding Radon measur $\mu_{\Phi_1 \otimes \Phi_2}$ is a left Haar measure on the group $G_1 \times G_2$. But if G_1 or G_2 is not σ-compact then the product measure $\mu_1 \otimes \mu_2$ is not defined, and if G_1 or G_2 is not second-countable, then the σ-algebra associated with $\mu_{\Phi_1 \otimes \Phi_2}$ has more Borel subsets than the σ-algebra associated with $\mu_1 \otimes \mu_2$ (see Definition 5.23).

As an application of Proposition 8.37, since the Lebesgue measure μ_L on $\mathbb{R}$ is both left and right-invariant, we see that the product measure $\mu_{L,n}$ of n copies of μ_L is a left and a right Haar measure on $\mathbb{R}^n$. To simplify notation, we may write μ_n instead of $\mu_{L,n}$, and $\mathcal{L}^1(\mu_n)$ instead of $\mathcal{L}^1_{\mu_n}(\mathbb{R}^n, \mathcal{B}(\mathbb{R}^n), \mathbb{C})$.

As a Haar measure, μ_n is both inner and outer regular.

8.10 G-Invariant Measures on Homogeneous Spaces

Let X be a locally compact space and let G be a locally compact group. Suppose we have a continuous left action $\varphi\colon G \times X \to X$ of G on X (which means that the map φ is continuous, see Definition 8.5). As usual, we write $g \cdot x$ instead of $\varphi(g, x)$. We would like to generalize the notion of left-invariance of a measure on G to the notion of G-invariance of a measure μ on X. This is easily done by replacing multiplication in G by the action of G on X.

Definition 8.18. Let G be a locally compact group, let X be a locally compact space, and let $\cdot\colon G \times X \to X$ be a continuous left action of G on X. For every $s \in G$, define $L_s\colon X \to X$ by

$$L_s(x) = s \cdot x \quad \text{for all } x \in X.$$

For every subset A of X and every $s \in G$, let

$$s \cdot A = \{s \cdot a \mid a \in A\}.$$

For every function $f\colon X \to \mathbb{C}$, the function $\lambda_s(f)$ is given by

$$(\lambda_s(f))(x) = f(s^{-1} \cdot x) \quad \text{for all } x \in X \text{ and } s \in G.$$

For every Borel measure μ on $(X, \mathcal{B}(X))$, the measure $\lambda_s(\mu)$ given by

$$(\lambda_s(\mu))(A) = \mu(s^{-1} \cdot A) \quad \text{for all } s \in G \text{ and all } A \in \mathcal{B}(X).$$

For every Radon functional $\Phi\colon \mathcal{K}_{\mathbb{C}}(X) \to \mathbb{C}$, the Radon functional $\lambda_s(\Phi)$ given by

$$(\lambda_s(\Phi))(f) = \Phi(\lambda_{s^{-1}}(f)) \quad \text{for all } s \in G \text{ and all } f \in \mathcal{K}_{\mathbb{C}}(X).$$

It is immediately verified that

$$L_{st} = L_s \circ L_t, \quad \lambda_{st}(f) = \lambda_s(\lambda_t(f)),$$

and

$$\lambda_{st}(\mu) = \lambda_s(\lambda_t(\mu)), \quad \lambda_{st}(\Phi) = \lambda_s(\lambda_t(\Phi)).$$

The proof of Proposition 8.16 is immediately adapted to show that

$$\int_X (\lambda_{s^{-1}}(f))(x)\, d\mu(x) = \int_X f(s \cdot x)\, d\mu(x) = \int_X f(x)\, d\lambda_s(\mu)(x),$$

for every $s \in G$, every $f \in \mathcal{L}^1_\mu(X, \mathcal{B}, \mathbb{C})$, and every Borel measure μ on X.

Definition 8.19. Let G be a locally compact group, let X be a locally compact space, and let $\cdot\colon G \times X \to X$ be a continuous left action of G on X. A Borel measure μ on X is *G-invariant* if

$$\lambda_s(\mu) = \mu \quad \text{for all } s \in G.$$

A Radon functional $\Phi\colon \mathcal{K}_{\mathbb{C}}(X) \to \mathbb{C}$ on X is *G-invariant* if

$$\lambda_s(\Phi) = \Phi \quad \text{for all } s \in G.$$

If μ is G-invariant, then

$$\int_X f(s \cdot x)\, d\mu(x) = \int_X f(x)\, \mu(x),$$

for every $f \in \mathcal{L}^1_\mu(X, \mathcal{B}, \mathbb{C})$ and all $s \in G$. The proof of Proposition 8.16 is immediately adapted to show that if

$$\int_X \lambda_{s^{-1}}(f)(x)\, d\mu(x) = \int_X f(s \cdot x)\, d\mu(x) = \int_X f(x)\, d\mu(x),$$

for all $f \in \mathcal{K}_{\mathbb{C}}(X)$ and all $s \in G$, then μ is G-invariant.

Our goal is to find sufficient conditions to ensure that X has some G-invariant measure. We will consider the case where $X = G/H$, with the left action of G on G/H given by

$$a \cdot (bH) = abH, \qquad a, b \in G;$$

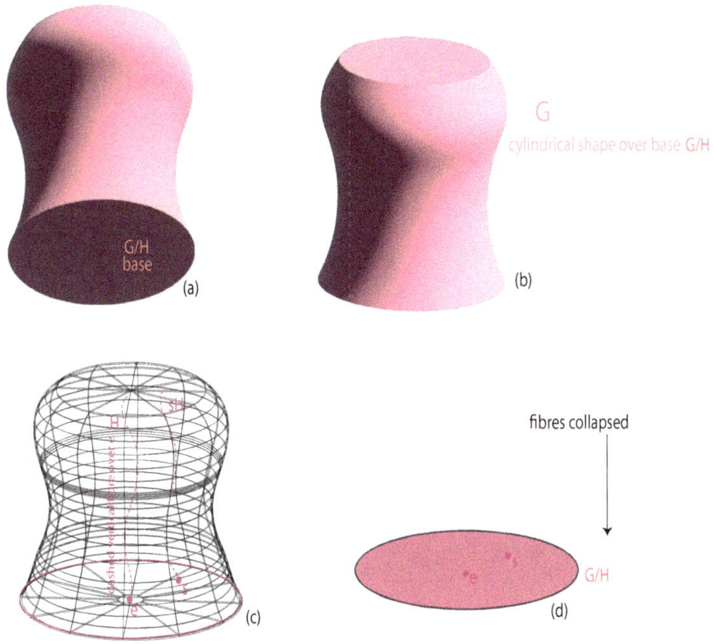

Fig. 8.17 Let G be the solid pink "cylindrical" shape; see Figures (a) and (b). The fibres sH are represented by wavy vertical lines over the circular base; see Figure (c). When these fibres are identified to the base point, we have effectively "collapsed" G to the circular base G/H; see Figure (d).

see Figure 8.17. In this case, by Proposition 8.6, the space X is also locally compact (and Hausdorff).

A G-invariant measure on G/H does not always exist. For example, if G is the affine "$ax + b$" group (with $a \neq 0$) and $X = \mathbb{R}$, obviously G acts transitively on $\mathbb{R}$ (see Example 8.6 for the definition of the action) and the stabilizer of 0 is $H = \mathbb{R}$. However, the only Borel measure on $\mathbb{R}$ invariant under translation is the Lebesgue measure, but it is not invariant under scaling transformations $x \mapsto ax$ with $a \neq 0, 1$.

It turns out that there is a necessary and sufficient condition for a G-invariant σ-Radon measure to exist on G/H in terms of Δ_G and Δ_H: Δ_H must the equal to the restriction of Δ_G on H. We proceed to explain this following Folland's exposition [Folland (1995)] (Chapter 2, Section 2.6).

Suppose μ is a left Haar measure on G and ξ is a left Haar measure on H. The group G is locally compact and σ-compact, and H is a closed subgroup of G. Denote the quotient map by $\pi\colon G \to G/H$. The first step is to define a map P from $\mathcal{K}_{\mathbb{C}}(G)$ to $\mathcal{K}_{\mathbb{C}}(G/H)$.

Definition 8.20. With (G,μ) and (H,ξ) as above, let $P\colon \mathcal{K}_{\mathbb{C}}(G) \to \mathcal{K}_{\mathbb{C}}(G/H)$ be the function defined as follows: for every $f \in \mathcal{K}_{\mathbb{C}}(G)$, for every $s \in G$, let

$$(P(f))(sH) = \int_H f(sh)\, d\xi(h);$$

see Figure 8.18.

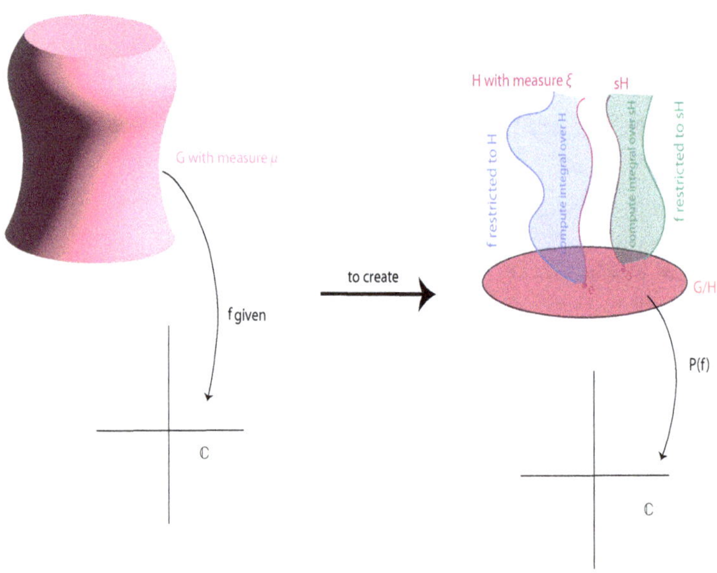

Fig. 8.18 Let G, H, and G/H be as in Figure 8.17. The right figure is a schematic interpretation of $P\colon \mathcal{K}_{\mathbb{C}}(G) \to \mathcal{K}_{\mathbb{C}}(G/H)$. For each $f \in \mathcal{K}_{\mathbb{C}}(G)$, restrict the domain of f to be over the fibre sH, and then integrate over that fibre using the measure ξ. This integral is represented as the shaded "area" between the restricted function image and the fibre.

We need to check that the map P is well-defined, that is, if $sH = tH$, then $(P(f))(sH) = (P(f))(tH)$, but this follows from the left-invariance of

ξ since

$$\begin{aligned}(P(f))(sH) &= \int_H f(sh)\,d\xi(h) = \int_H f(h)\,d\xi(h)\\ &= \int_H f(th)\,d\xi(h) = (P(f))(tH).\end{aligned}$$

Roughly speaking, $P(f)(sH)$ is obtained by averaging over H. The following properties are immediately verified.

Proposition 8.38. *The function $P\colon \mathcal{K}_{\mathbb{C}}(G) \to \mathcal{K}_{\mathbb{C}}(G/H)$ satisfies the following properties:*

(1) The function $P(f)$ is continuous.
(2) We have $\mathrm{supp}(P(f)) \subseteq \pi(\mathrm{supp}(f))$.
(3) For any $\varphi \in \mathcal{K}_{\mathbb{C}}(G/H)$, we have

$$P((\varphi \circ \pi)f) = \varphi\, P(f),$$

where $\varphi\, P(f)$ denotes the function defined by pointwise multiplication on G/H.

Our next goal is to show that P is surjective. The following technical result is needed.

Proposition 8.39. *For any compact subset F of G/H, there is a positive function $f \in \mathcal{K}_{\mathbb{C}}(G)$ such that $P(f) \equiv 1$ on F.*

Proof. Let E be a compact neighborhood of F in G/H. By Proposition 8.7 there is a compact subset K of G such that $\pi(K) = E$. Since G and G/H are locally compact, by Proposition A.39 we can find a positive function $g \in \mathcal{K}_{\mathbb{C}}(G)$ such that g is strictly positive on K and a function $\varphi \in \mathcal{K}_{\mathbb{C}}(G/H)$ such that $\varphi \equiv 1$ on F and $\mathrm{supp}(\varphi) \subseteq E$. Define f by

$$f(s) = \begin{cases} \dfrac{\varphi(\pi(s))}{(P(g))(\pi(s))}\, g(s) & \text{if } (P(g))(\pi(s)) \neq 0 \\ 0 & \text{if } (P(g))(\pi(s)) = 0. \end{cases}$$

Since $P(g) > 0$ on $\mathrm{supp}(\varphi)$, the function f is continuous, we have $\mathrm{supp}(f) \subseteq \mathrm{supp}(g)$, and by Proposition 8.38(3) applied to $\left(\frac{\varphi}{P(g)} \circ \pi\right) g$, we have $P(f) = (\varphi/P(g))P(g) = \varphi$, as desired. $\square$

Using Proposition 8.39, we obtain the surjectivity of P.

Proposition 8.40. *For any function $\varphi \in \mathcal{K}_{\mathbb{C}}(G/H)$, there is some function $f \in \mathcal{K}_{\mathbb{C}}(G)$ such that $P(f) = \varphi$. Furthermore, $\pi(\mathrm{supp}(f)) = \mathrm{supp}(\varphi)$, and if $\varphi \geq 0$, then $f \geq 0$.*

Proof. If $\varphi \in \mathcal{K}_{\mathbb{C}}(G/H)$, by Proposition 8.39 there is some function $g \geq 0$ in $\mathcal{K}_{\mathbb{C}}(G)$ such that $P(g) \equiv 1$ on $\mathrm{supp}(\varphi)$. Let $f = (\varphi \circ \pi)g$. Then by Proposition 8.38(3), we have $P(f) = \varphi P(g) = \varphi$ since $P(g) \equiv 1$ on $\mathrm{supp}(\varphi)$. The other properties are immediately verified. □

If G is a locally compact group and if H is a closed normal subgroup of G, then by Proposition 8.6, the group G/H is also a locally compact group. As application of the surjectivity of the map $P\colon \mathcal{K}_{\mathbb{C}}(G) \to \mathcal{K}_{\mathbb{C}}(G/H)$, the following proposition shows how to integrate on G by integrating on H and G/H.

Proposition 8.41. *Let G be a locally compact group and let H be a closed normal subgroup of G. If ξ is a left Haar measure on H and if γ is a left Haar measure on G/H, then the functional*

$$f \mapsto \int_{G/H} P(f)(sH)\, d\gamma(sH) = \int_{G/H}\int_H f(sh)\, d\xi(h) d\gamma(sH), \quad f \in \mathcal{K}_{\mathbb{C}}(G)$$

is a left Haar functional on G. Consequently, for any left Haar measure μ on G, by rescaling ξ or γ, we have

$$\int_G f(s) d\mu(s) = \int_{G/H}\int_H f(sh)\, d\xi(h) d\gamma(sH).$$

Proof skech. The verification that the given functional is a positive and left-invariant functional is left as an easy exercise. The fact that the functional is not the zero functional follows immediately from the surjectivity of P. By Theorem 7.8, there is a left Haar measure corresponding to this positive Radon functional, and by uniqueness of the left Haar measure up to a scalar, we can rescale ξ or γ as desired. □

We now come to our main theorem.

Theorem 8.42. *Let G be a locally compact group, and let H be closed subgroup of G. Suppose μ is a left Haar measure on G and ξ is a left Haar measure on H. There is a G-invariant σ-Radon measure γ on G/H if and only if Δ_H is equal to the restriction of Δ_G to H. In this case, γ is unique up to a scalar, and with a suitable choice of this factor, we have*

$$\int_G f(s) d\mu(s) = \int_{G/H} P(f)(sH)\, d\gamma(sH) = \int_{G/H}\int_H f(sh)\, d\xi(h)\, d\gamma(sH), \tag{†}$$

for all $f \in \mathcal{K}_{\mathbb{C}}(G)$; see Figure 8.19.

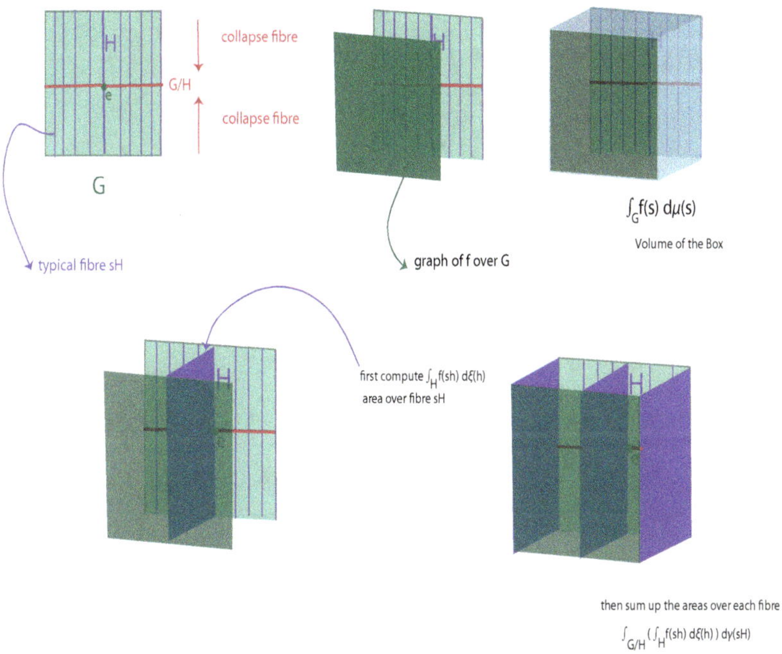

Fig. 8.19 The group G is represented by the green square, the fibres are the vertical purple lines, and G/H is represented by the horizontal red line. The graph of the function $f \in \mathcal{K}_{\mathbb{R}}(G)$ is represented by a second green square "floating" above G. The "volume" below the graph of f is computed by $\int_G f(s)d\mu(s)$. This volume can also be computed in an iterative manner by first compute the "area" over a fibre sH, and then "summing" up the areas by varying the fibre over G/H. Algebraically, this iterative process corresponds to calculating $\int_{G/H}\int_H f(sh)\, d\xi(h)\, d\gamma(sH)$.

Proof. First suppose that a G-invariant σ-Radon measure γ on G/H exists. The map $f \mapsto \int P(f)\, d\gamma$ is a nonzero left-invariant positive linear functional on $\mathcal{K}_{\mathbb{C}}(G)$, so by the uniqueness of Haar measure on G there is some $c > 0$ such that

$$\int P(f)(sH)\, d\gamma(sH) = c \int f(s) d\mu(s). \qquad (*)$$

By Radon-Riesz I (Theorem 7.8), the measure γ is uniquely determined by the functional $\varphi \mapsto \int \varphi(sH)\, d\gamma(sH)$ (with $\varphi \in \mathcal{K}_{\mathbb{C}}(G/H)$). By Proposition 8.40, since $P\colon \mathcal{K}_{\mathbb{C}}(G) \to \mathcal{K}_{\mathbb{C}}(G/H)$ is surjective, $(*)$ determines this functional completely, and thus γ is also completely determined, so γ is unique up to the scalar determining a left Haar measure. By replacing γ by $c^{-1}\gamma$, equation (†) holds. Then for any $\eta \in H$ and $f \in \mathcal{K}_{\mathbb{C}}(G)$, we have

$$\begin{aligned}
\Delta_G(\eta)\int_G f(s)d\mu(s) &= \int_G \rho_{\eta^{-1}}(f)(s)\,d\mu(s) \\
&\quad \text{by Proposition 8.22} \\
&= \int_{G/H}\int_H \rho_{\eta^{-1}}(f)(sh)\,d\xi(h)\,d\gamma(sH) \\
&\quad \text{by (†)} \\
&= \int_{G/H}\int_H f(sh\eta^{-1})\,d\xi(h)\,d\gamma(sH) \\
&\quad \text{by definition of } \rho_{\eta^{-1}}(f) \\
&= \Delta_H(\eta)\int_{G/H}\int_H f(sh)\,d\xi(h)\,d\gamma(sH) \\
&\quad \text{by Proposition 8.22} \\
&= \Delta_H(\eta)\int_G f(s)\,d\mu(s), \\
&\quad \text{by (†)}
\end{aligned}$$

which implies that $\Delta_G(\eta) = \Delta_H(\eta)$.

Conversely, assume that Δ_H is equal to the restriction of Δ_G to H. We claim that if $f \in \mathcal{K}_{\mathbb{C}}(G)$ and if $P(f) = 0$, then $\int f(s)d\mu(s) = 0$.

By Proposition 8.39 there is a positive function $\varphi \in \mathcal{K}_{\mathbb{C}}(G)$ such that $P(\varphi) \equiv 1$ on $\pi(\mathrm{supp}(f))$. Therefore we have

$$\begin{aligned}
0 = (P(f))(sH) &= \int_H f(sh)d\xi(h) && \text{by definition of } P(f) \\
&= \int_H f(sh^{-1})\Delta_H(h^{-1})d\xi(h) && \text{by Proposition 8.27} \\
&= \int_H f(sh^{-1})\Delta_G(h^{-1})d\xi(h), && \text{since } \Delta_H = \Delta_G \mid H
\end{aligned}$$

which implies

$$\begin{aligned}
0 &= \int_G\int_H \varphi(s)f(sh^{-1})\Delta_G(h^{-1})d\xi(h)\,d\mu(s) \\
&= \int_H\int_G \varphi(s)f(sh^{-1})\Delta_G(h^{-1})d\mu(s)\,d\xi(h) && \text{by Fubini} \\
&= \int_H \Delta_G(h^{-1})\int_G \varphi(s)f(sh^{-1})d\mu(s)\,d\xi(h) \\
&= \int_H\int_G \varphi(sh)f(s)d\mu(s)\,d\xi(h) && \text{by Proposition 8.22}
\end{aligned}$$

$$
\begin{aligned}
&= \int_G f(s) \int_H \varphi(sh) d\xi(h)\, d\mu(s) && \text{by Fubini} \\
&= \int_G (P(\varphi))(sH) f(s) d\mu(s) = \int f(s) d\mu(s) && P(\varphi) \equiv 1 \text{ on } \pi(\mathrm{supp}(f)).
\end{aligned}
$$

What we just showed implies that if $P(f) = P(g)$, then $\int f(s)d\mu(s) = \int g(s)d\mu(s)$. Since by Proposition 8.40 the map $P\colon \mathcal{K}_{\mathbb{C}}(G) \to \mathcal{K}_{\mathbb{C}}(G/H)$ is surjective, we define a functional Φ on $\mathcal{K}_{\mathbb{C}}(G/H)$ as follows: for every $\varphi \in \mathcal{K}_{\mathbb{C}}(G/H)$, let

$$
\Phi(\varphi) = \int_G f(s) d\mu(s) \quad \text{for any } f \in \mathcal{K}_{\mathbb{C}}(G) \text{ such that } P(f) = \varphi.
$$

Since $P(f) = P(g)$ implies that $\int f(s)d\mu(s) = \int g(s)d\mu(s)$, the functional Φ is well-defined, and it is immediately verified that Φ is a G-invariant positive linear functional on $\mathcal{K}_{\mathbb{C}}(G/H)$. By Radon–Riesz I (Theorem 7.8), this functional induces a G-invariant σ-Radon measure on G/H. ☐

If H is compact then by Proposition 8.25(2), we have $\Delta_H = \Delta_G \mid H = 1$, so we obtain the following useful corollary.

Proposition 8.43. *If G is a locally compact group, for any compact subgroup H of G, the space G/H admits a G-invariant σ-Radon measure (unique up to a scalar). In fact, if $\pi\colon G \to G/H$ is the quotient map, then for any left Haar measure μ on G, there is a unique G-invariant σ-Radon measure γ on G/H such that*

$$
\int_{G/H} f(x)\, d\gamma(x) = \int_G (f \circ \pi)(s)\, d\mu(s), \qquad \text{for all } f \in \mathcal{K}_{\mathbb{C}}(G/H).
$$

Proof. If $f \in \mathcal{K}_{\mathbb{C}}(G/H)$ has compact support K, then $f \circ \pi$ has support homeomorphic to $K \times H$, and since K and H are compact, it is compact. Thus $f \circ \pi \in \mathcal{K}_{\mathbb{C}}(G)$. The functional $\Phi\colon \pi(f) \mapsto \int_G (f \circ \pi)(s)\, d\mu(s)$ is well-defined, clearly a positive linear functional on $\mathcal{K}_{\mathbb{C}}(G/H)$, and since μ is left-invariant, it is G-invariant. By Radon–Riesz I, there is a unique G-invariant σ-Radon measure γ corresponding to Φ. ☐

Proposition 8.43 applies to the projective spaces and the Grassmannians with $G = \mathbf{SO}(n)$ and a suitable compact subgroup H. It also applies to $G = \mathbf{GL}(n, \mathbb{R})$ and $X = \mathbf{SPD}(n)$, where $\mathbf{SPD}(n)$ is the set of positive symmetric definite matrices discussed in Example C.11.

Example 8.8. Recall that the group $\mathbf{GL}(n) = \mathbf{GL}(n, \mathbb{R})$ acts on $\mathbf{SPD}(n)$ as follows: for all $A \in \mathbf{GL}(n)$ and all $S \in \mathbf{SPD}(n)$,

$$A \cdot S = ASA^{\top}.$$

This action is transitive, and in Section C.3, Example (d), we show that the stabilizer of I is $\mathbf{O}(n)$, so

$$\mathbf{GL}(n)/\mathbf{O}(n) = \mathbf{SPD}(n).$$

It can be shown that the unique (up to a scalar) $\mathbf{GL}(n, \mathbb{R})$-invariant measure on $\mathbf{SPD}(n)$ is given by

$$d\mu = (\det(H))^{-(n+1)/2} d\eta(H),$$

where η is the Haar measure on the additive group $\mathbf{S}(n)$ of real symmetric matrices. For a proof, see Bourbaki [Bourbaki (1965b)] (Chapter VII, Section 3, no. 3, Example 8).

When the condition $\Delta_H = \Delta_G \mid H$ fails, it is possible relax the notion of G-invariance and to obtain sufficient conditions for the existence of measures on G/H satisfying weaker invariance conditions. Such notions are relative invariance, quasi-invariance, and strong quasi-invariance. Strong quasi-invariance is discussed in Vol II, Section 6.6. For detailed expositions the interested reader is referred to Folland [Folland (1995)] (Chapter 2, Section 2.6) and Bourbaki [Bourbaki (1965b)] (Chapter VII, Section 2).

8.11 Convolution of Measures

Let G be a locally compact group equipped with a left Haar measure λ. Recall that $\mathcal{M}^1_{\mathrm{reg},\mathbb{C}}(G)$ denotes the Banach space of complex regular Borel measures on G (see Definition 7.22), and that $\mathrm{L}^1_\lambda(G, \mathcal{B}, \mathbb{C})$ denotes the space of integrable functions on the measure space $(G, \mathcal{B}, \lambda)$, where $\mathcal{B}$ is the σ-algebra of Borel sets of G. To simplify notation, from now on *we write* $\mathcal{M}^1(G)$ *for* $\mathcal{M}^1_{\mathrm{reg},\mathbb{C}}(G)$, *and* $\mathrm{L}^1(G)$ *for* $\mathrm{L}^1_\lambda(G, \mathcal{B}, \mathbb{C})$.

The vector space $\mathcal{M}^1(G)$ is a Banach space with the norm $\|\mu\| = |\mu|(G)$, and $\mathrm{L}^1(G)$ is a Banach space with the L^1-norm. There are three flavors of convolutions but we use mainly two of them:

(1) Convolutions of two measures $\mu, \nu \in \mathcal{M}^1(G)$. This makes $\mathcal{M}^1(G)$ into a Banach algebra with identity and with an involution.
(2) Convolution of two functions $f, g \in \mathrm{L}^1(G)$, which makes $\mathrm{L}^1(G)$ into a Banach algebra with involution, but without a multiplicative unit element, unless G is discrete.

(3) There is also a notion of convolution of a measure $\mu \in \mathcal{M}^1(G)$ and of a function $f \in \mathrm{L}^1(G)$, and of a function $f \in \mathrm{L}^1(G)$ and a measure $\mu \in \mathcal{M}^1(G)$.

Convolution applied to functions can be used as a regularization (or filtering) process. We begin with the convolution of measures.

Let $\mu, \nu \in \mathcal{M}^1(G)$ be two complex measures; then for any function $f \in \mathcal{C}_0(G, \mathbb{C})$ (recall that $\mathcal{C}_0(G, \mathbb{C})$ is the space of continuous functions that tend to zero at infinity), we have a linear functional $\Phi\colon \mathcal{C}_0(G, \mathbb{C}) \to \mathbb{C}$ given by

$$\Phi(f) = \iint f(st)\, d\mu(s)\, d\nu(t), \quad f \in \mathcal{C}_0(G; \mathbb{C}).$$

Observe that

$$|\Phi(f)| \leq \|f\|_\infty \|\mu\| \|\nu\|,$$

so Φ is a bounded linear functional. By Radon–Riesz III (Theorem 7.30), there is a unique measure $\mu * \nu \in \mathcal{M}^1(G)$ such that

$$\Phi(f) = \iint f(st)\, d\mu(s)\, d\nu(t) = \int f\, d(\mu * \nu), \quad f \in \mathcal{C}_0(G; \mathbb{C}),$$

with $\|\Phi\| = \|\mu * \nu\|$. Since

$$\|\Phi\| = \sup\{|\Phi(f)| \mid f \in \mathcal{C}_0(G; \mathbb{C}),\ \|f\|_\infty = 1\}$$

and $|\Phi(f)| \leq \|f\|_\infty \|\mu\| \|\nu\|$, we deduce that $\|\mu * \nu\| \leq \|\mu\| \|\nu\|$.

Definition 8.21. Let G be locally compact group. If $\mu, \nu \in \mathcal{M}^1(G)$ are two measures, then the measure $\mu * \nu$, called *convolution* of μ and ν, is the unique measure such that

$$\int f\, d(\mu * \nu) = \iint f(st)\, d\mu(s)\, d\nu(t) \quad \text{for all } f \in \mathcal{C}_0(G, \mathbb{C}).$$

We have $\|\mu * \nu\| \leq \|\mu\| \|\nu\|$.

Observe that by interchanging s and t, we have $\iint f(st)\, d\mu(s)\, d\nu(t) = \iint f(ts)\, d\mu(t)\, d\nu(s)$, and using Fubini's theorem, $\iint f(ts)\, d\mu(t)\, d\nu(s) = \iint f(ts)\, d\nu(s)\, d\mu(t)$, so we have

$$\iint f(st)\, d\mu(s)\, d\nu(t) = \iint f(ts)\, d\nu(s)\, d\mu(t).$$

Recall the Dirac measure δ_s given by $\delta_s(E) = 1$ iff $s \in E$, and $\delta_s(E) = 0$ otherwise (see Example 4.7).

Proposition 8.44. *Let G be locally compact group. The following properties hold.*

(1) Convolution is associative; that is, if $\mu, \nu, \sigma \in \mathcal{M}^1(G)$, then

$$(\mu * \nu) * \sigma = \mu * (\nu * \sigma)$$

(2) The measure δ_1 (where 1 is the identity element of G) is an identity for convolution; that is,

$$\mu * \delta_1 = \delta_1 * \mu = \mu \quad \text{for all } \mu \in \mathcal{M}^1(G).$$

We also have

$$\delta_s * \mu = \lambda_s(\mu), \qquad \mu * \delta_s = \rho_{s^{-1}}(\mu).$$

(3) For all $s, t \in G$, we have $\delta_s * \delta_t = \delta_{st}$, and convolution is commutative ($\mu * \nu = \nu * \mu$) if and only if G is abelian.

Most of Proposition 8.44 is proven in Folland [Folland (1995)] (Chapter 2, Section 2.5), and the other parts are proven using Proposition 8.16. See also Dieudonné [Dieudonné (2005)] (Chapter XIV, Section 6).

We need to define $\check{\mu}$ for complex measures. We simply use Definition 8.11.

Definition 8.22. Let G be locally compact group. For any complex measure $\mu \in \mathcal{M}^1(G)$, define $\check{\mu}$ by

$$\check{\mu}(A) = \mu(A^{-1}), \qquad \text{for all } A \in \mathcal{B}(G).$$

We also set $\mu^* = \overline{\check{\mu}}$ and call it the *adjoint* of μ.

The complex measure $\check{\mu}$ can be characterized by a property similar to the property of Proposition 8.27. Recall that the complex measure μ has a unique Jordan decomposition

$$\mu = \mu_1^+ - \mu_1^- + i(\mu_2^+ - \mu_2^-),$$

where the measures $\mu_1^+, \mu_1^-, \mu_2^+$, and μ_2^- are positive measures; see Theorem 7.22.

Proposition 8.45. *Let G be locally compact group. For any complex measure $\mu \in \mathcal{M}^1(G)$, for every function $f \in \mathcal{L}^1_{|\mu|}(X, \mathcal{A}, \mathbb{C})$, we have*

$$\int f d\check{\mu} = \int f(s^{-1}) d\mu(s).$$

Consequently, the complex measure $\check{\mu}$ is the unique measure in $\mathcal{M}^1(G)$ such that

$$\int \varphi d\check{\mu} = \int \varphi(s^{-1}) d\mu(s) \quad \text{for all } \varphi \in \mathcal{C}_0(G; \mathbb{C}).$$

Proof. We know that μ can be expressed uniquely as
$$\mu = \mu_1^+ - \mu_1^- + i(\mu_2^+ - \mu_2^-),$$
where the measures $\mu_1^+, \mu_1^-, \mu_2^+$, and μ_2^- are positive measures, and we have
$$\check{\mu} = \check{\mu_1^+} - \check{\mu_1^-} + i(\check{\mu_2^+} - \check{\mu_2^-}).$$
Then we have
$$\int f d\check{\mu} = \int f d\check{\mu_1^+} - \int f d\check{\mu_1^-} + i\int f d\check{\mu_2^+} - i\int f d\check{\mu_2^-}.$$
By Proposition 8.27,
$$\int f d\check{\mu_i^+} = \int f(s^{-1}) d\mu_i^+, \qquad \int f d\check{\mu_i^-} = \int f(s^{-1}) d\mu_i^-,$$
so we get
$$\begin{aligned}\int f d\check{\mu} &= \int f d\check{\mu_1^+} - \int f d\check{\mu_1^-} + i\int f d\check{\mu_2^+} - i\int f d\check{\mu_2^-} \\ &= \int f(s^{-1}) d\mu_1^+ - \int f(s^{-1}) d\mu_1^- + i\int f(s^{-1}) d\mu_2^+ - i\int f(s^{-1}) d\mu_2^- \\ &= \int f(s^{-1}) d\mu,\end{aligned}$$
as claimed. The second fact is an immediate consequence of Radon–Riesz III theorem. □

Recall from Proposition 7.24 that $\overline{\mu}$ is the unique measure in $\mathcal{M}^1(G)$ satisfying the equation
$$\int \varphi\, d\overline{\mu} = \overline{\int \overline{\varphi(s)}\, d\mu(s)}, \qquad \text{for all } \varphi \in \mathcal{C}_0(G; \mathbb{C}).$$

Since for any function φ, the function $\check{\varphi}$ is given by $\check{\varphi}(s) = \varphi(s^{-1})$ for all $s \in G$, if we define φ^* by $\varphi^*(s) = \overline{\varphi(s^{-1})} = \overline{\check{\varphi}}(s)$, observe that the measure $\mu^* = \overline{\check{\mu}}$ is characterized by the equation
$$\int \varphi\, d\mu^* = \overline{\int \varphi^*(s)\, d\mu(s)}, \qquad \text{for all } \varphi \in \mathcal{C}_0(G; \mathbb{C}). \qquad (*)$$
The verification of Equation $(*)$ is as follows:
$$\begin{aligned}\int \varphi(s)\, d\mu^*(s) &= \int \varphi(s)\overline{\check{\mu}}(s) \\ &= \overline{\int \overline{\varphi(s)}\, d\check{\mu}(s)}, && \text{by Proposition 7.24} \\ &= \overline{\int \overline{\varphi(s^{-1})}\, d\mu(s)}, && \text{by Proposition 8.45} \\ &= \overline{\int \varphi^*(s)\, d\mu(s)}.\end{aligned}$$

Proposition 8.46. *Let G be locally compact group. For any measures $\mu, \nu \in \mathcal{M}^1(G)$, we have*

$$\begin{aligned}
(\mu + \nu)^* &= \mu^* + \nu^* \\
(\alpha\mu)^* &= \overline{\alpha}\mu^* \quad (\alpha \in \mathbb{C}) \\
\overline{(\mu * \nu)} &= \overline{\mu} * \overline{\nu} \\
(\mu * \nu)\check{} &= \check{\nu} * \check{\mu} \\
(\mu * \nu)^* &= \nu^* * \mu^* \\
(\mu^*)^* &= \mu \\
\|\mu^*\| &= \|\mu\| \,.
\end{aligned}$$

Proof. We prove the second, third, and fourth equations, leaving the others as easy exercises. For every $\varphi \in \mathcal{C}_0(G;\mathbb{C})$ we have

$$\begin{aligned}
\int \varphi \, d(\alpha\mu)^* &= \overline{\int \overline{\varphi(s^{-1})} d(\alpha\mu)(s)}, && \text{by Equation } (*) \\
&= \overline{\alpha} \overline{\int \overline{\varphi(s^{-1})} d\mu(s)} \\
&= \overline{\alpha} \int \varphi \, d\mu^*,
\end{aligned}$$

which, by Radon–Riesz III, shows that $(\alpha\mu)^* = \overline{\alpha}\mu^*$. We also have

$$\begin{aligned}
\int \varphi \, d\overline{(\mu * \nu)} &= \overline{\int \overline{\varphi} \, d(\mu * \nu)}, && \text{by Proposition 7.24} \\
&= \overline{\int \left(\int \overline{\varphi(st)} \, d\mu(s) \right) d\nu(s)}, && \text{definition of } d(\mu * \nu) \\
&= \overline{\int \left(\overline{\int \varphi(st) \, d\overline{\mu}(s)} \right) d\nu(s)} \\
&= \int \left(\int \varphi(st) \, d\overline{\mu}(s) \right) d\overline{\nu}(s), && \text{by Proposition 7.24} \\
&= \int \varphi \, d(\overline{\mu} * \overline{\nu}),
\end{aligned}$$

which, by Radon–Riesz III, shows that $\overline{(\mu * \nu)} = \overline{\mu} * \overline{\nu}$. Finally, we have

$$\int \varphi \, d(\mu * \nu)\check{} = \int \varphi(s^{-1}) \, d(\mu * \nu), \qquad \text{by Proposition 8.45}$$

$$\begin{aligned}
&= \int\int \varphi((st)^{-1})\, d\mu(s)\, d\nu(t), && \text{definition of } d(\mu * \nu)\\
&= \int\Big(\int \varphi(t^{-1}s^{-1})\, d\nu(t)\Big)\, d\mu(s), && \text{Fubini's theorem}\\
&= \int\Big(\int \varphi(ts^{-1})\, d\check{\nu}(t)\Big)\, d\mu(s)\\
&= \int\Big(\int \varphi(ts)\, d\check{\nu}(t)\Big)\, d\check{\mu}(s), && \text{by Proposition 8.45}\\
&= \int \varphi\, d(\check{\nu} * \check{\mu}),
\end{aligned}$$

which, by Radon–Riesz III, shows that $(\mu * \nu)\check{} = \check{\nu} * \check{\mu}$. □

The identities of Proposition 8.46 show that $\mathcal{M}^1(G)$ is a *normed algebra with involution*; see Example 9.6. Furthermore, $\mathcal{M}^1(G)$ has an identity element δ_1 such that $\delta_1^* = \delta_1$, and it is complete. The algebra $\mathcal{M}^1(G)$ is called the *measure algebra* of G.

Remark: In general the identity

$$\|\mu^* * \mu\| = \|\mu\|^2$$

fails.

Observe that until now, we had no need for a Haar measure on G. Now we make use of the Haar measure.

8.12 Convolution and Cross-Correlation of Functions

We know from Proposition 7.32 that every $f \in \mathrm{L}^1(G)$ can be viewed as a complex measure $fd\lambda$ in $\mathcal{M}^1(G)$, where $fd\lambda$ is the unique complex regular Borel measure such that

$$\int fg\, d\lambda = \int g\, (fd\lambda) \quad \text{for all } g \in \mathcal{C}_0(G;\mathbb{C}).$$

Thus we can see what happens when we convolve two complex measures of the form $fd\lambda$ and $gd\lambda$, where $f, g \in \mathrm{L}^1(G)$. We need to figure out what

$$\Phi(h) = \int\int h(ts) f(t) g(s)\, d\lambda(s)\, d\lambda(t)$$

is for any $h \in \mathcal{C}_0(G;\mathbb{C})$, where we used the fact that

$$\int\int h(st)\, d\mu(s)\, d\nu(t) = \int\int h(ts)\, d\nu(s)\, d\mu(t),$$

with $\mu = fd\lambda$ and $\nu = gd\lambda$. Using the left-invariance of the Haar measure (changing s to $t^{-1}s$) and Fubini's theorem, we have

$$\begin{aligned}\Phi(h) &= \iint h(ts)f(t)g(s)\,d\lambda(s)\,d\lambda(t) = \int\Big(\int h(ts)f(t)g(s)\,d\lambda(s)\Big)\,d\lambda(t)\\ &= \int\Big(\int h(s)f(t)g(t^{-1}s)\,d\lambda(s)\Big)\,d\lambda(t)\\ &= \int\Big(\int h(s)f(t)g(t^{-1}s)\,d\lambda(t)\Big)\,d\lambda(s)\\ &= \int\Big(\int f(t)g(t^{-1}s)\,d\lambda(t)\Big)\,h(s)\,d\lambda(s).\end{aligned}$$

Fubini's theorem implies that the integral $\int f(t)g(t^{-1}s)\,d\lambda(t)$ is defined for almost all s. We also have

$$\left\|\int f(t)g(t^{-1}s)\,d\lambda(t)\right\|_1 \leq \|f\|_1\,\|g\|_1\,.$$

Indeed, by Fubini and by left-invariance of the Haar measure (changing s to ts), we have

$$\begin{aligned}\left\|\int f(t)g(t^{-1}s)\,d\lambda(t)\right\|_1 &= \int\left|\int f(t)g(t^{-1}s)\,d\lambda(t)\right|\,d\lambda(s)\\ &\leq \iint |f(t)g(t^{-1}s)|\,d\lambda(t)\,d\lambda(s)\\ &= \int |f(t)|\Big(\int |g(t^{-1}s)|\,d\lambda(s)\Big)\,d\lambda(t)\\ &= \int |f(t)|\Big(\int |g(s)|\,d\lambda(s)\Big)\,d\lambda(t)\\ &= \|f\|_1\,\|g\|_1\,.\end{aligned}$$

Then by Proposition 7.31, if the function $f * g$ is the function in $\mathrm{L}^1(G)$ given by

$$(f * g)(s) = \int f(t)g(t^{-1}s)\,d\lambda(t),$$

and since $h \in \mathcal{C}_0(G;\mathbb{C})$, by Proposition 7.31, the equation

$$\Phi(h) = \iint h(ts)f(t)g(s)\,d\lambda(s)\,d\lambda(t) = \int\Big(\int f(t)g(t^{-1}s)\,d\lambda(t)\Big)\,h(s)\,d\lambda(s)$$

shows that Φ is a bounded linear functional, and by Theorem 7.30, there is a unique complex measure $m = (f * g)d\lambda$ such that

$$\begin{aligned}\Phi(h) &= \int\Big(\int f(t)g(t^{-1}s)\,d\lambda(t)\Big)\,h(s)\,d\lambda(s) = \int (f * g)(s)h(s)\,d\lambda(s)\\ &= \int h(s)\,dm(s),\end{aligned}$$

which is the convolution of the measures $fd\lambda$ and $gd\lambda$. This suggests defining the convolution of functions as follows.

Definition 8.23. Let G be a locally compact group equipped with a left Haar measure λ. For any two functions $f, g \in \mathrm{L}^1(G)$, the function $f * g$ called the *convolution* of f and g is the function defined for all almost all s by

$$(f * g)(s) = \int f(t)g(t^{-1}s)\, d\lambda(t).$$

It satisfies the inequality $\|f * g\|_1 \leq \|f\|_1 \|g\|_1$.

The following expression for the convolution $f * g$ of two functions f and g may shed some light on what convolution does. Note that

$$\begin{aligned}(f * g)(s) &= \int_G f(t)g(t^{-1}s)\, d\lambda_G(t) = \int_G f(st)g(t^{-1})\, d\lambda_G(t) \\ &= \int_G (\lambda_{s^{-1}} f)(t)\check{g}(t)\, d\lambda_G(t).\end{aligned}$$

Think of $\check{g}$ as the function g flipped about the y-axis, which is what happens when $G = \mathbb{R}$, since in this case $\check{g}(t) = g(-t)$. Also think of $\lambda_{s^{-1}} f$ as the function f shifted along the x-axis by the amount s, which is what happens when $G = \mathbb{R}$, since $(\lambda_{s^{-1}} f)(t) = f(st) = f(t+s)$. Then $\int_G (\lambda_{s^{-1}} f)(t)\check{g}(t)\, d\lambda_G(t)$ is the "area" around s corresponding to overlapping the functions $\lambda_{s^{-1}} f$ and $\check{g}$ for all t. We can think of f as some kind of filter, and $f * g$ is the result of filtering, or smoothing g, using f. If f is tall and narrow and decays quickly near the origin, $(f * g)(s)$ is almost $g(s)$; see Figure 8.20. This is the idea behind the Dirac delta-function. If f is wider, it tends to perform a better smoothing effect.

In signal processing and computer vision, there is another interpretation of convolution obtained by viewing g as a function that performs some kind of template matching on the function f. The idea is to rewrite the convolution $f * g$ as an inner product of f with a "kernel" which is identified by the following computation. We have

$$\begin{aligned}(f * g)(s) &= \int_G f(t)g(t^{-1}s)\, d\lambda_G(t) \\ &= \int_G f(t)\overline{\bar{\check{g}}(s^{-1}t)}\, d\lambda_G(t) \\ &= \int_G f(t)\overline{\lambda_s(\bar{\check{g}})(t)}\, d\lambda_G(t) \\ &= \langle f, \lambda_s(\bar{\check{g}})\rangle.\end{aligned}$$

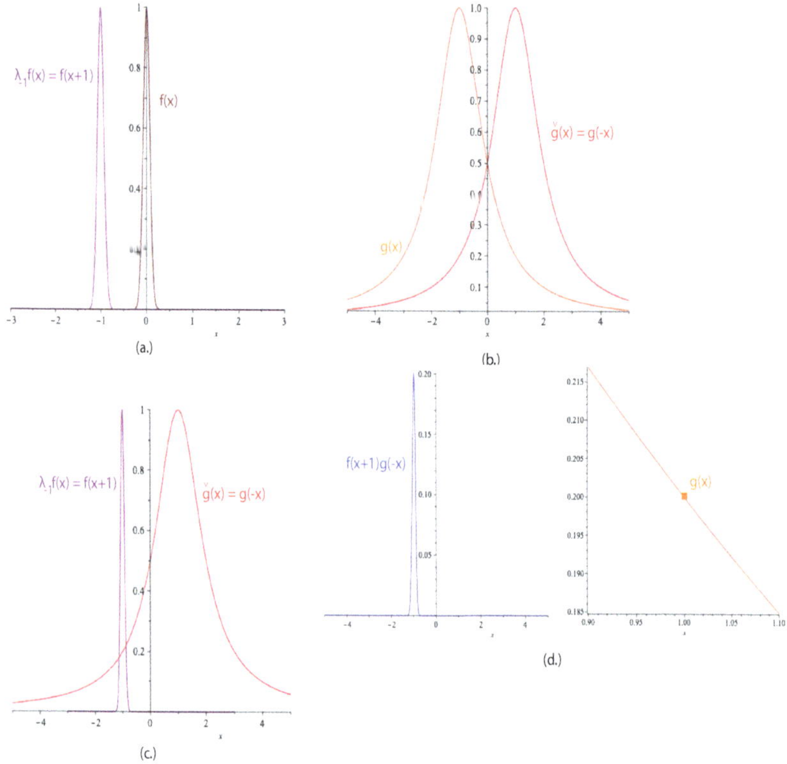

Fig. 8.20 Let $G = \mathbb{R}$, $f(x) = \exp(-100x^2)$, and $g(x) = \frac{1}{(x+1)^2+1}$. Figure (a) shows the graphs of $f(x)$ and $\lambda_{-1}f(x)$, while Figure (b) shows the graphs of $g(x)$ and $\check{g}(x)$. Figure (d) shows the graph of the integrand of $(f * g)(1)$ and the value of $g(1)$. The graph of the integrand (in blue) is a narrow peak whose apex has a y-value which is extremely close to the value of $g(1)$, which is denoted by the orange square in the second graph. Hence, the value of $(f * g)(1)$, which is the area under the blue curve, is almost $g(1)$.

The above is formally the inner product in $L^2_{\lambda_G}(G;\mathbb{C})$, but here we are not assuming that f and $\lambda_s(\overline{\check{g}})$ belong to $L^2_{\lambda_G}(G;\mathbb{C})$. We can think of the function f as a signal, say an image, with domain G, and of the function $k = \overline{\check{g}}$ as a template, a pattern to be identified in the image f. The function $\lambda_s(\overline{\check{g}})$ is the result of moving this template around using $s \in G$ according to the action of G on $L^1(G)$ (often called a *pose* of $\overline{\check{g}}$) and for each $t \in G$, the number $f(t)\overline{\lambda_s(\check{g})(t)}$ is a score of how well the pattern $\lambda_s(\overline{\check{g}})$ matches f at the location t. Then the integral $\langle f, \lambda_s(\overline{\check{g}})\rangle$ is a sort of average of all the scores $f(t)\overline{\lambda_s(\check{g})(t)}$ which indicates how much the pattern $\lambda_s(\overline{\check{g}})$ occurs in f. The number $\langle f, \lambda_s(\breve{\overline{g}})\rangle$ is called the *cross-correlation* of f and $\breve{\overline{g}}$ for $s \in G$. We are led to define the following definition.

Definition 8.24. Let G be a locally compact group with left Haar measure λ_G. For any two function k, f with $f \in \mathrm{L}^2(G)$ and $k \in \mathrm{L}^1(G)$ a function of compact support, the *cross-correlation operator*, or simply *correlation operator* of k and f, denoted $k \star f$, is defined by

$$(k \star f)(s) = \langle f, \lambda_s(k)\rangle = \int_G f(t)\overline{\lambda_s(k)(t)}\, d\lambda_G(t).$$

The function $k \in \mathrm{L}^1(G)$ is called the *correlation kernel.*

By a result of Folland [Folland (1995)] (Chapter 2, Proposition 2.39), $k \star f$ belongs to $\mathrm{L}^2(G)$.

The computation before Definition 8.24 shows that the cross-correlation $k \star f$ can be expressed as a convolution in terms of the following equation:

$$k \star f = f * \breve{\overline{k}}.$$

We can think of $\breve{\overline{k}}$ as a result of reflecting the kernel k. In many practical cases, k and f are real-valued functions and the conjugation can be omitted. Furthermore, if G is abelian, then $k \star f = \breve{k} * f$, which is the formula usually found in the computer vision literature.

The following result will be needed later and is easy to prove. It is a version of Proposition 8.46 functions in $\mathrm{L}^1(G)$.

Proposition 8.47. *The involution $\mu \mapsto \mu^*$ on $\mathcal{M}^1(G)$ yields an involution $f \mapsto f^*$ on $\mathrm{L}^1(G)$, with*

$$f^*(s) = \Delta(s^{-1})\overline{f(s^{-1})},$$

where Δ is the modular function of G. If G is unimodular, then we have the simpler formula

$$f^*(s) = \overline{f(s^{-1})}.$$

Furthermore, for any functions $f, g \in \mathrm{L}^1(G)$, we have

$$\begin{aligned}
(f+g)^* &= f^* + g^* \\
(\alpha f)^* &= \overline{\alpha} f^* \quad (\alpha \in \mathbb{C}) \\
\overline{(f * g)} &= \overline{f} * \overline{g} \\
(f * g)\check{} &= \check{g} * \check{f} \\
(f * g)^* &= g^* * f^* \\
(f^*)^* &= f \\
\|f^*\| &= \|f\| .
\end{aligned}$$

It is also easy to see that the convolution $f * g$ of two functions is given by the following equivalent equations:

$$\begin{aligned} f * g(s) &= \int f(t)g(t^{-1}s)\, d\lambda(t) \\ &= \int f(st)g(t^{-1})\, d\lambda(t) \\ &= \int f(t^{-1})g(ts)\Delta(t^{-1})\, d\lambda(t) \\ &= \int f(st^{-1})g(t)\Delta(t^{-1})\, d\lambda(t). \end{aligned}$$

We go from the first to the second equation using left-invariance by changing t to st. We go from the first to the third equation by changing t to t^{-1} and using the second equation of Proposition 8.27. We go from the second to the fourth equation by changing t to t^{-1} and using the second equation of Proposition 8.27.

Folland gives the following tips to remember how to arrange the variables:

(1) The variable of integration t appears as t in one factor and as t^{-1} in the other.
(2) The two occurrences of the variable of integration are *adjacent* to each other, not separated by the variable s.

When G is unimodular, the factor $\Delta(t^{-1})$ disappears. If G is abelian, it is customary to use an additive notation, and we have

$$f * g(x) = \int f(y)g(x-y)\, d\lambda(y) = \int f(x-y)g(y)\, d\lambda(y) = g * f(x).$$

If $G = \mathbb{R}$, then the function $x \mapsto f(x-y)$ is the function f translated along the x-axis by the amount y (and similarly the function $x \mapsto g(x-y)$ is the function g translated along the x-axis by the amount y). Thus we can think of $f * g(x)$ as a continuous superposition of translates of g, or as a continuous superposition of translates of f. We can interpret these continuous superpositions as moving weighted averages. For example, the convolution $f * g(x) = \int f(y)g(x-y)\, d\lambda(y)$ is the weighted average of f (on the whole line) with respect to the weight function $w(y) = g(x-y)$. In particular, if $g(x) = 0$ for all x such that $|x| > a$, then $g(x-y) = 0$ for all x such that $|x-y| > a$, and then $f * g(x)$ is a weighted average of f on the interval $[x-a, x+a]$; see Figure 8.21. In particular, if g is given by

$$g(x) = \begin{cases} \frac{1}{2a} & \text{if } -a < x < a; \\ 0 & \text{if } |x| \geq a, \end{cases} \qquad (*_1)$$

then

$$f * g(x) = \frac{1}{2a} \int_{x-a}^{x+a} f(y)\,dy.$$

If a is very small, then $f * g(x)$ is approximately $f(x)$. This corresponds to letting g be an approximation to the Dirac function.

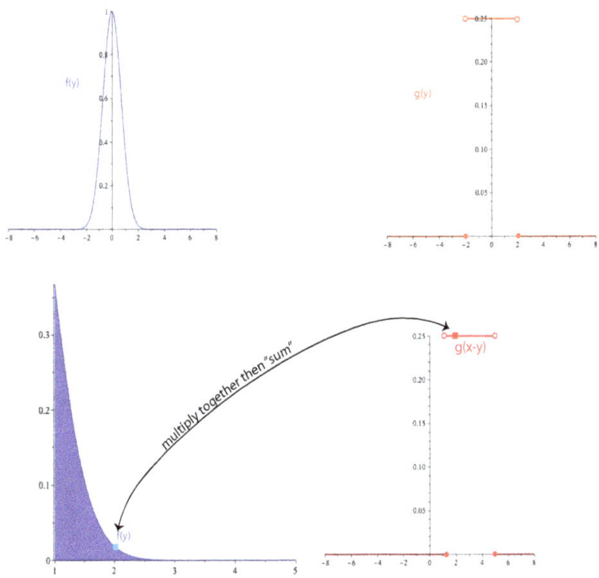

Fig. 8.21 Let $f(y) = \exp(-y^2)$, and in the definition of $g(y)$ provided by $(*_1)$ set $a = 2$. The graphs of these two functions are shown in the top row. The graph of $g(x-y)$ (with $x = 3$) is shown in the bottom row. This graph is obtained by reflecting the graph of $g(y)$ over the vertical axis and then shifting this reflection x units to the right. To compute $f*g(x)$, for any y such that $x - a \leq y \leq x+a$, multiply the values $f(y)g(x-y)$ and take the "infinite" sum of these values. For our particular case, since $f * g(x) = \frac{1}{2a}\int_{x-a}^{x+a} f(y)\,dy$, this "weighted" sum procedure is the area of the shaded blue region in the left figure of the bottom row multiplied by a scaling factor of $\frac{1}{2a}$.

With its involution operation, $\mathrm{L}^1(G)$ is a normed Banach algebra (with involution), but generally without a multiplicative unit. It is called the L^1 *group algebra of* G. When G is discrete or finite, $\mathrm{L}^1(G)$ is isomorphic to the algebra $\mathbb{C}[G]$ consisting of all *finite* formal linear combinations of the form

$$\sum_{s \in G} a_s s, \quad a_s \in \mathbb{C},$$

where $a_s = 0$ for all but finitely many $s \in G$, with the multiplication given by

$$\left(\sum_{s_1\in G} a_{s_1}s_1\right)\left(\sum_{s_2\in G} b_{s_2}s_2\right) = \sum_{s\in G}\left(\sum_{tu=s} a_t b_u\right)s = \sum_{s\in S}\left(\sum_{t\in G} a_t b_{t^{-1}s}\right)s,$$

where the last expression is the discrete convolution $a * b$ of $a = (a_{s_1})_{s_1\in G}$ and $b = (b_{s_2})_{s_2\in G}$, with

$$(a * b)_s = \sum_{t\in G} a_t b_{t^{-1}s}.$$

In particular, if $G = \mathbb{Z}$ and if $a = \sum_{i=0}^{m} a_i\, i$ and $b = \sum_{j=0}^{n} b_j\, j$, with no negative elements from $\mathbb{Z}$, we have

$$(a * b)_k = \sum_{i=0}^{k} a_i b_{k-i}, \quad 0 \le k \le m+n.$$

We can view $a = \sum_{i=0}^{m} a_i\, i$ as the polynomial $a(X) = \sum_{i=0}^{m} a_i X^i$ and $b = \sum_{j=0}^{n} b_j\, j$ as the polynomial $b(X) = \sum_{j=0}^{n} b_j X^j$.

Convolution can be extended from L^1 to the spaces L^2 and L^∞.

Proposition 8.48. *Let G be a locally compact group equipped with a left Haar measure λ. For any $f \in \mathrm{L}^1(G)$ and for any $g \in \mathrm{L}^p(G)$ with $p = 1, 2, \infty$, the following facts hold:*

(1) The integral $\int f(t)g(t^{-1}s)\, d\lambda(t)$ (defining $f * g$) converges absolutely for almost all s, and we have $f * g \in \mathrm{L}^p(G)$ and $\|f * g\|_p \le \|f\|_1 \|g\|_p$.
(2) If G is unimodular, then (1) holds with $f * g$ replaced by $g * f$.
(3) If G is not unimodular but f has compact support, then $g * f \in \mathrm{L}^p(G)$.
(4) When $p = \infty$, the function $f * g$ is continuous and under the conditions of (2) or (3), so is $g * f$.

Proposition 8.48 is proven in Folland [Folland (1995)] (Chapter 2, Section 2.5).

Observe that Proposition 8.48 *does not* say anything when $f \in \mathrm{L}^2(G)$. The following proposition takes care of this case.

Proposition 8.49. *Let G be a locally compact group equipped with a left Haar measure λ. For any $f, g \in \mathrm{L}^2(G)$, we have $f * g \in \mathcal{C}_0(G; \mathbb{C})$, and if G is unimodular, then $\|f * g\|_\infty \le \|f\|_2 \|g\|_2$.*

Proposition 8.49 is proven in Folland [Folland (1995)] (Chapter 2, Section 2.5). See also the variant in Dieudonné [Dieudonné (2005)] (Proposition 14.10.7).

8.13 Convolution of Measures and Functions

One of the applications of convolution is regularization. In order to prove that a complex measure can be approximated by complex measures with compact support we need to define the convolution of a complex measure μ and of a function g. The process of deriving a formula for $\mu * g$ is very similar to the process used to derive a formula for $f * g$, so we proceed quickly.

Let μ be a complex measure in $\mathcal{M}^1(G)$ and let g be a function in $\mathrm{L}^1(G)$. For any $h \in \mathcal{C}_0(G;\mathbb{C})$, we view g as the complex measure $gd\lambda$, and we have

$$\begin{aligned}\Phi(h) = \int\int h(ts)g(s)\,d\lambda(s)\,d\mu(t) &= \int\left(\int h(ts)g(s)\,d\lambda(s)\right)d\mu(t)\\ &= \int\left(\int h(s)g(t^{-1}s)\,d\lambda(s)\right)d\mu(t)\\ &= \int\left(\int h(s)g(t^{-1}s)\,d\mu(t)\right)d\lambda(s)\\ &= \int\left(\int g(t^{-1}s)\,d\mu(t)\right)h(s)\,d\lambda(s).\end{aligned}$$

Fubini's theorem implies that the integral $\int g(t^{-1}s)\,d\mu(t)$ is defined for almost all s. This leads to the following definition.

Definition 8.25. Let G be a locally compact group equipped with a left Haar measure λ. For any complex measure $\mu \in \mathcal{M}^1(G)$ and any function $g \in \mathrm{L}^1(G)$, the function $\mu * g$ called the *convolution function* of μ and g is the function defined for all almost all s by

$$(\mu * g)(s) = \int g(t^{-1}s)\,d\mu(t).$$

It satisfies the inequality $\|\mu * g\|_1 \leq \|\mu\|\,\|g\|_1$. The complex measure $(\mu * g)d\lambda$ is the *convolution* of μ and g. We have

$$\int\int h(ts)g(s)\,d\lambda(s)\,d\mu(t) = \int ((\mu * g)(s))\,h(s)\,d\lambda(s) \quad \text{for all } h \in \mathcal{C}_0(G;\mathbb{C}).$$

The convolution of a complex measure $\mu \in \mathcal{M}^1(G)$ and any function $g \in \mathrm{L}^2(G)$ is also defined by the same formula, and $\|\mu * g\|_2 \leq \|\mu\|\,\|g\|_2$. Observe that

$$(\delta_s * f)(t) = f(s^{-1}t) = (\lambda_s f)(t),$$

so

$$\delta_s * f = \lambda_s f. \qquad (*_{\lambda_s})$$

Similarly, let f be a function in $\mathrm{L}^1(G)$ and let μ be a complex measure in $\mathcal{M}^1(G)$. For any $h \in \mathcal{C}_0(G;\mathbb{C})$, we view f as the complex measure $fd\lambda$, and using Proposition 8.22 and Fubini's theorem, we have

$$\begin{aligned}\Phi(h) &= \int\int h(ts)f(t)\,d\lambda(t)\,d\mu(s) = \int\Big(\int h(ts)f(t)\,d\lambda(t)\Big)\,d\mu(s)\\ &= \int \Delta(s^{-1})\Big(\int h(t)f(ts^{-1})\,d\lambda(t)\Big)\,d\mu(s)\\ &= \int\Big(\int f(ts^{-1})\Delta(s^{-1})\,d\mu(s)\Big)\,h(t)\,d\lambda(t).\end{aligned}$$

Consequently, the convolution $f * \mu$ of a function $f \in \mathrm{L}^1(G)$ and a complex measure $\mu \in \mathcal{M}^1(G)$ is defined as follows.

Definition 8.26. Let G be a locally compact group equipped with a left Haar measure λ. For any function $f \in \mathrm{L}^1(G)$ and any complex measure $\mu \in \mathcal{M}^1(G)$, the function $f * \mu$ called the *convolution function* of f and μ is the function defined for all almost all s by

$$(f * \mu)(s) = \int f(st^{-1})\Delta(t^{-1})\,d\mu(t).$$

The complex measure $(f * \mu)d\lambda$ is the *convolution* of f and μ. The same definition applies if $f \in \mathrm{L}^2(G)$. The inequality $\|f * \mu\|_p \leq \|\mu\|\,\|f\|_p$ holds for $p = 1, 2$.

Observe that

$$(f * \delta_s)(t) = \Delta(s^{-1})f(ts^{-1}) = \Delta(s^{-1})(\rho_{s^{-1}}f)(t),$$

so

$$f * \delta_s = \Delta(s^{-1})\rho_{s^{-1}}f. \qquad (*_{\rho_{s^{-1}}})$$

8.14 Regularization

Given a function $g\colon G \to \mathbb{C}$, typically continuous, for some "well chosen" function f, the convolution $f * g$ might be more regular than g, for example, $f * g$ could become a polynomial function, or a sum of trigonometric functions, a C^2 functions, *etc.* In many cases there exists a sequence (f_n) of functions such that each $f_n * g$ is more "regular" than g, and the sequence $(f_n * g)$ converges uniformly to g, at least on every compact subset of G. It is argued in Folland that such sequences always exist; see Folland [Folland (1995)] (Chapter 2, Section 2.5).

Sequences of functions f_n as above can be thought of as approximations of the infamous Dirac δ function[1] in the sense that the sequence of integrals $\int |f_n| \, d\lambda$ is bounded, $\int f_n \, d\lambda$ tends to 1, and that for each open subset V containing 1, the integral $\int_{G-V} |f_n| \, d\lambda$ tends to zero. In fact, Lang calls such sequences *Dirac sequences*; see Lang [Lang (1996)] (Chapter VIII, Section 3).

There are various formulations of the regularization theorem. Here is a version due to Dieudonné; see [Dieudonné (2005)] (Chapter XIV, Section 11).

Proposition 8.50. *Let G be a locally compact group (with identity element e) equipped with a left Haar measure λ. Let (f_n) be a sequence of functions $f_n \in \mathcal{L}^1(G)$ whose supports are contained in a fixed compact subset K and which satisfy the following conditions:*

(1) The sequence of integrals $\int |f_n| \, d\lambda$ is bounded.
(2) The sequence of integrals $\int f_n \, d\lambda$ tends to 1.
(3) For each open subset V containing e, the sequence of integrals $\int_{G-V} |f_n| \, d\lambda$ tends to zero.

Then the following properties hold:

(i) For every bounded continuous function g on G, the sequence $(f_n * g)$ converges uniformly to g on every compact subset of G.
(ii) If $p = 1$ or 2, and if $g \in \mathcal{L}^p(G)$, the sequence of norms $\|(f_n * g) - g\|_p$ tends to 0 as n goes to infinity.

Proof. We follow Dieudonné's proof [Dieudonné (2005)] (Chapter XIV, Section 11, Theorem 14.11.1).

(i) For every $x \in G$ and for every compact neighborhood V of e, we have from the definitions

$$\begin{aligned} g(x) - (f_n * g)(x) = g(x) \left(1 - \int_V f_n(s) \, d\lambda(s)\right) \\ + \int_V f_n(s)(g(x) - g(s^{-1}x)) \, d\lambda(s) \\ - \int_{G-V} f_n(s) g(s^{-1}x) \, d\lambda(s). \end{aligned}$$

[1]The Dirac δ "function" is characterized by the following two properties: (1) $\delta(x) = 0$ for all $x \in \mathbb{R} - \{0\}$, $\delta(0) = +\infty$; (2) $\int \delta(x)dx = 1$. There is no such function. To make sense of it, one has to view δ as a distribution.

Next we specialize V. Let L be any compact subset of G, and let V_0 be a compact neighborhood of e. By continuity of the group operations, $V_0^{-1}L$ is a compact, so the restriction of g to $V_0^{-1}L$ is uniformly continuous. Hence, by Definition 8.4, for every $\epsilon > 0$, there is a compact neighborhood $V \subseteq V_0$ such that

$$|g(x) - g(s^{-1}x)| \leq \epsilon \qquad \text{for all } x \in L \text{ and all } s \in V. \tag{$*$}$$

Conditions (2) and (3) imply that we can pick n_0 such that

$$\int_{G-V} |f_n(s)|\, d\lambda(s) \leq \epsilon, \qquad \left|1 - \int f_n(s)\, d\lambda(s)\right| \leq \epsilon, \tag{$**$}$$

for all $n \geq n_0$. Since

$$\int |f_n(s)|\, d\lambda(s) = \int_V |f_n(s)|\, d\lambda(s) + \int_{G-V} |f_n(s)|\, d\lambda(s),$$

we have

$$1 - \int_V |f_n(s)|\, d\lambda(s) = 1 - \int |f_n(s)|\, d\lambda(s) + \int_{G-V} |f_n(s)|\, d\lambda(s),$$

so

$$\left|1 - \int_V f_n(s)\, d\lambda(s)\right| \leq \left|1 - \int f_n(s)\, d\lambda(s)\right| + \int_{G-V} |f_n(s)|\, d\lambda(s) \leq 2\epsilon, \tag{$\dagger$}$$

and therefore, since g is a bounded function, for all $x \in L$, Equations ($\dagger$), ($**$), and ($*$) imply that

$$\begin{aligned}
\left|g(x)\left(1 - \int_V f_n(s)\, d\lambda(s)\right)\right| &\leq \|g\|_\infty \left|1 - \int_V f_n(s)\, d\lambda(s)\right| \leq 2\,\|g\|_\infty\, \epsilon \\
\left|\int_{G-V} f_n(s) g(s^{-1}x)\, d\lambda(s)\right| &\leq \|g\|_\infty\, \epsilon \\
\left|\int_V f_n(s)(g(x) - g(s^{-1}x))\, d\lambda(s)\right| &\leq \int_V |f_n(s)||(g(x) - g(s^{-1}x)|\, d\lambda(s) \\
&\leq \epsilon \int_V |f_n(s)|\, d\lambda(s) \leq c\epsilon,
\end{aligned}$$

where

$$c = \sup_n \int |f_n(s)|\, d\lambda(s).$$

Note that Property (1) guarantees the existence of c. Consequently,

$$|g(x) - (f_n * g)(x)| \leq (c + 3\,\|g\|_\infty)\epsilon,$$

which proves the uniform convergence on the compact L.

(ii) By Theorem 7.10, since $\mathcal{K}_{\mathbb{C}}(G)$ is dense in $\mathcal{L}^p(G)$ for $p = 1, 2$, for every $\epsilon > 0$ there is some $h \in \mathcal{K}_{\mathbb{C}}(G)$ such that $\|g - h\|_p \leq \epsilon$. By Proposition 8.48(1), we have

$$\|(f_n * g) - (f_n * h)\|_p \leq \|f_n\|_1 \|g - h\|_p \leq c\epsilon,$$

since by Property (1) the $\|f_n\|_1$ are bounded by some constant $c > 0$. Since

$$\begin{aligned}\|f_n * g - g\|_p &= \|f_n * (g - h + h) - (g - h + h)\|_p \\ &= \|f_n * (g - h) - (g - h) + (f_n * h - h)\|_p \\ &\leq \|f_n * g - f_n * h\|_p + \|g - h\|_p + \|f_n * h - h\|_p \\ &\leq c\epsilon + \epsilon + \|f_n * h - h\|_p .\end{aligned}$$

Therefore we are reduced to proving (ii) for functions in $\mathcal{K}_{\mathbb{C}}(G)$. If $S = \mathrm{supp}(g)$, then it is easy to see that $\mathrm{supp}(f * g) \subseteq KS$, where K is a compact set such that $\mathrm{supp}(f_n) \subseteq K$ for all n, which exists by hypothesis. By (i), the sequence $(f_n * g)$ converges uniformly to g on $S \cup KS$ and vanishes (as does g) outside this set. To conclude that $\lim_{n \mapsto \infty} \|(f_n * g) - g\|_1 = 0$, we use Proposition 5.24(2), which says that

$$\left|\int_A f\, d\lambda\right| \leq \int_A |f|\, d\lambda \leq \|f\|_\infty \lambda(A).$$

Here set $f = (f_n * g) - g$ and $A = S \cup KS$, which is a compact subset, so $\lambda(A) < \infty$. By Proposition 8.48(1), if $g \in \mathcal{L}^1(G)$, then $f \in \mathcal{L}^1(G)$. When $g \in \mathcal{L}^2(G)$, we use Proposition 8.48(1) to conclude that $f \in \mathcal{L}^2(G)$. Then $\int_A |f|^2\, d\lambda$ is well defined with

$$\int_A |f|^2\, d\lambda \leq \|f\|_\infty^2 \lambda(A),$$

which implies that $\lim_{n \mapsto \infty} \|(f_n * g) - g\|_2 = 0$. The reader is also referred to Dieudonné [Dieudonné (2005)] (Chapter XIII, Section 12, Theorem 13.12.2.2). □

A sequence (f_n) of functions satisfying the conditions of Proposition 8.50 is called a *regularizing sequence*.

A neat application of Proposition 8.50 is a quick proof of the Weierstrass theorem on the uniform approximation of continuous functions on $[-1/2, 1/2]$ by polynomials.

Example 8.9. Let $G = \mathbb{R}$, with the Lebesgue measure, and let g be a continuous function with support in $[-1/2, 1/2]$. Consider the *Landau functions* f_n given by

$$f_n(x) = \begin{cases} \dfrac{1}{a_n}(1-x^2)^n & \text{if } |x| \leq 1 \\ 0 & \text{if } |x| > 1, \end{cases}$$

where $a_n = \int_{-1}^{1}(1-x^2)^n\,dx$; see Figure 8.22.

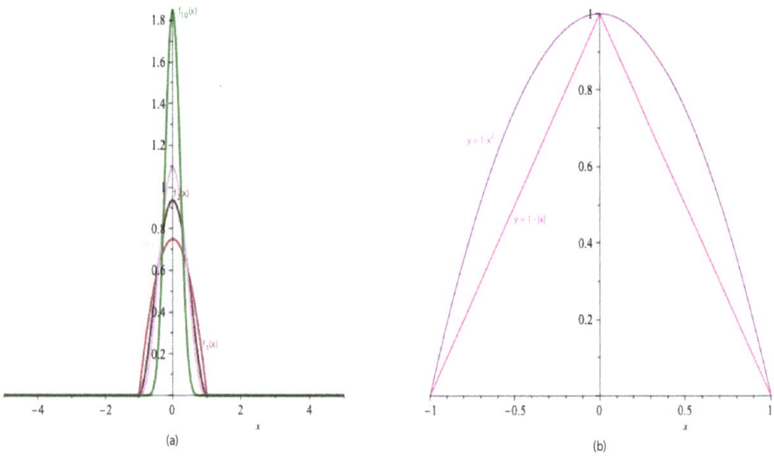

Fig. 8.22 Figure (a) shows the graphs of $f_1(x)$, $f_2(x)$, $f_3(x)$, and $f_{10}(x)$. As n increases the "peak" becomes higher and thinner. Figure (b) graphically shows that $1-x^2 \geq 1-|x|$ when $1 \leq x \leq -1$.

Since $1 - x^2 \geq 1 - |x|$ for $1 \leq x \leq -1$, we have

$$a_n \geq 2\int_0^1 (1-x)^n\,dx = \frac{2}{n+1},$$

which implies that

$$f_n(x) \leq (n+1)(1-x^2)^n$$

for all $x \in [-1, +1]$. Thus $f_n(x)$ tends to 0 uniformly on every compact interval not containing 0, which implies Property (3) of a regularizing sequence. Since by construction $\int |f_n(x)|dx = \int f_n(x)dx = 1$, (f_n) is a regularizing sequence. We have

$$f_n * g(x) = \frac{1}{a_n}\int_{-1/2}^{1/2} f_n(x-y)g(y)\,dy = \frac{1}{a_n}\int_{-1/2}^{1/2}(1-(x-y)^2)^n g(y)\,dy$$

and by using the binomial formula we see that $(1-(x-y)^2)^n$ is a polynomial in y,

$$(1-(x-y)^2)^n = \sum_{j=0}^{2n} u_j(x)y^j,$$

for some polynomials $u_j(x)$ in x, and so

$$\begin{aligned} f_n * g(x) &= \frac{1}{a_n}\int_{-1/2}^{1/2}(1-(x-y)^2)^n g(y)\,dy = \frac{1}{a_n}\int_{-1/2}^{1/2}\sum_{j=0}^{2n} u_j(x)y^j g(y)\,dy \\ &= \frac{1}{a_n}\sum_{j=0}^{2n} u_j(x)\int_{-1/2}^{1/2} y^j g(y)\,dy, \end{aligned}$$

a polynomial in x. Proposition 8.50 shows that the sequence of polynomials $(f_n * g)$ converges uniformly to g on the compact interval $[-1/2,+1/2]$, giving another proof of the Weierstrass approximation theorem.

In this example a continuous function g, which could be much more complicated than a polynomial, and in particular, could lack derivatives of order ≥ 1, becomes a polynomial when convolved with f_n. This is a perfect example of regularization.

In the next section we present another example from Lang [Lang (1996)] (Chapter VIII, Section 3), the *Cesàro summation* of Fourier series of continuous functions on the unit circle $\mathbb{T} = \mathbf{U}(1) = \{e^{i\theta} \mid -\pi \leq \theta < \pi\}$.

8.15 Dirichlet Kernels, Fejér Kernels, Poisson Kernels

Example 8.10. From now on, we will use the normalized Haar measure $dx/2\pi$ on $\mathbb{T}$ so that $\mathbb{T}$ has measure 1. With this normalized measure, the most important results come out cleaner (without an extra factor $1/2\pi$). With this measure, the integral of $f \in \mathrm{L}^1(\mathbb{T})$ is

$$\int_{-\pi}^{\pi} f(\theta)\,\frac{dx(\theta)}{2\pi},$$

and the convolution of two functions $f, g \in \mathrm{L}^1(\mathbb{T})$ is

$$(f*g)(\theta) = \int_{-\pi}^{\pi} f(\theta-\varphi)g(\varphi)\,\frac{dx(\varphi)}{2\pi} = \int_{-\pi}^{\pi} f(\varphi)g(\theta-\varphi)\,\frac{dx(\varphi)}{2\pi}.$$

Let g be a continuous periodic function (of period 2π) (equivalently, a function on $\mathbb{T} = \mathbf{U}(1)$). The nth partial sum $S_{n,g}$ of the *Fourier series* for g is given by

$$S_{n,g}(x) = \sum_{k=-n}^{n} c_k e^{ikx}, \qquad \text{with} \qquad c_k = \int_{-\pi}^{\pi} g(t)e^{-ikt}\,\frac{dx(t)}{2\pi},$$

where c_k is called the kth *Fourier coefficient* of g. Let $A_{n,g}$ be the average of these partial sums, that is,

$$A_{n,g} = \frac{1}{n}(S_{0,g} + \cdots + S_{n-1,g}).$$

The average sums $A_{n,g}$ are known as *Cesàro sums* (or *Cesàro means*). Since g is continuous and bounded, $g \in L^2(\mathbb{T})$, and although the partial sums $S_{n,g}$ converge to g in the $\| \|_2$-norm (see Theorem 6.2(3)), they may not converge pointwise to g; see Stein and Shakarchi [Stein and Shakarchi (2003)] (Chapter 3, Subsection 2.2), and Rudin [Rudin (1987)] (Chapters 4 and 5). On the other hand Fejér's theorem asserts that the sequence $(A_{n,g})$ of average sums converges uniformly to g (see Stein and Shakarchi [Stein and Shakarchi (2003)], Chapter 2, Section 5, Theorem 5.2).

This can be shown to be a consequence of Proposition 8.50 by defining the following regularizing functions D_n and K_n:

$$D_n(x) = \sum_{k=-n}^{n} e^{ikx}$$

$$K_n(x) = \frac{1}{n}\sum_{m=0}^{n-1}\sum_{k=-m}^{m} e^{ikx} = \frac{1}{n}(D_0(x) + \cdots + D_{n-1}(x)).$$

We leave it as an exercise to prove that

$$D_n(x) = \frac{\sin((2n+1)x/2)}{\sin(x/2)}$$

$$K_n(x) = \frac{1}{n}\left(\frac{\sin(nx/2)}{\sin(x/2)}\right)^2.$$

The functions D_n are known as *Dirichlet kernels*, and the functions K_n are *Fejér kernels*; see Stein and Shakarchi [Stein and Shakarchi (2003)] (Chapter 2). The graphs of various $D_n(x)$ and $K_n(x)$ were shown in Figures 6.4, and 6.5 respectively.

The Dirichlet kernels (D_n) do not form a regularizing sequence because they fail to satisfy Property (1) of Proposition 8.50. Indeed, we leave it as an exercise to prove that there is a constant $c > 0$ such that

$$\int_{-\pi}^{\pi} |D_n(x)|\,dx \geq c\log n, \quad \text{as } n \to \infty.$$

However, it is easy to check that (K_n) is a regularizing sequence, that $D_n * g = S_{n,g}$, and that $K_n * g = A_{n,g}$ (see immediately after Proposition 6.1, and also Stein and Shakarchi [Stein and Shakarchi (2003)], Chapter 2,

Section 5, Lemma 5.1). By Proposition 8.50, the sequence $(A_{n,g}) = (K_n * g)$ of averages of the partial sums of the Fourier series of g converge uniformly to g, which is Fejér's theorem (see Stein and Shakarchi [Stein and Shakarchi (2003)], Chapter 2, Section 5, Theorem 5.2).

Again, we have a very good example of regularization. After convolving a continuous periodic function g, (which could be much more complicated than a sum of complex exponentials), with D_n, we obtain a function $D_n * g$ which is a sum of complex exponentials.

It is possible to generalize regularizing sequences to families of functions parametrized by a continuous parameter, often called *kernels*.

Example 8.11. As in the previous example, we use the normalized Haar measure $dx/2\pi$ on $\mathbb{T}$ so that $\mathbb{T}$ has measure 1. The *Poisson kernel* on the unit disk is the family of functions $P_r(\theta)$, parametrized by $r \in [0, 1)$, and given by

$$P_r(\theta) = \sum_{n=-\infty}^{n=\infty} r^{|n|} e^{in\theta}.$$

To sum this series, using the formula for the sum of a geometric series, observe that

$$\begin{aligned}\sum_{n=-\infty}^{n=\infty} r^{|n|} e^{in\theta} &= \sum_{n=0}^{n=\infty} (re^{i\theta})^n + \sum_{p=1}^{p=\infty} (re^{-i\theta})^p \\ &= \frac{1}{1 - re^{i\theta}} + \frac{re^{-i\theta}}{1 - re^{-i\theta}} \\ &= \frac{1 - r^2}{1 - 2r\cos\theta + r^2}.\end{aligned}$$

Thus

$$P_r(\theta) = \frac{1 - r^2}{1 - 2r\cos\theta + r^2}.$$

Graphical interpretations of $P_r(\theta)$ were shown in Figures 6.2 and 6.3. Instead of being a sequence of functions indexed by natural numbers, the family (P_r) is a family of functions indexed by the continuous parameter $r \in [0, r)$, but it possesses properties analogous to the properties of regularizing functions, and Proposition 8.50 can be adapted to show that for a bounded periodic continuous function g on $\mathbb{R}$, the functions $(P_r * g)(\theta)$ converge to g uniformly as r tends to 1.

The functions P_r are harmonic for the Laplacian given in polar coordinates by

$$\Delta u = \frac{\partial^2 u}{\partial r^2} + \frac{1}{r}\frac{\partial u}{\partial r} + \frac{1}{r^2}\frac{\partial^2 u}{\partial \theta^2},$$

which means that

$$\Delta P_r = 0.$$

If we write $u(r, \theta) = (P_r * g)(\theta)$, then it can be shown that

$$\Delta u(r, \theta) = 0,$$

that is, $u(r, \theta)$ is harmonic. As a consequence of Proposition 8.50 (suitably generalized), when r tends to 1, we have $u(1, \theta) = g(\theta)$. This shows that $u(r, \theta)$ is a harmonic function solution of a boundary value problem, namely that

$$\Delta u(r, \theta) = 0,$$

with $u(1, \theta) = g(\theta)$ on the boundary, for a prescribed periodic function g.

Other kernels exist for solving partial differential equations, such as the heat equation for the Laplace operator; see Section 6.8. We refer the interested reader to Lang [Lang (1996)] (Chapter VIII, Section 3), Stein and Shakarchi [Stein and Shakarchi (2003)], and Folland [Folland (1992)], for more on this topic.

8.16 Regularization of Complex Measures

Regularization can also be used to prove various approximation results involving complex measures. Because we are dealing with locally compact spaces that may not be metrizable, we need the general machinery of filters to define convergence; see Section A.6. We begin with the following general result from Bourbaki [Bourbaki (1965b)] (Chapter VIII, Section 2, No. 7, Lemma 4).

Proposition 8.51. *Let X be a locally compact space, $a \in X$ a given point, M a subset of $\mathcal{M}^1(X, \mathcal{A})$, and $\mathcal{F}$ a filter on M. Suppose the following properties hold:*

(1) For every compact subset K of X, the set of numbers $\{|\mu|(K) \mid \mu \in M\}$ is bounded.
(2) For every compact subset K of $X - \{a\}$, we have $\lim_{\mu, \mathcal{F}} |\mu|(K) = 0$.
(3) There is some compact neighborhood V of a such that $\lim_{\mu, \mathcal{F}} |\mu|(V) = 1$.

Then the filter $\mathcal{F}$ converges to the Dirac measure δ_a in $\mathcal{M}^1(X, \mathcal{A})$ (in the norm topology).

Observe that the conditions of Proposition 8.51 are abstact versions of the conditions of Proposition 8.50.

Corollary 8.52. *Let X be a locally compact space, $a \in X$ a given point, M a subset of $\mathcal{M}^1(X, \mathcal{A})$, and $\mathcal{F}$ a filter on M. Suppose the properties of Proposition 8.51 hold and that there is a compact subset K_0 of X such that the complex measures in M have support in K_0. Then the filter $\mathcal{F}$ converges to the Dirac measure δ_a in the space $\mathcal{M}^1_c(X, \mathcal{A})$ of complex measures with compact support (in the norm topology).*

Corollary 8.52 is a more abstract version of Proposition 8.50(i).

Proposition 8.51 can be used to prove the following regularization result for complex measures from Bourbaki [Bourbaki (1965b)] (Chapter VIII, Section 4, No. 7, Proposition 19).

Proposition 8.53. *Let G be a locally compact group equipped with a left Haar measure λ. Let $\mathcal{B}$ be a filter basis of neighborhoods of the identity e, consisting of compact neighborhoods, and for every $V \in \mathcal{B}$, let f_V be a positive continuous functions with compact support contained in V, such that $\int f_V \, d\lambda = 1$. For any complex regular Borel measure $\mu \in \mathcal{M}^1(G)$, we have the family of measures $(\mu * f_V)d\lambda$, which has the structure of a filter base on $\mathcal{M}^1(G)$, by considering the subsets $S_V = \{(\mu * f_W)d\lambda \mid W \in \mathcal{B}, W \subseteq V\}$ (corresponding to the filter of sections of $\mathcal{B}$), and the filter base of subsets S_V converges to μ (in the norm topology).*

Since the measures $(\mu * f_V)d\lambda$ have compact support, Proposition 8.53 shows how the complex regular Borel measure μ can be approximated (in a suitable sense) by measures with compact support (and continuous density). Proposition 8.53 also has the following useful corollary from Bourbaki [Bourbaki (1965b)] (Chapter VIII, Section 4, No. 7, Corollary of Proposition 19). A slightly different version of this proposition is given in Folland [Folland (1995)] (Chapter 2, Proposition 2.42).

Proposition 8.54. *Let G be a locally compact group equipped with a left Haar measure λ. Let $\mathcal{B}$ be a filter basis of neighborhoods of the identity e, consisting of compact neighborhoods, and for every $V \in \mathcal{B}$, let f_V be a positive continuous functions with compact support contained in V, such that $\int f_V \, d\lambda = 1$. For any function $g \in \mathrm{L}^p(G)$, $p = 1, 2$, we have the family* *of functions* $(g * f_V)$*, which has the structure of a filter base on* $\mathrm{L}^p(G)$*, by considering the subsets* $S_V = \{(g * f_W) \mid W \in \mathcal{B},\, W \subseteq V\}$ *(corresponding to the filter of sections of* $\mathcal{B}$*), and the filter base of subsets* S_V *converges to* g *(in the norm topology* $\| \, \|_p$*).*

Proposition 8.54 is a more abstract version of Proposition 8.50(ii).

In particular, Proposition 8.54 implies that if μ and ν are two complex regular Borel measures, the identity of Definition 8.21 charactcrzing the convolution $\mu * \nu$ of μ and ν,

$$\int f\, d(\mu * \nu) = \int\int f(st)\, d\mu(s)\, d\nu(t)$$

holds not only for all functions $f \in \mathcal{C}_0(G, \mathbb{C})$, but also for all *bounded continuous* functions $f \in \mathcal{C}_b(G; \mathbb{C})$. This fact will be needed in Chapter 10.

Families of functions f_V as in Propositions 8.53 and 8.54 are easily constructed using continuous bump functions (since G is locally compact; see Proposition A.39).

8.17 Problems

Problem 8.1. Advanced Exercise: Prove Theorem 8.13. Hint: See Bourbaki [Bourbaki (1990b)], (Chapter IX, Section 5, Theorem 1).

Problem 8.2. Let $\varphi \in \mathcal{K}_{\mathbb{R}}(G)$ be a fixed nonzero positive function. For for every function $f \in \mathcal{K}_{\mathbb{R}}(G)$, recall that $(f : \varphi)$ is the greatest lower bound of the sums $\sum_{j=1}^{n} c_j$, over all sets $\{s_1, \ldots, s_n\}$ of elements of G and all finite sequences $(c_1, \ldots, c_n)$ of reals $c_j \in \mathbb{R}$ such that

$$f \leq \sum_{j=1}^{n} c_j \lambda_{s_j}(\varphi).$$

Show that the $(f : \varphi)$ satisfies the following properties:

$$\begin{aligned}
(f : \varphi) &= (\lambda_s(f) : \varphi) && \text{for all } s \in G \\
(f_1 + f_2 : \varphi) &\leq (f_1 : \varphi) + (f_2 : \varphi) && \text{for all } f_1, f_2 \in \mathcal{K}_{\mathbb{R}}(G) \\
(cf : \varphi) &= c(f : \varphi) && \text{for all } c \geq 0 \\
(f_1 : \varphi) &\leq (f_2 : \varphi) && \text{whenever } f_1 \leq f_2 \\
(f : \varphi) &\geq \|f\|_\infty / \|\varphi\|_\infty && \\
(f : \varphi) &\leq (f : \psi)(\psi : \varphi) && \text{for all positive } \psi \in \mathcal{K}_{\mathbb{R}}(G) \\
0 < \frac{1}{(f_0 : f)} \leq \frac{(f : \varphi)}{(f_0 : \varphi)} &\leq (f : f_0) && \text{for all positive } f, f_0 \in \mathcal{K}_{\mathbb{R}}(G).
\end{aligned}$$

Problem 8.3. Prove Proposition 8.18. Hint: See Folland [Folland (1995)] (Chapter 2, Lemma 2.18).

Problem 8.4. Let $G = \mathbf{GL}(n,\mathbb{R})$ be the group of invertible $n \times n$ real matrices. Show that a left (and right) Haar measure on $\mathbf{GL}(n,\mathbb{R})$ is given by

$$d\mu = \frac{dA}{|\det(A)|^n} = |\det(A)|^{-n} \bigotimes_{i,j} da_{ij}$$

with $A = (a_{ij})$, where da_{ij} is the Lebesgue measure on $\mathbb{R}$, and dA is the Lebesgue measure on $\mathbb{R}^{n^2}$.

Problem 8.5. Complete the details of the proof sketch of Proposition 8.23.

Problem 8.6. Prove Proposition 8.26. Hint: See Folland [Folland (1995)] (Chapter 2, Proposition 2.23).

Problem 8.7. Let $G = \mathbf{GA}(n,\mathbb{R})$, the affine group of $\mathbb{R}^n$, which consists of pairs (A, u) with $A \in \mathbf{GL}(n,\mathbb{R})$ and $u \in \mathbb{R}^n$, acting on $\mathbb{R}^n$ by $(A, u)(X) = Ax + u$.

(i) Show that

$$d\mu_L = |\det(A)|^{-n-1} \bigotimes_{i,j} da_{ij} \otimes \bigotimes_{i} du_i$$

with $A = (a_{ij})$, and $u = (u_i)$, where da_{ij} and du_i is the Lebesgue measure on $\mathbb{R}$, is a left Haar measure of G.

(ii) Show that

$$d\mu_R = |\det(A)|^{-n} \bigotimes_{i,j} da_{ij} \otimes \bigotimes_{i} du_i,$$

is a right Haar measure on G.

(iii) Prove that modular function is given by

$$\Delta((A, u)) = |\det(A)|^{-1}.$$

Hint: See Bourbaki [Bourbaki (1965b)] (Chapter VII, Section 2, no. 10, Proposition 14, and Section 3, no. 3, Example 2).

Problem 8.8. Let $G = \mathbf{T}(n,\mathbb{R})$ be the group of invertible upper triangular matrices.

(i) Show that a left Haar measure on $\mathbf{T}(n, \mathbb{R})$ is given by

$$d\mu_L = \prod_{i=1}^{n} |a_{ii}|^{i-n-1} \bigotimes_{i\leq j} da_{ij}$$

with $A = (a_{ij})$, and da_{ij} is the Lebesgue measure on $\mathbb{R}$.

(ii) Show that a right Haar measure is given by

$$d\mu_R = \prod_{i=1}^{n} |a_{ii}|^{-i} \bigotimes_{i\leq j} da_{ij}.$$

(iii) Show that the modular function is given by

$$\Delta(A) = \prod_{i=1}^{n} |a_{ii}|^{2i-n-1}.$$

Hint: See [Bourbaki (1965b)] (Chapter VII, Section 2, no. 10, Proposition 14, and Section 3, no. 3, Example 4).

Problem 8.9. Complete the details of the proof sketch provided for Proposition 8.32. Hint: See Dieudonné [Dieudonné (2005)] (Chapter XIV, Proposition 14.3.9.1).

Problem 8.10. Prove Proposition 8.35. Hint: See Dieudonné [Dieudonné (2005)] (Chapter XIV, Proposition 14.2.3).

Problem 8.11. Advanced Exercise: Complete the proof of Proposition 8.37. Hint: See Folland [Folland (1995)] (Chapter 2, Section 2.2).

Problem 8.12. Prove Proposition 8.38.

Problem 8.13. Complete the details of the proof sketch of Proposition 8.41.

Problem 8.14. Recall that the group $\mathbf{GL}(n) = \mathbf{GL}(n, \mathbb{R})$ acts on $\mathbf{SPD}(n)$ as follows: for all $A \in \mathbf{GL}(n)$ and all $S \in \mathbf{SPD}(n)$,

$$A \cdot S = ASA^{\top}.$$

This action is transitive, and the stabilizer of I is $\mathbf{O}(n)$, so

$$\mathbf{GL}(n)/\mathbf{O}(n) = \mathbf{SPD}(n).$$

Show that the unique (up to a scalar) $\mathbf{GL}(n, \mathbb{R})$-invariant measure on $\mathbf{SPD}(n)$ is given by

$$d\mu = (\det(H))^{-(n+1)/2} d\eta(H),$$

where η is the Haar measure on the additive group $\mathbf{S}(n)$ of real symmetric matrices. Hint: See Bourbaki [Bourbaki (1965b)] (Chapter VII, Section 3, no. 3, Example 8).

Problem 8.15. Prove Proposition 8.44. Hint: Use Proposition 8.16. Alternatively, see Folland [Folland (1995)] (Chapter 2, Section 2.5) or Dieudonné [Dieudonné (2005)] (Chapter XIV, Section 6).

Problem 8.16. Let G be locally compact group. For any measures $\mu, \nu \in \mathcal{M}^1(G)$, verify that

$$\begin{aligned}(\mu+\nu)^* &= \mu^* + \nu^* \\ (\mu^*)^* &= \mu \\ \|\mu^*\| &= \|\mu\| .\end{aligned}$$

Problem 8.17. Prove Proposition 8.47. Hint: See Proposition 8.46.

Problem 8.18. Prove Proposition 8.48. Hint: See Folland [Folland (1995)] (Chapter 2, Section 2.5).

Problem 8.19. Prove Proposition 8.49. Hint: See Folland [Folland (1995)] (Chapter 2, Section 2.5) or Dieudonné [Dieudonné (2005)] (Proposition 14.10.7).

Problem 8.20. Recall that

$$D_n(x) = \sum_{k=-n}^{n} e^{ikx}.$$

Prove that there is a constant $c > 0$ such that

$$\int_{-\pi}^{\pi} |D_n(x)|\, dx \geq c \log n, \quad \text{as } n \to \infty.$$

Problem 8.21. Recall that

$$\begin{aligned}D_n(x) &= \sum_{k=-n}^{n} e^{ikx} \\ K_n(x) &= \frac{1}{n}\sum_{m=0}^{n-1}\sum_{k=-m}^{m} e^{ikx} = \frac{1}{n}(D_0(x) + \cdots + D_{n-1}(x)).\end{aligned}$$

(i) Prove that

$$\begin{aligned}D_n(x) &= \frac{\sin((2n+1)x/2)}{\sin(x/2)} \\ K_n(x) &= \frac{1}{n}\left(\frac{\sin(nx/2)}{\sin(x/2)}\right)^2 .\end{aligned}$$

(ii) Prove that for any $g \in L^1(\mathbb{T})$, the sequence (K_n) is a regularizing sequence.

(iii) $D_n * g = S_{n,g}$, where

$$S_{n,g}(x) = \sum_{k=-n}^{n} c_k e^{ikx}, \qquad \text{with} \qquad c_k = \int_{-\pi}^{\pi} g(t)e^{-ikt}\, \frac{dx(t)}{2\pi}.$$

(iv) $K_n * g = A_{n,g}$, where

$$A_{n,g} = \frac{1}{n}(S_{0,g} + \cdots + S_{n-1,g}).$$

Problem 8.22. Recall that the family of functions on the unit disk $P_r(\theta)$, parametrized by $r \in [0, 1)$, is given by

$$P_r(\theta) = \sum_{n=-\infty}^{n=\infty} r^{|n|} e^{in\theta}.$$

(i) Show that P_r are harmonic for the Laplacian given in polar coordinates by

$$\Delta u = \frac{\partial^2 u}{\partial r^2} + \frac{1}{r}\frac{\partial u}{\partial r} + \frac{1}{r^2}\frac{\partial^2 u}{\partial \theta^2},$$

namely that

$$\Delta P_r = 0.$$

(ii) If $u(r, \theta) = (P_r * g)(\theta)$, show that

$$\Delta u(r, \theta) = 0.$$

Problem 8.23. Advanced Exercise: Prove Proposition 8.51, Hint: See Bourbaki [Bourbaki (1965b)] (Chapter VIII, Section 2, No. 7, Lemma 4).

Problem 8.24. Advanced Exercise: Prove Corollary 8.52. Hint: Use Proposition 8.51.

Problem 8.25. Advanced Exercise: Prove Proposition 8.53. Hint: Use Proposition 8.51. Alternatively, see Bourbaki [Bourbaki (1965b)] (Chapter VIII, Section 4, No. 7, Proposition 19).

Chapter 9

Normed Algebras and Spectral Theory

Let G be a locally compact abelian group. In order to define the notion of Fourier transform on $\mathrm{L}^1(G)$, one needs to figure out what is its domain. The answer is that the domain of the Fourier transform on $\mathrm{L}^1(G)$ is the group $\widehat{G}$ of (unitary) characters of G, the homomorphisms $\chi\colon G \to \mathbb{C}$ such that $|\chi(g)| = 1$ for all $g \in G$. Then one has to give $\widehat{G}$ a topology that makes it into a locally compact group. Doing this is not obvious, but it turns out that as a topological space, $\widehat{G}$ is homeomorphic to the space $\mathsf{X}(\mathrm{L}^1(G))$ of characters of the algebra $\mathrm{L}^1(G)$, the set of algebra homomorphisms $\chi\colon \mathrm{L}^1(G) \to \mathbb{C}$. Here $\mathrm{L}^1(G)$ is the commutative algebra whose multiplication operation is convolution, and it can be shown that $\mathsf{X}(\mathrm{L}^1(G))$ is locally compact.

There is an even deeper connection between the space $\mathsf{X}(\mathrm{L}^1(G))$ of algebra characters and the Fourier transform. Indeed the Fourier cotransform on the commutative algebra $\mathrm{L}^1(G)$ is the Gelfand transform on $\mathrm{L}^1(G)$. For any $f \in \mathrm{L}^1(G)$, the *Gelfand transform* $\mathcal{G}_f$ is a function defined on the set $\mathsf{X}(\mathrm{L}^1(G))$ of characters of $\mathrm{L}^1(G)$ by

$$\mathcal{G}_f(\zeta) = \zeta(f), \qquad \zeta \in \mathsf{X}(\mathrm{L}^1(G)).$$

In general, the algebra $\mathrm{L}^1(G)$ is a complete normed algebra (a Banach algebra), but it does not have a multiplicative unit. However, the space $\mathcal{M}^1_{\mathrm{reg},\mathbb{C}}(G)$ of complex regular Borel measures on G, simply denoted $\mathcal{M}^1(G)$, is a unital Banach algebra, with the norm $\|\mu\| = |\mu|(G)$ defined in Definition 7.10, with the convolution of measures as multiplication (see Definition 8.21), and with the Dirac measure δ_1 as multiplicative unit. Furthermore, $\mathrm{L}^1(G)$ can be identified with a subalgebra of $\mathcal{M}^1(G)$, using the embedding $f \mapsto f d\lambda$ given by Proposition 7.32.

Therefore we are led to the study of algebras and normed algebras, in particular to complete normed algebras, called Banach algebras. If an algebra A is commutative, then Gelfand had the idea to realize A as a set of complex-valued functions on the set of characters $\mathsf{X}(A)$ of A. Let $\mathbb{C}^{\mathsf{X}(A)}$ be the set of functions from $\mathsf{X}(A)$ to $\mathbb{C}$. The map $\mathcal{G}\colon A \to \mathbb{C}^{\mathsf{X}(A)}$, called *Gelfand transform*, is defined as follows: for every $a \in A$,

$$\mathcal{G}_a(\chi) = \chi(a), \qquad \chi \in \mathsf{X}(A).$$

If A is a commutative Banach algebra with multiplicative identity element e, then the range of the Gelfand transform $\mathcal{G}_a$, the set $\{\chi(a) \mid \chi \in \mathsf{X}(A)\}$, is equal to the spectrum $\sigma(a)$ of a, namely the set of complex numbers λ such that $\lambda e - a$ is not invertible in A. The spectrum $\sigma(a)$ of a is a generalization of the notion of eigenvalue of a linear map.

The study of algebras and normed algebras focuses on three concepts:

(1) The notion of *spectrum* $\sigma(a)$ of an element a of an algebra A.
(2) If A is a commutative algebra, the notion of *character*, and the space $\mathsf{X}(A)$ of characters of A.
(3) If A is a commutative algebra, the notion of *Gelfand transform*, $\mathcal{G}\colon A \to \mathcal{C}(\mathsf{X}(A); \mathbb{C})$.

In Section 9.1 we define algebras and algebras with a multiplicative unit, called unital algebras. We also define normed algebras (without a multiplicative unit), and Cauchy-complete algebras, called Banach algebras. In Section 9.2 we show that every nonunital K-algebra A can be embedded into a unital K-algebra $\widetilde{A}$. We also define the notion of quotient of a normed algebra by an ideal.

If E is an infinite-dimensional vector space, and if $f\colon E \to E$ is a linear map, the definition of an eigenvalue λ of a linear map f used in finite-dimension in terms of the existence of nonzero vector u such that $f(u) = \lambda u$ no longer works because a non-invertible linear map may still be injective. Consequently, there may be some complex number λ such that $\lambda\mathrm{id} - f$ is not invertible, yet there is no nonzero vector u such that $f(u) = \lambda u$. This suggests defining a spectral value as a complex number λ such that $\lambda\mathrm{id} - f$ is not invertible. Then it is an easy step to generalize this definition to any unital algebra A with multiplicative identity e. Given any $a \in A$, viewed as a sort of generalized linear map, a number $\lambda \in \mathbb{C}$ is a *spectral value* for a if $\lambda e - a$ is not invertible.

The notion of *spectrum* is defined and investigated in Section 9.3. The complement $\mathbb{C} - \sigma(a)$ of $\sigma(a)$ is called the *resolvent set* of a, and for any fixed $a \in A$, the function $R(a, \lambda)$ (defined on the set $\mathbb{C} - \sigma(a)$) given by $R(a, \lambda) = (\lambda e - a)^{-1}$ is called the *resolvent* of a. In general there is no guarantee that $\sigma(a)$ is nonempty, but if A is a unital Banach algebra, then $\sigma(a) \neq \emptyset$.

Next we define the notion of character of a commutative unital algebra A. This is a homomorphism $\chi\colon A \to \mathbb{C}$. The space of characters on A is denoted by $\mathsf{X}(A)$. A first connection between the notion of spectrum and the notion of character is that for any $a \in A$ and any $\chi \in \mathsf{X}(A)$, we have $\chi(a) \in \sigma(a)$.

The third key notion, the Gelfand transform, is then defined. The map $\mathcal{G}\colon A \to \mathbb{C}^{\mathsf{X}(A)}$, called *Gelfand transform*, is defined as follows: for every $a \in A$,

$$\mathcal{G}_a(\chi) = \chi(a), \qquad \chi \in \mathsf{X}(A).$$

The function $\mathcal{G}_a$ (or $\mathcal{G}(a)$) is called the *Gelfand transform* of a. If we give $\mathsf{X}(A)$ the topology of pointwise convergence, then $\mathcal{G}$ becomes a continuous map from A to the space $\mathcal{C}(\mathsf{X}(A); \mathbb{C})$ of continuous functions on $\mathsf{X}(A)$.

In order to obtain sharper results about spectra and characters, we consider unital Banach algebras in Section 9.5.

Theorem 9.13 states that if A is a unital Banach algebra, then for any $a \in A$, the spectrum $\sigma(a)$ is nonempty, compact, and contained in the closed ball of radius $\|a\|$. Furthermore, the map $\lambda \mapsto R(a, \lambda)$ is holomorphic.

We prove the Gelfand–Mazur theorem (Theorem 9.14), which says that if a unital Banach algebra A is a (possibly noncommutative) field, then A is isometrically isomorphic to $\mathbb{C}$.

Certain notions defined for complex matrices can be generalized to algebras and normed algebras. In particular, if A is a normed algebra, for any $a \in A$, the number $\rho(a) = \inf_n \|a^n\|^{1/n}$ converges and is called the *spectral radius* of a. Then if A is a unital Banach algebra, we have

$$\rho(a) = \sup\{|\lambda| \mid \lambda \in \sigma(a)\}.$$

If A is a unital Banach algebra, then $\mathsf{X}(A)$ is compact and Hausdorff. If A is a nonunital Banach algebra, then $\mathsf{X}(A)$ is locally compact.

Properties of the Gelfand transform holding for unital Banach algebras are proven in Section 9.7.

Let A be a commutative unital Banach algebra.

(1) For every $a \in A$, the range of $\mathcal{G}_a$ is equal to the spectrum $\sigma(a)$ of a; that is,

$$\mathcal{G}_a(\mathsf{X}(A)) = \{\chi(a) \mid \chi \in \mathsf{X}(A)\} = \sigma(a).$$

(2) The Gelfand transform $\mathcal{G}\colon A \to \mathcal{C}(\mathsf{X}(A); \mathbb{C})$ is a continuous homomorphism such that $\|\mathcal{G}_a\|_\infty = \rho(a) \leq \|a\|$, and $\mathcal{G}_e$ is the constant function 1.

(3) An element $a \in A$ is invertible iff $\mathcal{G}_a$ does not vanish on $\mathsf{X}(A)$.

If A is a commutative nonunital Banach algebra, then the Gelfand transform is a homomorphism $\mathcal{G}\colon A \to \mathcal{C}_0(\mathsf{X}(A); \mathbb{C})$.

We can characterize when the Gelfand transform is an isometry and when it is injective in terms of its radical. Given a commutative unital algebra A, the *radical* of A, $\operatorname{rad} A$, is the intersection of all maximal ideals in A.

Let A be a commutative unital Banach algebra. We have $\operatorname{Ker} \mathcal{G} = \operatorname{rad} A$, and the Gelfand transform $\mathcal{G}\colon A \to \mathcal{C}(\mathsf{X}(A); \mathbb{C})$ is injective iff the radical of A is trivial; that is, $\operatorname{rad} A = (0)$.

In Section 9.8, we consider special algebras equipped with an involution, in particular, C^*-algebras.

If A is an algebra an involution is a bijection $a \mapsto a^*$ that satisfies the equations of the conjugate-transpose $A^* = \overline{(A^\top)}$ on complex matrices.

If A is a normed algebra, then an *involutive normed algebra* is an algebra with an involution $a \mapsto a^*$ satisfying the extra axiom

$$\|a\| = \|a^*\|, \qquad \text{for all } a \in A. \tag{i}$$

A C^**-algebra* is a Banach algebra satisfying the axiom

$$\|a\|^2 = \|a^* a\|, \qquad \text{for all } a \in A. \tag{C^*}$$

A C^*-algebra automatically satisfies Axiom (i). The normed algebras $\mathcal{M}^1(G)$ and $\mathrm{L}^1(G)$ are involutive algebras, but in general, not C^*-algebras. The main example of a C^*-algebra is the algebra $\mathcal{L}(H)$ of continuous linear maps on a complex Hilbert space H.

If A is an involutive algebra, an element $a \in A$ is *hermitian* if $a = a^*$, *normal* if $aa^* = a^*a$, *unitary* if $aa^* = a^*a = e$. As in the case of matrices, if A is a unital C^*-algebra, for every $a \in A$, if a is hermitian, then $\sigma(a) \subseteq \mathbb{R}$, and if a is unitary then $\sigma(a) \subseteq \mathbb{T} = \mathbf{U}(1)$.

In Section 9.9 we consider characters and the Gelfand transform in a C^*-algebra. Let A be a commutative unital C^*-algebra. Then for any character $\chi \in \mathsf{X}(A)$, we have

$$\chi(a^*) = \overline{\chi(a)}, \qquad \text{for all } a \in A,$$

or equivalently $\chi(a) = \overline{\chi(a^*)}$. We say that the characters of A are *hermitian*.

The main theorem of the theory of commutative unital C^*-algebras, due to Gelfand and Naimark, states that every commutative unital C^*-algebra can be viewed as the algebra of continuous functions on a compact space, namely its space of characters $\mathsf{X}(A)$.

More precisely, let A be a commutative unital C^*-algebra. Then the Gelfand transform $\mathcal{G}\colon A \to \mathcal{C}(\mathsf{X}(A);\mathbb{C})$ is an isometric isomorphism between A and $\mathcal{C}(\mathsf{X}(A);\mathbb{C})$ (and so $\|\mathcal{G}_a\|_\infty = \|a\| = \rho(a)$ for all $a \in A$). Furthermore the Gelfand maps $\mathcal{G}_a$ are hermitian.

The Gelfand–Naimark theorem is used to prove the Plancherel–Godement theorem (see Vol II, Section 2.8, Theorem 2.41), and some representation theory results in harmonic analysis; see Dieudonné [Dieudonné (2003b)].

The spectral theory of C^*-algebras is the key machinery used to develop generalizations of the spectral theorems for normal matrices to bounded (and unbounded) operators of various kinds on a Hilbert space. A condensed presentation of these spectral theorems is given in Folland [Folland (1995)] (Chapter 1, Section 1.4). An extensive treatment of these spectral theorems is given in Rudin [Rudin (1991)], and in Lax [Lax (2002)].

In Section 9.10, given an involutive Banach algebra A, we construct a C^*-algebra $\mathrm{St}(A)$ and an involutive homomorphism $j\colon A \to \mathrm{St}(A)$ (see Definition 9.19) that satisfies a universal mapping condition with respect to homomorphisms of A into a C^*-algebra. For every involutive homomorphism $\varphi\colon A \to B$ of A into a C^*-algebra B, there is a unique involutive homomorphism $\overline{\varphi}\colon \mathrm{St}(A) \to B$ such that

$$\varphi = \overline{\varphi} \circ j,$$

as shown in the following commutative diagram:

$$\begin{array}{ccc} A & \xrightarrow{\;j\;} & \mathrm{St}(A) \\ & \searrow^{\varphi} & \downarrow{\scriptstyle\overline{\varphi}} \\ & & B. \end{array}$$

It is also possible to characterize the set of characters $\mathsf{X}(\mathrm{St}(A))$ of $\mathrm{St}(A)$. Let A be an involutive Banach algebra. Then the map $\mathsf{X}(j)\colon \mathsf{X}(\mathrm{St}(A)) \to \mathsf{X}(A)$ is a homeomorphism of the set of characters $\mathsf{X}(\mathrm{St}(A))$ onto the subspace H of hermitian characters in $\mathsf{X}(A)$; that is, the characters $\chi\colon A \to \mathbb{C}$ such that $\chi(a) = \overline{\chi(a^*)}$ for all $a \in A$.

The above result applies to the involutive Banach algebra $\mathrm{L}^1(G)$ associated with a locally compact group G. In general, $\mathrm{L}^1(G)$ is not a C^*-algebra. Thus we can form the enveloping C^*-algebra $\mathrm{St}(\mathrm{L}^1(G))$ of $\mathrm{L}^1(G)$, denoted $\mathrm{St}(G)$. Remarkably, the canonical map j is injective. As a consequence, it can be shown that there is a homeomorphism between $\mathsf{X}(\mathrm{St}(G))$ and $\mathsf{X}(\mathrm{L}^1(G))$.

More generally, the material of this chapter (spectra, algebra characters, the Gelfand transform), is covered in Bourbaki [Bourbaki (1967b)], Dieudonné [Dieudonné (2005)], Folland [Folland (1995)], Lang [Lang (1996)], Lax [Lax (2002)], Rudin [Rudin (1987, 1991)], and Schwartz [Schwartz (1991)], the most complete presentations being Rudin [Rudin (1987, 1991)] and Bourbaki [Bourbaki (1967b)].

9.1 Normed Algebras, Banach Algebras

Before defining normed algebras, let us recall the definition of an algebra over a field K. Since the algebras that we will be dealing with *do not always have a multiplicative unit*, for example $\mathrm{L}^1(G)$ with the convolution as product (where G is a locally compact group equipped with a left Haar measure), we use the following definition. A good reference for unital normed algebras is Rudin [Rudin (1987)] (Chapter 18).

Definition 9.1. Given a field K, a *K-algebra* is a K-vector space A together with a bilinear operation $\star\colon A \times A \to A$, called *multiplication*, which is associative; that is,

$$(a \star b) \star c = a \star (b \star c) \qquad \text{for all } a, b, c \in A.$$

A K-algebra A is *unital* if there is a multiplicative identity element $\mathbf{1} \neq 0$ so that

$$\mathbf{1} \star a = a \star \mathbf{1} = a, \qquad \text{for all } a \in A.$$

A K-algebra A is *commutative* if

$$a \star b = b \star a, \qquad \text{for all } a, b \in A.$$

If A is a unital algebra, an element $a \in A$ is *invertible* if there is some $b \in A$ such that $a \star b = b \star a = \mathbf{1}$.

Given two K-algebras A and B, a *K-algebra homomorphism* $h\colon A \to B$ is a linear map such that

$$h(a \star_A b) = h(a) \star_B h(b) \qquad \text{for all } a, b \in A.$$

If A and B are unital, we also require that $h(\mathbf{1}_A) = \mathbf{1}_B$.

For example, the ring $\mathrm{M}_n(K)$ of all $n \times n$ matrices over a field K is a unital K-algebra with multiplicative identity element $\mathbf{1} = I_n$.

There are obvious notions of *subalgebra* and *ideal* of a K-algebra.

Definition 9.2. A *subalgebra* B of a K-algebra A is linear subspace closed under multiplication; that is, $x \star y \in B$ for all $x, y \in B$. If A is unital, we require that $\mathbf{1} \in B$. A *left ideal* $\mathfrak{A} \subseteq A$ is a linear subspace of A such that $x \star a \in \mathfrak{A}$ for all $a \in \mathfrak{A}$ and all $x \in A$. A *right ideal* $\mathfrak{A} \subseteq A$ is a linear subspace of A such that $a \star y \in \mathfrak{A}$ for all $a \in \mathfrak{A}$ and all $y \in A$. An *ideal* (or *two-sided ideal*) $\mathfrak{A} \subseteq A$ is a linear subspace of A that is both a left ideal and a right ideal. A left ideal (right ideal) $\mathfrak{A}$ is a *proper left ideal* (a *proper right ideal*) if $\mathfrak{A} \neq A$. The same definition applies to a two-sided ideal. A left ideal (right ideal, two-sided ideal) $\mathfrak{A}$ is a *maximal left ideal* (*maximal right ideal*, *maximal two-sided ideal*) if it is proper and if there is no proper left ideal (proper right ideal, proper two-sided ideal) $\mathfrak{B}$ such that $\mathfrak{A} \subseteq \mathfrak{B}$ and $\mathfrak{A} \neq \mathfrak{B}$.

If A is nonunital, then an ideal is a subalgebra, but the converse is false in general. If A is unital with identity element $\mathbf{1}$, then an ideal $\mathfrak{A}$ is proper iff $\mathbf{1} \notin \mathfrak{A}$, iff $\mathfrak{A}$ contains no invertible element.

If $\mathfrak{A}$ is a two-sided ideal in a K-algebra A (not necessarily unital or commutative), not only is the quotient $A/\mathfrak{A}$ a K-vector space but it is also a K-algebra. Recall the construction of the multiplication operation. The quotient $A/\mathfrak{A}$ consists of the equivalence classes of the equivalence relation $\equiv$ on A defined by

$$x \equiv y \quad \text{iff} \quad x - y \in \mathfrak{A}, \qquad x, y \in A.$$

Let us denote the equivalence class $x + \mathfrak{A}$ of $x \in A$ by $[x]$. We define a multiplication operation on $A/\mathfrak{A}$ by

$$[x] \star [y] = [x \star y], \qquad x, y \in A.$$

This operation is well defined because if $x' \equiv x$ and $y' \equiv y$, then $x' = x + a_1$ and $y' = y + a_2$ for some $a_1, a_2 \in \mathfrak{A}$, so

$$x' \star y' = (x + a_1) \star (y + a_2) = x \star y + x \star a_2 + a_1 \star y + a_1 \star a_2,$$

and since $\mathfrak{A}$ is a two-sided ideal and $a_1, a_2 \in \mathfrak{A}$, we also have $a = x \star a_2 + a_1 \star y + a_1 \star a_2 \in \mathfrak{A}$, and thus $x' \star y' = x \star y + a$ with $a \in \mathfrak{A}$, that is, $[x' \star y'] = [x \star y]$. The verification that the multiplication axioms of an algebra are satisfied is left as an exercise.

In order to generalize certain results to nonunital algebra, in particular to define the notion of radical, we need the notion of regular ideal, as defined in the Appendix of Bourbaki [Bourbaki (1968a)].

Definition 9.3. Let A be an algebra, possibly nonunital. A left ideal $\mathfrak{A}$ is *regular* if there is some $u \in A$ such that $x \star u - x \in \mathfrak{A}$ for all $x \in A$. A right ideal $\mathfrak{A}$ is *regular* if there is some $u \in A$ such that $u \star x - x \in \mathfrak{A}$ for all $x \in A$.

If A is unital, then every left (right) ideal is regular (let $u = \mathbf{1}$). Observe that the element u used in Definition 9.3 is a right identity in $A/\mathfrak{A}$ if $\mathfrak{A}$ is a regular left ideal, and a left identity in $A/\mathfrak{A}$ if $\mathfrak{A}$ is a regular right ideal. The following result can be proven; see Bourbaki [Bourbaki (1968a)] (Appendix, Proposition 3).

Proposition 9.1. *Let A be a commutative not necessarily unital algebra. An ideal $\mathfrak{A}$ in A is a maximal regular ideal iff $A/\mathfrak{A}$ is a field.*

Using Zorn's lemma we obtain a generalization of another standard result.

Proposition 9.2. *Let A be an algebra, not necessarily unital or commutative. A regular left ideal (regular right ideal) $\mathfrak{A}$ in A distinct from A is contained in a maximal regular left ideal (maximal regular right ideal).*

From now on we assume that the field K is the field $\mathbb{C}$ of complex numbers.

Definition 9.4. A *normed algebra* is a $\mathbb{C}$-algebra A endowed with a norm $\| \; \| : A \to \mathbb{R}_+$ satisfying the inequality

$$\|x \star y\| \leq \|x\| \, \|y\| \qquad \text{for all } x, y \in A.$$

If A is unital with identity $\mathbf{1}$, then we require that

$$\|\mathbf{1}\| = 1.$$

If the underlying normed vector space of A is complete, then we say that A is a *Banach algebra.*

The inequality $\|x \star y\| \leq \|x\| \, \|y\|$ shows that the multiplication operation $\star$ is *continuous* (see Proposition A.68). It is a generalization of the inequality characterizing matrix norms. Intuitively, a normed algebra can be viewed as a sort of generalized space of matrices.

For simplicity of notation we write xy instead of $x \star y$, unless confusion arises. *We also use the notation e for the multiplicative identity of A instead of* $\mathbf{1}$.

Example 9.1.

(1) If E is a normed vector space, then the space $\mathcal{L}(E)$ of continuous linear maps $f\colon E \to E$ is a unital algebra under composition, with id_E as identity element. It is a normed algebra under the operator norm,

$$\|f\| = \sup\{\|f(x)\| \mid \|x\| = 1\}.$$

If E is a Banach space (that is, a complete normed vector space), then $\mathcal{L}(E)$ is a Banach algebra.

(2) Let X be a topological space. Then the space $\mathcal{C}_b(X;\mathbb{C})$ of bounded continuous functions on X is a commutative unital Banach algebra, with the norm $\| \, \|_\infty$; functions are multiplied pointwise; that is, $(fg)(x) = f(x)g(x)$ of all $x \in X$. The multiplicative identity is the constant function 1. If X is a (Hausdorff) locally compact space, then the space $\mathcal{C}_0(X;\mathbb{C})$ is a commutative Banach algebra and an ideal in $\mathcal{C}_b(X;\mathbb{C})$, but it not unital unless X is compact (recall that Definition 2.16 implies that $\mathcal{C}_0(X;\mathbb{C}) = \mathcal{K}_\mathbb{C}(X)$ if X is compact). The space of $\mathcal{K}_\mathbb{C}(X)$ of continuous functions with compact support is a commutative normed algebra and an ideal in $\mathcal{C}_b(X;\mathbb{C})$, but it is not complete (and not unital unless X is compact).

(3) Let G be a locally compact group. The space $\mathcal{M}^1_{\mathrm{reg},\mathbb{C}}(G)$ of complex regular Borel measures on G, simply denoted $\mathcal{M}^1(G)$, is a unital Banach algebra, with the norm $\|\mu\| = |\mu|(G)$ defined in Definition 7.10, with the convolution of measures as multiplication (see Definition 8.21), and with the Dirac measure δ_1 as multiplicative unit (see Section 8.11). This is the most important example of this book.

(4) Let G be a locally compact group equipped with a left Haar measure λ. The space $\mathrm{L}^1(G)$ (with the L^1-norm) can be identified with a subspace of $\mathcal{M}^1(G)$, using the norm-preserving embeding $f \mapsto f d\lambda$ given

by Proposition 7.32. The space $\mathrm{L}^1(G)$ is a Banach algebra with the convolution of functions as multiplication, but it is not unital unless G is discrete.

(5) As a special case of (4), let $G = \mathbb{Z}$, in which case $\mathrm{L}^1(G)$ is the set of all sequences $x = (x_m)_{m\in\mathbb{Z}}$ with $x_m \in \mathbb{C}$, such that $\sum_{m\in\mathbb{Z}} |x_m| < \infty$. This space is also denoted $l^1(\mathbb{Z})$. The convolution product $x * y$ of $x = (x_m)$ and $y = (y_m)$ is given by

$$(x * y)_m = \sum_{p\in\mathbb{Z}} x_p y_{m-p},$$

and the norm by $\|x\| = \sum_{m\in\mathbb{Z}} |x_m|$. This is a commutative unital Banach algebra with identity element e_0 such that $e_0(0) = 1$ and $e_0(m) = 0$ for all $m \neq 0$.

Define δ^m by $\delta^m(m) = 1$ and $\delta^m(k) = 0$ if $k \neq m$. It is easy to see that $\delta^0 = e_0$, and $\delta^m * \delta^n = \delta^{m+n}$, so $(\delta^m)^{-1} = \delta^{-m}$. Then we see that for any $x = (x_m)_{m\in\mathbb{Z}} \in l^1(\mathbb{Z})$, we have $x = \sum_{m\in\mathbb{Z}} x_m\delta^m$, which shows that $l^1(\mathbb{Z})$ is generated by δ^1 and $\delta^{-1} = (\delta^1)^{-1}$.

(6) For any $c = (c_m) \in l^1(\mathbb{Z})$, let $\varphi_c\colon \mathbb{T} \to \mathbb{C}$ be the function given by

$$\varphi_c(e^{i\theta}) = \sum_{m\in\mathbb{Z}} c_m e^{im\theta}, \qquad \theta \in \mathbb{R}/(2\pi\mathbb{Z}).$$

The series defining φ_c is absolutely and uniformly convergent. Thus we deduce that

$$c_m = \frac{1}{2\pi}\int_{-\pi}^{\pi} \varphi_c(e^{i\theta})e^{-im\theta}d\theta,$$

the mth Fourier coefficient of φ_c; so the map $\varphi\colon c \mapsto \varphi_c$ is an injection from $l^1(\mathbb{Z})$ into the commutative unital Banach algebra of absolutely and uniformly convergent Fourier series (under pointwise multiplication of functions in $\mathrm{L}^1(\mathbb{T})$). The norm on this space is the norm induced by φ from the norm on $l^1(\mathbb{Z})$. The identity is the constant function 1.

(7) Let G be a compact group equipped with a Haar measure. In this case $\mathrm{L}^2(G) \subseteq \mathrm{L}^1(G)$, and both are Banach algebras under convolution ($\mathrm{L}^2(G)$ with the L^2-norm and $\mathrm{L}^1(G)$ with the L^1-norm). These algebras are nonunital unless G is finite. The Banach algebra $\mathrm{L}^2(G)$ is also a Hilbert space. This yields even more structure on $\mathrm{L}^2(G)$ which turns out to be a *Hilbert algebra*, as discussed later. The significance of this fact is that there is a structure theorem about separable Hilbert algebras (they can be expressed as the Hilbert sum of special kinds of ideals), and this structure theorem is a key ingredient in the proof of

the Peter–Weyl theorem, the fundamental theorem about the structure of $\mathrm{L}^2(G)$ and of the irreducible unitary representations of the compact group G.

(8) Let $\mathcal{C}^n([0,1])$ be the algebra of functions $f\colon [0,1] \to \mathbb{C}$ having a continuous derivative $f^{(k)}$ for $k = 1, \ldots, n$, under pointwise addition and multiplication. If we let

$$\|f\| = \sum_{k=0}^{n} \frac{1}{k!} \sup_{0 \leq t \leq 1} |f^{(k)}(t)|,$$

then we can check that this is indeed a norm, and that $\|fg\| \leq \|f\| \, \|g\|$. With this norm, $\mathcal{C}^n([0,1])$ is a commutative unital Banach algebra. The identity is the constant function 1.

The following simple proposition is left as an exercise.

Proposition 9.3. *Let A be a normed algebra. The closure of a subalgebra of A is a subalgebra of A, and the closure of any ideal of A is an ideal in A.*

Part (2) of the following proposition implies that the set of invertible elements in a unital Banach algebra is open.

Proposition 9.4. *Let A be a unital Banach algebra.*

(1) For any element $a \in A$, if $\|a\| < 1$, then $e - a$ is invertible, the series $\sum_{n=0} a^n$ converges absolutely,

$$(e-a)^{-1} = \sum_{n=0} a^n = e + a + a^2 + \cdots + a^n + \cdots,$$

and

$$\left\|(e-a)^{-1} - e - a\right\| \leq \frac{\|a\|^2}{1 - \|a\|}.$$

(2) Let $a \in A$ be an invertible element. Then for any $h \in A$, if $\|h\| < (1/2)\left\|a^{-1}\right\|^{-1}$, then $a + h$ is invertible, and

$$\left\|(a+h)^{-1} - a^{-1} + a^{-1}ha^{-1}\right\| \leq 2\left\|a^{-1}\right\|^3 \|h\|^2.$$

Proof. (1) Since $\|xy\| \leq \|x\| \, \|y\|$, we have $\|a^n\| \leq \|a\|^n$, and so

$$1 + \|a\| + \cdots + \|a^n\| \leq 1 + \|a\| + \cdots + \|a\|^n = \frac{1 - \|a\|^{n+1}}{1 - \|a\|}.$$

Since $\|a\| < 1$, the geometric series converges, and so does the series $(\sum_n \|a^n\|)$. Let

$$S_n = e + a + \cdots + a^n.$$

For all $n \geq m$, we have

$$\|S_n - S_m\| = \left\|a^{m+1} + \cdots + a^n\right\| \leq \left\|a^{m+1}\right\| + \cdots + \|a^n\|.$$

Since the sequence $\sum_{k=0}^{m} \|a^m\|$ converges, it is a Cauchy sequence, thus for every $\epsilon > 0$, we can find $p > 0$ so that for all $m \geq n > p$ we have

$$\left\|a^{m+1}\right\| + \cdots + \|a^n\| \leq \epsilon,$$

thus the sequence (S_m) is also a Cauchy sequence. Since A is complete, this sequence has a limit, say b. Since

$$S_n(e - a) = (e - a)S_n = e - a^{n+1},$$

and since $\lim_{n \mapsto 0} a^{n+1} = 0$ (because $\left\|a^{n+1}\right\| \leq \|a\|^{n+1}$ and $\|a\| < 1$), by continuity of multiplication, we get

$$b(e - a) = (e - a)b = e,$$

which means that b is the inverse of $e - a$.

Since $b = \sum_{n=0}^{\infty} a^n$, we have

$$\|b - e - a\| = \left\|a^2 + \cdots + a^n + \cdots\right\| \leq \sum_{n=2}^{\infty} \|a\|^n = \frac{\|a\|^2}{1 - \|a\|}.$$

(2) Since a is invertible,

$$a + h = a(e + a^{-1}h),$$

and since $\|h\| < (1/2)\left\|a^{-1}\right\|^{-1}$, we get

$$\left\|a^{-1}h\right\| \leq \left\|a^{-1}\right\| \|h\| \leq \left\|a^{-1}\right\| (1/2) \left\|a^{-1}\right\|^{-1} = 1/2.$$

Applying (1) to $-a^{-1}h$, since $\left\|-a^{-1}h\right\| = \left\|a^{-1}h\right\| < 1/2$, we deduce that $e + a^{-1}h$ is invertible. Therefore $a + h = a(e + a^{-1}h)$ is invertible. We also have $(a + h)^{-1} = (a(e + a^{-1}h))^{-1} = (e + a^{-1}h)^{-1}a^{-1}$, and so

$$\begin{aligned}(a + h)^{-1} - a^{-1} + a^{-1}ha^{-1} &= ((e + a^{-1}h)^{-1} - e + a^{-1}h)a^{-1} \\ &= ((e - (-a^{-1}h))^{-1} - e - (-a^{-1}h))a^{-1},\end{aligned}$$

which by (1) and the fact that $\left\|-a^{-1}h\right\| < 1/2$ yields

$$\begin{aligned}\left\|(a + h)^{-1} - a^{-1} + a^{-1}ha^{-1}\right\| &= \left\|((e - (-a^{-1}h))^{-1} - e - (-a^{-1}h))a^{-1}\right\| \\ &\leq \frac{\left\|-a^{-1}h\right\|^2 \left\|a^{-1}\right\|}{1 - \left\|-a^{-1}h\right\|} \\ &\leq 2\left\|-a^{-1}h\right\|^2 \left\|a^{-1}\right\| \\ &\leq 2\left\|a^{-1}\right\|^3 \|h\|^2,\end{aligned}$$

as claimed. $\square$

As a corollary of Proposition 9.4, we obtain the following result.

Proposition 9.5. *Let A be a unital Banach algebra. The set $G(A)$ of invertible elements of A is open and contains the open ball of center e given by $\{a \in A \mid \|a - e\| < 1\}$. The inversion map $\iota\colon G(A) \to G(A)$ given by $\iota(a) = -a^{-1}$ is differentiable and its derivative is given by $d\iota(a)(h) = -a^{-1}ha^{-1}$ for all $a \in G(A)$ and all $h \in A$. Consequently, ι is continuous.*

9.2 Two Algebra Constructions

Let K be an arbitrary field. Every nonunital K-algebra A can be embedded into a unital K-algebra $\widetilde{A}$ as follows.

Definition 9.5. Consider the vector space $\widetilde{A} = K \times A$, and define a multiplication by

$$(\lambda, a)(\mu, b) = (\lambda\mu, \lambda b + \mu a + ab). \tag{$*$}$$

It is easily verified that $\widetilde{A}$ with this multiplication is a K-algebra, and that $e = (1, 0)$ is a multiplicative unit. The algebra A is embedded into $\widetilde{A}$ using the map $a \mapsto (0, a)$, and it is immediately verified that A is a left ideal in $\widetilde{A}$ (in fact, a maximal ideal). Since $(\lambda, a)(0, b) = (0, \lambda b + ab)$, and $e = (1, 0)$, no element in A is invertible.

Since $(\lambda, a) = \lambda(1, 0) + (0, a)$, we can write $(\lambda, a) = \lambda e + a$. With this notation the multiplication of (λ, a) and (μ, b) becomes

$$(\lambda e + a)(\mu e + b) = \lambda\mu e + \lambda b + \mu a + ab,$$

which shows that the formula $(*)$ defining multiplication is not so strange after all.

If A already has a multiplicative identity ϵ, we immediately check that $e - \epsilon = (1, -\epsilon)$ has the following properties:

$$(e - \epsilon)^2 = e - \epsilon, \quad (e - \epsilon)A = A(e - \epsilon) = \{(0, 0)\}.$$

Then $K(e - \epsilon)$ is a unital algebra with identity $e - \epsilon$ (in fact, a field). Let $(K(e-\epsilon)) \times A$ be the product algebra of $K(e-\epsilon)$ and A under componentwise addition and multiplication. It is a unital algebra with identity $(e - \epsilon, \epsilon)$. To avoid confusion between the two multiplications, for the rest of this paragraph, let use denote multiplication in $\widetilde{A}$ by $\star$. Define the map $\varphi\colon \widetilde{A} \to (K(e - \epsilon)) \times A$ by

$$\varphi(\lambda, a) = (\lambda(e - \epsilon), \lambda\epsilon + a).$$

We check immediately that φ is a linear map. We have

$$\begin{aligned}\varphi(\lambda, a)\varphi(\mu, b) &= (\lambda(e-\epsilon), \lambda\epsilon + a)(\mu(e-\epsilon), \mu\epsilon + b)\\ &= (\lambda\mu(e-\epsilon), (\lambda\epsilon + a)(\mu\epsilon + b))\\ &= (\lambda\mu(e-\epsilon), \lambda\mu\epsilon + \lambda b + \mu a + ab),\end{aligned}$$

and

$$\begin{aligned}\varphi((\lambda, a) \star (\mu, b)) &= \varphi(\lambda\mu, \lambda b + \mu a + ab)\\ &= (\lambda\mu(e-\epsilon), \lambda\mu\epsilon + \lambda b + \mu a + ab),\end{aligned}$$

so

$$\varphi((\lambda, a) \star (\mu, b)) = \varphi(\lambda, a)\varphi(\mu, b),$$

which shows that φ is an algebra homomorphism. The map φ is obviously injective, and it is surjective because

$$\varphi(\lambda, a - \lambda\epsilon) = (\lambda(e-\epsilon), \lambda\epsilon + a - \lambda\epsilon) = (\lambda(e-\epsilon), a).$$

In summary we have the following result.

Proposition 9.6. *If A is a unital K-algebra, then we have an algebra isomorphism $\varphi\colon \widetilde{A} \to (K(e-\epsilon)) \times A$.*

If A is a normed algebra, it is also easy to check that the map

$$\|(\lambda, a)\| = |\lambda| + \|a\|$$

makes $\widetilde{A}$ into a unital normed algebra, and if A is a Banach algebra, then so is $\widetilde{A}$.

Later on we will encounter special kinds of Banach algebras called C^*-algebras. In this case, in order to make $\widetilde{A}$ into a C^*-algebra, a different norm is needed. It turns out that

$$\|(\lambda, a)\| = \sup\{\|\lambda b + ab\| \mid \in A,\ \|b\| \leq 1\}$$

makes $\widetilde{A}$ into a C^*-algebra and agrees with the original norm on A. It is the only norm that does so; see Folland [Folland (1995)] (Chapter 1, Section 4), or Bourbaki [Bourbaki (1967b)] (Chapter 1, Section 6, No. 3).

In many concrete cases the general construction of $\widetilde{A}$ is not needed. For example, since $\mathrm{L}^1(G)$ is an algebra embedded in $\mathcal{M}^1(G)$, we can use $\mathrm{L}^1(G) \oplus \mathbb{C}\delta_1$ as the completion of $\mathrm{L}^1(G)$ into a unital algebra.

The following construction will be used in Section 9.5. If A is a normed algebra and if $\mathfrak{A}$ is a (topologically) closed ideal in A, then the quotient $A/\mathfrak{A}$ is an algebra which can be made into a normed algebra as follows. If $\pi\colon A \to A/\mathfrak{A}$ is the quotient map, then for any $a \in A$, let

$$\|\pi(a)\| = \inf\{\|a + z\| \mid z \in \mathfrak{A}\}.$$

Proposition 9.7. *Let A be a normed algebra and let $\mathfrak{A}$ be a closed ideal in A. The map $\| \ \| : A/\mathfrak{A} \to \mathbb{R}_+$ given by*

$$\|\pi(a)\| = \inf\{\|a+z\| \mid z \in \mathfrak{A}\}, \qquad a \in A,$$

is a norm on $A/\mathfrak{A}$. If A is a Banach algebra, then so is $A/\mathfrak{A}$, and if A is unital, then so is $A/\mathfrak{A}$.

Proof. Following Schwartz [Schwartz (1991)] (Chapter II, Sec. 2) we will prove that the map $\| \ \| : A/\mathfrak{A} \to \mathbb{R}_+$ is a norm if and only if $\mathfrak{A}$ is closed.

First observe that by definition of $\|\pi(a)\|$, we have $\|\pi(a)\| \leq \|a\|$, so π is continuous. Note that $\pi(a) = [a]$, so by definition of the multiplication on $A/\mathfrak{A}$ it is immediately verified that π is a homomorphism. Let us check the triangle inequality and the fact that if $\|\pi(a)\| = 0$, then $a \in \mathfrak{A}$, which means that $\pi(a) = 0$.

By definition of a greatest lower bound, given any $\epsilon > 0$, for any two elements $x, y \in A$, there exist some $u, v \in \mathfrak{A}$ such that

$$\begin{aligned} \|x+u\| &\leq \|\pi(x)\| + \epsilon/2 \\ \|y+v\| &\leq \|\pi(y)\| + \epsilon/2. \end{aligned}$$

But $\pi(x+y+u+v) = \pi(x+y)$, so

$$\|\pi(x+y)\| \leq \|x+y+u+v\| \leq \|x+u\| + \|y+v\| \leq \|\pi(x)\| + \|\pi(y)\| + \epsilon.$$

Since ϵ is arbitrary, we get

$$\|\pi(x+y)\| = \|\pi(x)+\pi(y)\| \leq \|\pi(x)\| + \|\pi(y)\|.$$

We leave the verification that

$$\|\lambda\pi(x)\| = |\lambda| \, \|\pi(x)\|$$

as an exercise.

Assume that $\|\pi(a)\| = 0$. We want to prove that $a \in \mathfrak{A}$. For every $n > 0$, there is some $u_n \in \mathfrak{A}$ such that $\|a + u_n\| < \|\pi(a)\| + 1/n = 1/n$. Consequently the sequence $(a + u_n)$ converges to 0, which implies that the sequence (u_n) converges to $-a$. Since $u_n \in \mathfrak{A}$ for all $n \geq 1$ and since $\mathfrak{A}$ is closed, $-a \in \mathfrak{A}$. But $\mathfrak{A}$ is a vector space so $a \in \mathfrak{A}$, that is, $\pi(a) = 0$, as claimed.

We also have

$$\|(x+u)(y+v)\| \leq \|x+u\| \, \|y+v\| \leq (\|\pi(x)\| + \epsilon/2)(\|\pi(y)\| + \epsilon/2).$$

Since $(x+u)(y+v) = xy+xv+uy+uv$ and $u, v \in \mathfrak{A}$, by definition $\|\pi(xy)\| \leq \|(x+u)(y+v)\|$. Since ϵ is arbitrary, since by definition $\|\pi(x)\pi(y)\| = \|\pi(xy)\|$, we deduce that

$$\|\pi(x)\pi(y)\| \leq \|\pi(x)\| \, \|\pi(y)\| \, .$$

If e is the unit of A, then $\pi(e)$ is the unit of $A/\mathfrak{A}$. By definition $\|\pi(e)\| \leq \|e\| = 1$. We also have

$$\|\pi(e)\| = \|(\pi(e))^2\| \leq (\|\pi(e)\|)^2,$$

which implies that $\|\pi(e)\| \geq 1$, so $\|\pi(e)\| = 1$.

It remains to prove that if A is complete, then $A/\mathfrak{A}$ is also complete. Let $(\pi(a_n))$ be a Cauchy sequence in $A/\mathfrak{A}$. Taking a subsequence we may assume that

$$\|\pi(a_n) - \pi(a_{n-1})\| < \frac{1}{2^n} \quad \text{for all } n \geq 1.$$

We construct inductively a sequence (x_n) in A such that $\pi(x_n) = \pi(a_n)$ and

$$\|x_n - x_{n-1}\| < \frac{1}{2^n} \quad \text{for all } n \geq 1.$$

Intitially $x_1 = a_1$, and if $x_1, \ldots, x_n$ have been defined, since

$$\|\pi(a_{n+1}) - \pi(a_n)\| = \|\pi(a_{n+1} - a_n)\| < \frac{1}{2^{n+1}}$$

there is some $y \in \mathfrak{A}$ such that

$$\|a_{n+1} - a_n + y\| < \frac{1}{2^{n+1}}.$$

Then we let $x_{n+1} = x_n + a_{n+1} - a_n + y$, which works since

$$\begin{aligned}\pi(x_n + a_{n+1} - a_n + y) &= \pi(x_n) + \pi(a_{n+1}) - \pi(a_n) + \pi(y) \\ &= \pi(a_n) + \pi(a_{n+1}) - \pi(a_n) = \pi(a_{n+1}).\end{aligned}$$

Using the triangle inequality and the fact that

$$\|x_n - x_{n-1}\| < \frac{1}{2^n} \quad \text{for all } n \geq 1,$$

for all $n, p \geq 1$, we have

$$\begin{aligned}\|x_{n+p} - x_n\| &\leq \|x_{n+p} - x_{n+p-1}\| + \|x_{n+p-1} - x_{n+p-2}\| + \cdots \\ &\quad + \|x_{n+1} - x_n\| \\ &< \frac{1}{2^{n+p}} + \frac{1}{2^{n+p-1}} + \cdots + \frac{1}{2^{n+1}} \\ &= \frac{1}{2^{n+1}} \left(1 + \frac{1}{2} + \cdots \frac{1}{2^{n+p-1}}\right) < \frac{1}{2^n}.\end{aligned}$$

Consequently, the sequence (x_n) is a Cauchy sequence in A, and since A is complete, this sequence converges to a limit $a \in A$. But π is continuous so the sequence $(\pi(a_n)) = (\pi(x_n))$ converges to $\pi(a)$, as desired. $\square$

9.3 Spectrum, I; For an Algebra

Let E be a finite-dimensional vector space over the field $\mathbb{C}$, and let $f\colon E \to E$ be a linear map. Recall that some $\lambda \in \mathbb{C}$ is an *eigenvalue* of f if there is some *nonzero* vector $u \in E$, called an *eigenvector* associated with λ, such that

$$f(u) = \lambda u. \tag{$*$}$$

Equation $(*)$ holds iff

$$(\lambda \mathrm{id} - f)(u) = 0$$

iff $\lambda \mathrm{id} - f$ is *not injective*. But since E is finite-dimensional, a linear map is injective iff it is invertible, thus $\lambda \mathrm{id} - f$ is not injective iff $\lambda \mathrm{id} - f$ is not invertible. Therefore, *$\lambda \in \mathbb{C}$ is an eigenvalue for f iff $\lambda \mathrm{id} - f$ is not invertible*, which we call the second definition of an eigenvalue. In turn, $\lambda \mathrm{id} - f$ is not invertible iff $\det(\lambda \mathrm{id} - f) = 0$, thus the eigenvalues of f are precisely the zeros of the polynomial $\det(\lambda \mathrm{id} - f) = 0$, viewed as a polynomial in λ.

If E is infinite-dimensional, the situation is more complicated because an injective linear map may not be surjective, so $\lambda \mathrm{id} - f$ could be noninvertible, yet injective for some λ, in which case there are no nonzero eigenvectors associated with λ. Here is an example illustrating this situation.

Example 9.2. Let H be the Hilbert space $\mathrm{L}^2([0,1])$, and let $E = \mathcal{L}(H)$, the Banach algebra of continuous linear maps from H to H. Let $T\colon H \to H$ be the operator in E given by

$$T(h)(x) = xh(x), \quad \text{for all } h \in H \text{ and all } x \in [0,1].$$

For any $\lambda \in [0,1]$, observe that

$$(\lambda \mathrm{id}_E - T)(h)(x) = \lambda h(x) - xh(x) = (\lambda - x)h(x),$$

so $(\lambda \mathrm{id}_E - T)(h)(\lambda) = 0$. This shows that for every $h \in H$, the function $(\lambda \mathrm{id}_E - T)(h)$ vanishes at λ. Thus $\lambda \mathrm{id}_E - T$ is not surjective, because the constant function 1 belongs to H, but it is not in the range of $\lambda \mathrm{id}_E - T$ since it does not vanish anywhere. Therefore $\lambda \mathrm{id}_E - T$ is not invertible, so λ satisfies the second definition for being an eigenvalue of T. However, there is no nonzero eigenvector $h \in H$ associated with λ. Indeed, such a function $h \in H$ would satisfy the equation

$$(\lambda \mathrm{id}_E - T)(h)(x) = 0 \quad \text{for all } x \in [0,1],$$

that is,

$$(\lambda - x)h(x) = 0 \quad \text{for all } x \in [0,1],$$

which implies that $h(x) = 0$ for all $x \in [0,1]$, except for $x = \lambda$. This function is equal to the zero function almost everywhere, so in $H = \mathrm{L}^2([0,1])$, it is the zero function. This shows that $\lambda \mathrm{id}_E - T$ is injective for all $\lambda \in [0,1]$, but we also showed earlier that $\lambda \mathrm{id}_E - T$ is not surjective.

In summary, the definition of an eigenvalue λ of a linear map f used in finite-dimension in terms of the existence of nonzero vector u such that $f(u) = \lambda u$ no longer works in infinite dimension. However, if we redefine an eigenvalue of T to be a complex number λ such that $\lambda \mathrm{id}_E - T$ is not invertible, then every $\lambda \in [0,1]$ is an eigenvalue of T, even though T has *no eigenvectors*.

Example 9.2 suggests a definition of the notion of eigenvalue for a linear map f defined on an infinite-dimensional space E: it is a number $\lambda \in \mathbb{C}$ such that $\lambda \mathrm{id} - f$ is *not* invertible. From this, it is an easy step to generalize this definition to any unital algebra A. Given any $a \in A$, viewed as a sort of generalized linear map, a number $\lambda \in \mathbb{C}$ is a *spectral value* for a if $\lambda e - a$ is not invertible.

Definition 9.6. Let A be complex unital algebra with multiplicative unit e ($e \neq 0$). For any $a \in A$, the *spectrum* $\sigma(a)$ of a is the set of all $\lambda \in \mathbb{C}$ such that $\lambda e - a$ is not invertible. The complement $\mathbb{C} - \sigma(a)$ of $\sigma(a)$ is called the *resolvent set* of a. For any fixed $a \in A$, the function $R(a, \lambda)$ with values in A defined on the set $\mathbb{C} - \sigma(a)$ and given by

$$R(a, \lambda) = (\lambda e - a)^{-1}$$

is called the *resolvant* of a.

If different algebras are involved, to avoid confusion, for any $a \in A$ we write $\sigma_A(a)$ for $\sigma(a)$. Note that there is no guarantee that the spectrum is nonempty for any $a \in A$. However, if A is a unital Banach algebra, we will see that $\sigma(a)$ is nonempty for all $a \in A$. More is true: each $\sigma(a)$ is compact.

Let us mention some simple properties of the spectrum, most of which are proven in Bourbaki [Bourbaki (1967b)] (Chapter 1, Section 1, No. 2). See also Rudin [Rudin (1987)] (Chapter 18).

Proposition 9.8. *The following properties of the spectrum hold.*

(1) *For all* $\lambda \in \mathbb{C}$, *we have* $\sigma(\lambda e) = \{\lambda\}$, *and* $\sigma(a + \lambda e) = \sigma(a) + \lambda$.
(2) *An element* $a \in A$ *is invertible iff* $0 \notin \sigma(a)$.
(3) *Let* $P(x) = \alpha_0 + \alpha_1 x + \cdots + \alpha_n x^n$ *be a polynomial of degree* $n \geq 1$, *with* $\alpha_n \neq 0$. *For any* $a \in A$, *let*

$$P(a) = \alpha_0 e + \alpha_1 a + \cdots + \alpha_n a^n.$$

Then we have

$$P(\sigma(a)) = \sigma(P(a)),$$

with $P(\sigma(a)) = \{P(\lambda) \mid \lambda \in \sigma(a)\}$.
(4) *If* $a \in A$ *is nilpotent (that is,* $a^n = 0$ *for some* $n \in \mathbb{N}$*), then* $\sigma(a) = \{0\}$.
(5) *Let* A *and* B *be two unital algebras, and let* $\varphi\colon A \to B$ *be a homomorphism between them; recall that* $\varphi(e_A) = e_B$. *Then for any* $a \in A$, *we have* $\sigma_B(\varphi(a)) \subseteq \sigma_A(a)$.

Definition 9.7. If A is not a unital algebra, for any $a \in A$ we define the spectrum $\sigma'(a)$ of a as the spectrum of a in $\widetilde{A}$.

Since no element of A is invertible, we always have $0 \in \sigma'(a)$. If A is already a unital algebra with multiplicative identity ϵ, then we saw in the previous section that $\widetilde{A}$ is isomorphic to the product algebra $(K(e-\epsilon)) \times A$. If (A_1, e_1) and (A_2, e_2) are unital algebras, then for any $(a_1, a_2) \in A_1 \times A_2$ we have

$$\sigma((a_1, a_2)) = \sigma(a_1) \cup \sigma(a_2),$$

because $\lambda(e_1, e_2) - (a_1, a_2) = (\lambda e_1 - a_1, \lambda e_2 - a_2)$ is not invertible iff either $\lambda e_1 - a_1$ is not invertible or $\lambda e_2 - a_2$ is not invertible. Therefore, by letting $A_1 = K(e - \epsilon)$ and $A_2 = A$, if A is a unital algebra, we obtain

$$\sigma'(a) = \sigma(a) \cup \{0\}.$$

In general, if A is a unital algebra, then as the following example demonstrates, $\sigma(ab) \neq \sigma(ba)$.

Example 9.3. Let H be a Hilbert space with a countable orthonormal basis $(e_1, e_2, \ldots, e_n, \ldots)$. Let $f\colon H \to H$ and $g\colon H \to H$ be the continuous linear maps defined by

$$\begin{aligned} f(e_n) &= e_{n+1}, && n \geq 1, \\ g(e_{n+1}) &= e_n, && n \geq 1, \qquad g(e_1) = 0. \end{aligned}$$

Then $g \circ f = \mathrm{id}_H$, but $(f \circ g)(e_{n+1}) = e_{n+1}$ and $(f \circ g)(e_1) = 0$, so $\sigma(g \circ f) = \{1\}$, and $\sigma(f \circ g) = \{0, 1\}$.

However, for any algebra (unital or not),

$$\sigma'(ab) = \sigma'(ba).$$

The above equation follows from the following proposition, which generalizes a well-known property of matrices; namely that if A is a $m \times n$ matrix and B is a $n \times m$ matrix, then AB and BA have the same nonzero eigenvalues.

Proposition 9.9. *Let A be a unital algebra. For any two elements $a, b \in A$ and any nonzero scalar $\lambda \in \mathbb{C}$) (actually, any field K), if $ab - \lambda e$ is invertible, then $ba - \lambda e$ is invertible.*

Proof. Let u be the inverse of $ab - \lambda e$. We have

$$e = (ab - \lambda e)u = abu - \lambda u,$$

so $abu = \lambda u + e$, and then

$$\begin{aligned}(ba - \lambda e)(bua - e) &= b(abu)a - ba - \lambda bua + \lambda e \\ &= b(\lambda u + e)a - ba - \lambda bua + \lambda e \\ &= \lambda e.\end{aligned}$$

We also have

$$e = u(ab - \lambda e) = uab - \lambda u,$$

so $uab = \lambda u + e$, and then

$$\begin{aligned}(bua - e)(ba - \lambda e) &= b(uab)a - ba - \lambda bua + \lambda e \\ &= b(\lambda u + e)a - ba - \lambda bua + \lambda e \\ &= \lambda e.\end{aligned}$$

Since $\lambda \neq 0$, the above shows that $\lambda^{-1}(bua - e)$ is the inverse of $ba - \lambda e$. ☐

9.4 Characters, Gelfand Transform, I; For an Algebra

The notion of character of an algebra plays a crucial role in harmonic analysis, as a technical tool to generalize the Fourier transform. Thus we introduce it right away.

Definition 9.8. Let A be a complex, *commutative*, unital algebra with multiplicative identity e. A *character* of A is any algebra homomorphism $\chi \colon A \to \mathbb{C}$. Thus it is a linear form such that

$$\chi(ab) = \chi(a)\chi(b), \qquad \text{for all } a, b \in A,$$

and

$$\chi(e) = 1.$$

The set of characters of A is denoted by $\mathsf{X}(A)$.

Note that even though A is commutative, we do not denote its multiplication by $+$, to avoid confusion with addition in A.

Remark: Definition 9.8 still makes sense if A if a noncommutative unital algebra. In fact, Rudin discusses properties of characters of noncommutative unital algebras in Chapter 10 of Rudin [Rudin (1991)] under the name of *complex homomorphisms*. However, certain results no longer hold. The main problem is that if A is noncommutative, although the Gelfand transform $\mathcal{G}\colon A \to \mathcal{C}(\mathsf{X}(A);\mathbb{C})$ (see Definition 9.11) is an algebra homomorphism, since the algebra $\mathcal{C}(\mathsf{X}(A);\mathbb{C})$ (under pointwise multiplication) is commutative, the Gelfand transform can't be injective.

As in all other sources known to us, we always assume when discussing the Gelfand transform that our algebras are commutative and unital.

Proposition 9.10. *Let A be a (commutative) unital algebra with multiplicative identity e. For any character χ, the condition $\chi(e) = 1$ is equivalent to the condition that χ is not identically 0. If so, χ is surjective.*

Proof. Obviously, if $\chi(e) = 1$, then χ is not identically 0. Conversely, if $\chi(a) \neq 0$ for some $a \in A$ then

$$\chi(a) = \chi(ae) = \chi(a)\chi(e),$$

so $\chi(a)(1 - \chi(e)) = 0$, and since $\chi(a) \neq 0$, then $\chi(e) = 1$. Since χ is linear, $\chi(\lambda e) = \lambda\chi(e) = \lambda$, so χ is surjective. ☐

Proposition 9.11. *Let A be a unital (commutative) algebra. For any character χ, if $a \in A$ is invertible, then $\chi(a) \neq 0$.*

Proof. We have

$$1 = \chi(e) = \chi(aa^{-1}) = \chi(a)\chi(a^{-1}),$$

which implies that $\chi(a) \neq 0$. ☐

Definition 9.9. Let $\varphi\colon A \to B$ be a homomorphism of (commutative) unital algebras. Then φ induces a map $\mathsf{X}(\varphi)\colon \mathsf{X}(B) \to \mathsf{X}(A)$ given by

$$\mathsf{X}(\varphi)(\chi) = \chi \circ \varphi, \quad \chi \in \mathsf{X}(B).$$

It is immediately verified that $\mathsf{X}(\psi \circ \varphi) = \mathsf{X}(\varphi) \circ \mathsf{X}(\psi)$ and $\mathsf{X}(\mathrm{id}_A) = \mathrm{id}_{\mathsf{X}(A)}$.

It is easy to show that if $\varphi\colon A \to B$ is surjective, then $\mathsf{X}(\varphi)$ is a bijection of $\mathsf{X}(B)$ onto the set of characters of A that vanish on $\operatorname{Ker}\varphi$.

The following proposition characterizes $\mathsf{X}(A)$ in terms of ideals of A and shows a connection between spectra and characters.

Proposition 9.12. *Let A be a unital commutative algebra.*

(1) If $\mathcal{Y}$ denotes the set of ideals of codimension 1 in A, then the map $\chi \mapsto \operatorname{Ker} \chi$ is a bijection between $\mathsf{X}(A)$ and $\mathcal{Y}$.
(2) If $a \in A$ and if $\chi \in \mathsf{X}(A)$, then $\chi(a) \in \sigma(a)$. This property also holds is A is noncommutative.

Proof. (1) Since $\mathbb{C}$ has dimension 1 and a character χ is surjective onto $\mathbb{C}$, the kernel of χ has codimension 1. If $\mathfrak{A} \in \mathcal{Y}$ is an ideal of codimension 1, then $A/\mathfrak{A}$ is an algebra of dimension 1, so it is isomorphic to $\mathbb{C}$. But since $A/\mathfrak{A}$ has an identity element ϵ (see Proposition 9.7), a homomorphism φ from $A/\mathfrak{A}$ to $\mathbb{C}$ is uniquely determined by its value on ϵ, so there is a unique isomorphism from $A/\mathfrak{A}$ to $\mathbb{C}$. If $\pi\colon A \to A/\mathfrak{A}$ is the canonical projection, then the homomorphism $\varphi \circ \pi$ from A to $\mathbb{C}$ is the unique character with kernel $\mathfrak{A}$.

(2) Since χ is a homomorphism, we have

$$\chi(\chi(a)e - a) = \chi(a)\chi(e) - \chi(a) = \chi(a)1 - \chi(a) = 0.$$

This implies that $\chi(a)e - a$ is not invertible, since otherwise by Proposition 9.11 we would have $\chi(\chi(a)e - a) \neq 0$. $\square$

Remark: When A is a (commutative) nonunital algebra, a character is still a homomorphism $\chi\colon A \to \mathbb{C}$, but this time, the trivial map with constant value 0 is a character.

Definition 9.10. We denote the set of characters of a nonunital algebra A by $\mathsf{X}'(A)$, and we let $\mathsf{X}(A) = \mathsf{X}'(A) - \{0\}$.

If A is already unital, then a character $\chi \in \mathsf{X}'(A)$ (which is a homomorphism not restricted to map e to 1) is nonzero iff $\chi(e) = 1$, which shows that $\mathsf{X}(A) = \mathsf{X}'(A) - \{0\}$ is equal to the set of characters of the unital algebra A. It is also easy to see that if A is nonunital, then there is a bijection between the set of characters $\mathsf{X}(\widetilde{A})$ and the set of characters $\mathsf{X}'(A)$.

The third fundamental concept in the theory of algebras is due to Gelfand. The idea is to realize a commutative algebra A as a set of complex-valued functions on the set of characters $\mathsf{X}(A)$ of A. Let $\mathbb{C}^{\mathsf{X}(A)}$ be the set of functions from $\mathsf{X}(A)$ to $\mathbb{C}$.

Definition 9.11. Let A be a commutative unital algebra. The map $\mathcal{G}\colon A \to \mathbb{C}^{\mathsf{X}(A)}$, called *Gelfand transform*, is defined as follows: for every

$a \in A$,

$$\mathcal{G}_a(\chi) = \chi(a), \qquad \chi \in \mathsf{X}(A).$$

The function $\mathcal{G}_a$ (or $\mathcal{G}(a)$) is called the *Gelfand transform* of a.

If necessary to avoid ambiguities, we write $\mathcal{G}_A$ instead of $\mathcal{G}$. Note that $\mathcal{G}_a(\chi)$ is just *evaluation of* χ *on* a.

The set $\mathbb{C}^{\mathsf{X}(A)}$ of functions from $\mathsf{X}(A)$ to $\mathbb{C}$ is a commutative unital algebra under pointwise multiplication. Observe that map $\mathcal{G}\colon A \to \mathbb{C}^{\mathsf{X}(A)}$ is a homomorphism. Indeed we have

$$\mathcal{G}_{ab}(\chi) = \chi(ab) = \chi(a)\chi(b) = \mathcal{G}_a(\chi)\mathcal{G}_b(\chi),$$

so $\mathcal{G}_{ab} = \mathcal{G}_a\mathcal{G}_b$. We also have $\mathcal{G}_e(\chi) = \chi(e) = 1$, so $\mathcal{G}_e$ is the multiplicative unit in $\mathbb{C}^{\mathsf{X}(A)}$.

Observe that the Gelfand transform is not necessarily injective. For example, if $a \in A$ is a nonzero nilpotent, then by Proposition 9.8(4), $\sigma(a) = \{0\}$, and since by Proposition 9.12(2), $\chi(a) \in \sigma(a)$, we obtain so $\chi(a) = 0$ for all characters χ, which means that $\mathcal{G}_a = 0$, even though $a \neq 0$.

Since $\mathsf{X}(A)$ consists of functions from A to $\mathbb{C}$, we can give it the topology of pointwise convergence; see Definition 2.2. Recall that in this topology, a subset of functions $f\colon A \to \mathbb{C}$ is open if it is the union of subsets U_Ω of functions for which there is a finite subset Ω of A and some open intervals $(-r_a, r_a)$ (with $r_a > 0$) for all $a \in \Omega$, such that $f(a) \in (-r_a, r_a)$ for all $a \in \Omega$ (and $f(a)$ is arbitrary for all $a \in A - \Omega$). The topology of pointwise convergence is also defined in terms of semi-norms; see Example 2.4 in Section 2.7. Because $\mathbb{C}$ is Hausdorff, the topology of pointwise convergence is Hausdorff. It is also easy to see that a sequence (f_n) of functions $f_n\colon A \to \mathbb{C}$ converges to a function f in this topology iff it converges pointwise to f (that is, for every $a \in A$, the sequence $(f_n(a))$ converges to $f(a)$); see Section 2.1 or Folland [Folland (1999)] (Chapter 4, Proposition 4.12). Then, by definition of the topology of pointwise convergence, each Gelfand map $\mathcal{G}_a\colon \chi \mapsto \chi(a)$ is continuous (for every open interval $(-r, r)$ of $\mathbb{C}$, we have $\mathcal{G}_a^{-1}((-r, r)) = \{\chi \in \mathsf{X}(A) \mid \mathcal{G}_a(\chi) = \chi(a) \in (-r, r)\}$, which is open in $\mathsf{X}(A)$, with $\Omega = \{a\}$, by definition of the pointwise topology on $\mathsf{X}(A)$). In fact, we leave it as an exercise to prove that the topology of pointwise convergence on $\mathsf{X}(A)$ is the weakest (coarsest) topology for which the Gelfand maps $\mathcal{G}_a$ are continuous. This topology is also called the *weak topology* on $\mathsf{X}(A)$. In summary, the Gelfand transform $\mathcal{G}$ maps A to the space $\mathcal{C}(\mathsf{X}(A); \mathbb{C})$ of continuous functions on $\mathsf{X}(A)$.

In order to obtain sharper results about spectra and characters, we now consider unital Banach algebras.

9.5 Spectrum, II; For a Unital Banach Algebra

If A is a unital Banach algebra, then we have a more precise characterization of the spectrum of an element $a \in A$. Part (3) of Theorem 9.13 is a nontrivial and deep fact.

Theorem 9.13. *Let A be a unital Banach algebra. The following properties hold.*

(1) *For every $a \in A$, the spectrum $\sigma(a)$ is a compact subset of $\mathbb{C}$ contained in the closed ball of radius $\|a\|$ (thus $|\lambda| \leq \|a\|$ for all $\lambda \in \sigma(a)$).*
(2) *For any fixed $a \in A$, the resolvent*

$$R(a, \lambda) = (\lambda e - a)^{-1}$$

is a holomorphic function $R(a, \lambda)\colon (\mathbb{C} - \sigma(a)) \to A$ (which means that $(d/d\lambda)(R(a, \lambda))$ exists for all $\lambda \in \mathbb{C} - \sigma(a)$), and tends to zero at infinity.
(3) *For every $a \in A$, the spectrum $\sigma(a)$ is nonempty.*

Proof sketch. Theorem 9.13 is proven in Dieudonné [Dieudonné (2005)] (Chapter XV, Section 2), Rudin [Rudin (1991)] (Chapter 10, Theorem 10.13), Bourbaki [Bourbaki (1967b)] (Chapter 1, Section 2, No. 5), and Folland [Folland (1995)] (Chapter 1, Section 1).

If $|\lambda| > \|a\|$, then by Proposition 9.4(1) the element $e - \lambda^{-1}a$ is invertible since $\left\|\lambda^{-1}a\right\| = |\lambda^{-1}| \, \|a\| < 1$. Since $\lambda \neq 0$, the element $\lambda e - a$ is also invertible, which shows that $\lambda \notin \sigma(a)$. Therefore, if $\lambda \in \sigma(a)$, then $|\lambda| \leq \|a\|$.

Define the map g by $g(\lambda) = \lambda e - a$. Then g is continuous and invertible on $\Omega = \mathbb{C} - \sigma(a)$. By Proposition 9.5, $G(A)$ is open, and since g is continuous, we conclude that $\Omega = g^{-1}(G(A))$ is open, where $G(A)$ is the set of invertible elements in A. Therefore $\sigma(a)$ is closed. As a closed subset of the compact ball of radius $\|a\|$, $\sigma(a)$ is compact.

Replace a by $\lambda e - a$ and h by $(\mu - \lambda)e$ in Proposition 9.4(2). If $\lambda \in \Omega$ and μ is close enough to λ we have

$$\left\|R(a, \mu) - R(a, \lambda) + (\mu - \lambda)R(a, \lambda)^2\right\| \leq 2 \|R(a, \lambda)\|^3 |\mu - \lambda|^2,$$

which shows that

$$\frac{d}{d\lambda}R(a,\lambda) = -R(a,\lambda)^2.$$

By induction we obtain

$$\frac{d^k}{d\lambda^k}R(a,\lambda) = (-1)^k k! R(a,\lambda)^{k+1}.$$

Therefore $R(a, \lambda)$ is holomorphic on Ω.

For λ such that $|\lambda| > \|a\|$ we know that $e - \lambda^{-1}a$ is invertible, and by Proposition 9.4(1) we get

$$R(a,\lambda) = (\lambda e - a)^{-1} = \lambda^{-1}(e - \lambda^{-1}a)^{-1} = \sum_{n=0}^{\infty} \lambda^{-(n+1)}a^n,$$

and when $|\lambda|$ tends to infinity, since $|\lambda^{-1}|$ tends to zero, the above series tends to 0.

As explained in Rudin [Rudin (1991)] (Chapter 3, pages 82–85), Cauchy's theorem and Liouville's theorem generalize to holomorphic functions from an open subset of $\mathbb{C}$ to a Banach space (even a Fréchet space). If $\sigma(a)$ was empty, then $R(a, \lambda)$ would be a holomorphic entire function which is bounded, so by Liouville's theorem it would be constant. Since $R(a, \lambda)$ tends to zero at infinity, this constant would be zero, which is absurd (see also Rudin [Rudin (1991)], Chapter 10, Theorem 10.13). □

It is shown in Bourbaki [Bourbaki (1967b)] (Chapter 1, Section 2, Corollary 1) that if A is *any* unital normed algebra ($A \neq (0)$), not necessarily complete, then the spectrum $\sigma(a)$ is nonempty for every $a \in A$.

Remark: It is easy to show that

$$R(a,\mu) - R(a,\lambda) = (\lambda - \mu)R(a,\lambda)R(a,\mu)$$

for all $(\lambda, \mu) \in (\mathbb{C} - \sigma(a)) \times (\mathbb{C} - \sigma(a))$. This shows that $R(a, \lambda)$ and $R(a, \mu)$ commute.

As a corollary of Theorem 9.13 we have the following theorem.

Theorem 9.14. *(Gelfand–Mazur) Let A be a unital Banach algebra. If A is a (possibly noncommutative) field, then A is isometrically isomorphic to $\mathbb{C}$.*

Proof. For any $a \in A$, since $\sigma(a)$ is nonempty there is some λ such that $\lambda e - a$ is not invertible. Since A is a field, every nonzero element is invertible, and we must have $\lambda e - a = 0$, so $a = \lambda e$. But if $\lambda_1 \neq \lambda_2$, then at most one of $\lambda_1 e - a$ and $\lambda_2 e - a$ is zero, so there is a unique $\lambda(a)$ such that $a = \lambda(a)e$, and so the map $a \mapsto \lambda(a)$ is an isomorphism. We have $|\lambda(a)| = \|\lambda(a)e\| = \lambda(a)$, so this map is an isometry. ☐

Proposition 9.15. *Let A be a unital Banach algebra. For any invertible element $a \in A$, if $\|a\| = \|a^{-1}\| = 1$, then $\sigma(a) \subseteq \mathbf{U}(1)$.*

Proof. We know by Theorem 9.13(1) that $\sigma(a) \subseteq \{z \in \mathbb{C} \mid |z| \leq 1\}$, and similarly $\sigma(a^{-1}) \subseteq \{z \in \mathbb{C} \mid |z| \leq 1\}$. However if a is invertible it is easy to show that $\sigma(a^{-1}) = (\sigma(a))^{-1} = \{\lambda^{-1} \mid \lambda \in \sigma(a)\}$ (or see Dieudonné [Dieudonné (2005)] (Chapter XV, Section 2), so $\sigma(a) \subseteq \mathbf{U}(1)$. ☐

Proposition 9.15 generalizes the fact that the eigenvalues of a unitary matrix belong to $\mathbf{U}(1)$.

We can improve the bound on the radius of the smallest closed disc containing $\sigma(a)$ by introducing the spectral radius of a.

Proposition 9.16. *Let A be a normed algebra. For any $a \in A$, the sequence $(\|a^n\|^{1/n})$ converges and its limit is $\inf_n \|a^n\|^{1/n}$.*

Proposition 9.16 is proven in Dieudonné [Dieudonné (2005)] (Chapter XV, Section 2) and Rudin [Rudin (1991)] (Chapter 10, Theorem 10.13).

Definition 9.12. Let A be a normed algebra. For any $a \in A$, the number $\rho(a) = \inf_n \|a^n\|^{1/n}$ is called the *spectral radius* of a.

By definition, $\rho(a) \leq \|a\|$.

Proposition 9.17. *Let A be a unital Banach algebra. For any $a \in A$, the spectral radius $\rho(a)$ of a is equal to the radius of the smallest closed disc containing the spectrum $\sigma(a)$ of a, that is, $\rho(a) = \sup\{|\lambda| \mid \lambda \in \sigma(a)\}$.*

A proof of Proposition 9.17 is given in in Dieudonné [Dieudonné (2005)] (Chapter XV, Section 2), Rudin [Rudin (1991)] (Chapter 10, Theorem 10.13), and Folland [Folland (1995)] (Chapter 1, Section 1.1). Proposition 9.17 is a generalization of a well-known fact about the spectral radius of a matrix A, which is the largest modulus of the eigenvalues of A.

Proposition 9.8(3) implies that $\sigma(a^n) = (\sigma(a))^n = \{\lambda^n \mid \lambda \in \sigma(a)\}$, so by Proposition 9.17, we have

$$\rho(a^n) = (\rho(a))^n.$$

If A and B are two unital Banach algebras, and if A is a subalgebra of B, for any element $a \in A$, if a has an inverse a^{-1} in A, then $a^{-1} \in B$ so a is also invertible in B, but a could have an inverse $a^{-1} \in B$ such that $a^{-1} \notin A$. The following proposition addresses this situation.

Proposition 9.18. *Let A and B be two unital Banach algebras, with A a closed subalgebra of B. For any $a \in A$, we have $\sigma_B(a) \subseteq \sigma_A(a)$. Every boundary point of $\sigma_A(a)$ belongs to $\sigma_B(a)$. Hence if $\sigma_A(a)$ has empty interior, then $\sigma_B(a) = \sigma_A(a)$.*

Proof. Since for any $a \in A$, if the element $\lambda e - a$ is not invertible in B then it is not invertible in A, we have $\sigma_B(a) \subseteq \sigma_A(a)$.

For the second statement, we need to show that if λ_0 belongs to the boundary of $\sigma_A(a)$, then $\lambda_0 e - a$ is not invertible in B. Since λ_0 belongs to the boundary of $\sigma_A(a)$, there is a sequence (λ_n) with $\lambda_n \in \mathbb{C} - \sigma_A(a)$ converging to λ_0. For every n, the inverse $(\lambda_n e - a)^{-1}$ exists in A, and thus $(\lambda_n e - a)^{-1} \in B$. If $\lambda_0 \notin \sigma_B(a)$, then $(\lambda_0 e - a)^{-1} \in B$, and by Theorem 9.13, since the resolvent $R(a, \lambda)$ is continuous, the sequence $((\lambda_n e - a)^{-1})$ would converge to $(\lambda_0 e - a)^{-1}$. Since A is closed in B, and since $(\lambda_n e - a)^{-1} \in A$, we would have $(\lambda_0 e - a)^{-1} \in A$, contradicting the hypothesis that $\lambda_0 \in \sigma_A(a)$. □

Let us now turn to the characters of a commutative unital Banach algebra.

9.6 Characters, II; Commutative Unital Banach Algebras

We will show in the next theorem that $\mathsf{X}(A)$ is contained in the unit ball B of the dual A' of A (the space of continuous linear forms on A under the operator norm induced by the norm on A). Unfortunately, the unit ball B in A' is generally not compact in A' (with the topology induced by the operator norm). However, if we consider a weaker topology on A', namely the topology of pointwise convergence on A', then by the Banach–Alaoglu theorem, B is compact in this topology (see Rudin [Rudin (1991)] (Chapter 3, Theorem 3.15, or Folland [Folland (1999)] (Chapter 5, Theorem

5.18)). Since $\mathbb{C}$ is Hausdorff, the topology of pointwise convergence on A' is Hausdorff (see the end of Section 2.1).

It turns out that $\mathsf{X}(A)$ is closed in B for the topology of pointwise convergence, and so $\mathsf{X}(A)$ is compact for the topology of pointwise convergence. This is the reason for dropping the norm topology on $\mathsf{X}(A)$ and adopting the topology of pointwise convergence. For historical reasons this topology is also known under another name.

Definition 9.13. We define the *weak***-topology* on A' as the topology of pointwise convergence on A'.

Theorem 9.19. *Let A be a commutative unital Banach algebra.*

(1) Every character $\chi \in \mathsf{X}(A)$ is a continuous map of norm ≤ 1 (by norm, we mean operator norm).
(2) The space $\mathsf{X}(A)$ is compact (and thus Hausdorff) in the topology of the pointwise convergence on A' restricted to $\mathsf{X}(A)$.

Proof sketch. Part (1) of Theorem 9.19 is easy. From Proposition 9.12, we have $\chi(a) \in \sigma(a)$ for all $\in A$, and by Proposition 9.17, $|\chi(a)| \leq \rho(a) \leq \|a\|$, which implies (1).

Part (2) is proven in Bourbaki [Bourbaki (1967b)] (Chapter 1, Section 3, No. 1) and Rudin [Rudin (1991)] (Chapter 11, Theorem 11.9). By Part (1), $\mathsf{X}(A)$ is a subset of the closed unit ball B in A'. As we explained earlier, the unit ball B in A' is compact in the weak*-topology on A'. It is not hard to show that $\mathsf{X}(A)$ is closed in A' in the weak*-topology on A', and the restriction of this topology to $\mathsf{X}(A)$ is the topology of pointwise convergence. Therefore, $\mathsf{X}(A)$ being closed in a compact subset is compact in the topology of pointwise convergence. $\square$

Remark: If the commutative unital Banach algebra A is also separable, then $\mathsf{X}(A)$ is metrizable; see Dieudonné [Dieudonné (2005)] (Chapter XV, Section 3, Theorem 15.3.2).

Proposition 9.20. *If A is a nonunital commutative Banach algebra, then $\mathsf{X}'(A)$ is compact, and $\mathsf{X}(A)$ is locally compact (in the topology of the pointwise convergence on A').*

For a proof, see Bourbaki [Bourbaki (1967b)] (Chapter 1, Section 3, No. 1) or Folland [Folland (1995)] (Chapter 1, Section 1.3, Theorem 1.30). Actually, the proof is almost the same as before, except that we prove that

$\mathsf{X}'(A)$ is closed in B in the weak *-topology on A'. Since $\mathsf{X}(A) = \mathsf{X}'(A) - \{0\}$, the result follows.

Theorem 9.21. *Let A be a nonunital commutative Banach algebra. The map $\chi \mapsto \operatorname{Ker} \chi$ is a bijection from $\mathsf{X}(A)$ to the set of maximal regular ideals in A. If A is a unital commutative Banach algebra, then the map $\chi \mapsto \operatorname{Ker} \chi$ is a bijection from $\mathsf{X}(A)$ to the set of maximal ideals in A.*

Proof. We only prove the second statement in which A is unital. The case where A is nonunital is dealt with in Bourbaki [Bourbaki (1967b)] (Chapter 1, Section 3, Theorem 2). Let $\mathfrak{A}$ be a maximal ideal in A. By Proposition 9.3, this ideal must be closed, since otherwise the closure of $\mathfrak{A}$ would be an ideal strictly containing $\mathfrak{A}$, and such an ideal is proper because $\mathfrak{A}$ is contained in the set of noninvertible elements of A (if $\mathfrak{A}$ contains any invertible element, then $\mathfrak{A} = A$, but a maximal ideal is properly contained in A), which is closed, contradicting the maximality of $\mathfrak{A}$. By Proposition 9.7, the quotient algebra $A/\mathfrak{A}$ is a unital Banach algebra. But since $\mathfrak{A}$ is a maximal ideal, $A/\mathfrak{A}$ is a field. By the Gelfand–Mazur theorem (Theorem 9.14), the Banach algebra $A/\mathfrak{A}$ is isomorphic to $\mathbb{C}$, which implies that $\mathfrak{A}$ has codimension 1. Obviously an ideal of codimension 1 is maximal. By Proposition 9.12, the map $\chi \mapsto \operatorname{Ker} \chi$ is a bijection from $\mathsf{X}(A)$ to the set of maximal ideals in A. □

As a corollary of Theorem 9.21 we have the following result proven in Bourbaki [Bourbaki (1967b)] (Chapter 1, Section 3, No. 2) and Dieudonné [Dieudonné (2005)] (Chapter 15, Example 15.3.7) showing that every compact space E is realized as the set of characters of some commutative unital Banach algebra of functions. Recall that for every $a \in E$, we have the linear functional (evaluation at a) given by $\delta_a(f) = f(a)$ for all $f \in \mathcal{K}_{\mathbb{C}}(E)$, called (with an abuse of language) a Dirac measure (see Example 7.1).

Proposition 9.22. *Let E be a compact space. The map $a \mapsto \delta_a$ is a homeomorphism from E to the set of characters $\mathsf{X}(\mathcal{C}_{\mathbb{C}}(E))$ of the unital Banach algebra $\mathcal{C}_{\mathbb{C}}(E)$. If E is a locally compact space, then the map $a \mapsto \delta_a$ is a homeomorphism from E to the set of characters $\mathsf{X}(\mathcal{C}_0(E;\mathbb{C}))$ of the unital Banach algebra $\mathcal{C}_0(E;\mathbb{C})$ of continuous functions that tend to zero at infinity.*

Proposition 9.22 implies that if E is compact then the set of characters of $\mathcal{C}_{\mathbb{C}}(E)$ is the set of Dirac measures. Similarly, if E is locally compact then the set of characters of $\mathcal{C}_0(E;\mathbb{C})$ is also the set of Dirac measures.

In both cases we have a bijection between the set of characters and E itself. Since the the characters are of the form δ_a for any $a \in E$, the Gelfand transform $\mathcal{G}_f$ of any function $f \in \mathcal{C}_0(E;\mathbb{C})$ is given by

$$\mathcal{G}_f(\delta_a) = \delta_a(f) = f(a).$$

Therefore, if we identify E and $\mathsf{X}(\mathcal{C}_0(E;\mathbb{C}))$ under the homeomorphism $a \mapsto \delta_a$, we see that the Gelfand map becomes the identity.

Using the fact that the unital commutative Banach algebra $l^1(\mathbb{Z})$ of Example 9.1(5) is generated by δ^1 and $(\delta^1)^{-1}$, it is shown in Proposition 9.24 that $\mathsf{X}(l^1(\mathbb{Z})) = \sigma(\delta^1)$.

Example 9.4. Let us show that $\sigma(\delta^1) = \mathbb{T}$. We follow Folland [Folland (1995)] (Chapter 1, Section 1.2). We need to figure out for which $\lambda \in \mathbb{C}$ is $\lambda\delta^0 - \delta^1$ is invertible. Suppose that $a \in l^1(\mathbb{Z})$ is an inverse of $\lambda\delta^0 - \delta^1$, so that

$$(\lambda\delta^0 - \delta^1) * a = \delta^0.$$

We leave it as an exercise to show that

$$[(\lambda\delta^0 - \delta^1) * a]_n = \lambda a_n - a_{n-1}.$$

Consequently, $(\lambda\delta^0 - \delta^1) * a = \delta^0$ iff

$$\lambda a_0 - a_{-1} = 1 \quad \text{and} \quad \lambda a_n - a_{n-1} = 0 \quad \text{for all } n \neq 0.$$

It is easy to solve these equations recursively and we obtain

$$\begin{aligned} a_{-1} &= \lambda a_0 - 1 \\ a_n &= \lambda^{-n} a_0, \quad n \geq 0 \\ a_{-n} &= \lambda^{n-1} a_{-1}, \quad n \geq 2. \end{aligned}$$

In order for a to belong to $l^1(\mathbb{Z})$ the condition $\sum_{n=-\infty}^{\infty} |a_n| < \infty$ must hold, which forces $a_0 = 0$ if $|\lambda| \leq 1$ and $a_{-1} = 0$ if $|\lambda| \geq 1$. These conditions imply that there is a unique inverse $a \in l^1(\mathbb{Z})$ iff $|\lambda| \neq 1$, namely

$$a = \begin{cases} -\sum_{n=1}^{\infty} \lambda^{n-1}\delta^{-n} & \text{if } |\lambda| < 1 \\ \sum_{n=0}^{\infty} \lambda^{-n-1}\delta^{n} & \text{if } |\lambda| > 1. \end{cases}$$

Therefore, $\lambda\delta^0 - \delta^1$ is not invertible iff $|\lambda| = 1$, which shows that $\sigma(\delta^1) = \mathbf{U}(1) = \mathbb{T}$.

If A is the unital commutative Banach algebra of absolutely convergent Fourier series of Example 9.1(6), it can be shown that again $\mathsf{X}(A) = \mathbb{T}$; see Folland [Folland (1995)] (Chapter 1, Section 1.2).

9.7 Gelfand Transform, II; For a Commutative Unital Banach Algebra

If A is a commutative unital algebra, we already know that for each $a \in A$ the Gelfand transform $\mathcal{G}_a$ is a continuous map $\mathcal{G}_a : \mathsf{X}(A) \to \mathbb{C}$ (where $\mathsf{X}(A)$ is given the topology of pointwise convergence). If A is a Banach algebra, then we have the following sharper result.

Theorem 9.23. *Let A be a commutative unital Banach algebra.*

(1) For every $a \in A$, the range of $\mathcal{G}_a$ is equal to the spectrum $\sigma(a)$ of a; that is,

$$\mathcal{G}_a(\mathsf{X}(A)) = \{\chi(a) \mid \chi \in \mathsf{X}(A)\} = \sigma(a).$$

(2) The Gelfand transform $\mathcal{G}\colon A \to \mathcal{C}(\mathsf{X}(A); \mathbb{C})$ is a continuous homomorphism such that $\|\mathcal{G}_a\|_\infty = \rho(a) \leq \|a\|$, and $\mathcal{G}_e$ is the constant function 1.

(3) An element $a \in A$ is invertible iff $\mathcal{G}_a$ does not vanish on $\mathsf{X}(A)$.

Proof. (1) We already know from Proposition 9.12 that $\chi(a) \in \sigma(a)$ for every $a \in A$, so we just have to prove that for every $\lambda \in \sigma(a)$, there is some $\chi \in \mathsf{X}(A)$ such that $\chi(a) = \lambda$. If $\lambda \in \sigma(a)$, then $\lambda e - a$ is not invertible. Since $\lambda e - a$ is not invertible, the ideal $A(\lambda e - a)$ generated by $\lambda e - a$ is distinct from A. Using Zorn's lemma, it is a standard argument to show that this ideal is contained in a maximal ideal $\mathfrak{A}$. By Theorem 9.21, there is some character χ such that $\mathfrak{A} = \operatorname{Ker} \chi$, so χ vanishes on $A(\lambda e - a)$, and in particular on $\lambda e - a$, which shows that $\chi(a) = \lambda$.

(2) We already showed that $\mathcal{G}$ is a homomorphism and that $\mathcal{G}_e = 1$. Since by (1) we have $\mathcal{G}_a(\mathsf{X}(A)) = \sigma(a)$ and since by Proposition 9.17, we have $\rho(a) = \sup\{|\lambda| \mid \lambda \in \sigma(a)\}$, we get

$$\|\mathcal{G}_a\|_\infty = \sup |\sigma(a)| = \rho(a).$$

We already know that $\rho(a) \leq \|a\|$, so $\|\mathcal{G}_a\| \leq \|a\|$, which shows that $\mathcal{G}$ is continuous.

(3) We know that a is invertible iff $0 \notin \sigma(a)$, and by (1), this is equivalent to the fact that $\mathcal{G}_a$ does not vanish on $\mathsf{X}(A)$. □

Remark: If A is a commutative nonunital Banach algebra, then the Gelfand transform is a homomorphism $\mathcal{G}\colon A \to \mathcal{C}_0(\mathsf{X}(A); \mathbb{C})$; see Bourbaki [Bourbaki (1967b)] (Chapter 1, Section 3, No. 3) or Folland [Folland (1995)] (Chapter 1, Section 1.3, Theorem 1.30).

As a corollary of Theorem 9.23 we have the following result.

Proposition 9.24. *Let A be a commutative unital Banach algebra. For any fixed $a \in A$, the Gelfand transform $\mathcal{G}_a$ is a homeomorphism from $\mathsf{X}(A)$ to $\sigma(a)$ in the following two cases:*

(1) The algebra A is generated by a and e.
(2) The algebra A is generated by a and a^{-1} (assuming that a is invertible).

Proof. By Theorem 9.23, the map $\mathcal{G}_a$ is continuous and surjective onto $\sigma(a)$. Since $\mathsf{X}(A)$ and $\sigma(a)$ are compact Hausdorff spaces, by the corollary to Proposition A.33, it suffices to show that this map is injective. But any character $\chi\colon A \to \mathbb{C}$ is uniquely determined by $\chi(a)$, which is trivial in (1), and in (2) follows from the fact that $\chi(a^{-1}) = (\chi(a))^{-1}$. If $\mathcal{G}_a(\chi_1) = \mathcal{G}_a(\chi_2)$, then $\chi_1(a) = \chi_2(a)$, and since χ_1 is completely determined by $\chi_1(a)$ and similarly χ_2 is completely determined by $\chi_2(a)$, we have $\chi_1 = \chi_2$, and $\mathcal{G}_a$ is injective. □

In particular, since $l^1(\mathbb{Z})$ is generated by δ^1 and $\delta^{-1} = (\delta^1)^{-1}$, Proposition 9.24 shows that $\mathsf{X}(l^1(\mathbb{Z}))$ is homeomorphic to the spectrum of δ^1 (recall Example 9.1(5)). As we said earlier, this spectrum is equal to $\mathbb{T}$, so $\mathsf{X}(l^1(\mathbb{Z})) \cong \mathbb{T}$.

Example 9.5. Proposition 9.24 allows us to figure out what the Gelfand transform on $l^1(\mathbb{Z})$ is. Indeed, for every spectral value $e^{i\theta} \in \mathbb{T}$, there is a unique character χ_θ such that $\chi_\theta(\delta^1) = e^{i\theta}$. Since every $c \in l^1(\mathbb{Z})$ is written uniquely as $c = \sum_{m\in\mathbb{Z}} c_m(\delta^1)^m$, we have

$$\chi_\theta(c) = \sum_{m\in\mathbb{Z}} c_i\chi_\theta((\delta^1)^m) = \sum_{m\in\mathbb{Z}} c_i(\chi_\theta(\delta^1))^m = \sum_{m\in\mathbb{Z}} c_i e^{im\theta},$$

the Fourier series associated with $c = (c_m)_{m\in\mathbb{Z}} \in l^1(\mathbb{Z})$. Since there is a bijection between $\mathsf{X}(l^1(\mathbb{Z}))$ and $\mathbb{T}$, we can identify χ_θ and $e^{i\theta}$, and we see that the Gelfand transform from $l^1(\mathbb{Z})$ to $\mathcal{C}(\mathsf{X}(l^1(\mathbb{Z})))$ is given by

$$\mathcal{G}_c(e^{i\theta}) = \chi_\theta(c) = \sum_{m\in\mathbb{Z}} c_i e^{im\theta}.$$

As a corollary of the above characterization of the Gelfand transform on $l^1(\mathbb{Z})$ and Theorem 9.23 we obtain the following nontrivial theorem of Wiener.

Proposition 9.25. *For any $c = (c_m)_{m\in\mathbb{Z}} \in l^1(\mathbb{Z})$, if the Fourier series $f(e^{i\theta}) = \sum_{m\in\mathbb{Z}} c_i e^{im\theta}$ does not vanish (does not take the value 0 for any*

$\theta \in \mathbb{R}/(2\pi\mathbb{Z})$), then $1/f$ is given by the absolutely convergent Fourier series $\sum_{m\in\mathbb{Z}} b_m e^{im\theta}$, with $b = c^{-1}$.

Proof. By Theorem 9.23(3), the element $c \in l^1(\mathbb{Z})$ is invertible iff the Fourier series $f = \mathcal{G}_c$ does not vanish. In this case, $\mathcal{G}_{c^{-1}}$ is the inverse $1/f$ of f. $\square$

The Gelfand transform on $l^1(\mathbb{Z})$ turns out to be the Fourier cotransform on $l^1(\mathbb{Z})$. More generally, if G is a commutative locally compact group equipped with a Haar measure λ, the Gelfand transform can be viewed as the Fourier cotransform on the commutative Banach algebra $\mathrm{L}^1(G)$. For any $f \in \mathrm{L}^1(G)$, the Gelfand transform $\mathcal{G}_f$ is a function defined on the set $\mathsf{X}(\mathrm{L}^1(G))$ of characters of $\mathrm{L}^1(G)$ by

$$\mathcal{G}_f(\zeta) = \zeta(f), \qquad \zeta \in \mathsf{X}(\mathrm{L}^1(G)).$$

However, it turns out that there is a homeomorphism between $\mathsf{X}(\mathrm{L}^1(G))$ and the dual group $\widehat{G}$ of G, which is the group of continuous homomorphisms $\chi\colon G \to \mathbf{U}(1)$; for example, see Theorem 10.6 or Folland [Folland (1995)] (Chapter 4, Section 1, Theorem 4.2). This homeomorphism from the dual group $\widehat{G}$ to $\mathsf{X}(\mathrm{L}^1(G))$ is given by the map $\chi \mapsto \zeta_\chi$, with

$$\mathcal{G}_f(\zeta_\chi) = \zeta_\chi(f) = \int \chi(s) f(s)\, d\lambda(s), \qquad \chi \in \widehat{G}.$$

Consequently we can view the Gelfand transform $\mathcal{G}_f$ of f as a map $\overline{\mathcal{F}}(f)$ defined on $\widehat{G}$ instead of $\mathsf{X}(\mathrm{L}^1(G))$, namely

$$\overline{\mathcal{F}}(f)(\chi) = \int \chi(s) f(s)\, d\lambda(s), \qquad \chi \in \widehat{G}.$$

The map $\overline{\mathcal{F}}(f)$ is the *Fourier cotransform* of f. For technical reasons this map is denoted as $\overline{\mathcal{F}}(f)$ instead of $\mathcal{F}(f)$. The *Fourier transform* of f is the map $\mathcal{F}(f)$ defined on $\widehat{G}$ by

$$\mathcal{F}(f)(\chi) = \int \overline{\chi(s)} f(s)\, d\lambda(s), \qquad \chi \in \widehat{G}.$$

Most authors define the Fourier transform with the conjugate term $\overline{\chi(s)}$ under the integral, but this convention is not universally adopted. As in Folland and Bourbaki, this is the convention that we adopt. The theory of Fourier transforms on a commutative locally compact group will be discussed thoroughly in Chapter 10.

Finally we can characterize when the Gelfand transform is an isometry and when it is injective.

Proposition 9.26. *Let A be a commutative unital Banach algebra. The Gelfand transform $\mathcal{G}\colon A \to \mathcal{C}(\mathsf{X}(A); \mathbb{C})$ is an isometry iff $\|a^2\| = \|a\|^2$ for all $a \in A$.*

Proof. First we prove that $\|\mathcal{G}_a\|_\infty = \|a\|$ iff $\left\|a^{2^k}\right\| = \|a\|^{2^k}$ for all $k \geq 1$.

Since by Theorem 9.23 we have $\|\mathcal{G}_a\|_\infty = \rho(a)$, and since

$$\rho(a) = \lim_{k \mapsto \infty} \left\|a^{2^k}\right\|^{1/2^k},$$

if $\left\|a^{2^k}\right\| = \|a\|^{2^k}$ for all $k \geq 1$, then $\|\mathcal{G}_a\|_\infty = \|a\|$.

Conversely, assume that $\|\mathcal{G}_a\|_\infty = \|a\|$. Then

$$\left\|a^{2^k}\right\| \leq \|a\|^{2^k} = \|\mathcal{G}_a\|_\infty^{2^k} = \rho(a)^{2^k} = \rho(a^{2^k}) \leq \left\|a^{2^k}\right\|,$$

which shows that $\left\|a^{2^k}\right\| = \|a\|^{2^k}$ for all $k \geq 1$. Now if $\|a^2\| = \|a\|^2$, then by induction $\left\|a^{2^k}\right\| = \|a\|^{2^k}$ for all $k \geq 1$, so $\|\mathcal{G}_a\|_\infty = \|a\|$. Conversely, we already proved that if $\|\mathcal{G}_a\|_\infty = \|a\|$, then $\left\|a^{2^k}\right\| = \|a\|^{2^k}$ for all $k \geq 1$; in particular, $\|a^2\| = \|a\|^2$. □

Definition 9.14. Given a commutative unital algebra A, the *radical* of A, $\mathrm{rad}\, A$, is the intersection of all maximal ideals in A.

Proposition 9.27. *Let A be a commutative unital Banach algebra. We have $\mathrm{Ker}\, \mathcal{G} = \mathrm{rad}\, A$, so the following statements are equivalent.*

(1) The Gelfand transform $\mathcal{G}\colon A \to \mathcal{C}(\mathsf{X}(A); \mathbb{C})$ is injective.
(2) The radical of A is trivial; that is, $\mathrm{rad}\, A = (0)$.
(3) The set $\{a \in A \mid \sigma(a) = \{0\}\}$ is reduced to $\{0\}$.
(4) The set $\{a \in A \mid \rho(a) = 0\}$ is reduced to $\{0\}$.

Proof. The Gelfand transform $\mathcal{G}\colon A \to \mathcal{C}(\mathsf{X}(A); \mathbb{C})$ is injective iff its kernel is (0). We have $a \in \mathrm{Ker}\, \mathcal{G}$ iff $\mathcal{G}_a = 0$, which means that $\chi(a) = 0$ for all $\chi \in \mathsf{X}(A)$, that is, $a \in \mathrm{Ker}\, \chi$ for all $\chi \in \mathsf{X}(A)$. Since $\mathrm{Ker}\, \chi$ is a maximal ideal this shows that $a \in \mathrm{rad}\, A$. Thus $\mathrm{Ker}\, \mathcal{G} \subseteq \mathrm{rad}\, A$. Since by Theorem 9.21, every maximal ideal is the kernel of some character, $\mathrm{rad}\, A \subseteq \mathrm{Ker}\, \mathcal{G}$, so we have $\mathrm{Ker}\, \mathcal{G} = \mathrm{rad}\, A$.

$\mathrm{Ker}\, \mathcal{G} = \mathrm{rad}\, A$ implies that (1) and (2) are equivalent.

Since $a \in \operatorname{Ker} \mathcal{G}$ iff $\sigma(a) = \mathcal{G}_a(\mathsf{X}(A)) = \{0\}$, we see that $\mathcal{G}$ is not injective iff there is some nonzero $a \in A$ such that $\sigma(a) = \{0\}$, so (1) and (3) are indeed equivalent.

Since $\sigma(a) = \{0\}$ iff $\rho(a) = 0$, (3) and (4) are equivalent. □

The notion of radical can also be defined for nonunital, noncommutative algebras using the notion of regular ideal (see Definition 9.3); see Bourbaki [Bourbaki (1968a)], Appendix, Section 3.

Definition 9.15. Let A be a nonunital, possibly noncommutative algebra. The *radical* rad A, is the intersection of all maximal regular left ideals in A.

It is shown in Bourbaki [Bourbaki (1968a)] (Appendix, Section 3, Proposition 5) that this more general notion of a radical rad A is a two-sided ideal which is isomorphic to the intersection of all maximal left ideals of $\widetilde{A}$, namely the radical of $\widetilde{A}$ (see Section 9.2 for the definition of $\widetilde{A}$). The following generalization of Proposition 9.27 holds; see Bourbaki [Bourbaki (1967b)] (Chapter 1, Section 3, Proposition 5).

Proposition 9.28. *If A be a commutative nonunital Banach algebra, then* $\operatorname{Ker} \mathcal{G} = \operatorname{rad} A$.

We will show later on that if G is a locally compact group, then $\mathrm{L}^1(G)$ has a trivial radical. Recall that in general $\mathrm{L}^1(G)$ is a Banach algebra which is neither commutative nor unital so Definition 9.15 is needed to define its radical.

9.8 Banach Algebras with Involution; C^*-Algebras

If A is a complex matrix, then we have its conjugate-transpose $A^* = \overline{(A^\top)} = (\overline{A})^\top$, and we know that it satisfies various identities such as

$$(A+B)^* = A^* + B^*, \quad (AB)^* = B^*A^*, \quad (A^*)^* = A,$$

and so on. It turns out to be fruitful to define algebras and normed algebras having an operation $a \mapsto a^*$ satisfying the most useful laws of conjugate-transposition.

Definition 9.16. Let A be an algebra over $\mathbb{C}$ (not necessarily unital). An *involution* on A is a bijection $a \mapsto a^*$ satisfying the following axioms:

$$(a^*)^* = a \qquad (a+b)^* = a^* + b^*$$
$$(\lambda a)^* = \overline{\lambda} a^* \qquad (ab)^* = b^* a^*,$$

for all $a, b \in A$ and all $\lambda \in \mathbb{C}$. The element a^* is called the *adjoint* of a. If $a = a^*$, then a is called *hermitian* (or *self-adjoint*). An algebra with an involution is called an *involutive algebra*.

If A is a normed algebra, then an *involutive normed algebra* is an algebra with an involution $a \mapsto a^*$ satisfying the extra axiom

$$\|a\| = \|a^*\|, \qquad \text{for all } a \in A. \tag{i}$$

A C^**-algebra* is a Banach algebra with an involution $a \mapsto a^*$ satisfying the axiom

$$\|a\|^2 = \|a^*a\|, \qquad \text{for all } a \in A. \tag{C^*}$$

Remark: Oddly, Rudin uses the term B^**-algebra* instead of C^*-algebra; see Rudin [Rudin (1991)], Definition 11.17. The term C^**-algebra* seems to be used predominantly.

Here are a few immediate consequences of the axioms.

(1) We have $0^* = 0$.

This is because $0^* = (0+0)^* = 0^* + 0^*$, so $0^* = 0$.

(2) If A is unital with identity e, then $e^* = e$.

This is because using the axioms, for all $a \in A$, we have

$$ae^* = (a^*)^*e^* = (ea^*)^* = (a^*)^* = a,$$

and similarly

$$e^*a = e^*(a^*)^* = (a^*e)^* = (a^*)^* = a,$$

so e^* is a multiplicative identity element, and by uniqueness of this element, $e^* = e$.

(3) If $a \in A$ is invertible, then so is a^* and $(a^*)^{-1} = (a^{-1})^*$.

We have

$$(a^{-1})^*a^* = (aa^{-1})^* = e^* = e,$$

and

$$a^*(a^{-1})^* = (a^{-1}a)^* = e^* = e,$$

so $(a^*)^{-1} = (a^{-1})^*$.

(4) For any $a \in A$, a is invertible iff a^* is invertible.

Since $(a^*)^* = a$, by applying (3) to a^* we see that if a^* is invertible, then a is invertible, and (3) gives the converse.

(5) For all $a \in A$ and all $\lambda \in \mathbb{C}$, we have $\lambda \in \sigma(a)$ iff $\overline{\lambda} \in \sigma(a^*)$.

This is because, by (4), $\lambda e - a$ is invertible iff $\overline{\lambda} e^* - a^* = \overline{\lambda} e - a^*$ is invertible.

(6) If A is a C^*-algebra, then Equation (i) holds. Therefore a C^*-algebra is an involutive algebra.

We already know that $0^* = 0$. For any $a \neq 0$,

$$\|a\|^2 = \|a^*a\| \leq \|a^*\| \, \|a\| \, ,$$

so $\|a\| \leq \|a^*\|$. Since $(a^*)^* = a$, we also get $\|a^*\| \leq \|a\|$, so $\|a^*\| = \|a\|$.

(7) If A is a C^*-algebra, then $\|aa^*\| = \|a\|^2$.

Using the fact that $(a^*)^* = a$, and $\|a\| = \|a^*\|$, substituting a^* for a in $\|a^*a\| = \|a\|^2$, we get $\|aa^*\| = \|a^*\|^2 = \|a\|^2$.

(8) If A is a unital C^*-algebra, then $\|e\| = 1$.

We have

$$\|e\|^2 = \|e^*e\| = \|e\| \, .$$

This implies that $\|e\| = 0, 1$, but since $e \neq 0$, we must have $\|e\| = 1$.

Example 9.6. The examples below are among the examples listed in Example 9.1.

(1) If E is a complex Hilbert space, then the space $\mathcal{L}(E)$ of continuous linear maps $f \colon E \to E$ is a C^*-algebra, with involution $h \mapsto h^*$.

(2) Let X be a topological space. Then the space $\mathcal{C}_b(X; \mathbb{C})$ of bounded continuous functions on X is a commutative C^* unital Banach algebra with involution $f \mapsto \overline{f}$. Let X be a compact topological space. Then the space $\mathcal{C}(X; \mathbb{C})$ of continuous functions on X is a commutative C^* unital Banach algebra with involution $f \mapsto \overline{f}$. If X is a locally compact space, then $\mathcal{C}_0(X; \mathbb{C})$ is a nonunital C^*-algebra with involution $f \mapsto \overline{f}$.

(3) Let G be a locally compact group. The space $\mathcal{M}^1_{\mathrm{reg},\mathbb{C}}(G)$ of complex regular Borel measures on G, simply denoted $\mathcal{M}^1(G)$, is a unital Banach algebra, with the norm $\|\mu\| = |\mu|(G)$ defined in Definition 7.10, with the convolution as multiplication, and with the Dirac measure δ_1 as multiplicative unit. By Proposition 8.46, the map $\mu \mapsto \mu^* = \overline{\check{\mu}}$ is an involution that makes $\mathcal{M}^1(G)$ into an involutive algebra. In general it is not a C^*-algebra.

(4) Let G be a locally compact group equipped with a left Haar measure λ. The space $\mathrm{L}^1(G)$ (with the L^1-norm) can be identified with a subspace

of $\mathcal{M}^1(G)$, using the embedding $f \mapsto f d\lambda$ given by Proposition 7.32. The space $\mathrm{L}^1(G)$ is a Banach algebra, but it is not unital unless G is discrete. If G is unimodular, then the map $f \mapsto f^*$ where $f^*(s) = \overline{f(s^{-1})}$ is an involution. If G is not unimodular, then we define f^* by $f^*(s) = \Delta(s^{-1})\overline{f(s^{-1})}$, and then the map $f \mapsto f^*$ is an involution such that the map $f \mapsto f\, d\lambda$ is an embedding of the involutive Banach algebra $\mathrm{L}^1(G)$ into the unital involutive Banach algebra $\mathcal{M}^1(G)$. In general $\mathrm{L}^1(G)$ is not a C^*-algebra.

(5) As a special case of (4), let $G = \mathbb{Z}$, in which case $\mathrm{L}^1(G)$ is the set of all sequences $x = (x_m)_{m\in\mathbb{Z}}$ with $x_m \in \mathbb{C}$, such that $\sum_{m\in\mathbb{Z}} |x_m| < \infty$. This space is also denoted $l^1(\mathbb{Z})$. The convolution product $x * y$ of $x = (x_m)$ and $y = (y_m)$ is given by

$$(x * y)_m = \sum_{p\in\mathbb{Z}} x_p y_{m-p},$$

and the norm by $\|x\| = \sum_{m\in\mathbb{Z}} |x_m|$. This is a commutative unital Banach algebra with identity element e_0 such that $e_0(0) = 1$ and $e_0(m) = 0$ for all $m \neq 0$. The map $x \mapsto x^*$ where

$$x_m^* = \overline{x}_{-m}$$

is an involution. The involutive algebra $l^1(\mathbb{Z})$ is not a C^*-algebra.

Definition 9.17. Let A be an involutive algebra. An element $a \in A$ is *hermitian* (or *self-adjoint*) if $a^* = a$.

Observe that for any $a \in A$, the elements $x_1 = (a + a^*)/2$ and $x_2 = (a - a^*)/(2i)$ are also hermitian, and so are aa^* and a^*a.

Proposition 9.29. *Let A be an involutive algebra. Every $a \in A$ can be written as $a = x_1 + ix_2$ for two unique hermitian elements x_1, x_2.*

Proof. The elements $x_1 = (a + a^*)/2$ and $x_2 = (a - a^*)/(2i)$ are hermitian, and we have $a = x_1 + ix_2$. Conversely, if $a = x_1 + ix_2$ with x_1, x_2 hermitian, then $a^* = x_1^* - ix_2^* = x_1 - ix_2$, so $x_1 = (a + a^*)/2$ and $x_2 = (a - a^*)/(2i)$ are uniquely determined hermitian elements. ☐

Definition 9.18. Let A be an involutive algebra. An element $a \in A$ is *normal* if $a^*a = aa^*$. If A is unital with identity element e, then $a \in A$ is *unitary* if $aa^* = a^*a = e$, that is, if a is invertible and if $a^{-1} = a^*$.

If A is a unital involutive algebra, then the unitary elements form a subgroup of A. If a is unitary, namely $a^{-1} = a^*$, then

$$(a^{-1})^* = (a^*)^* = a = (a^{-1})^{-1}$$

so a^{-1} is unitary. And if $a^{-1} = a^*$ and $b^{-1} = b^*$, then

$$(ab)^* = b^*a^* = b^{-1}a^{-1} = (ab)^{-1},$$

so ab is unitary.

As in the case of matrices, we have the following result about spectra.

Proposition 9.30. *Let A be a unital C^*-algebra. For every $a \in A$, if a is hermitian, then $\sigma(a) \subseteq \mathbb{R}$, and if a is unitary, then $\sigma(a) \subseteq \mathbb{T} = \mathbf{U}(1)$.*

Proof. Assume that $a^* = a$. If $\alpha + i\beta \in \sigma(a)$, with $\alpha, \beta \in \mathbb{R}$, then for every real number λ, we have $\alpha + i(\beta + \lambda) \in \sigma(a + i\lambda e)$, since $(\alpha + i(\beta + \lambda))e - (a + i\lambda e) = (\alpha + i\beta)e - a$. By Theorem 9.13, we have $|\mu| \leq \|b\|$ for every $\mu \in \sigma(b)$, and since λ is real $\overline{\lambda} = \lambda$, so we have

$$\begin{aligned} \alpha^2 + (\beta + \lambda)^2 &\leq \|a + i\lambda e\|^2 \\ &= \|(a + i\lambda e)^*(a + i\lambda e)\| \\ &= \left\|a^*a + i\lambda a^* - i\overline{\lambda}a + \lambda^2 e\right\| \\ &= \left\|a^*a + \lambda^2 e\right\| \\ &\leq \|a^*a\| + \lambda^2 \\ &= \|a\|^2 + \lambda^2, \end{aligned}$$

which yields

$$2\beta\lambda \leq \|a\|^2 - \alpha^2 - \beta^2.$$

Since the above holds for all $\lambda \in \mathbb{R}$, by picking λ of the same sign as β and $|\lambda|$ large enough we would violate the above inequality, so we must have $\beta = 0$.

If $aa^* = a^*a = e$, then

$$\|a\|^2 = \|a^*a\| = \|e\| = 1,$$

hence $\|a\| = 1$. Similarly, since a^{-1} is also unitary, we have $\left\|a^{-1}\right\| = 1$. By Proposition 9.15, we conclude that $\sigma(a) \subseteq \mathbf{U}(1)$. ☐

Remark: The first part of Proposition 9.30 also holds for a nonunital C^*-algebra A. For every $a \in A$, if a is hermitian then $\sigma'(a) \subseteq \mathbb{R}$; see Bourbaki [Bourbaki (1967b)] (Chapter I, §6, No. 3, Proposition 3).

Definition 9.19. Let A and B be two involutive algebras. A map $\varphi\colon A \to B$ is an *involutive homomorphism* if it is a homomorphism of algebras such that $\varphi(a^*) = \varphi(a)^*$ for all $a \in A$. As usual, if A and B are unital we require that $\varphi(e_A) = e_B$. We say that A is an *involutive subalgebra* of B if it is a subalgebra of B and if A is closed under the involution $a \mapsto a^*$; that is, if $a \in A$, then $a^* \in A$.

The following result is proven in L. Schwartz [Schwartz (1991)] (Chapter II, Section 14, page 374).

Proposition 9.31. *Let A be a C^*-algebra (not necessarily commutative). For every $a \in A$, if a is normal, then we have $\left\|a^2\right\| = \left\|a\right\|^2$. As a consequence, $\rho(a) = \|a\|$. In particular, if A is commutative then the above facts hold.*

Proof. Using the fact that $\|bb^*\| = \|b\|^2$, we get

$$\|aa^*(aa^*)^*\| = \|aa^*\|^2 = \|a\|^4 .$$

We also have $\|aa^*(aa^*)^*\| = \|aa^*(a^*)^*a^*\| = \|aa^*aa^*\|$, and since a is normal, $aa^* = a^*a$, so $\|aa^*aa^*\| = \left\|a^2(a^*)^2\right\|$. But $(a^2)^* = (aa)^* = a^*a^* = (a^*)^2$, so

$$\left\|a^2(a^*)^2\right\| = \left\|a^2(a^2)^*\right\| = \left\|a^2\right\|^2 .$$

Consequently $\left\|a^2\right\|^2 = \|a\|^4$, and so $\left\|a^2\right\| = \|a\|^2$. By induction, we get $\left\|a^{2^k}\right\| = \|a\|^{2^k}$. Since

$$\rho(a) = \lim_{k \mapsto \infty} \left\|a^{2^k}\right\|^{1/2^k},$$

we conclude that $\rho(a) = \|a\|$. ☐

Proposition 9.32. *Let A be a unital involutive Banach algebra and let B be a unital C^*-algebra. If $\varphi\colon A \to B$ is an involutive homomorphism, then $\|\varphi(a)\| \leq \|a\|$ for all $a \in A$. Thus φ is continuous.*

Proof. We know from Proposition 9.31 that $\rho(b) = \|b\|$ for every hermitian $b \in B$ (since a hermitian element is obviously normal). It can easily be shown that $\sigma_B(\varphi(a)) \subseteq \sigma_A(a)$, so by definition of $\rho(a)$ (see Definition 9.12), we have

$$\rho(\varphi(a)) \leq \rho(a) \leq \|a\| .$$

Since $\rho(b) = \|b\|$ for every hermitian $b \in B$, and since $\varphi(a^*a)$ is hermitian (because $\varphi(a^*a)^* = \varphi((a^*a)^*) = \varphi(a^*a)$), we get

$$\begin{aligned}\|\varphi(a)\|^2 &= \|\varphi(a)^*\varphi(a)\| = \|\varphi(a^*)\varphi(a)\| \\ &= \|\varphi(a^*a)\| = \rho(\varphi(a^*a)) \leq \|a^*a\| = \|a\|^2,\end{aligned}$$

which implies $\|\varphi(a)\| \leq \|a\|$. ☐

Proposition 9.32 also holds if A and B are not unital. This is because $\sigma'_B(\varphi(a)) \subseteq \sigma'_A(a)$; see Bourbaki [Bourbaki (1967b)] (Chapter I, §6, No. 3, Proposition 1).

Proposition 9.18 is sharpened as follows.

Proposition 9.33. *Let A and B be two unital C^*-algebra with A a closed involutive subalgebra of B. We have $\sigma_B(a) = \sigma_A(a)$ for all $a \in A$. If $a \in A$ is invertible in B, then $a^{-1} \in A$.*

Proof. If $a \in A$ is hermitian, we know from Proposition 9.30 that $\sigma_A(a) \subseteq \mathbb{R}$. Thus all points of $\sigma_A(a)$ are boundary points, so by Proposition 9.18 we have $\sigma_B(a) = \sigma_A(a)$.

Consider any $a \in A$ such that $a^{-1} \in B$. Then a^* is also invertible in B, so aa^* is invertible in B (similarly a^*a is invertible in B). Since $a \in A$ and A is an involutive subalgebra, $a^* \in A$, then $aa^* \in A$, and since aa^* is hermitian, by the fact we just proved above, $(aa^*)^{-1} \in A$. This implies that $aa^*(aa^*)^{-1} = e_A$, so a has a right inverse in A. A similar argument applied to a^*a shows that $(a^*a)^{-1} \in A$, so $(a^*a)^{-1}a^*a = e_A$ and a has a left inverse in A. Therefore $a^{-1} \in A$. This argument applied to $\lambda e - a$ (with $a \in A$) shows that $\sigma_B(a) = \sigma_A(a)$. ☐

Let A be an involutive algebra. For any linear form $f\colon A \to \mathbb{C}$, let f^* be the map given by

$$f^*(a) = \overline{f(a^*)}, \qquad a \in A.$$

We have

$$f^*(a + b) = \overline{f((a + b)^*)} = \overline{f(a^* + b^*)} = \overline{f(a^*)} + \overline{f(b^*)} = f^*(a) + f^*(b),$$

and

$$f^*(\lambda a) = \overline{f((\lambda a)^*)} = \overline{f(\overline{\lambda} a^*)} = \overline{\overline{\lambda} f(a^*)} = \lambda \overline{f(a^*)} = \lambda f^*(a),$$

so f^* is also a linear form. We verify immediately that

$$(f^*)^* = f, \qquad (f + g)^* = f^* + g^*, \qquad (\lambda f)^* = \overline{\lambda} f^*.$$

Definition 9.20. Let A be an involutive algebra. For any linear form $f: A \to \mathbb{C}$, the linear form f^* given by

$$f^*(a) = \overline{f(a^*)}, \qquad a \in A$$

is called the *adjoint* of f. We say that f is *hermitian* (or *self-adjoint*) if $f^* = f$.

9.9 Characters and Gelfand Transform in a C^*-Algebra

Interestingly, the characters of a commutative unital C^*-algebra are hermitian.

Proposition 9.34. *Let A be a commutative unital C^*-algebra. Then for any character $\chi \in \mathsf{X}(A)$, we have*

$$\chi(a^*) = \overline{\chi(a)}, \qquad \text{for all } a \in A,$$

or equivalently $\chi(a) = \overline{\chi(a^*)}$, which shows that the characters are hermitian. Consequently $\mathcal{G}_{a^*} = \overline{\mathcal{G}_a}$ for all $a \in A$, where $\mathcal{G}_a$ is the Gelfand transform of a.

Proof. First assume that $a \in A$ is hermitian. By Proposition 9.30 we have $\sigma(a) \subseteq \mathbb{R}$, and since by Proposition 9.12 we have $\chi(a) \in \sigma(a)$ for any $\chi \in \mathsf{X}(A)$, we have $\chi(a) \in \mathbb{R}$, and since a is hermitian $a^* = a$, so

$$\chi(a^*) = \chi(a) = \overline{\chi(a)}.$$

Any arbitrary $a \in A$ can be written as $a = x_1 + ix_2$ for two unique hermitian elements $x_1, x_2 \in A$ (see Proposition 9.29), and by the above fact

$$\chi(x_1) = \overline{\chi(x_1)}, \qquad \chi(x_2) = \overline{\chi(x_2)}$$

and since $a = x_1 + ix_2$ and $a^* = x_1^* - ix_2^* = x_1 - ix_2$ (because x_1 and x_2 are hermitian),

$$\begin{aligned}\chi(a^*) &= \chi(x_1 - ix_2) = \chi(x_1) - i\chi(x_2) = \overline{\chi(x_1)} - i\overline{\chi(x_1)} \\ &= \overline{\chi(x_1) + i\chi(x_2)} = \overline{\chi(a)}.\end{aligned}$$

Since by definition $\mathcal{G}_a(\chi) = \chi(a)$, we have

$$\mathcal{G}_{a^*}(\chi) = \chi(a^*) = \overline{\chi(a)} = \overline{\mathcal{G}_a(\chi)},$$

which means that $\mathcal{G}_{a^*} = \overline{\mathcal{G}_a}$, as claimed. ☐

Remark: Proposition 9.34 also holds if A is a noncommutative and nonunital C^*-algebra; see Bourbaki [Bourbaki (1967b)] (Chapter I, §6, No. 4, Theorem 1).

If A is a commutative unital C^*-algebra, we can make the following addition to Proposition 9.24.

Proposition 9.35. *Let A be a commutative unital C^*-algebra. For any fixed $a \in A$, if the algebra A is generated by a, a^*, and e, then the Gelfand transform $\mathcal{G}_a$ is a homeomorphism from $\mathsf{X}(A)$ to $\sigma(a)$.*

Proof. By Theorem 9.23, the map $\mathcal{G}_a$ is continuous and surjective onto $\sigma(a)$. Since $\mathsf{X}(A)$ and $\sigma(a)$ are compact Hausdorff spaces (by Theorem 9.19(2) and Theorem 9.13(1)), by the corollary to Proposition A.33, it suffices to show that this map is injective. But any character $\chi\colon A \to \mathbb{C}$ is uniquely determined by $\chi(a)$, since by Proposition 9.34, we have $\chi(a^*) = \overline{\chi(a)}$. If $\mathcal{G}_a(\chi_1) = \mathcal{G}_a(\chi_2)$, then $\chi_1(a) = \chi_2(a)$, and since χ_1 is completely determined by $\chi_1(a)$ and similarly χ_2 is completely determined by $\chi_2(a)$, we have $\chi_1 = \chi_2$, and $\mathcal{G}_a$ is injective. $\square$

As an application of Proposition 9.35, let H be a Hilbert space, and let T be a bounded normal operator on H (that is, $TT^* = T^*T$). Let $\mathcal{A}_T$ be the subalgebra of $\mathcal{L}(H)$ generated by T, T^* and I. Since T and T^* commute, $\mathcal{A}_T$ is a commutative unital C^*-algebra, and $\mathsf{X}(\mathcal{A}_T)$ is homeomorphic to the spectrum $\sigma(T)$ of $T \in \mathcal{L}(H)$. This is the first step in obtaining a spectral theorem for a bounded normal operator on a Hilbert space; see Folland [Folland (1995)] (Chapter 1, Section 1.4) and Dieudonné [Dieudonné (2005)] (Chapter XV, Section 11).

We are now ready to prove the main theorem of the theory of commutative unital C^*-algebras due to Gelfand and Naimark, namely that every commutative unital C^*-algebra can be viewed as the algebra of continuous functions on a compact space, namely its space of characters $\mathsf{X}(A)$. The proof makes use of the version of the Stone–Weierstrass theorem for complex-valued functions that we now recall.

Theorem 9.36. *(Stone–Weierstrass) Let X be a compact space, and let $\mathcal{C}(X;\mathbb{C})$ be the algebra of continuous functions on X. Let B be a subalgebra of $\mathcal{C}(X;\mathbb{C})$ satisfying the following properties:*

(1) The algebra B contains the constant functions.

(2) *The algebra B separates the points of X, which means that for any points $x, y \in X$, if $x \neq y$ then there is some function $f \in B$ such that $f(x) \neq f(y)$.*
(3) *The algebra B is stable under conjugation; that is, for any $f \in B$, the function $\overline{f}$ also belongs to B (where $\overline{f}(x) = \overline{f(x)}$ for all $x \in X$).*

Then B is dense in $\mathcal{C}(X;\mathbb{C})$ (with respect to the $\| \|_\infty$ norm); that is, for every function $f \in \mathcal{C}(X;\mathbb{C})$, there is a sequence (f_n) with $f_n \in B$ that converges uniformly to f.

Theorem 9.36 is a cornerstone of analysis. Its proof can be found in many books, including Schwartz [Schwartz (1991)], Folland [Folland (1999)] (Chapter 4, Theorem 4.15), and Rudin [Rudin (1991)] (Chapter 5, Theorem 5.7).

Theorem 9.37. *(Gelfand–Naimark) Let A be a commutative unital C^*-algebra. Then the Gelfand transform $\mathcal{G}\colon A \to \mathcal{C}(\mathsf{X}(A);\mathbb{C})$ is an isometric isomorphism between A and $\mathcal{C}(\mathsf{X}(A);\mathbb{C})$ (and so $\|\mathcal{G}_a\|_\infty = \|a\| = \rho(a)$ for all $a \in A$). Furthermore the Gelfand maps $\mathcal{G}_a$ are hermitian, which means that $\mathcal{G}_{a^*} = \overline{\mathcal{G}_a}$, for all $a \in A$.*

Proof. Since A is commutative, by Proposition 9.31, we have $\|a^2\| = \|a\|^2$ for all $a \in A$. Since A is a unital Banach algebra, by Proposition 9.26 the Gelfand transform $\mathcal{G}\colon A \to \mathcal{C}(\mathsf{X}(A);\mathbb{C})$ is an isometry. In particular it is injective. This implies that the image $\mathcal{G}(A)$ of A is closed in $\mathcal{C}(\mathsf{X}(A);\mathbb{C})$. Indeed, for any Cauchy sequence $(\mathcal{G}_{a_n})$ in $\mathcal{C}(\mathsf{X}(A);\mathbb{C})$, since $\|a_m - a_n\| = \|\mathcal{G}_{a_m} - \mathcal{G}_{a_n}\|_\infty$, the sequence (a_n) is a Cauchy sequence in A, and since A is a Banach space the sequence (a_n) has a limit $a \in A$. Since the Gelfand transform is continuous, $\mathcal{G}_a$ is the limit of the sequence $(\mathcal{G}_{a_n})$. Therefore $\mathcal{G}(A)$ is closed in $\mathcal{C}(\mathsf{X}(A);\mathbb{C})$. It remains to prove that $\mathcal{G}(A) = \mathcal{C}(\mathsf{X}(A);\mathbb{C})$.

For this we check that the hypotheses of the Stone–Weierstrass theorem (Theorem 9.36) are satisfied. Since A is algebra and $\mathcal{G}$ is a homomorphism, $B = \mathcal{G}(A)$ is a subalgebra of $\mathcal{C}(\mathsf{X}(A);\mathbb{C})$.

(1) The algebra B contains all the constant functions, since $\mathcal{G}_{\lambda e}$ is the constant function λ.
(2) The algebra B separates points. Indeed, if χ_1 and χ_2 are two distinct characters, then $\chi_1(a) \neq \chi_2(a)$ for some $a \in A$, and then $\mathcal{G}_a(\chi_1) = \chi_1(a) \neq \chi_2(a) = \mathcal{G}_a(\chi_2)$, so $\mathcal{G}_a$ separates χ_1 and χ_2.
(3) By Proposition 9.34, we have $\mathcal{G}_{a^*} = \overline{\mathcal{G}_a}$ for all $a \in A$, so B is stable under conjugation.

By the Stone–Weierstrass theorem, B is dense in $\mathcal{C}(\mathsf{X}(A);\mathbb{C})$. But B is closed in $\mathcal{C}(\mathsf{X}(A);\mathbb{C})$, so $\mathcal{G}(A) = B = \mathcal{C}(\mathsf{X}(A);\mathbb{C})$, proving that $\mathcal{G}$ is an isomorphism. $\square$

The Gelfand–Naimark theorem is used to prove the Plancherel–Godement theorem (see Vol II, Section 2.8, Theorem 2.41), and some representation theory results in harmonic analysis; see Dieudonné [Dieudonné (2003b)].

If A is a nonunital commutative C^*-algebra, then there is a version of the Gelfand–Naimark theorem in which the algebra $\mathcal{C}(\mathsf{X}(A);\mathbb{C})$ is replaced by the algebra $\mathcal{C}_0(\mathsf{X}(A);\mathbb{C})$ of continuous functions that tend to zero at infinity. Thus there is an isometric isomorphism $\mathcal{G}\colon A \to \mathcal{C}_0(\mathsf{X}(A);\mathbb{C})$; see Bourbaki [Bourbaki (1967b)] (Chapter 1, Section 6, No. 4) and Folland [Folland (1995)] (Chapter 1, Section 1.3).

There is also a version of the Gelfand–Naimark theorem for noncommutative C^*-algebras. Roughly speaking, a C^*-algebra is isometrically isomorphic to a C^*-subalgebra of the algebra of bounded operators on some Hilbert space; see Rudin [Rudin (1991)].

The spectral theory of C^*-algebras is the key machinery used to develop generalizations of the spectral theorems for normal matrices to bounded (and unbounded) operators of various kinds on a Hilbert space. A condensed presentation of these spectral theorems is given in Folland [Folland (1995)] (Chapter 1, Section 1.4) and in Dieudonné [Dieudonné (2005)] (Chapter XV, Section 11). An extensive treatment of spectral theorems is given in Rudin [Rudin (1991)], and in Lax [Lax (2002)].

Since the main goals of this book are to discuss harmonic analysis and representation theory, spectral theorems for families of operators on a Hilbert space are not a prime topic of interest. However, Theorem 9.37 and Proposition 9.35 yield an interesting preliminary version of the spectral theorem for bounded normal operators.

Let H be a Hilbert space, and let T be a bounded normal operator on H, which means that $TT^* = T^*T$. Let $\mathcal{A}_T$ be the subalgebra of $\mathcal{L}(H)$ generated by T, T^* and I. Since T and T^* commute, $\mathcal{A}_T$ is a commutative unital C^*-algebra. By Proposition 9.35, the Gelfand transform $\mathcal{G}_T\colon \mathsf{X}(\mathcal{A}_T) \to \sigma(T)$ (given by $\mathcal{G}_T(\chi) = \chi(T), \chi \in \mathsf{X}(\mathcal{A}_T)$) is a homeomorphism between $\mathsf{X}(\mathcal{A}_T)$ and the spectrum $\sigma(T)$ of $T \in \mathcal{A}_T$.

Actually, it is important to note that if $\mathcal{A}$ is any unital (not necessarily commutative) C^*-subalgebra of $\mathcal{L}(H)$, by Proposition 9.33, the spectrum of $T \in \mathcal{A}$ with respect to the algebra $\mathcal{A}$ is equal to the spectrum of T with repect to the algebra $\mathcal{L}(H)$. Thus from now on we will always assume that the spectrum $\sigma(T)$ of a map T in $\mathcal{A} \subseteq \mathcal{L}(H)$ is defined with respect to $\mathcal{L}(H)$.

Theorem 9.38. *Let H be a Hilbert space and let T be a bounded normal operator on H. There is an isometric isomorphism $G\colon \mathcal{A}_T \to \mathcal{C}(\sigma(T);\mathbb{C})$ such that*

$$G(T) = \mathrm{id}_{\sigma(T)}.$$

Proof. By Gelfand–Naimark (Theorem 9.37), the Gelfand transform $\mathcal{G}\colon \mathcal{A}_T \to \mathcal{C}(\mathsf{X}(\mathcal{A}_T), \mathbb{C})$ is an isometric isomorphism. The homeomorphism $\mathcal{G}_T\colon \mathsf{X}(\mathcal{A}_T) \to \sigma(T)$ has an inverse $\mathcal{G}_T^{-1}\colon \sigma(T) \to \mathsf{X}(\mathcal{A}_T)$, and the map $\mathcal{G}_T^{-1}$ induces a map $\theta\colon \mathcal{C}(\mathsf{X}(\mathcal{A}_T);\mathbb{C}) \to \mathcal{C}(\sigma(T);\mathbb{C})$ given by

$$\theta(f) = f \circ \mathcal{G}_T^{-1}, \quad f \in \mathcal{C}(\mathsf{X}(\mathcal{A}_T);\mathbb{C}).$$

We leave it as an exercise to check that θ is an isometric isomorphism. Let $G = \theta \circ \mathcal{G}$. Since both $\mathcal{G}$ and θ are isometric isomorphisms, G is an isometric isomorphism between $\mathcal{A}_T$ and $\mathcal{C}(\sigma(T);\mathbb{C})$, so it remains to prove that $G(T) = \mathrm{id}_{\sigma(T)}$. For $f = \mathcal{G}_T$, for every $\lambda \in \sigma(T)$, we have

$$\begin{aligned} G(T)(\lambda) &= (\theta(\mathcal{G}(T)))(\lambda) \\ &= (\theta(\mathcal{G}_T))(\lambda) \\ &= (\mathcal{G}_T \circ \mathcal{G}_T^{-1})(\lambda) = \lambda, \end{aligned}$$

so $G(T) = \mathrm{id}_{\sigma(T)}$, as claimed. □

It can be shown that the map G of Theorem 9.38 satisfying the property $G(T) = \mathrm{id}_{\sigma(T)}$ is unique; see Schwartz [Schwartz (1991)].

Observe that the inverse $G^{-1}\colon \mathcal{C}(\sigma(T);\mathbb{C}) \to \mathcal{A}_T$ of the isomorphism $G\colon \mathcal{A}_T \to \mathcal{C}(\sigma(T);\mathbb{C})$ is a bounded linear map on $\mathcal{C}(\sigma(T);\mathbb{C})$ taking its values in the Banach space $\mathcal{A}_T \subseteq \mathcal{L}(H)$. Thus G^{-1} is a vector-valued continuous Radon functional and we should expect that it can be defined as an integral with respect to some kind of measure. This can be indeed be done using H-projection-valued measures, as explained in Folland [Folland (1995)] (Chapter 1, Section 1.4) and in a slightly different formalism in Dieudonné [Dieudonné (2005)] (Chapter XV, Section 11). The key point is that for any two vectors $u, v \in H$, the map $\Phi_{u,v}$ defined on $\mathcal{C}(\sigma(T), \mathbb{C})$ by

$$\Phi_{u,v}(f) = \langle G^{-1}(f)(u), v\rangle$$

is a bounded Radon functional, so by Radon–Riesz III, it corresponds to a unique regular complex Borel measure $\mu_{u,v}$ such that

$$\langle G^{-1}(f)(u), v\rangle = \int_{\sigma(T)} f\, d\mu_{u,v} \quad \text{for all } f \in \mathcal{C}(\sigma(T); \mathbb{C}).$$

The next step is to define the projection-valued measures, but we will not do this here; we refer the reader to Folland [Folland (1995)] (Chapter 1, Section 1.4) and Dieudonné [Dieudonné (2005)] (Chapter XV, Section 11). Folland actually deals with the more general situation of an arbitary commutative C^*-subalgebra of $\mathcal{L}(H)$ containing I, whereas Dieudonné restricts his attention to the C^*-algebra $\mathcal{A}_T$. They both make the crucial observation that $G^{-1}\colon \mathcal{C}(\sigma(T); \mathbb{C}) \to \mathcal{A}_T$ is a C^*-algebra homomorphism whose range is a subspace of $\mathcal{L}(H)$, so it is a *representation* of the algebra $\mathcal{C}(\sigma(T); \mathbb{C})$ in the Hilbert space H, in the sense of Vol II, Definition 2.1 (in fact, it is a faithful representation). If H is a separable Hilbert space, representations of $\mathcal{C}(K; \mathbb{C})$ in H where K is a compact metrizable space can be completely classified, which is the approach followed by Dieudonné (recall that $\sigma(T)$ is compact). For more on this topic, see Vol II, Section 2.9.

9.10 Enveloping C^*-Algebra of an Involutive Banach Algebra

If A is an involutive Banach algebra, there is a C^*-algebra $\mathrm{St}(A)$ and an involutive homomorphism $j\colon A \to \mathrm{St}(A)$ that satisfy a universal mapping condition with respect to homomorphisms of A into a C^*-algebra. To construct $\mathrm{St}(A)$, first we establish the following result.

Proposition 9.39. *Let A be an involutive algebra (over $\mathbb{C}$) and let p be a semi-norm on A. The following conditions are equivalent.*

(1) We have $p(ab) \leq p(a)p(b)$, $p(a^*) = p(a)$, $(p(a))^2 = p(a^*a)$, for all $a, b \in A$.
(2) The set $\mathfrak{N}$ of all $a \in A$ such that $p(a) = 0$ is a self-adjoint ($\mathfrak{N}^* = \mathfrak{N}$) two-sided ideal of A, and the norm induced on $A/\mathfrak{N}$ (in Proposition 9.7) makes $A/\mathfrak{N}$ an involutive normed algebra whose completion is a C^*-algebra.
(3) There is a homomorphism φ of the involutive algebra A into a C^*-algebra such that $p(a) = \|\varphi(a)\|$ for all $a \in A$.

The proof of Proposition 9.39 is given in Bourbaki [Bourbaki (1967b)] (Chapter I, §6, No. 6, Lemma 1).

Definition 9.21. Let A be an involutive algebra (over $\mathbb{C}$). A semi-norm p satisfying the conditions of Proposition 9.39 is called a *stellar semi-norm* on A.

Let us now assume that A is an involutive Banach algebra. Let S be the set of stellar semi-norms on A. By Proposition 9.32, we have $p(a) \leq \|a\|$ for all $a \in A$ and all $p \in S$. Consequently the function $a \mapsto \|a\|_*$ given by

$$\|a\|_* = \sup_{p \in S} p(a)$$

is the largest stellar semi-norm in A.

Definition 9.22. Let A be an involutive Banach algebra, and let $\mathfrak{N}$ be the set of all elements $a \in A$ such that $\|a\|_* = 0$. The C^*-algebra obtained by completing the normed involutive algebra $A/\mathfrak{N}$ for the norm induced by $\| \|_*$ is called the *enveloping C^*-algebra* of A, and is denoted $\mathrm{St}(A)$. The map from A to the completion of $A/\mathfrak{N}$ induced by the canonical map from A to $A/\mathfrak{N}$ is denoted by j.

The enveloping C^*-algebra $\mathrm{St}(A)$ has the following universal mapping property.

Theorem 9.40. *Let A be an involutive Banach algebra. For every involutive homomorphism $\varphi\colon A \to B$ of A into a C^*-algebra B, there is a unique involutive homomorphism $\overline{\varphi}\colon \mathrm{St}(A) \to B$ such that*

$$\varphi = \overline{\varphi} \circ j,$$

as shown in the following commutative diagram:

$$\begin{array}{ccc} A & \xrightarrow{j} & \mathrm{St}(A) \\ & {\scriptstyle\varphi}\searrow & \downarrow{\scriptstyle\overline{\varphi}} \\ & & B. \end{array}$$

If A is commutative, then $\mathrm{St}(A)$ *is commutative, and if A is unital, then* $\mathrm{St}(A)$ *is unital.*

Proof sketch. The map $a \mapsto \|\varphi(a)\|$ is a stellar semi-norm on A. Thus $\|\varphi(a)\| \leq \|a\|_*$ for all $a \in A$, and it is easy to check that the homomorphism obtained by the quotient operation is continuous from $A/\mathfrak{N}$ to B, and extends uniquely to $\mathrm{St}(A)$. The uniqueness of $\overline{\varphi}$ is standard. $\square$

If A is abelian it is also possible to characterize the set of characters $\mathsf{X}(\mathrm{St}(A))$ of $\mathrm{St}(A)$.

Recall that given the homomorphism $j\colon A \to \mathrm{St}(A)$, the homomorphism $\mathsf{X}(j)\colon \mathsf{X}(\mathrm{St}(A)) \to \mathsf{X}(A)$ is given by

$$\mathsf{X}(j)(\chi) = \chi \circ j, \qquad \chi \in \mathsf{X}(\mathrm{St}(A)).$$

Proposition 9.41. *Let A be a commutative involutive Banach algebra. The map $\mathsf{X}(j)$ is a homeomorphism of the set of characters $\mathsf{X}(\mathrm{St}(A))$ onto the subspace H of hermitian characters in $\mathsf{X}(A)$; that is, the characters $\chi\colon A \to \mathbb{C}$ such that $\chi(a) = \overline{\chi(a^*)}$ for all $a \in A$.*

Proposition 9.41 is proven in Bourbaki [Bourbaki (1967b)] (Chapter I, §6, No. 6, corollary).

It is easy to show that if $a \in \mathrm{rad}\, A$ (the radical of A), then $j(a) = 0$.

If G is a locally compact group, then $\mathrm{L}^1(G)$ is an involutive Banach algebra, but in general it is not a C^*-algebra. Thus we can form the enveloping C^*-algebra $\mathrm{St}(\mathrm{L}^1(G))$ of $\mathrm{L}^1(G)$, denoted $\mathrm{St}(G)$.

Definition 9.23. If G is a locally compact group, then the enveloping C^*-algebra $\mathrm{St}(\mathrm{L}^1(G))$ of $\mathrm{L}^1(G)$ is denoted $\mathrm{St}(G)$.

Remarkably, the canonical map j is injective.

Proposition 9.42. *Let G be a locally compact group. The canonical map j from $\mathrm{L}^1(G)$ to its enveloping C^*-algebra $\mathrm{St}(G)$ is injective.*

Proposition 9.42 is proven in Bourbaki [Bourbaki (1967b)] (Chapter I, §6, No. 7, Proposition 12).

As a corollary, the following result is obtained; see Bourbaki [Bourbaki (1967b)] (Chapter 1, Section 7, No. 7).

Proposition 9.43. *If G is a locally compact group, then the radical of $\mathrm{L}^1(G)$ is the zero ideal.*

We will see in Section 10.1 that if G is an abelian locally compact group, then every character of $\mathrm{L}^1(G)$ is hermitian. By Proposition 9.41, there is a homeomorphism between $\mathsf{X}(\mathrm{St}(G))$ and $\mathsf{X}(\mathrm{L}^1(G))$ (see Proposition 10.8).

9.11 Problems

Problem 9.1. Let $\mathfrak{A}$ be a two-sided ideal in a K-algebra A (not necessarily unital or commutative). Show that $A/\mathfrak{A}$ a K-vector space which is also a

K-algebra. Hint: Recall that $A/\mathfrak{A}$ consists of the equivalence classes of the equivalence relation $\equiv$ on A defined by

$$x \equiv y \quad \text{iff} \quad x - y \in \mathfrak{A}, \qquad x, y \in A.$$

Problem 9.2. Prove Proposition 9.1. Hint: See Bourbaki [Bourbaki (1968a)] (Appendix, Proposition 3).

Problem 9.3. Prove Proposition 9.2. Hint: Use Zorn's lemma.

Problem 9.4. Let $\mathcal{C}^n([0,1])$ be the algebra of functions $f\colon [0,1] \to \mathbb{C}$ having a continuous derivative $f^{(k)}$ for $k = 1, \ldots, n$, under pointwise addition and multiplication. Let

$$\|f\| = \sum_{k=0}^{n} \frac{1}{k!} \sup_{0 \le t \le 1} |f^{(k)}(t)|.$$

(i) Check that the above equation defines a norm on $\mathcal{C}^n([0,1])$.
(ii) Prove that $\|fg\| \le \|f\| \, \|g\|$.
(iii) Prove that $\mathcal{C}^n([0,1])$ is a commutative unital Banach algebra, with identity as the constant function 1.

Problem 9.5. Prove Proposition 9.3.

Problem 9.6. Let K be an arbitrary field. Given a K-algebra (either unital or nonunital) A, define the K-algebra $\widetilde{A}$ as the vector space $\widetilde{A} = K \times A$, with multiplication given by

$$(\lambda, a)(\mu, b) = (\lambda\mu, \lambda b + \mu a + ab).$$

(i) Show $\widetilde{A}$ is a K-algebra, with multiplicative unit $e = (1, 0)$.
(ii) Show that A is a maximal left ideal in $\widetilde{A}$.

Now assume that A is unital with multiplicative identity ϵ.

(iii) Show that $K(e - \epsilon)$ is a unital algebra with identity $e - \epsilon$.
(iv) Show that $\varphi\colon \widetilde{A} \to (K(e - \epsilon)) \times A$ with

$$\varphi(\lambda, a) = (\lambda(e - \epsilon), \lambda\epsilon + a)$$

is a linear map.

Problem 9.7. Let A and $\tilde{A}$ be as defined in the previous problem.

(i) Show that the map

$$\|(\lambda, a)\| = |\lambda| + \|a\|$$

makes $\widetilde{A}$ into a unital normed algebra. Also show that if A is a Banach algebra, then so is $\widetilde{A}$.

(ii) Check that the norm defined as

$$\|(\lambda, a)\| = \sup\{\|\lambda b + ab\| \mid\in A,\ \|b\| \leq 1\}$$

makes $\widetilde{A}$ into a C^*-algebra. Furthermore show that this norm, when restricted to A, agrees with the original norm on A. Hint: See Folland [Folland (1995)] (Chapter 1, Section 4), or Bourbaki [Bourbaki (1967b)] (Chapter 1, Section 6, No. 3).

Problem 9.8. Let A be a normed algebra and let $\mathfrak{A}$ be a closed ideal in A. Define $\| \ \| : A/\mathfrak{A} \to \mathbb{R}_+$ as

$$\|\pi(a)\| = \inf\{\|a + z\| \mid z \in \mathfrak{A}\}, \qquad a \in A.$$

Show that

$$\|\lambda\pi(x)\| = |\lambda|\,\|\pi(x)\|\,.$$

Hint: See Proposition 9.7.

Problem 9.9. Prove Proposition 9.8. Hint: See Bourbaki [Bourbaki (1967b)] (Chapter 1, Section 1, No. 2) or Rudin [Rudin (1987)] (Chapter 18).

Problem 9.10. Let $\varphi\colon A \to B$ be a homomorphism of (commutative) unital algebras. Then φ induces a map $\mathsf{X}(\varphi)\colon \mathsf{X}(B) \to \mathsf{X}(A)$ given by

$$\mathsf{X}(\varphi)(\chi) = \chi \circ \varphi, \quad \chi \in \mathsf{X}(B).$$

(i) Verify that $\mathsf{X}(\psi \circ \varphi) = \mathsf{X}(\varphi) \circ \mathsf{X}(\psi)$ and $\mathsf{X}(\mathrm{id}_A) = \mathrm{id}_{\mathsf{X}(A)}$.
(ii) Show that if $\varphi\colon A \to B$ is surjective, then $\mathsf{X}(\varphi)$ is a bijection of $\mathsf{X}(B)$ onto the set of characters of A that vanish on $\operatorname{Ker}\varphi$.

Problem 9.11. Prove that the topology of pointwise convergence on $\mathsf{X}(A)$ is the weakest (coarsest) topology for which the Gelfand maps $\mathcal{G}_a$ are continuous.

Problem 9.12. Complete the proof sketch of Theorem 9.13. Hint: See Dieudonné [Dieudonné (2005)] (Chapter XV, Section 2), Rudin [Rudin (1991)] (Chapter 10, Theorem 10.13), Bourbaki [Bourbaki (1967b)] (Chapter 1, Section 2, No. 5), or Folland [Folland (1995)] (Chapter 1, Section 1).

Problem 9.13. Prove that

$$R(a, \mu) - R(a, \lambda) = (\lambda - \mu)R(a, \lambda)R(a, \mu)$$

for all $(\lambda, \mu) \in (\mathbb{C} - \sigma(a)) \times (\mathbb{C} - \sigma(a))$. Use this identity to deduce that $R(a, \lambda)$ and $R(a, \mu)$ commute.

Problem 9.14. Prove Proposition 9.16. Hint: See Dieudonné [Dieudonné (2005)] (Chapter XV, Section 2) or Rudin [Rudin (1991)] (Chapter 10, Theorem 10.13).

Problem 9.15. Prove Proposition 9.17. Hint: See Dieudonné [Dieudonné (2005)] (Chapter XV, Section 2), Rudin [Rudin (1991)] (Chapter 10, Theorem 10.13), or Folland [Folland (1995)] (Chapter 1, Section 1.1).

Problem 9.16. Complete the details in the proof sketch of Theorem 9.19. Hint: See Bourbaki [Bourbaki (1967b)] (Chapter 1, Section 3, No. 1) or Rudin [Rudin (1991)] (Chapter 11, Theorem 11.9).

Problem 9.17. Advanced Exercise: Prove that if the commutative unital Banach algebra A is separable, then $\mathsf{X}(A)$ is metrizable. Hint: See Dieudonné [Dieudonné (2005)] (Chapter XV, Section 3, Theorem 15.3.2).

Problem 9.18. Prove Proposition 9.20. Hint: Adapt the proof of Theorem 9.19. Alternatively, see Bourbaki [Bourbaki (1967b)] (Chapter 1, Section 3, No. 1) or Folland [Folland (1995)] (Chapter 1, Section 1.3, Theorem 1.30).

Problem 9.19. Let A be a nonunital commutative Banach algebra. Prove that the map $\chi \mapsto \operatorname{Ker} \chi$ is a bijection from $\mathsf{X}(A)$ to the set of maximal regular ideals in A. Hint: See Bourbaki [Bourbaki (1967b)] (Chapter 1, Section 3, Theorem 2).

Problem 9.20. Prove Proposition 9.22. Hint: Use Proposition 9.21. 9.21 Alternatively, see Bourbaki [Bourbaki (1967b)] (Chapter 1, Section 3, No. 2) or Dieudonné [Dieudonné (2005)] (Chapter 15, Example 15.3.7).

Problem 9.21. If A is a commutative nonunital Banach algebra, prove that the Gelfand transform is a homomorphism $\mathcal{G}\colon A \to \mathcal{C}_0(\mathsf{X}(A); \mathbb{C})$. Hint: See Bourbaki [Bourbaki (1967b)] (Chapter 1, Section 3, No. 3) or Folland [Folland (1995)] (Chapter 1, Section 1.3, Theorem 1.30).

Problem 9.22. Prove Proposition 9.28. Hint: See Bourbaki [Bourbaki (1967b)] (Chapter 1, Section 3, Proposition 5).

Problem 9.23. Prove that the conclusions of Proposition 9.34. hold when A is a noncommutative and nonunital C^*-algebra. Hint: See Bourbaki [Bourbaki (1967b)] (Chapter I, §6, No. 4, Theorem 1).

Problem 9.24. Advanced Exercise: Prove the Stone–Weierstrass theorem, Theorem 9.36. Hint: See Schwartz [Schwartz (1991)], Folland [Folland

(1999)] (Chapter 4, Theorem 4.15), or Rudin [Rudin (1991)] (Chapter 5, Theorem 5.7).

Problem 9.25. Advanced Exercise: If A is a nonunital commutative C^*-algebra, prove the version of the Gelfand–Naimark theorem in which the algebra $\mathcal{C}(\mathsf{X}(A);\mathbb{C})$ is replaced by the algebra $\mathcal{C}_0(\mathsf{X}(A);\mathbb{C})$ of continuous functions that tend to zero at infinity. Hint: See Bourbaki [Bourbaki (1967b)] (Chapter 1, Section 6, No. 4) or Folland [Folland (1995)] (Chapter 1, Section 1.3).

Problem 9.26. Let H be a Hilbert space and let T be a bounded normal operator on H. Show that isometric isomorphism $G\colon \mathcal{A}_T \to \mathcal{C}(\sigma(T);\mathbb{C})$

$$G(T) = \mathrm{id}_{\sigma(T)},$$

as defined in Theorem 9.38, is unique. Hint: See Schwartz [Schwartz (1991)].

Problem 9.27. Prove Proposition 9.39. Hint: See Bourbaki [Bourbaki (1967b)] (Chapter I, §6, No. 6, Lemma 1).

Problem 9.28. Complete the proof sketch of Theorem 9.40.

Problem 9.29. Prove Proposition 9.41. Hint: See Bourbaki [Bourbaki (1967b)] (Chapter I, §6, No. 6, corollary).

Problem 9.30. Given the homomorphism $j\colon A \to \mathrm{St}(A)$, where j is defined in Definition 9.22, prove that if $a \in \mathrm{rad}\, A$, then $j(a) = 0$.

Problem 9.31. Prove Proposition 9.42. Hint: See Bourbaki [Bourbaki (1967b)] (Chapter I, §6, No. 7, Proposition 12).

Problem 9.32. Prove Proposition 9.43. Hint: Bourbaki [Bourbaki (1967b)] (Chapter 1, Section 7, No. 7).

Chapter 10

Harmonic Analysis on Locally Compact Abelian Groups

In this chapter we generalize the various Fourier transforms and cotransforms defined in Chapter 6 to an arbitrary locally compact abelian group (often abbreviated as LCA groups).

The fact that the Fourier transform $\widehat{f}\colon \mathbb{R} \to \mathbb{C}$ of a function $f \in \mathrm{L}^1(\mathbb{R})$, given by

$$\widehat{f}(x) = \int e^{-iyx} f(y) \, \frac{dx(y)}{\sqrt{2\pi}}$$

is also a function defined on $\mathbb{R}$, is an accident (perhaps a convenient accident).

On the other hand, given a function $f \in \mathrm{L}^1(\mathbb{T})$ (a periodic function of period 2π), its Fourier transform $\widehat{f} = \mathcal{F}(f)$ is the *sequence* $\widehat{f} = (c_m)_{\in\mathbb{Z}}$ of *Fourier coefficients*

$$c_m = \int_{-\pi}^{\pi} e^{-imx} f(x) \frac{dx}{2\pi}.$$

We can view $\widehat{f} = (c_m)_{\in\mathbb{Z}}$ as a function $\widehat{f}\colon \mathbb{Z} \to \mathbb{C}$. The domain of the Fourier transform $\widehat{f}$ is $\mathbb{Z}$, which is completely different from $\mathbb{T}$.

If we now consider functions in $l^1(\mathbb{Z})$, which are sequences $c = (c_m)_{\in\mathbb{Z}}$ with $c_m \in \mathbb{C}$ such that $\sum_{m\in\mathbb{Z}} |c_m| < \infty$, then we can define the Fourier transform $\widehat{c} = \mathcal{F}(c)$ of c as the function $\mathcal{F}(c)\colon \mathbb{T} \to \mathbb{C}$ defined on $\mathbb{T}$ given by

$$\mathcal{F}(c)(e^{i\theta}) = \sum_{m\in\mathbb{Z}} c_m e^{-im\theta}.$$

The domain of this Fourier transform is $\mathbb{T}$, which is completely different from $\mathbb{Z}$. Because $\sum_{m\in\mathbb{Z}} |c_m| < \infty$, the above series is absolutely convergent, so $\mathcal{F}(c) \in \mathrm{L}^1(\mathbb{T})$.

Observe an asymmetry. If $f \in \mathrm{L}^1(\mathbb{T})$, then the Fourier transform $\widehat{f} = \mathcal{F}(f) = (c_m)_{\in\mathbb{Z}}$ may not belong to $l^1(\mathbb{Z})$. The same problem arises for the Fourier transform on $\mathrm{L}^1(\mathbb{R})$. In general, $\widehat{f} \notin \mathrm{L}^1(\mathbb{R})$. This is the problem of Fourier inversion. We would like to know when it is possible to recover a function f from its Fourier transform $\widehat{f}$. This is a difficult problem.

Remarkably, the Fourier transform on L^2 is better behaved. This is the content of the Plancherel theorem which shows that Fourier inversion is possible.

The question remains, what should be the domain of a Fourier transform?

Observe that the examples that we considered involve the Fourier transform of the L^1-functions on the abelian groups, $\mathbb{R}, \mathbb{T}$, and $\mathbb{Z}$. These groups are locally compact.

In the case of a commutative locally compact group G (equipped with a Haar measure λ), it turns out that a good solution is to define the domain of the Fourier transform as the dual group $\widehat{G}$ of G, which is a certain group of homomorphisms $\chi\colon G \to \mathbb{C}$, namely the continuous unitary homomorphisms $\chi\colon G \to \mathbf{U}(1)$.

The group $\widehat{G}$ of characters of G is defined in Section 10.1. Having defined a group structure on $\widehat{G}$, the next goal is to make $\widehat{G}$ into a topological group which is locally compact. Since $\widehat{G}$ consists of continuous functions from G to $\mathbf{U}(1)$, we can give $\widehat{G}$ the compact-open topology, but proving that the resulting space is locally compact is nontrivial. This can be done by proving that the spaces $\widehat{G}$ and $\mathsf{X}(\mathrm{L}^1(G))$ are homeomorphic; see Theorem 10.6. Since by Proposition 9.20, the space $\mathsf{X}(\mathrm{L}^1(G))$ is locally compact, we obtain the fact that $\widehat{G}$ is locally compact.

Actually, in Proposition 10.5, we prove that if G is a locally compact abelian group with a left Haar measure λ, then for every character $\chi \in \widehat{G}$, the map ζ_χ given by

$$\zeta_\chi(\mu) = \int \chi(a)\, d\mu(a) \qquad \text{for all } \mu \in \mathcal{M}^1(G)$$

is a hermitian character (an algebra homomorphism such that $\zeta_\chi(\check{\overline{\mu}}) = \overline{\zeta_\chi(\mu)}$) of the algebra $\mathcal{M}^1(G)$. By restriction to $\mathrm{L}^1(G)$, the map ζ_χ given by

$$\zeta_\chi(f) = \int \chi(a) f(a)\, d\lambda(a) \qquad \text{for all } f \in \mathrm{L}^1(G)$$

is a hermitian character of $\mathrm{L}^1(G)$ not equal to the zero function. Then the map $j\colon \widehat{G} \to \mathsf{X}(\mathrm{L}^1(G))$ given by

$$j(\chi)(f) = \zeta_\chi(f) = \int \chi(a) f(a)\, d\lambda(a), \qquad \chi \in \widehat{G},\ f \in \mathrm{L}^1(G),$$

is a homeomorphism of $\widehat{G}$ onto $\mathsf{X}(\mathrm{L}^1(G))$.

We also prove that, the spaces $\widehat{G}$, $\mathsf{X}(\mathrm{L}^1(G))$, and $\mathsf{X}(\mathrm{St}(G))$, are homeomorphic.

Next in Section 10.2 we determine the characters of the groups $\mathbb{Z}$, $\mathbb{T}$, $\mathbb{Z}/p\mathbb{Z}$, and $\mathbb{R}$. As a corollary, we obtain the isomorphisms

$$\widehat{\mathbb{R}^n} \cong \mathbb{R}^n, \quad \widehat{\mathbb{T}^n} \cong \mathbb{Z}^n, \quad \widehat{\mathbb{Z}^n} \cong \mathbb{T}^n.$$

We prove that if G is a finite locally compact abelian group, then $\widehat{G}$ is isomorphic to G.

If G is a compact abelian group of Haar measure 1, then its characters form an orthonormal set in $\mathrm{L}^2(G)$.

We conclude by showing that there is a natural injection of G into its double dual $\widehat{\widehat{G}}$.

Given any $a \in G$, define the map $\eta_a\colon \widehat{G} \to \mathbb{C}$ by

$$\eta_a(\chi) = \chi(a), \qquad \text{evaluation at } a.$$

The map $\eta\colon G \to \widehat{\widehat{G}}$ given by $\eta(a) = \eta_a$ is a continuous homomorphism from G to its double dual $\widehat{\widehat{G}}$.

Actually, η is an isomorphism, but this is much harder to prove (this is the Pontrjagin duality theorem).

Section 10.3 is devoted to the definition of the Fourier transform and the Fourier cotransform on an arbitrary locally compact abelian group G equipped with a Haar measure λ. For every function $f \in \mathrm{L}^1(G)$,

(1) The *Fourier transform* of f is the function $\mathcal{F}(f)\colon \widehat{G} \to \mathbb{C}$ given by

$$\mathcal{F}(f)(\chi) = \int \overline{\chi(a)} f(a)\, d\lambda(a), \qquad \chi \in \widehat{G}.$$

(2) The *Fourier cotransform* of f is the function $\overline{\mathcal{F}}(f)\colon \widehat{G} \to \mathbb{C}$ given by

$$\overline{\mathcal{F}}(f)(\chi) = \int \chi(a) f(a)\, d\lambda(a), \qquad \chi \in \widehat{G}.$$

These transforms are not independent. In fact, each one can be obtained from the other. For all $f \in \mathrm{L}^1(G)$, and all $\chi \in \widehat{G}$, we have

$$\overline{\mathcal{F}}(f)(\chi) = \mathcal{F}(f)(\chi^{-1}) = \mathcal{F}(\check{f})(\chi) = \overline{\mathcal{F}(\overline{f})(\chi)}.$$

We show that modulo the isomorphism $j\colon \widehat{G} \to \mathsf{X}(\mathrm{L}^1(G))$, the Fourier cotransform $\overline{\mathcal{F}}(f)$ is the Gelfand transform $\mathcal{G}_f$ from $\mathrm{L}^1(G)$ to $\mathsf{X}(\mathrm{L}^1(G))$.

Actually, it is possible to define the Fourier transform and the Fourier cotransform on the algebra $\mathcal{M}^1(G)$; see Definition 10.4.

We prove the main properties of the Fourier transform and of the Fourier cotransform. In particular, the Fourier transform $\mathcal{F}$ and the Fourier cotransform $\overline{\mathcal{F}}$ are injective involutive homomorphisms from the involutive Banach algebra $\mathrm{L}^1(G)$ to the involutive Banach algebra $\mathcal{C}_0(\widehat{G};\mathbb{C})$ of continuous functions on $\widehat{G}$ that tend to zero at infinity. In particular for any two functions $f, g \in \mathrm{L}^1(G)$, we have

$$\mathcal{F}(f * g) = \mathcal{F}(f)\mathcal{F}(g), \qquad \overline{\mathcal{F}}(f * g) = \overline{\mathcal{F}}(f)\overline{\mathcal{F}}(g),$$

and

$$\mathcal{F}(f^*) = (\mathcal{F}(f))^*, \qquad \overline{\mathcal{F}}(f^*) = (\overline{\mathcal{F}}(f))^*.$$

In Section 10.4 we discuss thoroughly the Fourier transform $\mathcal{F}$ on $\mathrm{L}^2(G)$ and the Fourier cotransform $\overline{\mathcal{F}}$ on $\mathrm{L}^2(\widehat{G})$ for a *finite* abelian group. In this case it is possible to work out directly Fourier inversion, the Plancherel theorem, and the convolution rule.

In Section 10.5 we make a brief excursion into number theory. If we consider the multiplicative group $G = (\mathbb{Z}/m\mathbb{Z})^*$ of units of the group $\mathbb{Z}/m\mathbb{Z}$, then its characters are the *Dirichlet characters*. They can be extended to $\mathbb{Z}$ and are called *Dirichlet characters* modulo m. It turns out the Fourier inversion formula for functions over $(\mathbb{Z}/m\mathbb{Z})^*$ is one of the steps in the proof of Dirichlet's famous theorem on arithmetic progressions of integers $mk + \ell$ with $\gcd(\ell, m) = 1$ and $k \in \mathbb{N}$, that says that such a sequence contains infinitely many primes. We briefly discuss this fascinating result.

If G is a finite abelian group, it is possible to formulate the Fourier transform $\mathcal{F}$ on $\mathrm{L}^2(G)$ and the Fourier cotransform $\overline{\mathcal{F}}$ on $\mathrm{L}^2(\widehat{G})$ in terms of matrices. This is achieved in Section 10.6. The details are a bit technical due to the appearance of various dual spaces. If we denote the vector space of functions from G to $\mathbb{C}$ as $[G \to \mathbb{C}]$ (which is equal to $\mathrm{L}^1(G)$ and $\mathrm{L}^2(G)$), then it turns out that the key is to extend $\mathcal{F}$ to a bilinear form on $[G \to \mathbb{C}]^*$, and to extend $\overline{\mathcal{F}}$ to a bilinear form on $[G \to \mathbb{C}]^{**}$. The matrices associated with these bilinear forms are called *Fourier matrices*, and they are mutual inverses.

In Section 10.7 we consider the special case of the group $G = \mathbb{Z}/n\mathbb{Z}$. We obtain the *discrete Fourier transform* and the *discrete Fourier cotransform* or *inverse discrete Fourier transform*. The Fourier matrix F is particularly interesting, as it is a Vandermonde matrix determined by the primitive nth root of unity $\omega = e^{-2\pi i/n}$ and its powers. Convolution of two sequences f and g in $\mathbb{C}^n$ can be expressed as $H(f)g$, where $H(f)$ is a *circulant matrix*. The matrix $H(f)$ has the remarkable property that its eigenvectors are the columns of the matrix $\overline{F}$, with corresponding eigenvalues the entries in the vector $n\widehat{f}$ (where $\widehat{f}$ is the discrete Fourier transform of f). As a consquence, we obtain another proof of the convolution rule.

Section 10.8 discusses Plancherel's theorem and Fourier inversion.

Let G be a locally compact abelian group equipped with a Haar measure λ. In general, given a function $f \in \mathrm{L}^1(G)$, its Fourier transform $\mathcal{F}(f)$ does not belong to $\mathrm{L}^1(\widehat{G})$.

Plancherel's theorem (Theorem 10.27) asserts that there is a Haar measure $\widehat{\lambda}$ on the dual group $\widehat{G}$ such that the map $f \mapsto \mathcal{F}(f)$ sends $\mathrm{L}^1(G) \cap \mathrm{L}^2(G)$ into $\mathrm{L}^2(\widehat{G})$, and has a unique extension which is an isometry from $\mathrm{L}^2(G)$ to $\mathrm{L}^2(\widehat{G})$.

One should realize that Theorem 10.27 does not say that the Fourier transform $\mathcal{F}$ (or the Fourier cotransform $\overline{\mathcal{F}}$) is defined on $\mathrm{L}^2(G)$, because in general the integral will not converge for f outside of $\mathrm{L}^1(G) \cap \mathrm{L}^2(G)$. What is happening is more subtle. It is always possible by using a limit process to define the Fourier transform of any $f \in \mathrm{L}^2(G)$, and this extension of $\mathcal{F}$ to $\mathrm{L}^2(G)$ is an isometry.

Plancherel's theorem has an interesting corollary when G is compact and abelian. If G is a compact abelian group endowed with a Haar measure λ normalized so that G has measure $\lambda(G) = 1$, then $\widehat{G}$ is a Hilbert basis for $\mathrm{L}^2(G)$ (it is orthonormal and dense in $\mathrm{L}^2(G)$).

The Pontrjagin duality theorem is presented in Section 10.9. This is the most important and most beautiful theorem in the theory of locally compact abelian groups.

Let G be a locally compact abelian group endowed with a Haar measure λ, let $\widehat{G}$ be its dual group endowed with the associated Haar measure $\widehat{\lambda}$ (see

Definition 10.19), and let $\widehat{\widehat{G}}$ be its double dual endowed with the associated measure $\widehat{\widehat{\lambda}}$. The Pontrjagin duality theorem asserts two facts:

(1) The map $\eta\colon G \to \widehat{\widehat{G}}$ is an isomorphism and a homeomorphism between the topological groups G and $\widehat{\widehat{G}}$.

(2) If we identify G and $\widehat{\widehat{G}}$ using the isomorphism η, then the extension $\mathcal{F}\colon \mathrm{L}^2(G) \to \mathrm{L}^2(\widehat{G})$ of the Fourier transform to $\mathrm{L}^2(G)$ and the extension $\overline{\mathcal{F}}\colon \mathrm{L}^2(\widehat{G}) \to \mathrm{L}^2(G)$ of the Fourier cotransform to $\mathrm{L}^2(\widehat{G})$ are mutual inverses. In particular, Fourier inversion holds; that is,

$$f = (\overline{\mathcal{F}} \circ \mathcal{F})(f) \circ \eta, \qquad \text{for all } f \in \mathrm{L}^2(G).$$

As a corollary of the Pontrjagin duality theorem we can show that Fourier inversion holds for an interesting class of functions. We define $B(G)$ as the set of functions

$$B(G) = \{f \in \mathrm{L}^1(G) \mid \mathcal{F}(f) \in \mathrm{L}^1(\widehat{G})\}.$$

The restriction of $\mathcal{F}$ to $B(G)$ is a bijection from $B(G)$ to $B(\widehat{G})$, whose inverse is the restriction of $\overline{\mathcal{F}}$ to $B(\widehat{G})$.

Another corollary is that for any locally compact abelian group G, the group G is discrete if and only if $\widehat{G}$ is compact (and by duality, G is compact if and only if $\widehat{G}$ is discrete).

The dual group $\widehat{G}$ was first defined by Pontrjagin (1934) and van Kampen (1935). Versions of the duality theorem were also first proven by Pontrjagin and van Kampen. The first proof of the general version of Pontrjagin duality appears to have been published by André Weil [Weil (1979)] (Chapter VI, Section 28). The definition of the Fourier transform on an arbitrary locally compact group is due to Weil [Weil (1979)] (Chapter VI, Section 30). In this same section Weil proves versions of Plancherel's theorem and of the Pontrjagin duality theorem.

Our exposition relies heavily on Folland [Folland (1995)] and Bourbaki [Bourbaki (1967b)], which is even more abstract than Folland. A more elementary presentation (dealing with σ-compact, metrizable, locally compact, abelian groups) can be found in Deitmar [Deitmar (2002)], which constitutes a very good warm up for the more general treatment given in this chapter.

10.1 Characters and The Dual Group

The dual of a commutative locally compact group G is defined in terms of certain homomorphisms $\chi\colon G \to \mathbf{U}(1)$ called characters. Even though G is commutative, we use a multiplicative notation for the group operation.

Definition 10.1. Let G be a commutative locally compact group (with identity element e). A *character*[1] is a continuous homomorphism $\chi\colon G \to \mathbf{U}(1)$, that is, we have

$$\begin{aligned} \chi(ab) &= \chi(a)\chi(b), && \text{for all } a, b \in G \\ |\chi(a)| &= 1, && \text{for all } a \in G. \end{aligned}$$

The set of characters of G is denoted by $\widehat{G}$.

The characters of G satisfy the following properties.

Proposition 10.1. *Let G be a commutative locally compact group. The following properties hold:*

(1) For every character $\chi \in \widehat{G}$, we have $\chi(e) = 1$.
(2) For every character $\chi \in \widehat{G}$, for every $a \in G$,

$$\chi(a^{-1}) = (\chi(a))^{-1}.$$

(3) For every character $\chi \in \widehat{G}$, for every $a \in G$,

$$\chi(a^{-1}) = \overline{\chi(a)}.$$

Proof. Since χ is a homomorphism, we have $\chi(e) = 1$, because $\chi(e) = \chi(ee) = \chi(e)\chi(e)$, and since $|\chi(e)| = 1$, we deduce that $\chi(e) = 1$. We also have

$$1 = \chi(aa^{-1}) = \chi(a)\chi(a^{-1})$$

and

$$1 = \chi(a^{-1}a) = \chi(a^{-1})\chi(a),$$

so $\chi(a^{-1}) = (\chi(a))^{-1}$, and since $\chi(a) \in \mathbb{C}$ and $|\chi(a)| = 1$, we have $(\chi(a))^{-1} = \overline{\chi(a)}$, so we get

$$\chi(a^{-1}) = \overline{\chi(a)}, \qquad \text{for all } a \in G. \qquad \square$$

The following fact will be needed in the proof of Theorem 10.6.

[1] Sometimes, to emphasize that their range is $\mathbf{U}(1)$, they are called *unitary characters*.

Proposition 10.2. *If a group homomorphism $\chi\colon G \to \mathbb{C}$ is bounded, which means that there is some $C > 0$ such that $|\chi(g)| \leq C$ for all $g \in G$, then $|\chi(g)| = 1$ for all $g \in G$, that is, $\chi\colon G \to \mathbf{U}(1)$.*

Proof. Suppose that $|\chi(g)| \neq 1$ for some $g \in G$. Since χ is a homomorphism, $\chi(g^{-1}) = (\chi(g))^{-1}$, so either $|\chi(g)| > 1$ or $|\chi(g^{-1})| > 1$, and we may assume that $|\chi(g)| > 1$. Since χ is a homomorphism, $\chi(g^n) = (\chi(g))^n$ for all $n \geq 0$, and since $|\chi(g)| > 1$, for n large enough we obtain $|\chi(g^n)| = |\chi(g)|^n > C$, contradicting the fact that χ is bounded. $\square$

Our next goal is to make the set of characters into a commutative locally compact group. The first step is to define a multiplication operation on characters. We proceed as follows.

Given two characters $\chi_1, \chi_2 \in \widehat{G}$, we define $\chi_1\chi_2$ by

$$(\chi_1\chi_2)(a) = \chi_1(a)\chi_2(a), \qquad a \in G.$$

Since $\mathbb{C}$ is commutative, we have

$$\begin{aligned}(\chi_1\chi_2)(ab) &= \chi_1(ab)\chi_2(ab)\\ &= \chi_1(a)\chi_1(b)\chi_2(a)\chi_2(b)\\ &= \chi_1(a)\chi_2(a)\chi_1(b)\chi_2(b)\\ &= (\chi_1\chi_2)(a)(\chi_1\chi_2)(b)\end{aligned}$$

so $\chi_1\chi_2$ is a homomorphism. Since $|\chi_1(a)| = |\chi_2(a)| = 1$, we also have

$$|(\chi_1\chi_2)(a)| = |\chi_1(a)\chi_2(a)| = |\chi_1(a)||\chi_2(a)| = 1.$$

Since χ_1 and χ_2 are continuous, and multiplication on $\mathbb{C}$ is continuous, the map $\chi_1\chi_2$ is continuous. Therefore $\chi_1\chi_2$ is a character.

If we define $\overline{\chi}$ by

$$\overline{\chi}(a) = \overline{\chi(a)}, \qquad a \in A,$$

then we have

$$\overline{\chi}(ab) = \overline{\chi(ab)} = \overline{\chi(a)\chi(b)} = \overline{\chi(a)}\,\overline{\chi(b)} = \overline{\chi}(a)\overline{\chi}(b),$$

and

$$|\overline{\chi}(a)| = |\overline{\chi(a)}| = |\chi(a)| = 1,$$

and since χ is continuous, and conjugation on $\mathbb{C}$ is continuous, $\overline{\chi}$ is obviously continuous. Thus $\overline{\chi}$ is a character. Finally, since $|\chi(a)| = 1$, Proposition 10.1(3) implies that

$$(\chi\overline{\chi})(a) = \chi(a)\overline{\chi(a)} = 1,$$

and that

$$(\overline{\chi}\chi)(a) = \overline{\chi(a)}\chi(a) = 1.$$

In summary, we proved the following result.

Proposition 10.3. *The set $\widehat{G}$ of characters with the multiplication operation $(\chi_1, \chi_2) \mapsto \chi_1\chi_2$ defined above is a commutative group with the constant function from G to $\mathbf{U}(1)$ with value 1 as identity. The inverse operation is $\chi \mapsto \overline{\chi}$.*

The next step is to give $\widehat{G}$ a topology that will make it a locally compact group. Although it is far from obvious why this works, since $\widehat{G}$ consists of continuous functions from G to $\mathbf{U}(1)$, we give it the compact-open topology (see Definition 2.11). A subbasis for this topology consists of the sets

$$S(K, U) = \{f \mid f \in \widehat{G},\ f(K) \subseteq U\},$$

where K is any compact subset of G and U is any open subset of $\mathbf{U}(1)$; see Definition 2.11. The group operations (multiplication and inversion) are continuous in this topology, although this not immediately obvious.

Since $\widehat{G}$ is a group, it suffices to show that for every open subset U of $\mathbf{U}(1)$ containing 1, for every open subset $S(K, U)$ containing the constant function 1, there is some subset $S(K_1, V_1)$ such that $S(K_1, V_1)S(K_1, V_1) \subseteq S(K, U)$, and that there is some subset $S(K_2, V_2)$ such that $(S(K_2, V_2))^{-1} \subseteq S(K, U)$. Since $\mathbf{U}(1)$ is a topological group, there is some open subset V_1 of U containing 1 such that $V_1V_1 \subseteq U$, and then $1 \in S(K, V_1)$ and $S(K, V_1)S(K, V_1) \subseteq S(K, U)$. Since inversion in G is continuous, there is also an open subset V_2 of $\mathbf{U}(1)$ such that $V_2^{-1} \subseteq U$, and then $1 \in S(K, V_2^{-1})$ and $(S(K, V_2^{-1}))^{-1} \subseteq S(K, U)$.

Since $\mathbf{U}(1)$ is Hausdorff, the compact-open topology is Hausdorff. Indeed, if $\chi_1, \chi_2 \in \widehat{G}$ and if $\chi_1 \neq \chi_2$, then there is some $a \in G$ such that $\chi_1(a) \neq \chi_2(a)$, and since $\mathbf{U}(1)$ is Hausdorff there exists two disjoint open subsets U_1, U_2 with $\chi_1(a) \in U_1$ and $\chi_2(a) \in U_2$. Then $S(\{a\}, U_2) \cap S(\{a\}, U_2) = \emptyset$, with $\chi_1 \in S(\{a\}, U_1)$ and $\chi_2 \in S(\{a\}, U_2)$.[2]

In summary, we proved the following result.

Proposition 10.4. *Let G be a locally compact abelian group. The set $\widehat{G}$ of characters of G with the multiplication $(\chi_1, \chi_2) \mapsto \chi_1\chi_2$ given by*

$$(\chi_1\chi_2)(a) = \chi_1(a)\chi_2(a), \qquad a \in G,$$

and endowed with the compact-open topology, is an abelian topological group.

---

[2]Recall that a compact subset K of G is a subset such that every open cover of K by open subsets in G contains a finite subfamily covering K. Obviously every finite subset is compact.

Definition 10.2. Let G be a locally compact abelian group. The topological abelian group $\widehat{G}$ of characters of G is called the *dual* (or *Pontrjagin dual*) of G.

Proving directly that $\widehat{G}$ is locally compact is not so easy. André Weil gives a clever proof in [Weil (1979)]. Another way to proceed is to use the fact that G is endowed with a Haar measure λ and to prove that $\widehat{G}$ is homeomorphic to $\mathsf{X}(\mathrm{L}^1(G))$, a remarkable result showing that the group characters of G and the algebra characters of $\mathrm{L}^1(G)$ are in some sense equivalent.

The proof that $\widehat{G}$ and $\mathsf{X}(\mathrm{L}^1(G))$ are homeomorphic is quite technical. Folland [Folland (1995)] (Chapter 4) gives the main idea, but does not prove that the map is injective, nor that its inverse is continuous. Bourbaki [Bourbaki (1967b)] (Chapter 2, Section 1) gives a complete, but terse proof. The details of the proof are not illuminating so we will only indicate its main ideas.

The crucial step is to show that every group character $\chi \in \widehat{G}$ induces an algebra character $\zeta_\chi \in \mathsf{X}(\mathrm{L}^1(G))$; namely, for every $f \in \mathrm{L}^1(G)$,

$$\zeta_\chi(f) = \int \chi(a) f(a)\, d\lambda(a).$$

Actually, every group character $\chi \in \widehat{G}$ induces an algebra character $\zeta_\chi \in \mathsf{X}(\mathcal{M}^1(G))$; namely, the map given by

$$\zeta_\chi(\mu) = \int \chi(a)\, d\mu(a)$$

for every complex measure $\mu \in \mathcal{M}^1(G)$ is a character of the algebra $\mathcal{M}^1(G)$. This more general fact will be needed.

Recall from Example 9.6(3) that $\mathcal{M}^1(G)$ is a unital Banach algebra under convolution, with involution given by $\mu^* = \overline{\check{\mu}}$. The identity element of $\mathcal{M}^1(G)$ is the Dirac measure δ_e. The algebra $\mathrm{L}^1(G)$ is also a Banach algebra under convolution, with involution given by $f \mapsto \overline{\check{f}}$, but unless G is discrete, it is nonunital. However, $\mathrm{L}^1(G)$ is embedded in $\mathcal{M}^1(G)$ as a closed Banach involutive subalgebra (via $f \mapsto f\, d\lambda$), so we can consider the unital Banach involutive subalgebra $\mathrm{L}^1(G) \oplus \mathbb{C}\delta_e$. Any character $\chi\colon \mathrm{L}^1(G) \to \mathbb{C}$ extends uniquely to a character $\chi'\colon (\mathrm{L}^1(G) \oplus \mathbb{C}\delta_e) \to \mathbb{C}$ by letting

$$\chi'(f\, d\lambda + \alpha\delta_e) = \chi(f) + \alpha.$$

Proposition 10.5. *Let G be any locally compact abelian group with a left Haar measure λ. For every character $\chi \in \widehat{G}$, the map ζ_χ given by*

$$\zeta_\chi(\mu) = \int \chi(a)\, d\mu(a) \qquad \text{for all } \mu \in \mathcal{M}^1(G)$$

is a hermitian character (an algebra homomorphism such that $\zeta_\chi(\overline{\check{\mu}}) = \overline{\zeta_\chi(\mu)}$) of the algebra $\mathcal{M}^1(G)$. By restriction to $\mathrm{L}^1(G)$, the map ζ_χ given by

$$\zeta_\chi(f) = \int \chi(a) f(a)\, d\lambda(a) \qquad \text{for all } f \in \mathrm{L}^1(G)$$

is a hermitian character of $\mathrm{L}^1(G)$ not equal to the zero function.

Proof. Let μ, ν be any two complex measures in $\mathcal{M}^1(G)$. The function $\chi \colon G \to \mathbb{C}$ is continuous and bounded, so by the corollary of Proposition 8.54, we have

$$\begin{aligned}
\zeta_\chi(\mu * \nu) &= \int \chi(a)\, d(\mu * \nu)(a) = \int\!\!\int \chi(ab)\, d\mu(a)\, d\nu(b) \\
&= \int\!\!\int \chi(a)\chi(b)\, d\mu(a)\, d\nu(b) \\
&= \left(\int \chi(a)\, d\mu(a)\right)\left(\int \chi(b)\, d\nu(b)\right) \\
&= \zeta_\chi(\mu)\zeta_\chi(\nu).
\end{aligned}$$

Recall from Proposition 7.24,

$$\int \varphi\, d\overline{\mu} = \overline{\int \overline{\varphi(s)}\, d\mu(s)},$$

and by Proposition 8.45,

$$\int \varphi\, d\check{\mu} = \int \check{\varphi}\, d\mu,$$

with $\check{\varphi}(a) = \varphi(a^{-1})$ for all $a \in G$. Thus we have

$$\begin{aligned}
\zeta_\chi(\overline{\check{\mu}}) &= \int \chi(a)\, d\overline{\check{\mu}} \\
&= \overline{\int \overline{\chi(a)}\, d\check{\mu}}, && \text{by Proposition 7.24} \\
&= \overline{\int \chi(a^{-1})\, d\check{\mu}}, && \text{by Proposition 10.1(3)} \\
&= \overline{\int \chi(a)\, d\mu(a)} = \overline{\zeta_\chi(\mu)}, && \text{by Proposition 8.45.}
\end{aligned}$$

Thus ζ_χ is a Hermitian character.

By restriction to $\mathrm{L}^1(G)$, we obtain an algebra homomorphism, with

$$\zeta_\chi(f) = \int f(a)\chi(a)\, d\lambda(a).$$

We need to prove that ζ_χ is not the zero function. The proof is not trivial. One method is to observe that this is a special case of the fundamental fact that there is a bijection between the set of unitary representations of the group G and the set of nondegenerate representations of the algebra $\mathrm{L}^1(G)$. This connection will be discussed in Vol II, Section 3.3. It is the method followed by Folland [Folland (1995)] (Chapter 3, Theorem 3.9, and Chapter 4, Section 4.1). The characters $\chi\colon G \to \mathbf{U}(1)$ are indeed unitary representations of G. The other method used by Bourbaki [Bourbaki (1967b)] (Chapter 2, Section 1, No. 1) is to use Corollary 8.52. We can choose a filter where the measures in $\mathcal{M}^1(G)$ are of the form $fd\lambda$ with $f \in \mathcal{K}_\mathbb{C}(G)$, so that if $fd\lambda$ tends to δ_e, then $\zeta_\chi(f)$, which is equal to $\zeta_\chi(fd\lambda)$, tends to $\zeta_\chi(\delta_e) = 1 \neq 0$. □

The following deep result is obtained.

Theorem 10.6. *Let G be a locally compact abelian group (equipped with a Haar measure λ). The map $j\colon \widehat{G} \to \mathsf{X}(\mathrm{L}^1(G))$ given by*

$$j(\chi)(f) = \zeta_\chi(f) = \int \chi(a)f(a)\, d\lambda(a), \qquad \chi \in \widehat{G},\ f \in \mathrm{L}^1(G),$$

is a homeomorphism of $\widehat{G}$ onto $\mathsf{X}(\mathrm{L}^1(G))$.

Proof sketch. We follow Bourbaki [Bourbaki (1967b)] (Chapter 2, Section 1, No. 1). In this proof we view every function in $\mathrm{L}^1(G)$ as a measure in $\mathcal{M}^1(G)$. Recall from Proposition 8.44(2) that $(\delta_a * f)(s) = (\lambda_a f)(s) = f(a^{-1}s)$. To show that j is injective, for any $\chi \in \widehat{G}$, we use the fact that for any $f \in \mathrm{L}^1(G)$ and any $a \in G$, we have

$$\zeta_\chi(\delta_a * f) = \zeta_\chi(\delta_a)\zeta_\chi(f) = \chi(a)\zeta_\chi(f),$$

because $\zeta_\chi(\delta_a) = \int \chi(b)\, d\delta_a(b) = \chi(a)$. If $\zeta_{\chi_1} = \zeta_{\chi_2}$, then the equation

$$\zeta_\chi(\delta_a * f) = \chi(a)\zeta_\chi(f), \quad \text{for all } a \in G \text{ and all } f \in \mathrm{L}^1(G)$$

applied to χ_1 and χ_2 shows that

$$\chi_1(a)\zeta_{\chi_1}(f) = \chi_2(a)\zeta_{\chi_2}(f), \quad \text{for all } a \in G \text{ and all } f \in \mathrm{L}^1(G).$$

Since by Proposition 10.5 there is some function $f \in \mathrm{L}^1(G)$ such that $\zeta_{\chi_1}(f) = \zeta_{\chi_2}(f) \neq 0$, we deduce that $\chi_1 = \chi_2$, so j is injective.

To prove surjectivity we use the following trick. For any $\zeta \in \mathsf{X}(\mathrm{L}^1(G))$ other than the zero function, pick some function $f \in \mathrm{L}^1(G)$ such that $\zeta(f) \neq 0$. Define $\chi \colon G \to \mathbb{C}$ by

$$\chi(a) = \zeta(\delta_a * f)/\zeta(f).$$

The goal is to show that χ is a character of G such that $\zeta = \zeta_\chi$.

It can be shown that the map $a \mapsto \delta_a * f$ is continuous, so χ is continuous. We also have

$$|\chi(a)| \leq \|\delta_a * f\|_1 / |\zeta(f)| = \|f\|_1 / |\zeta(f)|.$$

Therefore, χ is continuous and bounded. The next step is the most technical part of the proof. It can be shown that there is a filter base $\mathcal{B}$ of $e \in G$ consisting of compact subsets, such that, for every $V \in \mathcal{B}$, there is a continuous function g_V which is positive, zero outside of V, and with $\int g_V \, d\lambda = 1$, and such that

$$\delta_a * f = \lim \delta_a * g_V * f.$$

Then, since $\zeta(\delta_a * g_V * f) = \zeta(\delta_a * g_V)\zeta(f)$, and by definition $\chi(a) = \zeta(\delta_a * f)/\zeta(f)$, we get

$$\chi(a) = \lim \zeta(\delta_a * g_V),$$

and for any $h \in \mathrm{L}^1(G)$,

$$\zeta(\delta_a * h) = \lim \zeta(\delta_a * g_V * h) = (\lim \zeta(\delta_a * g_V))\zeta(h) = \chi(a)\zeta(h).$$

Using Proposition 8.44(3) ($\delta_{ab} = \delta_a * \delta_b$) and the above equation, (with $h = \delta_b * f$), for all $a, b \in G$, we have

$$\begin{aligned}\chi(ab) &= \zeta(\delta_a * \delta_b * f)/\zeta(f) = \zeta(\delta_a)\zeta(\delta_b * f)/\zeta(f) \\ &= \chi(a)\zeta(\delta_b * f)/\zeta(f) = \chi(a)\chi(b),\end{aligned}$$

which proves that χ is a homomorphism. But since χ is also bounded, by Proposition 10.2, it is a (unitary) character of G. Since for any $f \subset \mathrm{L}^1(G)$ we have $(\delta_a * f)(s) = f(a^{-1}s)$, (see just after Definition 8.25), we have

$$(g * f)(s) = \int g(a) f(a^{-1}s) \, d\lambda(a) = \int (\delta_a * f)(s) g(a) d\lambda(a). \qquad (\dagger)$$

(Alternatively, see Bourbaki [Bourbaki (1965b)] (Chapter VIII, Section 1, Proposition 7).

The equation

$$\zeta(g * f) = \int \zeta(\delta_a * f) g(a) \, d\lambda(a)$$

is needed to finish the proof of surjectivity. Since by Theorem 9.19(1), every character $\zeta \in \mathsf{X}(\mathrm{L}^1(G))$ is continuous, ζ is a continuous linear form on $\mathrm{L}^1(G)$, so $\zeta \in \mathrm{L}^1(G)'$, the dual of $\mathrm{L}^1(G)$. Theorem 5.51 asserts that $\mathrm{L}^\infty(G)$ and $\mathrm{L}^1(G)'$ are isomorphic, and more precisely that there is some (unique) $\varphi \in \mathrm{L}^\infty(G)$ such that

$$\zeta(f) = \int f(s)\varphi(s)\, d\lambda(s) \quad \text{for all } f \in \mathrm{L}^1(G). \tag{††}$$

Then using Fubini we have

$$\begin{aligned}
\zeta(g * f) &= \int (g * f)(s)\varphi(s)\, d\lambda(s) \\
&= \int\int (\delta_a * f)(s) g(a)\varphi(s)\, d\lambda(a) d\lambda(s) && \text{by (†)} \\
&= \int\left(\int (\delta_a * f)(s)\varphi(s)\, d\lambda(s)\right) g(a) d\lambda(a) \\
&= \int \zeta(\delta_a * f) g(a)\, d\lambda(a). && \text{by (††)}
\end{aligned}$$

Using the above fact, we get

$$\begin{aligned}
\zeta(g)\zeta(f) &= \zeta(g * f) \\
&= \int \zeta(\delta_a * f) g(a)\, d\lambda(a) \\
&= \zeta(f) \int \chi(a) g(a)\, d\lambda(a) \\
&= \zeta_\chi(g)\zeta(f).
\end{aligned}$$

If we pick f such that $\zeta(f) \neq 0$, we deduce that $\zeta = \zeta_\chi$, establishing the fact that j is surjective. Since j is injective and surjective, it is a bijection. It remains to prove that j is a homeomorphism. We skip this proof, referring the interested reader to Bourbaki [Bourbaki (1967b)] (Chapter 2, Section 1, No. 1). □

As a corollary of Theorem 10.6, since by Proposition 9.20 the space $\mathsf{X}(\mathrm{L}^1(G))$ is locally compact, we have the following fact.

Corollary 10.7. *Let G be a locally compact abelian group. The group $\widehat{G}$ of characters of G with the compact-open topology is locally compact.*

Theorem 10.6, Proposition 10.5, and Proposition 9.41, also imply the following result. Recall from Definition 9.23 that the enveloping C^*-algebra of $\mathrm{L}^1(G)$ is denoted by $\mathrm{St}(G)$.

Proposition 10.8. *Let G be a locally compact abelian group. The characters of the algebra $\mathrm{L}^1(G)$ are hermitian. There is a homeomorphism between the set of characters $\mathsf{X}(\mathrm{St}(G))$ of the enveloping C^*-algebra $\mathrm{St}(G)$ of $\mathrm{L}^1(G)$ and the set of characters $\mathsf{X}(\mathrm{L}^1(G))$ of $\mathrm{L}^1(G)$.*

In summary, the spaces $\widehat{G}$, $\mathsf{X}(\mathrm{L}^1(G))$, and $\mathsf{X}(\mathrm{St}(G))$, are homeomorphic.

It is instructive to figure out the duals of various familiar locally compact abelian groups.

10.2 Characters Groups of some LCA Groups

Proposition 10.9. *The locally compact abelian groups, $\mathbb{Z}, \mathbb{T}$, $\mathbb{Z}/n\mathbb{Z}$, and $\mathbb{R}$, have the following characters and dual groups.*

(1) For $\mathbb{Z}$, the homomorphisms $m \mapsto e^{im\theta} = (e^{i\theta})^m$, for any fixed $\theta \in \mathbb{R}/2\pi\mathbb{Z}$, and any $m \in \mathbb{Z}$. Therefore, the dual group $\widehat{\mathbb{Z}}$ of $\mathbb{Z}$ is isomorphic to $\mathbb{T}$.
(2) For $\mathbb{T}$, the homomorphisms $e^{i\theta} \mapsto e^{im\theta} = (e^{im})^\theta$, for any fixed $m \in \mathbb{Z}$, and any $\theta \in \mathbb{R}/2\pi\mathbb{Z}$. Therefore, the dual group $\widehat{\mathbb{T}}$ of $\mathbb{T}$ is isomorphic to $\mathbb{Z}$.
(3) For $\mathbb{Z}/n\mathbb{Z}$, the homomorphisms $m \mapsto e^{2\pi imk/n} = (e^{2\pi ik/n})^m$, for any fixed $k \in \mathbb{Z}/n\mathbb{Z}$, and any $m \in \mathbb{Z}/n\mathbb{Z}$. Therefore, the dual group $\widehat{\mathbb{Z}/n\mathbb{Z}}$ of $\mathbb{Z}/n\mathbb{Z}$ is isomorphic to $\mathbb{Z}/n\mathbb{Z}$ (itself).
(4) For $\mathbb{R}$, the homomorphisms $x \mapsto e^{iyx} = (e^{iy})^x$, for any fixed $y \in \mathbb{R}$, and all $x \in \mathbb{R}$. Therefore, the dual group $\widehat{\mathbb{R}}$ of $\mathbb{R}$ is isomorphic to $\mathbb{R}$ (itself).

Proof. (1) Since $\mathbb{Z}$ is a cyclic group generated by 1, every homomorphism $\varphi\colon \mathbb{Z} \to \mathbf{U}(1)$ satisfies the equation

$$\varphi(m) = (\varphi(1))^m, \qquad \text{for all } m \in \mathbb{Z}.$$

Thus φ is uniquely determined by picking $\varphi(1) = e^{i\theta}$ in $\mathbf{U}(1)$.

The characters of $\mathbb{T}$ are easily obtained from the characters of $\mathbb{R}$, so we consider (4) next.

(4) Folland has a particularly nice proof of (4). Any homomorphism $\varphi\colon \mathbb{R} \to \mathbf{U}(1)$ satisfies $\varphi(0) = 1$, and since φ is continuous, there is some $a > 0$ such that $\int_0^a \varphi(t)\, dt \neq 0$. Let $A = \int_0^a \varphi(t)\, dt$. Since $\varphi(x+t) = \varphi(x)\varphi(t)$, we have

$$\begin{aligned} A\varphi(x) &= \left(\int_0^a \varphi(t)\, dt \right) \varphi(x) = \int_0^a \varphi(t)\varphi(x)\, dt \\ &= \int_0^a \varphi(t+x)\, dt = \int_x^{a+x} \varphi(u)\, du. \end{aligned}$$

It follows that φ is differentiable, and we have

$$\varphi'(x) = A^{-1}(\varphi(a+x) - \varphi(x)) = A^{-1}(\varphi(a)\varphi(x) - \varphi(x)) = A^{-1}(\varphi(a) - 1)\varphi(x).$$

If we let $c = A^{-1}(\varphi(a) - 1)$, then (using the fact that $\varphi(0) = 1$) we deduce that

$$\varphi(x) = e^{cx}.$$

Since $|\varphi(x)| = 1$, c must be a pure imaginary number of the form $c = iy$, with $y \in \mathbb{R}$, so $\varphi(x) = e^{iyx}$.

(2) Recall that we have the surjective homomorphism $\sigma\colon \mathbb{R} \to \mathbb{T}$ given by $\sigma(\theta) = e^{i\theta}$, and that the kernel of σ is $2\pi\mathbb{Z}$. Therefore, there is a bijection between the continuous homomorphisms $\varphi\colon \mathbb{T} \to \mathbb{U}(1)$ and the continuous homomorphisms $\psi\colon \mathbb{R} \to \mathbb{U}(1)$ such that $\operatorname{Ker}\psi = 2\pi\mathbb{Z}$. By (4) the homomorphisms ψ are of the form $\psi(\theta) = (e^{iy})^\theta$ for some $y \in \mathbb{R}$, and for ψ to have kernel $2\pi\mathbb{Z}$, it must be the case that $\theta \equiv 0 \pmod{2\pi}$ implies that $y\theta \equiv 0 \pmod{2\pi}$, so $y = m \in \mathbb{Z}$, and $\varphi(e^{i\theta}) = (e^{im})^\theta$.

(3) We have the canonical surjective homomorphism $pr\colon \mathbb{Z} \to \mathbb{Z}/n\mathbb{Z}$ whose kernel is $\operatorname{Ker} pr = \mathbb{Z}/n\mathbb{Z}$. It follows that there is a bijection between the homomorphisms $\varphi\colon \mathbb{Z}/n\mathbb{Z} \to \mathbf{U}(1)$ and the homomorphisms $\psi\colon \mathbb{Z} \to \mathbf{U}(1)$ such that $\operatorname{Ker}\psi = \mathbb{Z}/n\mathbb{Z}$. By (1) the homomorphisms ψ are of the form $\psi(m) = e^{im\theta} = (e^{i\theta})^m$, and for ψ to have kernel $\mathbb{Z}/n\mathbb{Z}$, it must be the case that $m \equiv 0 \pmod{n}$ implies that $\theta m \equiv 0 \pmod{2\pi}$, so for $m = dn$ with $d \in \mathbb{Z}$, we must have $\theta dn \equiv 0 \pmod{2\pi}$, which implies that $\theta = 2\pi k/n$ with $k \in \mathbb{Z}/n\mathbb{Z}$, and then $\psi(m) = (e^{2\pi ik/n})^m$. $\square$

Remark: The proof of (4) shows that *any* continuous homomorphism $\varphi\colon \mathbb{R} \to \mathbb{C}^*$ is of the form $\varphi(x) = e^{cx}$ for some complex number $c \in \mathbb{C}$ (where $\mathbb{C}^*$ is the group of nonzero complex numbers under multiplication).

Given a locally compact abelian group G and its dual $\widehat{G}$, there is a canonical pairing $\langle -, - \rangle\colon G \times \widehat{G} \to \mathbb{T}$ given by evaluation,

$$\langle a, \chi \rangle = \chi(a), \qquad a \in G,\ \chi \in \widehat{G}.$$

In practice the dual group $\widehat{G}$ is only determined up to isomorphism. What this means is that if G_1 and G_2 are isomorphic to $\widehat{G}$, the two pairings are usually different. This issue comes up with the group $\mathbb{R}$. We figured out that one of the groups, $\widehat{\mathbb{R}}_1$, isomorphic to $\widehat{\mathbb{R}}$ consists of the characters $\chi^1(x) = (e^{iy})^x$, with $y \in \mathbb{R}$. The pairing is given by

$$\langle x, \chi_1 \rangle_1 = (e^{iy})^x.$$

However, another isomorphic copy $\widehat{\mathbb{R}}_2$ of $\widehat{\mathbb{R}}$ is often used, in which the characters are given by

$$\chi_2(x) = (e^{2\pi iy})^x,$$

with $x \in \mathbb{R}$, in which case the pairing is

$$\langle x, \chi_2 \rangle_2 = (e^{2\pi iy})^x.$$

A similar problem comes up with the groups $\widehat{\mathbb{Z}}$ and $\widehat{\mathbb{T}}$. The characters of $\mathbb{Z}$ can also be viewed as the homomorphisms $m \mapsto e^{2\pi im\theta} = (e^{2\pi i\theta})^m$, with $\theta \in \mathbb{R}/\mathbb{Z}$, and the characters of $\mathbb{T}$ can be viewed as the homomorphisms $e^{2\pi i\theta} \mapsto e^{2\pi im\theta} = (e^{2\pi im})^\theta$, with $\theta \in \mathbb{R}/\mathbb{Z}$.

This subtle issue is related to the choice of the normalization of the Haar measures on G and $\widehat{G}$ and will come up when we consider the inverse Fourier transform.

Products behave well with respect to characters.

Proposition 10.10. *If $G_1, \ldots, G_n$ are locally compact abelian groups, then*

$$(G_1 \times \cdots \times G_n)\widehat{} \cong \widehat{G_1} \times \cdots \times \widehat{G_n}.$$

Proof. Every tuple of characters $(\chi_1, \ldots, \chi_n)$ with $\chi_i \in \widehat{G_i}$ induces a character on $G_1 \times \cdots \times G_n$, by

$$(\chi_1, \ldots, \chi_n)(a_1, \ldots, a_n) = \chi_1(a_1) \cdots \chi_n(a_n)$$

for all $a_i \in G_i$, $i = 1, \ldots, n$. Conversely, every character $\chi \colon G_1 \times \cdots \times G_n \to \mathbb{T}$ can be written as

$$\chi(a_1, \ldots, a_n) = (\chi_1, \ldots, \chi_n)(a_1, \ldots, a_n) = \chi_1(a_1) \cdots \chi_n(a_n),$$

with χ_i given by

$$\chi_i(a_i) = \chi(1, \ldots, 1, a_i, 1, \ldots, 1).$$

This proves the isomorphism of the proposition. □

Propositions 10.9 and 10.10 imply the following facts.

Corollary 10.11. *We have the following isomorphisms:*

$$\widehat{\mathbb{R}^n} \cong \mathbb{R}^n, \quad \widehat{\mathbb{T}^n} \cong \mathbb{Z}^n, \quad \widehat{\mathbb{Z}^n} \cong \mathbb{T}^n.$$

In particular, the characters in $\widehat{\mathbb{R}^n}$ are the homomorphims from $\mathbb{R}^n$ to $\mathbb{T}$ given by

$$x \mapsto e^{iy\cdot x}, \qquad x, y \in \mathbb{R}^n,$$

where $y \cdot x$ is the Euclidean product in $\mathbb{R}^n$; that is, $y \cdot x = \sum_{k=1}^n y_k x_k$.

The characters in $\widehat{\mathbb{T}^n}$ are the homomorphims from $\mathbb{T}^n$ to $\mathbb{T}$ given by

$$(e^{i\theta_1}, \dots, e^{i\theta_n}) \mapsto e^{im\cdot\theta}, \qquad m \in \mathbb{Z}^n,\ \theta \in \mathbb{R}^n/2\pi\mathbb{Z}^n,$$

and the characters in $\widehat{\mathbb{Z}^n}$ are the homomorphims from $\mathbb{Z}^n$ to $\mathbb{T}$ given by

$$m \mapsto e^{im\cdot\theta}, \qquad \theta \in \mathbb{R}^n/2\pi\mathbb{Z}^n,\ m \in \mathbb{Z}^n.$$

As a corollary of Proposition 10.10, since by the structure theorem for finitely generated abelian groups, every *finite* abelian group is isomorphic to a product of cyclic groups $\mathbb{Z}/p\mathbb{Z}$, by Proposition 10.9 and Proposition 10.10, we see that every finite abelian group is isomorphic to its dual. Here we give G the discrete topology so it is automatically compact.

Proposition 10.12. *If G is a finite abelian group, then G is isomorphic to its dual $\widehat{G}$.*

This fact can also be shown more directly, and there is no canonical isomorphism; see Apostol [Apostol (1976)] (Chapter 6, Theorem 6.8).

If the abelian group is compact, then $\mathrm{L}^2(G)$ is a subspace of $\mathrm{L}^1(G)$, and a Hilbert space, with the hermitian inner product given by

$$\langle f, g\rangle = \int f(s)\overline{g(s)}\, d\lambda(s), \qquad \text{for all } f, g \in \mathrm{L}^2(G).$$

If we assume that $\int_G 1\, d\lambda = 1$, then it is remarkable that the set of characters is an orthonormal set in $\mathrm{L}^2(G)$.

Proposition 10.13. *Let G be a compact abelian group with a Haar measure normalized so that G has measure 1. Then for any character $\chi \in \widehat{G}$, we have $\langle \chi, \chi\rangle = 1$, and for any two distinct characters $\chi_1, \chi_2 \in \widehat{G}$, we have*

$$\langle \chi_1, \chi_2\rangle = 0;$$

that is, the characters form an orthonormal set in $\mathrm{L}^2(G)$.

Proof. We have

$$\langle \chi, \chi\rangle = \int \chi(s)\overline{\chi}(s)\, d\lambda(s) = \int 1\, d\lambda(s) = 1.$$

If $\chi_1 \neq \chi_2$, then there is some $a \in G$ such that $\chi_1(a) \neq \chi_2(a)$, which is equivalent to $(\chi_1\chi_2^{-1})(a) \neq 1$. Then by using properties of characters and left invariance of the Haar measure, we have

$$\begin{aligned}\langle \chi_1, \chi_2 \rangle &= \int \chi_1(s)\overline{\chi_2}(s)\, d\lambda(s)\\ &= \int (\chi_1\chi_2^{-1})(s)\, d\lambda(s)\\ &= \int (\chi_1\chi_2^{-1})(aa^{-1}s)\, d\lambda(s)\\ &= (\chi_1\chi_2^{-1})(a) \int (\chi_1\chi_2^{-1})(a^{-1}s)\, d\lambda(s)\\ &= (\chi_1\chi_2^{-1})(a) \int (\chi_1\chi_2^{-1})(s)\, d\lambda(s)\\ &= (\chi_1\chi_2^{-1})(a)\langle \chi_1, \chi_2 \rangle.\end{aligned}$$

Since $(\chi_1\chi_2^{-1})(a) \neq 1$, we conclude that $\langle \chi_1, \chi_2 \rangle = 0$. ☐

Proposition 10.13 implies the following fact.

Proposition 10.14. *For any compact abelian group G, for any character $\chi \in \widehat{G}$, if $\chi \neq 1$, that is, χ is not the trivial character with constant value 1, then $\int \chi(s)\, d\lambda(s) = 0$.*

Proof. Since $\chi \neq 1$, χ is orthogonal to the trivial character 1, so

$$0 = \langle \chi, 1 \rangle = \int \chi\overline{1}\, d\lambda = \int \chi d\lambda.$$ ☐

If G is a compact abelian group, it is remarkable that the orthonormal set of characters $\widehat{G}$ is a Hilbert basis of $\mathrm{L}^2(G)$. This means that the set $\widehat{G}$ is dense in $\mathrm{L}^2(G)$, so for every function $f \in \mathrm{L}^2(G)$ there is a sequence of linear combinations of characters converging to f in the L^2-norm. The proof is nontrivial, and relies on Plancherel's theorem; see Section 10.8 (it it also a corollary of the Peter–Weyl theorem; see Dieudonné [Dieudonné (2003a)] (Chapter XXI, Section 3)).

In particular, for $G = \mathbb{T}$, it turns out that the characters are the functions $e^{i\theta} \mapsto e^{im\theta}$, so we obtain a proof of the fact that $(e^{im\theta})_{m\in\mathbb{Z}}$ is an orthonormal system and that every function in $\mathrm{L}^1(\mathbb{T})$ (a periodic function) is given by a Fourier series.

In the case of a finite locally compact group G, we have $\mathrm{L}^1(G) = \mathrm{L}^2(G)$, the functions in this space are just finite sequences $x = (x_a)_{a\in G}$ of complex numbers indexed by G, and we can figure out explicitly what is integration, the inner product, and convolution.

Example 10.1. Let G be a locally compact abelian group. If G is finite, since G must be Hausdorff, it must have the discrete topology (because every singleton set must be closed, and since G is finite, by closure under finite unions of closed sets, every subset is closed). Thus G is actually compact. The Haar measure λ is just the counting measure, and the integral $\int x\, d\lambda$ is the sum $\sum_{a\in G} x_a$. Here we have a choice; if $|G| = n$, then we can normalize the Haar measure so that G has measure 1, or assume that it has measure n. Let us adopt the first choice, $\lambda(G) = 1$, which implies that

$$\int x_a\, d\lambda(a) = \frac{1}{|G|}\sum_{a\in G} x_a.$$

The inner product $\langle x, y\rangle$ of $x, y \in \mathrm{L}^2(G)$ is

$$\langle x, y\rangle = \int x_a \overline{y}_a\, d\lambda(a) = \frac{1}{|G|}\sum_{a\in G} x_a \overline{y_a}.$$

The convolution $x * y$ of $x, y \in \mathrm{L}^2(G)$ is given by

$$(x * y)_a = \frac{1}{|G|}\sum_{b\in G} x_b y_{b^{-1}a} = \frac{1}{|G|}\sum_{\substack{b,c\in G\\ b+c=a}} x_b y_c.$$

In the special case where $G = \mathbb{Z}/n\mathbb{Z}$, if $x = (x_0, \ldots, x_{n-1})$ and $y = (y_0, \ldots, y_{n-1})$, for $k = 0, \ldots, n-1$, we have

$$(x * y)_k = \frac{1}{n}\sum_{\substack{i,j\in\mathbb{Z}/n\mathbb{Z}\\ i+j\equiv k \pmod n}} x_i y_j.$$

For example, if $G = \mathbb{Z}/3\mathbb{Z}$, the convolution of $x = (x_0, x_1, x_2)$ and $y = (y_0, y_1, y_2)$ is the sequence

$$x * y = \frac{1}{3}(x_0y_0 + x_1y_2 + x_2y_1,\ x_0y_1 + x_1y_0 + x_2y_2,\ x_0y_2 + x_1y_1 + x_2y_0).$$

Observe that

$$\begin{pmatrix} x_0 & x_2 & x_1\\ x_1 & x_0 & x_2\\ x_2 & x_1 & x_0\end{pmatrix}\begin{pmatrix} y_0\\ y_1\\ y_2\end{pmatrix} = \begin{pmatrix} x_0y_0 + x_1y_2 + x_2y_1\\ x_0y_1 + x_1y_0 + x_2y_2\\ x_0y_2 + x_1y_1 + x_2y_0\end{pmatrix}.$$

The matrix

$$\begin{pmatrix} x_0 & x_2 & x_1\\ x_1 & x_0 & x_2\\ x_2 & x_1 & x_0\end{pmatrix}$$

is called a *circulant matrix*. It is obtained from a given column vector by repeatedly making cyclic permutations on the coordinates. The formula giving the convolution of two vectors in terms of a circulant matrix holds for any n.

Since the vector space $L^2(G)$ has dimension $|G|$ and since by Proposition 10.12 the dual group $\widehat{G}$ is isomorphic to G, by Proposition 10.13, the set of $\widehat{G}$ of characters of G is an orthonormal basis of $L^2(G)$.

Another remarkable property of the dual group is that there is a natural homomorphism from G to its double dual $\widehat{\widehat{G}}$. In fact, a fundamental (and famous) theorem due to Pontrjagin asserts that this map is an isomorphism (and a homeomorphism), which will allow us to define an inverse of the Fourier transform (considering functions in $L^2(G)$).

Proposition 10.15. *Let G be a locally compact abelian group. Given any $a \in G$, define the map $\eta_a \colon \widehat{G} \to \mathbb{C}$ by*

$$\eta_a(\chi) = \chi(a), \qquad \textit{evaluation at } a.$$

The map $\eta \colon G \to \widehat{\widehat{G}}$ given by $\eta(a) = \eta_a$ is a continuous homomorphism from G to its double dual $\widehat{\widehat{G}}$.

Proof. First let us check that η_a is a character of the group $\widehat{G}$. For any $a \in G$, for any two characters $\chi_1, \chi_2 \in \widehat{G}$, we have

$$\eta_a(\chi_1\chi_2) = \chi_1(a)\chi_2(a) = \eta_a(\chi_1)\eta_a(\chi_2),$$

so η_a is a homomorphism. Since the characters have range $\mathbf{U}(1)$ and $\eta_a(\chi) = \chi(a)$, we see that $\eta_a \colon \widehat{G} \to \mathbf{U}(1)$. Since G is locally compact, the map $(a, \chi) \mapsto \chi(a)$ from $G \times \widehat{G}$ to $\mathbb{C}$ is continuous, and this implies that η_a is continuous; for details, see Bourbaki [Bourbaki (1967b)] (Chapter 2, Section 1, No. 1).

Let us now check that η is a homomorphism. For all $a, b \in G$ and all $\chi \in \widehat{G}$, we have

$$\eta_{ab}(\chi) = \chi(ab) = \chi(a)\chi(b) = \eta_a(\chi)\eta_b(\chi),$$

showing that η is a homomorphism. The fact that η is continuous is a consequence of the continuity of the map $(a, \chi) \mapsto \chi(a)$; for details, see Bourbaki [Bourbaki (1967b)] (Chapter 2, Section 1, No. 1). ☐

Neither the injectivity nor the surjectivity of the map η is easy to prove.

10.3 The Fourier Transform and the Fourier Cotransform

Given a locally compact abelian group G equipped with a Haar measure λ, the Fourier transform $\mathcal{F}(f)$ of a function $f\colon G \to \mathbb{C}$ is a complex-valued function $\mathcal{F}(f)\colon \widehat{G} \to \mathbb{C}$ defined not on G, but on its dual group $\widehat{G}$ of characters, by the formula

$$\mathcal{F}(f)(\chi) = \int \overline{\chi(a)} f(a)\, d\lambda(a), \qquad \chi \in \widehat{G}. \tag{$*$}$$

The first issue it to determine when this integral converges. Since $|\chi|$ is a bounded continuous function $\chi\colon G \to \mathbb{T}$, by Proposition 5.36, the integral in $(*)$ is well-defined if $f \in \mathrm{L}^1(G)$. If G is not compact, in general, $\mathrm{L}^1(G)$ is not a subspace of $\mathrm{L}^2(G)$ and $\mathrm{L}^2(G)$ is not a subspace of $\mathrm{L}^1(G)$, so in general, the integral in $(*)$ does not converge if $f \in \mathrm{L}^2(G)$.

The second issue is to determine the class of functions on $\widehat{G}$ to which $\mathcal{F}(f)$ belongs.

We will see that if $f \in \mathrm{L}^1(G)$, then $\mathcal{F}(f) \in \mathcal{C}_0(\widehat{G};\mathbb{C})$, so in general $\mathcal{F}(f) \notin \mathrm{L}^1(\widehat{G})$. Remarkably, if $f \in \mathrm{L}^1(G) \cap \mathrm{L}^2(G)$, then $\mathcal{F}(f) \in \mathrm{L}^2(\widehat{G})$. In fact, because $\mathrm{L}^1(G) \cap \mathrm{L}^2(G)$ is dense in $\mathrm{L}^2(G)$, the Fourier transform $\mathcal{F}$ has a unique isometric extension from $\mathrm{L}^2(G)$ to $\mathrm{L}^2(\widehat{G})$. This is *Plancherel's theorem*.

If $\mathcal{F}(f) \in \mathrm{L}^1(\widehat{G})$ or if $\mathcal{F}(f) \in \mathrm{L}^2(\widehat{G})$, for some function $f\colon G \to \mathbb{C}$, then the question of finding an inverse of the Fourier transform arises. Is there a transform $\mathcal{G}$ defined on $\mathrm{L}^1(\widehat{G})$ or $\mathrm{L}^2(\widehat{G})$, or some subspace of them, such that

$$f = \mathcal{G}(\mathcal{F}(f))?$$

We call this equation a *Fourier inversion formula*. As stated, the problem does not make sense because $\mathcal{G}$, being defined on functions in $\mathrm{L}^1(\widehat{G})$ or $\mathrm{L}^2(\widehat{G})$, is a function defined on $\widehat{\widehat{G}}$, and not G. However, we showed in Proposition 10.15 that there is a homomorphism $\eta\colon G \to \widehat{\widehat{G}}$, and by Pontrjagin duality theorem, this map is an isomorphism, so the correct way to state Fourier inversion formula is so say that

$$f = (\mathcal{G} \circ \mathcal{F})(f) \circ \eta.$$

A bit less formally, since η is an isomorphism, by identifying G and $\widehat{\widehat{G}}$, we can drop η from the above formula. Then amazingly, the inverse $\mathcal{G}$ of $\mathcal{F}$

almost looks like $\mathcal{F}$, except that there is no conjugation on the character; that is, for every function $F \in L^1(\widehat{G})$, for every character $\zeta \in \widehat{\widehat{G}}$, we have

$$\mathcal{G}(F)(\zeta) = \int \zeta(\chi) F(\chi)\, d\widehat{\lambda}(\chi), \qquad \zeta \in \widehat{\widehat{G}}. \tag{**}$$

In the above formula, $\widehat{\lambda}$ is a Haar measure on the dual group $\widehat{G}$ (suitably normalized), and χ is any element in $\widehat{G}$ (so, χ is a character of G).

Following Bourbaki, it seems fair to define simultaneously two notions of Fourier transforms.

Definition 10.3. Let G be a locally compact abelian group equipped with a Haar measure λ. For every function $f \in L^1(G)$,

(1) The *Fourier transform* of f is the function $\mathcal{F}(f)\colon \widehat{G} \to \mathbb{C}$ given by

$$\mathcal{F}(f)(\chi) = \int \overline{\chi(a)} f(a)\, d\lambda(a), \qquad \chi \in \widehat{G}.$$

(2) The *Fourier cotransform* of f is the function $\overline{\mathcal{F}}(f)\colon \widehat{G} \to \mathbb{C}$ given by

$$\overline{\mathcal{F}}(f)(\chi) = \int \chi(a) f(a)\, d\lambda(a), \qquad \chi \in \widehat{G}.$$

Remark: We warn our readers that some authors define the Fourier transform as our notion of Fourier cotransform (and the notion of Fourier cotransform as our notion of Fourier transform). This is the convention adopted in Malliavin [Malliavian *et al.* (1995)].

These transforms are not independent. In fact, each one can be obtained from the other. Recall that for any function $f\colon G \to \mathbb{C}$, the function $\check{f}$ is given by $\check{f}(s) = f(s^{-1})$ for all $s \in G$; see Definition 8.11.

Proposition 10.16. *The Fourier transform $\mathcal{F}$ and the Fourier cotransform $\overline{\mathcal{F}}$ are related as follows: for all $f \in L^1(G)$ and all $\chi \in \widehat{G}$,*

$$\overline{\mathcal{F}}(f)(\chi) = \mathcal{F}(f)(\chi^{-1}) = \mathcal{F}(\check{f})(\chi) = \overline{\mathcal{F}(\overline{f})(\chi)}.$$

Proof. We have

$$\begin{aligned}\overline{\mathcal{F}}(f)(\chi) &= \int \chi(a) f(a)\, d\lambda(a)\\ &= \int \overline{\overline{\chi(a)}} f(a)\, d\lambda(a)\\ &= \int \overline{\chi^{-1}(a)} f(a)\, d\lambda(a) = \mathcal{F}(f)(\chi^{-1}),\end{aligned}$$

$$\int \overline{\overline{\chi(a)f(a)}}\, d\lambda(a) = \overline{\int \overline{\chi(a)\overline{f}(a)}\, d\lambda(a)} = \overline{\mathcal{F}(\overline{f})(\chi)},$$

and

$$\begin{aligned}\int \overline{\chi^{-1}(a)} f(a)\, d\lambda(a) &= \int \overline{\chi(a^{-1})} f(a)\, d\lambda(a)\\ &= \int \overline{\chi(a)} f(a^{-1})\, d\lambda(a) = \mathcal{F}(\check{f})(\chi),\end{aligned}$$

where we used the fact that a commutative locally compact group is unimodular (Proposition 8.25) and Proposition 8.27 to change a to a^{-1} in the second equation above, so $\overline{\mathcal{F}}(f)(\chi) = \mathcal{F}(f)(\chi^{-1}) = \mathcal{F}(\check{f})(\chi) = \overline{\mathcal{F}(\overline{f})(\chi)}$. □

With these definitions, the Fourier inversion formula is

$$f = (\overline{\mathcal{F}} \circ \mathcal{F})(f) \circ \eta. \tag{finv}$$

In the above formula, $\overline{\mathcal{F}}$ is the Fourier cotransform on functions defined on the dual group $\widehat{G}$. If we replace $\overline{\mathcal{F}}$ by $\mathcal{F}$, then we don't quite the right formula. By Proposition 10.16, we have

$$(\mathcal{F}(\mathcal{F}(f))(\zeta) = (\overline{\mathcal{F}}(\mathcal{F}))(\zeta^{-1}), \quad \zeta \in \widehat{\widehat{G}},$$

so with $\zeta = \eta(g)$ (with $g \in G$), since η is a homomorphism, we obtain

$$(\mathcal{F}(\mathcal{F}(f))(\eta(g)) = (\overline{\mathcal{F}}(\mathcal{F}))(\eta(g^{-1})),$$

which by (finv) yields

$$(\mathcal{F}(\mathcal{F}(f))(\eta(g)) = (\overline{\mathcal{F}}(\mathcal{F}))(\eta(g^{-1})) = f(g^{-1}).$$

Therefore, instead of (finv), we have

$$\check{f} = (\mathcal{F} \circ \mathcal{F})(f) \circ \eta.$$

Of course, in the above formula the leftmost occurrence of $\mathcal{F}$ is the Fourier transform on functions defined on the dual group $\widehat{G}$. But if we "apply $\mathcal{F}$ four times," we get

$$f = (\mathcal{F} \circ \mathcal{F} \circ \mathcal{F} \circ \mathcal{F})(f) \circ \eta.$$

This looks a little silly to us and seems another justification for considering $\overline{\mathcal{F}}$ on an equal footing with $\mathcal{F}$.

Actually, in view of the isomorphism $j\colon \widehat{G} \to \mathsf{X}(\mathrm{L}^1(G))$ given by Theorem 10.6, the Fourier cotransform $\overline{\mathcal{F}}$ can be viewed as the Gelfand transform from $\mathrm{L}^1(G)$ to $\mathsf{X}(\mathrm{L}^1(G))$. For any $f \in \mathrm{L}^1(G)$, the Gelfand transform $\mathcal{G}_f$ of f is given by

$$\mathcal{G}_f(\zeta) = \zeta(f), \qquad \zeta \in \mathsf{X}(\mathrm{L}^1(G)).$$

By Theorem 10.6, the map $j\colon \widehat{G} \to \mathsf{X}(\mathrm{L}^1(G))$ given by

$$j(\chi)(f) = \zeta_\chi(f) = \int \chi(a) f(a)\, d\lambda(a), \qquad \chi \in \widehat{G},\ f \in \mathrm{L}^1(G),$$

is a homeomorphism of $\widehat{G}$ onto $\mathsf{X}(\mathrm{L}^1(G))$. But

$$\overline{\mathcal{F}}(f)(\chi) = \int \chi(a) f(a)\, d\lambda(a),$$

so

$$j(\chi)(f) = \zeta_\chi(f) = \overline{\mathcal{F}}(f)(\chi),$$

and since $\mathcal{G}_f(\zeta_\chi) = \zeta_\chi(f)$ and $\zeta_\chi = j(\chi)$, we see that

$$\mathcal{G}_f(j(\chi)) = \mathcal{G}_f(\zeta_\chi) = \overline{\mathcal{F}}(f)(\chi). \tag{$\dagger$}$$

In summary, we proved the following result.

Proposition 10.17. *Modulo the isomorphism $j\colon \widehat{G} \to \mathsf{X}(\mathrm{L}^1(G))$, the Fourier cotransform $\overline{\mathcal{F}}(f)$ is the Gelfand transform $\mathcal{G}_f$ from $\mathrm{L}^1(G)$ to $\mathsf{X}(\mathrm{L}^1(G))$.*

This is good news because this shows that we can apply results known for the Gelfand transform to the Fourier cotransform, and thus to the Fourier transform.

The Fourier inversion formula turns out to hold in the following cases:

(1) If we identify G and $\widehat{\widehat{G}}$ using the Pontrjagin isomorphism theorem (and give suitably normalized Haar measures to G and $\widehat{G}$; see Section 10.9), then there is a unique extension of the Fourier tranform $\mathcal{F}$ to $\mathrm{L}^2(G)$ and a unique extension of the Fourier cotransform $\overline{\mathcal{F}}$ to $\mathrm{L}^2(\widehat{G})$, so that they are mutual inverses.

(2) If $A(G)$ is the subspace of $\mathrm{L}^1(G)$ spanned by all functions of the form $f*g$ with $f, g \in \mathrm{L}^1(G)\cap\mathrm{L}^2(G)$, then $A(G)$ is an ideal of $\mathrm{L}^1(G)$ contained in $\mathrm{L}^1(G) \cap \mathrm{L}^2(G)$; if $f \in A(G)$, then $\mathcal{F}(f) \in \mathrm{L}^1(\widehat{G})$ and the Fourier inversion formula holds.

(3) If $B(G) = \{f \in \mathrm{L}^1(G) \mid \mathcal{F}(f) \in \mathrm{L}^1(\widehat{G})\}$, then the restriction of $\mathcal{F}$ to $B(G)$ is a bijection onto $B(\widehat{G})$, and its inverse is the restriction of $\overline{\mathcal{F}}$ to $B(\widehat{G})$.

Thus it appears that the L^2 theory has the best behavior with respect to the Fourier transform. This had been observed for $G = \mathbb{R}$ long ago.

We now redefine the Fourier transform and the Fourier cotransform so that they apply to complex measures $\mu \in \mathcal{M}^1(G)$.

Definition 10.4. Let G be a locally compact abelian group equipped with a Haar measure λ. For every complex measure $\mu \in \mathcal{M}^1(G)$, the *Fourier transform* $\mathcal{F}(\mu)$ and the *Fourier cotransform* $\overline{\mathcal{F}}(\mu)$ of μ are the functions $\mathcal{F}(\mu)\colon \widehat{G} \to \mathbb{C}$ and $\overline{\mathcal{F}}(\mu)\colon \widehat{G} \to \mathbb{C}$ defined on the group $\widehat{G}$ by

$$\begin{aligned}\mathcal{F}(\mu)(\chi) &= \int \overline{\chi(a)}\, d\mu(a)\\ \overline{\mathcal{F}}(\mu)(\chi) &= \int \chi(a)\, d\mu(a),\end{aligned}$$

for all $\chi \in \widehat{G}$.

For every function $f \in \mathrm{L}^1(G)$, the *Fourier transform* $\mathcal{F}(f)$ and the *Fourier cotransform* $\overline{\mathcal{F}}(f)$ of f are the functions $\mathcal{F}(f)\colon \widehat{G} \to \mathbb{C}$ and $\overline{\mathcal{F}}(f)\colon \widehat{G} \to \mathbb{C}$ defined on the group $\widehat{G}$ by

$$\begin{aligned}\mathcal{F}(f)(\chi) &= \int \overline{\chi(a)} f(a)\, d\lambda(a)\\ \overline{\mathcal{F}}(f)(\chi) &= \int \chi(a) f(a)\, d\lambda(a),\end{aligned}$$

for all $\chi \in \widehat{G}$.

Remark: The Fourier cotransform is also called the *inverse Fourier transform* by some authors, including Hewitt and Ross.

As in the case of functions,

$$\overline{\mathcal{F}}(\mu)(\chi) = \mathcal{F}(\mu)(\chi^{-1}) = \mathcal{F}(\check{\mu})(\chi) = \overline{\mathcal{F}(\overline{\mu})(\chi)}.$$

Here is our first result. In particular, it gives the fundamental property of the Fourier transform (and cotransform), which is to convert a convolution into a (pointwise) product of functions. Recall that for a function $f\colon G \to \mathbb{C}$, we have $f^*(a) = \overline{f(a^{-1})}$.

Proposition 10.18.

(1) *The Fourier transform $\mathcal{F}$ and the Fourier cotransform $\overline{\mathcal{F}}$ are involutive homomorphisms from the unital involutive Banach algebra $\mathcal{M}^1(G)$ to the unital involutive Banach algebra $\mathcal{C}_b(\widehat{G};\mathbb{C})$ of continuous bounded*

functions on $\widehat{G}$. *In particular for any two complex measures* $\mu, \nu \in \mathcal{M}^1(G)$, *we have*

$$\mathcal{F}(\mu * \nu) = \mathcal{F}(\mu)\mathcal{F}(\nu), \qquad \overline{\mathcal{F}}(\mu * \nu) = \overline{\mathcal{F}}(\mu)\overline{\mathcal{F}}(\nu),$$

and

$$\mathcal{F}(\breve{\overline{\mu}}) = (\mathcal{F}(\mu))^*, \qquad \overline{\mathcal{F}}(\breve{\overline{\mu}}) = (\overline{\mathcal{F}}(\mu))^*.$$

(2) *The Fourier transform* $\mathcal{F}$ *and the Fourier cotransform* $\overline{\mathcal{F}}$ *are injective involutive homomorphisms from the involutive Banach algebra* $\mathrm{L}^1(G)$ *to the involutive Banach algebra* $\mathcal{C}_0(\widehat{G}; \mathbb{C})$ *of continuous functions on* $\widehat{G}$ *that tend to zero at infinity. In particular for any two functions* $f, g \in \mathrm{L}^1(G)$, *we have*

$$\mathcal{F}(f * g) = \mathcal{F}(f)\mathcal{F}(g), \qquad \overline{\mathcal{F}}(f * g) = \overline{\mathcal{F}}(f)\overline{\mathcal{F}}(g),$$

and

$$\mathcal{F}(f^*) = (\mathcal{F}(f))^*, \qquad \overline{\mathcal{F}}(f^*) = (\overline{\mathcal{F}}(f))^*.$$

Proof. (1) Observe that for any character $\chi \in \widehat{G}$, we have $\overline{\mathcal{F}}(\mu)(\chi) = \zeta_\chi(\mu)$, as defined in Proposition 10.5, and by this proposition, ζ_χ is a character of the algebra $\mathcal{M}^1(G)$, so $\overline{\mathcal{F}}$ is a homomorphism. Since $\mathcal{F}(\mu)(\chi) = \overline{\mathcal{F}}(\mu)(\chi^{-1})$, the Fourier transform $\mathcal{F}$ is also a homomorphism. Since $|\chi(a)| = 1$ for all $a \in G$, we have

$$|\mathcal{F}(\mu)(\chi)| = \left| \int \overline{\chi(a)}\, d\mu(a) \right| \leq \|\mu\|,$$

and the same argument applies to $\overline{\mathcal{F}}(\mu)$. Therefore $\mathcal{F}(\mu)$ and $\overline{\mathcal{F}}(\mu)$ are bounded functions on $\widehat{G}$.

Finally, by the definition of the compact-open topology on $\widehat{G}$, if χ_1 tends to χ_2 in $\widehat{G}$, then the function χ_1 defined on G tends to χ_2 uniformly on every compact set, while being bounded by 1. It follows that $\mathcal{F}(\mu)(\chi_1)$ tends to $\mathcal{F}(\mu)(\chi_2)$. Thus $\mathcal{F}(\mu)$ is continuous. A similar reasoning shows that $\overline{\mathcal{F}}(\mu)$ is continuous.

(2) Since $\mathrm{L}^1(G)$ is a subalgebra of $\mathcal{M}^1(G)$, we see immediately that $\mathcal{F}$ and $\overline{\mathcal{F}}$ are homomorphisms. Since $\widehat{G}$ and $\mathsf{X}(\mathrm{L}^1(G))$ are homeomorphic, $\overline{\mathcal{F}}(f)$ can be identified with the Gelfand transform $\mathcal{G}_f$, by the remark following Theorem 9.23, the Gelfand transform $\mathcal{G}$ maps $\mathrm{L}^1(G)$ into $\mathcal{C}_0(\mathsf{X}(\mathrm{L}^1(G)) \cong \mathcal{C}_0(\widehat{G})$, so $\overline{\mathcal{F}}$ maps $\mathrm{L}^1(G)$ to $\mathcal{C}_0(\widehat{G})$. Since $\mathcal{F}(\mu)(\chi) = \overline{\mathcal{F}}(\mu)(\chi^{-1})$, the Fourier transform $\mathcal{F}$ also maps $\mathrm{L}^1(G)$ to $\mathcal{C}_0(\widehat{G})$.

By Proposition 9.27, the Gelfand transform $\mathcal{G}$ is injective because it can be shown that $\mathrm{L}^1(G)$ has radical (0); see Proposition 9.43. As a consequence $\overline{\mathcal{F}}$ is injective, and since $\mathcal{F}(\mu)(\chi) = \overline{\mathcal{F}}(\mu)(\chi^{-1})$, the Fourier transform is also injective. □

Remarks:

(1) If $G = \mathbb{R}^n$, the fact that the Fourier transform $\mathcal{F}(f)$ is a continuous function that tends to zero at infinity is known as the *Riemann–Lebesgue lemma*; see Folland [Folland (1999)] (Chapter 8, Theorem 8.22).

(2) Since $\mathrm{L}^1(G)$ is a commutative Banach algebra, one may wonder whether a quicker proof of the injectivity of $\mathcal{G}$ could be obtained using the Gelfand-Naimark theorem (Theorem 9.37). Unfortunately, $\mathrm{L}^1(G)$ is *not* a unital C^*-algebra so this theorem does not apply. However, $\mathrm{L}^1(G)$ is dense in its enveloping C^*-algebra $\mathrm{St}(G)$ (see Definition 9.23), so $\mathcal{G}$ extends to an isomorphism between $\mathrm{St}(G)$ and $\mathcal{C}_0(\widehat{G})$.

Example 10.2.

(1) Let $G = \mathbb{R}$ and equip G with the Lebesgue measure dx. Recall from Proposition 10.9 that $\widehat{\mathbb{R}}$ is isomorphic to $\mathbb{R}$. If we choose the characters to be the maps $x \mapsto e^{iyx}$ for some $y \in \mathbb{R}$, then it turns out that for the Fourier inversion formula to come out right, $\widehat{\mathbb{R}}$ needs to be equipped with $dx/2\pi$. The Fourier transform on $\mathbb{R}$ is

$$\widehat{f}(x) = \mathcal{F}(f)(x) = \int f(y) e^{-ixy}\, dy,$$

and on $\widehat{\mathbb{R}}$ it is

$$\mathcal{F}(f)(x) = \int f(y) e^{-ixy}\, \frac{dy}{2\pi}.$$

The Fourier cotransform is obtained by changing the sign $-$ in the exponent to a $+$ sign. Then the inversion formula is

$$f(x) = \int \widehat{f}(y) e^{ixy}\, \frac{dy}{2\pi}.$$

This is the choice made in Malliavin [Malliavian *et al.* (1995)].

Another choice that works is to equip both $\mathbb{R}$ and $\widehat{\mathbb{R}}$ with the normalized Lebesgue measure $dx/\sqrt{2\pi}$, as in Rudin [Rudin (1987, 1991)], and in Chapter 6.

With the choice of characters $x \mapsto e^{2\pi i y x}$, the Lebesgue measure is self-dual; see Folland [Folland (1999)]. The Fourier transform on $\mathbb{R}$ and $\widehat{\mathbb{R}}$ is

$$\widehat{f}(x) = \mathcal{F}(f)(x) = \int f(y) e^{-2\pi i x y}\, dy.$$

The Fourier cotransform is obtained by changing the sign $-$ in the exponent to a $+$ sign, and the inversion formula is

$$f(x) = \int \widehat{f}(y) e^{2\pi i x y}\, dy.$$

The trick to pick the right normalization factor is that the Fourier transform $\widehat{g}$ of the function $g(x) = e^{-\pi x^2}$ should be g itself; $\widehat{g} = g$; see Folland [Folland (1999)] (Chapter 8, Proposition 8.24).

(2) Let $G = \mathbb{T}$, equipped with the Haar measure $d\nu_1/2\pi$ inherited from $\mathbb{R}$, by viewing $\mathbb{T}$ as $\mathbb{R}/2\pi\mathbb{Z}$, so that $\mathbb{T}$ has measure 1; see Example 8.4. The characters of $\mathbb{T}$ are the maps $e^{i\theta} \mapsto e^{im\theta}$, with $m \in \mathbb{Z}$. We equip $\mathbb{Z}$ with the counting measure, and the characters are the maps $m \mapsto e^{im\theta}$, with $\theta \in \mathbb{R}/2\pi\mathbb{Z}$. The Fourier transform $\widehat{f} = \mathcal{F}(f)$ of a function $f \in \mathrm{L}^1(\mathbb{T})$ is the $\mathbb{Z}$-indexed sequence whose mth element $\widehat{f}_m$, the *mth Fourier coefficient* of f, is given by

$$\widehat{f}_m = \mathcal{F}(f)(m) = \int_{-\pi}^{\pi} f(\theta) e^{-im\theta} \frac{d\theta}{2\pi}.$$

The Fourier transform $\mathcal{F}(c)$ of a sequence $c = (c_m)_{m\in\mathbb{Z}} \in \mathrm{L}^1(\mathbb{Z}) = l^1(\mathbb{Z})$ is the function

$$f(e^{i\theta}) = \mathcal{F}(c)(e^{i\theta}) = \sum_{m\in\mathbb{Z}} c_m e^{-im\theta}.$$

The Fourier cotransform is obtained by changing the sign $-$ in the exponent to a $+$ sign. This is traditionally called the *Fourier series* associated with $c = (c_m)_{m\in\mathbb{Z}}$. The inversion formula is

$$f(e^{i\theta}) = \sum_{m\in\mathbb{Z}} \widehat{f}_m e^{im\theta}.$$

(3) Consider the group $G = (\mathbb{R}_+^*, *)$, the group of positive reals under multiplication. This group comes up in the *Mellin transform*. Since we have the isomorphism $\varphi\colon \mathbb{R} \to \mathbb{R}_+^*$ given by $\varphi(x) = e^x$, where the group operation on $\mathbb{R}$ is addition, it is easy to figure out the characters of $\mathbb{R}_+^*$, which are homomorphisms $\chi\colon \mathbb{R}_+^* \to \mathbf{U}(1)$. Indeed, there is a bijection between the characters $\zeta \in \widehat{\mathbb{R}}$, which are homomorphisms

$\zeta\colon \mathbb{R} \to \mathbf{U}(1)$, and the characters $\chi \in \widehat{\mathbb{R}_+^*}$, with $\chi\colon \mathbb{R}_+^* \to \mathbf{U}(1)$, given by

$$\zeta = \chi \circ \varphi,$$

as illustrated by the following diagram.

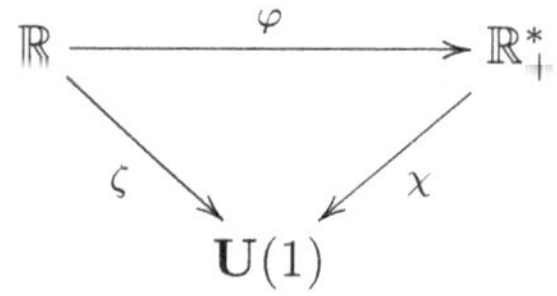

By Proposition 10.9, every character $\zeta \in \widehat{\mathbb{R}}$ is of the form $\zeta(z) = e^{ixz}$ for some $x \in \mathbb{R}$, and since $\chi(y) = \zeta(\varphi^{-1}(y)) = \zeta(\log y)$ for any $y > 0$, we get

$$\chi(y) = \zeta(\log y) = e^{ix \log y} = (e^{\log y})^{ix} = y^{ix}.$$

Therefore, the characters of $\mathbb{R}_+^*$ are of the form

$$y \mapsto y^{ix}, \quad y \in \mathbb{R}_+^*,$$

for some fixed $x \in \mathbb{R}$, which shows that

$$\widehat{\mathbb{R}_+^*} \cong \mathbb{R}.$$

We can also easily determine the characters of $\widehat{\mathbb{R}_+^*}$.

We know from Proposition 10.9 that every $\zeta \in \widehat{\mathbb{R}}$ is of the form $\zeta(z) = e^{ixz}$ for some $x \in \mathbb{R}$. If we let $y = e^x$, then $y > 0$, and

$$\zeta(z) = e^{ixz} = y^{iz}.$$

Therefore, every character of $\widehat{\mathbb{R}_+^*}$ is of the form

$$z \mapsto y^{iz}, \quad z \in \mathbb{R},$$

for some fixed $y \in \mathbb{R}_+^*$. Thus

$$\widehat{\widehat{\mathbb{R}_+^*}} \cong \mathbb{R}_+^*,$$

which confirms the general theory. If we pick the left-invariant Haar measure dy/y on $\mathbb{R}_+^*$ (see Example 8.3), then the Fourier transform on $\mathrm{L}^1(\mathbb{R}_+^*)$ is given by

$$\mathcal{F}(f)(x) = \int_0^\infty y^{-ix} f(y)\, \frac{dy}{y},$$

for any $f \in L^1(\mathbb{R}_+^*)$ and all $x \in \mathbb{R}$. This is one of the formulations of the *Mellin transform*, often denoted $\mathcal{M}(f)$.

If we give $\widehat{\mathbb{R}_+^*}$ the Haar measure $dx/2\pi$, then the Fourier cotransform on $L^1(\widehat{\mathbb{R}_+^*}) = L^1(\mathbb{R})$ is given by

$$\overline{\mathcal{F}}(g)(y) = \frac{1}{2\pi} \int_{\mathbb{R}} y^{ix} g(x)\, dx,$$

for any $g \in L^1(\mathbb{R})$ and all $y \in \mathbb{R}_+^*$. This is one of the formulations of the *inverse Mellin transform*, often denoted $\mathcal{M}^{-1}(g)$. The inversion formula is

$$f(y) = \frac{1}{2\pi} \int_{\mathbb{R}} y^{ix} \mathcal{M}(f)(x)\, dx, \quad y \in \mathbb{R}_+^*.$$

Observe that on $L^1(\mathbb{R}_+^*)$, convolution is given by

$$(f * g)(y) = \int_0^\infty f(z) g\left(\frac{y}{z}\right) \frac{dz}{z},$$

and by the general theory,

$$\mathcal{M}(f * g) = \mathcal{M}(f)\mathcal{M}(g).$$

Some useful properties of the Fourier transform are listed below.

Proposition 10.19. *Let G be locally compact abelian group. The Fourier transforms and the Fourier cotransforms satisfy the following equations: For all $f \in L^1(G)$, and all $\chi \in \widehat{G}$,*

(1)

$$\overline{\mathcal{F}}(f)(\chi) = \mathcal{F}(f)(\chi^{-1}) = \mathcal{F}(\check{f})(\chi) = \overline{\mathcal{F}(\overline{f})(\chi)}.$$

These equations also hold with f replaced by a complex measure $\mu \in \mathcal{M}^1(G)$. We also have

$$\mathcal{F}(\delta_a)(\chi) = \overline{\chi(a)}, \qquad \overline{\mathcal{F}}(\delta_a)(\chi) = \chi(a).$$

(2)

$$\|\mathcal{F}(f)\|_\infty = \left\|\overline{\mathcal{F}}(f)\right\|_\infty \le \|f\|_1 .$$

These equations also hold with f replaced by a complex measure $\mu \in \mathcal{M}^1(G)$.

(3)

$$\mathcal{F}(\lambda_a(f))(\chi) = \overline{\chi(a)}\mathcal{F}(f)(\chi)$$
$$\overline{\mathcal{F}}(\lambda_a(f))(\chi) = \chi(a)\overline{\mathcal{F}}(f)(\chi).$$

These equations also hold with f replaced by a complex measure $\mu \in \mathcal{M}^1(G)$.

(4) For all $f \in L^1(G)$, *and all* $\chi, \xi \in \widehat{G}$,

$$\mathcal{F}(\xi f)(\chi) = \mathcal{F}(f)(\xi^{-1}\chi) = \lambda_\xi(\mathcal{F}(f))(\chi)$$
$$\overline{\mathcal{F}}(\xi f)(\chi) = \overline{\mathcal{F}}(f)(\xi^{-1}\chi) = \lambda_\xi(\overline{\mathcal{F}}(f))(\chi).$$

These equations also hold with f *replaced by a complex measure* $\mu \in \mathcal{M}^1(G)$.

Proof. (1) We have already proven Equations (1) in Proposition 10.16. Part (2) is proven in the proof of Part (1) of Proposition 10.18.

(3) We have

$$\begin{aligned}\mathcal{F}(\lambda_a(f))(\chi) &= \int \overline{\chi(b)} f(a^{-1}b)\, d\lambda(b) \\ &= \int \overline{\chi(ab)} f(b)\, d\lambda(b) \\ &= \int \overline{\chi(a)\chi(b)} f(b)\, d\lambda(b) \\ &= \overline{\chi(a)} \int \overline{\chi(b)} f(b)\, d\lambda(b) = \overline{\chi(a)}\mathcal{F}(f)(\chi).\end{aligned}$$

The second equation is proven in a similar way.

(4) We have

$$\begin{aligned}\mathcal{F}(\xi f)(\chi) &= \int \overline{\chi(a)}\xi(a) f(a)\, d\lambda(a) \\ &= \int \overline{(\xi^{-1}\chi)(a)} f(a)\, d\lambda(a) \\ &= \mathcal{F}(f)(\xi^{-1}\chi) = \lambda_\xi(\mathcal{F}(f))(\chi).\end{aligned}$$

We leave the proof of the other equations as exercises. $\square$

Example 10.3. If $G = \mathbb{R}$, since the characters are of the form $x \mapsto e^{iyx}$ (with $y \in \mathbb{R}$), then, with a slight abuse of notation, Equations (3) yields the well-known formula

$$\mathcal{F}(\lambda_a(f))(x) = \mathcal{F}(f(x-a)) = e^{-iax}\mathcal{F}(f)(x),$$

and Equations (4) yields

$$\mathcal{F}(e^{iax} f(x)) = \mathcal{F}(f)(x-a);$$

see Rudin [Rudin (1987)] (Chapter 9) or Folland [Folland (1999)] (Chapter 8); recall that $\lambda_a(f)(x) = f(x-a)$.

If $G = \mathbb{T}$, since the characters are of the form $e^{i\theta} \mapsto e^{im\theta}$ (with $m \in \mathbb{Z}$), then, with a slight abuse of notation, Equations (3) yields the formula

$$\mathcal{F}(\lambda_{e^{i\varphi}}(f))(m) = e^{-im\varphi}\mathcal{F}(f)(m),$$

and Equations (4) yields

$$\mathcal{F}(e^{in\theta}f(e^{i\theta}))(m) = \mathcal{F}(f)(m-n).$$

Recall that $f\colon \mathbb{T} \to \mathbb{C}$, and that $\mathcal{F}(f)$ is a $\mathbb{Z}$-indexed sequence of complex numbers, namely the Fourier coefficients of f.

If $G = \mathbb{Z}$, since the characters are of the form $m \mapsto e^{im\theta}$ with $\theta \in \mathbb{R}/2\pi\mathbb{Z}$, Equations (3) yields the formula

$$\mathcal{F}(\lambda_n(f))(\theta) = e^{-in\theta}\mathcal{F}(f)(\theta),$$

and Equations (4) yields

$$\mathcal{F}(e^{im\varphi}f(m))(\theta) = \mathcal{F}(f)(\theta - \varphi).$$

Recall that $f\colon \mathbb{Z} \to \mathbb{C}$ is a $\mathbb{Z}$-indexed sequence, and that $\mathcal{F}(f)(\theta) = \sum_{m\in\mathbb{Z}} f(m)e^{-im\theta}$ is a Fourier series.

10.4 The Fourier Transform on a Finite Abelian Group

Let G be a finite locally compact abelian group, and as in Example 10.1, assume that the Haar measure is normalized so that $\lambda(G) = 1$. It is possible and very instructive to work out explicitly the Fourier transform on $\mathrm{L}^2(G)$ and the Fourier cotransform on $\mathrm{L}^2(\widehat{G})$. We will also prove directly that Fourier inversion holds.

Recall that the inner product $\langle x, y\rangle$ of $x, y \in \mathrm{L}^2(G)$ is given by

$$\langle x, y\rangle = \frac{1}{|G|}\sum_{a\in G} x_a\overline{y_a}.$$

Then the Fourier transform and the Fourier cotransform of $x = (x_a)_{a\in G} \in \mathrm{L}^2(G)$ are given by

$$\mathcal{F}(x)(\chi) = \frac{1}{|G|}\sum_{a\in G} x_a\overline{\chi(a)} \qquad (\dagger_1)$$

$$\overline{\mathcal{F}}(x)(\chi) = \frac{1}{|G|}\sum_{a\in G} x_a\chi(a), \qquad (\dagger_2)$$

where $\chi\colon G \to \mathbb{T}$ is a character of G.

Observe that

$$\mathcal{F}(x)(\chi) = \frac{1}{|G|} \sum_{a \in G} x_a \overline{\chi(a)} = \langle x, \chi \rangle.$$

Recall that $\widehat{G} \cong G$ by Proposition 10.12, so we can write $\widehat{G} = \{\chi_1, \ldots, \chi_n\}$, with $n = |G|$. We may call $\widehat{x}(\chi_i) = \mathcal{F}(x)(\chi_i) = \langle x, \chi_i \rangle$ the ith *Fourier coefficient* of x.

Proposition 10.20. *(Fourier inversion formula) Let G be a finite abelian group of order n, and let $\widehat{G} = \{\chi_1, \ldots, \chi_n\}$. For every $x \in \mathrm{L}^2(G)$, we have*

$$x = \sum_{j=1}^{n} \mathcal{F}(x)(\chi_j)\chi_j = \sum_{j=1}^{n} \widehat{x}(\chi_j)\chi_j,$$

with

$$\mathcal{F}(x)(\chi_j) = \langle x, \chi_j \rangle = \frac{1}{|G|} \sum_{a \in G} x_a \overline{\chi_j(a)}.$$

Proof. By Proposition 10.13 the characters $\{\chi_1, \ldots, \chi_n\}$ are orthonormal, and since $\mathrm{L}^2(G)$ is a vector space of dimension n, they form a basis. Consequently we can write

$$x = \sum_{j=1}^{n} c_j \chi_j$$

for some $c_j \in \mathbb{C}$. Taking the inner product with χ_j, we obtain

$$\langle x, \chi_j \rangle = c_j \langle \chi_j, \chi_j \rangle = c_j,$$

which concludes the proof. □

Let us equip $\widehat{G}$ with the counting measure $\widehat{\lambda}$ normalized so that $\widehat{\lambda}(\widehat{G}) = |\widehat{G}| = |G|$. Then the integral of a function $H \in \mathrm{L}^2(\widehat{G})$ is given by

$$\int H(\chi)\, d\widehat{\lambda}(\chi) = \sum_{\chi \in \widehat{G}} H_\chi,$$

the inner product of $F, H \in \mathrm{L}^2(\widehat{G})$ is given by

$$\langle F, H \rangle = \sum_{\chi \in \widehat{G}} F_\chi \overline{H_\chi},$$

and the Fourier transform and the Fourier cotransform of $H = (H_\chi)_{\chi \in \widehat{G}} \in \mathrm{L}^2(\widehat{G})$ are given by

$$\mathcal{F}(H)(\zeta) = \sum_{\chi \in \widehat{G}} H_\chi \overline{\zeta(\chi)} \tag{$\dagger_3$}$$

$$\overline{\mathcal{F}}(H)(\zeta) = \sum_{\chi \in \widehat{G}} H_\chi \zeta(\chi), \tag{$\dagger_4$}$$

where $\zeta \colon \widehat{G} \to \mathbb{T}$ is a character of $\widehat{G}$. Observe that the factor $1/|G|$ is missing.

Recall that $\eta_a \in \widehat{\widehat{G}} \subseteq \mathrm{L}^2(\widehat{G})$ is a character on $\widehat{G}$ such that $\eta_a(\chi) = \chi(a)$.

Proposition 10.21. *(Fourier inversion) For any $x \in \mathrm{L}^2(G)$ and any $a \in G$, we have*

$$(\overline{\mathcal{F}} \circ \mathcal{F})(x)(\eta_a) = x_a,$$

that is,

$$x = (\overline{\mathcal{F}} \circ \mathcal{F})(x) \circ \eta.$$

Proof. Let us compute $(\overline{\mathcal{F}} \circ \mathcal{F})(x)(\eta_a) = \overline{\mathcal{F}}(\mathcal{F}(x))(\eta_a)$, for any $x = (x_b)_{b \in G}$ and any $a \in G$. Since $\eta_a(\chi) = \chi(a)$, we have

$$\begin{aligned}
(\overline{\mathcal{F}} \circ \mathcal{F})(x)(\eta_a) &= \overline{\mathcal{F}}(\mathcal{F}(x))(\eta_a) \\
&= \sum_{\chi \in \widehat{G}} \mathcal{F}(x)_\chi \, \eta_a(\chi) \\
&= \frac{1}{|G|} \sum_{\chi \in \widehat{G}} \sum_{b \in G} x_b \overline{\chi(b)} \chi(a) \\
&= \frac{1}{|G|} \sum_{b \in G} x_b \sum_{\chi \in \widehat{G}} \chi(a) \overline{\chi(b)} \\
&= \frac{1}{|G|} \sum_{b \in G} x_b \sum_{\chi \in \widehat{G}} \eta_a(\chi) \overline{\eta_b}(\chi) \\
&= \frac{1}{|G|} \sum_{b \in G} x_b \langle \eta_a, \eta_b \rangle.
\end{aligned}$$

But $\eta_a, \eta_b \in \widehat{\widehat{G}} \subseteq \mathrm{L}^2(\widehat{G})$, so by Proposition 10.13 (applied to $\widehat{G}$), all terms $\langle \eta_a, \eta_b \rangle$ are zero if $a \neq b$, and $\langle \eta_a, \eta_a \rangle = |G|$, because the Haar measure on

$\widehat{G}$ is normalized so that $\widehat{\lambda}(\widehat{G}) = |\widehat{G}| = |G|$, so the factor $1/|G|$ is missing. Therefore,

$$(\overline{\mathcal{F}} \circ \mathcal{F})(x)(\eta_a) = \frac{1}{|G|} \sum_{b \in G} x_b \langle \eta_a, \eta_b \rangle = \frac{1}{|G|} x_a \langle \eta_a, \eta_a \rangle = \frac{1}{|G|} x_a |G| = x_a,$$

which proves that

$$x = (\overline{\mathcal{F}} \circ \mathcal{F})(x) \circ \eta,$$

namely, the Fourier inversion formula. □

Observe that in order for the inversion formula to be correct, the normalization factor of the Haar measure λ on G and the normalization factor of the Haar measure $\widehat{\lambda}$ on $\widehat{G}$ have to be chosen carefully. In our case, we chose $\lambda(G) = 1$ and $\widehat{\lambda}(\widehat{G}) = |G|$. Instead we could have chosen $\lambda(G) = |G|$ and $\widehat{\lambda}(\widehat{G}) = 1$. The reader should check that the self-dual choice $\lambda(G) = \widehat{\lambda}(\widehat{G}) = \sqrt{|G|}$ also works.

The reader should also check that if we use $\mathcal{F}$ instead of $\overline{\mathcal{F}}$ on $\mathrm{L}^2(\widehat{G})$, then we get

$$\breve{x} = (\mathcal{F} \circ \mathcal{F})(x) \circ \eta,$$

where $\breve{x}(a) = x_{a^{-1}}$.

Proposition 10.22. *(Plancherel theorem) For all $x, y \in \mathrm{L}^2(G)$, we have*

$$\langle x, y \rangle = \langle \mathcal{F}(x), \mathcal{F}(y) \rangle.$$

As a consequence, $\mathrm{L}^2(G)$ and $\mathrm{L}^2(\widehat{G})$ are isometric.

Proof. We have

$$\begin{aligned}
\langle \mathcal{F}(x), \mathcal{F}(y) \rangle &= \sum_{\chi \in \widehat{G}} \mathcal{F}(x)_\chi \overline{\mathcal{F}(y)_\chi} \\
&= \frac{1}{|G|^2} \sum_{\chi \in \widehat{G}} \sum_{a \in G} \sum_{b \in G} x_a \overline{\chi(a)} \overline{y_b} \chi(b) \\
&= \frac{1}{|G|^2} \sum_{a \in G} \sum_{b \in G} x_a \overline{y_b} \sum_{\chi \in \widehat{G}} \chi(b) \overline{\chi(a)} \\
&= \frac{1}{|G|^2} \sum_{a \in G} \sum_{b \in G} x_a \overline{y_b} \sum_{\chi \in \widehat{G}} \eta_b(\chi) \overline{\eta_a(\chi)}
\end{aligned}$$

$$
\begin{aligned}
&= \frac{1}{|G|^2} \sum_{a \in G} \sum_{b \in G} x_a \overline{y_b} \langle \eta_b, \eta_a \rangle \\
&= \frac{1}{|G|^2} \sum_{a \in G} x_a \overline{y_a} |G| \\
&= \langle x, y \rangle.
\end{aligned}
$$

Again, we used the fact that in $\mathrm{L}^2(\widehat{G})$, the measure $\widehat{\lambda}$ is normalized so that $\widehat{\lambda}(\widehat{G}) = |G|$, so $\langle \eta_a, \eta_a \rangle = |G|$ (and $\langle \eta_a, \eta_b \rangle = 0$ whenever $a \neq b$). $\square$

Proposition 10.22 is a special case of Plancherel theorem.

Proposition 10.23. *(Convolution rule) For all $x, y \in \mathrm{L}^2(G)$, we have*

$$\mathcal{F}(x * y) = \mathcal{F}(x)\mathcal{F}(y).$$

Proof. Recall from Example 10.1 that the convolution $x * y$ of $x, y \in \mathrm{L}^2(G)$ is given by

$$(x * y)_a = \frac{1}{|G|} \sum_{b \in G} x_b y_{b^{-1}a} = \frac{1}{|G|} \sum_{\substack{b, c \in G \\ b + c = a}} x_b y_c.$$

Thus we have

$$
\begin{aligned}
\mathcal{F}(x * y)(\chi) &= \frac{1}{|G|} \sum_{c \in G} (x * y)_c \overline{\chi(c)} \\
&= \frac{1}{|G|} \sum_{c \in G} \frac{1}{|G|} \sum_{d \in G} x_d y_{d^{-1}c} \overline{\chi(c)}.
\end{aligned}
$$

If we replace c by dc, since G is a group the sum does not change, and we get

$$
\begin{aligned}
\frac{1}{|G|} \sum_{c \in G} \frac{1}{|G|} \sum_{d \in G} x_d y_{d^{-1}c} \overline{\chi(c)} &= \frac{1}{|G|} \sum_{dc \in G} \frac{1}{|G|} \sum_{d \in G} x_d y_{d^{-1}dc} \overline{\chi(dc)} \\
&= \frac{1}{|G|} \sum_{d \in G} x_d \frac{1}{|G|} \sum_{c \in G} y_c \overline{\chi(d)}\, \overline{\chi(c)} \\
&= \frac{1}{|G|} \sum_{d \in G} x_d \overline{\chi(d)} \frac{1}{|G|} \sum_{c \in G} y_c \overline{\chi(c)} \\
&= \mathcal{F}(x)(\chi)\mathcal{F}(y)(\chi).
\end{aligned}
$$

Therefore, we proved that

$$\mathcal{F}(x * y) = \mathcal{F}(x)\mathcal{F}(y). \qquad \square$$

10.5 Dirichlet Characters

Let $G = (\mathbb{Z}/m\mathbb{Z})^*$ equipped with the counting measure. The group $(\mathbb{Z}/m\mathbb{Z})^*$ is the multiplicative group of units in $\mathbb{Z}/m\mathbb{Z}$, that is, those elements $a \in \mathbb{Z}/m\mathbb{Z}$ such that there is some $b \in \mathbb{Z}/m\mathbb{Z}$ and $ab = ba = 1 \pmod{m}$. It is well known that $a \in \mathbb{Z}/m\mathbb{Z}$ is a unit if and only if $\gcd(a, m) = 1$, and the group $(\mathbb{Z}/m\mathbb{Z})^*$ has $\varphi(m)$ elements, where φ is the Euler phi-function ($\varphi(m)$ is the number of integers a, with $1 \leq a \leq m$, such that $\gcd(a, m) = 1$). The group $(\mathbb{Z}/m\mathbb{Z})^*$ is not always cyclic, (for example, if $m = 2^k, k \geq 3$), but Gauss determined when this happens. However, when $(\mathbb{Z}/m\mathbb{Z})^*$ is cyclic, finding a generator for it is computationally hard.

Definition 10.5. The characters $\chi\colon (\mathbb{Z}/m\mathbb{Z})^* \to \mathbf{U}(1)$ are called *Dirichlet characters*. Every such character can be extended to a function $\chi\colon \mathbb{Z} \to \mathbb{C}$ given by

$$\chi(n) = \begin{cases} \chi(n \bmod m) & \text{if } \gcd(m, n) = 1 \\ 0 & \text{otherwise.} \end{cases}$$

It is immediately verified that such functions are multiplicative, which means that

$$\chi(rs) = \chi(r)\chi(s) \qquad \text{for all } r, s \in \mathbb{Z}.$$

They are also periodic with period m ($\chi(n+m) = \chi(n)$ for all $n \in \mathbb{Z}$). These functions are called *Dirichlet characters modulo m*. The trivial Dirichlet character is the Dirichlet character such that $\chi_0(m) = 1$ iff m and n are relatively prime, and 0 otherwise.

Definition 10.6. For every $\ell \in \mathbb{N} - \{0\}$, define the function $\delta_\ell\colon (\mathbb{Z}/m\mathbb{Z})^* \to \mathbb{C}$ given by

$$\delta_\ell(n) = \begin{cases} 1 & \text{if } n \equiv \ell \pmod{m} \\ 0 & \text{otherwise.} \end{cases}$$

Example 10.4. Let $G = (\mathbb{Z}/10\mathbb{Z})^* = \{1, 3, 7, 9\}$. The multiplication table for $(\mathbb{Z}/10\mathbb{Z})^*$ is shown below.

$\cdot$	1	3	7	9
1	1	3	7	9
3	3	9	1	7
7	7	1	9	3
9	9	7	3	1

Let $\ell = 13$. Since $13 \equiv 3 \pmod{10}$, we find that

$$\delta_{13}(1) = 0, \qquad \delta_{13}(3) = 1, \qquad \delta_{13}(7) = 0, \qquad \delta_{13}(9) = 0.$$

By the Fourier inversion formula (Proposition 10.20) we can write

$$\delta_\ell(n) = \sum_{\chi \in \widehat{G}} \mathcal{F}(\delta_\ell)(\chi)\chi(n),$$

with

$$\mathcal{F}(\delta_\ell)(\chi) = \frac{1}{\varphi(m)} \sum_{k \in G} \delta_\ell(k)\overline{\chi(k)} = \frac{1}{\varphi(m)} \overline{\chi(\ell)}.$$

Therefore,

$$\delta_\ell(n) = \frac{1}{\varphi(m)} \sum_{\chi \in \widehat{G}} \chi(n)\overline{\chi(\ell)}.$$

Like the characters, the functions δ_ℓ can be extended to $\mathbb{Z}$, by setting $\delta_\ell(n) = 0$ if m and n are not relatively prime. The above shows that

$$\delta_\ell(n) = \frac{1}{\varphi(m)} \sum_{\chi} \chi(n)\overline{\chi(\ell)}, \quad n \in \mathbb{Z},$$

where the sum is over the Dirichlet characters modulo m.

The above result is one of the steps in the proof of Dirichlet's theorem on arithmetic progressions of integers $mk + \ell$ with $\gcd(\ell, m) = 1$ and $k \in \mathbb{N}$, which says that such a sequence contains infinitely many primes.

Dirichlet's theorem is a consequence of the fact that the sum

$$\sum_{\substack{p \equiv \ell \pmod{m} \\ p \text{ prime}}} \frac{1}{p}$$

is infinite. This follows from the fact the the limit of the sum

$$\sum_{\substack{p \equiv \ell \pmod{m} \\ p \text{ prime}}} \frac{1}{p^s}$$

tends to infinity when $s > 1$ tends to 1. To simplify notation, write

$$\sum_{p \equiv \ell} \frac{1}{p^s},$$

where it is understood that p is prime and congruent to ℓ modulo m.

We can write

$$\sum_{p \equiv \ell} \frac{1}{p^s} = \sum_{p \text{ prime}} \frac{\delta_\ell(p)}{p^s} = \frac{1}{\varphi(m)} \sum_{\chi} \overline{\chi(\ell)} \sum_{p \text{ prime}} \frac{\chi(p)}{p^s}.$$

We can divide the above sum into two parts depending on whether or not χ is trivial, and we get

$$\begin{aligned} \sum_{p \equiv \ell} \frac{1}{p^s} &= \frac{1}{\varphi(m)} \sum_{p} \frac{\chi_0(p)}{p^s} + \frac{1}{\varphi(m)} \sum_{\chi \neq \chi_0} \overline{\chi(\ell)} \sum_{p} \frac{\chi(p)}{p^s} \\ &= \frac{1}{\varphi(m)} \sum_{p \nmid m} \frac{1}{p^s} + \frac{1}{\varphi(m)} \sum_{\chi \neq \chi_0} \overline{\chi(\ell)} \sum_{p} \frac{\chi(p)}{p^s}, \end{aligned}$$

where the sums on the right-hand side are over all primes p. Since there are only finitely many primes dividing m, the sum

$$\sum_{p \nmid m} \frac{1}{p^s}$$

tends to infinity when $s > 1$ tends to 1, because the series

$$\sum_{p \text{ prime}} \frac{1}{p}$$

diverges. This is a classical result of number theory going back to Euler (1737); see Stein and Shakarchi [Stein and Shakarchi (2003)] (Chapter 8), or Apostol [Apostol (1976)] (Chapter 1, Section 1.6).

If we could prove that the sum

$$\sum_{p \text{ prime}} \frac{\chi(p)}{p^s}$$

remains bounded when $s > 1$ tends to 1, we would be done, because then

$$\sum_{p \equiv \ell} \frac{1}{p^s}$$

tends to infinity as $s > 1$ tends to 1.

The above suggests studying the behavior of the functions $L(s, \chi)$ given by

$$L(s, \chi) = \sum_{n=1}^{\infty} \frac{\chi(n)}{n^s}, \qquad s > 1,$$

where χ is a Dirichlet character, called *L-functions*. The series $L(s, \chi)$ is absolutely convergent for $s > 1$. Actually, if $\chi \neq \chi_0$, the series $L(s, \chi)$

converges for $s > 0$ and is continuously differentiable for $s > 0$; see Stein and Shakarchi [Stein and Shakarchi (2003)] (Chapter 8). There is a also a remarkable product formula: For $s > 1$, we have

$$L(s, \chi) = \sum_{n=1}^{\infty} \frac{\chi(n)}{n^s} = \prod_{p \text{ prime}} \frac{1}{1 - \chi(p)p^{-s}}.$$

The above is a generalization of Euler's formula for expressing the zeta function

$$\zeta(s) = \sum_{n=1}^{\infty} \frac{1}{n^s}$$

as the infinite product

$$\prod_{p \text{ prime}} \frac{1}{1 - p^{-s}}.$$

In a tour de force Dirichlet proved that $L(1, \chi)$ is finite and that $L(1, \chi) \neq 0$ if χ is not the trivial character. By taking the logarithm of both sides of the product formula and using some properties of the log function one obtains

$$\log L(s, \chi) = \sum_{p \text{ prime}} \frac{\chi(p)}{p^s} + O(1),$$

so if $L(1, \chi)$ is nonzero, then the sum

$$\sum_{p \text{ prime}} \frac{\chi(p)}{p^s}$$

remains bounded when $s > 1$ tends to 1, and the proof of Dirichlet's theorem is completed.

The above sketch lacks rigor on several fronts, and a rigorous proof involves a lot of rather difficult technical details, the hardest step being the proof that $L(1, \chi) \neq 0$ if χ is not the trivial character. We refer the interested reader to Stein and Shakarchi [Stein and Shakarchi (2003)] (Chapter 8), Apostol [Apostol (1976)] (Chapters 6 and 7), or Serre [Serre (1973)] (Chapter VI).

10.6 Fourier Transform and Cotransform in Terms of Matrices

In this section we formulate the Fourier transform (and cotransform) on a finite abelian group in terms of matrices. Write $n = |G|$, and denote by

$[G \to \mathbb{C}]$ the vector space of all functions $x\colon G \to \mathbb{C}$. Since G is finite, we have $\mathrm{L}^1(G) = \mathrm{L}^2(G) = [G \to \mathbb{C}]$. Of course, the space $[G \to \mathbb{C}]$ is isomorphic to $\mathbb{C}^n$, but it is better to stick with $[G \to \mathbb{C}]$. Our goal is to extend the Fourier transform $\mathcal{F}$ on $\mathrm{L}^2(G) = [G \to \mathbb{C}]$ (defined on $\widehat{G}$) to a sesquilinear form on $[G \to \mathbb{C}]^*$, and to extend the Fourier transform $\overline{\mathcal{F}}$ on $\mathrm{L}^2(\widehat{G}) = [\widehat{G} \to \mathbb{C}]$ (defined on $\widehat{\widehat{G}}$) to a bilinear form on $[G \to \mathbb{C}]^{**}$. We will also prove the Fourier inversion theorem for these extensions.

For any $a \in G$, we denote by $e_a \in [G \to \mathbb{C}]$ the map given by

$$e_a(b) = \begin{cases} 1 & \text{if } b = a \\ 0 & \text{if } b \neq a. \end{cases}$$

If we order G as $(a_1, \ldots, a_n)$, then $(e_{a_1}, \ldots, e_{a_n})$ is a basis of $[G \to \mathbb{C}]$. Viewed as an element of $\mathbb{C}^n$, the vector e_{a_i} corresponds to the canonical ith basis vector e_i. For any function $x \in [G \to \mathbb{C}]$, we can write

$$x = \sum_{i=1}^{n} x(a_i) e_{a_i}.$$

For simplicity of notation, we write x_i (or x_{a_i}) instead of $x(a_i)$. To show that Fourier inversion holds, we need to view a function $x \in [G \to \mathbb{C}]$ as a linear form in $[G \to \mathbb{C}]^*$.

Definition 10.7. Let G be a finite abelian group and write $G = \{a_1, \ldots, a_n\}$. Every vector $x \in [G \to \mathbb{C}]$ determines uniquely the linear form $\widetilde{x} \in [G \to \mathbb{C}]^*$ defined on the basis $(e_{a_1}, \ldots, e_{a_n})$ of $[G \to \mathbb{C}]$ by

$$\widetilde{x}(e_{a_j}) = x(a_j).$$

In terms of the dual basis $(e^*_{a_1}, \ldots, e^*_{a_n})$ of the basis $(e_{a_1}, \ldots, e_{a_n})$, we have

$$\widetilde{x} = \sum_{j=1}^{n} x(a_j) e^*_{a_j}.$$

Since G is finite, we know that $\widehat{G}$ is isomorphic to G, and we order $\widehat{G}$ as $(\chi_1, \ldots, \chi_n)$. Every character $\chi_i \in \widehat{G}$ is a function $\chi_i\colon G \to \mathbb{C}$, so $\chi_i \in [G \to \mathbb{C}]$, and the orthogonality conditions of Proposition 10.13 imply that $(\chi_1, \ldots, \chi_n)$ are linearly independent.

As above, every character χ_i determines uniquely the linear form $\widetilde{\chi_i} \in [G \to \mathbb{C}]^*$ given by

$$\widetilde{\chi_i} = \sum_{j=1}^{n} \chi_i(a_j) e^*_{a_j}.$$

If there is a linear dependence

$$\sum_{j=1}^{n} \alpha_j \widetilde{\chi_j} = 0,$$

by applying the above linear form to the n vectors e_{a_k}, we obtain the n equations

$$\sum_{j=1}^{n} \alpha_j \widetilde{\chi_j}(e_{a_k}) = \sum_{j=1}^{n} \alpha_j \chi_j(a_k) = 0, \qquad k = 1, \ldots, n,$$

which are equivalent to

$$\sum_{j=1}^{n} \alpha_j (\chi_j(a_1), \ldots, \chi_j(a_n)) = 0,$$

that is

$$\sum_{j=1}^{n} \alpha_j \chi_j = 0,$$

and since $(\chi_1, \ldots, \chi_n)$ are linearly independent, we must have $\alpha_1 = \cdots = \alpha_n = 0$, and so the linear forms $(\widetilde{\chi_1}, \ldots, \widetilde{\chi_n})$ are also linearly independent. Thus they form a basis of $[G \to \mathbb{C}]^*$.

Our first goal is to extend the Fourier transform $\mathcal{F}$ on $\mathrm{L}^2(G) = [G \to \mathbb{C}]$ to a sesquilinear form on $[G \to \mathbb{C}]^*$. Given any $x \in [G \to \mathbb{C}]$, its Fourier transform is the map $\mathcal{F}(x) \colon \widehat{G} \to \mathbb{C}$ given by

$$\mathcal{F}(x)(\chi_i) = \frac{1}{n} \sum_{j=1}^{n} x(a_j) \overline{\chi_i(a_j)} = \frac{1}{n} \sum_{j=1}^{n} \widetilde{x}(e_{a_j}) \overline{\widetilde{\chi_i}(e_{a_j})}. \tag{$*$}$$

Definition 10.8. Let G be a finite abelian group and write $G = \{a_1, \ldots, a_n\}$. For all $f, \gamma \in [G \to \mathbb{C}]^*$ we define the *Fourier transform* $\mathcal{F}$ as the sequilinear form on $[G \to \mathbb{C}]^*$ given by

$$\mathcal{F}(f)(\gamma) = \frac{1}{n} \sum_{j=1}^{n} f(e_{a_j}) \overline{\gamma(e_{a_j})}.$$

If $f = \widetilde{x}$ and $\gamma = \widetilde{\chi_i}$, the right-hand side of the above definition is equal to the right-hand side of $(*)$, so this definition extends the definition of the Fourier transform on $[G \to \mathbb{C}]$ and $\widehat{G}$.

If we write $f = \sum_{i=1}^{n} x_i e_{a_i}^*$, then we have

$$\mathcal{F}(f)(\gamma) = \sum_{i=1}^{n} x_i \mathcal{F}(e_{a_i}^*)(\gamma) = \frac{1}{n} \sum_{i=1}^{n} \sum_{j=1}^{n} x_i e_{a_i}^*(e_{a_j}) \overline{\gamma(e_{a_j})} = \frac{1}{n} \sum_{j=1}^{n} x_j \overline{\gamma(e_{a_j})}.$$

Then if $\gamma = \sum_{i=1} w_i \widetilde{\chi_i}$, we get

$$\begin{aligned}\mathcal{F}(f)\left(\sum_{i=1}^{n} w_i \widetilde{\chi_i}\right) &= \sum_{i=1}^{n} \overline{w_i} \mathcal{F}(f)(\widetilde{\chi_i}) \\ &= \frac{1}{n} \sum_{i=1}^{n} \overline{w_i} \sum_{j=1}^{n} x_j \overline{\widetilde{\chi_i}(e_{a_j})} \\ &= \frac{1}{n} \sum_{i=1}^{n} \overline{w_i} \sum_{j=1}^{n} x_j \overline{\chi_i(a_j)} \\ &= \frac{1}{n} \sum_{i=1}^{n} \left(\sum_{j=1}^{n} \overline{\chi_i(a_j)} x_j\right) \overline{w_i}.\end{aligned}$$

Definition 10.9. Let G be a finite abelian group of order n, and write $G = \{a_1, \ldots, a_n\}$. Define the $n \times n$ matrix $F = (F_{ij})$, called the *Fourier matrix* of G with respect to the characters $\{\chi_1, \ldots, \chi_n\}$ of G, by

$$F_{ij} = \overline{\chi_i(a_j)}.$$

Example 10.5. As a concrete example of the preceding calculations set $n = 3$. Then

$$f = x_1 e^*_{a_1} + x_2 e^*_{a_2} + x_3 e^*_{a_3}, \qquad \gamma = w_1 \widetilde{\chi_1} + w_2 \widetilde{\chi_2} + w_3 \widetilde{\chi_3},$$

and

$$\widetilde{\chi_j} = \sum_{i=1}^{3} \chi_j(a_i) e^*_{a_i}, \qquad 1 \leq j \leq 3.$$

Since $f(e_{a_j}) = x_j$ whenever $1 \leq j \leq 3$, Definition 10.8 implies that

$$\mathcal{F}(f)(\gamma) = \frac{1}{3} \sum_{i=1}^{3} x_i \mathcal{F}(e^*_{a_i})(\gamma) = \frac{1}{3}\left[x_1 \overline{\gamma(e_{a_1})} + x_2 \overline{\gamma(e_{a_2})} + x_3 \overline{\gamma(e_{a_3})}\right].$$

Next we need to evaluate $\overline{\gamma(e_{a_j})}$ for $1 \leq j \leq 3$. We will demonstrate the evaluation of $\overline{\gamma(e_{a_1})}$ and leave the other two cases to the reader. By using the definition of γ and the fact each $\widetilde{\chi_i}$ can be expanded in terms of the dual basis $(e^*_{a_1}, e^*_{a_2}, e^*_{a_3})$, we find that

$$\begin{aligned}\gamma(e_{a_1}) &= w_1 \widetilde{\chi_1}(e_{a_1}) + w_2 \widetilde{\chi_2}(e_{a_1}) + w_3 \widetilde{\chi_3}(e_{a_1}) \\ &= w_1\left[\chi_1(a_1) e^*_{a_1}(e_{a_1}) + \chi_1(a_2) e^*_{a_2}(e_{a_1}) + \chi_1(a_3) e^*_{3_1}(e_{a_1})\right] \\ &\quad + w_2\left[\chi_2(a_1) e^*_{a_1}(e_{a_1}) + \chi_2(a_2) e^*_{a_2}(e_{a_1}) + \chi_2(a_3) e^*_{3_1}(e_{a_1})\right] \\ &\quad + w_3\left[\chi_3(a_1) e^*_{a_1}(e_{a_1}) + \chi_3(a_2) e^*_{a_2}(e_{a_1}) + \chi_3(a_3) e^*_{3_1}(e_{a_1})\right] \\ &= w_1 \chi_1(a_1) + w_2 \chi_2(a_1) + w_3 \chi_3(a_1).\end{aligned}$$

Hence

$$\overline{\gamma(e_{a_1})} = \overline{w_1}\overline{\chi_1(a_1)} + \overline{w_2}\overline{\chi_2(a_1)} + \overline{w_3}\overline{\chi_3(a_1)}.$$

Similar calculations show that

$$\overline{\gamma(e_{a_2})} = \overline{w_1}\overline{\chi_1(a_2)} + \overline{w_2}\overline{\chi_2(a_2)} + \overline{w_3}\overline{\chi_3(a_2)},$$
$$\overline{\gamma(e_{a_3})} = \overline{w_1}\overline{\chi_1(a_3)} + \overline{w_2}\overline{\chi_2(a_3)} + \overline{w_3}\overline{\chi_3(a_3)}.$$

Our calculations, when written in matrix form, become

$$\mathcal{F}(f)(\gamma) = \frac{1}{3}\begin{pmatrix} x_1 & x_2 & x_3 \end{pmatrix}\begin{pmatrix} \overline{\gamma(e_{a_1})} \\ \overline{\gamma(e_{a_2})} \\ \overline{\gamma(e_{a_3})} \end{pmatrix},$$

where

$$\begin{pmatrix} \overline{\gamma(e_{a_1})} \\ \overline{\gamma(e_{a_2})} \\ \overline{\gamma(e_{a_3})} \end{pmatrix} = \begin{pmatrix} \overline{\chi_1(a_1)} & \overline{\chi_2(a_1)} & \overline{\chi_3(a_1)} \\ \overline{\chi_1(a_2)} & \overline{\chi_2(a_2)} & \overline{\chi_2(a_3)} \\ \overline{\chi_1(a_3)} & \overline{\chi_2(a_3)} & \overline{\chi_3(a_3)} \end{pmatrix}\begin{pmatrix} \overline{w_1} \\ \overline{w_2} \\ \overline{w_3} \end{pmatrix} = F^\top \begin{pmatrix} \overline{w_1} \\ \overline{w_2} \\ \overline{w_3} \end{pmatrix},$$

with F being the corresponding Fourier matrix of Definition 10.9, namely

$$F = \begin{pmatrix} \overline{\chi_1(a_1)} & \overline{\chi_1(a_2)} & \overline{\chi_1(a_3)} \\ \overline{\chi_2(a_1)} & \overline{\chi_2(a_2)} & \overline{\chi_2(a_3)} \\ \overline{\chi_3(a_1)} & \overline{\chi_3(a_2)} & \overline{\chi_3(a_3)} \end{pmatrix}.$$

Hence

$$\mathcal{F}(f)(\gamma) = \frac{1}{3}\begin{pmatrix} x_1 & x_2 & x_3 \end{pmatrix} F^\top \begin{pmatrix} \overline{w_1} \\ \overline{w_2} \\ \overline{w_3} \end{pmatrix}.$$

In summary we obtained the following result.

Proposition 10.24. *If $f \in [G \to \mathbb{C}]^*$ is expressed over the basis $(e^*_{a_1}, \ldots, e^*_{a_n})$ as $f = \sum_{j=1}^n x_j e^*_{a_j}$, and if $\gamma \in [G \to \mathbb{C}]^*$ is expressed over the basis $(\widetilde{\chi_1}, \ldots, \widetilde{\chi_n})$ as $\gamma = \sum_{i=1}^n w_i \widetilde{\chi_i}$, then the value $\mathcal{F}(f)(\gamma)$ of the sequilinear form $\mathcal{F}$ is given by*

$$\mathcal{F}(f)(\gamma) = \frac{1}{n} w^* F x = \frac{1}{n} x^\top F^\top \overline{w}.$$

As a semilinear map from $[G \to \mathbb{C}]^*$ to $\mathbb{C}$, the matrix of $\mathcal{F}(f)$ over the basis $(\widetilde{\chi_1}, \ldots, \widetilde{\chi_n})$ is the row vector $\frac{1}{n} x^\top F^\top$. As a semilinear form on $[G \to \mathbb{C}]^*$, the semilinear form $\mathcal{F}(f)$ is represented by the column vector $\frac{1}{n} F x$ over the dual basis $(\widetilde{\chi_1}^*, \ldots, \widetilde{\chi_n}^*)$.

Our next goal is to extend the Fourier cotransform $\overline{\mathcal{F}}$ on the space $\mathrm{L}^2(\widehat{G}) = [\widehat{G} \to \mathbb{C}]$ to a bilinear form on $[G \to \mathbb{C}]^{**}$. Given any function $\xi \in [\widehat{G} \to \mathbb{C}]$, the Fourier cotransform $\overline{\mathcal{F}}(\xi)$ of ξ is the map $\overline{\mathcal{F}}(\xi) \colon \widehat{\widehat{G}} \to \mathbb{C}$ given by

$$\overline{\mathcal{F}}(\xi)(\zeta) = \sum_{j=1}^{n} \xi(\chi_j)\zeta(\chi_j), \qquad \zeta \in \widehat{\widehat{G}},$$

with $\widehat{G} = \{\chi_1, \ldots, \chi_n\}$.

Since every character in $\widehat{G}$ is a function in $[G \to \mathbb{C}]$, every function $\xi \in [\widehat{G} \to \mathbb{C}]$ can be viewed as a function in $[[G \to \mathbb{C}] \to \mathbb{C}]$. Similarly, every character $\zeta \in \widehat{\widehat{G}}$ is a function in $[[G \to \mathbb{C}] \to \mathbb{C}]$, and we know that $\widehat{\widehat{G}}$ is isomorphic to G.

Recall that every function $x \in [G \to \mathbb{C}]$ defines uniquely a linear form $\widetilde{x} \in [G \to \mathbb{C}]^*$. Similarly, every function $\xi \in [[G \to \mathbb{C}] \to \mathbb{C}]$ can be extended uniquely to the linear form $\widetilde{\xi} \in [G \to \mathbb{C}]^{**}$ defined on the basis $(\widetilde{\chi_1}, \ldots, \widetilde{\chi_n})$ of $[G \to \mathbb{C}]^*$ by

$$\widetilde{\xi}(\widetilde{\chi_i}) = \xi(\chi_i).$$

In terms of the dual basis $(\widetilde{\chi_1}^*, \ldots, \widetilde{\chi_n}^*)$ of the basis $(\widetilde{\chi_1}, \ldots, \widetilde{\chi_n})$, we have

$$\widetilde{\xi} = \sum_{i=1}^{n} \xi(\chi_i)\widetilde{\chi_i}^*.$$

The vector space $[\widehat{G} \to \mathbb{C}]$ is isomorphic to $[G \to \mathbb{C}]^{**}$. Using the basis $(\widetilde{\chi_1}^*, \ldots, \widetilde{\chi_n}^*)$ in $[G \to \mathbb{C}]^{**}$, any $\omega \in [G \to \mathbb{C}]^{**}$ can be expressed as $\omega = \sum_{i=1}^{n} y_i\widetilde{\chi_i}^*$.

Since $\overline{\mathcal{F}}(\xi)(\zeta)$ is defined as

$$\overline{\mathcal{F}}(\xi)(\zeta) = \sum_{j=1}^{n} \xi(\chi_j)\zeta(\chi_j),$$

we have

$$\overline{\mathcal{F}}(\xi)(\zeta) = \sum_{j=1}^{n} \xi(\chi_j)\zeta(\chi_j) = \sum_{j=1}^{n} \widetilde{\xi}(\widetilde{\chi_j})\widetilde{\zeta}(\widetilde{\chi_j}).$$

Definition 10.10. Let G be a finite abelian group, write $G = \{a_1, \ldots, a_n\}$, and let $\{\chi_1, \ldots, \chi_n\}$ be the characters of G. For all $\omega, \gamma \in [G \to \mathbb{C}]^{**}$, we define the *Fourier cotransform* $\overline{\mathcal{F}}$ as the bilinear form on $[G \to \mathbb{C}]^{**}$ given by

$$\overline{\mathcal{F}}(\omega)(\gamma) = \sum_{j=1}^{n} \omega(\widetilde{\chi_j})\gamma(\widetilde{\chi_j}).$$

If $\omega = \widetilde{\xi}$ and $\gamma = \widetilde{\zeta}$, then the right-hand side of the above equation is equal to $\overline{\mathcal{F}}(\xi)(\zeta)$, so this definition extends the Fourier cotransform on $[\widehat{G} \to \mathbb{C}]$ and $\widehat{\widehat{G}}$.

If we write

$$\omega = \sum_{i=1}^{n} y_i \widetilde{\chi_i}^*,$$

then we have

$$\overline{\mathcal{F}}(\omega)(\gamma) = \sum_{i=1}^{n} y_i \overline{\mathcal{F}}(\widetilde{\chi_i}^*)(\gamma) = \sum_{i=1}^{n}\sum_{j=1}^{n} y_i \widetilde{\chi_i}^*(\widetilde{\chi_j})\gamma(\widetilde{\chi_j}) = \sum_{j=1}^{n} y_j \gamma(\widetilde{\chi_j}).$$

We also have a natural isomorphism η from $[G \to \mathbb{C}]$ to $[G \to \mathbb{C}]^{**}$, defined such that the linear form $\eta_u \in [G \to \mathbb{C}]^{**}$ is given by

$$\eta_u(\chi) = \chi(u) \qquad \chi \in [G \to \mathbb{C}]^*,\ u \in [G \to \mathbb{C}].$$

Observe that

$$e_{a_i}^{**} = \eta_{e_{a_i}},$$

because on the basis $(e_{a_1}^*, \ldots, e_{a_n}^*)$ of $[G \to \mathbb{C}]^*$, we have

$$e_{a_i}^{**}(e_{a_j}^*) = \delta_{ij} = e_{a_j}^*(e_{a_i}) = \eta_{e_{a_i}}(e_{a_j}^*).$$

Then $(\eta_{e_{a_1}}, \ldots, \eta_{e_{a_n}}) = (e_{a_1}^{**}, \ldots, e_{a_n}^{**})$ is a basis of $[G \to \mathbb{C}]^{**}$, and if $\omega = \sum_{i=1}^{n} y_i \widetilde{\chi_i}^*$, then $\overline{\mathcal{F}}(\omega)$ is given by

$$\begin{aligned}
\overline{\mathcal{F}}(\omega)\left(\sum_{i=1}^{n} z_i \eta_{e_{a_i}}\right) &= \sum_{i=1}^{n} z_i \overline{\mathcal{F}}(\omega)(\eta_{e_{a_i}}) \\
&= \sum_{i=1}^{n} z_i \sum_{j=1}^{n} y_j \eta_{e_{a_i}}(\widetilde{\chi_j}) \\
&= \sum_{i=1}^{n} z_i \sum_{j=1}^{n} y_j \widetilde{\chi_j}(e_{a_i}) \\
&= \sum_{i=1}^{n} \left(\sum_{j=1}^{n} y_j \chi_j(a_i)\right) z_i \\
&= y^\top \overline{F} z = z^\top F^* y.
\end{aligned}$$

Example 10.6. Once again we provide a concrete demonstration of the preceding calculations. Set $n = 3$ and write Definition 10.10 as

$$\overline{\mathcal{F}}(\omega)(\gamma) = \omega(\widetilde{\chi_1})\gamma(\widetilde{\chi_1}) + \omega(\widetilde{\chi_2})\gamma(\widetilde{\chi_2}) + \omega(\widetilde{\chi_3})\gamma(\widetilde{\chi_3}).$$

Since $\omega = y_1\widetilde{\chi_1}^* + y_2\widetilde{\chi_2}^* + y_3\widetilde{\chi_3}^*$, a familiar calculation shows that preceding line becomes

$$\overline{\mathcal{F}}(\omega)(\gamma) = y_1\gamma(\widetilde{\chi_1}) + y_2\gamma(\widetilde{\chi_2}) + y_3\gamma(\widetilde{\chi_3}).$$

Now it is a matter of calculating $\gamma(\widetilde{\chi_i})$ for $1 \leq i \leq 3$. We will demonstrate in detail the calculation of $\gamma(\widetilde{\chi_1})$ and leave the other two cases to the reader. Since $\gamma = z_1 e^{**}_{a_1} + z_2 e^{**}_{a_2} + z_3 e^{**}_{a_3}$, we see that

$$\gamma(\widetilde{\chi_1}) = z_1 e^{**}_{a_1}(\widetilde{\chi_1}) + z_2 e^{**}_{a_2}(\widetilde{\chi_1}) + z_3 e^{**}_{a_3}(\widetilde{\chi_1}).$$

However, recall that

$$\widetilde{\chi_i} = \chi_i(a_1)e^*_{a_1} + \chi_i(a_2)e^*_{a_2} + \chi_i(a_3)e^*_{a_3}, \qquad 1 \leq i \leq 3.$$

Thus

$$\begin{aligned}\gamma(\widetilde{\chi_1}) &= z_1\left[\chi_1(a_1)e^{**}_{a_1}(e^*_{a_1}) + \chi_1(a_2)e^{**}_{a_1}(e^*_{a_2}) + \chi_1(a_3)e^{**}_{a_1}(e^*_{a_3})\right]\\ &\quad + z_2\left[\chi_1(a_1)e^{**}_{a_2}(e^*_{a_1}) + \chi_1(a_2)e^{**}_{a_2}(e^*_{a_2}) + \chi_1(a_3)e^{**}_{a_2}(e^*_{a_3})\right]\\ &\quad + z_3\left[\chi_1(a_1)e^{**}_{a_3}(e^*_{a_1}) + \chi_1(a_2)e^{**}_{a_3}(e^*_{a_2}) + \chi_1(a_3)e^{**}_{a_3}(e^*_{a_3})\right]\\ &= z_1\chi_1(a_1) + z_2\chi_1(a_2) + z_3\chi_1(a_3).\end{aligned}$$

Similar calculations show that

$$\begin{aligned}\gamma(\widetilde{\chi_2}) &= z_1\chi_2(a_1) + z_2\chi_2(a_2) + z_3\chi_2(a_3),\\ \gamma(\widetilde{\chi_3}) &= z_1\chi_3(a_1) + z_2\chi_3(a_2) + z_3\chi_3(a_3).\end{aligned}$$

It remains to write these calculations in matrix form. Observe that

$$\overline{\mathcal{F}}(\omega)(\gamma) = \begin{pmatrix} y_1 & y_2 & y_3\end{pmatrix}\begin{pmatrix}\gamma(\widetilde{\chi_1})\\ \gamma(\widetilde{\chi_2})\\ \gamma(\widetilde{\chi_3})\end{pmatrix},$$

with

$$\begin{pmatrix}\gamma(\widetilde{\chi_1})\\ \gamma(\widetilde{\chi_2})\\ \gamma(\widetilde{\chi_3})\end{pmatrix} = \begin{pmatrix}\chi_1(a_1) & \chi_1(a_2) & \chi_1(a_3)\\ \chi_2(a_1) & \chi_2(a_2) & \chi_2(a_3)\\ \chi_3(a_1) & \chi_3(a_2) & \chi_3(a_3)\end{pmatrix}\begin{pmatrix}z_1\\ z_2\\ z_3\end{pmatrix} = \overline{F}\begin{pmatrix}z_1\\ z_2\\ z_3\end{pmatrix},$$

where F is the Fourier matrix of Example 10.5. Thus we ultimately obtain

$$\overline{\mathcal{F}}(\omega)(\gamma) = \begin{pmatrix} y_1 & y_2 & y_3\end{pmatrix}\overline{F}\begin{pmatrix}z_1\\ z_2\\ z_3\end{pmatrix}.$$

In summary, we proved the following result.

Proposition 10.25. *If $\omega \in [G \to \mathbb{C}]^{**}$ is expressed over the basis $(\widetilde{\chi_1}^*, \ldots, \widetilde{\chi_n}^*)$ as $\omega = \sum_{i=1}^n y_i \widetilde{\chi_i}^*$, and if $\zeta \in [G \to \mathbb{C}]^{**}$ is expressed over the basis $(e_{a_1}^{**}, \ldots, e_{a_n}^{**})$ as $\zeta = \sum_{i=1}^n z_i e_{a_i}^{**}$, then the value $\overline{\mathcal{F}}(\omega)(\zeta)$ of the bilinear form $\overline{\mathcal{F}}$ is given by*

$$\overline{\mathcal{F}}(\omega)(\zeta) = y^\top \overline{F} z = z^\top F^* y.$$

As a linear map from $[G \to \mathbb{C}]^{**} \cong [G \to \mathbb{C}]$ to $\mathbb{C}$, the matrix of $\overline{\mathcal{F}}(\omega)$ over the basis $(e_{a_1}^{**}, \ldots, e_{a_n}^{**})$ is the row vector $y^\top \overline{F}$. As an element of $[G \to \mathbb{C}]^{***} \cong [G \to \mathbb{C}]^*$, the linear form $\overline{\mathcal{F}}(\omega)$ is represented by the column vector $F^* y$ over the dual basis $(e_{a_1}^{***}, \ldots, e_{a_n}^{***})$.

The orthogonality conditions of Proposition 10.13 imply that

$$\frac{1}{n} F F^* = \frac{1}{n} F^* F = I,$$

so F^* is the inverse of $\frac{1}{n} F$.

Now if $f \in [G \to \mathbb{C}]^*$ with $f = \sum_{j=1}^n x_j e_{a_j}^*$, then $\mathcal{F}(f) \in [G \to \mathbb{C}]^{**}$, and since $\mathcal{F}(f)$ is expressed over the basis $(\widetilde{\chi_1}^*, \ldots, \widetilde{\chi_n}^*)$ and $\overline{\mathcal{F}}$ is also defined over this basis, with $\omega = \mathcal{F}(f)$, we know that ω is represented by the column vector $y = Fx$ over the basis $(\widetilde{\chi_1}^*, \ldots, \widetilde{\chi_n}^*)$, and if $\zeta = \sum_{i=1}^n z_i e_{a_i}^{**} = \sum_{i=1}^n z_i \eta_{e_{a_i}}$, then $\overline{\mathcal{F}}(\omega)(\zeta)$ is given by

$$\overline{\mathcal{F}}(\omega)(\zeta) = z^\top F^* y,$$

so we get

$$(\overline{\mathcal{F}}(\mathcal{F}(f)))(\zeta) = z^\top \frac{1}{n} F^* F x = z^\top x = f\left(\sum_{i=1}^n z_i e_{a_i}\right).$$

But

$$\zeta = \sum_{i=1}^n z_i \eta_{e_{a_i}} = \eta_{\sum_{i=1}^n z_i e_{a_i}},$$

so the above equation shows that

$$(\overline{\mathcal{F}} \circ \mathcal{F})(f) \circ \eta_{\sum_{i=1}^n z_i e_{a_i}} = f\left(\sum_{i=1}^n z_i e_{a_i}\right),$$

that is,

$$f = (\overline{\mathcal{F}} \circ \mathcal{F})(f) \circ \eta.$$

In summary, we proved the following theorem.

Theorem 10.26. *Let G be a finite abelian group of order n. The Fourier transform $\mathcal{F}$ defined on $[G \to \mathbb{C}]^*$ in Definition 10.8 and the Fourier cotransform $\overline{\mathcal{F}}$ defined on $[G \to \mathbb{C}]^{**}$ in Definition 10.10 satisfy the Fourier inversion equation*

$$f = (\overline{\mathcal{F}} \circ \mathcal{F})(f) \circ \eta,$$

for all $f \in [G \to \mathbb{C}]^*$.

A nice feature of Definition 10.8 and Definition 10.10 is that they are intrinsic, that is, independent the choice of bases. A slight disadvantage is that the cotransform is defined on the double dual $[G \to \mathbb{C}]^{**}$ of $[G \to \mathbb{C}]$. But $[G \to \mathbb{C}]$ and $[G \to \mathbb{C}]^{**}$ are isomorphic (in fact, canonically), so there is an alternative method to define directly a Fourier cotransform on $[G \to \mathbb{C}]$. The slight complication is that we would like Fourier inversion to hold, and for this it appears that we need to use a noncanonical isomorphism, namely the isomorphism $\theta\colon [G \to \mathbb{C}] \to [G \to \mathbb{C}]^{**}$ defined as follows: if $c = \sum_{i=1}^n y_i e_{a_i}$, then

$$\theta_c = \sum_{i=1}^n y_i \widetilde{\chi_i}^*.$$

Definition 10.11. For any vector $c = \sum_{i=1}^n y_i e_{a_i} \in [G \to \mathbb{C}]$, the *Fourier cotransform* $\overline{\mathcal{F}}_2(c)$ of c is the linear form on $[G \to \mathbb{C}]$ defined such that for all $a \in [G \to \mathbb{C}]$, we have

$$\overline{\mathcal{F}}_2(c)(a) = \sum_{j=1}^n y_j \widetilde{\chi_j}(a).$$

Observe that as long as we express the vector $c \in [G \to \mathbb{C}]$ over the basis $(e_{a_1}, \ldots, e_{a_n})$, the map $\overline{\mathcal{F}}_2$ is bilinear. We may think of the components of c as the Fourier coefficients of some form $f \in [G \to \mathbb{C}]^*$, and

$$\overline{\mathcal{F}}_2(c) = \sum_{j=1}^n y_j \widetilde{\chi_j}$$

is the "Fourier series" associated with c (a linear form on $[G \to \mathbb{C}]$).

We can then repeat our familiar computation to prove that if $a = \sum_{j=1}^m z_j e_{a_j}$, then

$$\overline{\mathcal{F}}_2(c)(a) = y^\top \overline{F} z = z^\top F^* y.$$

As a linear map from $[G \to \mathbb{C}]$ to $\mathbb{C}$, the matrix of $\overline{\mathcal{F}}_2(c)$ over the basis $(e_{a_1}, \ldots, e_{a_n})$ is the row vector $y^\top \overline{F}$, and as an element of $[G \to \mathbb{C}]^*$, the linear form $\overline{\mathcal{F}}_2(c)$ is represented by the column vector $F^* y$.

In order to be able to compose $\mathcal{F}$ and $\overline{\mathcal{F}}_2$, we need to convert $\mathcal{F}(f) \in [G \to \mathbb{C}]^{**}$, the result of applying $\mathcal{F}$ to $f \in [G \to \mathbb{C}]^*$, to a vector in $[G \to \mathbb{C}]$, and we can do this by applying the isomorphism $\theta^{-1}\colon [G \to \mathbb{C}]^{**} \to [G \to \mathbb{C}]$. Then Fourier inversion becomes the identity

$$f = (\overline{\mathcal{F}}_2 \circ \theta^{-1} \circ \mathcal{F})(f),$$

for all $f \in [G \to \mathbb{C}]^*$, which is another way of stating Proposition 10.20.

Since the vector spaces $[G \to \mathbb{C}]$, $[G \to \mathbb{C}]^*$, $[G \to \mathbb{C}]^{**}$, and $[G \to \mathbb{C}]^{***}$, are all isomorphic (and isomorphic to $\mathbb{C}^n$, where $n = |G|$), if we are just interested in transformations on sequences of complex numbers of length n indexed by the elements of the group G, namely elements of $[G \to \mathbb{C}]$, we can formulate versions of the Fourier transform and of the Fourier cotransform in terms of the Fourier matrix defined in Definition 10.9.

Definition 10.12. Let G be a finite abelian group, $G = \{a_1, \ldots, a_n\}$, and let $\{\chi_1, \ldots, \chi_n\}$ be the characters of G. If $F = (\overline{\chi_i(a_j)})$ is the Fourier matrix of G (as in Definition 10.9), then for every sequence $x \in [G \to \mathbb{C}]$, the sequence

$$\widehat{x} = \mathcal{F}(x) = \frac{1}{n} Fx$$

is called the *Fourier transform* of x, and given any sequence $\xi \in [G \to \mathbb{C}]$, the sequence

$$\overline{\mathcal{F}}(\xi) = F^* \xi$$

is called the *inverse Fourier transform* or *Fourier cotransform* of ξ.

Recall that

$$\frac{1}{n} FF^* = \frac{1}{n} F^*F = I,$$

so the two transforms are mutual inverses.

Recall that we proved earlier that

$$\mathcal{F}(x * y) = \mathcal{F}(x)\mathcal{F}(y),$$

where $x * y$ is the convolution of x and y, given by

$$(x * y)_a = \frac{1}{|G|} \sum_{b \in G} x_b y_{b^{-1}a} = \frac{1}{|G|} \sum_{\substack{b,c \in G \\ b+c=a}} x_b y_c.$$

In matrix terms, $\mathcal{F}(x)\mathcal{F}(y) = \widehat{x}\,\widehat{y}$ is the vector whose ath entry $(\widehat{x}\,\widehat{y})_a$ is the product of the ath entry $\widehat{x}_a$ of the vector $\widehat{x}$ by the ath entry $\widehat{y}_a$ of the vector $\widehat{y}$. In matrix terms, it can be expressed as

$$\mathrm{diag}(\widehat{x})\widehat{y},$$

where $\mathrm{diag}(\widehat{x})$ is the diagonal matrix whose diagonal entries are the entries in the vector x.

Other aspects of harmonic analysis on finite abelian groups can be found in Terras [Terras (1999)]. In the next section we consider the special case where $G = \mathbb{Z}/n\mathbb{Z}$.

10.7 The Discrete Fourier Transform (on $\mathbb{Z}/n\mathbb{Z}$)

If $G = \mathbb{Z}/n\mathbb{Z}$, then we know from Proposition 10.9(3) that the characters of $\mathbb{Z}/n\mathbb{Z}$ are the n homomorphisms χ_k given by

$$m \mapsto e^{2\pi imk/n}, \quad k = 0, 1, \ldots, n-1, \quad m \in \mathbb{Z}/n\mathbb{Z}.$$

Observe that the characters are indexed by $0, 1, \ldots, n-1$ rather than $1, 2, \ldots, n$, but this is actually more convenient in what follows. The complex numbers

$$\{1, e^{2\pi i/n}, e^{2\pi i2/n}, \ldots, e^{2\pi im/n}, \ldots, e^{2\pi i(n-1)/n}\}$$

(with $0 \leq m \leq n-1$) are the *nth roots of unity* (because obviously, $(e^{2\pi ik/n})^n = e^{2\pi ik} = 1$). They form a subgroup of $\mathbb{C}$ denoted $\mu_n(\mathbb{C})$ isomorphic to $\mathbb{Z}/n\mathbb{Z}$ under the isomorphism $k \mapsto e^{2\pi ik/n}$, for $k \in \mathbb{Z}/n\mathbb{Z}$.

Definition 10.13. The Fourier matrix $F_n = \left(\overline{\chi_k(m)}\right)_{\substack{0\leq k\leq n-1\\ 0\leq m\leq n-1}}$ is given by

$$F_n = \left(e^{-2\pi ikm/n}\right)_{\substack{0\leq k\leq n-1\\ 0\leq m\leq n-1}}.$$

The first row ($k = 0$) consists of 1's, the second row ($k = 1$) consists of the consecutive inverse powers ζ^{-m} of $\zeta = e^{2\pi i/n}$ (a primitive nth root of unity) for $m = 0, \ldots, n-1$,

$$\begin{pmatrix}1 & e^{-2\pi i/n} & e^{-2\pi i2/n} & \cdots & e^{-2\pi im/n} & \cdots & e^{-2\pi i(n-1)/n}\end{pmatrix},$$

and the $(k+1)$th row ($0 \leq k \leq n-1$) consists of the kth powers of the entries in the second row,

$$\begin{pmatrix}1 & (e^{-2\pi i/n})^k & (e^{-2\pi i2/n})^k & \cdots & (e^{-2\pi im/n})^k & \cdots & (e^{-2\pi i(n-1)/n})^k\end{pmatrix}.$$

Observe that F_n is symmetric. It is also a Vandermonde matrix for the roots of unity

$$\{1, e^{-2\pi i/n}, e^{-2\pi i2/n}, \ldots, e^{-2\pi im/n}, \ldots, e^{-2\pi i(n-1)/n}\},$$

namely, with $\omega = \zeta^{-1} = e^{-2\pi i/n}$ (also a primitive nth root of unity), we have

$$F_n = \begin{pmatrix} 1 & 1 & \cdots & 1 & 1 \\ 1 & \omega & \cdots & \omega^{n-2} & \omega^{n-1} \\ 1 & \omega^2 & \cdots & \omega^{2(n-2)} & \omega^{2(n-1)} \\ 1 & \omega^3 & \cdots & \omega^{3(n-2)} & \omega^{3(n-1)} \\ \vdots & \vdots & \ddots & \vdots & \vdots \\ 1 & \omega^{n-2} & \cdots & \omega^{(n-2)^2} & \omega^{(n-2)(n-1)} \\ 1 & \omega^{n-1} & \cdots & \omega^{(n-1)(n-2)} & \omega^{(n-1)^2} \end{pmatrix}.$$

For example, if $n = 5$, we have

$$F_5 = \begin{pmatrix} 1 & 1 & 1 & 1 & 1 \\ 1 & e^{-2\pi i/5} & e^{-2\pi i2/5} & e^{-2\pi i3/5} & e^{-2\pi i4/5} \\ 1 & e^{-2\pi i2/5} & e^{-2\pi i4/5} & e^{-2\pi i6/5} & e^{-2\pi i8/5} \\ 1 & e^{-2\pi i3/5} & e^{-2\pi i6/5} & e^{-2\pi i9/5} & e^{-2\pi i12/5} \\ 1 & e^{-2\pi i4/5} & e^{-2\pi i8/5} & e^{-2\pi i12/5} & e^{-2\pi i16/5} \end{pmatrix}$$

$$= \begin{pmatrix} 1 & 1 & 1 & 1 & 1 \\ 1 & e^{-2\pi i/5} & e^{-2\pi i2/5} & e^{-2\pi i3/5} & e^{-2\pi i4/5} \\ 1 & e^{-2\pi i2/5} & e^{-2\pi i4/5} & e^{-2\pi i/5} & e^{-2\pi i3/5} \\ 1 & e^{-2\pi i3/5} & e^{-2\pi i/5} & e^{-2\pi i4/5} & e^{-2\pi i2/5} \\ 1 & e^{-2\pi i4/5} & e^{-2\pi i3/5} & e^{-2\pi i2/5} & e^{-2\pi i/5} \end{pmatrix}.$$

Definition 10.14. Given a sequence $x = (x_0, \ldots, x_{n-1}) \in \mathbb{C}^n$, its Fourier transform, also called *discrete Fourier transform*, is

$$\widehat{x} = \frac{1}{n} F_n x.$$

We can think of $c = \widehat{x}$ as the sequence of Fourier coefficients of x.

Definition 10.15. Similarly, given a sequence $c = (c_0, \ldots, c_{n-1}) \in \mathbb{C}^n$, its *discrete inverse Fourier transform* (or *discrete Fourier cotransform*) is

$$\overline{\mathcal{F}}(c) = F_n^* c = \overline{F_n} c.$$

Definition 10.16. Every sequence $c = (c_0, \ldots, c_{n-1}) \in \mathbb{C}^n$ of "Fourier coefficients" determines a periodic function $f_c \colon \mathbb{R} \to \mathbb{C}$ (of period 2π) known as *discrete Fourier series, or phase polynomial*, defined such that

$$f_c(\theta) = c_0 + c_1 e^{i\theta} + \cdots + c_{n-1} e^{i(n-1)\theta} = \sum_{k=0}^{n-1} c_k e^{ik\theta}.$$

Then given any sequence $f = (f_0, \ldots, f_{n-1})$ of data points, it is desirable to find the "Fourier coefficients" $c = (c_0, \ldots, c_{n-1})$ of the discrete Fourier series f_c such that

$$f_c(2\pi k/n) = f_k,$$

for every k, $0 \leq k \leq n-1$.

The problem amounts to solving the linear system of n equations

$$c_0 + c_1 e^{2\pi ik/n} + \cdots + c_{n-1} e^{2\pi ik(n-1)/n} = f_k, \quad k = 0, \ldots, n-1,$$

which is just the system

$$\overline{F_n} c = f.$$

Since

$$\frac{1}{n} F_n \overline{F_n} = \frac{1}{n} \overline{F_n} F_n = I,$$

we see that c is given by

$$c = \frac{1}{n} F_n f = \widehat{f},$$

the discrete Fourier transform of f, so

$$c_k = \frac{1}{n} \sum_{j=0}^{n-1} f_j e^{-2\pi ijk/n}, \qquad k = 0, \ldots, n-1.$$

Example 10.7. Let us see how to obtain the "Fourier coefficients" $c = (c_0, c_1, c_2)$ when $n = 3$ and $f = (f_0, f_1, f_2)$. The condition $f_c(2\pi k/n) = f_k$ for $0 \leq k \leq 2$ translates into three equations

$$f_c(0) = f_0, \qquad f_c(2\pi/3) = f_1, \qquad f_c(4\pi/3) = f_2.$$

After expanding via the definition $f_c(\theta) = \sum_{k=0}^{n-1} c_k e^{ik\theta}$, the above three equations become

$$\begin{aligned} c_0 + c_1 + c_2 &= f_0 \\ c_0 + c_1 e^{2\pi i/3} + c_2 e^{4\pi i/3} &= f_1 \\ c_0 + c_1 e^{4\pi i/3} + c_2 e^{2\pi i/3} &= f_2. \end{aligned}$$

These three equations can be written as

$$\begin{pmatrix} 1 & 1 & 1 \\ 1 & e^{2\pi i/3} & e^{4\pi i/3} \\ 1 & e^{4\pi i/3} & e^{2\pi i/3} \end{pmatrix} \begin{pmatrix} c_0 \\ c_1 \\ c_2 \end{pmatrix} = \begin{pmatrix} f_0 \\ f_1 \\ f_2 \end{pmatrix}.$$

Since

$$F_3 = \begin{pmatrix} 1 & 1 & 1 \\ 1 & e^{-2\pi i/3} & e^{-4\pi i/3} \\ 1 & e^{-4\pi i/3} & e^{-2\pi i/3} \end{pmatrix}$$

the above matrix system is equivalent to

$$\overline{F}_3 \begin{pmatrix} c_0 \\ c_1 \\ c_2 \end{pmatrix} = \begin{pmatrix} f_0 \\ f_1 \\ f_2 \end{pmatrix},$$

which implies that

$$\begin{pmatrix} c_0 \\ c_1 \\ c_2 \end{pmatrix} = \frac{1}{3} F_3 \begin{pmatrix} f_0 \\ f_1 \\ f_2 \end{pmatrix}.$$

Note the analogy with the case of $\mathbb{T}$ and $\mathbb{Z}$, where the Fourier cotransform $\overline{\mathcal{F}}(c)$ of the sequence $(c_m)_{m\in\mathbb{Z}}$ is given by

$$f(\theta) = \sum_{k=-\infty}^{\infty} c_k e^{ik\theta},$$

and the Fourier coefficients of the function $f\colon \mathbb{T} \to \mathbb{C}$ are given by the formulae

$$c_k = \frac{1}{2\pi} \int_{-\pi}^{\pi} f(x) e^{-ikx} dx.$$

In $\mathbb{Z}/n\mathbb{Z}$, the convolution of two sequences $f = (f_0, \ldots, f_{n-1})$ and $g = (g_0, \ldots, g_{n-1})$ is given by

$$(f * g)_k = \frac{1}{n} \sum_{\substack{i,j\in\mathbb{Z}/n\mathbb{Z} \\ i+j\equiv k \pmod n}} f_i g_j, \qquad k = 0, \ldots, n-1.$$

It is remarkable that the convolution $f * g$ can be expressed in matrix form as

$$f * g = \frac{1}{n} H(f) g$$

for some matrix $H(f)$. The matrix $H(f)$ is a *circulant matrix*.

Definition 10.17. The *circular shift matrix* S_n *(of order* n*)* is defined as the matrix

$$S_n = \begin{pmatrix} 0 & 0 & 0 & 0 & \cdots & 0 & 1 \\ 1 & 0 & 0 & 0 & \cdots & 0 & 0 \\ 0 & 1 & 0 & 0 & \cdots & 0 & 0 \\ 0 & 0 & 1 & 0 & \cdots & 0 & 0 \\ \vdots & \vdots & \vdots & \vdots & \vdots & \vdots & \vdots \\ 0 & 0 & 0 & 0 & \cdots & 1 & 0 \end{pmatrix}$$

consisting of cyclic permutations of its first column. For any sequence $f = (f_0, \ldots, f_{n-1}) \in \mathbb{C}^n$, we define the *circulant matrix* $H(f)$ as

$$H(f) = \sum_{j=0}^{n-1} f_j S_n^j,$$

where $S_n^0 = I_n$, as usual.

For example, the circulant matrix associated with the sequence $f = (a, b, c, d)$ is

$$\begin{pmatrix} a & d & c & b \\ b & a & d & c \\ c & b & a & d \\ d & c & b & a \end{pmatrix}.$$

It is not hard to prove that the convolution $f * g$ of two sequences $f = (f_0, \ldots, f_{n-1})$ and $g = (g_0, \ldots, g_{n-1})$ is given by

$$f * g = \frac{1}{n} H(f)\, g,$$

viewing f and g as column vectors. Then the miracle (which is not too hard to prove!), is that we have

$$H(f)\overline{F_n} = \overline{F_n}\,\mathrm{diag}(n\widehat{f}), \tag{$\dagger$}$$

where $\mathrm{diag}(n\widehat{f})$ is the diagonal matrix whose diagonal entries are the elements of the vector $n\widehat{f}$, which means that the columns of the Fourier matrix $\overline{F_n}$ are the eigenvectors of the circulant matrix $H(f)$, and that the eigenvalue associated with the kth eigenvector is $(n\widehat{f})_k$, that is, n times the kth component of the Fourier transform $\widehat{f}$ of f (counting from 0).

To prove ($\dagger$), we first prove that the eigenvectors u_k of the circular shift matrix S_n (indexing from 0 to $n-1$) are the columns of $\overline{F_n}$, where the column of index k whose entries are

$$\begin{pmatrix} 1 & (e^{2\pi i/n})^k & (e^{2\pi i2/n})^k & \cdots & (e^{2\pi im/n})^k & \cdots & (e^{2\pi i(n-1)/n})^k \end{pmatrix}$$

is associated with the eigenvalue $e^{-2\pi ik/n}$.

Indeed, applying S_n to u_k, the last entry $e^{2\pi i(n-1)k/n} = e^{-2\pi ik/n}$ in u_k becomes the first entry in $S_n u_k$, so the corresponding eigenvalue is $e^{-2\pi ik/n}$. For example, if $n = 4$, since $(1, e^{2\pi i/4}, e^{2\pi i2/4}, e^{2\pi i3/4}) = (1, i, -1, -i)$ and $(1, e^{-2\pi i/4}, e^{-2\pi i2/4}, e^{-2\pi i3/4}) = (1, -i, -1, i)$, we have

$$\begin{pmatrix} 0 & 0 & 0 & 1 \\ 1 & 0 & 0 & 0 \\ 0 & 1 & 0 & 0 \\ 0 & 0 & 1 & 0 \end{pmatrix} \begin{pmatrix} 1 & 1 & 1 & 1 \\ 1 & i & -1 & -i \\ 1 & -1 & 1 & -1 \\ 1 & -i & -1 & i \end{pmatrix} = \begin{pmatrix} 1 & 1 & 1 & 1 \\ 1 & i & -1 & -i \\ 1 & -1 & 1 & -1 \\ 1 & -i & -1 & i \end{pmatrix} \begin{pmatrix} 1 & 0 & 0 & 0 \\ 0 & -i & 0 & 0 \\ 0 & 0 & -1 & 0 \\ 0 & 0 & 0 & i \end{pmatrix}.$$

Since $H(f) = \sum_{j=0}^{n-1} f_j S_n^j$, the eigenvectors remain the same, and it is easy to see that the kth eigenvalue of $H(f)$ (indexing from 0) is $\sum_{j=0}^{n-1} f_j e^{-2\pi ijk/n} = n\widehat{f}_k$.

If we recall that $F_n\overline{F_n} = \overline{F_n}F_n = n\,I_n$, multiplying the equation $H(f)\overline{F_n} = \overline{F_n}\,\mathrm{diag}(n\widehat{f})$ both on the left and on the right by F_n, we get

$$F_n H(f)(n\,I_n) = (n\,I_n)\,\mathrm{diag}(n\widehat{f})\,F_n,$$

that is,

$$F_n H(f) = \mathrm{diag}(n\widehat{f})\,F_n.$$

If we apply both sides to any sequence $g \in \mathbb{C}^n$, we get

$$F_n H(f) g = \mathrm{diag}(n\widehat{f})\,F_n g.$$

Since $\widehat{f} = \frac{1}{n}F_n f$, $\widehat{g} = \frac{1}{n}F_n g$, $f * g = \frac{1}{n}H(f)g$, and $\widehat{f * g} = \frac{1}{n}F_n(f * g)$, multiplying both sides by $1/n^2$, the above equation yields

$$\frac{1}{n}F_n \frac{1}{n}H(f)g = \mathrm{diag}(\widehat{f})\,\frac{1}{n}F_n g,$$

which means that

$$\widehat{f * g} = \mathrm{diag}(\widehat{f})\,\widehat{g} = \widehat{f}\,\widehat{g},$$

where $\widehat{f}\,\widehat{g}$ is the column vector obtained by pointwise multiplication $((\widehat{f}\,\widehat{g})(j) = \widehat{f}(j)\widehat{g}(f))$.

Therefore, we have given another proof of the convolution rule.

10.8 Plancherel's Theorem and Fourier Inversion

Let G be a locally compact abelian group equipped with a Haar measure λ. In general, given a function $f \in \mathrm{L}^1(G)$, its Fourier transform $\mathcal{F}(f)$ does not belong to $\mathrm{L}^1(\widehat{G})$.

Plancherel's theorem (Theorem 10.27) asserts that there is a Haar measure $\widehat{\lambda}$ on the dual group $\widehat{G}$ such that the map $f \mapsto \mathcal{F}(f)$ sends $\mathrm{L}^1(G) \cap \mathrm{L}^2(G)$ into $\mathrm{L}^2(\widehat{G})$ and has a unique extension which is an isometry from $\mathrm{L}^2(G)$ to $\mathrm{L}^2(\widehat{G})$.

We will follow the proof in Bourbaki [Bourbaki (1967b)] (Chapter 2, Section 1, No. 3). The crucial step is to define a subspace $A(G)$ of $\mathrm{L}^1(G) \cap \mathrm{L}^2(G)$ which is dense in both $\mathrm{L}^1(G)$ and $\mathrm{L}^2(G)$, and to show that there is a Haar measure ν on $\widehat{G}$ such that

$$\int_{\widehat{G}} |\mathcal{F}(f)|^2 \, d\nu = \int_G |f|^2 \, d\lambda$$

for all $f \in A(G)$. There are many technical details so we will focus on the main ideas.

Definition 10.18. Let $A(G)$ be the subspace of $\mathrm{L}^1(G)$ spanned by the set of functions of the form $f * g$, with $f, g \in \mathrm{L}^1(G) \cap \mathrm{L}^2(G)$.

It is immediately verified that $A(G)$ is an ideal of $\mathrm{L}^1(G)$ contained in $\mathrm{L}^1(G) \cap \mathrm{L}^2(G)$.

The first step is to show that there is a filter base $\mathcal{B}$ defined on $A(G) \cap \mathcal{K}_{\mathbb{C}}(G)$, where the functions in this filter base approximate the Dirac measure, so that

(1) $\delta_e = \lim_{\mathcal{B}} \varphi d\lambda$ for all φ in any subset in $\mathcal{B}$.
(2) $\lim_{\mathcal{B}} \mathcal{F}(\varphi) = 1$, and $\|\mathcal{F}(\varphi)\|_\infty \leq 1$, for all φ in any subset in $\mathcal{B}$ (recall from Proposition 10.18 that $\mathcal{F}(\varphi)$ is bounded).
(3) $\lim_{\mathcal{B}} \varphi * f = f$, for all $f \in \mathrm{L}^p(G)$, $p = 1, 2$.

Condition (3) implies that $A(G)$ is dense in both $\mathrm{L}^1(G)$ and $\mathrm{L}^2(G)$, and it can also be shown that $\mathcal{F}(A(G))$ is dense in $\mathcal{C}_0(\widehat{G})$.

The second step is to show that there is a Haar measure ν on $\widehat{G}$ such that

$$\int_{\widehat{G}} |\mathcal{F}(f)|^2 \, d\nu = \int_G |f|^2 \, d\lambda \tag{$*$}$$

for all $f \in A(G)$. This goes as follows.

It can be shown that for every $f \in A(G)$, there is a unique positive measure μ_f on $\widehat{G}$ such that

$$(g * f)(e) = \int_{\widehat{G}} \mathcal{F}(g) \, d\mu_f, \qquad \text{for all } g \in A(G).$$

Then for every $f \in A(G)$, define Ω_f as the open subset of $\widehat{G}$ given by

$$\Omega_f = \{\chi \in \widehat{G} \mid \mathcal{F}(f)(\chi) \neq 0\}.$$

It can be shown that the subsets Ω_f form an open cover of $\widehat{G}$ when f ranges over $A(G)$. For every Ω_f, let ν_f be the positive measure on Ω_f associated with the Radon functional given by $\Phi_{\nu_f} = 1/(\mathcal{F}(f))\,\Phi_{\mu_f}$, where Φ_{μ_f} is the Radon functional associated with the measure μ_f. It can be shown that the local measures ν_f patch to a global Haar measure ν on $\widehat{G}$, and that $(*)$ is satisfied.

Definition 10.19. Let G be a locally compact abelian group. For every Haar measure λ on G, the Haar measure $\widehat{\lambda} = \nu$ given by the previous construction is called the *measure associated with* λ or the *dual measure*.

Theorem 10.27. *(Plancherel) Let G be a locally compact abelian group equipped with a Haar measure λ. There is a Haar measure $\widehat{\lambda}$ on the dual $\widehat{G}$ such that for any $f \in \mathrm{L}^1(G) \cap \mathrm{L}^2(G)$, we have $\mathcal{F}(f) \in \mathrm{L}^2(\widehat{G})$. Furthermore, the map $f \mapsto \mathcal{F}(f)$ from $\mathrm{L}^1(G) \cap \mathrm{L}^2(G)$ to $\mathrm{L}^2(\widehat{G})$ has a unique extension which is an isometry from $\mathrm{L}^2(G)$ to $\mathrm{L}^2(\widehat{G})$.*

Proof sketch. By Property (3) of the filter base, $A(G)$ is dense in $\mathrm{L}^2(G)$. By $(*)$, the Fourier transform is an isometry from $A(G) \subseteq \mathrm{L}^2(G)$ to a subspace of $\mathrm{L}^2(\widehat{G})$, which is complete Therefore, by Proposition A.61, it has a unique extension Φ to $\mathrm{L}^2(G)$ (an isometry is uniformly continuous). To finish the proof, is suffices to show that $\Phi(L^2(G)) = \mathrm{L}^2(\widehat{G})$ and that $\Phi(f) = \mathcal{F}(f)$ for all $f \in \mathrm{L}^1(G) \cap \mathrm{L}^2(G)$.

Since $\mathcal{F}$ is an isometry between $A(G) \subseteq \mathrm{L}^2(G)$ and the subspace $\mathcal{F}(A(G))$ of $\mathrm{L}^2(\widehat{G})$, a sequence $(\mathcal{F}(f_n))_n$ of functions in $\mathcal{F}(A(G))$ is a Cauchy sequence iff $(f_n)_n$ is a Cauchy sequence in $A(G)$, and since $\mathrm{L}^2(G)$ and $\mathrm{L}^2(\widehat{G})$ are complete and Φ is continuous, the sequence $(\mathcal{F}(f_n))_n$ converges to a function $g \in \mathrm{L}^2(\widehat{G})$ iff the sequence $(f_n)_n$ converges to a function $f \in \mathrm{L}^2(G)$ such that $\Phi(f) = g$. Therefore, to prove that $\Phi(L^2(G)) = \mathrm{L}^2(\widehat{G})$, it suffices to show that $\mathcal{F}(A(G))$ is dense in $\mathrm{L}^2(\widehat{G})$.

Assume that some $h \in \mathrm{L}^2(\widehat{G})$ is orthogonal to $\mathcal{F}(A(G))$. For all $f, g \in A(G)$, we have $\mathcal{F}(f)\mathcal{F}(g) = \mathcal{F}(f * g) \in \mathcal{F}(A(G))$. Since $\check{f}(a) = f(a^{-1})$ and $f^*(a) = \overline{f(a^{-1})} = \overline{f}(a^{-1})$, we have $f^* = \check{\overline{f}}$, so by Proposition 10.19(1), the equation

$$\mathcal{F}(\check{f})(\chi) = \overline{\mathcal{F}(\overline{f})(\chi)}$$

implies that $\mathcal{F}(f^*)(\chi) = \overline{\mathcal{F}(f)(\chi)}$. Then

$$\begin{aligned}\langle h\mathcal{F}(f), \mathcal{F}(g)\rangle &= \int h(\chi)\mathcal{F}(f)(\chi)\overline{\mathcal{F}(g)(\chi)}\, d\widehat{\lambda}(\chi) \\ &= \int h(\chi)\overline{\overline{\mathcal{F}(f)(\chi)}\, \mathcal{F}(g)(\chi)}\, d\widehat{\lambda}(\chi) \\ &= \int h(\chi)\overline{\mathcal{F}(f^*)(\chi)\, \mathcal{F}(g)(\chi)}\, d\widehat{\lambda}(\chi) \\ &= \int h(\chi)\overline{\mathcal{F}(f^* * g)(\chi)}\, d\widehat{\lambda}(\chi) \\ &= \langle h, \mathcal{F}(f^* * g)\rangle = 0,\end{aligned}$$

where Proposition 10.18 was used in the next to the last equation, and because $f^* * g \in A(G)$ and h is orthogonal to $\mathcal{F}(A(G))$. The above shows that $h\mathcal{F}(f)$ is orthogonal to $\mathcal{F}(A(G))$. Then it can be shown that this implies that $h = 0$. By a well known fact of Hilbert space theory, $\mathcal{F}(A(G))$ is dense in $\mathrm{L}^2(\widehat{G})$. The last step is to show that $\Phi(f) = \mathcal{F}(f)$ for all $f \in \mathrm{L}^1(G) \cap \mathrm{L}^2(G)$. The details are a bit involved, so we refer the reader to Bourbaki [Bourbaki (1967b)] (Chapter 2, Section 1, No. 3, Theorem 1). Another proof of Plancherel's theorem is given in Folland [Folland (1995)] (Chapter 4, Section 2, Theorem 4.25). It follows similar lines but uses a class of functions $\mathcal{B}^1$ different from $A(G)$. □

The unique extension of $\mathcal{F}$ is also denoted $\mathcal{F}$. By using the same techniques as above, it is easy to see that $\overline{\mathcal{F}}$ also has a unique extension to $\mathrm{L}^2(G)$, and that it is an isometry.

On should realize that Theorem 10.27 does not say that the Fourier transform $\mathcal{F}$ (or the Fourier cotransform $\overline{\mathcal{F}}$) is defined on $\mathrm{L}^2(G)$, because in general the integral will not converge for f outside of $\mathrm{L}^1(G) \cap \mathrm{L}^2(G)$. What is happening is more subtle. It is always possible by using a limit process to define the Fourier transform of any $f \in \mathrm{L}^2(G)$, and this extension of $\mathcal{F}$ to $\mathrm{L}^2(G)$ is an isometry. This is still quite remarkable because there is no such result for $\mathrm{L}^1(G)$. Since the extension of $\mathcal{F}$ to $\mathrm{L}^2(G)$ is an isometry, it has an inverse, but it is far from obvious that this inverse has any relation to the Fourier cotransform on $\mathrm{L}^2(\widehat{G})$. In fact it does, but this requires proving Gelfand's duality theorem, that G and its double dual $\widehat{\widehat{G}}$ are isomorphic.

As a corollary of Theorem 10.27 and Proposition 10.13, we can prove the fact announced just after Proposition 10.13.

Proposition 10.28. *If G is a compact abelian group endowed with a Haar measure λ normalized so that G has measure $\lambda(G) = 1$, then $\widehat{G}$ is a Hilbert basis for $\mathrm{L}^2(G)$ (it is orthonormal and dense in $\mathrm{L}^2(G)$).*

Proof. By a well known fact of Hilbert space theory, it suffices to show that there is no nonzero function $f \in \mathrm{L}^2(G)$ orthogonal to every character $\chi \in \widehat{G}$. Assume $f \in \mathrm{L}^2(G)$ is orthogonal to every character $\chi \in \widehat{G}$. This means that $\int f(a)\overline{\chi(a)}\, d\lambda(a) = 0$, and since

$$\int f(a)\overline{\chi(a)}\, d\lambda(a) = \mathcal{F}(f)(\chi),$$

we get

$$\mathcal{F}(f)(\chi) = 0, \qquad \text{for all } \chi \in \widehat{G}.$$

Since Plancherel's theorem asserts that $\mathcal{F}$ is an isometry, it is injective, so $f = 0$, establishing our result. $\square$

We now turn to the issue of Fourier inversion. If $f, g \in \mathrm{L}^2(G)$, the convolution $f * g$ is given by

$$(f * g)(s) = \int f(t) g(t^{-1}s)\, d\lambda(t)$$

so for $s = e$,

$$\begin{aligned}(f * g)(e) &= \int f(t) g(t^{-1})\, d\lambda(s) = \int f(t)\check{g}(t)\, d\lambda(s) \\ &= \int f(t)\overline{\overline{\check{g}}(t)}\, d\lambda(s) = \int f(t)\overline{g^*(t)}\, d\lambda(s).\end{aligned}$$

We also have

$$\langle f, g\rangle = \int f(t)\overline{g(t)}\, d\lambda(t),$$

so we deduce

$$(f * g^*)(e) = \langle f, g\rangle. \tag{$\dagger_1$}$$

But Plancherel's theorem implies that

$$\langle f, g\rangle = \langle \mathcal{F}(f), \mathcal{F}(g)\rangle = \int \mathcal{F}(f)(\chi)\overline{\mathcal{F}(g)(\chi)}\, d\widehat{\lambda}(\chi),$$

and we conclude that

$$(f * g^*)(e) = \int \mathcal{F}(f)(\chi)\overline{\mathcal{F}(g)(\chi)}\, d\widehat{\lambda}(\chi). \tag{$\dagger_2$}$$

Proposition 10.29. *(Fourier inversion for $A(G)$) If $f \in A(G)$, then $\mathcal{F}(f) \in \mathrm{L}^1(\widehat{G})$, and*

$$f(a) = \int_{\widehat{G}} \chi(a)\mathcal{F}(f)(\chi)\, d\widehat{\lambda}(\chi) \qquad \text{for all } a \in G.$$

Equivalently,

$$f = (\overline{\mathcal{F}} \circ \mathcal{F})(f) \circ \eta, \qquad \text{for all } f \in A(G),$$

where $\eta\colon G \to \widehat{\widehat{G}}$ is the canonical map.

Sketch of proof. Using functions g that approximate the Dirac measure δ_e, Formula ($\dagger_2$) yields

$$f(e) = \int \mathcal{F}(f)(\chi)\, d\widehat{\lambda}(\chi),$$

which is the result of Proposition 10.29 for $a = e$ since $\chi(e) = 1$. For any arbitrary $a \in G$, replace f by $\lambda_{a^{-1}} f$ and use Proposition 10.19(3). See Bourbaki [Bourbaki (1967b)] (Chapter 2, Section 1, No. 4). $\square$

Since the ideal $A(G)$ of Definition 10.18 is contained in $\mathrm{L}^1(G) \cap \mathrm{L}^2(G)$, unlike the situation in Plancherel's theorem, there is no need to extend $\mathcal{F}$ (and $\overline{\mathcal{F}}$ on $\mathrm{L}^1(\widehat{G})$).

It is also shown in Bourbaki that the inversion formula holds for all $f \in \mathrm{L}^2(G)$ such that $\mathcal{F}(f) \in \mathrm{L}^1(\widehat{G})$.

In order to proceed any further, we need Pontrjagin's duality theorem asserting that η is an isomorphism.

10.9 Pontrjagin Duality and Fourier Inversion

The Pontrjagin duality theorem is one of the most important and most beautiful theorems of the theory of locally compact abelian groups. Recall that we have a canonical map $\eta\colon G \to \widehat{\widehat{G}}$ given by

$$\eta_a(\chi) = \chi(a), \qquad a \in G,\ \chi \in \widehat{G},$$

which is a homomorphism.

Theorem 10.30. *(Pontrjagin duality theorem) Let G be a locally compact abelian group endowed with a Haar measure λ, let $\widehat{G}$ be its dual group endowed with the associated Haar measure $\widehat{\lambda}$ (see Definition 10.19), and let $\widehat{\widehat{G}}$ be its double dual endowed with the associated measure $\widehat{\widehat{\lambda}}$. The map*

$\eta\colon G \to \widehat{\widehat{G}}$ is an isomorphism and a homeomorphism between the topological groups G and $\widehat{\widehat{G}}$ that maps the measure λ to the measure $\widehat{\widehat{\lambda}}$, which means that $\lambda = \eta^{-1}(\widehat{\widehat{\lambda}})$, as in Definition 8.14. If we identify G and $\widehat{\widehat{G}}$ using the isomorphism η, then the extension $\mathcal{F}\colon \mathrm{L}^2(G) \to \mathrm{L}^2(\widehat{G})$ of the Fourier transform to $\mathrm{L}^2(G)$ and the extension $\overline{\mathcal{F}}\colon \mathrm{L}^2(\widehat{G}) \to \mathrm{L}^2(G)$ of the Fourier cotransform to $\mathrm{L}^2(\widehat{G})$ are mutual inverses. In particular, Fourier inversion holds; that is,

$$f = (\overline{\mathcal{F}} \circ \mathcal{F})(f) \circ \eta, \qquad \textit{for all } f \in \mathrm{L}^2(G).$$

Proof idea. The proof of Theorem 10.30 is too technical to be presented in full detail here. A proof can found in Bourbaki [Bourbaki (1967b)] (Chapter 2, Section 1, No. 5, Theorem 2) and in Folland [Folland (1995)] (Chapter 4, Section 4.3, Theorem 4.31).

The first part of the proof of Pontrjagin duality establishes the fact that η is injective and a homeomorphism onto its image, which is closed in $\widehat{\widehat{G}}$. To prove that η is injective and that η^{-1} is continuous, it suffices to prove that for every neighborhood U of e in G there is some neighborhood W of $\widehat{\widehat{e}}$ in $\widehat{\widehat{G}}$, such that $\eta^{-1}(W) \subseteq U$.

We can find a compact symmetric neighborhood V of e in G such that $V^2 \subseteq U$, and some positive function $f \in \mathcal{K}_{\mathbb{C}}(G)$ whose support is contained in V. If we let $g = f^* * f$, then we see that $g \in A(G)$, $\mathrm{supp}(g) \subseteq U$, and $g(e) > 0$, which follows from Equation ($\dagger_1$) of the previous section. Since $\widehat{G}$ has the compact open topology, which is equivalent to the topology of pointwise convergence on $\mathrm{L}^1(\widehat{G})$ (by Theorem 10.6 applied to $\widehat{G}$), there is a neighborhood W of the identity $\widehat{\widehat{e}}$ in $\widehat{\widehat{G}}$ such that

$$|\overline{\mathcal{F}}(\mathcal{F}(g))(\zeta) - \overline{\mathcal{F}}(\mathcal{F}(g))(\widehat{e})| < \frac{1}{2}g(e), \qquad \zeta \in W,$$

so if $a \in \eta^{-1}(W)$, since $g \in A(G)$, by Proposition 10.29 we have $g = (\overline{\mathcal{F}} \circ \mathcal{F})(g) \circ \eta$, and we obtain

$$|g(a) - g(e)| < \frac{1}{2}g(e).$$

Therefore, $g(a) \neq 0$, and since $\mathrm{supp}(g) \subseteq U$, we have $a \in U$, which shows that $\eta^{-1}(W) \subseteq U$, as desired.

The second part is to prove that η is surjective. Bourbaki's proof uses the following fact which shows the existence of certain kinds of bump functions on $\mathrm{L}^1(G)$.

Proposition 10.31. *Given any closed subset P of $\widehat{G}$ and any $\chi \in \widehat{G}$, if $\chi \notin P$, then there exists some $f \in \mathrm{L}^1(G)$ such that $\mathcal{F}(f)(\chi) = 1$ and $\mathcal{F}(f)$ vanishes on P.*

Assume that there is some $\zeta \in \widehat{\widehat{G}}$ such that $\zeta \notin \eta(G)$. By Proposition 10.31 applied to $\widehat{\widehat{G}}$, since we proved in the first part of the proof that $\eta(G)$ is closed, there is some nonzero function $f \in \mathrm{L}^1(\widehat{G})$ such that $\mathcal{F}(f)$ vanishes on $\eta(G)$, that is (since $\eta_a(\chi) = \chi(a)$),

$$\mathcal{F}(f)(\eta_a) = \int f(\chi)\overline{\eta_a(\chi)}\, d\widehat{\lambda}(\chi) = \int f(\chi)\overline{\chi(a)}\, d\widehat{\lambda}(\chi) = 0, \qquad \text{for all } a \in G.$$

Using Fubini's theorem, for any $g \in \mathrm{L}^1(G)$, we have

$$\begin{aligned}\int f(\chi)\mathcal{F}(g)(\chi)\, d\widehat{\lambda}(\chi) &= \int f(\chi)\left(\int g(a)\overline{\chi(a)}\, d\lambda(a)\right) d\widehat{\lambda}(\chi)\\ &= \int\left(\int f(\chi)\overline{\chi(a)}\, d\widehat{\lambda}(\chi)\right) g(a)\, d\lambda(a) = 0.\end{aligned}$$

By the second remark just after Proposition 10.18, $\mathcal{F}(\mathrm{L}^1(G))$ is dense in $\mathcal{C}_0(\widehat{G})$, so

$$\int f(\chi)\mathcal{F}(g)(\chi)\, d\widehat{\lambda}(\chi) = 0$$

for all $g \in \mathrm{L}^1(G)$ implies that $f \equiv 0$, a contradiction. Therefore, η is surjective, thus an isomorphism. Then using Plancherel's Theorem (Theorem 10.27) and Proposition 10.29, we can shown that $\overline{\mathcal{F}}$ (defined on $\mathrm{L}^2(\widehat{G})$) is an isometry between $\mathrm{L}^2(\widehat{G})$ and $\mathrm{L}^2(\widehat{\widehat{G}})$, and that $\lambda = \eta^{-1}(\widehat{\widehat{\lambda}})$. ☐

From now on we identify G and $\widehat{\widehat{G}}$ unless specified otherwise.

We will now show that there is another class of functions for which $\mathcal{F}$ and $\overline{\mathcal{F}}$ are mutual inverses. For this we need the following result.

Proposition 10.32. *For every $f \in \mathrm{L}^1(G)$ and every $H \in \mathrm{L}^1(\widehat{G})$, we have*

$$\int_G f(a)\mathcal{F}(H)(a)\, d\lambda(a) = \int_{\widehat{G}} \mathcal{F}(f)(\chi)H(\chi)\, d\widehat{\lambda}(\chi).$$

The proof of Proposition 10.32 is an application of Fubini's theorem.

Definition 10.20. Let G be a locally compact abelian group. We define $B(G)$ as the set of functions

$$B(G) = \{f \in \mathrm{L}^1(G) \mid \mathcal{F}(f) \in \mathrm{L}^1(\widehat{G})\}.$$

Recall that we are identifying G and $\widehat{\widehat{G}}$ so that we view $\mathcal{F}$ and $\overline{\mathcal{F}}$ as mutual inverses. Then $B(\widehat{G})$ is defined by

$$B(\widehat{G}) = \{f \in \mathrm{L}^1(\widehat{G}) \mid \overline{\mathcal{F}}(f) \in \mathrm{L}^1(G)\}.$$

Theorem 10.33. *The restriction of $\mathcal{F}$ to $B(G)$ is a bijection from $B(G)$ to $B(\widehat{G})$, whose inverse is the restriction of $\overline{\mathcal{F}}$ to $B(\widehat{G})$.*

Proof. If $f \in B(G)$, then $\mathcal{F}(f) \in \mathrm{L}^1(\widehat{G}) \cap \mathcal{C}_0(\widehat{G}) \subseteq \mathrm{L}^1(\widehat{G}) \cap \mathrm{L}^2(\widehat{G})$. Let $h = (\overline{\mathcal{F}} \circ \mathcal{F})(f) \in \mathrm{L}^2(G)$. For every $g \in \mathcal{K}_{\mathbb{C}}(\widehat{G})$, by Theorem 10.30, $\overline{\mathcal{F}}$ is an isometry, $\mathcal{F}$ and $\overline{\mathcal{F}}$ are mutual inverses, so that $\mathcal{F}(h) = \mathcal{F}(f)$, and by Proposition 10.32, we have

$$\begin{aligned}
\int_G h(a)\mathcal{F}(g)(a)\,d\lambda &= \langle \mathcal{F}(g), \overline{h}\rangle \\
&= \langle g, \overline{\mathcal{F}}(\overline{h})\rangle && \mathcal{F} \text{ is an isometry} \\
&= \langle g, \overline{\mathcal{F}(h)}\rangle && \text{by Proposition 10.16} \\
&= \int_{\widehat{G}} \mathcal{F}(h)(\chi)g(\chi)\,d\widehat{\lambda}(\chi) \\
&= \int_{\widehat{G}} \mathcal{F}(f)(\chi)g(\chi)\,d\widehat{\lambda}(\chi) && \mathcal{F}(h) = \mathcal{F}(f) \\
&= \int_G f(a)\mathcal{F}(g)(a)\,d\lambda(a) && \text{by Proposition 10.32.}
\end{aligned}$$

It follows that

$$\int_G h(a)\mathcal{F}(g)(a)\,d\lambda = \int_G f(a)\mathcal{F}(g)(a)\,d\lambda(a) \qquad \text{for all } g \in \mathcal{K}_{\mathbb{C}}(\widehat{G}),$$

and we deduce that $h = f \in \mathrm{L}^1(G)$. Since $h = (\overline{\mathcal{F}} \circ \mathcal{F})(f)$, we conclude that $\mathcal{F}(f) \in B(\widehat{G})$. We also proved that $\overline{\mathcal{F}} \circ \mathcal{F}$ is the identity on $B(G)$. By exchanging the roles of G and $\widehat{G}$, we can show that $\mathcal{F} \circ \overline{\mathcal{F}}$ is the identity on $B(\widehat{G})$. The theorem follows immediately. ☐

Note that unlike the situation in Theorem 10.30, in Theorem 10.33 there is no need to extend $\mathcal{F}$ (and $\overline{\mathcal{F}}$ on $B(\widehat{G})$).

Remark: The space $B(G)$ is an algebra for the pointwise product and the convolution product. It is also easy to see that $\mathcal{F}(fg) = \mathcal{F}(f) * \mathcal{F}(g)$, in addition to $\mathcal{F}(f * g) = \mathcal{F}(f)\mathcal{F}(g)$.

The property $\mathcal{F}(fg) = \mathcal{F}(f) * \mathcal{F}(g)$ is also satisfied by $\mathrm{L}^2(G)$.

Proposition 10.34. *For any two functions $f, g \in \mathrm{L}^2(G)$, we have*

$$\mathcal{F}(fg) = \mathcal{F}(f) * \mathcal{F}(g).$$

Proposition 10.34 is proven in Bourbaki [Bourbaki (1967b)] (Chapter 2, Section 1, No. 6, Theorem 3), and Folland [Folland (1995)] (Chapter 4, Section 4.3, Proposition 4.36).

The following proposition explains a phenomenon that we have already observed for $G = \mathbb{T}$ and $G = \mathbb{Z}$.

Proposition 10.35. *For any locally compact abelian group G, the group G is discrete if and only if $\widehat{G}$ is compact (and by duality, G is compact if and only if $\widehat{G}$ is discrete). Furthermore, if G is compact and endowed with the Haar measure λ normalized so that $\lambda(G) = 1$, then the associated measure $\widehat{\lambda}$ on $\widehat{G}$ is the counting measure. If G is discrete and endowed with the counting measure, then the associated measure $\widehat{\lambda}$ on $\widehat{G}$ is normalized so that $\widehat{\lambda}(\widehat{G}) = 1$.*

Proof. Assume that G is discrete. If so $\mathrm{L}^1(G)$ is unital with identity δ_e. By Theorem 9.19, the algebra $\mathsf{X}(\mathrm{L}^1(G))$ is compact, and since $\widehat{G}$ is homeomorphic to $\mathsf{X}(\mathrm{L}^1(G))$, we deduce that $\widehat{G}$ is compact.

Assume now that $\widehat{G}$ is compact. Since the characters are uniformly continuous, there is an open subset V containing the identity in $\widehat{G}$ such that for all $\chi \in V$, for all $a \in G$, we have $|\chi(a) - 1| \leq 1$. Since $a \in G$ is arbitrary, we can replace it by a^n for any $n \in \mathbb{Z}$, and since $\chi(a^n) = \chi(a)^n$, we get

$$|\chi(a)^n - 1| \leq 1, \qquad \text{for all } a \in G \text{ and all } n \in \mathbb{Z}.$$

Since $\chi(a)$ is a complex number of unit length, say $\chi(a) = \cos\theta + i\sin\theta$, with $0 \leq \theta < 2\pi$, we have $\chi(a)^n = \cos n\theta + i\sin n\theta$, and

$$\begin{aligned} |\cos n\theta - 1 + i\sin n\theta|^2 &= (\cos n\theta - 1)^2 + \sin^2 n\theta \\ &= \cos^2 n\theta - 2\cos n\theta + 1 + \sin^2 n\theta = 2(1 - \cos n\theta). \end{aligned}$$

Unless $\theta = 0$, we can find some n so that $\cos n\theta < 0$, and we get a contradiction to the inequality $|\chi(a)^n - 1| \leq 1$. Therefore, $\chi(a) = 1$ for all $a \in G$, which implies that $V = \{e\}$ (since the other characters are not the constant character 1). Thus we proved that $\{e\}$ is an open subset of $\widehat{G}$, so every singleton subset $\{\chi\}$ is open, which means that $\widehat{G}$ is discrete. The second part of the proposition is proven in Folland [Folland (1995)] (Chapter 4, Section 2, Proposition 4.24). $\square$

We haven't discussed functions of positive type yet. They play an important role in the theory of unitary representations of a locally compact group. A function $\varphi \in \mathrm{L}^\infty(G)$ is of *positive type* if

$$\int (f^* * f)\varphi \, d\lambda \geq 0, \qquad \text{for all } f \in \mathrm{L}^1(G).$$

Let $\mathcal{P}_+(G)$ be the space of functions of positive type. There is a connection with the dual group $\widehat{G}$. Indeed, for any measure $\mu \in \mathcal{M}(\widehat{G})$, define φ_μ by

$$\varphi_\mu(a) = \int_{\widehat{G}} \chi(a) d\mu(\chi).$$

Then a theorem of Bochner states that for any function of positive type $\varphi \in \mathcal{P}_+(G)$, there is a *unique positive measure* $\mu \in \mathcal{M}(\widehat{G})$ such that $\varphi = \varphi_\mu$. We will return to positive functions in Vol II, Chapter 3 (Section 3.5) and Chapter 9, and refer the reader to Folland for a discussion of this topic; see [Folland (1995)] Chapters 3 and Chapter 4, Theorem 4.18.

10.10 Problems

Problem 10.1. Complete the proof sketch of Theorem 10.6. In particular show that if G is a locally compact abelian group (equipped with a Haar measure λ), the bijection $j\colon \widehat{G} \to \mathsf{X}(\mathrm{L}^1(G))$ given by

$$j(\chi)(f) = \zeta_\chi(f) = \int \chi(a) f(a) \, d\lambda(a), \qquad \chi \in \widehat{G},\ f \in \mathrm{L}^1(G),$$

is actually a homeomorphism. Hint: See Bourbaki [Bourbaki (1967b)] (Chapter 2, Section 1, No. 1).

Problem 10.2. Let G be a locally compact abelian group. Given any $a \in G$, define the map $\eta_a \colon \widehat{G} \to \mathbb{C}$ by

$$\eta_a(\chi) = \chi(a), \qquad \text{evaluation at } a.$$

Show that the homomorphism $\eta\colon G \to \widehat{\widehat{G}}$ given by given by $\eta(a) = \eta_a$ is continuous. Hint: Show that the map $(a, \chi) \mapsto \chi(a)$ from $G \times \widehat{G}$ to $\mathbb{C}$ is continuous. Alternatively, see Bourbaki [Bourbaki (1967b)] (Chapter 2, Section 1, No. 1).

Problem 10.3. Verify that the identities of Proposition 10.19 hold when f is replaced by a complex measure $\mu \in \mathcal{M}^1(G)$.

Problem 10.4. Let G be a finite locally compact abelian group whose Haar measure is normalized so that $\lambda(G) = 1$. Recall that the Fourier transform of $x = (x_a)_{a \in G} \in \mathrm{L}^2(G)$ is given by

$$\mathcal{F}(x)(\chi) = \frac{1}{|G|} \sum_{a \in G} x_a \overline{\chi(a)},$$

where $\chi\colon G \to \mathbb{T}$ is a character of G. Prove that

$$\check{x} = (\mathcal{F} \circ \mathcal{F})(x) \circ \eta,$$

where $\check{x}(a) = x_{a^{-1}}$.

Problem 10.5. Recall that the characters $\chi\colon (\mathbb{Z}/m\mathbb{Z})^* \to \mathbf{U}(1)$ can be extended to a function $\chi\colon \mathbb{Z} \to \mathbb{C}$ given by

$$\chi(n) = \begin{cases} \chi(n \bmod m) & \text{if } \gcd(m, n) = 1 \\ 0 & \text{otherwise.} \end{cases}$$

(i) Verify that these extended functions are multiplicative, which means that

$$\chi(rs) = \chi(r)\chi(s) \qquad \text{for all } r, s \in \mathbb{Z}.$$

(ii) Verify these extended functions are also periodic with period m ($\chi(n + m) = \chi(n)$ for all $n \in \mathbb{Z}$).

Problem 10.6. Complete the calculations left to the reader in Example 10.5 and Example 10.6.

Problem 10.7. Recall the following notational conventions of Section 10.6:

(i) $(e_{a_1}, \ldots, e_{a_n})$ is a basis of $[G \to \mathbb{C}]$;
(ii) $(e^*_{a_1}, \ldots, e^*_{a_n})$ is the dual of the basis $(e_{a_1}, \ldots, e_{a_n})$;
(iii) $(\widetilde{\chi_1}, \ldots, \widetilde{\chi_n})$ is the dual character basis of $[G \to \mathbb{C}]^*$.

These conventions imply that for all $f, \gamma \in [G \to \mathbb{C}]^*$, the Fourier transform $\mathcal{F}$ is the sequilinear form on $[G \to \mathbb{C}]^*$ given by

$$\mathcal{F}(f)(\gamma) = \frac{1}{n} \sum_{j=1}^{n} f(e_{a_j}) \overline{\gamma(e_{a_j})};$$

Next recall the noncanonical isomorphism $\theta\colon [G \to \mathbb{C}] \to [G \to \mathbb{C}]^{**}$ as follows: if $c = \sum_{i=1}^{n} y_i e_{a_i}$, then

$$\theta_c = \sum_{i=1}^{n} y_i \widetilde{\chi_i}^*.$$

By using this noncanonical isomorphism, we define for any vector $c = \sum_{i=1}^{n} y_i e_{a_i} \in [G \to \mathbb{C}]$, the *Fourier cotransform* $\overline{\mathcal{F}}_2(c)$ of c as the linear form on $[G \to \mathbb{C}]$ such that for all $a \in [G \to \mathbb{C}]$, we have

$$\overline{\mathcal{F}}_2(c)(a) = \sum_{j=1}^{n} y_j \widetilde{\chi_j}(a).$$

Prove that if $a = \sum_{j=1}^{m} z_j e_{a_j}$, then

$$\overline{\mathcal{F}}_2(c)(a) = y^\top \overline{F} z = z^\top F^* y.$$

Then show that the Fourier inversion becomes the identity

$$f = (\overline{\mathcal{F}}_2 \circ \theta^{-1} \circ \mathcal{F})(f),$$

for all $f \in [G \to \mathbb{C}]^*$.

Problem 10.8. Given $f = (f_0, \dots, f_{n-1})$ and $g = (g_0, \dots, g_{n-1})$ in $\mathbb{Z}/n\mathbb{Z}$, prove that $f * g = \frac{1}{n} H(f) g$, where $H(f)$ is the circulant matrix of Definition 10.17.

Problem 10.9. Recall that $A(G)$ is the subspace of $\mathrm{L}^1(G)$ spanned by the set of functions of the form $f * g$, with $f, g \in \mathrm{L}^1(G) \cap \mathrm{L}^2(G)$. Verify that $A(G)$ is an ideal of $\mathrm{L}^1(G)$ contained in $\mathrm{L}^1(G) \cap \mathrm{L}^2(G)$.

Problem 10.10. Advanced Exercise: Show that for every $f \in A(G)$, there is a unique positive measure μ_f on $\widehat{G}$ such that

$$(g * f)(e) = \int_{\widehat{G}} \mathcal{F}(g)\, d\mu_f, \qquad \text{for all } g \in A(G).$$

Hint: See Bourbaki [Bourbaki (1967b)] (Chapter 2, Section 1, No. 3).

Problem 10.11. Advanced Exercise: For every $f \in A(G)$, define Ω_f as the open subset of $\widehat{G}$ given by

$$\Omega_f = \{\chi \in \widehat{G} \mid \mathcal{F}(f)(\chi) \neq 0\}.$$

Show that the subsets Ω_f form an open cover of $\widehat{G}$ when f ranges over $A(G)$. Hint: See Bourbaki [Bourbaki (1967b)] (Chapter 2, Section 1, No. 3).

Problem 10.12. Advanced Exercise: Complete the proof sketch of Plancherel's theorem, Theorem 10.27. Hint: See Bourbaki [Bourbaki (1967b)] (Chapter 2, Section 1, No. 3, Theorem 1) or Folland [Folland (1995)] (Chapter 4, Section 2, Theorem 4.25).

Problem 10.13. Complete the proof sketch of Proposition 10.29. Hint: See Bourbaki [Bourbaki (1967b)] (Chapter 2, Section 1, No. 4).

Problem 10.14. Prove Proposition 10.31. Hint: See Bourbaki [Bourbaki (1967b)] (Chapter 2, Section 1, No. 5, Theorem 2).

Problem 10.15. Prove Proposition 10.32. Hint: Use Fubini's theorem.

Problem 10.16. Let G be a locally compact abelian group. Recall that $B(G)$ is the set of functions

$$B(G) = \{f \in \mathrm{L}^1(G) \mid \mathcal{F}(f) \in \mathrm{L}^1(\widehat{G})\}.$$

Show $B(G)$ is an algebra for the pointwise product and the convolution product. Also show that for $f, g \in B(G)$ $\mathcal{F}(fg) = \mathcal{F}(f) * \mathcal{F}(g)$, and that $\mathcal{F}(f * g) = \mathcal{F}(f)\mathcal{F}(g)$.

Problem 10.17. Prove Proposition 10.34. Hint: See Bourbaki [Bourbaki (1967b)] (Chapter 2, Section 1, No. 6, Theorem 3) or Folland [Folland (1995)] (Chapter 4, Section 4.3, Proposition 4.36).

Appendix A

Topology

A.1 Metric Spaces and Normed Vector Spaces

This chapter contains a review of basic topological concepts. First metric spaces are defined. Next normed vector spaces are defined. Closed and open sets are defined, and their basic properties are stated. The general concept of a topological space is defined. The closure and the interior of a subset are defined. The subspace topology and the product topology are defined. Continuous maps and homeomorphisms are defined. Limits of sequences are defined. Continuous linear maps and multilinear maps are defined and studied briefly.

Most spaces considered in this book have a topological structure given by a metric or a norm, and we first review these notions. We begin with metric spaces. Recall that $\mathbb{R}_+ = \{x \in \mathbb{R} \mid x \geq 0\}$.

Definition A.1. A *metric space* is a set E together with a function $d\colon E \times E \to \mathbb{R}_+$, called a *metric, or distance*, assigning a nonnegative real number $d(x, y)$ to any two points $x, y \in E$, and satisfying the following conditions for all $x, y, z \in E$:

(D1) $d(x, y) = d(y, x)$. (symmetry)
(D2) $d(x, y) \geq 0$, and $d(x, y) = 0$ iff $x = y$. (positivity)
(D3) $d(x, z) \leq d(x, y) + d(y, z)$. (triangle inequality)

Geometrically, Condition (D3) expresses the fact that in a triangle with vertices x, y, z, the length of any side is bounded by the sum of the lengths of the other two sides. From (D3), we immediately get

$$|d(x, y) - d(y, z)| \leq d(x, z).$$

Let us give some examples of metric spaces. Recall that the *absolute value* $|x|$ of a real number $x \in \mathbb{R}$ is defined such that $|x| = x$ if $x \geq 0$, $|x| = -x$ if $x < 0$, and for a complex number $x = a + ib$, by $|x| = \sqrt{a^2 + b^2}$.

Example A.1.

(1) Let $E = \mathbb{R}$, and $d(x, y) = |x - y|$, the absolute value of $x - y$. This is the so-called natural metric on $\mathbb{R}$.

(2) Let $E = \mathbb{R}^n$ (or $E = \mathbb{C}^n$). We have the *Euclidean metric*

$$d_2(x, y) = \left(|x_1 - y_1|^2 + \cdots + |x_n - y_n|^2\right)^{\frac{1}{2}},$$

the distance between the points $(x_1, \ldots, x_n)$ and $(y_1, \ldots, y_n)$.

(3) For every set E, we can define the *discrete metric*, defined such that $d(x, y) = 1$ iff $x \neq y$, and $d(x, x) = 0$.

(4) For any $a, b \in \mathbb{R}$ such that $a < b$, we define the following sets:

$[a, b] = \{x \in \mathbb{R} \mid a \leq x \leq b\}$, (closed interval)

$(a, b) = \{x \in \mathbb{R} \mid a < x < b\}$, (open interval)

$[a, b) = \{x \in \mathbb{R} \mid a \leq x < b\}$, (interval closed on the left, open on the right)

$(a, b] = \{x \in \mathbb{R} \mid a < x \leq b\}$, (interval open on the left, closed on the right)

Let $E = [a, b]$, and $d(x, y) = |x - y|$. Then $([a, b], d)$ is a metric space.

We will need to define the notion of proximity in order to define convergence of limits and continuity of functions. For this we introduce some standard "small neighborhoods."

Definition A.2. Given a metric space E with metric d, for every $a \in E$, for every $\rho \in \mathbb{R}$, with $\rho > 0$, the set

$$B(a, \rho) = \{x \in E \mid d(a, x) \leq \rho\}$$

is called the *closed ball of center a and radius ρ*, the set

$$B_0(a, \rho) = \{x \in E \mid d(a, x) < \rho\}$$

is called the *open ball of center a and radius ρ*, and the set

$$S(a, \rho) = \{x \in E \mid d(a, x) = \rho\}$$

is called the *sphere of center a and radius ρ*. It should be noted that ρ is finite (i.e., not $+\infty$). A subset X of a metric space E is *bounded* if there is a closed ball $B(a, \rho)$ such that $X \subseteq B(a, \rho)$.

Clearly, $B(a, \rho) = B_0(a, \rho) \cup S(a, \rho)$.

Example A.2.

(1) In $E = \mathbb{R}$ with the distance $|x - y|$, an open ball of center a and radius ρ is the open interval $(a - \rho, a + \rho)$.
(2) In $E = \mathbb{R}^2$ with the Euclidean metric, an open ball of center a and radius ρ is the set of points inside the disk of center a and radius ρ, excluding the boundary points on the circle.
(3) In $E = \mathbb{R}^3$ with the Euclidean metric, an open ball of center a and radius ρ is the set of points inside the sphere of center a and radius ρ, excluding the boundary points on the sphere.

One should be aware that intuition can be misleading in forming a geometric image of a closed (or open) ball. For example, if d is the discrete metric, a closed ball of center a and radius $\rho < 1$ consists only of its center a, and a closed ball of center a and radius $\rho \geq 1$ consists of the entire space!

If $E = [a, b]$, and $d(x, y) = |x - y|$, as in Example A.1, an open ball $B_0(a, \rho)$, with $\rho < b - a$, is in fact the interval $[a, a + \rho)$, which is closed on the left.

We now consider a very important special case of metric spaces, normed vector spaces. Normed vector spaces have already been defined in Chapter B (Definition B.1) but for the reader's convenience we repeat the definition.

Definition A.3. Let E be a vector space over a field K, where K is either the field $\mathbb{R}$ of reals, or the field $\mathbb{C}$ of complex numbers. A *norm on E* is a function $\| \, \| \colon E \to \mathbb{R}_+$, assigning a nonnegative real number $\|u\|$ to any vector $u \in E$, and satisfying the following conditions for all $x, y \in E$:

(N1) $\|x\| \geq 0$, and $\|x\| = 0$ iff $x = 0$. (positivity)
(N2) $\|\lambda x\| = |\lambda| \, \|x\|$. (scaling)
(N3) $\|x + y\| \leq \|x\| + \|y\|$. (triangle inequality)

A vector space E together with a norm $\| \, \|$ is called a *normed vector space*. A function $\| \, \| \colon E \to \mathbb{R}_+$ satisfying only properties (N2) and (N3) is called a *semi-norm*.

From (N3), we easily get

$$|\|x\| - \|y\|| \leq \|x - y\|.$$

Given a normed vector space E, if we define d such that

$$d(x, y) = \|x - y\|,$$

it is easily seen that d is a metric. Thus, every normed vector space is immediately a metric space. Note that the metric associated with a norm is invariant under translation, that is,

$$d(x+u,\, y+u) = d(x,\, y).$$

For this reason, we can restrict ourselves to open or closed balls of center 0.

If $\| \; \|\colon E \to \mathbb{R}_+$ is a semi-norm, then $\|x\| = 0$ does not necessarily imply that $x = 0$. However by setting $\lambda = 0$ and $x = 0$ in (N2), we see that $\|0\| = 0$. If we let $\mathcal{N} = \{x \in E \mid \|x\| = 0\}$, then $\mathcal{N}$ is a subspace of E. Indeed, $0 \in \mathcal{N}$, and if $\|x\| = \|y\| = 0$, then by (N2) and (N3) we have

$$\|\lambda x + \mu y\| \le \|\lambda x\| + \|\mu y\| = |\lambda|\, \|x\| + |\mu|\, \|y\| = 0 + 0 = 0,$$

so $\lambda x + \mu y \in \mathcal{N}$. We can form the quotient space $E/\mathcal{N}$, and then it is easy to see that the semi-norm $\| \; \|$ induces a norm on $X/\mathcal{N}$.

Natural examples of semi-norms arise in integration theory; see Chapter 5.

Examples of normed vector spaces are given in Example B.1. We mention the most important examples.

Example A.3. Let $E = \mathbb{R}^n$ (or $E = \mathbb{C}^n$). There are three standard norms. For every $(x_1, \ldots, x_n) \in E$, we have the norm $\|x\|_1$, defined such that,

$$\|x\|_1 = |x_1| + \cdots + |x_n|,$$

we have the *Euclidean norm* $\|x\|_2$, defined such that,

$$\|x\|_2 = \left(|x_1|^2 + \cdots + |x_n|^2\right)^{\frac{1}{2}},$$

and the *sup*-norm $\|x\|_\infty$, defined such that,

$$\|x\|_\infty = \max\{|x_i| \mid 1 \le i \le n\}.$$

More generally, we define the *ℓ^p-norm* (for $p \ge 1$) by

$$\|x\|_p = (|x_1|^p + \cdots + |x_n|^p)^{1/p}.$$

We prove in Proposition B.1 that the ℓ^p-norms are indeed norms. The closed unit balls centered at $(0, 0)$ for $\| \; \|_1$, $\| \; \|_2$, and $\| \; \|_\infty$, along with the containment relationships, are shown in Figures A.1 and A.2. Figures A.3 and A.4 illustrate the situation in $\mathbb{R}^3$.

In a normed vector space we define a closed ball or an open ball of radius ρ as a closed ball or an open ball of center 0. We may use the notation $B(\rho)$ and $B_0(\rho)$.

We will now define the crucial notions of open sets and closed sets, and of a topological space.

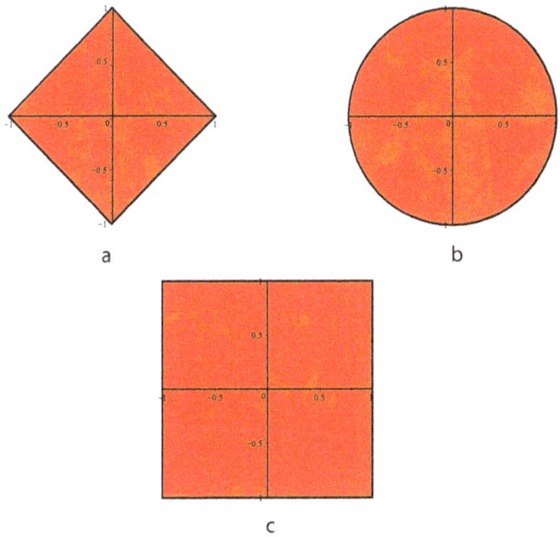

Fig. A.1 Figure (a) shows the diamond shaped closed ball associated with $\| \|_1$. Figure (b) shows the closed unit disk associated with $\| \|_2$, while Figure (c) illustrates the closed unit ball associated with $\| \|_\infty$.

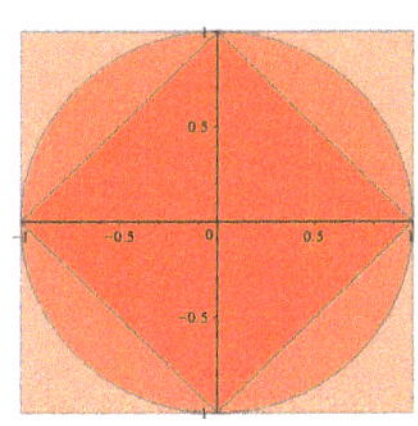

Fig. A.2 The relationship between the closed unit balls centered at $(0, 0)$.

Definition A.4. Let (E, d) be a metric space. A subset $U \subseteq E$ is an *open set* in E if either $U = \emptyset$, or for every $a \in U$, there is some open ball $B_0(a, \rho)$ such that, $B_0(a, \rho) \subseteq U$.[3] A subset $F \subseteq E$ is a *closed set* in E if its complement $E - F$ is open in E. See Figure A.5.

The set E itself is open, since for every $a \in E$, every open ball of center a is contained in E. In $E = \mathbb{R}^n$, given n intervals $[a_i, b_i]$, with $a_i < b_i$, it is easy to show that the open n-cube

$$\{(x_1, \ldots, x_n) \in E \mid a_i < x_i < b_i,\, 1 \leq i \leq n\}$$

[3]Recall that $\rho > 0$.

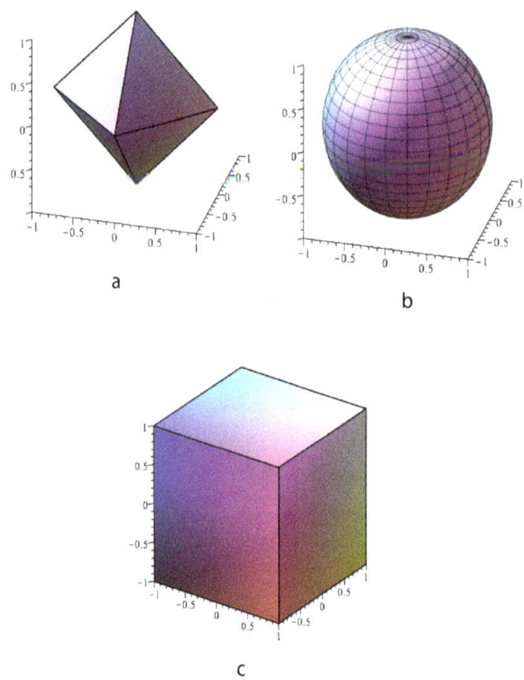

Fig. A.3 Figure (a) shows the octahedral shaped closed ball associated with $\| \|_1$. Figure (b) shows the closed spherical associated with $\| \|_2$, while Figure (c) illustrates the closed unit ball associated with $\| \|_\infty$.

is an open set. In fact, it is possible to find a metric for which such open n-cubes are open balls! Similarly, we can define the closed n-cube

$$\{(x_1, \ldots, x_n) \in E \mid a_i \leq x_i \leq b_i,\ 1 \leq i \leq n\},$$

which is a closed set.

The open sets satisfy some important properties that lead to the definition of a topological space.

Proposition A.1. *Given a metric space E with metric d, the family $\mathcal{O}$ of all open sets defined in Definition A.4 satisfies the following properties:*

(O1) For every finite family $(U_i)_{1 \leq i \leq n}$ of sets $U_i \in \mathcal{O}$, we have $U_1 \cap \cdots \cap U_n \in \mathcal{O}$, i.e., $\mathcal{O}$ is closed under finite intersections.

(O2) For every arbitrary family $(U_i)_{i \in I}$ of sets $U_i \in \mathcal{O}$, we have $\bigcup_{i \in I} U_i \in \mathcal{O}$, i.e., $\mathcal{O}$ is closed under arbitrary unions.

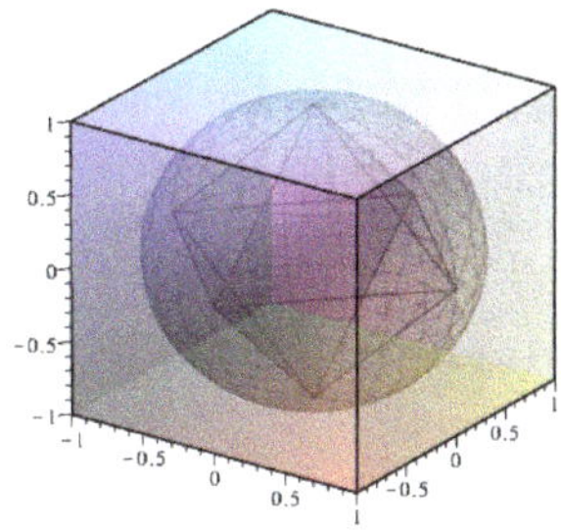

Fig. A.4 The relationship between the closed unit balls centered at $(0, 0, 0)$.

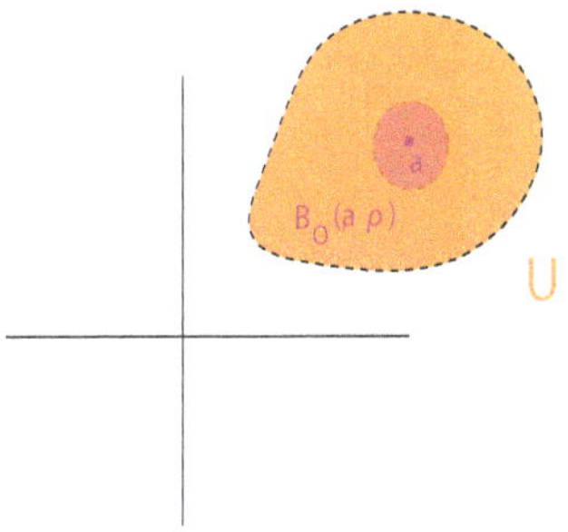

Fig. A.5 An open set U in $E = \mathbb{R}^2$ under the standard Euclidean metric. Any point in the peach set U is surrounded by a small raspberry open set which lies within U.

(O3) $\emptyset \in \mathcal{O}$, *and* $E \in \mathcal{O}$, *i.e.,* $\emptyset$ *and* E *belong to* $\mathcal{O}$.

Furthermore, for any two distinct points $a \neq b$ *in* E, *there exist two open sets* U_a *and* U_b *such that,* $a \in U_a$, $b \in U_b$, *and* $U_a \cap U_b = \emptyset$.

Proof. It is straightforward. For the last point, letting $\rho = d(a, b)/3$ (in fact $\rho = d(a, b)/2$ works too), we can pick $U_a = B_0(a, \rho)$ and $U_b = B_0(b, \rho)$. By the triangle inequality, we must have $U_a \cap U_b = \emptyset$. □

The above proposition leads to the very general concept of a topological space.

One should be careful that, in general, the family of open sets is not closed under infinite intersections. For example, in $\mathbb{R}$ under the metric

$|x - y|$, letting $U_n = (-1/n, +1/n)$, each U_n is open, but $\bigcap_n U_n = \{0\}$, which is not open.

Later on, given any nonempty subset A of a metric space (E, d), we will need to know that certain special sets containing A are open.

Definition A.5. Let (E, d) be a metric space. For any nonempty subset A of E and any $x \in E$, let

$$d(x, A) = \inf_{a \in A} d(x, a).$$

Proposition A.2. *Let (E, d) be a metric space. For any nonempty subset A of E and for any two points $x, y \in E$, we have*

$$|d(x, A) - d(y, A)| \leq d(x, y).$$

Proof. For all $a \in A$ we have

$$d(x, a) \leq d(x, y) + d(y, a),$$

which implies

$$\begin{aligned} d(x, A) &= \inf_{a \in A} d(x, a) \\ &\leq \inf_{a \in A} (d(x, y) + d(y, a)) \\ &= d(x, y) + \inf_{a \in A} d(y, a) \\ &= d(x, y) + d(y, A). \end{aligned}$$

By symmetry, we also obtain $d(y, A) \leq d(x, y) + d(x, A)$, and thus

$$|d(x, A) - d(y, A)| \leq d(x, y),$$

as claimed. ☐

Definition A.6. Let (E, d) be a metric space. For any nonempty subset A of E, and any $r > 0$, let

$$V_r(A) = \{x \in E \mid d(x, A) < r\}.$$

Proposition A.3. *Let (E, d) be a metric space. For any nonempty subset A of E, and any $r > 0$, the set $V_r(A)$ is an open set containing A.*

Proof. For any $y \in E$ such that $d(x, y) < r - d(x, A)$, by Proposition A.2 we have

$$d(y, A) \leq d(x, A) + d(x, y) \leq d(x, A) + r - d(x, A) = r,$$

so $V_r(A)$ contains the open ball $B_0(x, r - d(x, A))$, which means that it is open. Obviously, $A \subseteq V_r(A)$. ☐

A.2 Topological Spaces

Motivated by Proposition A.1, a topological space is defined in terms of a family of sets satisfying the properties of open sets stated in that proposition.

Definition A.7. Given a set E, a *topology on E (or a topological structure on E)*, is defined as a family $\mathcal{O}$ of subsets of E called *open sets*, and satisfying the following three properties:

(1) For every finite family $(U_i)_{1 \leq i \leq n}$ of sets $U_i \in \mathcal{O}$, we have $U_1 \cap \cdots \cap U_n \in \mathcal{O}$, i.e., $\mathcal{O}$ is closed under finite intersections.
(2) For every arbitrary family $(U_i)_{i \in I}$ of sets $U_i \in \mathcal{O}$, we have $\bigcup_{i \in I} U_i \in \mathcal{O}$, i.e., $\mathcal{O}$ is closed under arbitrary unions.
(3) $\emptyset \in \mathcal{O}$, and $E \in \mathcal{O}$, i.e., $\emptyset$ and E belong to $\mathcal{O}$.

A set E together with a topology $\mathcal{O}$ on E is called a *topological space*. Given a topological space $(E, \mathcal{O})$, a subset F of E is a *closed set* if $F = E - U$ for some open set $U \in \mathcal{O}$, i.e., F is the complement of some open set.

It is possible that an open set is also a closed set. For example, $\emptyset$ and E are both open and closed. When a topological space contains a proper nonempty subset U which is both open and closed, the space E is said to be *disconnected*.

Definition A.8. A topological space $(E, \mathcal{O})$ is said to satisfy the *Hausdorff separation axiom (or T_2-separation axiom)* if for any two distinct points $a \neq b$ in E, there exist two open sets U_a and U_b such that, $a \in U_a$, $b \in U_b$, and $U_a \cap U_b = \emptyset$. When the T_2-separation axiom is satisfied, we also say that $(E, \mathcal{O})$ is a *Hausdorff space*.

As shown by Proposition A.1, any metric space is a topological Hausdorff space, the family of open sets being in fact the family of arbitrary unions of open balls. Similarly, any normed vector space is a topological Hausdorff space, the family of open sets being the family of arbitrary unions of open balls. The topology $\mathcal{O}$ consisting of all subsets of E is called the *discrete topology*.

Remark: Most (if not all) spaces used in analysis are Hausdorff spaces. Intuitively, the Hausdorff separation axiom says that there are enough "small" open sets. Without this axiom, some counter-intuitive behaviors may arise.

For example, a sequence may have more than one limit point (or a compact set may not be closed). Nevertheless, non-Hausdorff topological spaces arise naturally in algebraic geometry. But even there, some substitute for separation is used.

One of the reasons why topological spaces are important is that the definition of a topology only involves a certain family $\mathcal{O}$ of sets, and not **how** such family is generated from a metric or a norm. For example, different metrics or different norms can define the same family of open sets. Many topological properties only depend on the family $\mathcal{O}$ and not on the specific metric or norm. But the fact that a topology is definable from a metric or a norm is important, because it usually implies nice properties of a space. All our examples will be spaces whose topology is defined by a metric or a norm.

Definition A.9. A topological space $(E, \mathcal{O})$ is *metrizable* if there is a distance on E defining the topology $\mathcal{O}$.

Note that in a metric space (E, d), the metric d is *explicitly given*. However, in general, the topology of a metrizable space $(E, \mathcal{O})$ is not specified by an explicitly given metric, but *some metric* defining the topology $\mathcal{O}$ exists. Obviously, a metrizable topological space must be Hausdorff. Actually, a stronger separation property holds, a metrizable space is normal; see Definition A.30.

Remark: By taking complements we can state properties of the closed sets dual to those of Definition A.7. Thus, $\emptyset$ and E are closed sets, and the closed sets are closed under finite unions and arbitrary intersections.

It is also worth noting that the Hausdorff separation axiom implies that for every $a \in E$, the set $\{a\}$ is closed. Indeed, if $x \in E - \{a\}$, then $x \neq a$, and so there exist open sets U_a and U_x such that $a \in U_a$, $x \in U_x$, and $U_a \cap U_x = \emptyset$. See Figure A.6. Thus, for every $x \in E - \{a\}$, there is an open set U_x containing x and contained in $E - \{a\}$, showing by (O3) that $E - \{a\}$ is open, and thus that the set $\{a\}$ is closed.

Given a topological space $(E, \mathcal{O})$, given any subset A of E, since $E \in \mathcal{O}$ and E is a closed set, the family $\mathcal{C}_A = \{F \mid A \subseteq F,\ F \text{ a closed set}\}$ of closed sets containing A is nonempty, and since any arbitrary intersection of closed sets is a closed set, the intersection $\bigcap \mathcal{C}_A$ of the sets in the family $\mathcal{C}_A$ is the smallest closed set containing A. By a similar reasoning, the union of all the open subsets contained in A is the largest open set contained in A.

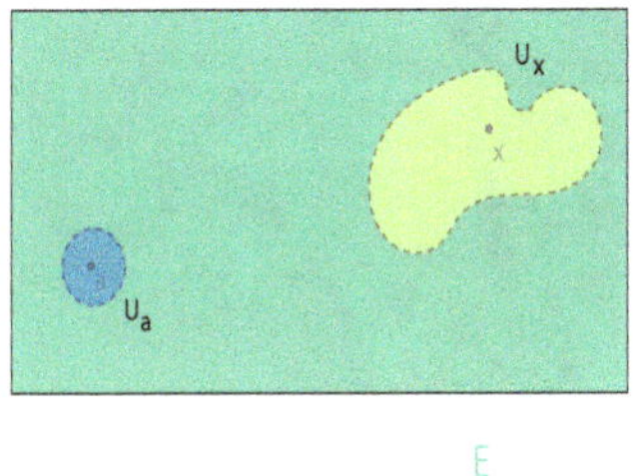

Fig. A.6 A schematic illustration of the Hausdorff separation property.

Definition A.10. Given a topological space $(E, \mathcal{O})$, given any subset A of E, the smallest closed set containing A is denoted by $\overline{A}$, and is called the *closure, or adherence of* A. See Figure A.7. A subset A of E is *dense in* E if $\overline{A} = E$. The largest open set contained in A is denoted by $\overset{\circ}{A}$ and is called the *interior of* A. See Figure A.8. The set $\text{Fr}\, A = \overline{A} \cap \overline{E - A}$ is called the *boundary (or frontier) of* A. We also denote the boundary of A by ∂A. See Figure A.9.

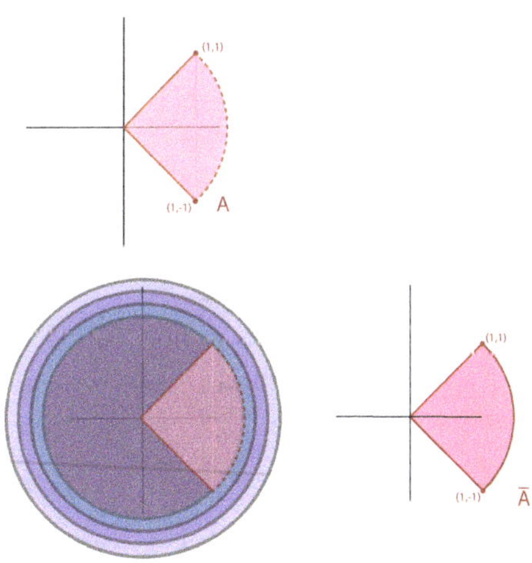

Fig. A.7 The topological space $(E, \mathcal{O})$ is $\mathbb{R}^2$ with topology induced by the Euclidean metric. The subset A is the section $B_0(1)$ in the first and fourth quadrants bound by the lines $y = x$ and $y = -x$. The closure of A is obtained by the intersection of A with the closed unit ball.

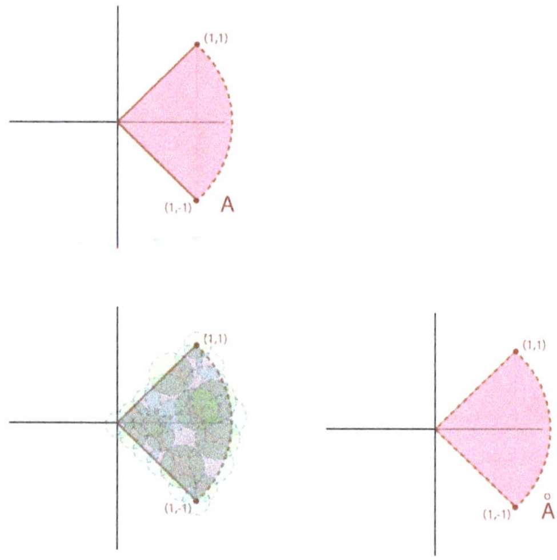

Fig. A.8 The topological space $(E, \mathcal{O})$ is $\mathbb{R}^2$ with topology induced by the Euclidean metric. The subset A is the section $B_0(1)$ in the first and fourth quadrants bound by the lines $y = x$ and $y = -x$. The interior of A is obtained by the covering A with small open balls.

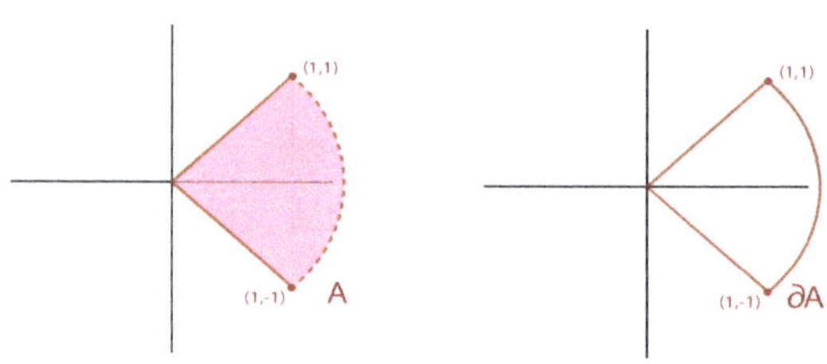

Fig. A.9 The topological space $(E, \mathcal{O})$ is $\mathbb{R}^2$ with topology induced by the Euclidean metric. The subset A is the section $B_0(1)$ in the first and fourth quadrants bound by the lines $y = x$ and $y = -x$. The boundary of A is $\overline{A} - \overset{\circ}{A}$.

Remark: The notation $\overline{A}$ for the closure of a subset A of E is somewhat unfortunate, since $\overline{A}$ is often used to denote the set complement of A in E. Still, we prefer it to more cumbersome notations such as $\text{clo}(A)$, and we denote the complement of A in E by $E - A$ (or sometimes, A^c).

By definition, it is clear that a subset A of E is closed iff $A = \overline{A}$. The set $\mathbb{Q}$ of rationals is dense in $\mathbb{R}$. It is easily shown that $\overline{A} = \overset{\circ}{A} \cup \partial A$ and $\overset{\circ}{A} \cap \partial A = \emptyset$. Another useful characterization of $\overline{A}$ is given by the following proposition.

Proposition A.4. *Given a topological space $(E, \mathcal{O})$, given any subset A of E, the closure $\overline{A}$ of A is the set of all points $x \in E$ such that for every open set U containing x, then $U \cap A \neq \emptyset$. See Figure A.10.*

Proof. If $A = \emptyset$, since $\emptyset$ is closed, the proposition holds trivially. Thus, assume that $A \neq \emptyset$. First assume that $x \in \overline{A}$. Let U be any open set such that $x \in U$. If $U \cap A = \emptyset$, since U is open, then $E - U$ is a closed set containing A, and since $\overline{A}$ is the intersection of all closed sets containing A, we must have $x \in E - U$, which is impossible. Conversely, assume that $x \in E$ is a point such that for every open set U containing x, then $U \cap A \neq \emptyset$. Let F be any closed subset containing A. If $x \notin F$, since F is closed, then $U = E - F$ is an open set such that $x \in U$, and $U \cap A = \emptyset$, a contradiction. Thus, we have $x \in F$ for every closed set containing A, that is, $x \in \overline{A}$. $\square$

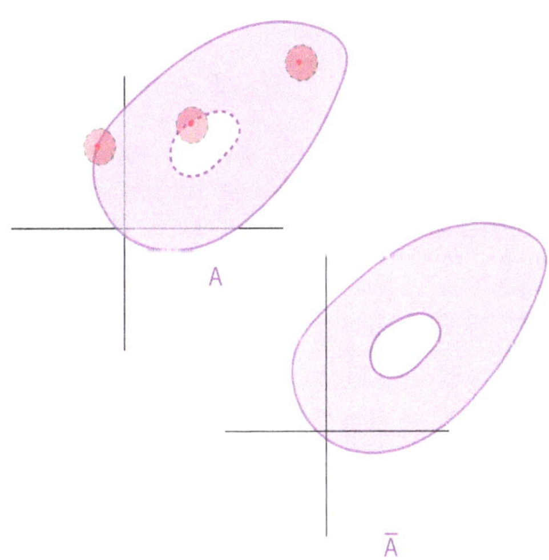

Fig. A.10 The topological space $(E, \mathcal{O})$ is $\mathbb{R}^2$ with topology induced by the Euclidean metric. The purple subset A is illustrated with three red points, each in its closure since the open ball centered at each point has nontrivial intersection with A.

Often it is necessary to consider a subset A of a topological space E, and to view the subset A as a topological space. The following proposition shows how to define a topology on a subset.

Proposition A.5. *Given a topological space $(E, \mathcal{O})$, given any subset A of E, let*

$$\mathcal{U} = \{U \cap A \mid U \in \mathcal{O}\}$$

be the family of all subsets of A obtained as the intersection of any open set in $\mathcal{O}$ with A. The following properties hold.

(1) The space $(A, \mathcal{U})$ is a topological space.
(2) If E is a metric space with metric d, then the restriction $d_A \colon A \times A \to \mathbb{R}_+$ of the metric d to A defines a metric space. Furthermore, the topology induced by the metric d_A agrees with the topology defined by $\mathcal{U}$, as above.

Proof. Left as an exercise. □

Proposition A.5 suggests the following definition.

Definition A.11. Given a topological space $(E, \mathcal{O})$, given any subset A of E, the *subspace topology on A induced by $\mathcal{O}$* is the family $\mathcal{U}$ of open sets defined such that

$$\mathcal{U} = \{U \cap A \mid U \in \mathcal{O}\}$$

is the family of all subsets of A obtained as the intersection of any open set in $\mathcal{O}$ with A. We say that $(A, \mathcal{U})$ has the *subspace topology*. If (E, d) is a metric space, the restriction $d_A \colon A \times A \to \mathbb{R}_+$ of the metric d to A is called the *subspace metric*.

For example, if $E = \mathbb{R}^n$ and d is the Euclidean metric, we obtain the subspace topology on the closed n-cube

$$\{(x_1, \ldots, x_n) \in E \mid a_i \leq x_i \leq b_i,\, 1 \leq i \leq n\}.$$

See Figure A.11.

One should realize that every open set $U \in \mathcal{O}$ which is entirely contained in A is also in the family $\mathcal{U}$, but $\mathcal{U}$ may contain open sets that are not in $\mathcal{O}$. For example, if $E = \mathbb{R}$ with $|x - y|$, and $A = [a, b]$, then sets of the form $[a, c)$, with $a < c < b$ belong to $\mathcal{U}$, but they are not open sets for $\mathbb{R}$ under $|x - y|$. However, there is agreement in the following situation.

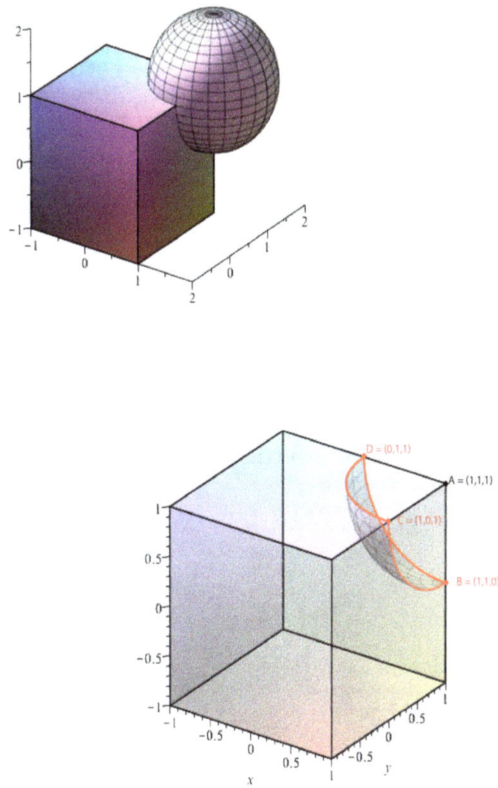

Fig. A.11 An example of an open set in the subspace topology for $\{(x, y, z) \in \mathbb{R}^3 \mid -1 \leq x \leq 1, -1 \leq y \leq 1, -1 \leq z \leq 1\}$. The open set is the corner region $ABCD$ and is obtained by intersecting with the cube $B_0((1, 1, 1), 1)$.

Proposition A.6. *Given a topological space $(E, \mathcal{O})$, given any subset A of E, if $\mathcal{U}$ is the subspace topology, then the following properties hold.*

(1) If A is an open set $A \in \mathcal{O}$, then every open set $U \in \mathcal{U}$ is an open set $U \in \mathcal{O}$.
(2) If A is a closed set in E, then every closed set w.r.t. the subspace topology is a closed set w.r.t. $\mathcal{O}$.

Proof. Left as an exercise. $\square$

The concept of product topology is also useful. We have the following proposition.

Proposition A.7. *Given n topological spaces $(E_i, \mathcal{O}_i)$, let $\mathcal{B}$ be the family of subsets of $E_1 \times \cdots \times E_n$ defined as follows:*

$$\mathcal{B} = \{U_1 \times \cdots \times U_n \mid U_i \in \mathcal{O}_i,\, 1 \leq i \leq n\},$$

and let $\mathcal{P}$ be the family consisting of arbitrary unions of sets in $\mathcal{B}$, including $\emptyset$. Then $\mathcal{P}$ is a topology on $E_1 \times \cdots \times E_n$.

Proof. Left as an exercise. □

Definition A.12. Given n topological spaces $(E_i, \mathcal{O}_i)$, the *product topology on* $E_1 \times \cdots \times E_n$ is the family $\mathcal{P}$ of subsets of $E_1 \times \cdots \times E_n$ defined as follows: if

$$\mathcal{B} = \{U_1 \times \cdots \times U_n \mid U_i \in \mathcal{O}_i,\, 1 \leq i \leq n\},$$

then $\mathcal{P}$ is the family consisting of arbitrary unions of sets in $\mathcal{B}$, including $\emptyset$. See Figure A.12.

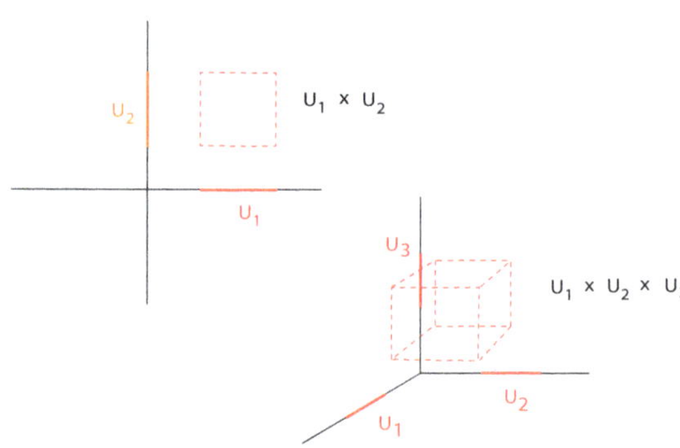

Fig. A.12 Examples of open sets in the product topology for $\mathbb{R}^2$ and $\mathbb{R}^3$ induced by the Euclidean metric.

If each (E_i, d_{E_i}) is a metric space, there are three natural metrics that can be defined on $E_1 \times \cdots \times E_n$:

$$d_1((x_1,\ldots,x_n),(y_1,\ldots,y_n)) = d_{E_1}(x_1,y_1) + \cdots + d_{E_n}(x_n,y_n),$$

$$d_2((x_1,\ldots,x_n),(y_1,\ldots,y_n)) = \left((d_{E_1}(x_1,y_1))^2 + \cdots + (d_{E_n}(x_n,y_n))^2\right)^{\frac{1}{2}},$$

$$d_\infty((x_1,\ldots,x_n),(y_1,\ldots,y_n)) = \max\{d_{E_1}(x_1,y_1),\ldots,d_{E_n}(x_n,y_n)\}.$$

It is easy to show that

$$\begin{aligned} d_\infty((x_1,\ldots,x_n),(y_1,\ldots,y_n)) &\le d_2((x_1,\ldots,x_n),(y_1,\ldots,y_n)) \\ &\le d_1((x_1,\ldots,x_n),(y_1,\ldots,y_n)) \\ &\le n d_\infty((x_1,\ldots,x_n),(y_1,\ldots,y_n)), \end{aligned}$$

so these distances define the same topology, which is the product topology.

If each $(E_i, \| \|_{E_i})$ is a normed vector space, there are three natural norms that can be defined on $E_1 \times \cdots \times E_n$:

$$\begin{aligned} \|(x_1,\ldots,x_n)\|_1 &= \|x_1\|_{E_1} + \cdots + \|x_n\|_{E_n}, \\ \|(x_1,\ldots,x_n)\|_2 &= \left(\|x_1\|_{E_1}^2 + \cdots + \|x_n\|_{E_n}^2\right)^{\frac{1}{2}}, \\ \|(x_1,\ldots,x_n)\|_\infty &= \max\left\{\|x_1\|_{E_1},\ldots,\|x_n\|_{E_n}\right\}. \end{aligned}$$

It is easy to show that

$$\|(x_1,\ldots,x_n)\|_\infty \le \|(x_1,\ldots,x_n)\|_2 \le \|(x_1,\ldots,x_n)\|_1 \le n\|(x_1,\ldots,x_n)\|_\infty,$$

so these norms define the same topology, which is the product topology. It can also be verified that when $E_i = \mathbb{R}$, with the standard topology induced by $|x - y|$, the topology product on $\mathbb{R}^n$ is the standard topology induced by the Euclidean norm.

Definition A.13. Two metrics d and d' on a space E are *equivalent* if they induce the same topology $\mathcal{O}$ on E (i.e., they define the same family $\mathcal{O}$ of open sets). Similarly, two norms $\| \|$ and $\| \|'$ on a space E are *equivalent* if they induce the same topology $\mathcal{O}$ on E.

Given a topological space $(E, \mathcal{O})$, it is often useful, as in Proposition A.7, to define the topology $\mathcal{O}$ in terms of a subfamily $\mathcal{B}$ of subsets of E.

Definition A.14. We say that a family $\mathcal{B}$ of subsets of E is a *basis for the topology* $\mathcal{O}$, if $\mathcal{B}$ is a subset of $\mathcal{O}$, and if every open set U in $\mathcal{O}$ can be obtained as some union (possibly infinite) of sets in $\mathcal{B}$ (agreeing that the empty union is the empty set).

For example, given any metric space (E, d), $\mathcal{B} = \{B_0(a, \rho) \mid a \in E, \rho > 0\}$. In particular, if $d = \| \|_2$, the open intervals form a basis for $\mathbb{R}$, while the open disks form a basis for $\mathbb{R}^2$. The open rectangles also form a basis for $\mathbb{R}^2$ with the standard topology.

It is immediately verified that if a family $\mathcal{B} = (U_i)_{i \in I}$ is a basis for the topology of $(E, \mathcal{O})$, then $E = \bigcup_{i \in I} U_i$, and the intersection of any two sets $U_i, U_j \in \mathcal{B}$ is the union of some sets in the family $\mathcal{B}$ (again, agreeing that the empty union is the empty set). Conversely, a family $\mathcal{B}$ with these properties is the basis of the topology obtained by forming arbitrary unions of sets in $\mathcal{B}$.

Definition A.15. A *subbasis for* $\mathcal{O}$ is a family $\mathcal{S}$ of subsets of E, such that the family $\mathcal{B}$ of all finite intersections of sets in $\mathcal{S}$ (including E itself, in case of the empty intersection) is a basis of $\mathcal{O}$. See Figure A.13.

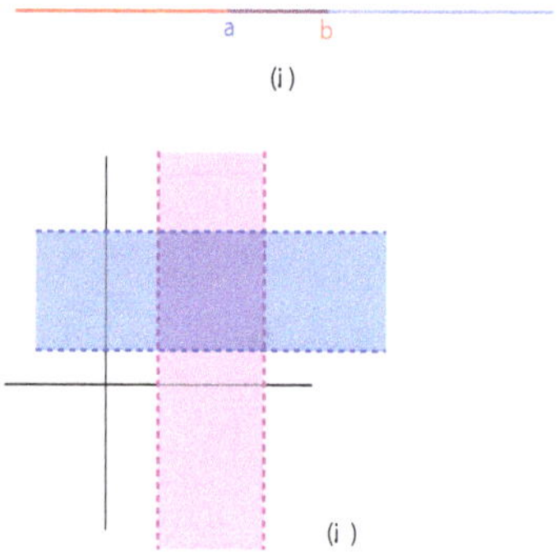

Fig. A.13 Figure (*i.*) shows that the set of infinite open intervals forms a subbasis for $\mathbb{R}$. Figure (*ii.*) shows that the infinite open strips form a subbasis for $\mathbb{R}^2$.

The following proposition gives useful criteria for determining whether a family of open subsets is a basis of a topological space.

Proposition A.8. *Given a topological space* $(E, \mathcal{O})$ *and a family* $\mathcal{B}$ *of open subsets in* $\mathcal{O}$ *the following properties hold:*

(1) *The family* $\mathcal{B}$ *is a basis for the topology* $\mathcal{O}$ *iff for every open set* $U \in \mathcal{O}$ *and every* $x \in U$*, there is some* $B \in \mathcal{B}$ *such that* $x \in B$ *and* $B \subseteq U$*. See Figure A.14.*

(2) *The family* $\mathcal{B}$ *is a basis for the topology* $\mathcal{O}$ *iff*

  (a) *For every* $x \in E$*, there is some* $B \in \mathcal{B}$ *such that* $x \in B$*.*

  (b) *For any two open subsets,* $B_1, B_2 \in \mathcal{B}$*, for every* $x \in E$*, if* $x \in B_1 \cap B_2$*, then there is some* $B_3 \in \mathcal{B}$ *such that* $x \in B_3$ *and* $B_3 \subseteq B_1 \cap B_2$*. See Figure A.15.*

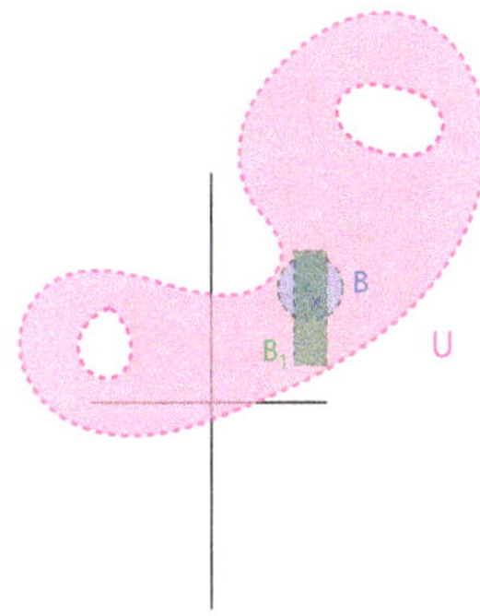

Fig. A.14 Given an open subset U of $\mathbb{R}^2$ and $x \in U$, there exists an open ball B containing x with $B \subset U$. There also exists an open rectangle B_1 containing x with $B_1 \subset U$.

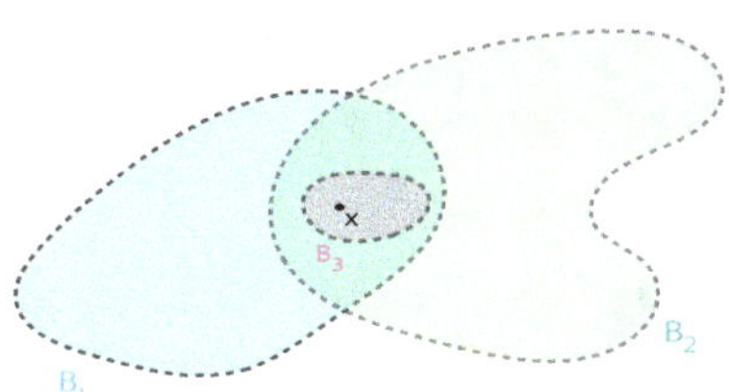

Fig. A.15 A schematic illustration of Condition (b) in Proposition A.8.

We now consider the fundamental property of continuity.

A.3 Continuous Functions, Limits

Definition A.16. Let $(E, \mathcal{O}_E)$ and $(F, \mathcal{O}_F)$ be topological spaces, and let $f\colon E \to F$ be a function. For every $a \in E$, we say that f *is continuous at* a, if for every open set $V \in \mathcal{O}_F$ containing $f(a)$ there is some open set $U \in \mathcal{O}_E$ containing a, such that $f(U) \subseteq V$. See Figure A.16. We say that f *is continuous* if it is continuous at every $a \in E$.

Define a *neighborhood of* $a \in E$ as any subset N of E containing some open set $O \in \mathcal{O}$ such that $a \in O$. If f is continuous at a and N is any neighborhood of $f(a)$, there is some open set $V \subseteq N$ containing $f(a)$, and since f is continuous at a, there is some open set U containing a,

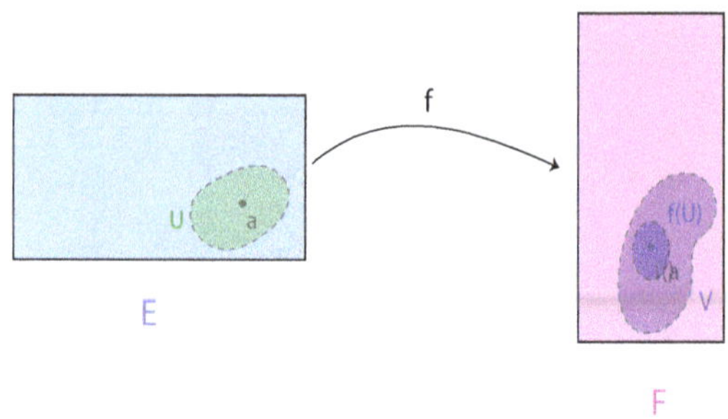

Fig. A.16 A schematic illustration of Definition A.16.

such that $f(U) \subseteq V$. Since $V \subseteq N$, the open set U is a subset of $f^{-1}(N)$ containing a, and $f^{-1}(N)$ is a neighborhood of a. Conversely, if $f^{-1}(N)$ is a neighborhood of a whenever N is any neighborhood of $f(a)$, it is immediate that f is continuous at a. See Figure A.17.

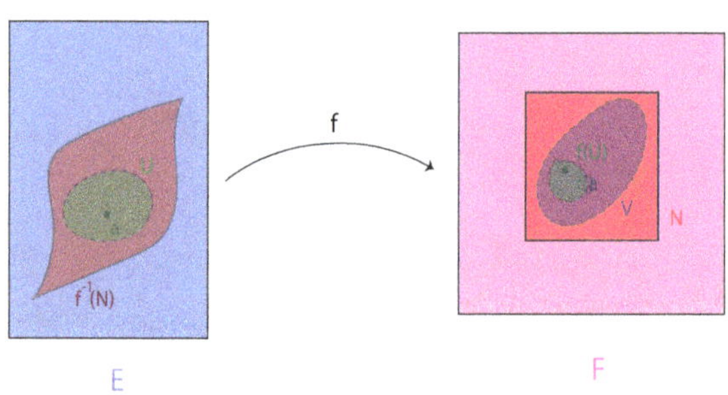

Fig. A.17 A schematic illustration of the neighborhood condition.

It is easy to see that Definition A.16 is equivalent to the following statements.

Proposition A.9. *Let $(E, \mathcal{O}_E)$ and $(F, \mathcal{O}_F)$ be topological spaces, and let $f \colon E \to F$ be a function. For every $a \in E$, the function f is continuous at $a \in E$ iff for every neighborhood N of $f(a) \in F$, then $f^{-1}(N)$ is a neighborhood of a. The function f is continuous on E iff $f^{-1}(V)$ is an open set in $\mathcal{O}_E$ for every open set $V \in \mathcal{O}_F$.*

If E and F are metric spaces defined by metrics d_E and d_F, we can show easily that f is continuous at a iff

for every $\epsilon > 0$, there is some $\eta > 0$, such that, for every $x \in E$,

$$\text{if } d_E(a, x) \leq \eta, \text{ then } d_F(f(a), f(x)) \leq \epsilon.$$

Similarly, if E and F are normed vector spaces defined by norms $\| \, \|_E$ and $\| \, \|_F$, we can show easily that f is continuous at a iff

for every $\epsilon > 0$, there is some $\eta > 0$, such that, for every $x \in E$,

$$\text{if } \|x - a\|_E \leq \eta, \text{ then } \|f(x) - f(a)\|_F \leq \epsilon.$$

It is worth noting that continuity is a topological notion, in the sense that equivalent metrics (or equivalent norms) define exactly the same notion of continuity.

Definition A.17. If $(E, \mathcal{O}_E)$ and $(F, \mathcal{O}_F)$ are topological spaces and $f \colon E \to F$ is a function, for every nonempty subset $A \subseteq E$ of E, we say that f *is continuous on* A if the restriction of f to A is continuous with respect to $(A, \mathcal{U})$ and $(F, \mathcal{O}_F)$, where $\mathcal{U}$ is the subspace topology induced by $\mathcal{O}_E$ on A.

Given a product $E_1 \times \cdots \times E_n$ of topological spaces, as usual, we let $\pi_i \colon E_1 \times \cdots \times E_n \to E_i$ be the projection function such that, $\pi_i(x_1, \ldots, x_n) = x_i$. It is immediately verified that each π_i is continuous.

Given a topological space $(E, \mathcal{O})$, we say that a point $a \in E$ is *isolated* if $\{a\}$ is an open set in $\mathcal{O}$. Then if $(E, \mathcal{O}_E)$ and $(F, \mathcal{O}_F)$ are topological spaces, any function $f \colon E \to F$ is continuous at every isolated point $a \in E$. In the discrete topology, every point is isolated.

In a nontrivial normed vector space $(E, \| \, \|)$ (with $E \neq \{0\}$), no point is isolated. To show this, we show that every open ball $B_0(u, \rho,)$ contains some vectors different from u. Indeed, since E is nontrivial, there is some $v \in E$ such that $v \neq 0$, and thus $\lambda = \|v\| > 0$ (by (N1)). Let

$$w = u + \frac{\rho}{\lambda + 1} v.$$

Since $v \neq 0$ and $\rho > 0$, we have $w \neq u$. Then,

$$\|w - u\| = \left\| \frac{\rho}{\lambda + 1} v \right\| = \frac{\rho \lambda}{\lambda + 1} < \rho,$$

which shows that $\|w - u\| < \rho$, for $w \neq u$.

The following proposition is easily shown.

Proposition A.10. *Given topological spaces $(E, \mathcal{O}_E)$, $(F, \mathcal{O}_F)$, and $(G, \mathcal{O}_G)$, and two functions $f\colon E \to F$ and $g\colon F \to G$, if f is continuous at $a \in E$ and g is continuous at $f(a) \in F$, then $g \circ f\colon E \to G$ is continuous at $a \in E$. Given n topological spaces $(F_i, \mathcal{O}_i)$, for every function $f\colon E \to F_1 \times \cdots \times F_n$, then f is continuous at $a \in E$ iff every $f_i\colon E \to F_i$ is continuous at a, where $f_i = \pi_i \circ f$.*

One can also show that in a metric space (E, d), the distance $d\colon E \times E \to \mathbb{R}$ is continuous, where $E \times E$ has the product topology. By the triangle inequality, we have

$$d(x, y) \leq d(x, x_0) + d(x_0, y_0) + d(y_0, y) = d(x_0, y_0) + d(x_0, x) + d(y_0, y)$$

and

$$d(x_0, y_0) \leq d(x_0, x) + d(x, y) + d(y, y_0) = d(x, y) + d(x_0, x) + d(y_0, y).$$

Consequently,

$$|d(x, y) - d(x_0, y_0)| \leq d(x_0, x) + d(y_0, y),$$

which proves that d is continuous at (x_0, y_0). In fact this shows that d is uniformly continuous; see Definition A.45.

Given any nonempty subset A of E, by Proposition A.2, the map $x \mapsto d(x, A)$ is continuous (in fact, uniformy continuous).

Similarly, for a normed vector space $(E, \| \|)$, the norm $\| \|\colon E \to \mathbb{R}$ is (uniformly) continuous.

Given a function $f\colon E_1 \times \cdots \times E_n \to F$, we can fix $n-1$ of the arguments, say $a_1, \ldots, a_{i-1}, a_{i+1}, \ldots, a_n$, and view f as a function of the remaining argument,

$$x_i \mapsto f(a_1, \ldots, a_{i-1}, x_i, a_{i+1}, \ldots, a_n),$$

where $x_i \in E_i$. If f is continuous, it is clear that each f_i is continuous.

One should be careful that the converse is false! For example, consider the function $f\colon \mathbb{R} \times \mathbb{R} \to \mathbb{R}$, defined such that,

$$f(x, y) = \frac{xy}{x^2 + y^2} \quad \text{if } (x, y) \neq (0, 0), \quad \text{and} \quad f(0, 0) = 0.$$

The function f is continuous on $\mathbb{R} \times \mathbb{R} - \{(0, 0)\}$, but on the line $y = mx$, with $m \neq 0$, we have $f(x, y) = \frac{m}{1+m^2} \neq 0$, and thus, on this line, $f(x, y)$ does not approach 0 when (x, y) approaches $(0, 0)$. See Figure A.18.

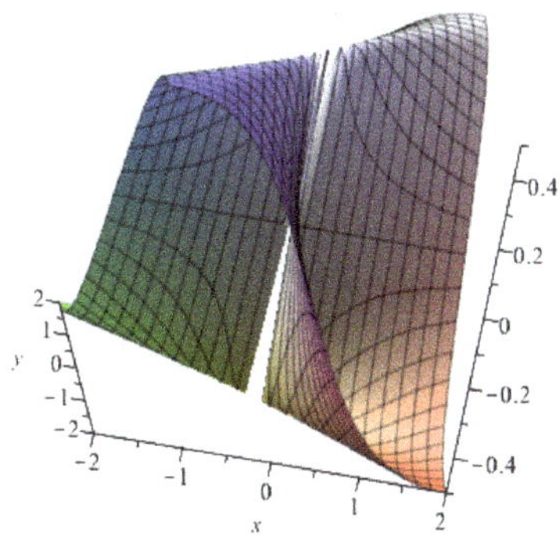

Fig. A.18 The graph of $f(x, y) = \frac{xy}{x^2+y^2}$ for $(x, y) \neq (0, 0)$. The bottom of this graph, which shows the approach along the line $y = -x$, does not have a z value of 0.

The following proposition is useful for showing that real-valued functions are continuous.

Proposition A.11. *If E is a topological space, and $(\mathbb{R}, |x - y|)$ the reals under the standard topology, for any two functions $f\colon E \to \mathbb{R}$ and $g\colon E \to \mathbb{R}$, for any $a \in E$, for any $\lambda \in \mathbb{R}$, if f and g are continuous at a, then $f + g$, λf, $f \cdot g$, are continuous at a, and f/g is continuous at a if $g(a) \neq 0$.*

Proof. Left as an exercise. □

Using Proposition A.11, we can show easily that every real polynomial function is continuous.

The notion of isomorphism of topological spaces is defined as follows.

Definition A.18. Let $(E, \mathcal{O}_E)$ and $(F, \mathcal{O}_F)$ be topological spaces, and let $f\colon E \to F$ be a function. We say that f *is a homeomorphism between* E *and* F if f is bijective, and both $f\colon E \to F$ and $f^{-1}\colon F \to E$ are continuous.

One should be careful that a bijective continuous function $f\colon E \to F$ is not necessarily a homeomorphism. For example, if $E = \mathbb{R}$ with the discrete topology, and $F = \mathbb{R}$ with the standard topology, the identity is not a homeomorphism. Another interesting example involving a parametric curve is given below. Let $L\colon \mathbb{R} \to \mathbb{R}^2$ be the function, defined such that,

$$L_1(t) = \frac{t(1+t^2)}{1+t^4},$$
$$L_2(t) = \frac{t(1-t^2)}{1+t^4}.$$

If we think of $(x(t), y(t)) = (L_1(t), L_2(t))$ as a geometric point in $\mathbb{R}^2$, the set of points $(x(t), y(t))$ obtained by letting t vary in $\mathbb{R}$ from $-\infty$ to $+\infty$, defines a curve having the shape of a "figure eight," with self-intersection at the origin, called the "lemniscate of Bernoulli." See Figure A.19. The map L is continuous, and in fact bijective, but its inverse L^{-1} is not continuous. Indeed, when we approach the origin on the branch of the curve in the upper left quadrant (i.e., points such that, $x \leq 0$, $y \geq 0$), then t goes to $-\infty$, and when we approach the origin on the branch of the curve in the lower right quadrant (i.e., points such that, $x \geq 0$, $y \leq 0$), then t goes to $+\infty$.

Fig. A.19 The lemniscate of Bernoulli.

We also review the concept of limit of a sequence. Given any set E, a *sequence* is any function $x\colon \mathbb{N} \to E$, usually denoted by $(x_n)_{n\in\mathbb{N}}$, or $(x_n)_{n\geq 0}$, or even by (x_n).

Definition A.19. Given a topological space $(E, \mathcal{O})$, we say that *a sequence* $(x_n)_{n\in\mathbb{N}}$ *converges to some* $a \in E$ if for every open set U containing a, there is some $n_0 \geq 0$, such that, $x_n \in U$, for all $n \geq n_0$. We also say that *a is a limit of* $(x_n)_{n\in\mathbb{N}}$. See Figure A.20.

When E is a metric space with metric d, it is easy to show that this is equivalent to the fact that,

for every $\epsilon > 0$, there is some $n_0 \geq 0$, such that $d(x_n, a) \leq \epsilon$, for all $n \geq n_0$.

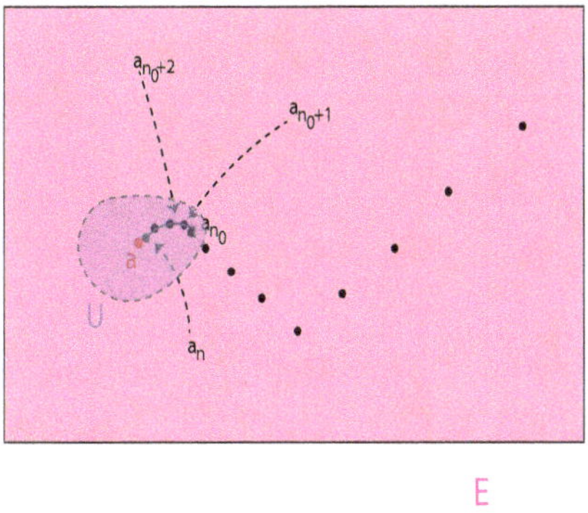

Fig. A.20 A schematic illustration of Definition A.19.

When E is a normed vector space with norm $\| \ \|$, it is easy to show that this is equivalent to the fact that,

for every $\epsilon > 0$, there is some $n_0 \geq 0$, such that $\|x_n - a\| \leq \epsilon$, for all $n \geq n_0$.

The following proposition shows the importance of the Hausdorff separation axiom.

Proposition A.12. *Given a topological space $(E, \mathcal{O})$, if the Hausdorff separation axiom holds, then every sequence has at most one limit.*

Proof. Left as an exercise. □

It is worth noting that the notion of limit is topological, in the sense that a sequence converge to a limit b iff it converges to the same limit b in any equivalent metric (and similarly for equivalent norms).

If E is a metric space and if A is a subset of E, there is a convenient way of showing that a point $x \in E$ belongs to the closure $\overline{A}$ of A in terms of sequences.

Proposition A.13. *Given any metric space (E, d), for any subset A of E and any point $x \in E$, we have $x \in \overline{A}$ iff there is a sequence (a_n) of points $a_n \in A$ converging to x.*

Proof. If the sequence (a_n) of points $a_n \in A$ converges to x, then for every open subset U of E containing x, there is some n_0 such that $a_n \in U$ for all $n \geq n_0$, so $U \cap A \neq \emptyset$, and Proposition A.4 implies that $x \in \overline{A}$.

Conversely, assume that $x \in \overline{A}$. Then for every $n \geq 1$, consider the open ball $B_0(x, 1/n)$. By Proposition A.4, we have $B_0(x, 1/n) \cap A \neq \emptyset$, so we can pick some $a_n \in B_0(x, 1/n) \cap A$. This way, we define a sequence (a_n) of points in A, and by construction $d(x, a_n) < 1/n$ for all $n \geq 1$, so the sequence (a_n) converges to x. $\square$

We still need one more concept of limit for functions.

Definition A.20. Let $(E, \mathcal{O}_E)$ and $(F, \mathcal{O}_F)$ be topological spaces, let A be some nonempty subset of E, and let $f \colon A \to F$ be a function. For any $a \in \overline{A}$ and any $b \in F$, we say that $f(x)$ *approaches* b *as* x *approaches* a *with values in* A if for every open set $V \in \mathcal{O}_F$ containing b, there is some open set $U \in \mathcal{O}_E$ containing a such that $f(U \cap A) \subseteq V$. See Figure A.21. This is denoted by

$$\lim_{x \to a, x \in A} f(x) = b.$$

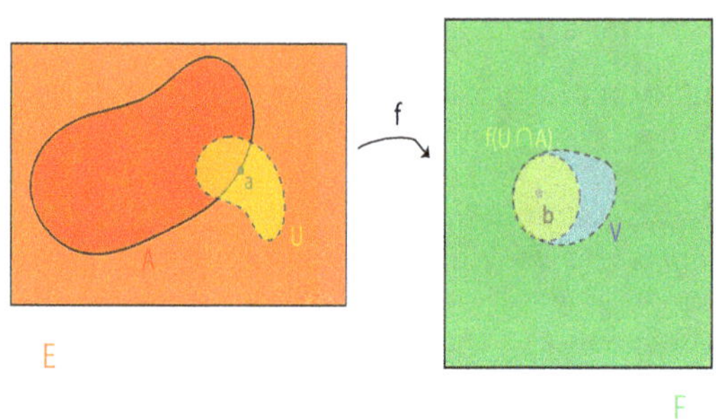

Fig. A.21 A schematic illustration of Definition A.20.

First, note that by Proposition A.4, since $a \in \overline{A}$, for every open set U containing a, we have $U \cap A \neq \emptyset$, and the definition is nontrivial. Also, even if $a \in A$, the value $f(a)$ of f at a plays no role in this definition. When E and F are metric space with metrics d_E and d_F, it can be shown easily that the definition can be stated as follows:

For every $\epsilon > 0$, there is some $\eta > 0$, such that for every $x \in A$,

$$\text{if } d_E(x, a) \leq \eta, \text{ then } d_F(f(x), b) \leq \epsilon.$$

When E and F are normed vector spaces with norms $\| \, \|_E$ and $\| \, \|_F$, it can be shown easily that the definition can be stated as follows:

For every $\epsilon > 0$, there is some $\eta > 0$, such that for every $x \in A$,

$$\text{if } \|x - a\|_E \leq \eta, \text{ then } \|f(x) - b\|_F \leq \epsilon.$$

We have the following result relating continuity at a point and the previous notion.

Proposition A.14. *Let $(E, \mathcal{O}_E)$ and $(F, \mathcal{O}_F)$ be two topological spaces, and let $f: E \to F$ be a function. For any $a \in E$, the function f is continuous at a iff $f(x)$ approaches $f(a)$ when x approaches a (with values in E).*

Proof. Left as a trivial exercise. ☐

Another important proposition relating the notion of convergence of a sequence to continuity, is stated without proof.

Proposition A.15. *Let $(E, \mathcal{O}_E)$ and $(F, \mathcal{O}_F)$ be two topological spaces, and let $f: E \to F$ be a function.*

(1) If f is continuous, then for every sequence $(x_n)_{n \in \mathbb{N}}$ in E, if (x_n) converges to a, then $(f(x_n))$ converges to $f(a)$.
(2) If E is a metric space and $(f(x_n))$ converges to $f(a)$ whenever (x_n) converges to a, for every sequence $(x_n)_{n \in \mathbb{N}}$ in E, then f is continuous.

A special case of Definition A.20 will be used when E and F are (nontrivial) normed vector spaces with norms $\| \, \|_E$ and $\| \, \|_F$. Let U be any nonempty open subset of E. We showed earlier that E has no isolated points and that every set $\{v\}$ is closed, for every $v \in E$. Since E is nontrivial, for every $v \in U$, there is a nontrivial open ball contained in U (an open ball not reduced to its center). Then for every $v \in U$, $A = U - \{v\}$ is open and nonempty, and clearly, $v \in \overline{A}$. For any $v \in U$, if $f(x)$ approaches b when x approaches v with values in $A = U - \{v\}$, we say that $f(x)$ *approaches b when x approaches v with values $\neq v$ in U*. This is denoted by

$$\lim_{x \to v, x \in U, x \neq v} f(x) = b.$$

Remark: Variations of the above case show up in the following case: $E = \mathbb{R}$, and F is some arbitrary topological space. Let A be some nonempty subset of $\mathbb{R}$, and let $f\colon A \to F$ be some function. For any $a \in A$, we say that f *is continuous on the right at* a if

$$\lim_{x \to a, x \in A \cap [a, +\infty)} f(x) = f(a).$$

We can define *continuity on the left* at a in a similar fashion.

Let us consider another variation. Let A be some nonempty subset of $\mathbb{R}$, and let $f\colon A \to F$ be some function. For any $a \in A$, we say that f *has a discontinuity of the first kind at* a if

$$\lim_{x \to a, x \in A \cap\, (-\infty, a)} f(x) = f(a_-)$$

and

$$\lim_{x \to a, x \in A \cap\, (a, +\infty)} f(x) = f(a_+)$$

both exist, and either $f(a_-) \neq f(a)$, or $f(a_+) \neq f(a)$.

Note that it is possible that $f(a_-) = f(a_+)$, but f is still discontinuous at a if this common value differs from $f(a)$. Functions defined on a nonempty subset of $\mathbb{R}$, and that are continuous, except for some points of discontinuity of the first kind, play an important role in analysis.

We now turn to connectivity properties of topological spaces.

A.4 Connected Sets

Connectivity properties of topological spaces play a very important role in understanding the topology of surfaces. This section gathers the facts needed to have a good understanding of the classification theorem for compact surfaces (with boundary). The main references are Ahlfors and Sario [Ahlfors and Sario (1960)] and Massey [Massey (1987, 1991)]. For general background on topology, geometry, and algebraic topology, we also highly recommend Bredon [Bredon (1993)] and Fulton [Fulton (1995)].

Definition A.21. A topological space $(E, \mathcal{O})$ is *connected* if the only subsets of E that are both open and closed are the empty set and E itself. Equivalently, $(E, \mathcal{O})$ is connected if E cannot be written as the union $E = U \cup V$ of two disjoint nonempty open sets U, V, or if E cannot be written as the union $E = U \cup V$ of two disjoint nonempty closed sets. A subset $S \subseteq E$ is *connected* if it is connected in the subspace topology on S induced by $(E, \mathcal{O})$. See Figure A.22. A connected open set is called a *region*, and a closed set is a *closed region* if its interior is a connected (open) set.

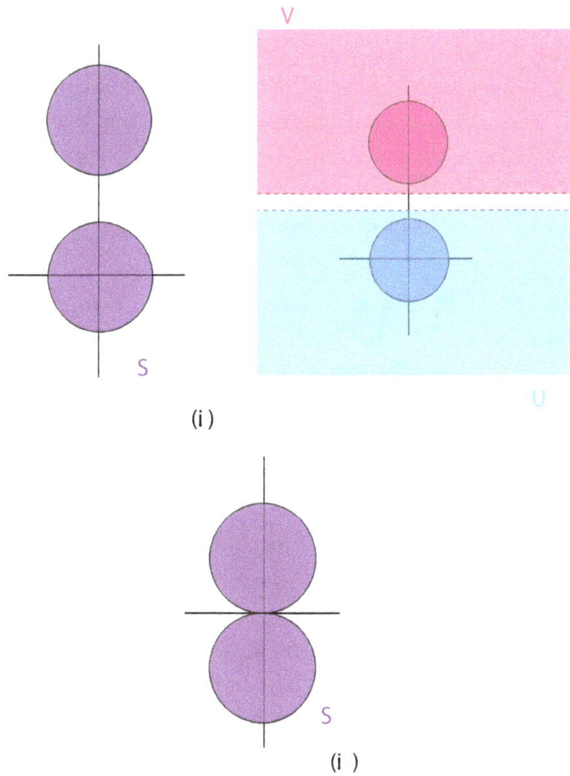

Fig. A.22 Figure (i) shows that the union of two disjoint disks in $\mathbb{R}^2$ is a disconnected set since each circle can be separated by open half regions. Figure (ii) is an example of a connected subset of $\mathbb{R}^2$ since the two disks can not separated by open sets.

The definition of connectivity is meant to capture the fact that a connected space S is "one piece." Given the metric space $(\mathbb{R}^n, \| \|_2)$, the quintessential examples of connected spaces are $B_0(a, \rho)$ and $B(a, \rho)$. In particular, the following standard proposition characterizing the connected subsets of $\mathbb{R}$ can be found in most topology texts (for example, Munkres [Munkres (2000)], Schwartz [Schwartz (1991)]). For the sake of completeness, we give a proof.

Proposition A.16. *A subset of the real line $\mathbb{R}$ is connected iff it is an interval, i.e., of the form $[a, b]$, $(a, b]$, where $a = -\infty$ is possible, $[a, b)$, where $b = +\infty$ is possible, or (a, b), where $a = -\infty$ or $b = +\infty$ is possible.*

Proof. Assume that A is a connected nonempty subset of $\mathbb{R}$. The cases where $A = \emptyset$ or A consists of a single point are trivial. Otherwise, we show that whenever $a, b \in A$, $a < b$, then the entire interval $[a, b]$ is a subset of A. Indeed, if this was not the case, there would be some $c \in (a, b)$ such that $c \notin A$, and then we could write $A = ((-\infty, c) \cap A) \cup ((c, +\infty) \cap A)$, where $(-\infty, c) \cap A$ and $(c, +\infty) \cap A$ are nonempty and disjoint open subsets of A, contradicting the fact that A is connected. It follows easily that A must be an interval.

Conversely, we show that an interval I must be connected. Let A be any nonempty subset of I which is both open and closed in I. We show that $I = A$. Fix any $x \in A$ and consider the set R_x of all y such that $[x, y] \subseteq A$. If the set R_x is unbounded, then $R_x = [x, +\infty)$. Otherwise, if this set is bounded, let b be its least upper bound. We claim that b is the right boundary of the interval I. Because A is closed in I, unless I is open on the right and b is its right boundary, we must have $b \in A$. In the first case, $A \cap [x, b) = I \cap [x, b) = [x, b)$. In the second case, because A is also open in I, unless b is the right boundary of the interval I (closed on the right), there is some open set $(b - \eta, b + \eta)$ contained in A, which implies that $[x, b + \eta/2] \subseteq A$, contradicting the fact that b is the least upper bound of the set R_x. Thus, b must be the right boundary of the interval I (closed on the right). A similar argument applies to the set L_y of all x such that $[x, y] \subseteq A$ and either L_y is unbounded, or its greatest lower bound a is the left boundary of I (open or closed on the left). In all cases, we showed that $A = I$, and the interval must be connected. $\square$

Intuitively, if a space is not connected, it is possible to define a continuous function which is constant on disjoint "connected components" and which takes possibly distinct values on disjoint components. This can be stated in terms of the concept of a locally constant function.

Definition A.22. Given two topological spaces X, Y, a function $f : X \to Y$ is *locally constant* if for every $x \in X$, there is an open set $U \subseteq X$ such that $x \in U$ and f is constant on U.

We claim that a locally constant function is continuous. In fact, we will prove that $f^{-1}(V)$ is open for every subset, $V \subseteq Y$ (not just for an open set V). It is enough to show that $f^{-1}(y)$ is open for every $y \in Y$, since for every subset $V \subseteq Y$,

$$f^{-1}(V) = \bigcup_{y \in V} f^{-1}(y),$$

and open sets are closed under arbitrary unions. However, either $f^{-1}(y) = \emptyset$ if $y \in Y - f(X)$ or f is constant on $U = f^{-1}(y)$ if $y \in f(X)$ (with value y), and since f is locally constant, for every $x \in U$, there is some open set, $W \subseteq X$, such that $x \in W$ and f is constant on W, which implies that $f(w) = y$ for all $w \in W$ and thus, that $W \subseteq U$, showing that U is a union of open sets and thus, is open. The following proposition shows that a space is connected iff every locally constant function is constant:

Proposition A.17. *A topological space is connected iff every locally constant function is constant. See Figure A.23.*

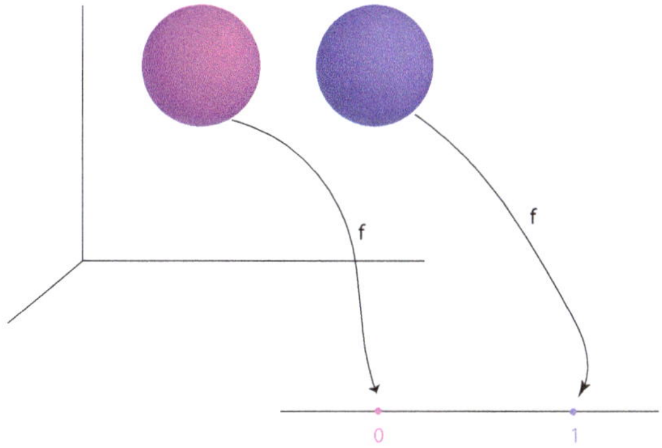

Fig. A.23 An example of a locally constant, but not constant, real-valued function f over the disconnected set consisting of the disjoint union of the two solid balls. On the pink ball, f is 0, while on the purple ball, f is 1.

Proof. First assume that X is connected. Let $f\colon X \to Y$ be a locally constant function to some space Y and assume that f is not constant. Pick any $y \in f(X)$. Since f is not constant, $U_1 = f^{-1}(y) \neq X$, and of course, $U_1 \neq \emptyset$. We proved just before Proposition A.17 that $f^{-1}(V)$ is open for every subset $V \subseteq Y$, and thus $U_1 = f^{-1}(y) = f^{-1}(\{y\})$ and $U_2 = f^{-1}(Y - \{y\})$ are both open, nonempty, and clearly $X = U_1 \cup U_2$ with U_1 and U_2 disjoint. This contradicts the fact that X is connected so f must be constant.

Assume that every locally constant function $f\colon X \to Y$ is constant. If X is not connected, we can write $X = U_1 \cup U_2$, where both U_1, U_2 are open, disjoint, and nonempty. We can define the function $f\colon X \to \mathbb{R}$ such that

$f(x) = 1$ on U_1 and $f(x) = 0$ on U_2. Since U_1 and U_2 are open, the function f is locally constant, and yet not constant, a contradiction. $\square$

A characterization on the connected subsets of $\mathbb{R}^n$ is harder and requires the notion of arcwise connectedness. One of the most important properties of connected sets is that they are preserved by continuous maps.

Proposition A.18. *Given any continuous map $f: E \to F$, if $A \subseteq E$ is connected, then $f(A)$ is connected.*

Proof. If $f(A)$ is not connected, then there exist some nonempty open sets, U, V in F such that $f(A) \cap U$ and $f(A) \cap V$ are nonempty and disjoint, and

$$f(A) = (f(A) \cap U) \cup (f(A) \cap V).$$

Then $f^{-1}(U)$ and $f^{-1}(V)$ are nonempty and open since f is continuous and

$$A = (A \cap f^{-1}(U)) \cup (A \cap f^{-1}(V)),$$

with $A \cap f^{-1}(U)$ and $A \cap f^{-1}(V)$ nonempty, disjoint, and open in A, contradicting the fact that A is connected. $\square$

An important corollary of Proposition A.18 is that for every continuous function $f: E \to \mathbb{R}$, where E is a connected space, $f(E)$ is an interval. Indeed, this follows from Proposition A.16. Thus, if f takes the values a and b where $a < b$, then f takes all values $c \in [a, b]$. This is a very important property known as the intermediate value theorem.

Even if a topological space is not connected, it turns out that it is the disjoint union of maximal connected subsets and these connected components are closed in E. In order to obtain this result, we need a few lemmas.

Lemma A.19. *Given a topological space E, for any family $(A_i)_{i \in I}$ of (nonempty) connected subsets of E, if $A_i \cap A_j \neq \emptyset$ for all $i, j \in I$, then the union $A = \bigcup_{i \in I} A_i$ of the family $(A_i)_{i \in I}$ is also connected.*

Proof. Assume that $\bigcup_{i \in I} A_i$ is not connected. There exists two nonempty open subsets U and V of E such that $A \cap U$ and $A \cap V$ are disjoint and nonempty and such that

$$A = (A \cap U) \cup (A \cap V).$$

Now, for every $i \in I$, we can write

$$A_i = (A_i \cap U) \cup (A_i \cap V),$$

where $A_i \cap U$ and $A_i \cap V$ are disjoint, since $A_i \subseteq A$ and $A \cap U$ and $A \cap V$ are disjoint. Since A_i is connected, either $A_i \cap U = \emptyset$ or $A_i \cap V = \emptyset$. This implies that either $A_i \subseteq A \cap U$ or $A_i \subseteq A \cap V$. However, by assumption, $A_i \cap A_j \neq \emptyset$, for all $i, j \in I$, and thus, either both $A_i \subseteq A \cap U$ and $A_j \subseteq A \cap U$, or both $A_i \subseteq A \cap V$ and $A_j \subseteq A \cap V$, since $A \cap U$ and $A \cap V$ are disjoint. Thus, we conclude that either $A_i \subseteq A \cap U$ for all $i \in I$, or $A_i \subseteq A \cap V$ for all $i \in I$. But this proves that either

$$A = \bigcup_{i \in I} A_i \subseteq A \cap U,$$

or

$$A = \bigcup_{i \in I} A_i \subseteq A \cap V,$$

contradicting the fact that both $A \cap U$ and $A \cap V$ are disjoint and nonempty. Thus, A must be connected. □

In particular, the above lemma applies when the connected sets in a family $(A_i)_{i \in I}$ have a point in common.

Lemma A.20. *If A is a connected subset of a topological space E, then for every subset B such that $A \subseteq B \subseteq \overline{A}$, where $\overline{A}$ is the closure of A in E, the set B is connected.*

Proof. If B is not connected, then there are two nonempty open subsets U, V of E such that $B \cap U$ and $B \cap V$ are disjoint and nonempty, and

$$B = (B \cap U) \cup (B \cap V).$$

Since $A \subseteq B$, the above implies that

$$A = (A \cap U) \cup (A \cap V),$$

and since A is connected, either $A \cap U = \emptyset$, or $A \cap V = \emptyset$. Without loss of generality, assume that $A \cap V = \emptyset$, which implies that $A \subseteq A \cap U \subseteq B \cap U$. However, $B \cap U$ is closed in the subspace topology for B and since $B \subseteq \overline{A}$ and $\overline{A}$ is closed in E, the closure of A in B w.r.t. the subspace topology of B is clearly $B \cap \overline{A} = B$, which implies that $B \subseteq B \cap U$ (since the closure is the smallest closed set containing the given set). Thus, $B \cap V = \emptyset$, a contradiction. □

In particular, Lemma A.20 shows that if A is a connected subset, then its closure $\overline{A}$ is also connected. We are now ready to introduce the connected components of a space.

Definition A.23. Given a topological space $(E, \mathcal{O})$, we say that two points $a, b \in E$ are *connected* if there is some connected subset A of E such that $a \in A$ and $b \in A$.

It is immediately verified that the relation "a and b are connected in E" is an equivalence relation. Only transitivity is not obvious, but it follows immediately as a special case of Lemma A.19. Thus, the above equivalence relation defines a partition of E into nonempty disjoint *connected components*. The following proposition is easily proved using Lemma A.19 and Lemma A.20:

Proposition A.21. *Given any topological space E, for any $a \in E$, the connected component containing a is the largest connected set containing a. The connected components of E are closed.*

The notion of a locally connected space is also useful.

Definition A.24. A topological space $(E, \mathcal{O})$ is *locally connected* if for every $a \in E$, for every neighborhood V of a, there is a connected neighborhood U of a such that $U \subseteq V$. See Figure A.24.

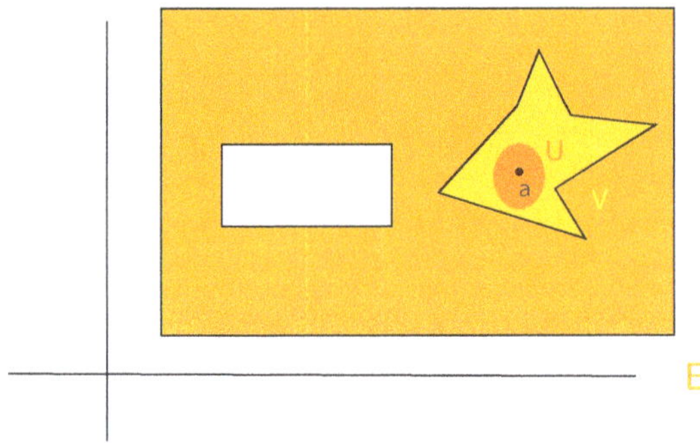

Fig. A.24 The topological space E, which is homeomorphic to an annulus, is locally connected since each point is surrounded by a small disk contained in E.

As we shall see in a moment, it would be equivalent to require that E has a basis of connected open sets.

There are connected spaces that are not locally connected and there are locally connected spaces that are not connected. The two properties are independent. For example, the subspace of $\mathbb{R}^2$ $S = \{(x, \sin(1/x)), \mid x > 0\} \cup \{(0, y) \mid -1 \leq y \leq 1\}$ is connected but not locally connected. See Figure A.25. The subspace S of $\mathbb{R}$ consisting $[0, 1] \cup [2, 3]$ is locally connected but not connected.

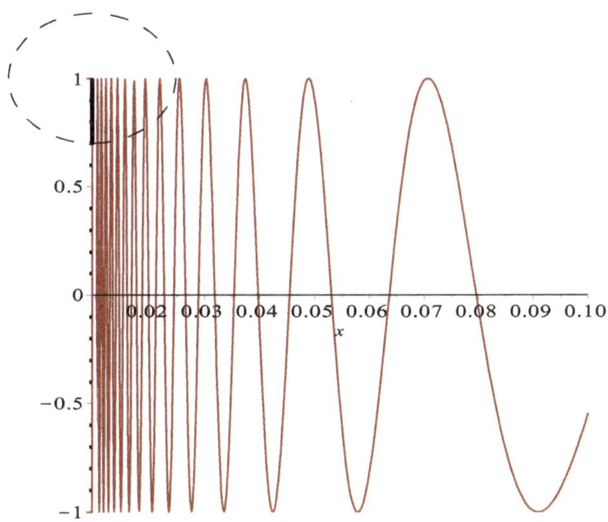

Fig. A.25 Let S be the graph of $f(x) = \sin(1/x)$ union the y-axis between -1 and 1. This space is connected, but not locally connected.

Proposition A.22. *A topological space E is locally connected iff for every open subset A of E, the connected components of A are open.*

Proof. Assume that E is locally connected. Let A be any open subset of E and let C be one of the connected components of A. For any $a \in C \subseteq A$, there is some connected neighborhood U of a such that $U \subseteq A$ and since C is a connected component of A containing a, we must have $U \subseteq C$. This shows that for every $a \in C$, there is some open subset containing a contained in C, so C is open.

Conversely, assume that for every open subset A of E, the connected components of A are open. Then, for every $a \in E$ and every neighborhood

U of a, since U contains some open set A containing a, the interior $\overset{\circ}{U}$ of U is an open set containing a and its connected components are open. In particular, the connected component C containing a is a connected open set containing a and contained in U. $\square$

Proposition A.22 shows that in a locally connected space, the connected open sets form a basis for the topology. It is easily seen that $\mathbb{R}^n$ is locally connected. Another very important property of surfaces and more generally, manifolds, is to be arcwise connected. The intuition is that any two points can be joined by a continuous arc of curve. This is formalized as follows.

Definition A.25. Given a topological space $(E, \mathcal{O})$, an *arc (or path)* is a continuous map $\gamma\colon [a, b] \to E$, where $[a, b]$ is a closed interval of the real line $\mathbb{R}$. The point $\gamma(a)$ is the *initial point* of the arc, and the point $\gamma(b)$ is the *terminal point* of the arc. We say that γ *is an arc joining* $\gamma(a)$ *and* $\gamma(b)$. See Figure A.26. An arc is a *closed curve* if $\gamma(a) = \gamma(b)$. The set $\gamma([a, b])$ is the *trace* of the arc γ.

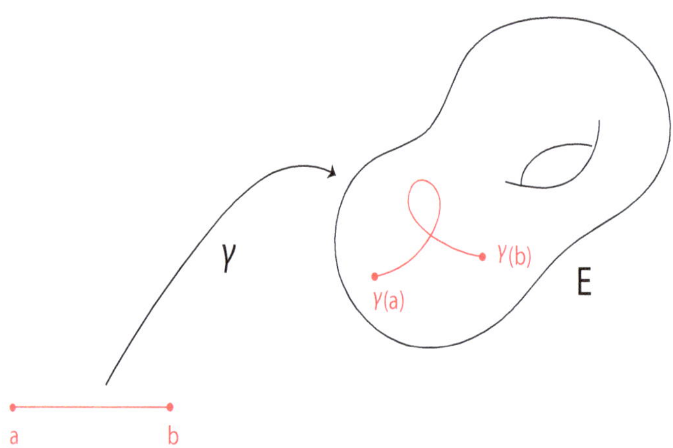

Fig. A.26 Let E be the torus with subspace topology induced from $\mathbb{R}^3$ with red arc $\gamma([a, b])$. The torus is both arcwise connected and locally arcwise connected.

Typically, $a = 0$ and $b = 1$.

One should not confuse an arc $\gamma\colon [a,b] \to E$ with its trace. For example, γ could be constant, and thus, its trace reduced to a single point.

An arc is a *Jordan arc* if γ is a homeomorphism onto its trace. An arc $\gamma\colon [a,b] \to E$ is a *Jordan curve* if $\gamma(a) = \gamma(b)$ and γ is injective on $[a,b)$. Since $[a,b]$ is connected, by Proposition A.18, the trace $\gamma([a,b])$ of an arc is a connected subset of E.

Given two arcs $\gamma\colon [0,1] \to E$ and $\delta\colon [0,1] \to E$ such that $\gamma(1) = \delta(0)$, we can form a new arc defined as follows:

Definition A.26. Given two arcs $\gamma\colon [0,1] \to E$ and $\delta\colon [0,1] \to E$ such that $\gamma(1) = \delta(0)$, we can form their *composition (or product)* $\gamma\delta$, defined such that

$$\gamma\delta(t) = \begin{cases} \gamma(2t) & \text{if } 0 \le t \le 1/2; \\ \delta(2t-1) & \text{if } 1/2 \le t \le 1. \end{cases}$$

The *inverse* γ^{-1} *of the arc* γ is the arc defined such that $\gamma^{-1}(t) = \gamma(1-t)$, for all $t \in [0,1]$.

It is trivially verified that Definition A.26 yields continuous arcs.

Definition A.27. A topological space E is *arcwise connected* if for any two points $a, b \in E$, there is an arc $\gamma\colon [0,1] \to E$ joining a and b, i.e., such that $\gamma(0) = a$ and $\gamma(1) = b$. A topological space E is *locally arcwise connected* if for every $a \in E$, for every neighborhood V of a, there is an arcwise connected neighborhood U of a such that $U \subseteq V$. See Figure A.26.

The space $\mathbb{R}^n$ is locally arcwise connected, since for any open ball, any two points in this ball are joined by a line segment. Manifolds and surfaces are also locally arcwise connected. Proposition A.18 also applies to arcwise connectedness (this is a simple exercise). The following theorem is crucial to the theory of manifolds and surfaces:

Theorem A.23. *If a topological space E is arcwise connected, then it is connected. If a topological space E is connected and locally arcwise connected, then E is arcwise connected.*

Proof. First assume that E is arcwise connected. Pick any point a in E. Since E is arcwise connected, for every $b \in E$, there is a path $\gamma_b\colon [0,1] \to E$ from a to b and so,

$$E = \bigcup_{b \in E} \gamma_b([0,1])$$

a union of connected subsets all containing a. By Lemma A.19, E is connected.

Now assume that E is connected and locally arcwise connected. For any point $a \in E$, let F_a be the set of all points b such that there is an arc $\gamma_b \colon [0,1] \to E$ from a to b. Clearly, F_a contains a. We show that F_a is both open and closed. For any $b \in F_a$, since E is locally arcwise connected, there is an arcwise connected neighborhood U containing b (because E is a neighborhood of b). Thus, b can be joined to every point $c \in U$ by an arc, and since by the definition of F_a, there is an arc from a to b, the composition of these two arcs yields an arc from a to c, which shows that $c \in F_a$. But then $U \subseteq F_a$ and thus, F_a is open. See Figure A.27(i.). Now assume that b is in the complement of F_a. As in the previous case, there is some arcwise connected neighborhood U containing b. Thus, every point $c \in U$ can be joined to b by an arc. If there was an arc joining a to c, we would get an arc from a to b, contradicting the fact that b is in the complement of F_a. Thus, every point $c \in U$ is in the complement of F_a, which shows that U is contained in the complement of F_a, and thus, that the complement of F_a is open. See Figure A.27(ii.). Consequently, we have shown that F_a is both open and closed and since it is nonempty, we must have $E = F_a$, which shows that E is arcwise connected. □

If E is locally arcwise connected, the above argument shows that the connected components of E are arcwise connected.

It is not true that a connected space is arcwise connected. For example, the space consisting of the graph of the function

$$f(x) = \sin(1/x),$$

where $x > 0$, together with the portion of the y-axis, for which $-1 \le y \le 1$, is connected, but not arcwise connected. See Figure A.25.

A trivial modification of the proof of Theorem A.23 shows that in a normed vector space E a connected open set is arcwise connected by polygonal lines (i.e., arcs consisting of line segments). This is because in every open ball, any two points are connected by a line segment. Furthermore, if E is finite dimensional, these polygonal lines can be forced to be parallel to basis vectors.

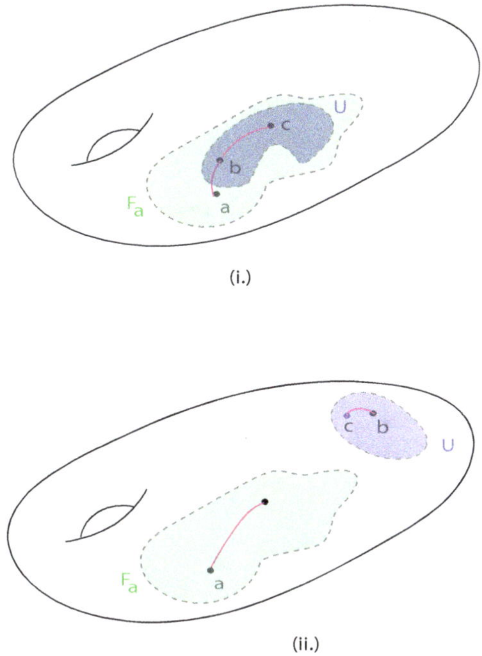

Fig. A.27 Schematic illustrations of the proof techniques that show F_a is both open and closed.

We now consider compactness.

A.5 Compact Sets and Locally Compact Spaces

The property of compactness is very important in topology and analysis. We provide a quick review geared towards the study of manifolds, and for details we refer the reader to Munkres [Munkres (2000)], Schwartz [Schwartz (1991)]. In this section we will need to assume that the topological spaces are Hausdorff spaces. This is not a luxury, as many of the results are false otherwise.

We begin this section by providing the definition of compactness and describing a collection of compact spaces in $\mathbb{R}$. There are various equivalent ways of defining compactness. For our purposes, the most convenient way involves the notion of open cover.

Definition A.28. Given a topological space E, for any subset A of E, an *open cover* $(U_i)_{i\in I}$ *of* A is a family of open subsets of E such that $A \subseteq \bigcup_{i\in I} U_i$. An *open subcover* of an open cover $(U_i)_{i\in I}$ of A is any subfamily $(U_j)_{j\in J}$ which is an open cover of A, with $J \subseteq I$. An open cover $(U_i)_{i\in I}$ of A is *finite* if I is finite. See Figure A.28. The topological space E is *compact* if it is Hausdorff and for every open cover $(U_i)_{i\in I}$ of E, there is a finite open subcover $(U_j)_{j\in J}$ of E. Given any subset A of E, we say that A is *compact* if it is compact with respect to the subspace topology. We say that A is *relatively compact* if its closure $\overline{A}$ is compact.

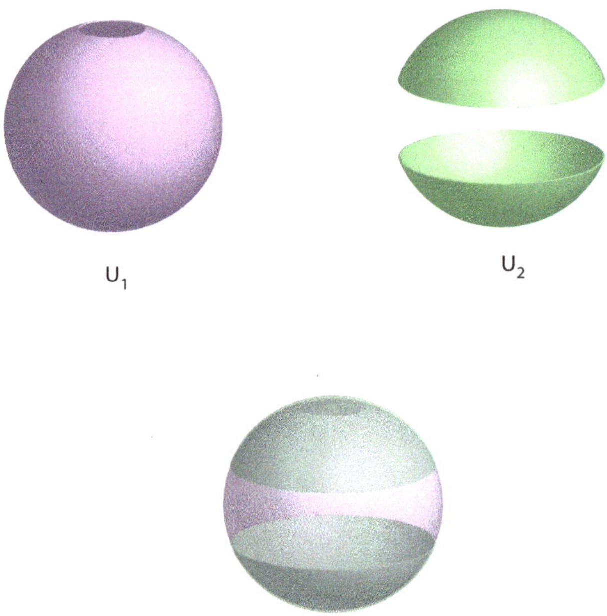

Fig. A.28 An open cover of S^2 using two open sets induced by the Euclidean topology of $\mathbb{R}^3$.

It is immediately verified that a subset A of E is compact in the subspace topology relative to A iff for every open cover $(U_i)_{i\in I}$ of A by open subsets of E, there is a finite open subcover $(U_j)_{j\in J}$ of A. The property that every open cover contains a finite open subcover is often called the *Heine-Borel-Lebesgue* property. By considering complements, a Hausdorff space is compact iff for every family $(F_i)_{i\in I}$ of closed sets, if $\bigcap_{i\in I} F_i = \emptyset$, then $\bigcap_{j\in J} F_j = \emptyset$ for some finite subset J of I.

Definition A.28 requires that a compact space be Hausdorff. There are books in which a compact space is not necessarily required to be Hausdorff. Following Schwartz, we prefer calling such a space *quasi-compact*.

Another equivalent and useful characterization can be given in terms of families having the finite intersection property.

Definition A.29. A family $(F_i)_{i \in I}$ of sets has the *finite intersection property* if $\bigcap_{j \in J} F_j \neq \emptyset$ for every finite subset J of I.

Proposition A.24. *A topological Hausdorff space E is compact iff for every family $(F_i)_{i \in I}$ of closed sets having the finite intersection property, then $\bigcap_{i \in I} F_i \neq \emptyset$.*

Proof. If E is compact and $(F_i)_{i \in I}$ is a family of closed sets having the finite intersection property, then $\bigcap_{i \in I} F_i$ cannot be empty, since otherwise we would have $\bigcap_{j \in J} F_j = \emptyset$ for some finite subset J of I, a contradiction. The converse is equally obvious. $\square$

Another useful consequence of compactness is as follows. For any family $(F_i)_{i \in I}$ of closed sets such that $F_{i+1} \subseteq F_i$ for all $i \in I$, if $\bigcap_{i \in I} F_i = \emptyset$, then $F_i = \emptyset$ for some $i \in I$. Indeed, there must be some finite subset J of I such that $\bigcap_{j \in J} F_j = \emptyset$, and since $F_{i+1} \subseteq F_i$ for all $i \in I$, we must have $F_j = \emptyset$ for the smallest F_j in $(F_j)_{j \in J}$. Using this fact, we note that $\mathbb{R}$ is *not* compact. Indeed, the family of closed sets, $([n, +\infty))_{n \geq 0}$, is decreasing and has an empty intersection.

It is immediately verified that every finite union of compact subsets is compact. Similarly, every finite union of relatively compact subsets is relatively compact (use the fact that $\overline{A \cup B} = \overline{A} \cap \overline{B}$).

Given a metric space, if we define a *bounded subset* to be a subset that can be enclosed in some closed ball (of finite radius), then any nonbounded subset of a metric space is not compact. However, a closed interval $[a, b]$ of the real line is compact.

Proposition A.25. *Every closed interval $[a, b]$ of the real line is compact.*

Proof. We proceed by contradiction. Let $(U_i)_{i \in I}$ be any open cover of $[a, b]$ and assume that there is no finite open subcover. Let $c = (a + b)/2$. If both $[a, c]$ and $[c, b]$ had some finite open subcover, so would $[a, b]$, and thus, either $[a, c]$ does not have any finite subcover, or $[c, b]$ does not have any finite open subcover. Let $[a_1, b_1]$ be such a bad subinterval. The same argument applies, and we split $[a_1, b_1]$ into two equal subintervals, one of which must be bad. Thus, having defined $[a_n, b_n]$ of length $(b-a)/2^n$ as an interval having no finite open subcover and splitting $[a_n, b_n]$ into two equal intervals, we know that at least one of the two has no finite open subcover and we denote such a bad interval by $[a_{n+1}, b_{n+1}]$. See Figure A.29. The sequence (a_n) is nondecreasing and bounded from above by b, and thus, by a fundamental property of the real line, it converges to its least upper bound α. Similarly, the sequence (b_n) is nonincreasing and bounded from below by a, and thus it converges to its greatest lowest bound β. Since $[a_n, b_n]$ has length $(b-a)/2^n$, we must have $\alpha = \beta$. However, the common limit $\alpha = \beta$ of the sequences (a_n) and (b_n) must belong to some open set U_i of the open cover and since U_i is open, it must contain some interval $[c, d]$ containing α. Then because α is the common limit of the sequences (a_n) and (b_n), there is some N such that the intervals $[a_n, b_n]$ are all contained in the interval $[c, d]$ for all $n \geq N$, which contradicts the fact that none of the intervals $[a_n, b_n]$ has a finite open subcover. Thus, $[a, b]$ is indeed compact. ☐

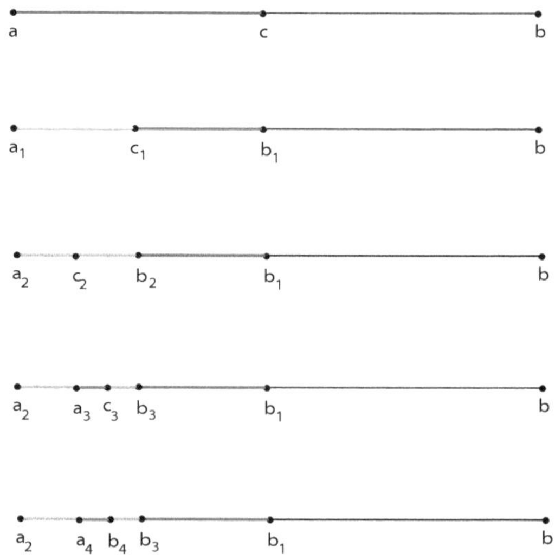

Fig. A.29 The first four stages of the nested interval construction utilized in the proof of Proposition A.25.

The argument of Proposition A.25 can be adapted to show that in $\mathbb{R}^m$, every closed set $[a_1, b_1] \times \cdots \times [a_m, b_m]$ is compact. At every stage we need to divide into 2^m subpieces instead of 2.

We next discuss some important properties of compact spaces. We begin with two separations axioms which only hold for Hausdorff spaces:

Proposition A.26. *Given a topological Hausdorff space E, for every compact subset A and every point b not in A, there exist disjoint open sets U and V such that $A \subseteq U$ and $b \in V$. See Figure A.30. As a consequence, every compact subset is closed.*

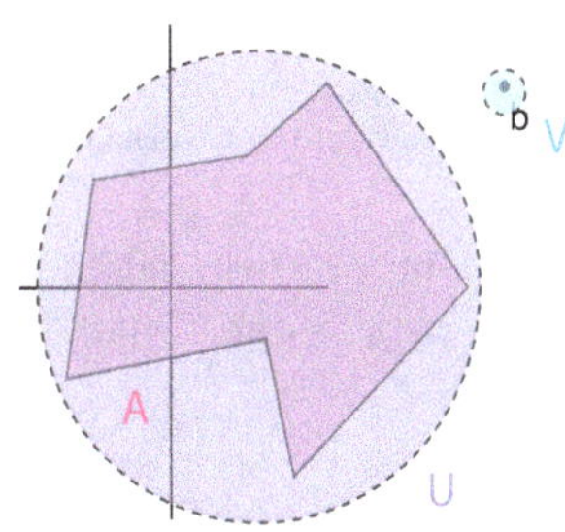

Fig. A.30 The compact set of $\mathbb{R}^2$, A, is separated by any point in its complement.

Proof. Since E is Hausdorff, for every $a \in A$, there are some disjoint open sets U_a and V_a containing a and b respectively. Thus, the family $(U_a)_{a \in A}$ forms an open cover of A. Since A is compact there is a finite open subcover $(U_j)_{j \in J}$ of A, where $J \subseteq A$, and then $\bigcup_{j \in J} U_j$ is an open set containing A disjoint from the open set $\bigcap_{j \in J} V_j$ containing b. This shows that every point, b, in the complement of A belongs to some open set in this complement and thus, that the complement is open, i.e., that A is closed. See Figure A.31. ☐

Actually, the proof of Proposition A.26 can be used to show the following useful property.

Proposition A.27. *Given a topological Hausdorff space E, for every pair of compact disjoint subsets A and B, there exist disjoint open sets U and V such that $A \subseteq U$ and $B \subseteq V$.*

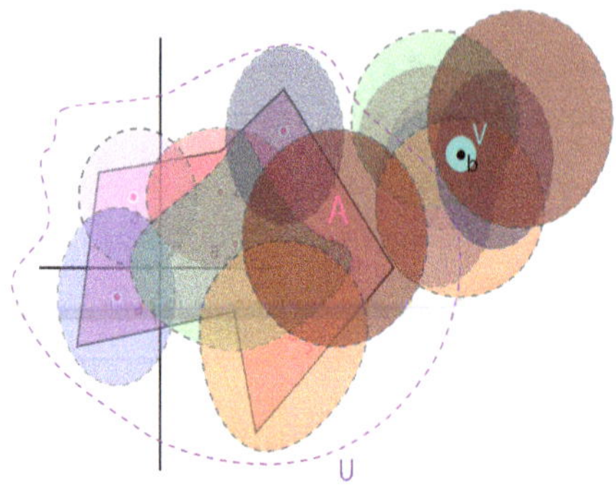

Fig. A.31 For the pink compact set A, U is the union of the seven disks which cover A, while V is the intersection of the seven open sets containing b.

Proof. We repeat the argument of Proposition A.26 with B playing the role of b and use Proposition A.26 to find disjoint open sets U_a containing $a \in A$, and V_a containing B. $\square$

The following proposition shows that in a compact topological space, every closed set is compact:

Proposition A.28. *Given a compact topological space E, every closed set is compact.*

Proof. Since A is closed, $E - A$ is open and from any open cover $(U_i)_{i \in I}$ of A, we can form an open cover of E by adding $E - A$ to $(U_i)_{i \in I}$. Since E is compact, a finite subcover $(U_j)_{j \in J} \cup \{E - A\}$ of E can be extracted such that $(U_j)_{j \in J}$ is a finite subcover of A. See Figure A.32. $\square$

Remark: Proposition A.28 also holds for quasi-compact spaces, i.e., the Hausdorff separation property is not needed.

Putting Proposition A.27 and Proposition A.28 together, we note that if X is compact, then for every pair of disjoint closed sets A and B, there exist disjoint open sets U and V such that $A \subseteq U$ and $B \subseteq V$.

Definition A.30. A topological space E is *normal* if every one-point set is closed, and for every pair of disjoint closed sets A and B, there exists disjoint open sets U and V such that $A \subseteq U$ and $B \subseteq V$. A topological space E is *regular* if every one-point set is closed, and for every point $a \in E$

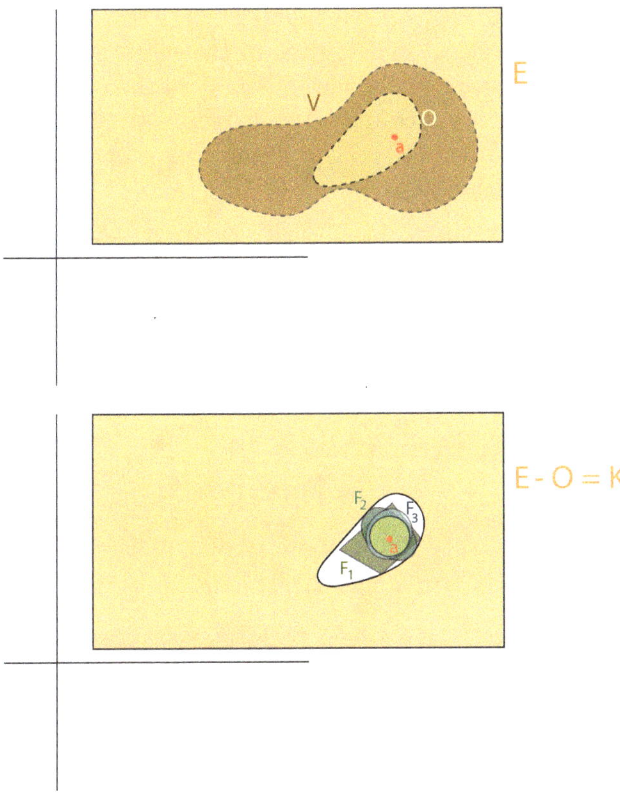

Fig. A.32 An illustration of the proof of Proposition A.28. Both E and A are closed squares in $\mathbb{R}^2$. Note that an open cover of A, namely the green circles, when combined with the yellow square annulus $E - A$ covers all of the yellow square E.

and every closed subset B of E, if $a \notin B$, there exists disjoint open sets U and V such that $a \in U$ and $B \subseteq V$.

It is clear that a normal space is regular, and a regular space is Hausdorff. There are examples of Hausdorff spaces that are not regular and of regular spaces that are not normal.

We just observed that a compact space is normal, and this is worth recording as a proposition.

Proposition A.29. *Every (Hausdorff) compact space is normal.*

An important property of metrizable spaces is that they are normal.

Proposition A.30. *Every metrizable space E is normal.*

Proof. Assume the topology of E is given by the metric d. Since B is closed and $A \cap B = \emptyset$, for every $a \in A$ since $a \notin \overline{B} = B$, there is some open ball $B_0(a, \epsilon_a)$ of radius $\epsilon_a > 0$ such that $B_0(a, \epsilon_a) \cap B = \emptyset$. Similarly, since A is closed and $A \cap B = \emptyset$, for every $b \in B$ there is some open ball $B_0(b, \epsilon_b)$ of radius $\epsilon_b > 0$ such that $B_0(b, \epsilon_b) \cap A = \emptyset$. Let

$$U = \bigcup_{a \in A} B_0(a, \epsilon_a/2), \quad V = \bigcup_{b \in B} B_0(b, \epsilon_b/2).$$

Then A and B are open sets such that $A \subseteq U$ and $B \subseteq V$, and we claim that $U \cap V = \emptyset$.

If not, then there is some $z \in U \cap V$, which implies that for some $a \in A$ and some $b \in B$, we have

$$z \in B_0(a, \epsilon_a/2) \cap B_0(b, \epsilon_b/2).$$

It follows that

$$d(a, b) \leq d(a, z) + d(z, b) < (\epsilon_a + \epsilon_b)/2.$$

If $\epsilon_a \leq \epsilon_b$, then $d(a, b) < \epsilon_b$, so $a \in B_0(b, \epsilon_b)$, contradicting the fact that $B_0(b, \epsilon_b) \cap A = \emptyset$. If $\epsilon_b \leq \epsilon_a$, then $d(a, b) < \epsilon_a$, so $b \in B_0(a, \epsilon_a)$, contradicting the fact that $B_0(a, \epsilon_a) \cap B = \emptyset$. □

Normal spaces have a strong separation property regarding disjoint closed subsets A and B. Actually, this separation property can be stated as the existence of a certain continuous function $f\colon E \to [0, 1]$ taking the value 1 on A and the value 0 on B. This result is known as *Urysohn lemma*. It is an important tool in topology and analysis.

Theorem A.31. *(Urysohn Lemma) Let E be a normal space. For any two closed disjoint subsets A and B, there is a continuous function $f\colon E \to [0, 1]$ such that $f(x) = 1$ for all $x \in A$ and $f(x) = 0$ for all $x \in B$.*

A proof of Theorem A.31 can be found in Munkres [Munkres (2000)] (Chapter 4, Section 33, Theorem 33.1). Theorem A.31 is one of the ingredients in the Urysohn metrization theorem (Theorem A.48).

Compact spaces also have the following property.

Proposition A.32. *Given a compact topological space E, for every $a \in E$, for every neighborhood V of a, there exists a compact neighborhood U of a such that $U \subseteq V$. See Figure A.33.*

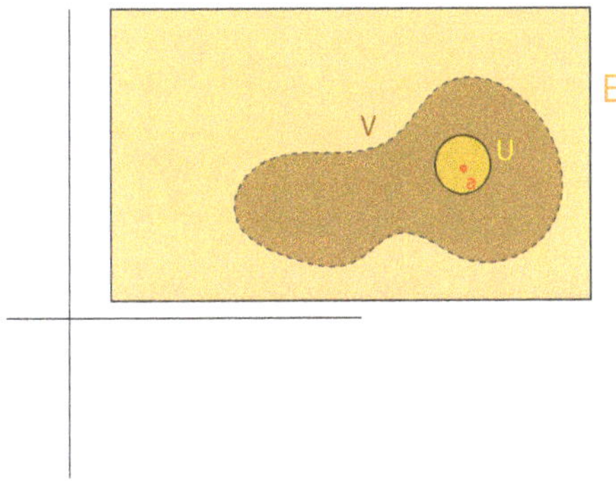

Fig. A.33 Let E be the peach square of $\mathbb{R}^2$. Each point of E is contained in a compact neighborhood U, in this case the small closed yellow disk.

Proof. Since V is a neighborhood of a, there is some open subset O of V containing a. Then the complement $K = E - O$ of O is closed and since E is compact, by Proposition A.28, K is compact. Now, if we consider the family of all closed sets of the form $K \cap F$, where F is any closed neighborhood of a, since $a \notin K$, this family has an empty intersection and thus, there is a finite number of closed neighborhoods $F_1, \ldots, F_n$ of a such that $K \cap F_1 \cap \cdots \cap F_n = \emptyset$. Then $U = F_1 \cap \cdots \cap F_n$ is closed and hence by Proposition A.28, a compact neighborhood of a contained in $O \subseteq V$. See Figure A.34. □

It can be shown that in a normed vector space of finite dimension, a subset is compact iff it is closed and bounded. For $\mathbb{R}^n$ the proof is simple.

In a normed vector space of infinite dimension, there are closed and bounded sets that are not compact!

More could be said about compactness in metric spaces but we will only need the notion of Lebesgue number, which will be discussed a little later. Another crucial property of compactness is that it is preserved under continuity.

Proposition A.33. *Let E be a topological space, and let F be a topological Hausdorff space. For every compact subset A of E, for every continuous map $f \colon E \to F$, the subspace $f(A)$ is compact.*

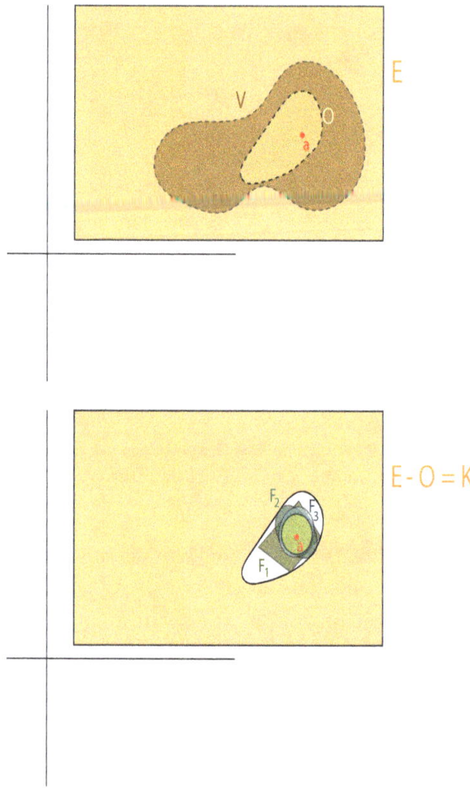

Fig. A.34 Let E be the peach square of $\mathbb{R}^2$. The compact neighborhood of a, U, is the intersection of the closed sets F_1, F_2, F_3, each of which are contained in the complement of K.

Proof. Let $(U_i)_{i\in I}$ be an open cover of $f(A)$. We claim that $(f^{-1}(U_i))_{i\in I}$ is an open cover of A, which is easily checked. Since A is compact, there is a finite open subcover $(f^{-1}(U_j))_{j\in J}$ of A, and thus, $(U_j)_{j\in J}$ is an open subcover of $f(A)$. □

As a corollary of Proposition A.33, if E is compact, F is Hausdorff, and $f\colon E \to F$ is continuous and bijective, then f is a homeomorphism. Indeed, it is enough to show that f^{-1} is continuous, which is equivalent to showing that f maps closed sets to closed sets. However, closed sets are compact and Proposition A.33 shows that compact sets are mapped to compact sets, which, by Proposition A.26, are closed.

Another important corollary of Proposition A.33 is the following result.

Proposition A.34. *If E is a compact nonempty topological space and if $f\colon E \to \mathbb{R}$ is a continuous function, then there are points $a, b \in E$ such that $f(a)$ is the minimum of $f(E)$ and $f(b)$ is the maximum of $f(E)$.*

Proof. The set $f(E)$ is a compact subset of $\mathbb{R}$ and thus, a closed and bounded set which contains its greatest lower bound and its least upper bound. □

The following property also holds.

Proposition A.35. *Let (E, d) be a metric space. For any nonempty subset A of E, if A is compact, then for every open subset U such that $A \subseteq U$, there is some $r > 0$ such that $V_r(A) \subseteq U$.*

Proof. The function $x \mapsto d(x, E - U)$ is continuous and $d(x, E - U) > 0$ for $x \in A$ (since $A \subseteq U$). By Proposition A.34, there is some $a \in A$ such that

$$d(a, E - U) = \inf_{x \in A} d(x, E - U).$$

But $d(a, E - U) = r > 0$, which implies that $V_r(A) \subseteq U$. □

Another useful notion is that of local compactness. Indeed manifolds and surfaces are locally compact.

Definition A.31. A topological space E is *locally compact* if it is Hausdorff and for every $a \in E$, there is some compact neighborhood K of a. See Figure A.33.

From Proposition A.32, every compact space is locally compact but the converse is false. For example, $\mathbb{R}$ is locally compact but not compact. In fact it can be shown that a normed vector space of finite dimension is locally compact.

Proposition A.36. *Given a locally compact topological space E, for every $a \in E$, for every neighborhood N of a, there exists a compact neighborhood U of a such that $U \subseteq N$.*

Proof. For any $a \in E$, there is some compact neighborhood V of a. By Proposition A.32, every neighborhood of a relative to V contains some compact neighborhood U of a relative to V. But every neighborhood of a relative to V is a neighborhood of a relative to E, and every neighborhood N of a in E yields a neighborhood $V \cap N$ of a in V and thus for every neighborhood N of a, there exists a compact neighborhood U of a such that $U \subseteq N$. □

When E is a metric space, the subsets $V_r(A)$ defined in Definition A.6 have the following property.

Proposition A.37. *Let (E, d) be a metric space. If E is locally compact, then for any nonempty compact subset A of E, there is some $r > 0$ such that $\overline{V_r(A)}$ is compact.*

Proof. Since E is locally compact, for every $x \in A$, there is some compact subset V_x whose interior $\overset{\circ}{V}_x$ contains x. The family of open subsets $\overset{\circ}{V}_x$ is an open cover A, and since A is compact, it has a finite subcover $\{\overset{\circ}{V}_{x_1}, \ldots, \overset{\circ}{V}_{x_n}\}$. Then $U = V_{x_1} \cup \cdots \cup V_{x_n}$ is compact (as a finite union of compact subsets), and it contains an open subset containing A (the union of the $\overset{\circ}{V}_{x_i}$). By Proposition A.35, there is some $r > 0$ such that $\overline{V_r(A)} \subseteq \overset{\circ}{U}$, and thus $\overline{V_r(A)} \subseteq U$. Since U is compact and $\overline{V_r(A)}$ is closed, $\overline{V_r(A)}$ is compact. □

Another very important property of locally compact spaces is the Proposition A.39 below. This result implies the existence of continuous partitions of unity for a finite open cover of a compact subset. Such partitions of unity are used in proving that Radon functionals correspond to certain Borel measures. First we have the following proposition.

Proposition A.38. *Let E be a locally compact (Hausdorff) space. For every compact subset K and every open subset V, if $K \subseteq V$, then there is an open set W with compact closure such that $K \subseteq W \subseteq \overline{W} \subseteq V$.*

A proof of Proposition A.38 can be found in Rudin [Rudin (1987)] (Chapter 2, Theorem 2.7). The following proposition shows the existence of continuous "bump functions" in a locally compact space. It is sometimes called Urysohn lemma (which is a bit confusing since there is already a Urysohn lemma (Proposition A.31).

Proposition A.39. *Let E be a locally compact (Hausdorff) space. For every compact subset K and every open subset V of E, if $K \subseteq V$, there is a continuous function $f \colon E \to [0, 1]$ such that $f(x) = 1$ for all $x \in K$, and such that* $\mathrm{supp}(f)$ *is compact and* $\mathrm{supp}(f) \subseteq V$*, where* $\mathrm{supp}(f)$ *is the closure of the subset $\{x \in E \mid f(x) \neq 0\}$, called the support of f.*

Proof. Theorem A.39 follows easily from the Urysohn lemma (Theorem A.31). Since E is locally compact, by Proposition A.38 we can find some open subset W with compact closure $\overline{W}$ such that $K \subseteq W \subseteq \overline{W} \subseteq V$. Since $\overline{W}$ is compact, it is normal, so we can apply Theorem A.31 to find a continuous function $f\colon \overline{W} \to [0,1]$ such that $f(x) = 1$ for all $x \in K$ and $f(x) = 0$ for all $x \in \overline{W} - W$ (the boundary of W). Then we extend f to E by setting to 0 outside $\overline{W}$. Since the support of f is contained in $\overline{W}$, this function is continuous. □

As a corollary of Proposition A.39 we obtain the existence of continuous partitions of unity for a finite open cover of a compact subset.

Proposition A.40. *Let E be a locally compact (Hausdorff) space. For any compact subset K of E and any finite open cover $(U_1, \dots, U_n)$ of K (that is, $K \subseteq \bigcup_{i=1}^n U_i$), there exist n continuous functions $f_i\colon E \to [0,1]$ such that f_i has compact support* $\mathrm{supp}(f_i) \subseteq U_i$, *and*

$$\sum_{i=1}^{n} f_i(x) = 1 \quad \textit{for all } x \in K.$$

A proof of Proposition A.40 is not difficult. It can be found in Rudin [Rudin (1987)] (Chapter 2, Theorem 2.13) and Lang [Lang (1996)] (Chapter IX, §2). A family $(f_1, \dots, f_n)$ satisfying the properties of Proposition A.40 is called a *partition of unity on K subordinate to the cover* $(U_1, \dots, U_n)$.

It is much harder to deal with noncompact manifolds than it is to deal with compact manifolds. However, manifolds are locally compact and it turns out that there are various ways of embedding a locally compact Hausdorff space into a compact Hausdorff space. The most economical construction consists in adding just one point. This construction, known as the *Alexandroff compactification*, is technically useful, and we now describe it and sketch the proof that it achieves its goal.

To help the reader's intuition, let us consider the case of the plane $\mathbb{R}^2$. If we view the plane $\mathbb{R}^2$, as embedded in 3-space $\mathbb{R}^3$, say as the xy plane of equation $z = 0$, we can consider the sphere Σ of radius 1 centered on the z-axis at the point $(0,0,1)$ and tangent to the xOy plane at the origin (sphere of equation $x^2 + y^2 + (z-1)^2 = 1$). If N denotes the north pole on the sphere, i.e., the point of coordinates $(0,0,2)$, then any line D passing through the north pole and not tangent to the sphere, (i.e., not parallel to the xOy plane), intersects the xOy plane in a unique point M and the sphere in a unique point P, other than the north pole N. This way we obtain a bijection between the xOy plane and the punctured sphere Σ, i.e., the sphere with the north pole N deleted. This bijection is called a *stereographic projection*. See Figure A.35.

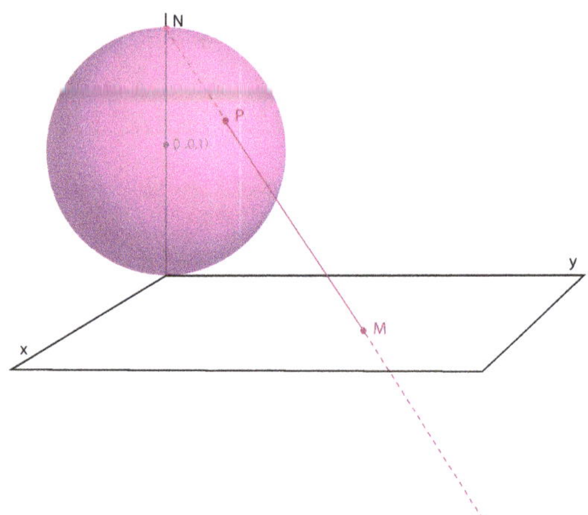

Fig. A.35 The stereographic projections of $x^2 + y^2 + (z-1)^2 = 1$ onto the xy-plane.

The Alexandroff compactification of the plane puts the north pole back on the sphere, which amounts to adding a single point at infinity ∞ to the plane. Intuitively, as we travel away from the origin O towards infinity (in any direction!), we tend towards an ideal point at infinity ∞. Imagine that we "bend" the plane so that it gets wrapped around the sphere, according to stereographic projection. See Figure A.36. A simpler example takes a line and gets a circle as its compactification. The Alexandroff compactification is a generalization of these simple constructions.

Definition A.32. Let $(E, \mathcal{O})$ be a locally compact space. Let ω be any point not in E, and let $E_\omega = E \cup \{\omega\}$. Define the family $\mathcal{O}_\omega$ as follows:

$$\mathcal{O}_\omega = \mathcal{O} \cup \{(E - K) \cup \{\omega\} \mid K \text{ compact in } E\}.$$

The pair $(E_\omega, \mathcal{O}_\omega)$ is called the *Alexandroff compactification (or one point compactification) of* $(E, \mathcal{O})$. See Figure A.37.

The following theorem shows that $(E_\omega, \mathcal{O}_\omega)$ is indeed a topological space, and that it is compact.

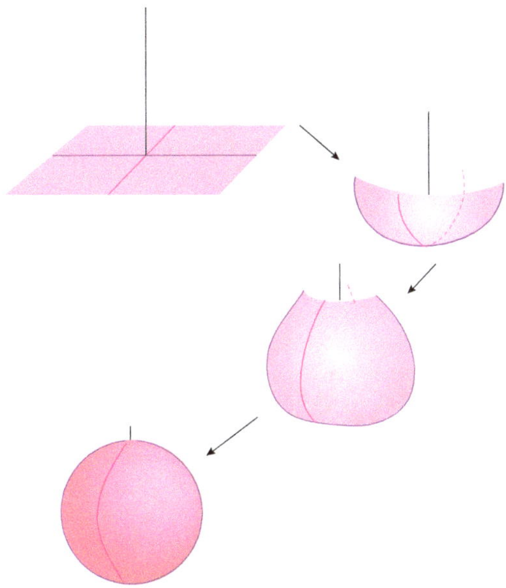

Fig. A.36 A four stage illustration of how the xy-plane is wrapped around the unit sphere centered at $(0, 0, 1)$. When finished all of the sphere is covered except the point $(0, 0, 2)$.

Theorem A.41. *Let E be a locally compact topological space. The Alexandroff compactification E_ω of E is a compact space such that E is a subspace of E_ω, and if E is not compact, then $\overline{E} = E_\omega$.*

Proof. The verification that $\mathcal{O}_\omega$ is a family of open sets is not difficult but a bit tedious. Details can be found in Munkres [Munkres (2000)] or Schwartz [Schwartz (1991)]. Let us show that E_ω is compact. For every open cover $(U_i)_{i\in I}$ of E_ω, since ω must be covered, there is some U_{i_0} of the form

$$U_{i_0} = (E - K_0) \cup \{\omega\}$$

where K_0 is compact in E. Consider the family $(V_i)_{i\in I}$ defined as follows:

$$\begin{aligned} V_i &= U_i \quad \text{if} \quad U_i \in \mathcal{O}, \\ V_i &= E - K \quad \text{if} \quad U_i = (E - K) \cup \{\omega\}, \end{aligned}$$

where K is compact in E. Then, because each K is compact and thus closed in E (since E is Hausdorff), $E - K$ is open, and every V_i is an open subset of E. Furthermore, the family $(V_i)_{i\in(I-\{i_0\})}$ is an open cover of K_0.

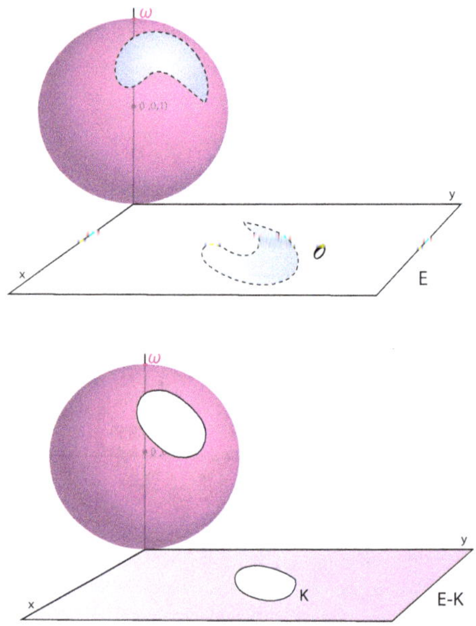

Fig. A.37 The two types of open sets associated with the Alexandroff compactification of the xy-plane. The first type of open set does not include ω, i.e. the north pole, while the second type of open set contains ω.

Since K_0 is compact, there is a finite open subcover $(V_j)_{j\in J}$ of K_0, and thus, $(U_j)_{j\in J\cup\{i_0\}}$ is a finite open cover of E_ω.

Let us show that E_ω is Hausdorff. Given any two points $a, b \in E_\omega$, if both $a, b \in E$, since E is Hausdorff and every open set in $\mathcal{O}$ is an open set in $\mathcal{O}_\omega$, there exist disjoint open sets U, V (in $\mathcal{O}$) such that $a \in U$ and $b \in V$. If $b = \omega$, since E is locally compact, there is some compact set K containing an open set U containing a, and then U and $V = (E - K) \cup \{\omega\}$ are disjoint open sets (in $\mathcal{O}_\omega$) such that $a \in U$ and $b \in V$.

The space E is a subspace of E_ω because for every open set U in $\mathcal{O}_\omega$, either $U \in \mathcal{O}$ and $E \cap U = U$ is open in E, or $U = (E - K) \cup \{\omega\}$, where K is compact in E, and thus, $U \cap E = E - K$, which is open in E, since K is compact in E and thus, closed (since E is Hausdorff). Finally, if E is not compact, for every compact subset K of E, $E - K$ is nonempty and thus, for every open set $U = (E - K) \cup \{\omega\}$ containing ω, we have $U \cap E \neq \emptyset$, which shows that $\omega \in \overline{E}$ and thus, that $\overline{E} = E_\omega$. $\square$

A.6 Neighborhood Bases and Filters

When dealing with convolution we will need a notion of convergence more general than the notion of convergence of a sequence. There are two equivalent definitions of such a general notion of convergence. One in terms of nets, and the other in terms of filters. For our purposes, the definition in terms of filters is more convenient.

First let us review the notion of neighborhood and neighborhood base.

Definition A.33. Let X be a topological space whose topology is specified by a set $\mathcal{O}$ of open sets. For any subset $A \subseteq X$, a *neighborhood* of A is any subset N containing some open subset U containing A; in short, there is some $U \in \mathcal{O}$ such that $A \subseteq U \subseteq N$; see Figure A.38. If $A = \{x\}$, a neighborhood of $\{x\}$ is called simply a *neighborhood of* x.

A *neighborhood base* of a point x (resp. of a subset A) is a family $\mathcal{N}$ of neighborhoods of x (resp. of neighborhoods of A), such for every neighborhood V of x (resp. neighborhood of A), there is some $N \in \mathcal{N}$ such that $N \subseteq V$; see Figure A.38.

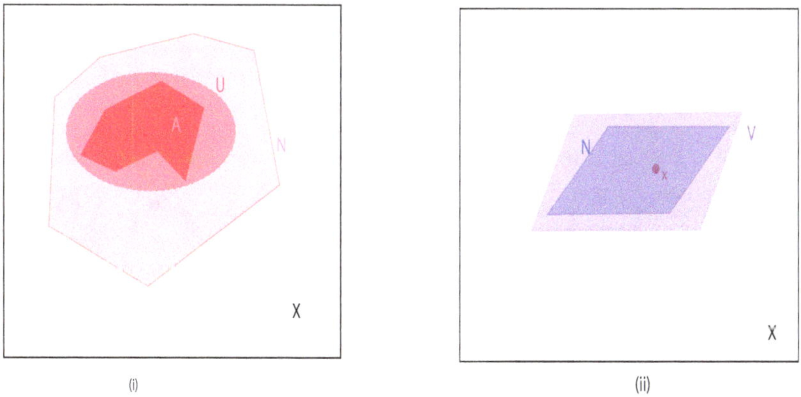

Fig. A.38 Figure (i) illustrates a neighborhood of A, while Figure (ii) illustrates a neighborhood base of x.

In many cases a neighborhood base consists of open sets. Let us now define the notion of filter and filter base. This notion is defined for any set, not just for a topological space.

Definition A.34. Let X be any set. A *filter* $\mathcal{F}$ on X is a family of subsets of X satisfying the following properties.

(1) For any two subset A, B of X, if $A \in \mathcal{F}$ and if $A \subseteq B$, then $B \in \mathcal{F}$ ($\mathcal{F}$ is upward-closed).
(2) For any two subsets A, B of X, if $A \in \mathcal{F}$ and $B \in \mathcal{F}$, then $A \cap B \in \mathcal{F}$ (closure under intersection).
(3) We have $X \in \mathcal{F}$.
(4) The empty set *does not* belong to $\mathcal{F}$.

The axioms of a filter show that filters only exist on nonempty sets. In particular, Axiom (4) prevents $\mathcal{F} = 2^X$ from being a filter.

Example A.4.

(1) If $X \neq \emptyset$, for any nonempty subset A of X, the family of all subsets of X containing A is a filter.
(2) If $(X, \mathcal{O})$ is a topological space, then for any $x \in X$ (resp. any nonempty subset A of X), the family of neighborhoods of x (resp. A) is a filter; see Figure A.39.
(3) If X is an infinite set, the family of complements of finite subsets of X is a filter. If $X = \mathbb{N}$, then the filter of complements of finite subsets of $\mathbb{N}$ is called the *Fréchet filter*.
(4) Let $\mathcal{F}$ be a filter on X. For any $A \in \mathcal{F}$, let $S(A)$ be the family

$$S(A) = \{B \in \mathcal{F} \mid B \subseteq A\},$$

called a *section*. It is easy to check that the family of sections $S(A)$ (for all $A \in \mathcal{F}$) is a filter on the set $\mathcal{F}$, called the *filter of sections of* $\mathcal{F}$.

Filters are compared as follows.

Definition A.35. Let X be any nonempty set. Given two filters $\mathcal{F}$ and $\mathcal{F}'$ on X, we say that $\mathcal{F}'$ is *finer* than $\mathcal{F}$ if $\mathcal{F} \subseteq \mathcal{F}'$.

A convenient way to generate a filter is to use a filter base.

Definition A.36. Let X be any nonempty set. A *filter base* $\mathcal{B}$ on X is a family of subsets of X satisfying the following properties.

(1) For any two subsets A, B of X, if $A \in \mathcal{B}$ and $B \in \mathcal{B}$, then there is some $C \in \mathcal{B}$ such that $C \subseteq A \cap B$.
(2) The family $\mathcal{B}$ is nonempty.

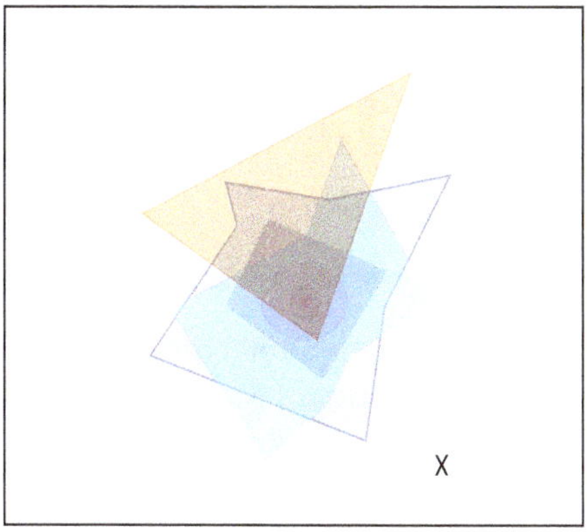

Fig. A.39 An illustration of six elements of the (canonical) neighborhood filter of x described by Example A.4, (2).

(3) The empty set *does not* belong to $\mathcal{B}$.

It is immediately verified that if $\mathcal{B}$ is a filter base on X, then the family of subsets of X containing some subset in $\mathcal{B}$ is a filter called the *filter generated by* $\mathcal{B}$.

If $(X, \mathcal{O})$ is a topological space, for any $x \in X$, the filter bases of neighborhoods of x are exactly the neighborhood bases of x.

The main reason for introducing filters is to define the following general notion of convergence.

Definition A.37. Let X be a topological space whose topology is specified by a set $\mathcal{O}$ of open sets. For any $x \in X$, a filter $\mathcal{F}$ *converges* to x, or x is a *limit* of the filter $\mathcal{F}$, if every neighborhood N of x belongs to $\mathcal{F}$; equivalently, the filter $\mathcal{B}(x)$ of neighborhoods of x is a subset of the filter $\mathcal{F}$; that is the filter $\mathcal{F}$ is finer than the filter $\mathcal{B}(x)$. A filter base $\mathcal{B}$ *converges* to x if the filter generated by $\mathcal{B}$ converges to x.

The following proposition is an immediate consequence of the definition.

Proposition A.42. *Let X be a topological space whose topology is specified by a set $\mathcal{O}$ of open sets. For any $x \in X$, a filter base $\mathcal{B}$ converges to x iff every neighborhood base of x contains some set in $\mathcal{B}$.*

Intuitively, x is a limit of a filter base $\mathcal{B}$ if there are sets in $\mathcal{B}$ as close to x as desired; see Figure A.40.

The limit of a sequence $(x_n)_{n\geq 0}$ of points $x_n \in X$ is a special case of Definition A.37; see Figure A.41. Indeed, if we define for every $n \geq 0$ the set S_n given by

$$S_n = \{x_p \mid p > n\},$$

then the family of sets S_n forms a filter base, and an element $y \in X$ is a limit of the sequence (x_n) iff the filter base $\{S_n\}$ converges to y. Indeed, by Proposition A.42, the filter base $\{S_n\}$ converges to y iff for every neighborhood V of y, there is some $S_n \subseteq V$, in other words, there is some $n \geq 1$ such that $x_p \in V$ for all $p \geq n$, which is the standard definition of convergence of a sequence.

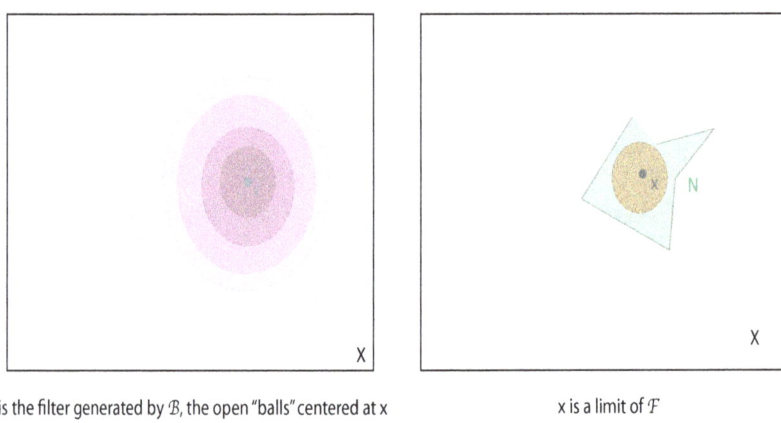

Fig. A.40 Let X be a metric space, say $X = \mathbb{R}^2$. Let $\mathcal{F}$ be the filter generated $\mathcal{B}$, where an element of $\mathcal{B}$ is an open ball centered at x. Then by Proposition A.42, x is a limit of $\mathcal{F}$.

We can also define the notion of limit of a function. Let $f\colon X \to Y$ be a function where X is any nonempty set and Y is a topological space. Then if $\mathcal{F}$ is any filter on X, it is immediately verified that the family of sets $f(U)$, with $U \in \mathcal{F}$, is a filter base on Y.

Definition A.38. Let $f\colon X \to Y$ be a function where X is any nonempty set and Y is a topological space. For any filter $\mathcal{F}$ on X, and for any $y \in Y$, we say that y *is a limit of* f *according to* $\mathcal{F}$ (or simply that y *is a limit of*

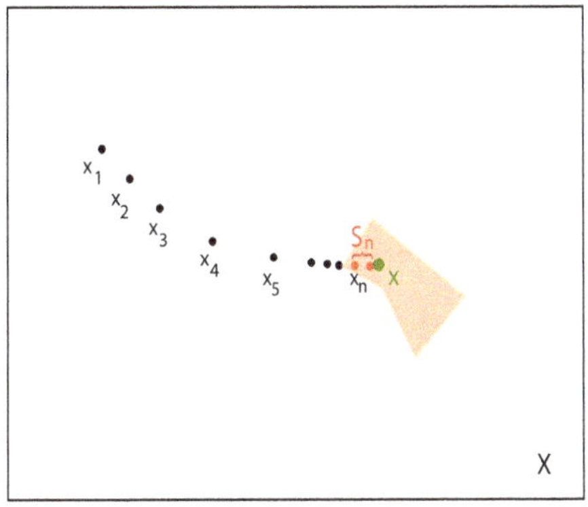

Fig. A.41 The convergence of a sequence $(x_n)_{n\geq 0}$ reinterpreted in terms of Definition A.37.

$\mathcal{F}$) if the filter basis $f(\mathcal{F})$ (consisting of the subsets $f(U)$ of Y with $U \in \mathcal{F}$) converges to y. We write

$$\lim_{x,\mathcal{F}} f(x) = y.$$

If we view a sequence $(x_n)_{n\geq 0}$ of points in a topological space X as a function $x\colon \mathbb{N} \to X$, then (x_n) converges to y in the traditional sense iff x converges to y according to the Fréchet filter on $\mathbb{N}$ (the family of subsets of $\mathbb{N}$ having a finite complement); see Figure A.42.

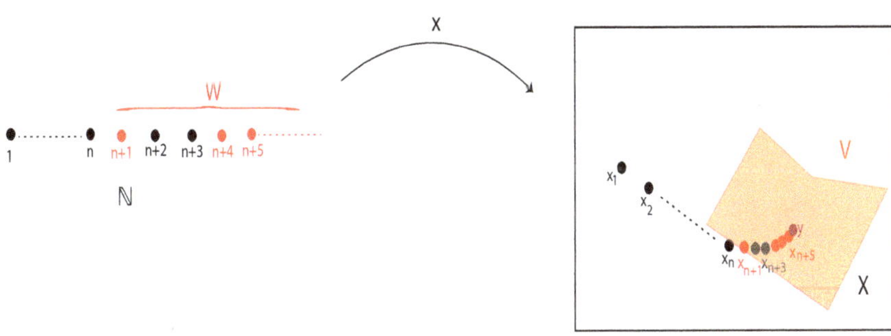

Fig. A.42 The convergence of a sequence $(x_n)_{n\geq 0}$ reinterpreted in terms of Definition A.38 and the Fréchet filter on $\mathbb{N}$. For this illustration $W = \mathbb{N} - \{1, 2, \ldots, n, n+2, n+3\}$.

The following useful characterization of a limit of a filter is immediate from the definitions.

Proposition A.43. *Let $f\colon X \to Y$ be a function where X is any nonempty set and Y is a topological space. A point $y \in Y$ is a limit of a filter $\mathcal{F}$ on X if and only if for every neighborhood V of y, there is some $W \in \mathcal{F}$ such that $f(W) \subseteq V$, or equivalently, $f^{-1}(V) \in \mathcal{F}$ for every neighborhood V of y.*

Filters also provide a useful characterization of the notion of compactness. First we define ultrafilters.

Definition A.39. Let X be any nonempty set. A filter $\mathcal{F}$ on X is an *ultrafilter* if it is a maximal filter; that is, there is no filter different from $\mathcal{F}$ and finer than $\mathcal{F}$.

For example, for any $x \in X$, the filter of subsets containing x is an ultrafilter. The following important result shows that there are many ultrafilters, but they are very nonconstructive in nature. The proof uses Zorn's lemma.

Theorem A.44. *Let X be any nonempty set. Every filter $\mathcal{F}$ on X is contained in a finer ultrafilter.*

Observe that an ultrafilter $\mathcal{F}$ has the following completeness property: for any subset A of X, either $A \in \mathcal{F}$ or its complement $X - A \in \mathcal{F}$, but not both.

Indeed, if $A \notin \mathcal{F}$ and $X - A \notin \mathcal{F}$, then it is easy to see that the family $\mathcal{G}$ of subsets of X given by

$$\mathcal{G} = \{B \subseteq X \mid A \cup B \in \mathcal{F}\}$$

is a filter finer than $\mathcal{F}$ and containing $X - A$, thus strictly finer than $\mathcal{F}$, contradicting the maximality of $\mathcal{F}$. This property of ultrafilters is used in logic to prove completeness results.

We also need to define a notion weaker than the notion of limit of a filter.

Definition A.40. Let X be a topological space whose topology is specified by a set $\mathcal{O}$ of open sets. A point $x \in X$ is a *cluster point* (or *cluster*) of a filter base $\mathcal{B}$ if every neighborhood of x has a nonempty intersection with every set in $\mathcal{B}$ (equivalently, if $x \in \bigcap_{V \in \mathcal{B}} \overline{V}$).

A limit x of a filter is a cluster point, but the converse is false in general; see Figure A.43.

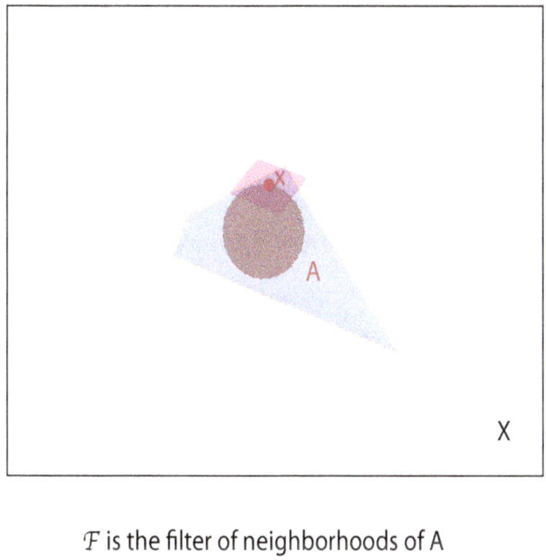

Fig. A.43 Let A be an open disk and x any point on the boundary of X. Such an x is a cluster point of $\mathcal{F}$, the filter of neighborhoods of A, since any pink neighborhood of x has an intersection with any blue neighborhood of A. However this x is not a limit of $\mathcal{F}$ since $\mathcal{F}$ is not finer than the filter of neighborhood of x.

We see immediately that x is a cluster point of a filter $\mathcal{F}$ iff there is a filter $\mathcal{G}$ finer than $\mathcal{F}$ and the filter $\mathcal{G}$ converges to x. An ultrafilter $\mathcal{F}$ converges to a limit x iff x is a cluster point of $\mathcal{F}$.

Finally, we have the following characterizations of compactness.

Theorem A.45. *Let X be a topological space whose topology is specified by a set $\mathcal{O}$ of open sets. The following properties are equivalent.*

(1) Every filter $\mathcal{F}$ on X has some cluster point.
(2) Every ultrafilter $\mathcal{F}$ on X converges to some limit.
(3) Every open cover $(U_\alpha)_{\alpha\in I}$ of X contains some finite subcover; that is, if $\bigcup_{\alpha\in I} U_\alpha = X$, then there is a finite subset J of I such that $\bigcup_{\alpha\in J} U_\alpha = X$.
(4) For every family $(F_\alpha)_{\alpha\in I}$ of closed subsets of X, if $\bigcap_{\alpha\in I} F_\alpha = \emptyset$, then there is a finite subset J of I such that $\bigcap_{\alpha\in J} F_\alpha = \emptyset$.

Let us also mention that a topological space X is Hausdorff if and only if every filter has at most one limit.

The theory of filters and their use in topology is discussed quite extensively in Bourbaki [Bourbaki (1990a)] (Chapter 1).

A.7 Second-Countable and Separable Spaces

In studying surfaces and manifolds, an important property is the existence of a countable basis for the topology. Indeed this property, among other things, guarantees the existence of triangulations of manifolds, and the fact that a manifold is metrizable.

Definition A.41. A topological space E is called *second-countable* if there is a countable basis for its topology, i.e., if there is a countable family, $(U_i)_{i\geq 0}$, of open sets such that every open set of E is a union of open sets U_i.

It is easily seen that $\mathbb{R}^n$ is second-countable and more generally, that every normed vector space of finite dimension is second-countable. More generally, a metric space is second-countable if and only if it is separable, a very useful property that holds for all of the spaces that we will consider in practice.

Definition A.42. A topological space E is *separable* if it contains some countable subset S which is dense in X, that is, $\overline{S} = E$.

Observe that by Proposition A.4, a subset S of E is dense in E iff every nonempty open subset of E contains some element of S.

The (metric) space $\mathbb{R}$ is separable because $\mathbb{Q}$ is a countable dense subset of $\mathbb{R}$. Similarly, $\mathbb{C}$ is separable. In general, $\mathbb{Q}^n$ is dense in $\mathbb{R}^n$, so $\mathbb{R}^n$ is separable, and similarly, every finite-dimensional normed vector space over $\mathbb{R}$ (or $\mathbb{C}$) is separable. For metric spaces, we have the following useful result.

Proposition A.46. *If E is a metric space, then E is second-countable if and only if E is separable.*

Proof. If $\mathcal{B} = (B_n)$ is a countable basis for the topology of E, then for any set S obtained by picking some point s_n in B_n, since every nonempty open subset U of E is the union of some of the B_n, the intersection $U \cap S$ is nonempty, and so S is dense in E.

Conversely, assume that there is a countable subset $S = (s_n)$ of E which is dense in E. We claim that the countable family $\mathcal{B}$ of open balls $B_0(s_n, 1/m)$ $(m \in \mathbb{N}, m > 0)$ is a basis for the topology of E. For every $x \in E$ and every $r > 0$, there is some $m > 0$ such that $1/m < r/2$, and some n such that $s_n \in B_0(x, 1/m)$. It follows that $x \in B_0(s_n, 1/m)$. For all $y \in B_0(s_n, 1/m)$, we have

$$d(x, y) \le d(x, s_n) + d(s_n, y) \le 2/m < r,$$

thus $B_0(s_n, 1/m) \subseteq B_0(x, r)$, which by Proposition A.8(a) implies that $\mathcal{B}$ is a basis for the topology of E. □

Proposition A.47. *If E is a compact metric space, then E is separable.*

Proof. For every $n > 0$, the family of open balls of radius $1/n$ forms an open cover of E, and since E is compact, there is a finite subset A_n of E such that $E = \bigcup_{a_i \in A_n} B_0(a_i, 1/n)$. It is easy to see that this is equivalent to the condition $d(x, A_n) < 1/n$ for all $x \in E$. Let $A = \bigcup_{n \ge 1} A_n$. Then A is countable, and for evey $x \in E$, we have

$$d(x, A) \le d(x, A_n) < \frac{1}{n}, \quad \text{for all } n \ge 1,$$

which implies that $d(x, A) = 0$; that is, A is dense in E. □

The following theorem due to Uryshon gives a very useful sufficient condition for a topological space to be metrizable.

Theorem A.48. *(Urysohn metrization theorem) If a topological space E is regular and second-countable, then it is metrizable.*

The proof of Theorem A.48 can be found in Munkres [Munkres (2000)] (Chapter 4, Theorem 34.1). As a corollary of Theorem A.48, every (second-countable) manifold, and thus every Lie group is metrizable.

The following technical result shows that a locally compact metrizable space which is also separable can be expressed as the union of a countable monotonic sequence of compact subsets. This gives us a method for generalizing various properties of compact metric spaces to locally compact metric spaces of the above kind.

Proposition A.49. *Let E be a locally compact metrizable space. The following properties are equivalent:*

(1) There is a sequence $(U_n)_{n \ge 0}$ of open subsets such that for all $n \in \mathbb{N}$, $U_n \subseteq U_{n+1}$, $\overline{U_n}$ is compact, $\overline{U_n} \subseteq U_{n+1}$, and $E = \bigcup_{n \ge 0} U_n = \bigcup_{n \ge 0} \overline{U_n}$.
(2) The space E is the union of a countable family of compact subsets of E.
(3) The space E is separable.

Proof. We show (1) implies (2), (2) implies (3), and (3) implies (1). Obviously, (1) implies (2) since the $\overline{U_n}$ are compact.

If (2) holds, then $E = \bigcup_{n\geq 0} K_n$, for some compact subsets K_n. By Proposition A.47, each compact subset K_n is separable, so let S_n be a countable dense subset of K_n. Then $S = \bigcup_{n\geq 0} S_n$ is a countable dense subset of E, since

$$E = \bigcup_{n\geq 0} K_n \subseteq \bigcup_{n\geq 0} \overline{S_n} \subseteq \overline{S} \subseteq E.$$

Consequently (3) holds.

If (3) holds, let $S = \{s_n\}$ be a countable dense subset of E. By Proposition A.46, the space E has a countable basis $\mathcal{B}$ of open sets O_n. Since E is locally compact, for every $x \in E$, there is some compact neighborhood W_x containing x, and by Proposition A.8, there some index $n(x)$ such that $x \in O_{n(x)} \subseteq W_x$. Since W_x is a compact neighborhood, we deduce that $\overline{O_{n(x)}}$ is compact. Consequently, there is a subfamily of $\mathcal{B}$ consisting of open subsets O_i such that $\overline{O_i}$ is compact, which is a countable basis for the topology of E, so we may assume that we restrict our attention to this basis. We define the sequence $(U_n)_{n\geq 1}$ of open subsets of E by induction as follows. Set $U_1 = O_1$, and let

$$U_{n+1} = O_{n+1} \cup V_r(\overline{U_n}),$$

where $r > 0$ is chosen so that $\overline{V_r(\overline{U_n})}$ is compact, which is possible by Proposition A.37. We immediately check that the U_n satisfy (1) of Proposition A.49. ☐

Definition A.43. Given a topological space E, a subset A of E is *σ-compact* (or *countable at infinity*) if A is the union of countably many compact subsets.

Note that Proposition A.49 implies that a locally compact metrizable space is separable iff it is σ-compact.

It can also be shown that if E is a locally compact space that has a countable basis, then E_ω also has a countable basis (and in fact, is metrizable).

We also have the following property.

Proposition A.50. *Given a second-countable topological space E, every open cover $(U_i)_{i\in I}$ of E contains some countable subcover.*

Proof. Let $(O_n)_{n\geq 0}$ be a countable basis for the topology. Then all sets O_n contained in some U_i can be arranged into a countable subsequence, $(\Omega_m)_{m\geq 0}$, of $(O_n)_{n\geq 0}$ and for every Ω_m, there is some U_{i_m} such that $\Omega_m \subseteq U_{i_m}$. Furthermore, every U_i is some union of sets Ω_j, and thus, every $a \in E$ belongs to some Ω_j, which shows that $(\Omega_m)_{m\geq 0}$ is a countable open subcover of $(U_i)_{i\in I}$. ☐

As an immediate corollary of Proposition A.50, a locally connected second-countable space has countably many connected components.

A.8 Sequential Compactness

For a general topological Hausdorff space E, the definition of compactness relies on the existence of finite cover. However, when E has a countable basis or is a metric space, we may define the notion of compactness in terms of sequences. To understand how this is done, we need to first define accumulation points.

Definition A.44. Given a topological Hausdorff space E, given any sequence (x_n) of points in E, a point $l \in E$ is an *accumulation point (or cluster point)* of the sequence (x_n) if every open set U containing l contains x_n for infinitely many n. See Figure A.44.

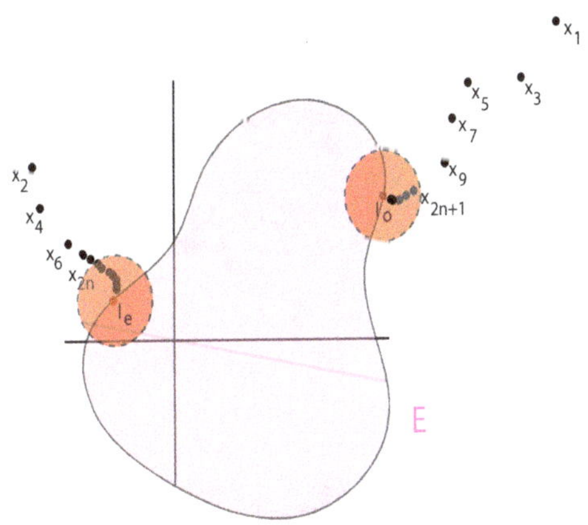

Fig. A.44 The space E is the closed, bounded pink subset of $\mathbb{R}^2$. The sequence (x_n) has two accumulation points, one for the subsequence (x_{2n+1}) and one for (x_{2n}).

Clearly, if l is a limit of the sequence (x_n), then it is an accumulation point, since every open set U containing a contains all x_n except for finitely many n.

For second-countable spaces we are able to give another characterization of accumulation points.

Proposition A.51. *Given a second-countable topological Hausdorff space E, a point l is an accumulation point of the sequence (x_n) iff l is the limit of some subsequence (x_{n_k}) of (x_n).*

Proof. Clearly, if l is the limit of some subsequence (x_{n_k}) of (x_n), it is an accumulation point of (x_n).

Conversely, let $(U_k)_{k\geq 0}$ be the sequence of open sets containing l, where each U_k belongs to a countable basis of E, and let $V_k = U_1 \cap \cdots \cap U_k$. For every $k \geq 1$, we can find some $n_k > n_{k-1}$ such that $x_{n_k} \in V_k$, since l is an accumulation point of (x_n). Now, since every open set containing l contains some U_{k_0} and since $x_{n_k} \in U_{k_0}$ for all $k \geq 0$, the sequence (x_{n_k}) has limit l. $\square$

Remark: Proposition A.51 also holds for metric spaces.

As an illustration of Proposition A.51 let (x_n) be the sequence $(1, -1, 1, -1, \dots)$. This sequence has two accumulation points, namely 1 and -1 since $(x_{2n+1}) = (1)$ and $(x_{2n}) = (-1)$.

In second-countable Hausdorff spaces, compactness can be characterized in terms of accumulation points (this is also true for metric spaces).

Proposition A.52. *A second-countable topological Hausdorff space E is compact iff every sequence (x_n) of E has some accumulation point in E.*

Proof. Assume that every sequence (x_n) has some accumulation point. Let $(U_i)_{i\in I}$ be some open cover of E. By Proposition A.50, there is a countable open subcover $(O_n)_{n\geq 0}$ for E. Now if E is not covered by any finite subcover of $(O_n)_{n\geq 0}$, we can define a sequence (x_m) by induction as follows.

Let x_0 be arbitrary and for every $m \geq 1$, let x_m be some point in E not in $O_1 \cup \cdots \cup O_m$, which exists since $O_1 \cup \cdots \cup O_m$ is not an open cover of E. We claim that the sequence (x_m) does not have any accumulation point. Indeed, for every $l \in E$, since $(O_n)_{n\geq 0}$ is an open cover of E, there is some O_m such that $l \in O_m$, and by construction, every x_n with $n \geq m+1$ does not belong to O_m, which means that $x_n \in O_m$ for only finitely many n and l is not an accumulation point. See Figure A.45.

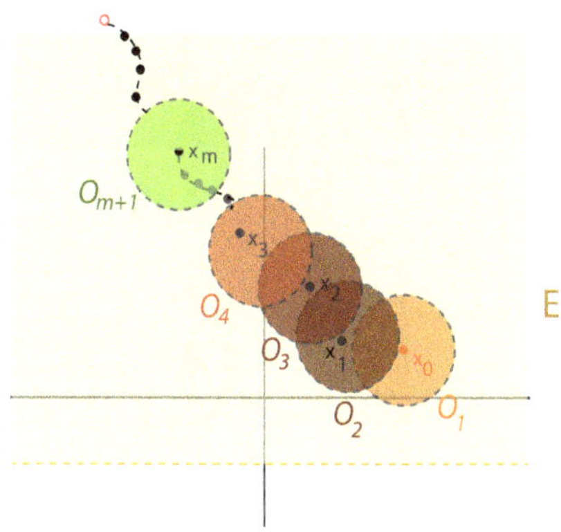

Fig. A.45 The space E is the open half plane above the line $y = -1$. S ince E is not compact, we inductively build a sequence, (x_n) that will have no accumulation point in E. Note the y coordinate of x_n approaches infinity.

Conversely, assume that E is compact, and let (x_n) be any sequence . If $l \in E$ is not an accumulation point of the sequence, then there is some open set U_l such that $l \in U_l$ and $x_n \in U_l$ for only finitely many n. Thus, if (x_n) does not have any accumulation point, the family $(U_l)_{l \in E}$ is an open cover of E and since E is compact, it has some finite open subcover $(U_l)_{l \in J}$, where J is a finite subset of E. But every U_l with $l \in J$ is such that $x_n \in U_l$ for only finitely many n, and since J is finite, $x_n \in \bigcup_{l \in J} U_l$ for only finitely many n, which contradicts the fact that $(U_l)_{l \in J}$ is an open cover of E, and thus contains all the x_n. Thus, (x_n) has some accumulation point. See Figure A.46. $\square$

Remarks:

(1) By combining Propositions A.51 and A.52, we have observed that a second-countable Hausdorff space E is compact iff every sequence (x_n) has a convergent subsequence (x_{n_k}). In other words, we say a second-countable Hausdorff space E is compact iff it is *sequentially compact.*

(2) It should be noted that the proof showing that if E is compact, then

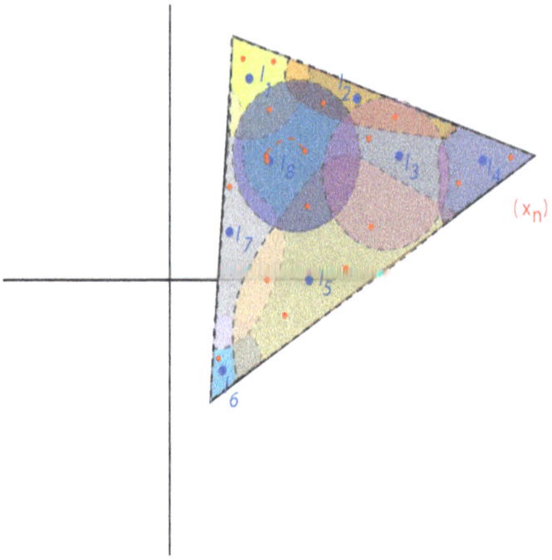

Fig. A.46 The space E the closed triangular region of $\mathbb{R}^2$. Given a sequence (x_n) of red points in E, if the sequence has no accumulation points, then each l_i for $1 \leq i \leq 8$, is not an accumulation point. But as implied by the illustration, l_8 actually is an accumulation point of (x_n).

every sequence has some accumulation point, holds for any arbitrary compact space (the proof does not use a countable basis for the topology). The converse also holds for metric spaces. We will prove this converse since it is a major property of metric spaces.

Given a metric space in which every sequence has some accumulation point, we first prove the existence of a *Lebesgue number*.

Lemma A.53. *Given a metric space E, if every sequence (x_n) has an accumulation point, for every open cover $(U_i)_{i\in I}$ of E, there is some $\delta > 0$ (a Lebesgue number for $(U_i)_{i\in I}$) such that for every open ball $B_0(a, \epsilon)$ of radius $\epsilon \leq \delta$, there is some open subset U_i such that $B_0(a, \epsilon) \subseteq U_i$. See Figure A.47*

Proof. If there was no δ with the above property, then for every natural number n, there would be some open ball $B_0(a_n, 1/n)$ which is not contained in any open set U_i of the open cover $(U_i)_{i\in I}$. However, the sequence (a_n) has some accumulation point a, and since $(U_i)_{i\in I}$ is an open cover of E, there is some U_i such that $a \in U_i$. Since U_i is open, there is some open ball of center a and radius ϵ contained in U_i. Now, since a is an accumulation

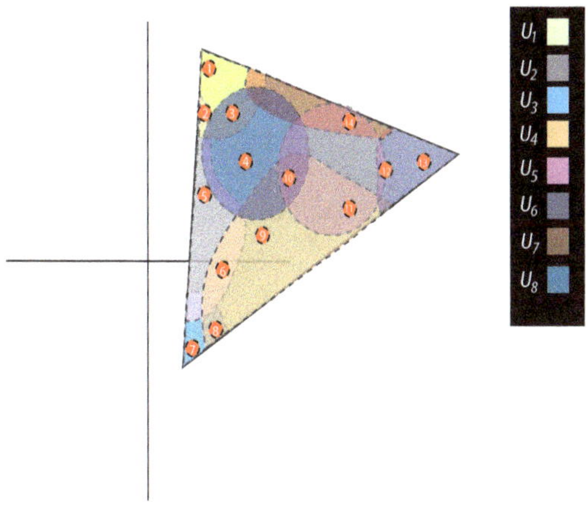

Fig. A.47 The space E the closed triangular region of $\mathbb{R}^2$. It's open cover is $(U_i)_{i=1}^8$. The Lebesque number is the radius of the small orange balls labelled 1 through 14. Each open ball of this radius entirely contained within at least one U_i. For example, Ball 2 is contained in both U_1 and U_2.

point of the sequence (a_n), every open set containing a contains a_n for infinitely many n and thus, there is some n large enough so that

$$1/n \leq \epsilon/2 \quad \text{and} \quad a_n \in B_0(a, \epsilon/2),$$

which implies that

$$B_0(a_n, 1/n) \subseteq B_0(a, \epsilon) \subseteq U_i,$$

a contradiction. □

By a previous remark, since the proof of Proposition A.52 implies that in a compact topological space, every sequence has some accumulation point, by Lemma A.53, in a compact metric space, every open cover has a Lebesgue number. This fact can be used to prove another important property of compact metric spaces, the uniform continuity theorem.

Definition A.45. Given two metric spaces (E, d_E) and (F, d_F), a function $f\colon E \to F$ is *uniformly continuous* if for every $\epsilon > 0$, there is some $\eta > 0$, such that for all $a, b \in E$,

$$\text{if} \quad d_E(a, b) \leq \eta, \quad \text{then} \quad d_F(f(a), f(b)) \leq \epsilon.$$

See Figures A.48 and A.49.

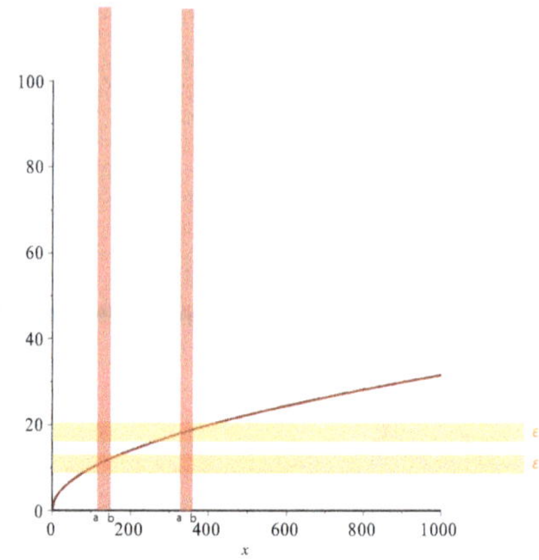

Fig. A.48 The real valued function $f(x) = \sqrt{x}$ is uniformly continuous over $(0, \infty)$. Fix ϵ. If the x values lie within the rose colored η strip, the y values always lie within the peach ϵ strip.

As we saw earlier, the metric on a metric space is uniformly continuous, and the norm on a normed metric space is uniformly continuous.

The *uniform continuity theorem* can be stated as follows.

Theorem A.54. *Given two metric spaces (E, d_E) and (F, d_F), if E is compact and if $f\colon E \to F$ is a continuous function, then f is uniformly continuous.*

Proof. Consider any $\epsilon > 0$ and let $(B_0(y, \epsilon/2))_{y \in F}$ be the open cover of F consisting of open balls of radius $\epsilon/2$. Since f is continuous, the family

$$(f^{-1}(B_0(y, \epsilon/2)))_{y \in F},$$

is an open cover of E. Since E is compact, by Lemma A.53, there is a Lebesgue number δ such that for every open ball $B_0(a, \eta)$ of radius $\eta \leq \delta$, then $B_0(a, \eta) \subseteq f^{-1}(B_0(y, \epsilon/2))$, for some $y \in F$. In particular, for any $a, b \in E$ such that $d_E(a, b) \leq \eta = \delta/2$, we have $a, b \in B_0(a, \delta)$ and thus, $a, b \in f^{-1}(B_0(y, \epsilon/2))$, which implies that $f(a), f(b) \in B_0(y, \epsilon/2)$. But then $d_F(f(a), f(b)) \leq \epsilon$, as desired. $\square$

We now prove another lemma needed to obtain the characterization of compactness in metric spaces in terms of accumulation points.

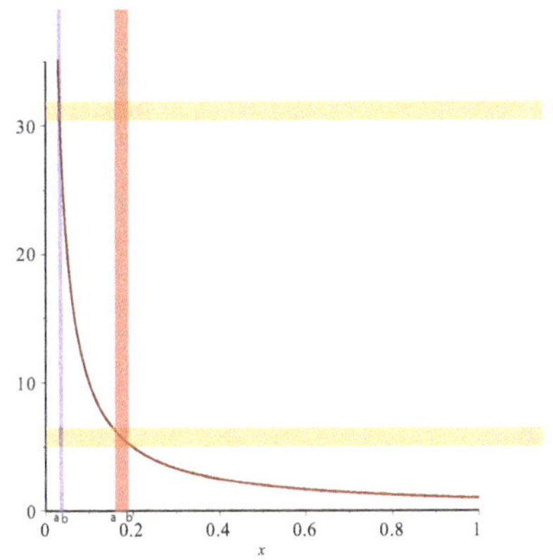

Fig. A.49 The real valued function $f(x) = 1/x$ is not uniformly continuous over $(0, \infty)$. Fix ϵ. In order for the y values to lie within the peach epsilon strip, the widths of the eta strips decrease as $x \to 0$.

Lemma A.55. *Given a metric space E, if every sequence (x_n) has an accumulation point, then for every $\epsilon > 0$, there is a finite open cover $B_0(a_0, \epsilon) \cup \cdots \cup B_0(a_n, \epsilon)$ of E by open balls of radius ϵ.*

Proof. Let a_0 be any point in E. If $B_0(a_0, \epsilon) = E$, then the lemma is proven. Otherwise, assume that a sequence $(a_0, a_1, \ldots, a_n)$ has been defined such that $B_0(a_0, \epsilon) \cup \cdots \cup B_0(a_n, \epsilon)$ does not cover E. Then there is some a_{n+1} not in $B_0(a_0, \epsilon) \cup \cdots \cup B_0(a_n, \epsilon)$ and either

$$B_0(a_0, \epsilon) \cup \cdots \cup B_0(a_{n+1}, \epsilon) = E,$$

in which case the lemma is proven, or we obtain a sequence $(a_0, a_1, \ldots, a_{n+1})$ such that $B_0(a_0, \epsilon) \cup \cdots \cup B_0(a_{n+1}, \epsilon)$ does not cover E. If this process goes on forever, we obtain an infinite sequence (a_n) such that $d(a_m, a_n) > \epsilon$ for all $m \neq n$. Since every sequence in E has some accumulation point, the sequence (a_n) has some accumulation point a. Then for infinitely many n, we must have $d(a_n, a) \leq \epsilon/3$ and thus, for at least two distinct natural numbers p, q, we must have $d(a_p, a) \leq \epsilon/3$ and $d(a_q, a) \leq \epsilon/3$, which implies $d(a_p, a_q) \leq d(a_p, a) + d(a_q, a) \leq 2\epsilon/3$, contradicting the fact that $d(a_m, a_n) > \epsilon$ for all $m \neq n$. See Figure A.50. Thus, there must be some n such that

$$B_0(a_0, \epsilon) \cup \cdots \cup B_0(a_n, \epsilon) = E. \qquad \square$$

Definition A.46. A metric space E is said to be *precompact* (or *totally bounded*) if for every $\epsilon > 0$, there is a finite open cover $B_0(a_0, \epsilon) \cup \cdots \cup B_0(a_n, \epsilon)$ of E by open balls of radius ϵ.

We now obtain the *Weierstrass–Bolzano* property.

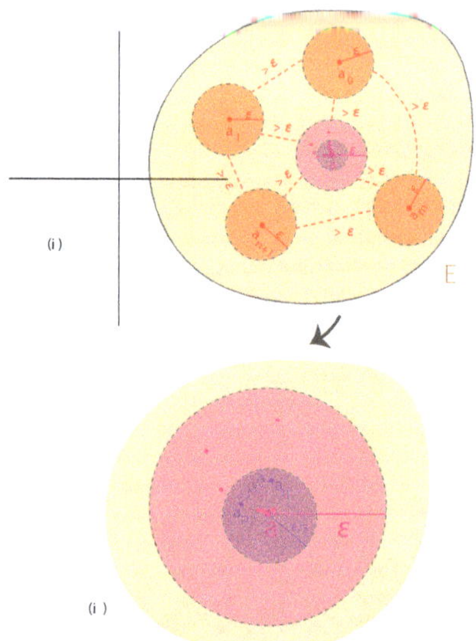

Fig. A.50 Let E be the peach region of $\mathbb{R}^2$. If E is not covered by a finite collection of orange balls with radius ϵ, the points of the sequence (a_n) are separated by a distance of at least ϵ. This contradicts the fact that a is the accumulation point of a, as evidenced by the enlargement of the plum disk in Figure (ii).

Theorem A.56. *A metric space E is compact iff every sequence (x_n) has an accumulation point.*

Proof. We already observed that the proof of Proposition A.52 shows that for any compact space (not necessarily metric), every sequence (x_n) has an accumulation point. Conversely, let E be a metric space, and assume that every sequence (x_n) has an accumulation point. Given any open cover $(U_i)_{i \in I}$ for E, we must find a finite open subcover of E. By Lemma A.53, there is some $\delta > 0$ (a Lebesgue number for $(U_i)_{i \in I}$) such that for every open ball $B_0(a, \epsilon)$ of radius $\epsilon \leq \delta$, there is some open subset U_j such that $B_0(a, \epsilon) \subseteq U_j$. By Lemma A.55, for every $\delta > 0$, there is a finite open cover

$B_0(a_0, \delta) \cup \cdots \cup B_0(a_n, \delta)$ of E by open balls of radius δ. But from the previous statement, every open ball $B_0(a_i, \delta)$ is contained in some open set U_{j_i}, and thus, $\{U_{j_1}, \ldots, U_{j_n}\}$ is an open cover of E. □

A.9 Complete Metric Spaces and Compactness

Another very useful characterization of compact metric spaces is obtained in terms of Cauchy sequences. Such a characterization is quite useful in fractal geometry (and elsewhere). First recall the definition of a Cauchy sequence and of a complete metric space.

Definition A.47. Given a metric space (E, d), a sequence $(x_n)_{n \in \mathbb{N}}$ in E is a *Cauchy sequence* if the following condition holds:

for every $\epsilon > 0$, there is some $p \geq 0$, such that for all $m, n \geq p$, $d(x_m, x_n) \leq \epsilon$.

If every Cauchy sequence in (E, d) converges we say that (E, d) is a *complete metric space.*

First let us show the following proposition:

Proposition A.57. *Given a metric space E, if a Cauchy sequence (x_n) has some accumulation point a, then a is the limit of the sequence (x_n).*

Proof. Since (x_n) is a Cauchy sequence, for every $\epsilon > 0$, there is some $p \geq 0$, such that for all $m, n \geq p$, $d(x_m, x_n) \leq \epsilon/2$. Since a is an accumulation point for (x_n), for infinitely many n, we have $d(x_n, a) \leq \epsilon/2$, and thus, for at least some $n \geq p$, we have $d(x_n, a) \leq \epsilon/2$. Then for all $m \geq p$,

$$d(x_m, a) \leq d(x_m, x_n) + d(x_n, a) \leq \epsilon,$$

which shows that a is the limit of the sequence (x_n). □

We can now prove the following theorem.

Theorem A.58. *A metric space E is compact iff it is precompact and complete.*

Proof. Let E be compact. For every $\epsilon > 0$, the family of all open balls of radius ϵ is an open cover for E and since E is compact, there is a finite subcover $B_0(a_0, \epsilon) \cup \cdots \cup B_0(a_n, \epsilon)$ of E by open balls of radius ϵ. Thus E is precompact. Since E is compact, by Theorem A.56, every sequence (x_n) has some accumulation point. Thus every Cauchy sequence (x_n) has some accumulation point a, and by Proposition A.57, a is the limit of (x_n). Thus, E is complete.

Now assume that E is precompact and complete. We prove that every sequence (x_n) has an accumulation point. By the other direction of Theorem A.56, this shows that E is compact. Given any sequence (x_n), we construct a Cauchy subsequence (y_n) of (x_n) as follows. Since E is precompact, letting $\epsilon = 1$, there exists a finite cover $\mathcal{U}_1$ of E by open balls of radius 1. Thus some open ball B_o^0 in the cover $\mathcal{U}_1$ contains infinitely many elements from the sequence (x_n). Let y_0 be any element of (x_n) in B_o^0. By induction, assume that a sequence of open balls $(B_o^i)_{1\leq i\leq m}$ has been defined such that every ball B_o^i has radius $\frac{1}{2^i}$, contains infinitely many elements from the sequence (x_n), and contains some y_i from (x_n) such that

$$d(y_i, y_{i+1}) \leq \frac{1}{2^i},$$

for all i, $0 \leq i \leq m-1$. See Figure A.51. Then letting $\epsilon = \frac{1}{2^{m+1}}$, because E is precompact, there is some finite cover $\mathcal{U}_{m+1}$ of E by open balls of radius ϵ and thus, of the open ball B_o^m. Thus, some open ball B_o^{m+1} in the cover $\mathcal{U}_{m+1}$ contains infinitely many elements from the sequence (x_n), and we let y_{m+1} be any element of (x_n) in B_o^{m+1}. Thus, we have defined by induction a sequence (y_n) which is a subsequence of (x_n), and such that

$$d(y_i, y_{i+1}) \leq \frac{1}{2^i},$$

for all i. However, for all $m, n \geq 1$, we have

$$d(y_m, y_n) \leq d(y_m, y_{m+1}) + \cdots + d(y_{n-1}, y_n) \leq \sum_{i=m}^{n} \frac{1}{2^i} \leq \frac{1}{2^{m-1}},$$

and thus, (y_n) is a Cauchy sequence. Since E is complete, the sequence (y_n) has a limit, and since it is a subsequence of (x_n), the sequence (x_n) has some accumulation point. □

Another useful property of a complete metric space is that a subset is closed iff it is complete. This is shown in the following two propositions.

Proposition A.59. *Let (E, d) be a metric space, and let A be a subset of E. If A is complete (which means that every Cauchy sequence of elements in A converges to some point of A), then A is closed in E.*

Proof. Assume $x \in \overline{A}$. By Proposition A.13, there is some sequence (a_n) of points $a_n \in A$ which converges to x. Consequently (a_n) is a Cauchy

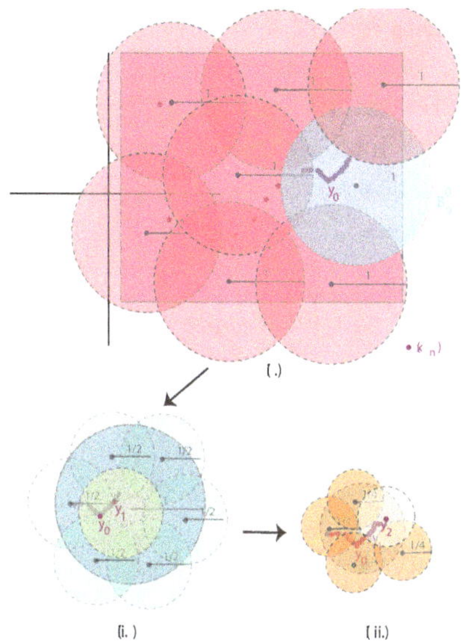

Fig. A.51 The first three stages of the construction of the Cauchy sequence (y_n), where E is the pink square region of $\mathbb{R}^2$. The original sequence (x_n) is illustrated with plum colored dots. Figure ($i.$) covers E with ball of radius 1 and shows the selection of B_o^0 and y_0. Figure ($ii.$) covers B_o^0 with balls of radius 1/2 and selects the yellow ball as B_o^1 with point y_1. Figure ($iii.$) covers B_o^1 with balls of radius 1/4 and selects the pale peach ball as B_o^2 with point y_2.

sequence in E, and thus a Cauchy sequence in A (since $a_n \in A$ for all n). Since A is complete, the sequence (a_n) has a limit $a \in A$, but since E is a metric space it is Hausdorff, so $a = x$, which shows that $x \in A$; that is, A is closed. □

Proposition A.60. *Let (E, d) be a metric space, and let A be a subset of E. If E is complete and if A is closed in E, then A is complete.*

Proof. Let (a_n) be a Cauchy sequence in A. The sequence (a_n) is also a Cauchy sequence in E, and since E is complete, it has a limit $x \in E$. But $a_n \in A$ for all n, so by Proposition A.13 we must have $x \in \overline{A}$. Since A is closed, actually $x \in A$, which proves that A is complete. □

An arbitrary metric space (E, d) is not necessarily complete, but there is a construction of a metric space $(\widehat{E}, \widehat{d})$ such that $\hat{E}$ is complete, and there is a continuous (injective) distance-preserving map $\varphi\colon E \to \widehat{E}$ such that $\varphi(E)$

is dense in $\widehat{E}$. This is a generalization of the construction of the set $\mathbb{R}$ of real numbers from the set $\mathbb{Q}$ of rational numbers in terms of Cauchy sequences. This construction can be immediately adapted to a normed vector space $(E, \| \|)$ to embed $(E, \| \|)$ into a complete normed vector space $(\widehat{E}, \| \|_{\widehat{E}})$ (a Banach space). This construction is used heavily in integration theory, where E is a set of functions.

A.10 Completion of a Metric Space

In order to prove a kind of uniqueness result for the completion $(\widehat{E}, \widehat{d})$ of a metric space (E, d), we need the following result about extending a uniformly continuous function.

Recall that E_0 is dense in E iff $\overline{E_0} = E$. Since E is a metric space, by Proposition A.13, this means that for every $x \in E$, there is some sequence (x_n) converging to x, with $x_n \in E_0$.

Theorem A.61. *Let E and F be two metric spaces, let E_0 be a dense subspace of E, and let $f_0 \colon E_0 \to F$ be a continuous function. If f_0 is uniformly continuous and if F is complete, then there is a unique uniformly continuous function $f \colon E \to F$ extending f_0.*

Proof. We follow Schwartz's proof; see Schwartz [Schwartz (1980)] (Chapter XI, Section 3, Theorem 1).

Step 1. We begin by constructing a function $f \colon E \to F$ extending f_0. Since E_0 is dense in E, for every $x \in E$, there is some sequence (x_n) converging to x, with $x_n \in E_0$. Then the sequence (x_n) is a Cauchy sequence in E. We claim that $(f_0(x_n))$ is a Cauchy sequence in F.

Proof of the claim. For every $\epsilon > 0$, since f_0 is uniformly continuous, there is some $\eta > 0$ such that for all $(y, z) \in E_0$, if $d(y, z) \leq \eta$, then $d(f_0(y), f_0(z)) \leq \epsilon$. Since (x_n) is a Cauchy sequence with $x_n \in E_0$, there is some integer $p > 0$ such that if $m, n \geq p$, then $d(x_m, x_n) \leq \eta$, thus $d(f_0(x_m), f_0(x_n)) \leq \epsilon$, which proves that $(f_0(x_n))$ is a Cauchy sequence in F. ☐

Since F is complete and $(f_0(x_n))$ is a Cauchy sequence in F, the sequence $(f_0(x_n))$ converges to some element of F; denote this element by $f(x)$.

Step 2. Let us now show that $f(x)$ does not depend on the sequence (x_n) converging to x. Suppose that (x'_n) and (x''_n) are two sequences of elements in E_0 converging to x. Then the mixed sequence

$$x'_0, x''_0, x'_1, x''_1, \ldots, x'_n, x''_n, \ldots,$$

also converges to x. It follows that the sequence

$$f_0(x'_0), f_0(x''_0), f_0(x'_1), f_0(x''_1), \ldots, f_0(x'_n), f_0(x''_n), \ldots,$$

is a Cauchy sequence in F, and since F is complete, it converges to some element of F, which implies that the sequences $(f_0(x'_n))$ and $(f_0(x''_n))$ converge to the same limit.

As a summary, we have defined a function $f \colon E \to F$ by

$$f(x) = \lim_{n \mapsto \infty} f_0(x_n),$$

for any sequence (x_n) converging to x, with $x_n \in E_0$. See Figure A.52.

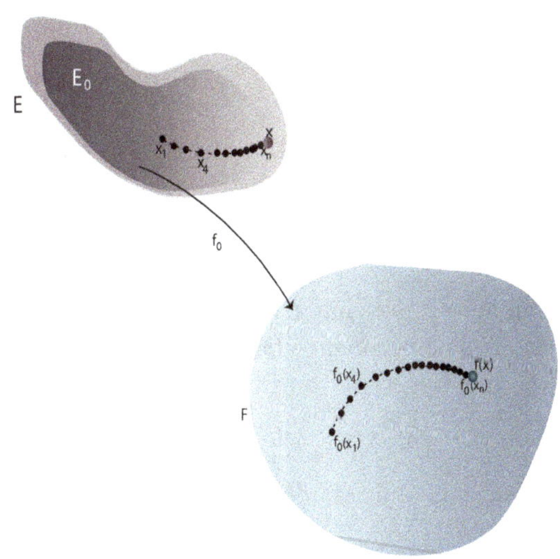

Fig. A.52 A schematic illustration of the construction of $f \colon E \to F$ where $f(x) = \lim_{n \mapsto \infty} f_0(x_n)$ for any sequence (x_n) converging to x, with $x_n \in E_0$.

Step 3. The function f extends f_0. Since every element $x \in E_0$ is the limit of the constant sequence (x_n) with $x_n = x$ for all $n \geq 0$, by definition

$f(x)$ is the limit of the sequence $(f_0(x_n))$, which is the constant sequence with value $f_0(x)$, so $f(x) = f_0(x)$; that is, f extends f_0.

Step 4. We now prove that f is uniformly continuous. Since f_0 is uniformly continuous, for every $\epsilon > 0$, there is some $\eta > 0$ such that if $a, b \in E_0$ and $d(a, b) \leq \eta$, then $d(f_0(a), f_0(b)) \leq \epsilon$. Consider any two points $x, y \in E$ such that $d(x, y) \leq \eta/2$. We claim that $d(f(x), f(y)) \leq \epsilon$, which shows that f is uniformly continuous.

Let (x_n) be a sequence of points in E_0 converging to x, and let (y_n) be a sequence of points in E_0 converging to y. By the triangle inequality,

$$d(x_n, y_n) \leq d(x_n, x) + d(x, y) + d(y, y_n) = d(x, y) + d(x_n, x) + d(y_n, y),$$

and since (x_n) converges to x and (y_n) converges to y, there is some integer $p > 0$ such that for all $n \geq p$, we have $d(x_n, x) \leq \eta/4$ and $d(y_n, y) \leq \eta/4$, and thus

$$d(x_n, y_n) \leq d(x, y) + \frac{\eta}{2}.$$

Since we assumed that $d(x, y) \leq \eta/2$, we get $d(x_n, y_n) \leq \eta$ for all $n \geq p$, and by uniform continuity of f_0, we get

$$d(f_0(x_n), f_0(y_n)) \leq \epsilon$$

for all $n \geq p$. Since the distance function on F is also continuous, and since $(f_0(x_n))$ converges to $f(x)$ and $(f_0(y_n))$ converges to $f(y)$, we deduce that the sequence $(d(f_0(x_n), f_0(y_n)))$ converges to $d(f(x), f(y))$. This implies that $d(f(x), f(y)) \leq \epsilon$, as desired.

Step 5. It remains to prove that f is unique. Since E_0 is dense in E, for every $x \in E$, there is some sequence (x_n) converging to x, with $x_n \in E_0$. Since f extends f_0 and since f is continuous, we get

$$f(x) = \lim_{n \mapsto \infty} f_0(x_n),$$

which only depends on f_0 and x and shows that f is unique. □

Remark: It can be shown that the theorem no longer holds if we either omit the hypothesis that F is complete or omit that f_0 is uniformly continuous.

For example, if $E_0 \neq E$ and if we let $F = E_0$ and f_0 be the identity function, it is easy to see that f_0 cannot be extended to a continuous function from E to E_0 (for any $x \in E - E_0$, any continuous extension f of f_0 would satisfy $f(x) = x$, which is absurd since $x \notin E_0$).

If f_0 is continuous but not uniformly continuous, a counter-example can be given by using $E = \overline{\mathbb{R}} = \mathbb{R} \cup \{\infty\}$ made into a metric space, $E_0 = \mathbb{R}$, $F = \mathbb{R}$, and f_0 the identity function; for details, see Schwartz [Schwartz (1980)] (Chapter XI, Section 3, page 134).

Definition A.48. If (E, d_E) and (F, d_F) are two metric spaces, then a function $f\colon E \to F$ is *distance-preserving*, or an *isometry*, if

$$d_F(f(x), f(y)) = d_E(x, y), \quad \text{for all for all } x, y \in E.$$

Observe that an isometry must be injective, because if $f(x) = f(y)$, then $d_F(f(x), f(y)) = 0$, and since $d_F(f(x), f(y)) = d_E(x, y)$, we get $d_E(x, y) = 0$, but $d_E(x, y) = 0$ implies that $x = y$. Also, an isometry is uniformly continuous (since we can pick $\eta = \epsilon$ to satisfy the condition of uniform continuity). However, an isometry is not necessarily surjective.

We now give a construction of the completion of a metric space. This construction is just a generalization of the classical construction of $\mathbb{R}$ from $\mathbb{Q}$ using Cauchy sequences.

Theorem A.62. *Let (E, d) be any metric space. There is a complete metric space $(\widehat{E}, \widehat{d})$ called a completion of (E, d), and a distance-preserving (uniformly continuous) map $\varphi\colon E \to \widehat{E}$ such that $\varphi(E)$ is dense in $\widehat{E}$, and the following extension property holds: for every complete metric space F and for every uniformly continuous function $f\colon E \to F$, there is a unique uniformly continuous function $\widehat{f}\colon \widehat{E} \to F$ such that*

$$f = \widehat{f} \circ \varphi,$$

as illustrated in the following diagram.

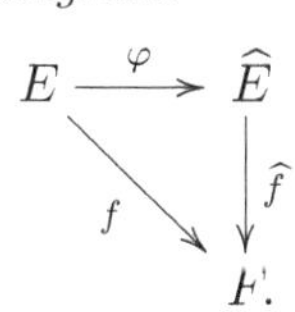

As a consequence, for any two completions $(\widehat{E}_1, \widehat{d}_1)$ and $(\widehat{E}_2, \widehat{d}_2)$ of (E, d), there is a unique bijective isometry between $(\widehat{E}_1, \widehat{d}_1)$ and $(\widehat{E}_2, \widehat{d}_2)$.

Proof. Consider the set $\mathcal{E}$ of all Cauchy sequences (x_n) in E, and define the relation $\sim$ on $\mathcal{E}$ as follows:

$$(x_n) \sim (y_n) \quad \text{iff} \quad \lim_{n \mapsto \infty} d(x_n, y_n) = 0.$$

It is easy to check that $\sim$ is an equivalence relation on $\mathcal{E}$, and let $\widehat{E} = \mathcal{E}/\sim$ be the quotient set, that is, the set of equivalence classes modulo $\sim$. Our goal is to show that we can endow $\widehat{E}$ with a distance that makes it into a complete metric space satisfying the conditions of the theorem. We proceed in several steps.

Step 1. First let us construct the function $\varphi\colon E \to \widehat{E}$. For every $a \in E$, we have the constant sequence (a_n) such that $a_n = a$ for all $n \geq 0$, which is obviously a Cauchy sequence. Let $\varphi(a) \in \widehat{E}$ be the equivalence class $[(a_n)]$ of the constant sequence (a_n) with $a_n = a$ for all n. By definition of $\sim$, the equivalence class $\varphi(a)$ is also the equivalence class of all sequences converging to a. The map $a \mapsto \varphi(a)$ is injective because a metric space is Hausdorff, so if $a \neq b$, then a sequence converging to a does not converge to b. After having defined a distance on $\widehat{E}$, we will check that φ is an isometry.

Step 2. Let us now define a distance on $\widehat{E}$. Let $\alpha = [(a_n)]$ and $\beta = [(b_n)]$ be two equivalence classes of Cauchy sequences in E. The triangle inequality implies that

$$\begin{aligned} d(a_m, b_m) &\leq d(a_m, a_n) + d(a_n, b_n) + d(b_n, b_m) \\ &= d(a_n, b_n) + d(a_m, a_n) + d(b_m, b_n) \end{aligned}$$

and

$$\begin{aligned} d(a_n, b_n) &\leq d(a_n, a_m) + d(a_m, b_m) + d(b_m, b_n) \\ &= d(a_m, b_m) + d(a_m, a_n) + d(b_m, b_n), \end{aligned}$$

which implies that

$$|d(a_m, b_m) - d(a_n, b_n)| \leq d(a_m, a_n) + d(b_m, b_n).$$

Since (a_n) and (b_n) are Cauchy sequences, the above inequality shows that $(d(a_n, b_n))$ is a Cauchy sequence of nonnegative reals. Since $\mathbb{R}$ is complete, the sequence $(d(a_n, b_n))$ has a limit, which we denote by $\widehat{d}(\alpha, \beta)$; that is, we set

$$\widehat{d}(\alpha, \beta) = \lim_{n \mapsto \infty} d(a_n, b_n), \quad \alpha = [(a_n)],\ \beta = [(b_n)].$$

See Figure A.53.

Step 3. Let us check that $\widehat{d}(\alpha, \beta)$ does not depend on the Cauchy sequences (a_n) and (b_n) chosen in the equivalence classes α and β.

If $(a_n) \sim (a'_n)$ and $(b_n) \sim (b'_n)$, then $\lim_{n \mapsto \infty} d(a_n, a'_n) = 0$ and $\lim_{n \mapsto \infty} d(b_n, b'_n) = 0$, and since

$$d(a'_n, b'_n) \leq d(a'_n, a_n) + d(a_n, b_n) + d(b_n, b'_n) = d(a_n, b_n) + d(a_n, a'_n) + d(b_n, b'_n),$$

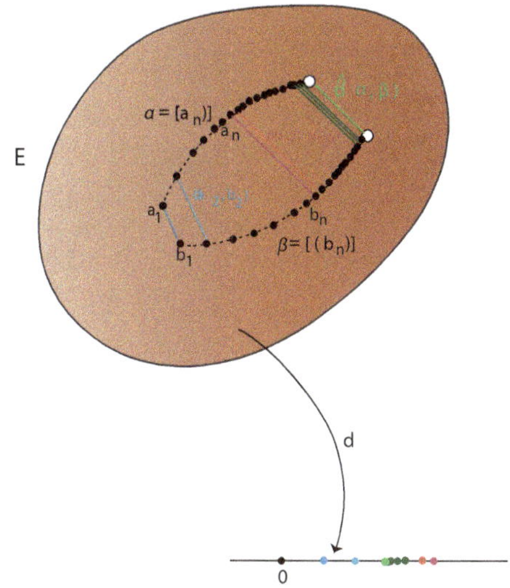

Fig. A.53 A schematic illustration of $\widehat{d}(\alpha, \beta)$ from the Cauchy sequence $(d(a_n, b_n))$.

and

$$\begin{aligned} d(a_n, b_n) &\leq d(a_n, a_n') + d(a_n', b_n') + d(b_n', b_n) \\ &= d(a_n', b_n') + d(a_n, a_n') + d(b_n, b_n'), \end{aligned}$$

we have

$$|d(a_n, b_n) - d(a_n', b_n')| \leq d(a_n, a_n') + d(b_n, b_n'),$$

so we have $\lim_{n \mapsto \infty} d(a_n', b_n') = \lim_{n \mapsto \infty} d(a_n, b_n) = \widehat{d}(\alpha, \beta)$. Therefore, $\widehat{d}(\alpha, \beta)$ is indeed well defined.

Step 4. Let us check that φ is indeed an isometry.

Given any two elements $\varphi(a)$ and $\varphi(b)$ in $\widehat{E}$, since they are the equivalence classes of the constant sequences (a_n) and (b_n) such that $a_n = a$ and $b_n = b$ for all n, the constant sequence $(d(a_n, b_n))$ with $d(a_n, b_n) = d(a, b)$ for all n converges to $d(a, b)$, so by definition $\widehat{d}(\varphi(a), \varphi(b)) = \lim_{n \mapsto \infty} d(a_n, b_n) = d(a, b)$, which shows that φ is an isometry.

Step 5. Let us verify that $\widehat{d}$ is a metric on $\widehat{E}$. By definition it is obvious that $\widehat{d}(\alpha, \beta) = \widehat{d}(\beta, \alpha)$. If α and β are two distinct equivalence classes, then for any Cauchy sequence (a_n) in the equivalence class α and for any Cauchy sequence (b_n) in the equivalence class β, the sequences (a_n) and (b_n) are inequivalent, which means that $\lim_{n\mapsto\infty} d(a_n, b_n) \neq 0$, that is, $\widehat{d}(\alpha, \beta) \neq 0$. Obviously, $\widehat{d}(\alpha, \alpha) = 0$.

For any equivalence classes $\alpha = [(a_n)]$, $\beta = [(b_n)]$, and $\gamma = [(c_n)]$, we have the triangle inequality

$$d(a_n, c_n) \leq d(a_n, b_n) + d(b_n, c_n),$$

so by continuity of the distance function, by passing to the limit, we obtain

$$\widehat{d}(\alpha, \gamma) \leq \widehat{d}(\alpha, \beta) + \widehat{d}(\beta, \gamma),$$

which is the triangle inequality for $\widehat{d}$. Therefore, $\widehat{d}$ is a distance on $\widehat{E}$.

Step 6. Let us prove that $\varphi(E)$ is dense in $\widehat{E}$. For any $\alpha = [(a_n)]$, let (x_n) be the constant sequence such that $x_k = a_n$ for all $k \geq 0$, so that $\varphi(a_n) = [(x_n)]$. Then we have

$$\widehat{d}(\alpha, \varphi(a_n)) = \lim_{m\mapsto\infty} d(a_m, a_n) \leq \sup_{p,q\geq n} d(a_p, a_q).$$

Since (a_n) is a Cauchy sequence, $\sup_{p,q\geq n} d(a_p, a_q)$ tends to 0 as n goes to infinity, so

$$\lim_{n\mapsto\infty} d(\alpha, \varphi(a_n)) = 0,$$

which means that the sequence $(\varphi(a_n))$ converge to α, and $\varphi(E)$ is indeed dense in $\widehat{E}$.

Step 7. Finally let us prove that the metric space $\widehat{E}$ is complete.

Let (α_n) be a Cauchy sequence in $\widehat{E}$. Since $\varphi(E)$ is dense in $\widehat{E}$, for every $n > 0$, there some $a_n \in E$ such that

$$\widehat{d}(\alpha_n, \varphi(a_n)) \leq \frac{1}{n}.$$

Since

$$\begin{aligned}\widehat{d}(\varphi(a_m), \varphi(a_n)) &\leq \widehat{d}(\varphi(a_m), \alpha_m) + \widehat{d}(\alpha_m, \alpha_n) + \widehat{d}(\alpha_n, \varphi(a_n))\\ &\leq \widehat{d}(\alpha_m, \alpha_n) + \frac{1}{m} + \frac{1}{n},\end{aligned}$$

and since (α_m) is a Cauchy sequence, so is $(\varphi(a_n))$, and as φ is an isometry, the sequence (a_n) is a Cauchy sequence in E. Let $\alpha \in \widehat{E}$ be the equivalence class of (a_n). Since

$$\widehat{d}(\alpha, \varphi(a_n)) = \lim_{m\mapsto\infty} d(a_m, a_n)$$

and (a_n) is a Cauchy sequence, we deduce that the sequence $(\varphi(a_n))$ converges to α, and since $d(\alpha_n, \varphi(a_n)) \leq 1/n$ for all $n > 0$, the sequence (α_n) also converges to α.

Step 8. Let us prove the extension property. Let F be any complete metric space and let $f\colon E \to F$ be any uniformly continuous function. The function $\varphi\colon E \to \widehat{E}$ is an isometry and a bijection between E and its image $\varphi(E)$, so its inverse $\varphi^{-1}\colon \varphi(E) \to E$ is also an isometry, and thus is uniformly continuous. If we let $g = f \circ \varphi^{-1}$, then $g\colon \varphi(E) \to F$ is a uniformly continuous function, and $\varphi(E)$ is dense in $\widehat{E}$, so by Theorem A.61 there is a unique uniformly continuous function $\widehat{f}\colon \widehat{E} \to F$ extending $g = f \circ \varphi^{-1}$; see the diagram below:

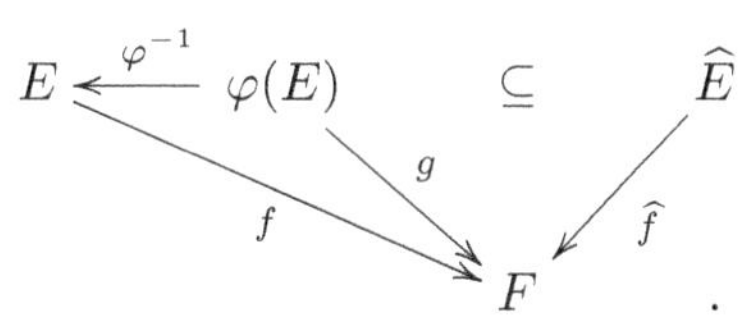

This means that

$$\widehat{f}|\varphi(E) = f \circ \varphi^{-1},$$

which implies that

$$(\widehat{f}|\varphi(E)) \circ \varphi = f,$$

that is, $f = \widehat{f} \circ \varphi$, as illustrated in the diagram below:

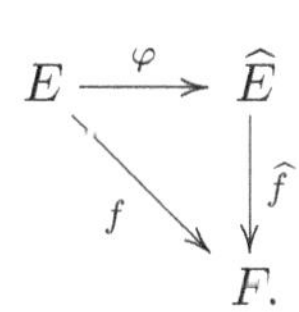

If $h\colon \widehat{E} \to F$ is any other uniformly continuous function such that $f = h \circ \varphi$, then $g = f \circ \varphi^{-1} = h|\varphi(E)$, so h is a uniformly continuous function extending g, and by Theorem A.61, we have $h = \widehat{f}$, so $\widehat{f}$ is indeed unique.

Step 9. Uniqueness of the completion $(\widehat{E}, \widehat{d})$ up to a bijective isometry.

Let $(\widehat{E}_1, \widehat{d}_1)$ and $(\widehat{E}_2, \widehat{d}_2)$ be any two completions of (E, d). Then we have two uniformly continuous isometries $\varphi_1\colon E \to \widehat{E}_1$ and $\varphi_2\colon E \to \widehat{E}_2$, so by the unique extension property, there exist unique uniformly continuous maps $\widehat{\varphi_2}\colon \widehat{E}_1 \to \widehat{E}_2$ and $\widehat{\varphi_1}\colon \widehat{E}_2 \to \widehat{E}_1$ such that the following diagrams commute:

$$\begin{array}{ccc} E & \xrightarrow{\varphi_1} & \widehat{E}_1 \\ & \searrow^{\varphi_2} & \downarrow \widehat{\varphi_2} \\ & & \widehat{E}_2 \end{array} \qquad \begin{array}{ccc} E & \xrightarrow{\varphi_2} & \widehat{E}_2 \\ & \searrow^{\varphi_1} & \downarrow \widehat{\varphi_1} \\ & & \widehat{E}_1 \end{array}$$

Consequently we have the following commutative diagrams:

$$\begin{array}{ccc} & & \widehat{E}_2 \\ & \nearrow^{\varphi_2} & \downarrow \widehat{\varphi_1} \\ E & \xrightarrow{\varphi_1} & \widehat{E}_1 \\ & \searrow^{\varphi_2} & \downarrow \widehat{\varphi_2} \\ & & \widehat{E}_2 \end{array} \qquad \begin{array}{ccc} & & \widehat{E}_1 \\ & \nearrow^{\varphi_1} & \downarrow \widehat{\varphi_2} \\ E & \xrightarrow{\varphi_2} & \widehat{E}_2 \\ & \searrow^{\varphi_1} & \downarrow \widehat{\varphi_1} \\ & & \widehat{E}_1. \end{array}$$

However, $\mathrm{id}_{\widehat{E}_1}$ and $\mathrm{id}_{\widehat{E}_2}$ are uniformly continuous functions making the following diagrams commute

$$\begin{array}{ccc} E & \xrightarrow{\varphi_1} & \widehat{E}_1 \\ & \searrow^{\varphi_1} & \downarrow \mathrm{id}_{\widehat{E}_1} \\ & & \widehat{E}_1 \end{array} \qquad \begin{array}{ccc} E & \xrightarrow{\varphi_2} & \widehat{E}_2 \\ & \searrow^{\varphi_2} & \downarrow \mathrm{id}_{\widehat{E}_2} \\ & & \widehat{E}_2, \end{array}$$

so by the uniqueness of extensions we must have

$$\widehat{\varphi_1} \circ \widehat{\varphi_2} = \mathrm{id}_{\widehat{E}_1} \quad \text{and} \quad \widehat{\varphi_2} \circ \widehat{\varphi_1} = \mathrm{id}_{\widehat{E}_2}.$$

This proves that $\widehat{\varphi_1}$ and $\widehat{\varphi_2}$ are mutual inverses. Now since $\varphi_2 = \widehat{\varphi_2} \circ \varphi_1$, we have

$$\widehat{\varphi_2}|\varphi_1(E) = \varphi_2 \circ \varphi_1^{-1},$$

and since φ_1^{-1} and φ_2 are isometries, so is $\widehat{\varphi_2}|\varphi_1(E)$. But we showed in Step 8 that $\widehat{\varphi_2}$ is the uniform continuous extension of $\widehat{\varphi_2}|\varphi_1(E)$ and $\varphi_1(E)$ is dense in $\widehat{E}_1$, so for any two elements $\alpha, \beta \in \widehat{E}_1$, if (a_n) and (b_n) are sequences in $\varphi_1(E)$ converging to α and β, we have

$$\widehat{d}_2((\widehat{\varphi_2}|\varphi_1(E))(a_n), ((\widehat{\varphi_2}|\varphi_1(E))(b_n)) = \widehat{d}_1(a_n, b_n),$$

and by passing to the limit we get

$$\widehat{d}_2(\widehat{\varphi_2}(\alpha), \widehat{\varphi_2}(\beta)) = \widehat{d}_1(\alpha, \beta),$$

which shows that $\widehat{\varphi_2}$ is an isometry (similarly, $\widehat{\varphi_1}$ is an isometry). $\square$

Remarks:

(1) Except for Step 8 and Step 9, the proof of Theorem A.62 is the proof given in Schwartz [Schwartz (1980)] (Chapter XI, Section 4, Theorem 1), and Kolmogorov and Fomin [Kolmogorov and Fomin (1975)] (Chapter 2, Section 7, Theorem 4).

(2) The construction of $\widehat{E}$ relies on the completeness of $\mathbb{R}$, and so it cannot be used to construct $\mathbb{R}$ from $\mathbb{Q}$. However, this construction can be modified to yield a construction of $\mathbb{R}$ from $\mathbb{Q}$.

We show in Section A.13 that Theorem A.62 yields a construction of the completion of a normed vector space.

A.11 The Contraction Mapping Theorem

If (E, d) is a nonempty complete metric space, every map $f \colon E \to E$ for which there is some k such that $0 \leq k < 1$ and

$$d(f(x), f(y)) \leq k d(x, y)$$

for all $x, y \in E$, has the very important property that it has a unique fixed point, that is, there is a unique $a \in E$ such that $f(a) = a$. A map as above is called a *contraction mapping*. Furthermore, the fixed point of a contraction mapping can be computed as the limit of a fast converging sequence.

The fixed point property of contraction mappings is used to show some important theorems of analysis, such as the implicit function theorem and the existence of solutions to certain differential equations. It can also be used to show the existence of fractal sets defined in terms of iterated function systems. Since the proof is quite simple, we prove the fixed point property of contraction mappings. First, observe that a contraction mapping is (uniformly) continuous.

Proposition A.63. *If (E, d) is a nonempty complete metric space, every contraction mapping $f \colon E \to E$ has a unique fixed point. Furthermore, for every $x_0 \in E$, defining the sequence (x_n) such that $x_{n+1} = f(x_n)$, the sequence (x_n) converges to the unique fixed point of f.*

Proof. First we prove that f has at most one fixed point. Indeed, if $f(a) = a$ and $f(b) = b$, since

$$d(a, b) = d(f(a), f(b)) \leq k d(a, b)$$

and $0 \leq k < 1$, we must have $d(a, b) = 0$, that is, $a = b$.

Next, we prove that (x_n) is a Cauchy sequence. Observe that

$$\begin{aligned} d(x_2, x_1) &\leq kd(x_1, x_0), \\ d(x_3, x_2) &\leq kd(x_2, x_1) \leq k^2 d(x_1, x_0), \\ &\vdots \qquad\quad \vdots \\ d(x_{n+1}, x_n) &\leq kd(x_n, x_{n-1}) \leq \cdots \leq k^n d(x_1, x_0). \end{aligned}$$

Thus, we have

$$\begin{aligned} d(x_{n+p}, x_n) &\leq d(x_{n+p}, x_{n+p-1}) + d(x_{n+p-1}, x_{n+p-2}) + \cdots + d(x_{n+1}, x_n) \\ &\leq (k^{p-1} + k^{p-2} + \cdots + k + 1)k^n d(x_1, x_0) \\ &\leq \frac{k^n}{1-k} d(x_1, x_0). \end{aligned}$$

We conclude that $d(x_{n+p}, x_n)$ converges to 0 when n goes to infinity, which shows that (x_n) is a Cauchy sequence. Since E is complete, the sequence (x_n) has a limit a. Since f is continuous, the sequence $(f(x_n))$ converges to $f(a)$. But $x_{n+1} = f(x_n)$ converges to a and so $f(a) = a$, the unique fixed point of f. □

Note that no matter how the starting point x_0 of the sequence (x_n) is chosen, (x_n) converges to the unique fixed point of f. Also, the convergence is fast, since

$$d(x_n, a) \leq \frac{k^n}{1-k} d(x_1, x_0).$$

The Hausdorff distance between compact subsets of a metric space provides a very nice illustration of some of the theorems on complete and compact metric spaces just presented.

Definition A.49. Given a metric space (X, d), for any subset $A \subseteq X$, for any $\epsilon \geq 0$, define the *ϵ-hull* of A as the set

$$V_\epsilon(A) = \{x \in X,\ \exists a \in A \mid d(a, x) \leq \epsilon\}.$$

See Figure A.54. Given any two nonempty bounded subsets A, B of X, define $D(A, B)$, *the Hausdorff distance between A and B*, by

$$D(A, B) = \inf\{\epsilon \geq 0 \mid A \subseteq V_\epsilon(B) \text{ and } B \subseteq V_\epsilon(A)\}.$$

Note that since we are considering nonempty bounded subsets, $D(A, B)$ is well defined (i.e., not infinite). However, D is not necessarily a distance function. It is a distance function if we restrict our attention to nonempty

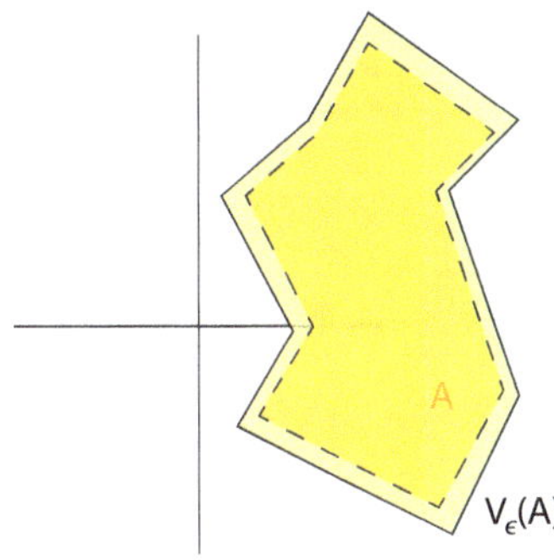

Fig. A.54 The ϵ-hull of a polygonal region A of $\mathbb{R}^2$.

compact subsets of X (actually, it is also a metric on closed and bounded subsets). We let $\mathcal{K}(X)$ denote the set of all nonempty compact subsets of X. The remarkable fact is that D is a distance on $\mathcal{K}(X)$ and that if X is complete or compact, then so is $\mathcal{K}(X)$. The following theorem is taken from Edgar [Edgar (1992)].

Theorem A.64. *If (X, d) is a metric space, then the Hausdorff distance D on the set $\mathcal{K}(X)$ of nonempty compact subsets of X is a distance. If (X, d) is complete, then $(\mathcal{K}(X), D)$ is complete and if (X, d) is compact, then $(\mathcal{K}(X), D)$ is compact.*

Proof. Since (nonempty) compact sets are bounded, $D(A, B)$ is well defined. Clearly D is symmetric. Assume that $D(A, B) = 0$. Then for every $\epsilon > 0$, $A \subseteq V_\epsilon(B)$, which means that for every $a \in A$, there is some $b \in B$ such that $d(a, b) \leq \epsilon$, and thus, that $A \subseteq \overline{B}$. Since Proposition A.26 implies that B is closed, $\overline{B} = B$, and we have $A \subseteq B$. Similarly, $B \subseteq A$, and thus, $A = B$. Clearly, if $A = B$, we have $D(A, B) = 0$. It remains to prove the triangle inequality. Assume that $D(A, B) \leq \epsilon_1$ and that $D(B, C) \leq \epsilon_2$. We must show that $D(A, C) \leq \epsilon_1 + \epsilon_2$. This will be accomplished if we can show that $C \subseteq V_{\epsilon_1+\epsilon_2}(A)$ and $A \subseteq V_{\epsilon_1+\epsilon_2}(C)$. By assumption and definition of D, $B \subseteq V_{\epsilon_1}(A)$ and $C \subseteq V_{\epsilon_2}(B)$. Then

$$V_{\epsilon_2}(B) \subseteq V_{\epsilon_2}(V_{\epsilon_1}(A)),$$

and since a basic application of the triangle inequality implies that

$$V_{\epsilon_2}(V_{\epsilon_1}(A)) \subseteq V_{\epsilon_1+\epsilon_2}(A),$$

we get

$$C \subseteq V_{\epsilon_2}(B) \subseteq V_{\epsilon_1+\epsilon_2}(A).$$

See Figure A.55.

Similarly, the conditions $(A, B) \leq \epsilon_1$ and $D(B, C) \leq \epsilon_2$ imply that

$$A \subseteq V_{\epsilon_1}(B), \qquad B \subseteq V_{\epsilon_2}(C).$$

Hence

$$A \subseteq V_{\epsilon_1}(B) \subseteq V_{\epsilon_1}(V_{\epsilon_2}(C)) \subseteq V_{\epsilon_1+\epsilon_2}(C),$$

and thus the triangle inequality follows.

Next we need to prove that if (X, d) is complete, then $(\mathcal{K}(X), D)$ is also complete. First we show that if (A_n) is a sequence of nonempty compact sets converging to a nonempty compact set A in the Hausdorff metric, then

$$A = \{x \in X \mid \text{there is a sequence } (x_n), \text{ with } x_n \in A_n, \text{ converging to } x\}.$$

Indeed, if (x_n) is a sequence with $x_n \in A_n$ converging to x and (A_n) con-

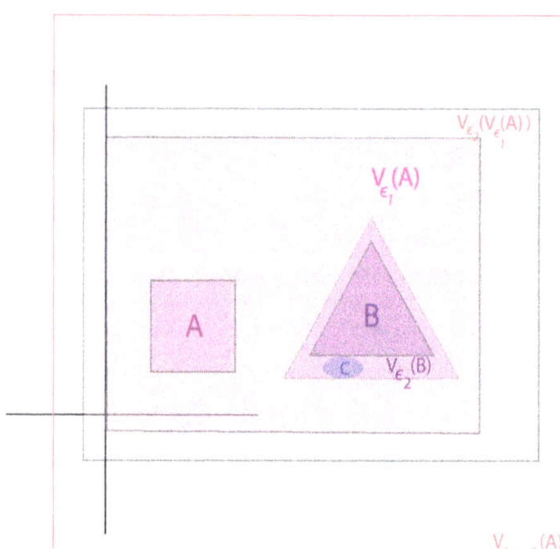

Fig. A.55 Let A be the small pink square and B be the small purple triangle in $\mathbb{R}^2$. The periwinkle oval C is contained in $V_{\epsilon_1+\epsilon_2}(A)$.

verges to A, then for every $\epsilon > 0$, there is some x_n such that $d(x_n, x) \leq \epsilon/2$, and there is some $a_n \in A$ such that $d(a_n, x_n) \leq \epsilon/2$ and thus, $d(a_n, x) \leq \epsilon$,

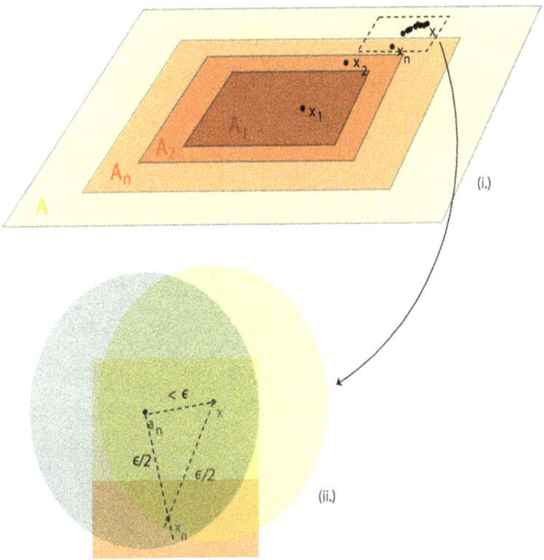

Fig. A.56 Let (A_n) be the sequence of parallelograms converging to A, the large pale yellow parallelogram. Figure $(ii.)$ expands the dashed region and shows why $d(a_n, x) < \epsilon$.

which shows that $x \in \overline{A}$. Since A is compact, it is closed, and $x \in A$. See Figure A.56.

Conversely, since (A_n) converges to A, for every $x \in A$, for every $n \geq 1$, there is some $x_n \in A_n$ such that $d(x_n, x) \leq 1/n$ and the sequence (x_n) converges to x.

Now let (A_n) be a Cauchy sequence in $\mathcal{K}(X)$. It can be proven that (A_n) converges to the set

$$A = \{x \in X \mid \text{there is a sequence } (x_n), \text{ with } x_n \in A_n, \text{ converging to } x\},$$

and that A is nonempty and compact. To prove that A is compact, one proves that it is totally bounded and complete. Details are given in Edgar [Edgar (1992)].

Finally we need to prove that if (X, d) is compact, then $(\mathcal{K}(X), D)$ is compact. Since we already know that $(\mathcal{K}(X), D)$ is complete if (X, d) is, it is enough to prove that $(\mathcal{K}(X), D)$ is totally bounded if (X, d) is, which is not hard. $\square$

In view of Theorem A.64 and Theorem A.63, it is possible to define some nonempty compact subsets of X in terms of fixed points of contraction maps. This can be done in terms of iterated function systems, yielding a large class of fractals. However, we will omit this topic and instead refer the reader to Edgar [Edgar (1992)].

A.12 Continuous Linear and Multilinear Maps

If E and F are normed vector spaces, we first characterize when a linear map $f\colon E \to F$ is continuous.

Proposition A.65. *Given two normed vector spaces E and F, for any linear map $f\colon E \to F$, the following conditions are equivalent:*

(1) The function f is continuous at 0.
(2) There is a constant $k \geq 0$ such that,

$$\|f(u)\| \leq k, \text{ for every } u \in E \text{ such that } \|u\| \leq 1.$$

(3) There is a constant $k \geq 0$ such that,

$$\|f(u)\| \leq k\|u\|, \text{ for every } u \in E.$$

(4) The function f is continuous at every point of E.

Proof. Assume (1). Then for every $\epsilon > 0$, there is some $\eta > 0$ such that, for every $u \in E$, if $\|u\| \leq \eta$, then $\|f(u)\| \leq \epsilon$. Pick $\epsilon = 1$, so that there is some $\eta > 0$ such that, if $\|u\| \leq \eta$, then $\|f(u)\| \leq 1$. If $\|u\| \leq 1$, then $\|\eta u\| \leq \eta\|u\| \leq \eta$, and so, $\|f(\eta u)\| \leq 1$, that is, $\eta\|f(u)\| \leq 1$, which implies $\|f(u)\| \leq \eta^{-1}$. Thus, (2) holds with $k = \eta^{-1}$.

Assume that (2) holds. If $u = 0$, then by linearity, $f(0) = 0$, and thus $\|f(0)\| \leq k\|0\|$ holds trivially for all $k \geq 0$. If $u \neq 0$, then $\|u\| > 0$, and since

$$\left\|\frac{u}{\|u\|}\right\| = 1,$$

we have

$$\left\|f\left(\frac{u}{\|u\|}\right)\right\| \leq k,$$

which implies that

$$\|f(u)\| \leq k\|u\|.$$

Thus, (3) holds.

If (3) holds, then for all $u, v \in E$, we have

$$\|f(v) - f(u)\| = \|f(v - u)\| \leq k\|v - u\|.$$

If $k = 0$, then f is the zero function, and continuity is obvious. Otherwise, if $k > 0$, for every $\epsilon > 0$, if $\|v - u\| \leq \frac{\epsilon}{k}$, then $\|f(v - u)\| \leq \epsilon$, which shows continuity at every $u \in E$. Finally, it is obvious that (4) implies (1). $\square$

Among other things, Proposition A.65 shows that a linear map is continuous iff the image of the unit (closed) ball is bounded. Since a continuous linear map satisfies the condition $\|f(u)\| \leq k\|u\|$ (for some $k \geq 0$), it is also uniformly continuous.

If E and F are normed vector spaces, the set of all continuous linear maps $f \colon E \to F$ is denoted by $\mathcal{L}(E; F)$.

Using Proposition A.65, we can define a norm on $\mathcal{L}(E; F)$ which makes it into a normed vector space.

Definition A.50. Given two normed vector spaces E and F, for every continuous linear map $f \colon E \to F$, we define the *operator norm* $\|f\|$ *of* f as

$$\begin{aligned}\|f\| &= \inf\{k \geq 0 \mid \|f(x)\| \leq k\|x\|, \text{ for all } x \in E\} \\ &= \sup\{\|f(x)\| \mid \|x\| \leq 1\} \\ &= \sup\{\|f(x)\| \mid \|x\| = 1\}.\end{aligned}$$

From Definition A.50, for every continuous linear map $f \in \mathcal{L}(E; F)$, we have

$$\|f(x)\| \leq \|f\|\|x\|,$$

for every $x \in E$. It is easy to verify that $\mathcal{L}(E; F)$ is a normed vector space under the norm of Definition A.50. Furthermore, if E, F, G, are normed vector spaces, and $f \colon E \to F$ and $g \colon F \to G$ are continuous linear maps, we have

$$\|g \circ f\| \leq \|g\|\|f\|.$$

We can now show that when $E = \mathbb{R}^n$ or $E = \mathbb{C}^n$, with any of the norms $\| \ \|_1$, $\| \ \|_2$, or $\| \ \|_\infty$, then every linear map $f \colon E \to F$ is continuous.

Proposition A.66. *If $E = \mathbb{R}^n$ or $E = \mathbb{C}^n$, with any of the norms $\| \ \|_1$, $\| \ \|_2$, or $\| \ \|_\infty$, and F is any normed vector space, then every linear map $f \colon E \to F$ is continuous.*

Proof. Let $(e_1, \ldots, e_n)$ be the standard basis of $\mathbb{R}^n$ (a similar proof applies to $\mathbb{C}^n$). In view of Proposition B.2, it is enough to prove the proposition for the norm

$$\|x\|_\infty = \max\{|x_i| \mid 1 \leq i \leq n\}.$$

We have,

$$\begin{aligned}\|f(v) - f(u)\| = \|f(v-u)\| &= \left\| f\Big(\sum_{1\leq i\leq n} (v_i - u_i)e_i\Big)\right\| \\ &= \left\| \sum_{1\leq i\leq n} (v_i - u_i)f(e_i)\right\|,\end{aligned}$$

and so,

$$\begin{aligned}\|f(v) - f(u)\| &\leq \Big(\sum_{1\leq i\leq n} \|f(e_i)\|\Big) \max_{1\leq i\leq n} |v_i - u_i| \\ &= \Big(\sum_{1\leq i\leq n} \|f(e_i)\|\Big) \|v - u\|_\infty.\end{aligned}$$

By the argument used in Proposition A.65 to prove that (3) implies (4), f is continuous. ☐

Actually, we prove in Theorem B.3 that if E is a vector space of finite dimension, then any two norms are equivalent, so that they define the same topology. This fact together with Proposition A.66 proves the following.

Theorem A.67. *If E is a vector space of finite dimension (over $\mathbb{R}$ or $\mathbb{C}$), then all norms are equivalent (define the same topology). Furthermore, for any normed vector space F, every linear map $f \colon E \to F$ is continuous.*

If E is a normed vector space of infinite dimension, a linear map $f \colon E \to F$ may not be continuous. As an example, let E be the infinite vector space of all polynomials over $\mathbb{R}$. Let

$$\|P(X)\| = \max_{0\leq x\leq 1} |P(x)|.$$

We leave as an exercise to show that this is indeed a norm. Let $F = \mathbb{R}$, and let $f \colon E \to F$ be the map defined such that $f(P(X)) = P(3)$. It is clear that f is linear. Consider the sequence of polynomials

$$P_n(X) = \left(\frac{X}{2}\right)^n.$$

It is clear that $\|P_n\| = \left(\frac{1}{2}\right)^n$, and thus, the sequence P_n has the null polynomial as a limit. However, we have

$$f(P_n(X)) = P_n(3) = \left(\frac{3}{2}\right)^n,$$

and the sequence $f(P_n(X))$ diverges to $+\infty$. Consequently, in view of Proposition A.15 (1), f is not continuous.

We now consider the continuity of multilinear maps. We treat explicitly bilinear maps, the general case being a straightforward extension.

Proposition A.68. *Given normed vector spaces E, F and G, for any bilinear map $f\colon E \times F \to G$, the following conditions are equivalent:*

(1) The function f is continuous at $\langle 0, 0\rangle$.
(2) There is a constant $k \geq 0$ such that,

$$\|f(u,v)\| \leq k, \text{ for all } u \in E, v \in F \text{ such that } \|u\|, \|v\| \leq 1.$$

(3) There is a constant $k \geq 0$ such that,

$$\|f(u,v)\| \leq k\|u\|\|v\|, \text{ for all } u \in E, v \in F.$$

(4) The function f is continuous at every point of $E \times F$.

Proof. It is similar to that of Proposition A.65, with a small subtlety in proving that (3) implies (4), namely that two different η's that are not independent are needed. □

In contrast to continuous linear maps, which must be uniformly continuous, nonzero continuous bilinear maps are **not** uniformly continuous. Let $f\colon E \times F \to G$ be a continuous bilinear map such that $f(a, b) \neq 0$ for some $a \in E$ and some $b \in F$. Consider the sequences (u_n) and (v_n) (with $n \geq 1$) given by

$$\begin{aligned} u_n &= (x_n, y_n) = (na, nb) \\ v_n &= (x'_n, y'_n) = \left(\left(n + \frac{1}{n}\right) a, \left(n + \frac{1}{n}\right) b\right). \end{aligned}$$

Obviously

$$\|v_n - u_n\| \leq \frac{1}{n}(\|a\| + \|b\|),$$

so $\lim_{n\mapsto\infty} \|v_n - u_n\| = 0$. On the other hand

$$f(x'_n, y'_n) - f(x_n, y_n) = \left(2 + \frac{1}{n^2}\right) f(a, b),$$

and thus $\lim_{n\mapsto\infty} \|f(x'_n, y'_n) - f(x_n, y_n)\| = 2\|f(a,b)\| \neq 0$, which shows that f is not uniformly continuous, because if this was the case, this limit would be zero.

If E, F, and G are normed vector spaces, we denote the set of all continuous bilinear maps $f\colon E\times F \to G$ by $\mathcal{L}_2(E, F; G)$. Using Proposition A.68, we can define a norm on $\mathcal{L}_2(E, F; G)$ which makes it into a normed vector space.

Definition A.51. Given normed vector spaces E, F, and G, for every continuous bilinear map $f\colon E \times F \to G$, we define the *norm* $\|f\|$ *of* f as

$$\begin{aligned}\|f\| &= \inf\{k \geq 0 \mid \|f(x,y)\| \leq k\|x\|\|y\|, \text{ for all } x \in E, y \in F\} \\ &= \sup\{\|f(x,y)\| \mid \|x\|, \|y\| \leq 1\}.\end{aligned}$$

From Definition A.50, for every continuous bilinear map $f \in \mathcal{L}_2(E, F; G)$, we have

$$\|f(x,y)\| \leq \|f\|\|x\|\|y\|,$$

for all $x \in E, y \in F$. It is easy to verify that $\mathcal{L}_2(E, F; G)$ is a normed vector space under the norm of Definition A.51.

Given a bilinear map $f\colon E\times F \to G$, for every $u \in E$, we obtain a linear map denoted $fu\colon F \to G$, defined such that $fu(v) = f(u,v)$. Furthermore, since

$$\|f(x,y)\| \leq \|f\|\|x\|\|y\|,$$

it is clear that fu is continuous. We can then consider the map $\varphi\colon E \to \mathcal{L}(F;G)$, defined such that $\varphi(u) = fu$ for any $u \in E$, or equivalently, such that,

$$\varphi(u)(v) = f(u,v).$$

Actually, it is easy to show that φ is linear and continuous, and that $\|\varphi\| = \|f\|$. Thus, $f \mapsto \varphi$ defines a map from $\mathcal{L}_2(E, F; G)$ to $\mathcal{L}(E; \mathcal{L}(F; G))$. We can also go back from $\mathcal{L}(E; \mathcal{L}(F; G))$ to $\mathcal{L}_2(E, F; G)$. We summarize all this in the following proposition.

Proposition A.69. *Let E, F, G be three normed vector spaces. The map $f \mapsto \varphi$, from $\mathcal{L}_2(E, F; G)$ to $\mathcal{L}(E; \mathcal{L}(F; G))$, defined such that for every $f \in \mathcal{L}_2(E, F; G)$,*

$$\varphi(u)(v) = f(u,v),$$

is an isomorphism of vector spaces, and furthermore, $\|\varphi\| = \|f\|$.

As a corollary of Proposition A.69, we get the following proposition which will be useful when we define second-order derivatives.

Proposition A.70. *Let E, F be normed vector spaces. The map app from $\mathcal{L}(E; F) \times E$ to F, defined such that, for every $f \in \mathcal{L}(E; F)$, for every $u \in E$,*

$$app(f, u) = f(u),$$

is a continuous bilinear map.

Remark: If E and F are nontrivial, it can be shown that $\|\text{app}\| = 1$. It can also be shown that composition

$$\circ\colon \mathcal{L}(E; F) \times \mathcal{L}(F; G) \to \mathcal{L}(E; G),$$

is bilinear and continuous.

The above propositions and definition generalize to arbitrary n-multilinear maps, with $n \geq 2$. Proposition A.68 extends in the obvious way to any n-multilinear map $f\colon E_1 \times \cdots \times E_n \to F$, but condition (3) becomes:

(3) There is a constant $k \geq 0$ such that,

$$\|f(u_1, \ldots, u_n)\| \leq k\|u_1\| \cdots \|u_n\|, \text{ for all } u_1 \in E_1, \ldots, u_n \in E_n.$$

Definition A.51 also extends easily to

$$\begin{aligned} \|f\| &= \inf \{k \geq 0 \mid \|f(x_1, \ldots, x_n)\| \leq k\|x_1\| \cdots \|x_n\|,\ x_i \in E_i, 1 \leq i \leq n\} \\ &= \sup \{\|f(x_1, \ldots, x_n)\| \mid \|x_n\|, \ldots, \|x_n\| \leq 1\}. \end{aligned}$$

Proposition A.69 is also easily extended, and we get an isomorphism between continuous n-multilinear maps in $\mathcal{L}_n(E_1, \ldots, E_n; F)$, and continuous linear maps in

$$\mathcal{L}(E_1; \mathcal{L}(E_2; \ldots; \mathcal{L}(E_n; F))).$$

An obvious extension of Proposition A.70 also holds.

Definition A.52. A normed vector space $(E, \| \,\|)$ over $\mathbb{R}$ (or $\mathbb{C}$) which is a complete metric space for the distance $d(u, v) = \|v - u\|$, is called a *Banach space*.

It can be shown that every normed vector space of finite dimension is a Banach space (is complete). This is because $\mathbb{R}$ (and $\mathbb{C}$) are complete. The following theorem is a key result of the theory of Banach spaces worth proving.

Theorem A.71. *If E and F are normed vector spaces, and if F is a Banach space, then $\mathcal{L}(E; F)$ is a Banach space (with the operator norm).*

Proof. Let $(f)_{n\geq 1}$ be a Cauchy sequence of continuous linear maps $f_n : E \to F$. We proceed in several steps.

Step 1. Define the pointwise limit $f : E \to F$ of the sequence $(f_n)_{n\geq 1}$.

Since $(f)_{n\geq 1}$ is a Cauchy sequence, for every $\epsilon > 0$, there is some $N > 0$ such that $\|f_m - f_n\| < \epsilon$ for all $m, n \geq N$. Since $\| \ \|$ is the operator norm, we deduce that for any $u \in E$, we have

$$\|f_m(u) - f_n(u)\| = \|(f_m - f_n)(u)\| \leq \|f_m - f_n\| \, \|u\| \leq \epsilon \, \|u\|$$

for all $m, n \geq N$, that is,

$$\|f_m(u) - f_n(u)\| \leq \epsilon \, \|u\| \tag{$*_1$}$$

for all $m, n \geq N$. If $u = 0$, then $f_m(0) = f_n(0) = 0$ for all m, n, so the sequence $(f_n(0))$ is a Cauchy sequence in F converging to 0. If $u \neq 0$, by replacing ϵ by $\epsilon / \|u\|$, we see that the sequence $(f_n(u))$ is a Cauchy sequence in F. Since F is complete, the sequence $(f_n(u))$ has a limit which we denote by $f(u)$. This defines our candidate limit function f by

$$f(u) = \lim_{n \mapsto \infty} f_n(u).$$

It remains to prove that

(1) f is linear.
(2) f is continous.
(3) f is the limit of (f_n) for the operator norm.

Step 2. The function f is linear.

Recall that in a normed vector space, addition and multiplication by a fixed scalar are continuous (since $\|u + v\| \leq \|u\| + \|v\|$ and $\|\lambda u\| \leq |\lambda| \, \|u\|$). Thus by definition of f and since the f_n are linear we have

$$\begin{aligned} f(u + v) &= \lim_{n \mapsto \infty} f_n(u + v) && \text{by definition of } f \\ &= \lim_{n \mapsto \infty} (f_n(u) + f_n(v)) && \text{by linearity of } f_n \\ &= \lim_{n \mapsto \infty} f_n(u) + \lim_{n \mapsto \infty} f_n(v) && \text{since + is continuous} \\ &= f(u) + f(v) && \text{by definition of } f. \end{aligned}$$

Similarly,

$$\begin{aligned}
f(\lambda u) &= \lim_{n\mapsto\infty} f_n(\lambda u) && \text{by definition of } f\\
&= \lim_{n\mapsto\infty} \lambda f_n(u) && \text{by linearity of } f_n\\
&= \lambda \lim_{n\mapsto\infty} f_n(u) && \text{by continuity of scalar multiplication}\\
&= \lambda f(u) && \text{by definition of } f.
\end{aligned}$$

Therefore, f is linear.

Step 3. The function f is continuous.

Since $(f_n)_{n\geq 1}$ is a Cauchy sequence, for every $\epsilon > 0$, there is some $N > 0$ such that $\|f_m - f_n\| < \epsilon$ for all $m, n \geq N$. Since $f_m = f_n + f_m - f_n$, we get $\|f_m\| \leq \|f_n\| + \|f_m - f_n\|$, which implies that

$$\|f_m\| \leq \|f_n\| + \epsilon \quad \text{for all } m, n \geq N. \tag{$*_2$}$$

Using $(*_2)$, we also have

$$\|f_m(u)\| \leq \|f_m\| \|u\| \leq (\|f_n\| + \epsilon) \|u\| \quad \text{for all } m, n \geq N,$$

that is,

$$\|f_m(u)\| \leq (\|f_n\| + \epsilon) \|u\| \quad \text{for all } m, n \geq N. \tag{$*_3$}$$

Hold $n \geq N$ fixed and let m tend to $+\infty$ in $(*_3)$. Since the norm is continuous, we get

$$\|f(u)\| \leq (\|f_n\| + \epsilon) \|u\|,$$

which shows that f is continuous.

Step 4. The function f is the limit of (f_n) for the operator norm.

Recall $(*_1)$:

$$\|f_m(u) - f_n(u)\| \leq \epsilon \|u\| \quad \text{for all } m, n \geq N. \tag{$*_1$}$$

Hold $n \geq N$ fixed but this time let m tend to $+\infty$ in $(*_1)$. By continuity of the norm we get

$$\|f(u) - f_n(u)\| = \|(f - f_n)(u)\| \leq \epsilon \|u\|.$$

By definition of the operator norm,

$$\|f - f_n\| = \sup\{\|(f - f_n)(u)\| \mid \|u\| = 1\} \leq \epsilon \quad \text{for all } n \geq N,$$

which proves that f_n converges to f for the operator norm. ☐

As a special case of Theorem A.71, if we let $F = \mathbb{R}$ (or $F = \mathbb{C}$ in the case of complex vector spaces) we see that $E' = \mathcal{L}(E;\mathbb{R})$ (or $E' = \mathcal{L}(E;\mathbb{C})$) is complete (since $\mathbb{R}$ and $\mathbb{C}$ are complete). The space E' of continuous linear forms on E is called the *dual* of E. It is a subspace of the *algebraic dual* E^* of E which consists of *all* linear forms on E, not necessarily continuous.

It can also be shown that if E, F and G are normed vector spaces, and if G is a Banach space, then $\mathcal{L}_2(E, F; G)$ is a Banach space. The proof is essentially identical.

A.13 Completion of a Normed Vector Space

An easy corollary of Theorem A.62 and Theorem A.61 is that every normed vector space can be embedded in a complete normed vector space, that is, a Banach space.

Theorem A.72. *If $(E, \| \,\|)$ is a normed vector space, then its completion $(\widehat{E}, \widehat{d})$ as a metric space (where E is given the metric $d(x, y) = \|x - y\|$) can be given a unique vector space structure extending the vector space structure on E, and a norm $\| \,\|_{\widehat{E}}$, so that $(\widehat{E}, \| \,\|_{\widehat{E}})$ is a Banach space, and the metric $\widehat{d}$ is associated with the norm $\| \,\|_{\widehat{E}}$. Furthermore, the isometry $\varphi\colon E \to \widehat{E}$ given by Theorem A.62 is a linear isometry, and $\varphi(E)$ is dense in $\widehat{E}$.*

Proof. The addition operation $+\colon E \times E \to E$ is uniformly continuous because

$$\|(u' + v') - (u'' + v'')\| \leq \|u' - u''\| + \|v' - v''\|.$$

It is not hard to show that $\widehat{E} \times \widehat{E}$ is a complete metric space and that $E \times E$ is dense in $\widehat{E} \times \widehat{E}$. Then, by Theorem A.61, the uniformly continuous function $+$ has a unique continuous extension $+\colon \widehat{E} \times \widehat{E} \to \widehat{E}$.

The map $\cdot\colon \mathbb{R} \times E \to E$ is not uniformly continuous, but for any fixed $\lambda \in \mathbb{R}$, the map $L_\lambda\colon E \to E$ given by $L_\lambda(u) = \lambda \cdot u$ is uniformly continuous, so by Theorem A.61 the function L_λ has a unique continuous extension $L_\lambda\colon \widehat{E} \to \widehat{E}$, which we use to define the scalar multiplication $\cdot\colon \mathbb{R} \times \widehat{E} \to \widehat{E}$. It is easily checked that with the above addition and scalar multiplication, $\widehat{E}$ is a vector space.

Since the norm $\| \,\|$ on E is uniformly continuous, it has a unique continuous extension $\| \,\|_{\widehat{E}}\colon \widehat{E} \to \mathbb{R}_+$. The identities $\|u + v\| \leq \|u\| + \|v\|$ and $\|\lambda u\| \leq |\lambda| \, \|u\|$ extend to $\widehat{E}$ by continuity. The equation

$$d(u, v) = \|u - v\|$$

also extends to $\widehat{E}$ by continuity and yields

$$\widehat{d}(\alpha, \beta) = \|\alpha - \beta\|_{\widehat{E}},$$

which shows that $\| \, \|_{\widehat{E}}$ is indeed a norm, and that the metric $\widehat{d}$ is associated to it. Finally, it is easy to verify that the map φ is linear. The uniqueness of the structure of normed vector space follows from the uniqueness of continuous extensions in Theorem A.61. □

Theorem A.72 and Theorem A.61 will be used to show that every Hermitian space can be embedded in a Hilbert space; see Theorem D.1.

The following version of Theorem A.61 for normed vector spaces will be needed in the theory of integration.

Theorem A.73. *Let E and F be two normed vector spaces, let E_0 be a dense subspace of E, and let $f_0 \colon E_0 \to F$ be a continuous function. If f_0 is uniformly continuous and if F is complete, then there is a unique uniformly continuous function $f \colon E \to F$ extending f_0. Furthermore, if f_0 is a continuous linear map, then f is also a linear continuous map, and $\|f\| = \|f_0\|$.*

Proof. We only need to prove the second statement. Given any two vectors $x, y \in E$, since E_0 is dense on E we can pick sequences (x_n) and (y_n) of vectors $x_n, y_n \in E_0$ such that $x = \lim_{n\mapsto\infty} x_n$ and $y = \lim_{n\mapsto\infty} y_n$. Since addition and scalar multiplication are continuous, we get

$$\begin{aligned} x + y &= \lim_{n\mapsto\infty} (x_n + y_n) \\ \lambda x &= \lim_{n\mapsto\infty} (\lambda x_n) \end{aligned}$$

for any $\lambda \in \mathbb{R}$ (or $\lambda \in \mathbb{C}$). Since $f(x)$ is defined by

$$f(x) = \lim_{n\mapsto\infty} f_0(x_n)$$

independently of the sequence (x_n) converging to x, and similarly for $f(y)$ and $f(x + y)$, since f_0 is linear, we have

$$\begin{aligned} f(x + y) &= \lim_{n\mapsto\infty} f_0(x_n + y_n) \\ &= \lim_{n\mapsto\infty} (f_0(x_n) + f_0(y_n)) \\ &= \lim_{n\mapsto\infty} f_0(x_n) + \lim_{n\mapsto\infty} f_0(y_n) \\ &= f(x) + f(y). \end{aligned}$$

Similarly,

$$\begin{aligned} f(\lambda x) &= \lim_{n \mapsto \infty} f_0(\lambda x_n) \\ &= \lim_{n \mapsto \infty} \lambda f_0(x_n) \\ &= \lambda \lim_{n \mapsto \infty} f_0(x_n) \\ &= \lambda f(x). \end{aligned}$$

Therefore, f is linear. Since the norm is continuous, we have

$$\|f(x)\| = \left\| \lim_{n \mapsto \infty} f_0(x_n) \right\| = \lim_{n \mapsto \infty} \|f_0(x_n)\|,$$

and since f_0 is continuous

$$\|f_0(x_n)\| \leq \|f_0\| \, \|x_n\| \quad \text{for all } n \geq 1,$$

so we get

$$\lim_{n \mapsto \infty} \|f_0(x_n)\| \leq \lim_{n \mapsto \infty} \|f_0\| \, \|x_n\| \quad \text{for all } n \geq 1,$$

that is,

$$\|f(x)\| \leq \|f_0\| \, \|x\|.$$

Since

$$\|f\| = \sup_{\|x\|=1,\, x \in E} \|f(x)\|,$$

we deduce that $\|f\| \leq \|f_0\|$. But since $E_0 \subseteq E$ and f agrees with f_0 on E_0, we also have

$$\begin{aligned} \|f_0\| = \sup_{\|x\|=1,\, x \in E_0} \|f_0(x)\| &= \sup_{\|x\|=1,\, x \in E_0} \|f(x)\| \\ \leq \sup_{\|x\|=1,\, x \in E} \|f(x)\| &= \|f\|, \end{aligned}$$

and thus $\|f\| = \|f_0\|$. □

A.14 Further Readings

A thorough treatment of general topology can be found in Munkres [Munkres (2000, 1991)], Dixmier [Dixmier (1984)], Lang [Lang (1997, 1996)], Schwartz [Schwartz (1991, 1980)], and Bredon [Bredon (1993)].

Appendix B

Vector Norms and Matrix Norms

B.1 Normed Vector Spaces

In order to define how close two vectors or two matrices are, and in order to define the convergence of sequences of vectors or matrices, we can use the notion of a norm. Recall that $\mathbb{R}_+ = \{x \in \mathbb{R} \mid x \geq 0\}$. Also recall that if $z = a + ib \in \mathbb{C}$ is a complex number, with $a, b \in \mathbb{R}$, then $\overline{z} = a - ib$ and $|z| = \sqrt{z\overline{z}} = \sqrt{a^2 + b^2}$ ($|z|$ is the *modulus* of z).

Definition B.1. Let E be a vector space over a field K, where K is either the field $\mathbb{R}$ of reals, or the field $\mathbb{C}$ of complex numbers. A *norm* on E is a function $\| \; \|\colon E \to \mathbb{R}_+$, assigning a nonnegative real number $\|u\|$ to any vector $u \in E$, and satisfying the following conditions for all $x, y \in E$:

(N1) $\|x\| \geq 0$, and $\|x\| = 0$ iff $x = 0$. (positivity)
(N2) $\|\lambda x\| = |\lambda| \, \|x\|$. (homogeneity (or scaling))
(N3) $\|x + y\| \leq \|x\| + \|y\|$. (triangle inequality)

A vector space E together with a norm $\| \; \|$ is called a *normed vector space*.

By (N2), setting $\lambda = -1$, we obtain

$$\|-x\| = \|(-1)x\| = |-1| \, \|x\| = \|x\| \, ;$$

that is, $\|-x\| = \|x\|$. From (N3), we have

$$\|x\| = \|x - y + y\| \leq \|x - y\| + \|y\| \, ,$$

which implies that

$$\|x\| - \|y\| \leq \|x - y\| \, .$$

By exchanging x and y and using the fact that by (N2),

$$\|y - x\| = \|-(x - y)\| = \|x - y\| \, ,$$

we also have

$$\|y\| - \|x\| \leq \|x - y\|.$$

Therefore,

$$|\|x\| - \|y\|| \leq \|x - y\|, \quad \text{for all } x, y \in E. \tag{$*$}$$

Observe that setting $\lambda = 0$ in (N2), we deduce that $\|0\| = 0$ without assuming (N1). Then, by setting $y = 0$ in $(*)$, we obtain

$$|\|x\|| \leq \|x\|, \quad \text{for all } x \in E.$$

Therefore, the condition $\|x\| \geq 0$ in (N1) follows from (N2) and (N3), and (N1) can be replaced by the weaker condition

(N1') For all $x \in E$, if $\|x\| = 0$ then $x = 0$,

A function $\| \ \| : E \to \mathbb{R}$ satisfying axioms (N2) and (N3) is called a *semi-norm*. From the above discussion, a semi-norm also has the properties

$\|x\| \geq 0$ for all $x \in E$, and $\|0\| = 0$.

However, there may be nonzero vectors $x \in E$ such that $\|x\| = 0$. Let us give some examples of normed vector spaces.

Example B.1.

(1) Let $E = \mathbb{R}$, and $\|x\| = |x|$, the absolute value of x.
(2) Let $E = \mathbb{C}$, and $\|z\| = |z|$, the modulus of z.
(3) Let $E = \mathbb{R}^n$ (or $E = \mathbb{C}^n$). There are three standard norms. For every $(x_1, \ldots, x_n) \in E$, we have the norm $\|x\|_1$, defined such that,

$$\|x\|_1 = |x_1| + \cdots + |x_n|,$$

we have the *Euclidean norm* $\|x\|_2$, defined such that,

$$\|x\|_2 = \left(|x_1|^2 + \cdots + |x_n|^2\right)^{\frac{1}{2}},$$

and the *sup*-norm $\|x\|_\infty$, defined such that,

$$\|x\|_\infty = \max\{|x_i| \mid 1 \leq i \leq n\}.$$

More generally, we define the ℓ^p-*norm* (for $p \geq 1$) by

$$\|x\|_p = (|x_1|^p + \cdots + |x_n|^p)^{1/p}.$$

There are other norms besides the ℓ^p-norms. Here are some examples.

(1) For $E = \mathbb{R}^2$,

$$\|(u_1, u_2)\| = |u_1| + 2|u_2|.$$

(2) For $E = \mathbb{R}^2$,

$$\|(u_1, u_2)\| = \left((u_1 + u_2)^2 + u_1^2\right)^{1/2}.$$

(3) For $E = \mathbb{C}^2$,

$$\|(u_1, u_2)\| = |u_1 + iu_2| + |u_1 - iu_2|.$$

The reader should check that they satisfy all the axioms of a norm.

Some work is required to show the triangle inequality for the ℓ^p-norm.

Proposition B.1. *If E is a finite-dimensional vector space over $\mathbb{R}$ or $\mathbb{C}$, for every real number $p \geq 1$, the ℓ^p-norm is indeed a norm.*

Proof. The cases $p = 1$ and $p = \infty$ are easy and left to the reader. If $p > 1$, then let $q > 1$ such that

$$\frac{1}{p} + \frac{1}{q} = 1.$$

We will make use of the following fact: for all $\alpha, \beta \in \mathbb{R}$, if $\alpha, \beta \geq 0$, then

$$\alpha\beta \leq \frac{\alpha^p}{p} + \frac{\beta^q}{q}. \tag{$*$}$$

To prove the above inequality, we use the fact that the exponential function $t \mapsto e^t$ satisfies the following convexity inequality:

$$e^{\theta x + (1-\theta)y} \leq \theta e^x + (1 - \theta)e^y,$$

for all $x, y \in \mathbb{R}$ and all θ with $0 \leq \theta \leq 1$.

Since the case $\alpha\beta = 0$ is trivial, let us assume that $\alpha > 0$ and $\beta > 0$. If we replace θ by $1/p$, x by $p \log \alpha$ and y by $q \log \beta$, then we get

$$e^{\frac{1}{p} p \log \alpha + \frac{1}{q} q \log \beta} \leq \frac{1}{p} e^{p \log \alpha} + \frac{1}{q} e^{q \log \beta},$$

which simplifies to

$$\alpha\beta \leq \frac{\alpha^p}{p} + \frac{\beta^q}{q},$$

as claimed.

We will now prove that for any two vectors $u, v \in E$, we have

$$\sum_{i=1}^{n} |u_i v_i| \leq \|u\|_p \|v\|_q. \tag{$**$}$$

Since the above is trivial if $u = 0$ or $v = 0$, let us assume that $u \neq 0$ and $v \neq 0$. Then, the inequality (∗) with $\alpha = |u_i| / \|u\|_p$ and $\beta = |v_i| / \|v\|_q$ yields

$$\frac{|u_i v_i|}{\|u\|_p \|v\|_q} \leq \frac{|u_i|^p}{p \|u\|_p^p} + \frac{|v_i|^q}{q \|u\|_q^q},$$

for $i = 1, \ldots, n$, and by summing up these inequalities, we get

$$\sum_{i=1}^{n} |u_i v_i| \leq \|u\|_p \|v\|_q ,$$

as claimed. To finish the proof, we simply have to prove that property (N3) holds, since (N1) and (N2) are clear. Now, for $i = 1, \ldots, n$, we can write

$$(|u_i| + |v_i|)^p = |u_i|(|u_i| + |v_i|)^{p-1} + |v_i|(|u_i| + |v_i|)^{p-1},$$

so that by summing up these equations we get

$$\sum_{i=1}^{n}(|u_i| + |v_i|)^p = \sum_{i=1}^{n} |u_i|(|u_i| + |v_i|)^{p-1} + \sum_{i=1}^{n} |v_i|(|u_i| + |v_i|)^{p-1},$$

and using the inequality (∗∗), we get

$$\sum_{i=1}^{n}(|u_i| + |v_i|)^p \leq (\|u\|_p + \|v\|_p)\left(\sum_{i=1}^{n}(|u_i| + |v_i|)^{(p-1)q}\right)^{1/q}.$$

However, $1/p + 1/q = 1$ implies $pq = p + q$, that is, $(p-1)q = p$, so we have

$$\sum_{i=1}^{n}(|u_i| + |v_i|)^p \leq (\|u\|_p + \|v\|_p)\left(\sum_{i=1}^{n}(|u_i| + |v_i|)^p\right)^{1/q},$$

which yields

$$\left(\sum_{i=1}^{n}(|u_i| + |v_i|)^p\right)^{1/p} \leq \|u\|_p + \|v\|_p .$$

Since $|u_i + v_i| \leq |u_i| + |v_i|$, the above implies the triangle inequality $\|u + v\|_p \leq \|u\|_p + \|v\|_p$, as claimed. □

For $p > 1$ and $1/p + 1/q = 1$, the inequality

$$\sum_{i=1}^{n} |u_i v_i| \leq \left(\sum_{i=1}^{n} |u_i|^p\right)^{1/p} \left(\sum_{i=1}^{n} |v_i|^q\right)^{1/q}$$

is known as *Hölder's inequality*. For $p = 2$, it is the *Cauchy–Schwarz inequality*.

Actually, if we define the *Hermitian inner product* $\langle -, - \rangle$ on $\mathbb{C}^n$ by

$$\langle u, v \rangle = \sum_{i=1}^{n} u_i \overline{v}_i,$$

where $u = (u_1, \ldots, u_n)$ and $v = (v_1, \ldots, v_n)$, then

$$|\langle u, v \rangle| \leq \sum_{i=1}^{n} |u_i \overline{v}_i| = \sum_{i=1}^{n} |u_i v_i|,$$

so Hölder's inequality implies the inequality

$$|\langle u, v \rangle| \leq \|u\|_p \|v\|_q$$

also called *Hölder's inequality*, which, for $p = 2$ is the standard Cauchy–Schwarz inequality. The triangle inequality for the ℓ^p-norm,

$$\left(\sum_{i=1}^{n} (|u_i + v_i|)^p \right)^{1/p} \leq \left(\sum_{i=1}^{n} |u_i|^p \right)^{1/p} + \left(\sum_{i=1}^{n} |v_i|^p \right)^{1/p},$$

is known as *Minkowski's inequality*.

When we restrict the Hermitian inner product to real vectors, $u, v \in \mathbb{R}^n$, we get the *Euclidean inner product*

$$\langle u, v \rangle = \sum_{i=1}^{n} u_i v_i.$$

It is very useful to observe that if we represent (as usual) $u = (u_1, \ldots, u_n)$ and $v = (v_1, \ldots, v_n)$ (in $\mathbb{R}^n$) by column vectors, then their Euclidean inner product is given by

$$\langle u, v \rangle = u^\top v = v^\top u,$$

and when $u, v \in \mathbb{C}^n$, their Hermitian inner product is given by

$$\langle u, v \rangle = v^* u = \overline{u^* v}.$$

In particular, when $u = v$, in the complex case we get

$$\|u\|_2^2 = u^* u,$$

and in the real case, this becomes

$$\|u\|_2^2 = u^\top u.$$

As convenient as these notations are, we still recommend that you do not abuse them; the notation $\langle u, v \rangle$ is more intrinsic and still "works" when our vector space is infinite dimensional.

The following proposition is easy to show.

Proposition B.2. *The following inequalities hold for all $x \in \mathbb{R}^n$ (or $x \in \mathbb{C}^n$):*

$$\begin{aligned}\|x\|_\infty &\leq \|x\|_1 \leq n\|x\|_\infty,\\ \|x\|_\infty &\leq \|x\|_2 \leq \sqrt{n}\|x\|_\infty,\\ \|x\|_2 &\leq \|x\|_1 \leq \sqrt{n}\|x\|_2.\end{aligned}$$

Proposition B.2 is a special case of a very important result: in a finite-dimensional vector space, any two norms are equivalent.

Definition B.2. Given any (real or complex) vector space E, two norms $\| \|_a$ and $\| \|_b$ are *equivalent* iff there exists some positive reals $C_1, C_2 > 0$, such that

$$\|u\|_a \leq C_1 \|u\|_b \quad \text{and} \quad \|u\|_b \leq C_2 \|u\|_a, \text{ for all } u \in E.$$

Given any norm $\| \|$ on a vector space of dimension n, for any basis $(e_1, \ldots, e_n)$ of E, observe that for any vector $x = x_1e_1 + \cdots + x_ne_n$, we have

$$\begin{aligned}\|x\| = \|x_1e_1 + \cdots + x_ne_n\| &\leq |x_1| \|e_1\| + \cdots + |x_n| \|e_n\|\\ &\leq C(|x_1| + \cdots + |x_n|) = C \|x\|_1,\end{aligned}$$

with $C = \max_{1 \leq i \leq n} \|e_i\|$ and

$$\|x\|_1 = \|x_1e_1 + \cdots + x_ne_n\| = |x_1| + \cdots + |x_n|.$$

The above implies that

$$| \|u\| - \|v\| | \leq \|u - v\| \leq C \|u - v\|_1,$$

which means that the map $u \mapsto \|u\|$ is *continuous* with respect to the norm $\| \|_1$.

Let S_1^{n-1} be the unit sphere with respect to the norm $\| \|_1$, namely

$$S_1^{n-1} = \{x \in E \mid \|x\|_1 = 1\}.$$

Now, S_1^{n-1} is a closed and bounded subset of a finite-dimensional vector space, so by Heine–Borel (or equivalently, by Bolzano–Weiertrass), S_1^{n-1} is compact. On the other hand, it is a well known result of analysis that any continuous real-valued function on a nonempty compact set has a minimum and a maximum, and that they are achieved. Using these facts, we can prove the following important theorem:

Theorem B.3. *If E is any real or complex vector space of finite dimension, then any two norms on E are equivalent.*

Proof. It is enough to prove that any norm $\| \ \|$ is equivalent to the 1-norm. We already proved that the function $x \mapsto \|x\|$ is continuous with respect to the norm $\| \ \|_1$ and we observed that the unit sphere S_1^{n-1} is compact. Now, we just recalled that because the function $f\colon x \mapsto \|x\|$ is continuous and because S_1^{n-1} is compact, the function f has a minimum m and a maximum M, and because $\|x\|$ is never zero on S_1^{n-1}, we must have $m > 0$. Consequently, we just proved that if $\|x\|_1 = 1$, then

$$0 < m \leq \|x\| \leq M,$$

so for any $x \in E$ with $x \neq 0$, we get

$$m \leq \|x/\|x\|_1\| \leq M,$$

which implies

$$m \|x\|_1 \leq \|x\| \leq M \|x\|_1 .$$

Since the above inequality holds trivially if $x = 0$, we just proved that $\| \ \|$ and $\| \ \|_1$ are equivalent, as claimed. ☐

Next, we will consider norms on matrices.

B.2 Matrix Norms

For simplicity of exposition, we will consider the vector spaces $\mathrm{M}_n(\mathbb{R})$ and $\mathrm{M}_n(\mathbb{C})$ of square $n \times n$ matrices. Most results also hold for the spaces $\mathrm{M}_{m,n}(\mathbb{R})$ and $\mathrm{M}_{m,n}(\mathbb{C})$ of rectangular $m \times n$ matrices. Since $n \times n$ matrices can be multiplied, the idea behind matrix norms is that they should behave "well" with respect to matrix multiplication.

Definition B.3. A *matrix norm* $\| \ \|$ on the space of square $n \times n$ matrices in $\mathrm{M}_n(K)$, with $K = \mathbb{R}$ or $K = \mathbb{C}$, is a norm on the vector space $\mathrm{M}_n(K)$, with the additional property called *submultiplicativity* that

$$\|AB\| \leq \|A\| \|B\| ,$$

for all $A, B \in \mathrm{M}_n(K)$. A norm on matrices satisfying the above property is often called a *submultiplicative* matrix norm.

Since $I^2 = I$, from $\|I\| = \|I^2\| \leq \|I\|^2$, we get $\|I\| \geq 1$, for every matrix norm.

Before giving examples of matrix norms, we need to review some basic definitions about matrices. Given any matrix $A = (a_{ij}) \in \mathrm{M}_{m,n}(\mathbb{C})$, the *conjugate* $\overline{A}$ of A is the matrix such that

$$\overline{A}_{ij} = \overline{a_{ij}}, \quad 1 \leq i \leq m, \ 1 \leq j \leq n.$$

The *transpose* of A is the $n \times m$ matrix $A^\top$ such that

$$A^\top_{ij} = a_{ji}, \quad 1 \leq i \leq m,\ 1 \leq j \leq n.$$

The *adjoint* of A is the $n \times m$ matrix A^* such that

$$A^* = \overline{(A^\top)} = (\overline{A})^\top.$$

When A is a real matrix, $A^* = A^\top$. A matrix $A \in \mathrm{M}_n(\mathbb{C})$ is *Hermitian* if

$$A^* = A.$$

If A is a real matrix ($A \in \mathrm{M}_n(\mathbb{R})$), we say that A is *symmetric* if

$$A^\top = A.$$

A matrix $A \in \mathrm{M}_n(\mathbb{C})$ is *normal* if

$$AA^* = A^*A,$$

and if A is a real matrix, it is *normal* if

$$AA^\top = A^\top A.$$

A matrix $U \in \mathrm{M}_n(\mathbb{C})$ is *unitary* if

$$UU^* = U^*U = I.$$

A real matrix $Q \in \mathrm{M}_n(\mathbb{R})$ is *orthogonal* if

$$QQ^\top = Q^\top Q = I.$$

Given any matrix $A = (a_{ij}) \in \mathrm{M}_n(\mathbb{C})$, the *trace* $\mathrm{tr}(A)$ of A is the sum of its diagonal elements

$$\mathrm{tr}(A) = a_{11} + \cdots + a_{nn}.$$

It is easy to show that the trace is a linear map, so that

$$\mathrm{tr}(\lambda A) = \lambda \mathrm{tr}(A)$$

and

$$\mathrm{tr}(A + B) = \mathrm{tr}(A) + \mathrm{tr}(B).$$

Moreover, if A is an $m \times n$ matrix and B is an $n \times m$ matrix, it is not hard to show that

$$\mathrm{tr}(AB) = \mathrm{tr}(BA).$$

We also review eigenvalues and eigenvectors. We content ourselves with definition involving matrices.

Definition B.4. Given any square matrix $A \in \mathrm{M}_n(\mathbb{C})$, a complex number $\lambda \in \mathbb{C}$ is an *eigenvalue* of A if there is some *nonzero* vector $u \in \mathbb{C}^n$, such that

$$Au = \lambda u.$$

If λ is an eigenvalue of A, then the *nonzero* vectors $u \in \mathbb{C}^n$ such that $Au = \lambda u$ are called *eigenvectors of A associated with λ*; together with the zero vector, these eigenvectors form a subspace of $\mathbb{C}^n$ denoted by $E_\lambda(A)$, and called the *eigenspace associated with λ*.

Remark: Note that Definition B.4 *requires an eigenvector to be nonzero.* A somewhat unfortunate consequence of this requirement is that the set of eigenvectors is *not* a subspace, since the zero vector is missing! On the positive side, whenever eigenvectors are involved, there is no need to say that they are nonzero. In contrast, even if we allow 0 to be an eigenvector, in order for a scalar λ to be an eigenvalue, there must be a *nonzero vector* u such that $Au = \lambda u$. Without this restriction, since $A0 = \lambda 0 = 0$ for all λ, every scalar would be an eigenvector, which would make the definition of an eigenvalue trivial and useless. The fact that eigenvectors are nonzero is implicitly used in all the arguments involving them, so it seems preferable (but perhaps not as elegant) to stipulate that eigenvectors should be nonzero.

If A is a square real matrix $A \in \mathrm{M}_n(\mathbb{R})$, then we restrict Definition B.4 to real eigenvalues $\lambda \in \mathbb{R}$ and real eigenvectors. However, it should be noted that although every complex matrix always has at least some complex eigenvalue, a real matrix may not have any real eigenvalues. For example, the matrix

$$A = \begin{pmatrix} 0 & 1 \\ 1 & 0 \end{pmatrix}$$

has the complex eigenvalues i and $-i$, but no real eigenvalues. Thus, typically even for real matrices, we consider complex eigenvalues.

Observe that $\lambda \in \mathbb{C}$ is an eigenvalue of A

- iff $Au = \lambda u$ for some nonzero vector $u \in \mathbb{C}^n$
- iff $(\lambda I - A)u = 0$
- iff the matrix $\lambda I - A$ defines a linear map which has a nonzero kernel, that is,
- iff $\lambda I - A$ not invertible.

However, it is a standard fact of linear algebra that $\lambda I - A$ is not invertible iff

$$\det(\lambda I - A) = 0.$$

Now $\det(\lambda I - A)$ is a polynomial of degree n in the indeterminate λ, in fact, of the form

$$\lambda^n - \operatorname{tr}(A)\lambda^{n-1} + \cdots + (-1)^n \det(A).$$

Thus we see that the eigenvalues of A are the zeros (also called *roots*) of the above polynomial. Since every complex polynomial of degree n has exactly n roots, counted with their multiplicity, we have the following definition:

Definition B.5. Given any square $n \times n$ matrix $A \in \mathrm{M}_n(\mathbb{C})$, the polynomial

$$\det(\lambda I - A) = \lambda^n - \operatorname{tr}(A)\lambda^{n-1} + \cdots + (-1)^n \det(A)$$

is called the *characteristic polynomial* of A. The n (not necessarily distinct) roots $\lambda_1, \ldots, \lambda_n$ of the characteristic polynomial are all the *eigenvalues* of A and constitute the *spectrum* of A. We let

$$\rho(A) = \max_{1 \leq i \leq n} |\lambda_i|$$

be the largest modulus of the eigenvalues of A, called the *spectral radius* of A.

Since the eigenvalues $\lambda_1, \ldots, \lambda_n$ of A are the zeros of the polynomial

$$\det(\lambda I - A) = \lambda^n - \operatorname{tr}(A)\lambda^{n-1} + \cdots + (-1)^n \det(A),$$

we deduce that

$$\begin{aligned} \operatorname{tr}(A) &= \lambda_1 + \cdots + \lambda_n \\ \det(A) &= \lambda_1 \cdots \lambda_n. \end{aligned}$$

Proposition B.4. *For any matrix norm $\| \ \|$ on $\mathrm{M}_n(\mathbb{C})$ and for any square $n \times n$ matrix $A \in \mathrm{M}_n(\mathbb{C})$, we have*

$$\rho(A) \leq \|A\|.$$

Proof. Let λ be some eigenvalue of A for which $|\lambda|$ is maximum, that is, such that $|\lambda| = \rho(A)$. If u ($\neq 0$) is any eigenvector associated with λ and if U is the $n \times n$ matrix whose columns are all u, then $Au = \lambda u$ implies

$$AU = \lambda U,$$

and since

$$|\lambda| \, \|U\| = \|\lambda U\| = \|AU\| \leq \|A\| \, \|U\|$$

and $U \neq 0$, we have $\|U\| \neq 0$, and get

$$\rho(A) = |\lambda| \leq \|A\|,$$

as claimed. $\square$

Proposition B.4 also holds for any real matrix norm $\| \, \|$ on $\mathrm{M}_n(\mathbb{R})$ but the proof is more subtle and requires the notion of induced norm. We prove it after giving Definition B.7.

It turns out that if A is a real $n \times n$ symmetric matrix, then the eigenvalues of A are all real and there is some orthogonal matrix Q such that

$$A = Q\mathrm{diag}(\lambda_1, \ldots, \lambda_n)Q^\top,$$

where $\mathrm{diag}(\lambda_1, \ldots, \lambda_n)$ denotes the matrix whose only nonzero entries (if any) are its diagonal entries, which are the (real) eigenvalues of A. Similarly, if A is a complex $n \times n$ Hermitian matrix, then the eigenvalues of A are all real and there is some unitary matrix U such that

$$A = U\mathrm{diag}(\lambda_1, \ldots, \lambda_n)U^*,$$

where $\mathrm{diag}(\lambda_1, \ldots, \lambda_n)$ denotes the matrix whose only nonzero entries (if any) are its diagonal entries, which are the (real) eigenvalues of A.

We now return to matrix norms. We begin with the so-called *Frobenius norm*, which is just the norm $\| \, \|_2$ on $\mathbb{C}^{n^2}$, where the $n \times n$ matrix A is viewed as the vector obtained by concatenating together the rows (or the columns) of A. The reader should check that for any $n \times n$ complex matrix $A = (a_{ij})$,

$$\left(\sum_{i,j=1}^{n} |a_{ij}|^2 \right)^{1/2} = \sqrt{\mathrm{tr}(A^*A)} = \sqrt{\mathrm{tr}(AA^*)}.$$

Definition B.6. The *Frobenius norm* $\| \, \|_F$ is defined so that for every square $n \times n$ matrix $A \in \mathrm{M}_n(\mathbb{C})$,

$$\|A\|_F = \left(\sum_{i,j-1}^{n} |a_{ij}|^2 \right)^{1/2} = \sqrt{\mathrm{tr}(AA^*)} = \sqrt{\mathrm{tr}(A^*A)}.$$

The following proposition show that the Frobenius norm is a matrix norm satisfying other nice properties.

Proposition B.5. *The Frobenius norm* $\| \, \|_F$ *on* $\mathrm{M}_n(\mathbb{C})$ *satisfies the following properties:*

(1) *It is a matrix norm; that is,* $\|AB\|_F \leq \|A\|_F \|B\|_F$, *for all* $A, B \in \mathrm{M}_n(\mathbb{C})$.
(2) *It is unitarily invariant, which means that for all unitary matrices* U, V, *we have*

$$\|A\|_F = \|UA\|_F = \|AV\|_F = \|UAV\|_F.$$

(3) $\sqrt{\rho(A^*A)} \leq \|A\|_F \leq \sqrt{n}\sqrt{\rho(A^*A)}$, *for all* $A \in \mathrm{M}_n(\mathbb{C})$.

Proof. (1) The only property that requires a proof is the fact $\|AB\|_F \leq \|A\|_F \|B\|_F$. This follows from the Cauchy–Schwarz inequality:

$$\begin{aligned} \|AB\|_F^2 &= \sum_{i,j=1}^n \left| \sum_{k=1}^n a_{ik} b_{kj} \right|^2 \\ &\leq \sum_{i,j=1}^n \left(\sum_{h=1}^n |a_{ih}|^2 \right) \left(\sum_{k=1}^n |b_{kj}|^2 \right) \\ &= \left(\sum_{i,h=1}^n |a_{ih}|^2 \right) \left(\sum_{k,j=1}^n |b_{kj}|^2 \right) = \|A\|_F^2 \|B\|_F^2 . \end{aligned}$$

(2) We have

$$\|A\|_F^2 = \mathrm{tr}(AA^*) = \mathrm{tr}(AVV^*A^*) = \mathrm{tr}(AV(AV)^*) = \|AV\|_F^2 ,$$

and

$$\|A\|_F^2 = \mathrm{tr}(A^*A) = \mathrm{tr}(A^*U^*UA) = \|UA\|_F^2 .$$

The identity

$$\|A\|_F = \|UAV\|_F$$

follows from the previous two.

(3) It is known by linear algebra that the trace of a matrix is equal to the sum of its eigenvalues. Furthermore, A^*A is symmetric positive semidefinite (which means that its eigenvalues are nonnegative), so $\rho(A^*A)$ is the largest eigenvalue of A^*A and

$$\rho(A^*A) \leq \mathrm{tr}(A^*A) \leq n\rho(A^*A),$$

which yields (3) by taking square roots. □

Remark: The Frobenius norm is also known as the *Hilbert-Schmidt norm* or the *Schur norm*. So many famous names associated with such a simple thing!

B.3 Subordinate Norms

We now give another method for obtaining matrix norms using subordinate norms. First we need a proposition that shows that in a finite-dimensional space, the linear map induced by a matrix is bounded, and thus continuous.

Proposition B.6. *For every norm $\| \ \|$ on $\mathbb{C}^n$ (or $\mathbb{R}^n$), for every matrix $A \in \mathrm{M}_n(\mathbb{C})$ (or $A \in \mathrm{M}_n(\mathbb{R})$), there is a real constant $C_A \geq 0$, such that*

$$\|Au\| \leq C_A \|u\|,$$

for every vector $u \in \mathbb{C}^n$ (or $u \in \mathbb{R}^n$ if A is real).

Proof. For every basis $(e_1, \ldots, e_n)$ of $\mathbb{C}^n$ (or $\mathbb{R}^n$), for every vector $u = u_1e_1 + \cdots + u_ne_n$, we have

$$\begin{aligned}\|Au\| &= \|u_1A(e_1) + \cdots + u_nA(e_n)\| \\ &\leq |u_1| \|A(e_1)\| + \cdots + |u_n| \|A(e_n)\| \\ &\leq C_1(|u_1| + \cdots + |u_n|) = C_1 \|u\|_1,\end{aligned}$$

where $C_1 = \max_{1 \leq i \leq n} \|A(e_i)\|$. By Theorem B.3, the norms $\| \ \|$ and $\| \ \|_1$ are equivalent, so there is some constant $C_2 > 0$ so that $\|u\|_1 \leq C_2 \|u\|$ for all u, which implies that

$$\|Au\| \leq C_A \|u\|,$$

where $C_A = C_1C_2$. ∎

Proposition B.6 says that every linear map on a finite-dimensional space is *bounded*. This implies that every linear map on a finite-dimensional space is continuous. Actually, it is not hard to show that a linear map on a normed vector space E is bounded iff it is continuous, regardless of the dimension of E.

Proposition B.6 implies that for every matrix $A \in \mathrm{M}_n(\mathbb{C})$ (or $A \in \mathrm{M}_n(\mathbb{R})$),

$$\sup_{\substack{x \in \mathbb{C}^n \\ x \neq 0}} \frac{\|Ax\|}{\|x\|} \leq C_A.$$

Since $\|\lambda u\| = |\lambda| \|u\|$, for every nonzero vector x, we have

$$\frac{\|Ax\|}{\|x\|} = \frac{\|x\| \|A(x/\|x\|)\|}{\|x\|} = \|A(x/\|x\|)\|,$$

which implies that

$$\sup_{\substack{x \in \mathbb{C}^n \\ x \neq 0}} \frac{\|Ax\|}{\|x\|} = \sup_{\substack{x \in \mathbb{C}^n \\ \|x\|=1}} \|Ax\|.$$

Similarly

$$\sup_{\substack{x \in \mathbb{R}^n \\ x \neq 0}} \frac{\|Ax\|}{\|x\|} = \sup_{\substack{x \in \mathbb{R}^n \\ \|x\|=1}} \|Ax\|.$$

The above considerations justify the following definition.

Definition B.7. If $\| \; \|$ is any norm on $\mathbb{C}^n$, we define the function $\| \; \|_{\text{op}}$ on $\mathrm{M}_n(\mathbb{C})$ by

$$\|A\|_{\text{op}} = \sup_{\substack{x \in \mathbb{C}^n \\ x \neq 0}} \frac{\|Ax\|}{\|x\|} = \sup_{\substack{x \in \mathbb{C}^n \\ \|x\|=1}} \|Ax\| .$$

The function $A \mapsto \|A\|_{\text{op}}$ is called the *subordinate matrix norm* or *operator norm* induced by the norm $\| \; \|$.

Another notation for the operator norm of a matrix A (in particular, used by Horn and Johnson [Horn and Johnson (1990)]) is $|\!|\!| A |\!|\!|$.

It is easy to check that the function $A \mapsto \|A\|_{\text{op}}$ is indeed a norm, and by definition, it satisfies the property

$$\|Ax\| \leq \|A\|_{\text{op}} \|x\| , \quad \text{for all } x \in \mathbb{C}^n.$$

A norm $\| \; \|_{\text{op}}$ on $\mathrm{M}_n(\mathbb{C})$ satisfying the above property is said to be *subordinate* to the vector norm $\| \; \|$ on $\mathbb{C}^n$. As a consequence of the above inequality, we have

$$\|ABx\| \leq \|A\|_{\text{op}} \|Bx\| \leq \|A\|_{\text{op}} \|B\|_{\text{op}} \|x\| ,$$

for all $x \in \mathbb{C}^n$, which implies that

$$\|AB\|_{\text{op}} \leq \|A\|_{\text{op}} \|B\|_{\text{op}} \quad \text{for all } A, B \in \mathrm{M}_n(\mathbb{C}),$$

showing that $A \mapsto \|A\|_{\text{op}}$ is a matrix norm (it is submultiplicative).

Observe that the operator norm is also defined by

$$\|A\|_{\text{op}} = \inf\{\lambda \in \mathbb{R} \mid \|Ax\| \leq \lambda \|x\| , \text{ for all } x \in \mathbb{C}^n\}.$$

Since the function $x \mapsto \|Ax\|$ is continuous (because $| \, \|Ay\| - \|Ax\| \, | \leq \|Ay - Ax\| \leq C_A \|x - y\|$) and the unit sphere $S^{n-1} = \{x \in \mathbb{C}^n \mid \|x\| = 1\}$ is compact, there is some $x \in \mathbb{C}^n$ such that $\|x\| = 1$ and

$$\|Ax\| = \|A\|_{\text{op}} .$$

Equivalently, there is some $x \in \mathbb{C}^n$ such that $x \neq 0$ and

$$\|Ax\| = \|A\|_{\text{op}} \|x\| .$$

Consequently we can replace sup by max in the definition of $\|A\|_{\text{op}}$ (and inf by min), namely

$$\|A\|_{\text{op}} = \max_{\substack{x \in \mathbb{C}^n \\ \|x\|=1}} \|Ax\| .$$

The definition of an operator norm also implies that

$$\|I\|_{\mathrm{op}} = 1.$$

The above shows that the Frobenius norm is not a subordinate matrix norm for $n \geq 2$ (why?).

If $\| \; \|$ is a vector norm on $\mathbb{C}^n$, the operator norm $\| \; \|_{\mathrm{op}}$ that it induces applies to matrices in $\mathrm{M}_n(\mathbb{C})$. If we are careful to denote vectors and matrices so that no confusion arises, for example, by using lower case letters for vectors and upper case letters for matrices, it should be clear that $\|A\|_{\mathrm{op}}$ is the operator norm of the matrix A and that $\|x\|$ is the vector norm of x. Consequently, following common practice to alleviate notation, we will drop the subscript "op" and simply write $\|A\|$ instead of $\|A\|_{\mathrm{op}}$.

The notion of subordinate norm can be slightly generalized.

Definition B.8. If $K = \mathbb{R}$ or $K = \mathbb{C}$, for any norm $\| \; \|$ on $\mathrm{M}_{m,n}(K)$, and for any two norms $\| \; \|_a$ on K^n and $\| \; \|_b$ on K^m, we say that the norm $\| \; \|$ is *subordinate* to the norms $\| \; \|_a$ and $\| \; \|_b$ if

$$\|Ax\|_b \leq \|A\| \, \|x\|_a \quad \text{for all } A \in \mathrm{M}_{m,n}(K) \text{ and all } x \in K^n.$$

Remark: For any norm $\| \; \|$ on $\mathbb{C}^n$, we can define the function $\| \; \|_{\mathbb{R}}$ on $\mathrm{M}_n(\mathbb{R})$ by

$$\|A\|_{\mathbb{R}} = \sup_{\substack{x \in \mathbb{R}^n \\ x \neq 0}} \frac{\|Ax\|}{\|x\|} = \sup_{\substack{x \in \mathbb{R}^n \\ \|x\| = 1}} \|Ax\|.$$

The function $A \mapsto \|A\|_{\mathbb{R}}$ is a matrix norm on $\mathrm{M}_n(\mathbb{R})$, and

$$\|A\|_{\mathbb{R}} \leq \|A\|,$$

for all real matrices $A \in \mathrm{M}_n(\mathbb{R})$. However, it is possible to construct vector norms $\| \; \|$ on $\mathbb{C}^n$ and *real* matrices A such that

$$\|A\|_{\mathbb{R}} < \|A\|.$$

In order to avoid this kind of difficulties, we define subordinate matrix norms over $\mathrm{M}_n(\mathbb{C})$. Luckily, it turns out that $\|A\|_{\mathbb{R}} = \|A\|$ for the vector norms, $\| \; \|_1$, $\| \; \|_2$, and $\| \; \|_\infty$.

We now prove Proposition B.4 for real matrix norms.

Proposition B.7. *For any matrix norm $\| \; \|$ on $\mathrm{M}_n(\mathbb{R})$ and for any square $n \times n$ matrix $A \in \mathrm{M}_n(\mathbb{R})$, we have*

$$\rho(A) \leq \|A\|.$$

Proof. We follow the proof in Denis Serre's book [Serre (2010)]. If A is a real matrix, the problem is that the eigenvectors associated with the eigenvalue of maximum modulus may be complex. We use a trick based on the fact that for every matrix A (real or complex),

$$\rho(A^k) = (\rho(A))^k,$$

which is left as an exercise.

Pick any complex matrix norm $\|\ \|_c$ on $\mathbb{C}^n$ (for example, the Frobenius norm, or any subordinate matrix norm induced by a norm on $\mathbb{C}^n$). The restriction of $\|\ \|_c$ to real matrices is a real norm that we also denote by $\|\ \|_c$. Now by Theorem B.3, since $\mathrm{M}_n(\mathbb{R})$ has finite dimension n^2, there is some constant $C > 0$ so that

$$\|B\|_c \leq C\,\|B\|, \quad \text{for all} \quad B \in \mathrm{M}_n(\mathbb{R}).$$

Furthermore, for every $k \geq 1$ and for every real $n \times n$ matrix A, by Proposition B.4, $\rho(A^k) \leq \|A^k\|_c$, and because $\|\ \|$ is a matrix norm, $\|A^k\| \leq \|A\|^k$, so we have

$$(\rho(A))^k = \rho(A^k) \leq \|A^k\|_c \leq C\,\|A^k\| \leq C\,\|A\|^k,$$

for all $k \geq 1$. It follows that

$$\rho(A) \leq C^{1/k}\,\|A\|, \quad \text{for all} \quad k \geq 1.$$

However because $C > 0$, we have $\lim_{k\mapsto\infty} C^{1/k} = 1$ (we have $\lim_{k\mapsto\infty} \frac{1}{k}\log(C) = 0$). Therefore, we conclude that

$$\rho(A) \leq \|A\|,$$

as desired. ☐

We now determine explicitly what are the subordinate matrix norms associated with the vector norms $\|\ \|_1$, $\|\ \|_2$, and $\|\ \|_\infty$.

Proposition B.8. *For every square matrix $A = (a_{ij}) \in \mathrm{M}_n(\mathbb{C})$, we have*

$$\|A\|_1 = \sup_{\substack{x\in\mathbb{C}^n \\ \|x\|_1=1}} \|Ax\|_1 = \max_j \sum_{i=1}^n |a_{ij}|$$

$$\|A\|_\infty = \sup_{\substack{x\in\mathbb{C}^n \\ \|x\|_\infty=1}} \|Ax\|_\infty = \max_i \sum_{j=1}^n |a_{ij}|$$

$$\|A\|_2 = \sup_{\substack{x\in\mathbb{C}^n \\ \|x\|_2=1}} \|Ax\|_2 = \sqrt{\rho(A^*A)} = \sqrt{\rho(AA^*)}.$$

Note that $\|A\|_1$ *is the maximum of the* ℓ^1*-norms of the columns of* A *and* $\|A\|_\infty$ *is the maximum of the* ℓ^1*-norms of the rows of* A*. Furthermore,* $\|A^*\|_2 = \|A\|_2$*, the norm* $\| \ \|_2$ *is unitarily invariant, which means that*

$$\|A\|_2 = \|UAV\|_2$$

for all unitary matrices U, V*, and if* A *is a normal matrix, then* $\|A\|_2 = \rho(A)$.

Proof. For every vector u, we have

$$\|Au\|_1 = \sum_i \left| \sum_j a_{ij} u_j \right| \leq \sum_j |u_j| \sum_i |a_{ij}| \leq \left(\max_j \sum_i |a_{ij}| \right) \|u\|_1 ,$$

which implies that

$$\|A\|_1 \leq \max_j \sum_{i=1}^n |a_{ij}|.$$

It remains to show that equality can be achieved. For this let j_0 be some index such that

$$\max_j \sum_i |a_{ij}| = \sum_i |a_{ij_0}|,$$

and let $u_i = 0$ for all $i \neq j_0$ and $u_{j_0} = 1$.

In a similar way, we have

$$\|Au\|_\infty = \max_i \left| \sum_j a_{ij} u_j \right| \leq \left(\max_i \sum_j |a_{ij}| \right) \|u\|_\infty ,$$

which implies that

$$\|A\|_\infty \leq \max_i \sum_{j=1}^n |a_{ij}|.$$

To achieve equality, let i_0 be some index such that

$$\max_i \sum_j |a_{ij}| = \sum_j |a_{i_0 j}|.$$

The reader should check that the vector given by

$$u_j = \begin{cases} \frac{\overline{a}_{i_0 j}}{|a_{i_0 j}|} & \text{if } a_{i_0 j} \neq 0 \\ 1 & \text{if } a_{i_0 j} = 0 \end{cases}$$

works.

We have

$$\|A\|_2^2 = \sup_{\substack{x\in\mathbb{C}^n \\ x^*x=1}} \|Ax\|_2^2 = \sup_{\substack{x\in\mathbb{C}^n \\ x^*x=1}} x^*A^*Ax.$$

Since the matrix A^*A is symmetric, it has real eigenvalues and it can be diagonalized with respect to a unitary matrix. These facts can be used to prove that the function $x \mapsto x^*A^*Ax$ has a maximum on the sphere $x^*x = 1$ equal to the largest eigenvalue of A^*A, namely, $\rho(A^*A)$. We postpone the proof until we discuss optimizing quadratic functions. Therefore,

$$\|A\|_2 = \sqrt{\rho(A^*A)}.$$

Let use now prove that $\rho(A^*A) = \rho(AA^*)$. First assume that $\rho(A^*A) > 0$. In this case, there is some eigenvector $u\,(\neq 0)$ such that

$$A^*Au = \rho(A^*A)u,$$

and since $\rho(A^*A) > 0$, we must have $Au \neq 0$. Since $Au \neq 0$,

$$AA^*(Au) = A(A^*Au) = \rho(A^*A)Au$$

which means that $\rho(A^*A)$ is an eigenvalue of AA^*, and thus

$$\rho(A^*A) \leq \rho(AA^*).$$

Because $(A^*)^* = A$, by replacing A by A^*, we get

$$\rho(AA^*) \leq \rho(A^*A),$$

and so $\rho(A^*A) = \rho(AA^*)$.

If $\rho(A^*A) = 0$, then we must have $\rho(AA^*) = 0$, since otherwise by the previous reasoning we would have $\rho(A^*A) = \rho(AA^*) > 0$. Hence, in all case

$$\|A\|_2^2 = \rho(A^*A) = \rho(AA^*) = \|A^*\|_2^2.$$

For any unitary matrices U and V, it is an easy exercise to prove that V^*A^*AV and A^*A have the same eigenvalues, so

$$\|A\|_2^2 = \rho(A^*A) = \rho(V^*A^*AV) = \|AV\|_2^2,$$

and also

$$\|A\|_2^2 = \rho(A^*A) = \rho(A^*U^*UA) = \|UA\|_2^2.$$

Finally, if A is a normal matrix ($AA^* = A^*A$), it can be shown that there is some unitary matrix U so that

$$A = UDU^*,$$

where $D = \mathrm{diag}(\lambda_1, \ldots, \lambda_n)$ is a diagonal matrix consisting of the eigenvalues of A, and thus

$$A^*A = (UDU^*)^*UDU^* = UD^*U^*UDU^* = UD^*DU^*.$$

However, $D^*D = \mathrm{diag}(|\lambda_1|^2, \ldots, |\lambda_n|^2)$, which proves that

$$\rho(A^*A) = \rho(D^*D) = \max_i |\lambda_i|^2 = (\rho(A))^2,$$

so that $\|A\|_2 = \rho(A)$. $\square$

Definition B.9. For $A = (a_{ij}) \in \mathrm{M}_n(\mathbb{C})$, the norm $\|A\|_2$ is often called the *spectral norm*.

Observe that Property (3) of Proposition B.5 says that

$$\|A\|_2 \leq \|A\|_F \leq \sqrt{n}\, \|A\|_2\,,$$

which shows that the Frobenius norm is an upper bound on the spectral norm. The Frobenius norm is much easier to compute than the spectral norm.

The reader will check that the above proof still holds if the matrix A is real (change unitary to orthogonal), confirming the fact that $\|A\|_{\mathbb{R}} = \|A\|$ for the vector norms $\| \|_1$, $\| \|_2$, and $\| \|_\infty$. It is also easy to verify that the proof goes through for *rectangular* $m \times n$ matrices, with the same formulae. Similarly, the Frobenius norm given by

$$\|A\|_F = \left(\sum_{i=1}^m \sum_{j=1}^n |a_{ij}|^2 \right)^{1/2} = \sqrt{\mathrm{tr}(A^*A)} = \sqrt{\mathrm{tr}(AA^*)}$$

is also a norm on rectangular matrices. For these norms, whenever AB makes sense, we have

$$\|AB\| \leq \|A\|\, \|B\|\,.$$

Remark: It can be shown that for any two real numbers $p, q \geq 1$ such that $\dfrac{1}{p} + \dfrac{1}{q} = 1$, we have

$$\begin{aligned} \|A^*\|_q = \|A\|_p &= \sup\{\Re(y^*Ax) \mid \|x\|_p = 1, \|y\|_q = 1\} \\ &= \sup\{|\langle Ax, y\rangle| \mid \|x\|_p = 1, \|y\|_q = 1\}, \end{aligned}$$

where $\|A^*\|_q$ and $\|A\|_p$ are the operator norms.

Remark: Let $(E, \| \|)$ and $(F, \| \|)$ be two normed vector spaces (for simplicity of notation, we use the same symbol $\| \|$ for the norms on E and F;

this should not cause any confusion). Recall that a function $f\colon E \to F$ is *continuous* if for every $a \in E$, for every $\epsilon > 0$, there is some $\eta > 0$ such that for all $x \in E$,

$$\text{if} \quad \|x - a\| \leq \eta \quad \text{then} \quad \|f(x) - f(a)\| \leq \epsilon.$$

It is not hard to show that a *linear map* $f\colon E \to F$ is continuous iff there is some constant $C \geq 0$ such that

$$\|f(x)\| \leq C \|x\| \text{ for all } x \in E.$$

If so, we say that f is *bounded* (or a *linear bounded operator*). We let $\mathcal{L}(E; F)$ denote the set of all continuous (equivalently, bounded) linear maps from E to F. Then we can define the *operator norm* (or *subordinate norm*) $\| \ \|$ on $\mathcal{L}(E; F)$ as follows: for every $f \in \mathcal{L}(E; F)$,

$$\|f\| = \sup_{\substack{x \in E \\ x \neq 0}} \frac{\|f(x)\|}{\|x\|} = \sup_{\substack{x \in E \\ \|x\|=1}} \|f(x)\|,$$

or equivalently by

$$\|f\| = \inf\{\lambda \in \mathbb{R} \mid \|f(x)\| \leq \lambda \|x\|, \text{ for all } x \in E\}.$$

Here because E may be infinite-dimensional, sup can't be replaced by max and inf can't be replaced by min. It is not hard to show that the map $f \mapsto \|f\|$ is a norm on $\mathcal{L}(E; F)$ satisfying the property

$$\|f(x)\| \leq \|f\| \|x\|$$

for all $x \in E$, and that if $f \in \mathcal{L}(E; F)$ and $g \in \mathcal{L}(F; G)$, then

$$\|g \circ f\| \leq \|g\| \|f\|.$$

Operator norms play an important role in functional analysis, especially when the spaces E and F are *complete*.

Appendix C

Basics of Groups and Group Actions

This chapter gathers basics of the theory of groups and group actions.

C.1 Groups, Subgroups, Cosets

Definition C.1. A *group* is a set G equipped with a binary operation $\cdot\colon G \times G \to G$ that associates an element $a \cdot b \in G$ to every pair of elements $a, b \in G$, and having the following properties: $\cdot$ is associative, has an identity element $e \in G$, and every element in G is invertible (w.r.t. the group operation $\cdot$). More explicitly, this means that the following equations hold for all $a, b, c \in G$:

(G1) $a \cdot (b \cdot c) = (a \cdot b) \cdot c.$ (associativity);
(G2) $a \cdot e = e \cdot a = a.$ (identity);
(G3) For every $a \in G$, there is some $a^{-1} \in G$ such that $a \cdot a^{-1} = a^{-1} \cdot a = e.$ (inverse).

A group G is *abelian* (or *commutative*) if

$$a \cdot b = b \cdot a \quad \text{for all } a, b \in G.$$

A set M together with an operation $\cdot\colon M \times M \to M$ and an element e satisfying only Conditions (G1) and (G2) is called a *monoid*. For example, the set $\mathbb{N} = \{0, 1, \ldots, n, \ldots\}$ of natural numbers is a (commutative) monoid under addition. However, it is not a group.

Some examples of groups are given below.

Example C.1.

(1) The set $\mathbb{Z} = \{\ldots, -n, \ldots, -1, 0, 1, \ldots, n, \ldots\}$ of integers is an abelian group under addition, with identity element 0. However, $\mathbb{Z}^* = \mathbb{Z} - \{0\}$ is not a group under multiplication.

(2) The set $\mathbb{Q}$ of rational numbers (fractions p/q with $p, q \in \mathbb{Z}$ and $q \neq 0$) is an abelian group under addition, with identity element 0. The set $\mathbb{Q}^* = \mathbb{Q} - \{0\}$ is also an abelian group under multiplication, with identity element 1.

(3) Given any nonempty set S, the set of bijections $f\colon S \to S$, also called *permutations of* S, is a group under function composition (i.e., the multiplication of f and g is the composition $g \circ f$), with identity element the identity function id_S. This group is not abelian as soon as S has more than two elements. The permutation group of the set $S = \{1, \ldots, n\}$ is often denoted $\mathfrak{S}_n$ and called the *symmetric group* on n elements.

(4) For any positive integer $p \in \mathbb{N}$, define a relation on $\mathbb{Z}$, denoted $m \equiv n \pmod{p}$, as follows:

$$m \equiv n \pmod{p} \quad \text{iff} \quad m - n = kp \quad \text{for some } k \in \mathbb{Z}.$$

The reader will easily check that this is an equivalence relation, and, moreover, it is compatible with respect to addition and multiplication, which means that if $m_1 \equiv n_1 \pmod{p}$ and $m_2 \equiv n_2 \pmod{p}$, then $m_1 + m_2 \equiv n_1 + n_2 \pmod{p}$ and $m_1 m_2 \equiv n_1 n_2 \pmod{p}$. Consequently, we can define an addition operation and a multiplication operation of the set of equivalence classes $\pmod{p}$:

$$[m] + [n] = [m + n]$$

and

$$[m] \cdot [n] = [mn].$$

The reader will easily check that addition of residue classes $\pmod{p}$ induces an abelian group structure with $[0]$ as zero. This group is denoted $\mathbb{Z}/p\mathbb{Z}$.

(5) The set of $n \times n$ invertible matrices with real (or complex) coefficients is a group under matrix multiplication, with identity element the identity matrix I_n. This group is called the *general linear group* and is usually denoted by $\mathbf{GL}(n, \mathbb{R})$ (or $\mathbf{GL}(n, \mathbb{C})$).

(6) The set of $n \times n$ invertible matrices A with real (or complex) coefficients such that $\det(A) = 1$ is a group under matrix multiplication, with identity element the identity matrix I_n. This group is called the *special linear group* and is usually denoted by $\mathbf{SL}(n, \mathbb{R})$ (or $\mathbf{SL}(n, \mathbb{C})$).

(7) The set of $n \times n$ matrices Q with real coefficients such that

$$QQ^\top = Q^\top Q = I_n$$

is a group under matrix multiplication, with identity element the identity matrix I_n; we have $Q^{-1} = Q^\top$. This group is called the *orthogonal group* and is usually denoted by $\mathbf{O}(n)$.

(8) The set of $n \times n$ invertible matrices Q with real coefficients such that

$$QQ^\top = Q^\top Q = I_n \quad \text{and} \quad \det(Q) = 1$$

is a group under matrix multiplication, with identity element the identity matrix I_n; as in (6), we have $Q^{-1} = Q^\top$. This group is called the *special orthogonal group* or *rotation group* and is usually denoted by $\mathbf{SO}(n)$.

The groups in (5)–(8) are nonabelian for $n \geq 2$, except for $\mathbf{SO}(2)$ which is abelian (but $\mathbf{O}(2)$ is not abelian).

It is customary to denote the operation of an abelian group G by $+$, in which case the inverse a^{-1} of an element $a \in G$ is denoted by $-a$.

The identity element of a group is *unique*. In fact, we can prove a more general fact:

Proposition C.1. *If a binary operation* $\cdot\colon M \times M \to M$ *is associative and if* $e' \in M$ *is a left identity and* $e'' \in M$ *is a right identity, which means that*

$$e' \cdot a = a \quad \text{for all} \quad a \in M \tag{G2l}$$

and

$$a \cdot e'' = a \quad \text{for all} \quad a \in M, \tag{G2r}$$

then $e' = e''$.

Proof. If we let $a = e''$ in equation (G2l), we get

$$e' \cdot e'' = e'',$$

and if we let $a = e'$ in equation (G2r), we get

$$e' \cdot e'' = e',$$

and thus

$$e' = e' \cdot e'' = e'',$$

as claimed. ☐

Proposition C.1 implies that the identity element of a monoid is unique, and since every group is a monoid, the identity element of a group is unique. Furthermore, every element in a group has a *unique inverse*. This is a consequence of a slightly more general fact:

Proposition C.2. *In a monoid M with identity element e, if some element $a \in M$ has some left inverse $a' \in M$ and some right inverse $a'' \in M$, which means that*

$$a' \cdot a = e \tag{G3l}$$

and

$$a \cdot a'' = e, \tag{G3r}$$

then $a' = a''$.

Proof. Using (G3l) and the fact that e is an identity element, we have

$$(a' \cdot a) \cdot a'' = e \cdot a'' = a''.$$

Similarly, Using (G3r) and the fact that e is an identity element, we have

$$a' \cdot (a \cdot a'') = a' \cdot e = a'.$$

However, since M is monoid, the operation $\cdot$ is associative, so

$$a' = a' \cdot (a \cdot a'') = (a' \cdot a) \cdot a'' = a'',$$

as claimed. ☐

Remark: Axioms (G2) and (G3) can be weakened a bit by requiring only (G2r) (the existence of a right identity) and (G3r) (the existence of a right inverse for every element) (or (G2l) and (G3l)). It is a good exercise to prove that the group axioms (G2) and (G3) follow from (G2r) and (G3r).

Definition C.2. If a group G has a finite number n of elements, we say that G is a group of *order* n. If G is infinite, we say that G has *infinite order*. The order of a group is usually denoted by $|G|$ (if G is finite).

Given a group G, for any two subsets $R, S \subseteq G$, we let

$$RS = \{r \cdot s \mid r \in R,\ s \in S\}.$$

In particular, for any $g \in G$, if $R = \{g\}$, we write

$$gS = \{g \cdot s \mid s \in S\},$$

and similarly, if $S = \{g\}$, we write

$$Rg = \{r \cdot g \mid r \in R\}.$$

From now on, we will drop the multiplication sign and write $g_1 g_2$ for $g_1 \cdot g_2$.

Definition C.3. Let G be a group. For any $g \in G$, define L_g, the *left translation by* g, by $L_g(a) = ga$, for all $a \in G$, and R_g, the *right translation by* g, by $R_g(a) = ag$, for all $a \in G$.

The following simple fact is often used.

Proposition C.3. *Given a group* G*, the translations* L_g *and* R_g *are bijections.*

Proof. We show this for L_g, the proof for R_g being similar. If $L_g(a) = L_g(b)$, then $ga = gb$, and multiplying on the left by g^{-1}, we get $a = b$, so L_g injective. For any $b \in G$, we have $L_g(g^{-1}b) = gg^{-1}b = b$, so L_g is surjective. Therefore, L_g is bijective. □

Definition C.4. Given a group G, a subset H of G is a *subgroup of* G iff

(1) The identity element e of G also belongs to H ($e \in H$);
(2) For all $h_1, h_2 \in H$, we have $h_1 h_2 \in H$;
(3) For all $h \in H$, we have $h^{-1} \in H$.

The proof of the following proposition is left as an exercise.

Proposition C.4. *Given a group* G*, a subset* $H \subseteq G$ *is a subgroup of* G *iff* H *is nonempty and whenever* $h_1, h_2 \in H$*, then* $h_1 h_2^{-1} \in H$.

If the group G is finite, then the following criterion can be used.

Proposition C.5. *Given a finite group* G*, a subset* $H \subseteq G$ *is a subgroup of* G *iff*

(1) $e \in H$*;*
(2) H *is closed under multiplication.*

Proof. We just have to prove that Condition (3) of Definition C.4 holds. For any $a \in H$, since the left translation L_a is bijective, its restriction to H is injective, and since H is finite, it is also bijective. Since $e \in H$, there is a unique $b \in H$ such that $L_a(b) = ab = e$. However, if a^{-1} is the inverse of a in G, we also have $L_a(a^{-1}) = aa^{-1} = e$, and by injectivity of L_a, we have $a^{-1} = b \in H$. □

Example C.2.

(1) For any integer $n \in \mathbb{Z}$, the set $n\mathbb{Z} = \{nk \mid k \in \mathbb{Z}\}$ is a subgroup of the group $\mathbb{Z}$.

(2) The set of matrices

$$\mathbf{GL}^+(n, \mathbb{R}) = \{A \in \mathbf{GL}(n, \mathbb{R}) \mid \det(A) > 0\}$$

is a subgroup of the group $\mathbf{GL}(n, \mathbb{R})$.

(3) The group $\mathbf{SL}(n, \mathbb{R})$ is a subgroup of the group $\mathbf{GL}(n, \mathbb{R})$.

(4) The group $\mathbf{O}(n)$ is a subgroup of the group $\mathbf{GL}(n, \mathbb{R})$.

(5) The group $\mathbf{SO}(n)$ is a subgroup of the group $\mathbf{O}(n)$, and a subgroup of the group $\mathbf{SL}(n, \mathbb{R})$.

(6) It is not hard to show that every 2×2 rotation matrix $R \in \mathbf{SO}(2)$ can be written as

$$R = \begin{pmatrix} \cos\theta & -\sin\theta \\ \sin\theta & \cos\theta \end{pmatrix}, \quad \text{with} \quad 0 \leq \theta < 2\pi.$$

Then $\mathbf{SO}(2)$ can be considered as a subgroup of $\mathbf{SO}(3)$ by viewing the matrix

$$R = \begin{pmatrix} \cos\theta & -\sin\theta \\ \sin\theta & \cos\theta \end{pmatrix}$$

as the matrix

$$Q = \begin{pmatrix} \cos\theta & -\sin\theta & 0 \\ \sin\theta & \cos\theta & 0 \\ 0 & 0 & 1 \end{pmatrix}.$$

(7) The set of 2×2 upper-triangular matrices of the form

$$\begin{pmatrix} a & b \\ 0 & c \end{pmatrix} \quad a, b, c \in \mathbb{R},\ a, c \neq 0$$

is a subgroup of the group $\mathbf{GL}(2, \mathbb{R})$.

(8) The set V consisting of the four matrices

$$\begin{pmatrix} \pm 1 & 0 \\ 0 & \pm 1 \end{pmatrix}$$

is a subgroup of the group $\mathbf{GL}(2, \mathbb{R})$ called the *Klein four-group*.

Definition C.5. If H is a subgroup of G and $g \in G$ is any element, the sets of the form gH are called *left cosets of H in G* and the sets of the form Hg are called *right cosets of H in G*. The left cosets (resp. right cosets) of H induce an equivalence relation $\sim$ defined as follows: For all $g_1, g_2 \in G$,

$$g_1 \sim g_2 \quad \text{iff} \quad g_1 H = g_2 H$$

(resp. $g_1 \sim g_2$ iff $Hg_1 = Hg_2$). Obviously, $\sim$ is an equivalence relation.

Now, we claim the following fact:

Proposition C.6. *Given a group G and any subgroup H of G, we have $g_1H = g_2H$ iff $g_2^{-1}g_1H = H$ iff $g_2^{-1}g_1 \in H$, for all $g_1, g_2 \in G$.*

Proof. If we apply the bijection $L_{g_2^{-1}}$ to both g_1H and g_2H we get $L_{g_2^{-1}}(g_1H) = g_2^{-1}g_1H$ and $L_{g_2^{-1}}(g_2H) = H$, so $g_1H = g_2H$ iff $g_2^{-1}g_1H = H$. If $g_2^{-1}g_1H = H$, since $1 \in H$, we get $g_2^{-1}g_1 \in H$. Conversely, if $g_2^{-1}g_1 \in H$, since H is a group, the left translation $L_{g_2^{-1}g_1}$ is a bijection of H, so $g_2^{-1}g_1H = H$. Thus, $g_2^{-1}g_1H = H$ iff $g_2^{-1}g_1 \in H$. ☐

It follows that the equivalence class of an element $g \in G$ is the coset gH (resp. Hg). Since L_g is a bijection between H and gH, the cosets gH all have the same cardinality. The map $L_{g^{-1}} \circ R_g$ is a bijection between the left coset gH and the right coset Hg, so they also have the same cardinality. Since the distinct cosets gH form a partition of G, we obtain the following fact:

Proposition C.7. *(Lagrange) For any finite group G and any subgroup H of G, the order h of H divides the order n of G.*

Definition C.6. Given a finite group G and a subgroup H of G, if $n = |G|$ and $h = |H|$, then the ratio n/h is denoted by $(G : H)$ and is called the *index of H in G*.

The index $(G : H)$ is the number of left (and right) cosets of H in G. Proposition C.7 can be stated as

$$|G| = (G : H)|H|.$$

The set of left cosets of H in G (which, in general, is **not** a group) is denoted G/H. The "points" of G/H are obtained by "collapsing" all the elements in a coset into a single element.

Example C.3.

(1) Let n be any positive integer, and consider the subgroup $n\mathbb{Z}$ of $\mathbb{Z}$ (under addition). The coset of 0 is the set $\{0\}$, and the coset of any nonzero integer $m \in \mathbb{Z}$ is

$$m + n\mathbb{Z} = \{m + nk \mid k \in \mathbb{Z}\}.$$

By dividing m by n, we have $m = nq + r$ for some unique r such that $0 \leq r \leq n - 1$. But then we see that r is the smallest positive element of the coset $m + n\mathbb{Z}$. This implies that there is a bijection betwen the cosets of the subgroup $n\mathbb{Z}$ of $\mathbb{Z}$ and the set of residues $\{0, 1, \ldots, n-1\}$ modulo n, or equivalently a bijection with $\mathbb{Z}/n\mathbb{Z}$.

(2) The cosets of $\mathbf{SL}(n, \mathbb{R})$ in $\mathbf{GL}(n, \mathbb{R})$ are the sets of matrices

$$A\,\mathbf{SL}(n, \mathbb{R}) = \{AB \mid B \in \mathbf{SL}(n, \mathbb{R})\},\ A \in \mathbf{GL}(n, \mathbb{R}).$$

Since A is invertible, $\det(A) \neq 0$, and we can write

$$A = (\det(A))^{1/n}((\det(A))^{-1/n}A) \text{ if } \det(A) > 0$$

and

$$A = (-\det(A))^{1/n}((-\det(A))^{-1/n}A) \text{ if } \det(A) < 0.$$

But we have $(\det(A))^{-1/n}A \in \mathbf{SL}(n, \mathbb{R})$ if $\det(A) > 0$ and $-(-\det(A))^{-1/n}A \in \mathbf{SL}(n, \mathbb{R})$ if $\det(A) < 0$, so the coset $A\,\mathbf{SL}(n, \mathbb{R})$ contains the matrix

$$(\det(A))^{1/n}I_n \text{ if } \det(A) > 0, \quad -(-\det(A))^{1/n}I_n \quad \text{if } \det(A) < 0.$$

It follows that there is a bijection between the cosets of $\mathbf{SL}(n, \mathbb{R})$ in $\mathbf{GL}(n, \mathbb{R})$ and $\mathbb{R}$.

(3) The cosets of $\mathbf{SO}(n)$ in $\mathbf{GL}^+(n, \mathbb{R})$ are the sets of matrices

$$A\,\mathbf{SO}(n) = \{AQ \mid Q \in \mathbf{SO}(n)\},\ A \in \mathbf{GL}^+(n, \mathbb{R}).$$

It can be shown (using the polar form for matrices) that there is a bijection between the cosets of $\mathbf{SO}(n)$ in $\mathbf{GL}^+(n, \mathbb{R})$ and the set of $n \times n$ symmetric, positive, definite matrices; these are the symmetric matrices whose eigenvalues are strictly positive.

(4) The cosets of $\mathbf{SO}(2)$ in $\mathbf{SO}(3)$ are the sets of matrices

$$Q\,\mathbf{SO}(2) = \{QR \mid R \in \mathbf{SO}(2)\},\ Q \in \mathbf{SO}(3).$$

The group $\mathbf{SO}(3)$ moves the points on the sphere S^2 in $\mathbb{R}^3$, namely for any $x \in S^2$,

$$x \mapsto Qx \quad \text{for any rotation } Q \in \mathbf{SO}(3).$$

Here,

$$S^2 = \{(x, y, z) \in \mathbb{R}^3 \mid x^2 + y^2 + z^2 = 1\}.$$

Let $N = (0, 0, 1)$ be the north pole on the sphere S^2. Then it is not hard to show that $\mathbf{SO}(2)$ is precisely the subgroup of $\mathbf{SO}(3)$ that leaves N fixed. As a consequence, all rotations QR in the coset $Q\,\mathbf{SO}(2)$ map N to the same point $QN \in S^2$, and it can be shown that there is a bijection between the cosets of $\mathbf{SO}(2)$ in $\mathbf{SO}(3)$ and the points on S^2. The surjectivity of this map has to do with the fact that the action of $\mathbf{SO}(3)$ on S^2 is transitive, which means that for any point $x \in S^2$, there is some rotation $Q \in \mathbf{SO}(3)$ such that $QN = x$.

It is tempting to define a multiplication operation on left cosets (or right cosets) by setting

$$(g_1H)(g_2H) = (g_1g_2)H,$$

but this operation is not well defined in general, unless the subgroup H possesses a special property. In Example C.3, it is possible to define multiplication of cosets in (1), but it is not possible in (2) and (3).

The property of the subgroup H that allows defining a multiplication operation on left cosets is typical of the kernels of group homomorphisms, so we are led to the following definition.

Definition C.7. Given any two groups G and G', a function $\varphi\colon G \to G'$ is a *homomorphism* iff

$$\varphi(g_1g_2) = \varphi(g_1)\varphi(g_2), \quad \text{for all } g_1, g_2 \in G.$$

Taking $g_1 = g_2 = e$ (in G), we see that

$$\varphi(e) = e',$$

and taking $g_1 = g$ and $g_2 = g^{-1}$, we see that

$$\varphi(g^{-1}) = (\varphi(g))^{-1}.$$

Example C.4.

(1) The map $\varphi\colon \mathbb{Z} \to \mathbb{Z}/n\mathbb{Z}$ given by $\varphi(m) = m \bmod n$ for all $m \in \mathbb{Z}$ is a homomorphism.
(2) The map $\det\colon \mathbf{GL}(n, \mathbb{R}) \to \mathbb{R}$ is a homomorphism because $\det(AB) = \det(A)\det(B)$ for any two matrices A, B. Similarly, the map $\det\colon \mathbf{O}(n) \to \mathbb{R}$ is a homomorphism.

If $\varphi\colon G \to G'$ and $\psi\colon G' \to G''$ are group homomorphisms, then $\psi \circ \varphi\colon G \to G''$ is also a homomorphism. If $\varphi\colon G \to G'$ is a homomorphism of groups, and if $H \subseteq G$, $H' \subseteq G'$ are two subgroups, then it is easily checked that

$$\text{Im}\ \varphi = \varphi(H) = \{\varphi(g) \mid g \in H\}$$

is a subgroup of G' and

$$\varphi^{-1}(H') = \{g \in G \mid \varphi(g) \in H'\}$$

is a subgroup of G. In particular, when $H' = \{e'\}$, we obtain the *kernel*, $\text{Ker}\ \varphi$, of φ.

Definition C.8. If $\varphi\colon G \to G'$ is a homomorphism of groups, and if $H \subseteq G$ is a subgroup of G, then the subgroup of G',

$$\text{Im } \varphi = \varphi(H) = \{\varphi(g) \mid g \in H\},$$

is called the *image of* H *by* φ, and the subgroup of G,

$$\text{Ker } \varphi = \{g \in G \mid \varphi(g) = e'\},$$

is called the *kernel* of φ.

Example C.5.

(1) The kernel of the homomorphism $\varphi\colon \mathbb{Z} \to \mathbb{Z}/n\mathbb{Z}$ is $n\mathbb{Z}$.
(2) The kernel of the homomorphism $\det\colon \mathbf{GL}(n, \mathbb{R}) \to \mathbb{R}$ is $\mathbf{SL}(n, \mathbb{R})$. Similarly, the kernel of the homomorphism $\det\colon \mathbf{O}(n) \to \mathbb{R}$ is $\mathbf{SO}(n)$.

The following characterization of the injectivity of a group homomorphism is used all the time.

Proposition C.8. *If $\varphi\colon G \to G'$ is a homomorphism of groups, then $\varphi\colon G \to G'$ is injective iff* Ker $\varphi = \{e\}$. *(We also write* Ker $\varphi = (0)$.*)*

Proof. Assume φ is injective. Since $\varphi(e) = e'$, if $\varphi(g) = e'$, then $\varphi(g) = \varphi(e)$, and by injectivity of φ we must have $g = e$, so Ker $\varphi = \{e\}$.

Conversely, assume that Ker $\varphi = \{e\}$. If $\varphi(g_1) = \varphi(g_2)$, then by multiplication on the left by $(\varphi(g_1))^{-1}$ we get

$$e' = (\varphi(g_1))^{-1}\varphi(g_1) = (\varphi(g_1))^{-1}\varphi(g_2),$$

and since φ is a homomorphism $(\varphi(g_1))^{-1} = \varphi(g_1^{-1})$, so

$$e' = (\varphi(g_1))^{-1}\varphi(g_2) = \varphi(g_1^{-1})\varphi(g_2) = \varphi(g_1^{-1}g_2).$$

This shows that $g_1^{-1}g_2 \in$ Ker φ, but since Ker $\varphi = \{e\}$ we have $g_1^{-1}g_2 = e$, and thus $g_2 = g_1$, proving that φ is injective. $\square$

Definition C.9. We say that a group homomorphism $\varphi\colon G \to G'$ is an *isomorphism* if there is a homomorphism $\psi\colon G' \to G$, so that

$$\psi \circ \varphi = \text{id}_G \quad \text{and} \quad \varphi \circ \psi = \text{id}_{G'}. \qquad (\dagger)$$

If φ is an isomorphism we say that the groups G and G' are *isomorphic*. When $G' = G$, a group isomorphism is called an *automorphism*.

The reasoning used in the proof of Proposition C.2 shows that if a a group homomorphism $\varphi\colon G \to G'$ is an isomorphism, then the homomorphism $\psi\colon G' \to G$ satisfying Condition (†) is unique. This homomorphism is denoted φ^{-1}.

The left translations L_g and the right translations R_g are automorphisms of G.

Suppose $\varphi\colon G \to G'$ is a bijective homomorphism, and let φ^{-1} be the inverse of φ (as a function). Then for all $a, b \in G$, we have

$$\varphi(\varphi^{-1}(a)\varphi^{-1}(b)) = \varphi(\varphi^{-1}(a))\varphi(\varphi^{-1}(b)) = ab,$$

and so

$$\varphi^{-1}(ab) = \varphi^{-1}(a)\varphi^{-1}(b),$$

which proves that φ^{-1} is a homomorphism. Therefore, we proved the following fact.

Proposition C.9. *A bijective group homomorphism $\varphi\colon G \to G'$ is an isomorphism.*

Observe that the property

$$gH = Hg, \quad \text{for all } g \in G \tag{$*$}$$

is equivalent by multiplication on the right by g^{-1} to

$$gHg^{-1} = H, \quad \text{for all } g \in G,$$

and the above is equivalent to

$$gHg^{-1} \subseteq H, \quad \text{for all } g \in G. \tag{$**$}$$

This is because $gHg^{-1} \subseteq H$ implies $H \subseteq g^{-1}Hg$, and this for all $g \in G$.

Proposition C.10. *Let $\varphi\colon G \to G'$ be a group homomorphism. Then $H = \mathrm{Ker}\,\varphi$ satisfies Property ($**$), and thus Property ($*$).*

Proof. We have

$$\varphi(ghg^{-1}) = \varphi(g)\varphi(h)\varphi(g^{-1}) = \varphi(g)e'\varphi(g)^{-1} = \varphi(g)\varphi(g)^{-1} = e',$$

for all $h \in H = \mathrm{Ker}\,\varphi$ and all $g \in G$. Thus, by definition of $H = \mathrm{Ker}\,\varphi$, we have $gHg^{-1} \subseteq H$. $\square$

Definition C.10. For any group G, a subgroup N of G is a *normal subgroup* of G iff

$$gNg^{-1} = N, \quad \text{for all } g \in G.$$

This is denoted by $N \lhd G$.

Proposition C.10 shows that the kernel $\mathrm{Ker}\,\varphi$ of a homomorphism $\varphi\colon G \to G'$ is a normal subgroup of G.

Observe that if G is abelian, then *every* subgroup of G is normal.

Consider Example C.2. Let $R \in \mathbf{SO}(2)$ and $A \in \mathbf{SL}(2,\mathbb{R})$ be the matrices

$$R = \begin{pmatrix} 0 & -1 \\ 1 & 0 \end{pmatrix}, \quad A = \begin{pmatrix} 1 & 1 \\ 0 & 1 \end{pmatrix}.$$

Then

$$A^{-1} = \begin{pmatrix} 1 & -1 \\ 0 & 1 \end{pmatrix},$$

and we have

$$ARA^{-1} = \begin{pmatrix} 1 & 1 \\ 0 & 1 \end{pmatrix} \begin{pmatrix} 0 & -1 \\ 1 & 0 \end{pmatrix} \begin{pmatrix} 1 & -1 \\ 0 & 1 \end{pmatrix} = \begin{pmatrix} 1 & -1 \\ 1 & 0 \end{pmatrix} \begin{pmatrix} 1 & -1 \\ 0 & 1 \end{pmatrix} = \begin{pmatrix} 1 & -2 \\ 1 & -1 \end{pmatrix},$$

and clearly $ARA^{-1} \notin \mathbf{SO}(2)$. Therefore $\mathbf{SO}(2)$ is not a normal subgroup of $\mathbf{SL}(2,\mathbb{R})$. The same counter-example shows that $\mathbf{O}(2)$ is not a normal subgroup of $\mathbf{GL}(2,\mathbb{R})$.

Let $R \in \mathbf{SO}(2)$ and $Q \in \mathbf{SO}(3)$ be the matrices

$$R = \begin{pmatrix} 0 & -1 & 0 \\ 1 & 0 & 0 \\ 0 & 0 & 1 \end{pmatrix}, \quad Q = \begin{pmatrix} 1 & 0 & 0 \\ 0 & 0 & -1 \\ 0 & 1 & 0 \end{pmatrix}.$$

Then

$$Q^{-1} = Q^\top = \begin{pmatrix} 1 & 0 & 0 \\ 0 & 0 & 1 \\ 0 & -1 & 0 \end{pmatrix}$$

and we have

$$\begin{aligned} QRQ^{-1} &= \begin{pmatrix} 1 & 0 & 0 \\ 0 & 0 & -1 \\ 0 & 1 & 0 \end{pmatrix} \begin{pmatrix} 0 & -1 & 0 \\ 1 & 0 & 0 \\ 0 & 0 & 1 \end{pmatrix} \begin{pmatrix} 1 & 0 & 0 \\ 0 & 0 & 1 \\ 0 & -1 & 0 \end{pmatrix} = \begin{pmatrix} 0 & -1 & 0 \\ 0 & 0 & -1 \\ 1 & 0 & 0 \end{pmatrix} \begin{pmatrix} 1 & 0 & 0 \\ 0 & 0 & 1 \\ 0 & -1 & 0 \end{pmatrix} \\ &= \begin{pmatrix} 0 & 0 & -1 \\ 0 & 1 & 0 \\ 1 & 0 & 0 \end{pmatrix}. \end{aligned}$$

Observe that $QRQ^{-1} \notin \mathbf{SO}(2)$, so $\mathbf{SO}(2)$ is not a normal subgroup of $\mathbf{SO}(3)$.

Let T and $A \in \mathbf{GL}(2,\mathbb{R})$ be the following matrices

$$T = \begin{pmatrix} 1 & 1 \\ 0 & 1 \end{pmatrix}, \quad A = \begin{pmatrix} 0 & 1 \\ 1 & 0 \end{pmatrix}.$$

We have

$$A^{-1} = \begin{pmatrix} 0 & 1 \\ 1 & 0 \end{pmatrix} = A,$$

and

$$ATA^{-1} = \begin{pmatrix} 0 & 1 \\ 1 & 0 \end{pmatrix}\begin{pmatrix} 1 & 1 \\ 0 & 1 \end{pmatrix}\begin{pmatrix} 0 & 1 \\ 1 & 0 \end{pmatrix} = \begin{pmatrix} 0 & 1 \\ 1 & 1 \end{pmatrix}\begin{pmatrix} 0 & 1 \\ 1 & 0 \end{pmatrix} = \begin{pmatrix} 1 & 0 \\ 1 & 1 \end{pmatrix}.$$

The matrix T is upper triangular, but ATA^{-1} is not, so the group of 2×2 upper triangular matrices is not a normal subgroup of $\mathbf{GL}(2,\mathbb{R})$.

Let $Q \in V$ and $A \in \mathbb{GL}(2,\mathbb{R})$ be the following matrices

$$Q = \begin{pmatrix} 1 & 0 \\ 0 & -1 \end{pmatrix}, \quad A = \begin{pmatrix} 1 & 1 \\ 0 & 1 \end{pmatrix}.$$

We have

$$A^{-1} = \begin{pmatrix} 1 & -1 \\ 0 & 1 \end{pmatrix}$$

and

$$AQA^{-1} = \begin{pmatrix} 1 & 1 \\ 0 & 1 \end{pmatrix}\begin{pmatrix} 1 & 0 \\ 0 & -1 \end{pmatrix}\begin{pmatrix} 1 & -1 \\ 0 & 1 \end{pmatrix} = \begin{pmatrix} 1 & -1 \\ 0 & -1 \end{pmatrix}\begin{pmatrix} 1 & -1 \\ 0 & 1 \end{pmatrix} = \begin{pmatrix} 1 & -2 \\ 0 & -1 \end{pmatrix}.$$

Clearly $AQA^{-1} \notin V$, which shows that the Klein four group is not a normal subgroup of $\mathbf{GL}(2,\mathbb{R})$.

The reader should check that the subgroups $n\mathbb{Z}$, $\mathbf{GL}^+(n,\mathbb{R})$, $\mathbf{SL}(n,\mathbb{R})$, and $\mathbf{SO}(n,\mathbb{R})$ as a subgroup of $\mathbf{O}(n,\mathbb{R})$, are normal subgroups.

If N is a normal subgroup of G, the equivalence relation $\sim$ induced by left cosets (see Definition C.5) is the same as the equivalence induced by right cosets. Furthermore, this equivalence relation is a *congruence*, which means that: For all $g_1, g_2, g_1', g_2' \in G$,

(1) If $g_1N = g_1'N$ and $g_2N = g_2'N$, then $g_1g_2N = g_1'g_2'N$, and
(2) If $g_1N = g_2N$, then $g_1^{-1}N = g_2^{-1}N$.

As a consequence, we can define a group structure on the set $G/\sim$ of equivalence classes modulo $\sim$, by setting

$$(g_1N)(g_2N) = (g_1g_2)N.$$

Definition C.11. Let G be a group and N be a normal subgroup of G. The group obtained by defining the multiplication of (left) cosets by

$$(g_1N)(g_2N) = (g_1g_2)N, \quad g_1, g_2 \in G$$

is denoted G/N, and called the *quotient of* G *by* N. The equivalence class gN of an element $g \in G$ is also denoted $\overline{g}$ (or $[g]$). The map $\pi\colon G \to G/N$ given by

$$\pi(g) = \overline{g} = gN$$

is a group homomorphism called the *canonical projection*.

Since the kernel of a homomorphism is a normal subgroup, we obtain the following very useful result.

Proposition C.11. *Given a homomorphism of groups* $\varphi\colon G \to G'$*, the groups* $G/\mathrm{Ker}\ \varphi$ *and* $\mathrm{Im}\ \varphi = \varphi(G)$ *are isomorphic.*

Proof. Since φ is surjective onto its image, we may assume that φ is surjective, so that $G' = \mathrm{Im}\ \varphi$. We define a map $\overline{\varphi}\colon G/\mathrm{Ker}\ \varphi \to G'$ as follows:

$$\overline{\varphi}(\overline{g}) = \varphi(g), \quad g \in G.$$

We need to check that the definition of this map does not depend on the representative chosen in the coset $\overline{g} = g\,\mathrm{Ker}\ \varphi$, and that it is a homomorphism. If g' is another element in the coset $g\,\mathrm{Ker}\ \varphi$, which means that $g' = gh$ for some $h \in \mathrm{Ker}\,\varphi$, then

$$\varphi(g') = \varphi(gh) = \varphi(g)\varphi(h) = \varphi(g)e' = \varphi(g),$$

since $\varphi(h) = e'$ as $h \in \mathrm{Ker}\ \varphi$. This shows that

$$\overline{\varphi}(\overline{g'}) = \varphi(g') = \varphi(g) = \overline{\varphi}(\overline{g}),$$

so the map $\overline{\varphi}$ is well defined. It is a homomorphism because

$$\begin{aligned}
\overline{\varphi}(\overline{g}\overline{g'}) &= \overline{\varphi}(\overline{gg'}) \\
&= \varphi(gg') \\
&= \varphi(g)\varphi(g') \\
&= \overline{\varphi}(\overline{g})\overline{\varphi}(\overline{g'}).
\end{aligned}$$

The map $\overline{\varphi}$ is injective because $\overline{\varphi}(\overline{g}) = e'$ iff $\varphi(g) = e'$ iff $g \in \mathrm{Ker}\ \varphi$, iff $\overline{g} = \overline{e}$. The map $\overline{\varphi}$ is surjective because φ is surjective. Therefore $\overline{\varphi}$ is a bijective homomorphism, and thus an isomorphism, as claimed. $\square$

Proposition C.11 is called the *first isomorphism theorem*.

A useful way to construct groups is the *direct product* construction.

Definition C.12. Given two groups G an H, we let $G \times H$ be the Cartesian product of the sets G and H with the multiplication operation $\cdot$ given by

$$(g_1, h_1) \cdot (g_2, h_2) = (g_1 g_2, h_1 h_2).$$

It is immediately verified that $G \times H$ is a group called the *direct product* of G and H.

Similarly, given any n groups $G_1, \ldots, G_n$, we can define the direct product $G_1 \times \cdots \times G_n$ is a similar way.

If G is an abelian group and $H_1, \ldots, H_n$ are subgroups of G, the situation is simpler. Consider the map

$$a \colon H_1 \times \cdots \times H_n \to G$$

given by

$$a(h_1, \ldots, h_n) = h_1 + \cdots + h_n,$$

using $+$ for the operation of the group G. It is easy to verify that a is a group homomorphism, so its image is a subgroup of G denoted by $H_1 + \cdots + H_n$, and called the *sum* of the groups H_i. The following proposition will be needed.

Proposition C.12. *Given an abelian group G, if H_1 and H_2 are any subgroups of G such that $H_1 \cap H_2 = \{0\}$, then the map a is an isomorphism*

$$a \colon H_1 \times H_2 \to H_1 + H_2.$$

Proof. The map is surjective by definition, so we just have to check that it is injective. For this, we show that $\operatorname{Ker} a = \{(0,0)\}$. We have $a(a_1, a_2) = 0$ iff $a_1 + a_2 = 0$ iff $a_1 = -a_2$. Since $a_1 \in H_1$ and $a_2 \in H_2$, we see that $a_1, a_2 \in H_1 \cap H_2 = \{0\}$, so $a_1 = a_2 = 0$, which proves that $\operatorname{Ker} a = \{(0,0)\}$. ☐

Under the conditions of Proposition C.12, namely $H_1 \cap H_2 = \{0\}$, the group $H_1 + H_2$ is called the *direct sum* of H_1 and H_2; it is denoted by $H_1 \oplus H_2$, and we have an isomorphism $H_1 \times H_2 \cong H_1 \oplus H_2$.

C.2 Group Actions: Part I, Definition and Examples

If X is a set (usually some kind of geometric space, for example, the sphere in $\mathbb{R}^3$, the upper half-plane, etc.), the "symmetries" of X are often captured by the action of a group G on X. In fact, if G is a Lie group and the action satisfies some simple properties, the set X can be given a manifold structure which makes it a projection (quotient) of G, a so-called "homogeneous space."

Definition C.13. Given a set X and a group G, a *left action of* G *on* X (for short, an *action of* G *on* X) is a function $\varphi\colon G \times X \to X$, such that:

(1) For all $g, h \in G$ and all $x \in X$,
$$\varphi(g, \varphi(h, x)) = \varphi(gh, x).$$

(2) For all $x \in X$,
$$\varphi(1, x) = x,$$
where $1 \in G$ is the identity element of G.

To alleviate the notation, we usually write $g \cdot x$ or even gx for $\varphi(g, x)$, in which case the above axioms read:

(1) For all $g, h \in G$ and all $x \in X$,
$$g \cdot (h \cdot x) = gh \cdot x.$$

(2) For all $x \in X$,
$$1 \cdot x = x.$$

The set X is called a *(left) G-set*. The action φ is *faithful* or *effective* iff for every g, if $g \cdot x = x$ for all $x \in X$, then $g = 1$. Faithful means that if the action of some element g behaves like the identity, then g must be the identity element. The action φ is *transitive* iff for any two elements $x, y \in X$, there is some $g \in G$ so that $g \cdot x = y$.

Given an action $\varphi\colon G \times X \to X$, for every $g \in G$, we have a function $\varphi_g\colon X \to X$ defined by
$$\varphi_g(x) = g \cdot x, \quad \text{for all } x \in X.$$
Observe that φ_g has $\varphi_{g^{-1}}$ as inverse, since
$$\varphi_{g^{-1}}(\varphi_g(x)) = \varphi_{g^{-1}}(g \cdot x) = g^{-1} \cdot (g \cdot x) = (g^{-1}g) \cdot x = 1 \cdot x = x,$$

and similarly, $\varphi_g \circ \varphi_{g^{-1}} = \mathrm{id}$. Therefore, φ_g is a bijection of X; that is, φ_g is a permutation of X. Moreover, we check immediately that

$$\varphi_g \circ \varphi_h = \varphi_{gh},$$

so the map $g \mapsto \varphi_g$ is a group homomorphism from G to $\mathfrak{S}_X$, the group of permutations of X. With a slight abuse of notation, this group homomorphism $G \longrightarrow \mathfrak{S}_X$ is also denoted φ.

Conversely, it is easy to see that any group homomorphism $\varphi\colon G \to \mathfrak{S}_X$ yields a group action $\cdot\colon G \times X \longrightarrow X$, by setting

$$g \cdot x = \varphi(g)(x).$$

Observe that an action φ is faithful iff the group homomorphism $\varphi\colon G \to \mathfrak{S}_X$ is injective, i.e. iff φ has a trivial kernel. Also, we have $g \cdot x = y$ iff $g^{-1} \cdot y = x$, since $(gh) \cdot x = g \cdot (h \cdot x)$ and $1 \cdot x = x$, for all $g, h \in G$ and all $x \in X$.

Definition C.14. Given two G-sets X and Y, a function $f\colon X \to Y$ is said to be *equivariant*, or a *G-map*, iff for all $x \in X$ and all $g \in G$, we have

$$f(g \cdot x) = g \cdot f(x).$$

Equivalently, if the G-actions are denoted by $\varphi\colon G \times X \to X$ and $\psi\colon G \times Y \to Y$, we have the following commutative diagram for all $g \in G$:

$$\begin{array}{ccc} X & \xrightarrow{\varphi_g} & X \\ {\scriptstyle f}\downarrow & & \downarrow{\scriptstyle f} \\ Y & \xrightarrow[\psi_g]{} & Y. \end{array}$$

Remark: We can also define a *right action* $\cdot\colon X \times G \to X$ of a group G on a set X as a map satisfying the conditions

(1) For all $g, h \in G$ and all $x \in X$,

$$(x \cdot g) \cdot h = x \cdot gh.$$

(2) For all $x \in X$,

$$x \cdot 1 = x.$$

Every notion defined for left actions is also defined for right actions in the obvious way.

However, one change is necessary. For every $g \in G$, the map $\varphi_g \colon X \to X$ must be defined as

$$\varphi_g(x) = x \cdot g^{-1},$$

in order for the map $g \mapsto \varphi_g$ from G to $\mathfrak{S}_X$ to be a homomorphism ($\varphi_g \circ \varphi_h = \varphi_{gh}$). Conversely, given a homomorphism $\varphi \colon G \to \mathfrak{S}_X$, we get a right action $\cdot \colon X \times G \longrightarrow X$ by setting

$$x \cdot g = \varphi(g^{-1})(x).$$

Here are some examples of (left) group actions.

Example C.6. The unit sphere S^2 (more generally, S^{n-1}).

Recall that for any $n \geq 1$, the *(real) unit sphere* S^{n-1} is the set of points in $\mathbb{R}^n$ given by

$$S^{n-1} = \{(x_1, \dots, x_n) \in \mathbb{R}^n \mid x_1^2 + \cdots + x_n^2 = 1\}.$$

In particular, S^2 is the usual sphere in $\mathbb{R}^3$. Since the group $\mathbf{SO}(3) = \mathbf{SO}(3, \mathbb{R})$ consists of (orientation preserving) linear isometries, i.e., *linear* maps that are distance preserving (and of determinant $+1$), and every linear map leaves the origin fixed, we see that any rotation maps S^2 into itself.

Beware that this would be false if we considered the group of *affine* isometries $\mathbf{SE}(3)$ of $\mathbb{E}^3$. For example, a screw motion does *not* map S^2 into itself, even though it is distance preserving, because the origin is translated.

Thus, for $X = S^2$ and $G = \mathbf{SO}(3)$, we have an action $\cdot \colon \mathbf{SO}(3) \times S^2 \to S^2$, given by the matrix multiplication

$$R \cdot x = Rx.$$

The verification that the above is indeed an action is trivial. This action is transitive. This is because, for any two points x, y on the sphere S^2, there is a rotation whose axis is perpendicular to the plane containing x, y and the center O of the sphere (this plane is not unique when x and y are antipodal, i.e., on a diameter) mapping x to y. See Figure C.1.

Similarly, for any $n \geq 1$, let $X = S^{n-1}$ and $G = \mathbf{SO}(n)$ and define the action $\cdot \colon \mathbf{SO}(n) \times S^{n-1} \to S^{n-1}$ as $R \cdot x = Rx$. It is easy to show that this action is transitive.

Analogously, we can define the *(complex) unit sphere* Σ^{n-1}, as the set of points in $\mathbb{C}^n$ given by

$$\Sigma^{n-1} = \{(z_1, \dots, z_n) \in \mathbb{C}^n \mid z_1\overline{z}_1 + \cdots + z_n\overline{z}_n = 1\}.$$

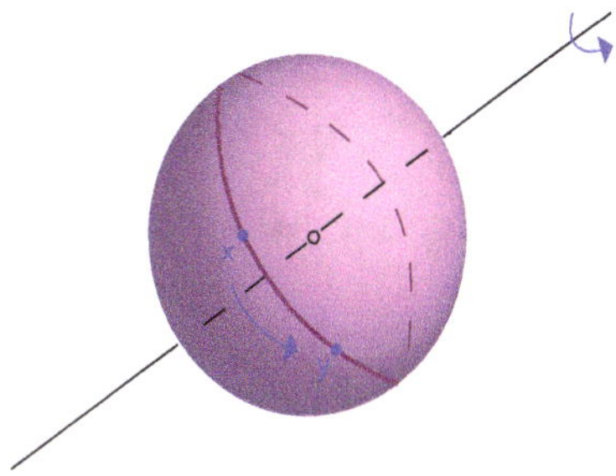

Fig. C.1 The rotation which maps x to y.

If we write $z_j = x_j + iy_j$, with $x_j, y_j \in \mathbb{R}$, then

$$\Sigma^{n-1} = \{(x_1, \dots, x_n, y_1, \dots, y_n) \in \mathbb{R}^{2n} \mid x_1^2 + \cdots + x_n^2 + y_1^2 + \cdots + y_n^2 = 1\}.$$

Therefore, we can view the complex sphere Σ^{n-1} (in $\mathbb{C}^n$) as the real sphere S^{2n-1} (in $\mathbb{R}^{2n}$). By analogy with the real case, we can define for $X = \Sigma^{n-1}$ and $G = \mathbf{SU}(n)$ an action $\cdot\colon \mathbf{SU}(n) \times \Sigma^{n-1} \to \Sigma^{n-1}$ of the group $\mathbf{SU}(n)$ of *linear* maps of $\mathbb{C}^n$ preserving the Hermitian inner product (and the origin, as all linear maps do), and this action is transitive.

One should not confuse the unit sphere Σ^{n-1} with the hypersurface $S_{\mathbb{C}}^{n-1}$, given by

$$S_{\mathbb{C}}^{n-1} = \{(z_1, \dots, z_n) \in \mathbb{C}^n \mid z_1^2 + \cdots + z_n^2 = 1\}.$$

For instance, one should check that a line L through the origin intersects Σ^{n-1} in a circle, whereas it intersects $S_{\mathbb{C}}^{n-1}$ in exactly two points! Recall for a fixed $u = (x_1, \dots x_n, y_1, \dots y_n) \in \mathbb{C}^n$, that $L = \{\gamma u \mid \gamma \in \mathbb{C}\}$. Since $\gamma = \rho(\cos\theta + i\sin\theta)$, we deduce that L is actually the two dimensional subspace through the origin spanned by the orthogonal vectors $(x_1, \dots x_n, y_1, \dots y_n)$ and $(-y_1, \cdots - y_n, x_1, \dots x_n)$.

Example C.7. The upper half-plane.

The *upper half-plane* H is the open subset of $\mathbb{R}^2$ consisting of all points $(x, y) \in \mathbb{R}^2$, with $y > 0$. It is convenient to identify H with the set of complex numbers $z \in \mathbb{C}$ such that $\Im z > 0$. Then we can let $X = H$ and $G = \mathbf{SL}(2, \mathbb{R})$ and define an action $\cdot\colon \mathbf{SL}(2, \mathbb{R}) \times H \to H$ of the group $\mathbf{SL}(2, \mathbb{R})$ on H, as follows: For any $z \in H$, for any $A \in \mathbf{SL}(2, \mathbb{R})$,

$$A \cdot z = \frac{az + b}{cz + d},$$

where

$$A = \begin{pmatrix} a & b \\ c & d \end{pmatrix}$$

with $ad - bc = 1$.

It is easily verified that $A \cdot z$ is indeed always well defined and in H when $z \in H$ (check this). To see why this action is transitive, let z and w be two arbitrary points of H where $z = x + iy$ and $w = u + iv$ with $x, u \in \mathbb{R}$ and $y, v \in \mathbb{R}^+$ (i.e. y and v are positive real numbers). Define

$$A = \begin{pmatrix} \sqrt{\frac{v}{y}} & \frac{uy - vx}{\sqrt{yv}} \\ 0 & \sqrt{\frac{y}{v}} \end{pmatrix}.$$

Note that $A \in \mathbf{SL}(2, \mathbb{R})$. A routine calculation shows that $A \cdot z = w$.

Before introducing Example C.8, we need to define the groups of Möbius transformations and the Riemann sphere. Maps of the form

$$z \mapsto \frac{az + b}{cz + d},$$

where $z \in \mathbb{C}$ and $ad - bc = 1$, are called *Möbius transformations*. Here, $a, b, c, d \in \mathbb{R}$, but in general, we allow $a, b, c, d \in \mathbb{C}$. Actually, these transformations are not necessarily defined everywhere on $\mathbb{C}$, for example, for $z = -d/c$ if $c \neq 0$. To fix this problem, we add a "point at infinity" ∞ to $\mathbb{C}$, and define Möbius transformations as functions $\mathbb{C} \cup \{\infty\} \longrightarrow \mathbb{C} \cup \{\infty\}$. If $c = 0$, the Möbius transformation sends ∞ to itself, otherwise, $-d/c \mapsto \infty$ and $\infty \mapsto a/c$.

The space $\mathbb{C} \cup \{\infty\}$ can be viewed as the plane $\mathbb{R}^2$ extended with a point at infinity. Using a stereographic projection from the sphere S^2 to the plane (say from the north pole to the equatorial plane), we see that there is a bijection between the sphere S^2 and $\mathbb{C} \cup \{\infty\}$. More precisely, the *stereographic projection* σ_N of the sphere S^2 from the north pole $N = (0, 0, 1)$ to the plane $z = 0$ (extended with the point at infinity ∞) is given by

$$(x, y, z) \in S^2 - \{(0, 0, 1)\} \mapsto \left(\frac{x}{1 - z}, \frac{y}{1 - z}\right) = \frac{x + iy}{1 - z} \in \mathbb{C}, \;\; (0, 0, 1) \mapsto \infty.$$

The inverse stereographic projection σ_N^{-1} is given by

$$(u,v) \mapsto \left(\frac{2u}{u^2+v^2+1}, \frac{2v}{u^2+v^2+1}, \frac{u^2+v^2-1}{u^2+v^2+1}\right), \quad \text{with} \quad \infty \mapsto (0,0,1).$$

Intuitively, the inverse stereographic projection "wraps" the equatorial plane around the sphere. See Figure C.2.

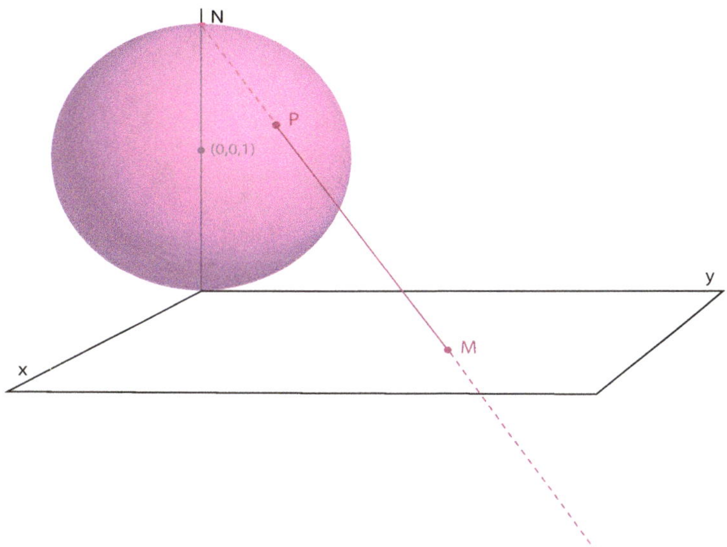

Fig. C.2 Inverse stereographic projection.

The space $\mathbb{C}\cup\{\infty\}$ is known as the *Riemann sphere*. We will see shortly that $\mathbb{C}\cup\{\infty\} \cong S^2$ is also the complex projective line $\mathbb{CP}^1$. In summary, Möbius transformations are bijections of the Riemann sphere. It is easy to check that these transformations form a group under composition for all $a,b,c,d \in \mathbb{C}$, with $ad-bc=1$. This is the *Möbius group*, denoted $\mathbf{M\ddot{o}b}^+$. The Möbius transformations corresponding to the case $a,b,c,d \in \mathbb{R}$, with $ad-bc=1$ form a subgroup of $\mathbf{M\ddot{o}b}^+$ denoted $\mathbf{M\ddot{o}b}^+_{\mathbb{R}}$.

The map from $\mathbf{SL}(2,\mathbb{C})$ to $\mathbf{M\ddot{o}b}^+$ that sends $A \in \mathbf{SL}(2,\mathbb{C})$ to the corresponding Möbius transformation is a surjective group homomorphism, and one checks easily that its kernel is $\{-I, I\}$ (where I is the 2×2 identity matrix). Therefore, the Möbius group $\mathbf{M\ddot{o}b}^+$ is isomorphic to the quotient group $\mathbf{SL}(2,\mathbb{C})/\{-I,I\}$, denoted $\mathbf{PSL}(2,\mathbb{C})$. This latter group turns out to be the group of projective transformations of the projective space $\mathbb{CP}^1$. The same reasoning shows that the subgroup $\mathbf{M\ddot{o}b}^+_{\mathbb{R}}$ is isomorphic to $\mathbf{SL}(2,\mathbb{R})/\{-I,I\}$, denoted $\mathbf{PSL}(2,\mathbb{R})$.

Example C.8. The Riemann sphere $\mathbb{C}\cup\{\infty\}$.

Let $X=\mathbb{C}\cup\{\infty\}$ and $G=\mathbf{SL}(2,\mathbb{C})$. The group $\mathbf{SL}(2,\mathbb{C})$ acts on $\mathbb{C}\cup\{\infty\}\cong S^2$ the same way that $\mathbf{SL}(2,\mathbb{R})$ acts on H, namely: For any $A\in\mathbf{SL}(2,\mathbb{C})$, for any $z\in\mathbb{C}\cup\{\infty\}$,

$$A\cdot z=\frac{az+b}{cz+d},$$

where

$$A=\begin{pmatrix} a & b\\ c & d\end{pmatrix}\quad\text{with}\quad ad-bc=1.$$

This action is transitive, an exercise we leave for the reader.

Example C.9. The unit disk.

One may recall from complex analysis that the scaled (complex) Möbius transformation

$$z\mapsto\frac{z-i}{z+i}$$

is a biholomorphic or analytic isomorphism between the upper half plane H and the open unit disk

$$D=\{z\in\mathbb{C}\mid |z|<1\}.$$

As a consequence, it is possible to define a transitive action of $\mathbf{SL}(2,\mathbb{R})$ on D. This can be done in a more direct fashion, using a group isomorphic to $\mathbf{SL}(2,\mathbb{R})$, namely, $\mathbf{SU}(1,1)$ (a group of complex matrices), but we don't want to do this right now.

Example C.10. The unit Riemann sphere revisited.

Another interesting action is the action of $\mathbf{SU}(2)$ on the extended plane $\mathbb{C}\cup\{\infty\}$. Recall that the group $\mathbf{SU}(2)$ consists of all complex matrices of the form

$$A=\begin{pmatrix} \alpha & \beta\\ -\overline{\beta} & \overline{\alpha}\end{pmatrix}\qquad \alpha,\beta\in\mathbb{C},\qquad \alpha\overline{\alpha}+\beta\overline{\beta}=1.$$

Let $X=\mathbb{C}\cup\{\infty\}$ and $G=\mathbf{SU}(2)$. The action $\cdot\colon \mathbf{SU}(2)\times(\mathbb{C}\cup\{\infty\})\to\mathbb{C}\cup\{\infty\}$ is given by

$$A\cdot w=\frac{\alpha w+\beta}{-\overline{\beta}w+\overline{\alpha}},\quad w\in\mathbb{C}\cup\{\infty\}.$$

Let us denote this action by $\Phi_{\mathbb{C}}$. The action $\Phi_{\mathbb{C}}$ is transitive. The proof relies on the fact that there is another action $\Phi_{S^2}\colon \mathbf{SU}(2) \times S^2 \to S^2$ such that $(\Phi_{S^2})_A \in \mathbf{SO}(3)$ for all $A \in \mathbf{SU}(2)$, that

$$(\Phi_{\mathbb{C}})_A = \sigma_N \circ (\Phi_{S^2})_A \circ \sigma_N^{-1},$$

and the fact that the map $\rho\colon \mathbf{SU}(2) \to \mathbf{SO}(3)$ defined by $\rho(A) = (\Phi_{S^2})_A$ is a surjective group homomorphism. Then we use the transitivity of the action of $\mathbf{SO}(3)$ on S^2. Let us provide details.

Using the stereographic projection σ_N from S^2 onto $\mathbb{C} \cup \{\infty\}$ and its inverse σ_N^{-1}, we define an action $\Phi_{S^2}\colon \mathbf{SU}(2) \times S^2 \to S^2$ of $\mathbf{SU}(2)$ on S^2 by

$$(\Phi_{S^2})_A(x, y, z) = \sigma_N^{-1}((\Phi_{\mathbb{C}})_A(\sigma_N(x, y, z))), \quad (x, y, z) \in S^2.$$

By definition we have

$$(\Phi_{S^2})_A = \sigma_N^{-1} \circ (\Phi_{\mathbb{C}})_A \circ \sigma_N,$$

and so

$$(\Phi_{\mathbb{C}})_A = \sigma_N \circ (\Phi_{S^2})_A \circ \sigma_N^{-1}.$$

Although this is not immediately obvious, it turns out that the map $(\Phi_{S^2})_A$ resulting from the action of $\mathbf{SU}(2)$ on S^2 is the restriction of a linear map $\rho(A)$ to S^2, and since this linear map preserves S^2, it is an orthogonal transformations. Thus, we obtain a continuous (in fact, smooth) group homomorphism

$$\rho\colon \mathbf{SU}(2) \to \mathbf{O}(3),$$

where $\rho(A)$ is the orthogonal transformation associated with $(\Phi_{S^2})_A$. Since $\mathbf{SU}(2)$ is connected and ρ is continuous, the image of $\mathbf{SU}(2)$ is contained in the connected component of I in $\mathbf{O}(3)$, namely $\mathbf{SO}(3)$, so ρ is a homomorphism

$$\rho\colon \mathbf{SU}(2) \to \mathbf{SO}(3).$$

We will see that this homomorphism is surjective and that its kernel is $\{I, -I\}$. The upshot is that we have an isomorphism

$$\mathbf{SO}(3) \cong \mathbf{SU}(2)/\{I, -I\}.$$

The fact that the action $\Phi_{\mathbb{C}}$ is transitive follows the surjectivity of $\rho\colon \mathbf{SU}(2) \to \mathbf{SO}(3)$, the fact that

$$(\Phi_{\mathbb{C}})_A = \sigma_N \circ \rho(A) \circ \sigma_N^{-1},$$

and the transitivity of the action of $\mathbf{SO}(3)$ on S^2. More precisely, take $z, w \in \mathbb{C}\cup\{\infty\}$, use the inverse stereographic projection to obtain two points on S^2, namely $\sigma_N^{-1}(z)$ and $\sigma_N^{-1}(w)$, and then apply some appropriate rotation $R \in \mathbf{SO}(3)$ to map $\sigma_N^{-1}(z)$ onto $\sigma_N^{-1}(w)$, that is, $\sigma_N^{-1}(w) = R(\sigma_N^{-1}(z))$. Such a rotation exists by the argument presented in Example C.6. Since $\rho\colon \mathbf{SU}(2) \to \mathbf{SO}(3)$ is surjective (see below), there must exist $A \in \mathbf{SU}(2)$ such that $\rho(A) = R$ and then

$$(\Phi_{\mathbb{C}})_A(z) = (\sigma_N \circ \rho(A) \circ \sigma_N^{-1})(z) = \sigma_N(R(\sigma_N^{-1})(z)) = \sigma_N(\sigma_N^{-1})(w) = w.$$

The homomorphism ρ is a way of describing how a unit quaternion (any element of $\mathbf{SU}(2)$) induces a rotation, *via* the stereographic projection and its inverse. If we write $\alpha = a + ib$ and $\beta = c + id$, a rather tedious computation yields

$$\rho(A) = \begin{pmatrix} a^2 - b^2 - c^2 + d^2 & -2ab - 2cd & -2ac + 2bd \\ 2ab - 2cd & a^2 - b^2 + c^2 - d^2 & -2ad - 2bc \\ 2ac + 2bd & 2ad - 2bc & a^2 + b^2 - c^2 - d^2 \end{pmatrix}.$$

One can check that $\rho(A)$ is indeed a rotation matrix which represents the rotation whose axis is the line determined by the vector $(d, -c, b)$ and whose angle $\theta \in [0, 2\pi)$ is determined by

$$\cos\frac{\theta}{2} = a.$$

Remark: If we use the *right* action of $\mathbf{SU}(2)$ on $\mathbb{C} \cup \{\infty\}$ given by

$$A^{\top} \cdot w = \frac{\alpha w - \overline{\beta}}{\beta w + \overline{\alpha}}, \quad w \in \mathbb{C} \cup \{\infty\},$$

the effect is to change b to $-b$ and then $\rho(A)$ is the rotation of axis (d, c, b) and same angle $\theta \in [0, 2\pi)$ as before.

We can also compute the derivative $d\rho_I\colon \mathfrak{su}(2) \to \mathfrak{so}(3)$ of ρ at I as follows. Recall that $\mathfrak{su}(2)$ consists of all complex matrices of the form

$$\begin{pmatrix} ib & c + id \\ -c + id & -ib \end{pmatrix}, \quad b, c, d \in \mathbb{R},$$

so pick the following basis for $\mathfrak{su}(2)$,

$$X_1 = \begin{pmatrix} i & 0 \\ 0 & -i \end{pmatrix}, \quad X_2 = \begin{pmatrix} 0 & 1 \\ -1 & 0 \end{pmatrix}, \quad X_3 = \begin{pmatrix} 0 & i \\ i & 0 \end{pmatrix},$$

and define the curves in $\mathbf{SU}(2)$ through I given by

$$c_1(t) = \begin{pmatrix} e^{it} & 0 \\ 0 & e^{-it} \end{pmatrix}, \quad c_2(t) = \begin{pmatrix} \cos t & \sin t \\ -\sin t & \cos t \end{pmatrix}, \quad c_3(t) = \begin{pmatrix} \cos t & i\sin t \\ i\sin t & \cos t \end{pmatrix}.$$

It is easy to check that $c_i'(0) = X_i$ for $i = 1, 2, 3$, and that

$$d\rho_I(X_1) = 2\begin{pmatrix} 0 & -1 & 0 \\ 1 & 0 & 0 \\ 0 & 0 & 0 \end{pmatrix}, d\rho_I(X_2) = 2\begin{pmatrix} 0 & 0 & -1 \\ 0 & 0 & 0 \\ 1 & 0 & 0 \end{pmatrix}, d\rho_I(X_3) = 2\begin{pmatrix} 0 & 0 & 0 \\ 0 & 0 & -1 \\ 0 & 1 & 0 \end{pmatrix}.$$

Thus we have

$$d\rho_I(X_1) = 2E_3, \quad d\rho_I(X_2) = -2E_2, \quad d\rho_I(X_3) = 2E_1,$$

where (E_1, E_2, E_3) is the basis of $\mathfrak{so}(3)$ given by

$$\left(E_1 = \begin{pmatrix} 0 & 0 & 0 \\ 0 & 0 & -1 \\ 0 & 1 & 0 \end{pmatrix}, \; E_2 = \begin{pmatrix} 0 & 0 & 1 \\ 0 & 0 & 0 \\ -1 & 0 & 0 \end{pmatrix}, \; E_3 = \begin{pmatrix} 0 & -1 & 0 \\ 1 & 0 & 0 \\ 0 & 0 & 0 \end{pmatrix} \right),$$

which means that $d\rho_I$ is an isomorphism between the Lie algebras $\mathfrak{su}(2)$ and $\mathfrak{so}(3)$.

Recall that we have the commutative diagram

$$\begin{array}{ccc} \mathbf{SU}(2) & \xrightarrow{\rho} & \mathbf{SO}(3) \\ {\scriptstyle \exp}\uparrow & & \uparrow{\scriptstyle \exp} \\ \mathfrak{su}(2) & \xrightarrow[d\rho_I]{} & \mathfrak{so}(3) \, . \end{array}$$

Since $d\rho_I$ is surjective and the exponential map $\exp\colon \mathfrak{so}(3) \to \mathbf{SO}(3)$ is surjective, we conclude that ρ is surjective. (We also know that $\exp\colon \mathfrak{su}(2) \to \mathbf{SU}(2)$ is surjective.) Observe that $\rho(-A) = \rho(A)$, and it is easy to check that $\operatorname{Ker} \rho = \{I, -I\}$.

Example C.11. The set of $n \times n$ symmetric, positive, definite matrices, $\mathbf{SPD}(n)$.

Let $X = \mathbf{SPD}(n)$ and $G = \mathbf{GL}(n)$. The group $\mathbf{GL}(n) = \mathbf{GL}(n, \mathbb{R})$ acts on $\mathbf{SPD}(n)$ as follows: for all $A \in \mathbf{GL}(n)$ and all $S \in \mathbf{SPD}(n)$,

$$A \cdot S = ASA^{\top}.$$

It is easily checked that $ASA^{\top}$ is in $\mathbf{SPD}(n)$ if S is in $\mathbf{SPD}(n)$. First observe that $ASA^{\top}$ is symmetric since

$$(ASA^{\top})^{\top} = AS^{\top}A^{\top} = ASA^{\top}.$$

Next recall the following characterization of positive definite matrix, namely

$$y^{\top}Sy > 0, \qquad \text{whenever } y \neq 0.$$

We want to show $x^\top(A^\top SA)x > 0$ for all $x \neq 0$. Since A is invertible, we have $x = A^{-1}y$ for some nonzero y, and hence

$$\begin{aligned} x^\top(A^\top SA)x &= y^\top(A^{-1})^\top A^\top SAA^{-1}y \\ &= y^\top Sy > 0. \end{aligned}$$

Hence $A^\top SA$ is positive definite. This action is transitive because every SPD matrix S can be written as $S = AA^\top$, for some invertible matrix A (prove this as an exercise). Given any two SPD matrices $S_1 = A_1A_1^\top$ and $S_2 = A_2A_2^\top$ with A_1 and A_2 invertible, if $A = A_2A_1^{-1}$, we have

$$\begin{aligned} A \cdot S_1 &= A_2A_1^{-1}S_1(A_2A_1^{-1})^\top = A_2A_1^{-1}S_1(A_1^\top)^{-1}A_2^\top \\ &= A_2A_1^{-1}A_1A_1^\top(A_1^\top)^{-1}A_2^\top = A_2A_2^\top = S_2. \end{aligned}$$

Example C.12. The projective spaces $\mathbb{RP}^n$ and $\mathbb{CP}^n$.

The *(real) projective space* $\mathbb{RP}^n$ is the set of all lines through the origin in $\mathbb{R}^{n+1}$; that is, the set of one-dimensional subspaces of $\mathbb{R}^{n+1}$ (where $n \geq 0$). Since a one-dimensional subspace $L \subseteq \mathbb{R}^{n+1}$ is spanned by any nonzero vector $u \in L$, we can view $\mathbb{RP}^n$ as the set of equivalence classes of nonzero vectors in $\mathbb{R}^{n+1} - \{0\}$ modulo the equivalence relation

$$u \sim v \quad \text{iff} \quad v = \lambda u, \quad \text{for some} \quad \lambda \in \mathbb{R},\ \lambda \neq 0.$$

In terms of this definition, there is a projection $pr\colon (\mathbb{R}^{n+1} - \{0\}) \to \mathbb{RP}^n$, given by $pr(u) = [u]_\sim$, the equivalence class of u modulo $\sim$. Write $[u]$ for the line defined by the nonzero vector u. Since every line L in $\mathbb{R}^{n+1}$ intersects the sphere S^n in two antipodal points, we can view $\mathbb{RP}^n$ as the quotient of the sphere S^n by identification of antipodal points. See Figures C.3 and C.4.

Let $X = \mathbb{RP}^n$ and $G = \mathbf{SO}(n+1)$. We define an action of $\mathbf{SO}(n+1)$ on $\mathbb{RP}^n$ as follows.

For any line $L = [u]$, for any $R \in \mathbf{SO}(n+1)$,

$$R \cdot L = [Ru].$$

Since R is linear, the line $[Ru]$ is well defined; that is, does not depend on the choice of $u \in L$. The reader can show that this action is transitive.

The *(complex) projective space* $\mathbb{CP}^n$ is defined analogously as the set of all lines through the origin in $\mathbb{C}^{n+1}$; that is, the set of one-dimensional subspaces of $\mathbb{C}^{n+1}$ (where $n \geq 0$). This time, we can view $\mathbb{CP}^n$ as the set of equivalence classes of vectors in $\mathbb{C}^{n+1} - \{0\}$ modulo the equivalence relation

$$u \sim v \quad \text{iff} \quad v = \lambda u, \quad \text{for some} \quad \lambda \neq 0 \in \mathbb{C}.$$

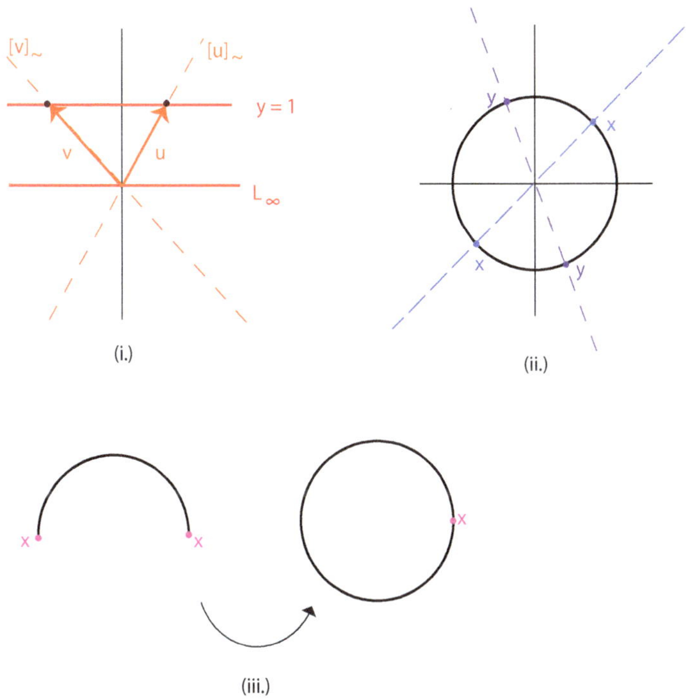

Fig. C.3 Three constructions for $\mathbb{RP}^1 \cong S^1$. Illustration (i.) applies the equivalence relation. Since any line through the origin, excluding the x-axis, intersects the line $y = 1$, its equivalence class is represented by its point of intersection on $y = 1$. Hence, $\mathbb{RP}^n$ is the disjoint union of the line $y = 1$ and the point of infinity given by the x-axis. Illustration (ii.) represents $\mathbb{RP}^1$ as the quotient of the circle S^1 by identification of antipodal points. Illustration (iii.) is a variation which glues the equatorial points of the upper semicircle.

We have the projection $pr\colon \mathbb{C}^{n+1} - \{0\} \to \mathbb{CP}^n$, given by $pr(u) = [u]_\sim$, the equivalence class of u modulo $\sim$. Again, write $[u]$ for the line defined by the nonzero vector u. Let $X = \mathbb{CP}^n$ and $G = \mathbf{SU}(n+1)$.

We define an action of $\mathbf{SU}(n+1)$ on $\mathbb{CP}^n$ as follows: For any line $L = [u]$, for any $R \in \mathbf{SU}(n+1)$,

$$R \cdot L = [Ru].$$

Again, this action is well defined and it is transitive. (Check this.)

Before progressing to our final example of group actions, we take a moment to construct $\mathbb{CP}^n$ as a quotient space of S^{2n+1}. Recall that $\Sigma^n \subseteq$

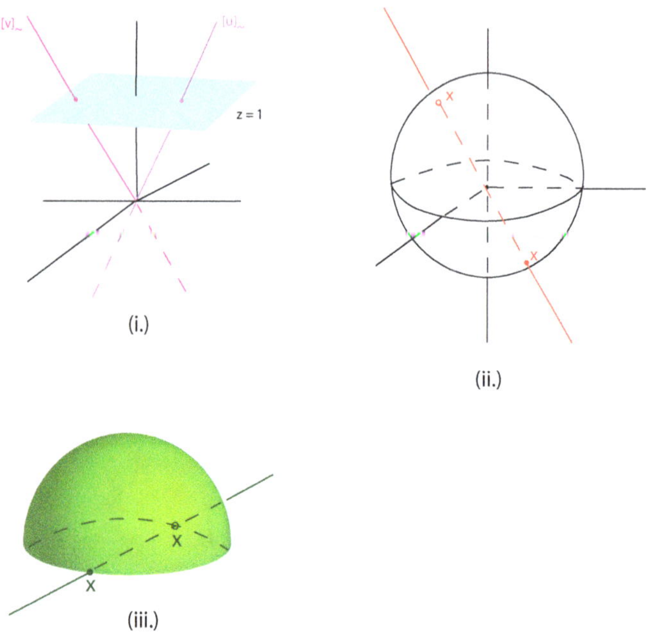

Fig. C.4 Three constructions for $\mathbb{RP}^2$. Illustration (i.) applies the equivalence relation. Since any line through the origin which is not contained in the xy-plane intersects the plane $z = 1$, its equivalence class is represented by its point of intersection on $z = 1$. Hence, $\mathbb{RP}^2$ is the disjoint union of the plane $z = 1$ and the copy of $\mathbb{RP}^1$ provided by the xy-plane. Illustration (ii.) represents $\mathbb{RP}^2$ as the quotient of the sphere S^2 by identification of antipodal points. Illustration (iii.) is a variation which glues the antipodal points on boundary of the unit disk, which is represented as as the upper hemisphere.

$\mathbb{C}^{n+1}$, the unit sphere in $\mathbb{C}^{n+1}$, is defined by

$$\Sigma^n = \{(z_1, \ldots, z_{n+1}) \in \mathbb{C}^{n+1} \mid z_1\overline{z}_1 + \cdots + z_{n+1}\overline{z}_{n+1} = 1\}.$$

For any line $L = [u]$, where $u \in \mathbb{C}^{n+1}$ is a nonzero vector, writing $u = (u_1, \ldots, u_{n+1})$, a point $z \in \mathbb{C}^{n+1}$ belongs to L iff $z = \lambda(u_1, \ldots, u_{n+1})$, for some $\lambda \in \mathbb{C}$. Therefore, the intersection $L \cap \Sigma^n$ of the line L and the sphere Σ^n is given by

$$L \cap \Sigma^n = \{\lambda(u_1, \ldots, u_{n+1}) \in \mathbb{C}^{n+1} \mid \lambda \in \mathbb{C}, \lambda\overline{\lambda}(u_1\overline{u}_1 + \cdots + u_{n+1}\overline{u}_{n+1}) = 1\},$$

i.e.,

$$L \cap \Sigma^n = \left\{ \lambda(u_1, \ldots, u_{n+1}) \in \mathbb{C}^{n+1} \;\middle|\; \lambda \in \mathbb{C}, |\lambda| = \frac{1}{\sqrt{|u_1|^2 + \cdots + |u_{n+1}|^2}} \right\}.$$

Thus, we see that there is a bijection between $L \cap \Sigma^n$ and the circle S^1; that is, geometrically $L \cap \Sigma^n$ is a circle. Moreover, since any line L through the origin is determined by just one other point, we see that for any two lines L_1 and L_2 through the origin,

$$L_1 \neq L_2 \quad \text{iff} \quad (L_1 \cap \Sigma^n) \cap (L_2 \cap \Sigma^n) = \emptyset.$$

However, Σ^n is the sphere S^{2n+1} in $\mathbb{R}^{2n+2}$. It follows that $\mathbb{CP}^n$ is the quotient of S^{2n+1} by the equivalence relation $\sim$ defined such that

$$y \sim z \quad \text{iff} \quad y, z \in L \cap \Sigma^n, \quad \text{for some line, } L, \text{ through the origin.}$$

Therefore, we can write

$$S^{2n+1}/S^1 \cong \mathbb{CP}^n.$$

The case $n = 1$ is particularly interesting, as it turns out that

$$S^3/S^1 \cong S^2.$$

This is the famous *Hopf fibration*. To show this, proceed as follows: As

$$S^3 \cong \Sigma^1 = \{(z, z') \in \mathbb{C}^2 \mid |z|^2 + |z'|^2 = 1\},$$

define a map, $\mathrm{HF}\colon S^3 \to S^2$, by

$$\mathrm{HF}((z, z')) = (2z\overline{z'}, |z|^2 - |z'|^2).$$

We leave as a homework exercise to prove that this map has range S^2 and that

$$\mathrm{HF}((z_1, z_1')) = \mathrm{HF}((z_2, z_2')) \text{ iff } (z_1, z_1') = \lambda(z_2, z_2'), \text{ for some } \lambda \text{ with } |\lambda| = 1.$$

In other words, for any point, $p \in S^2$, the inverse image $\mathrm{HF}^{-1}(p)$ (also called *fibre* over p) is a circle on S^3. Consequently, S^3 can be viewed as the union of a family of disjoint circles. This is the *Hopf fibration*. It is possible to visualize the Hopf fibration using the stereographic projection from S^3 onto $\mathbb{R}^3$. This is a beautiful and puzzling picture. For example, see Berger [Berger (1990)]. Therefore, HF induces a bijection from $\mathbb{CP}^1$ to S^2, and it is a homeomorphism.

Example C.13. Affine spaces.

Let X be a set and E a real vector space. A transitive and faithful action $\cdot\colon E \times X \to X$ of the additive group of E on X makes X into an *affine space*. The intuition is that the members of E are translations.

Those familiar with affine spaces as in Gallier [Gallier (2011)] (Chapter 2) or Berger [Berger (1990)] will point out that if X is an affine space, then not only is the action of E on X transitive, but more is true: For any two points $a, b \in E$, there is a *unique* vector $u \in E$, such that $u \cdot a = b$. By the way, the action of E on X is usually considered to be a right action and is written additively, so $u \cdot a$ is written $a + u$ (the result of translating a by u). Thus, it would seem that we have to require more of our action. However, this is not necessary because E (under addition) is *abelian*. More precisely, we have the proposition

Proposition C.13. *If G is an abelian group acting on a set X and the action $\cdot\colon G \times X \to X$ is transitive and faithful, then for any two elements $x, y \in X$, there is a unique $g \in G$ so that $g \cdot x = y$ (the action is simply transitive).*

Proof. Since our action is transitive, there is at least some $g \in G$ so that $g \cdot x = y$. Assume that we have $g_1, g_2 \in G$ with

$$g_1 \cdot x = g_2 \cdot x = y.$$

We shall prove that

$$g_1 \cdot z = g_2 \cdot z, \quad \text{for all } z \in X.$$

This implies that

$$g_1 g_2^{-1} \cdot z = z, \quad \text{for all } z \in X.$$

As our action is faithful, $g_1 g_2^{-1} = 1$, and we must have $g_1 = g_2$, which proves our proposition.

Pick any $z \in X$. As our action is transitive, there is some $h \in G$ so that $z = h \cdot x$. Then, we have

$$\begin{aligned}
g_1 \cdot z &= g_1 \cdot (h \cdot x) \\
&= (g_1 h) \cdot x \\
&= (h g_1) \cdot x \qquad \text{(since } G \text{ is abelian)} \\
&= h \cdot (g_1 \cdot x) \\
&= h \cdot (g_2 \cdot x) \qquad \text{(since} \quad g_1 \cdot x = g_2 \cdot x) \\
&= (h g_2) \cdot x \\
&= (g_2 h) \cdot x \qquad \text{(since } G \text{ is abelian)} \\
&= g_2 \cdot (h \cdot x) \\
&= g_2 \cdot z.
\end{aligned}$$

Therefore, $g_1 \cdot z = g_2 \cdot z$ for all $z \in X$, as claimed. ☐

C.3 Group Actions: Part II, Stabilizers and Homogeneous Spaces

Now that we have an understanding of how a group G acts on a set X, we may use this action to form new topological spaces, namely homogeneous spaces. In the construction of homogeneous spaces, the subset of group elements that leaves some given element $x \in X$ fixed plays an important role.

Definition C.15. Given an action $\cdot\colon G \times X \to X$ of a group G on a set X, for any $x \in X$, the group G_x (also denoted $\mathrm{Stab}_G(x)$), called the *stabilizer* of x or *isotropy group at* x, is given by

$$G_x = \{g \in G \mid g \cdot x = x\}.$$

We have to verify that G_x is indeed a subgroup of G, but this is easy. Indeed, if $g \cdot x = x$ and $h \cdot x = x$, then we also have $h^{-1} \cdot x = x$ and so, we get $gh^{-1} \cdot x = x$, proving that G_x is a subgroup of G. In general, G_x is **not** a normal subgroup.

Observe that

$$G_{g \cdot x} = gG_xg^{-1},$$

for all $g \in G$ and all $x \in X$. Indeed,

$$\begin{aligned} G_{g\cdot x} &= \{h \in G \mid h \cdot (g \cdot x) = g \cdot x\} \\ &= \{h \in G \mid hg \cdot x = g \cdot x\} \\ &= \{h \in G \mid g^{-1}hg \cdot x = x\}, \end{aligned}$$

which shows $g^{-1}G_{g\cdot x}g \subseteq G_x$, or equivalently that $G_{g\cdot x} \subseteq gG_xg^{-1}$. It remains to show that $gG_xg^{-1} \subseteq G_{g\cdot x}$. Take an element of gG_xg^{-1}, which has the form ghg^{-1} with $h \cdot x = x$. Since $h \cdot x = x$, we have $(ghg^{-1}) \cdot gx = gx$, which shows that $ghg^{-1} \in G_{g\cdot x}$.

Because $G_{g\cdot x} = gG_xg^{-1}$, the stabilizers of x and $g \cdot x$ are conjugate of each other.

When the action of G on X is transitive, for any fixed $x \in G$, the set X is a quotient (as a set, not as group) of G by G_x. Indeed, we can define the map, $\pi_x\colon G \to X$, by

$$\pi_x(g) = g \cdot x, \quad \text{for all } g \in G.$$

Observe that

$$\pi_x(gG_x) = (gG_x) \cdot x = g \cdot (G_x \cdot x) = g \cdot x = \pi_x(g).$$

This shows that $\pi_x \colon G \to X$ induces a quotient map $\overline{\pi}_x \colon G/G_x \to X$, from the set G/G_x of (left) cosets of G_x to X, defined by

$$\overline{\pi}_x(gG_x) = g \cdot x.$$

Since

$$\begin{aligned} &\pi_x(g) = \pi_x(h) \quad \text{iff} \quad g \cdot x = h \cdot x \quad \text{iff} \quad g^{-1}h \cdot x = x \\ &\text{iff} \quad g^{-1}h \in G_x \quad \text{iff} \quad gG_x = hG_x, \end{aligned}$$

we deduce that $\overline{\pi}_x \colon G/G_x \to X$ is injective. However, since our action is transitive, for every $y \in X$, there is some $g \in G$ so that $g \cdot x = y$, and so $\overline{\pi}_x(gG_x) = g \cdot x = y$; that is, the map $\overline{\pi}_x$ is also surjective. Therefore, the map $\overline{\pi}_x \colon G/G_x \to X$ is a bijection (of sets, not groups). The map $\pi_x \colon G \to X$ is also surjective. Let us record this important fact as

Proposition C.14. *If $\cdot \colon G \times X \to X$ is a transitive action of a group G on a set X, for every fixed $x \in X$, the surjection $\pi_x \colon G \to X$ given by*

$$\pi_x(g) = g \cdot x$$

induces a bijection

$$\overline{\pi}_x \colon G/G_x \to X,$$

where G_x is the stabilizer of x. See Figure C.5.

The map $\pi_x \colon G \to X$ (corresponding to a fixed $x \in X$) is sometimes called a *projection* of G onto X. Proposition C.14 shows that for every $y \in X$, the subset $\pi_x^{-1}(y)$ of G (called the *fibre above* y) is equal to some coset gG_x of G, and thus is in bijection with the group G_x itself. We can think of G as a moving family of fibres G_x parametrized by X. This point of view of viewing a space as a moving family of simpler spaces is typical in (algebraic) geometry, and underlies the notion of (principal) fibre bundle.

Note that if the action $\cdot \colon G \times X \to X$ is transitive, then the stabilizers G_x and G_y of any two elements $x, y \in X$ are isomorphic, as they as conjugates. Thus, in this case, it is enough to compute one of these stabilizers for a "convenient" x.

As the situation of Proposition C.14 is of particular interest, we make the following definition:

Definition C.16. A set X is said to be a *homogeneous space* if there is a transitive action $\cdot \colon G \times X \to X$ of some group G on X.

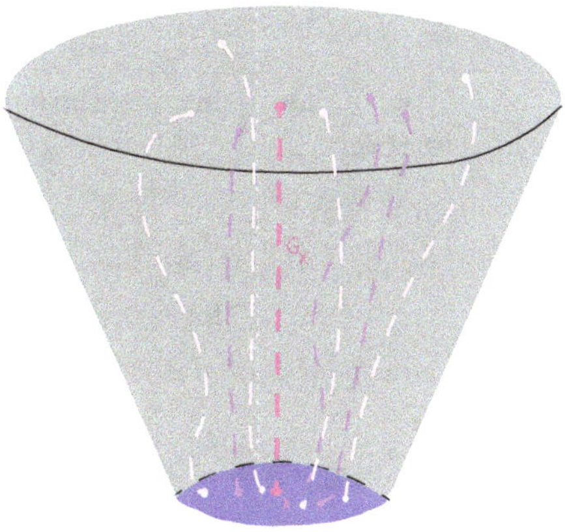

Fig. C.5 A schematic representation of $G/G_x \cong X$, where G is the gray solid, X is its purple circular base, and G_x is the pink vertical strand. The dotted strands are the fibres gG_x.

We see that all the spaces of Examples C.6–C.13, are homogeneous spaces. Another example that will play an important role when we deal with Lie groups is the situation where we have a group G, a subgroup H of G (not necessarily normal), and where $X = G/H$, the set of left cosets of G modulo H. The group G acts on G/H by left multiplication:

$$a \cdot (gH) = (ag)H,$$

where $a, g \in G$. This action is clearly transitive and one checks that the stabilizer of gH is gHg^{-1}. If G is a topological group and H is a closed subgroup of G (see later for an explanation), it turns out that G/H is Hausdorff. If G is a Lie group, we obtain a manifold.

Even if G and X are topological spaces and the action $\cdot\colon G \times X \to X$ is continuous, in general, the space G/G_x under the quotient topology is **not** homeomorphic to X.

We will give later sufficient conditions that insure that X is indeed a topological space or even a manifold. In particular, X will be a manifold when G is a Lie group.

In general, an action $\cdot\colon G \times X \to X$ is not transitive on X, but for every $x \in X$, it is transitive on the set

$$O(x) = G \cdot x = \{g \cdot x \mid g \in G\}.$$

Such a set is called the *orbit* of x. The orbits are the equivalence classes of the following equivalence relation:

Definition C.17. Given an action $\cdot\colon G \times X \to X$ of some group G on X, the equivalence relation $\sim$ on X is defined so that, for all $x, y \in X$,

$$x \sim y \quad \text{iff} \quad y = g \cdot x, \quad \text{for some } g \in G.$$

For every $x \in X$, the equivalence class of x is the *orbit of* x, denoted $O(x)$ or $G \cdot x$, with

$$G \cdot x = O(x) = \{g \cdot x \mid g \in G\}.$$

The set of orbits is denoted X/G.

We warn the reader that some authors use the notation $G\backslash X$ for the the set of orbits $G \cdot x$, by analogy with right cosets Hg of a subgroup H of G.

The orbit space X/G is obtained from X by an identification (or merging) process: For every orbit, all points in that orbit are merged into a single point. This is akin to the process of forming the identification topology. For example, if $X = S^2$ and G is the group consisting of the restrictions of the two linear maps I and $-I$ of $\mathbb{R}^3$ to S^2 (where $(-I)(x) = -x$ for all $x \in \mathbb{R}^3$), then

$$X/G = S^2/\{I, -I\} \cong \mathbb{RP}^2.$$

See Figure C.4. More generally, if S^n is the n-sphere in $\mathbb{R}^{n+1}$, then we have a bijection between the orbit space $S^n/\{I, -I\}$ and $\mathbb{RP}^n$:

$$S^n/\{I, -I\} \cong \mathbb{RP}^n.$$

Many manifolds can be obtained in this fashion, including the torus, the Klein bottle, the Möbius band, *etc.*

Since the action of G is transitive on $O(x)$, by Proposition C.14, we see that for every $x \in X$, we have a bijection

$$O(x) \cong G/G_x.$$

As a corollary, if both X and G are finite, for any set $A \subseteq X$ of representatives from every orbit, we have the *orbit formula*:

$$|X| = \sum_{a \in A} [G\colon G_a] = \sum_{a \in A} |G|/|G_a|.$$

Even if a group action $\cdot\colon G \times X \to X$ is not transitive, when X is a manifold, we can consider the set of orbits X/G, and if the action of G on X satisfies certain conditions, X/G is actually a manifold. Manifolds arising in this fashion are often called *orbifolds*. In summary, we see that manifolds arise in at least two ways from a group action:

(1) As homogeneous spaces G/G_x, if the action is transitive.
(2) As orbifolds X/G (under certain conditions on the action).

Of course, in both cases, the action must satisfy some additional properties.

For the rest of this section, we reconsider Examples C.6–C.13 in the context of homogeneous space by determining some stabilizers for those actions.

(a) Consider the action $\cdot\colon \mathbf{SO}(n) \times S^{n-1} \to S^{n-1}$ of $\mathbf{SO}(n)$ on the sphere S^{n-1} $(n \geq 1)$ defined in Example C.6. Since this action is transitive, we can determine the stabilizer of any convenient element of S^{n-1}, say $e_1 = (1, 0, \ldots, 0)$. In order for any $R \in \mathbf{SO}(n)$ to leave e_1 fixed, the first column of R must be e_1, so R is an orthogonal matrix of the form

$$R = \begin{pmatrix} 1 & U \\ 0 & S \end{pmatrix}, \quad \text{with} \quad \det(S) = 1,$$

where U is a $1 \times (n-1)$ row vector. As the rows of R must be unit vectors, we see that $U = 0$ and $S \in \mathbf{SO}(n-1)$. Therefore, the stabilizer of e_1 is isomorphic to $\mathbf{SO}(n-1)$, and we deduce the bijection

$$\mathbf{SO}(n)/\mathbf{SO}(n-1) \cong S^{n-1}.$$

Strictly speaking, $\mathbf{SO}(n-1)$ is not a subgroup of $\mathbf{SO}(n)$, and in all rigor, we should consider the subgroup $\widetilde{\mathbf{SO}}(n-1)$ of $\mathbf{SO}(n)$ consisting of all matrices of the form

$$\begin{pmatrix} 1 & 0 \\ 0 & S \end{pmatrix}, \quad \text{with} \quad \det(S) = 1,$$

and write

$$\mathbf{SO}(n)/\widetilde{\mathbf{SO}}(n-1) \cong S^{n-1}.$$

However, it is common practice to identify $\mathbf{SO}(n-1)$ with $\widetilde{\mathbf{SO}}(n-1)$.

When $n = 2$, as $\mathbf{SO}(1) = \{1\}$, we find that $\mathbf{SO}(2) \cong S^1$, a circle, a fact that we already knew. When $n = 3$, we find that $\mathbf{SO}(3)/\mathbf{SO}(2) \cong S^2$. This says that $\mathbf{SO}(3)$ is somehow the result of glueing circles to the surface of a sphere (in $\mathbb{R}^3$), in such a way that these circles do not intersect. This is hard to visualize!

A similar argument for the complex unit sphere Σ^{n-1} shows that

$$\mathbf{SU}(n)/\mathbf{SU}(n-1) \cong \Sigma^{n-1} \cong S^{2n-1}.$$

Again, we identify $\mathbf{SU}(n-1)$ with a subgroup of $\mathbf{SU}(n)$, as in the real case. In particular, when $n = 2$, as $\mathbf{SU}(1) = \{1\}$, we find that

$$\mathbf{SU}(2) \cong S^3;$$

that is, the group $\mathbf{SU}(2)$ is topologically the sphere S^3! Actually, this is not surprising if we remember that $\mathbf{SU}(2)$ is in fact the group of unit quaternions.

(b) We saw in Example C.7 that the action $\cdot\colon \mathbf{SL}(2,\mathbb{R}) \times H \to H$ of the group $\mathbf{SL}(2,\mathbb{R})$ on the upper half plane is transitive. Let us find out what the stabilizer of $z = i$ is. We should have

$$\frac{ai+b}{ci+d} = i,$$

that is, $ai + b = -c + di$, i.e.,

$$(d-a)i = b + c.$$

Since a, b, c, d are real, we must have $d = a$ and $b = -c$. Moreover, $ad - bc = 1$, so we get $a^2 + b^2 = 1$. We conclude that a matrix in $\mathbf{SL}(2,\mathbb{R})$ fixes i iff it is of the form

$$\begin{pmatrix} a & -b \\ b & a \end{pmatrix}, \quad \text{with} \quad a^2 + b^2 = 1.$$

Clearly, these are the rotation matrices in $\mathbf{SO}(2)$, and so the stabilizer of i is $\mathbf{SO}(2)$. We conclude that

$$\mathbf{SL}(2,\mathbb{R})/\mathbf{SO}(2) \cong H.$$

This time we can view $\mathbf{SL}(2,\mathbb{R})$ as the result of glueing circles to the upper half plane. This is not so easy to visualize. There is a better way to visualize the topology of $\mathbf{SL}(2,\mathbb{R})$ by making it act on the open disk D. We will return to this action in a little while.

(c) Now consider the action of $\mathbf{SL}(2,\mathbb{C})$ on $\mathbb{C} \cup \{\infty\} \cong S^2$ given in Example C.8. As it is transitive, let us find the stabilizer of $z = 0$. We must have

$$\frac{b}{d} = 0,$$

and as $ad - bc = 1$, we must have $b = 0$ and $ad = 1$. Thus the stabilizer of 0 is the subgroup $\mathbf{SL}(2,\mathbb{C})_0$ of $\mathbf{SL}(2,\mathbb{C})$ consisting of all matrices of the form

$$\begin{pmatrix} a & 0 \\ c & a^{-1} \end{pmatrix}, \quad \text{where} \quad a \in \mathbb{C} - \{0\} \quad \text{and} \quad c \in \mathbb{C}.$$

We get

$$\mathbf{SL}(2,\mathbb{C})/\mathbf{SL}(2,\mathbb{C})_0 \cong \mathbb{C} \cup \{\infty\} \cong S^2,$$

but this is not very illuminating.

(d) In Example C.11 we considered the action $\cdot\colon \mathbf{GL}(n) \times \mathbf{SPD}(n) \to \mathbf{SPD}(n)$ of $\mathbf{GL}(n)$ on $\mathbf{SPD}(n)$, the set of symmetric positive definite matrices. As this action is transitive, let us find the stabilizer of I. For any $A \in \mathbf{GL}(n)$, the matrix A stabilizes I iff

$$AIA^\top = AA^\top = I.$$

Therefore the stabilizer of I is $\mathbf{O}(n)$, and we find that

$$\mathbf{GL}(n)/\mathbf{O}(n) = \mathbf{SPD}(n).$$

Observe that if $\mathbf{GL}^+(n)$ denotes the subgroup of $\mathbf{GL}(n)$ consisting of all matrices with a strictly positive determinant, then we have an action $\cdot\colon \mathbf{GL}^+(n) \times \mathbf{SPD}(n) \to \mathbf{SPD}(n)$ of $\mathbf{GL}^+(n)$ on $\mathbf{SPD}(n)$. This action is transitive and we find that the stabilizer of I is $\mathbf{SO}(n)$; consequently, we get

$$\mathbf{GL}^+(n)/\mathbf{SO}(n) = \mathbf{SPD}(n).$$

(e) In Example C.12 we considered the action $\cdot\colon \mathbf{SO}(n+1) \times \mathbb{RP}^n \to \mathbb{RP}^n$ of $\mathbf{SO}(n+1)$ on the (real) projective space $\mathbb{RP}^n$. As this action is transitive, let us find the stabilizer of the line $L = [e_1]$, where $e_1 = (1, 0, \ldots, 0)$. For any $R \in \mathbf{SO}(n+1)$, the line L is fixed iff either $R(e_1) = e_1$ or $R(e_1) = -e_1$, since e_1 and $-e_1$ define the same line. As R is orthogonal with $\det(R) = 1$, this means that R is of the form

$$R = \begin{pmatrix} \alpha & 0 \\ 0 & S \end{pmatrix}, \quad \text{with} \quad \alpha = \pm 1 \quad \text{and} \quad \det(S) = \alpha.$$

But, S must be orthogonal, so we conclude $S \in \mathbf{O}(n)$. Therefore, the stabilizer of $L = [e_1]$ is isomorphic to the group $\mathbf{O}(n)$, and we find that

$$\mathbf{SO}(n+1)/\mathbf{O}(n) \cong \mathbb{RP}^n.$$

Strictly speaking, $\mathbf{O}(n)$ is not a subgroup of $\mathbf{SO}(n+1)$, so the above equation does not make sense. We should write

$$\mathbf{SO}(n+1)/\widetilde{\mathbf{O}}(n) \cong \mathbb{RP}^n,$$

where $\widetilde{\mathbf{O}}(n)$ is the subgroup of $\mathbf{SO}(n+1)$ consisting of all matrices of the form

$$\begin{pmatrix} \alpha & 0 \\ 0 & S \end{pmatrix}, \quad \text{with} \quad S \in \mathbf{O}(n),\ \alpha = \pm 1 \quad \text{and} \quad \det(S) = \alpha.$$

This group is also denoted $S(\mathbf{O}(1) \times \mathbf{O}(n))$. However, the common practice is to write $\mathbf{O}(n)$ instead of $S(\mathbf{O}(1) \times \mathbf{O}(n))$.

We should mention that $\mathbb{RP}^3$ and $\mathbf{SO}(3)$ are homeomorphic spaces. This is shown using the quaternions; for example, see Gallier [Gallier (2011)], Chapter 8.

A similar argument applies to the action $\cdot\colon \mathbf{SU}(n+1) \times \mathbb{CP}^n \to \mathbb{CP}^n$ of $\mathbf{SU}(n+1)$ on the (complex) projective space $\mathbb{CP}^n$. We find that

$$\mathbf{SU}(n+1)/\mathbf{U}(n) \cong \mathbb{CP}^n.$$

Again, the above is a bit sloppy as $\mathbf{U}(n)$ is not a subgroup of $\mathbf{SU}(n+1)$. To be rigorous, we should use the subgroup $\widetilde{\mathbf{U}}(n)$ consisting of all matrices of the form

$$\begin{pmatrix} \alpha & 0 \\ 0 & S \end{pmatrix}, \quad \text{with} \quad S \in \mathbf{U}(n),\ |\alpha| = 1 \quad \text{and} \quad \det(S) = \overline{\alpha}.$$

This group is also denoted $S(\mathbf{U}(1) \times \mathbf{U}(n))$. The common practice is to write $\mathbf{U}(n)$ instead of $S(\mathbf{U}(1) \times \mathbf{U}(n))$. In particular, when $n = 1$, we find that

$$\mathbf{SU}(2)/\mathbf{U}(1) \cong \mathbb{CP}^1.$$

But, we know that $\mathbf{SU}(2) \cong S^3$, and clearly $\mathbf{U}(1) \cong S^1$. So, again, we find that $S^3/S^1 \cong \mathbb{CP}^1$ (we know more, namely, $S^3/S^1 \cong S^2 \cong \mathbb{CP}^1$.)

Observe that $\mathbb{CP}^n$ can also be viewed as the orbit space of the action $\cdot\colon S^1 \times S^{2n+1} \to S^{2n+1}$ given by

$$\lambda \cdot (z_1, \ldots, z_{n+1}) = (\lambda z_1, \ldots, \lambda z_{n+1}),$$

where $S^1 = \mathbf{U}(1)$ (the group of complex numbers of modulus 1) and S^{2n+1} is identified with Σ^n.

We now return to Case (b) to give a better picture of $\mathbf{SL}(2, \mathbb{R})$. Instead of having $\mathbf{SL}(2, \mathbb{R})$ act on the upper half plane, we define an action of

$\mathbf{SL}(2,\mathbb{R})$ on the open unit disk D as we did in Example C.9. Technically, it is easier to consider the group $\mathbf{SU}(1,1)$, which is isomorphic to $\mathbf{SL}(2,\mathbb{R})$, and to make $\mathbf{SU}(1,1)$ act on D. The group $\mathbf{SU}(1,1)$ is the group of 2×2 complex matrices of the form

$$\begin{pmatrix} a & b \\ \overline{b} & \overline{a} \end{pmatrix}, \quad \text{with} \quad a\overline{a} - b\overline{b} = 1.$$

The reader should check that if we let

$$g = \begin{pmatrix} 1 & -i \\ 1 & i \end{pmatrix},$$

then the map from $\mathbf{SL}(2,\mathbb{R})$ to $\mathbf{SU}(1,1)$ given by

$$A \mapsto gAg^{-1}$$

is an isomorphism. Observe that the scaled Möbius transformation associated with g is

$$z \mapsto \frac{z-i}{z+i},$$

which is the holomorphic isomorphism mapping H to D mentionned earlier! We can define a bijection between $\mathbf{SU}(1,1)$ and $S^1 \times D$ given by

$$\begin{pmatrix} a & b \\ \overline{b} & \overline{a} \end{pmatrix} \mapsto (a/|a|, b/a).$$

We conclude that $\mathbf{SL}(2,\mathbb{R}) \cong \mathbf{SU}(1,1)$ is topologically an open solid torus (i.e., with the surface of the torus removed). It is possible to further classify the elements of $\mathbf{SL}(2,\mathbb{R})$ into three categories and to have geometric interpretations of these as certain regions of the torus. For details, the reader should consult Carter, Segal and Macdonald [Carter *et al.* (1995)] or Duistermatt and Kolk [Duistermaat and Kolk (2000)] (Chapter 1, Section 1.2).

The group $\mathbf{SU}(1,1)$ acts on D by interpreting any matrix in $\mathbf{SU}(1,1)$ as a Möbius tranformation; that is,

$$\begin{pmatrix} a & b \\ \overline{b} & \overline{a} \end{pmatrix} \mapsto \left(z \mapsto \frac{az+b}{\overline{b}z+\overline{a}} \right).$$

The reader should check that these transformations preserve D.

Both the upper half-plane and the open disk are models of Lobachevsky's non-Euclidean geometry (where the parallel postulate fails). They are also models of hyperbolic spaces (Riemannian manifolds with constant negative curvature, see Gallot, Hulin and Lafontaine [Gallot *et al.* (1993)], Chapter III).

According to Dubrovin, Fomenko, and Novikov [Dubrovin *et al.* (1985)] (Chapter 2, Section 13.2), the open disk model is due to Poincaré and the upper half-plane model to Klein, although Poincaré was the first to realize that the upper half-plane is a hyperbolic space.

C.4 The Grassmann and Stiefel Manifolds

In this section we introduce two very important homogeneous manifolds, the Grassmann manifolds and the Stiefel manifolds. The Grassmann manifolds are generalizations of projective spaces (real and complex), while the Stiefel manifold are generalizations of $\mathbf{O}(n)$. Both of these manifolds are examples of reductive homogeneous spaces. We begin by defining the Grassmann manifolds $G(k,n)$.

First consider the real case.

Definition C.18. Given any $n \geq 1$, for any k with $0 \leq k \leq n$, the set $G(k,n)$ of all linear k-dimensional subspaces of $\mathbb{R}^n$ (also called *k-planes*) is called a *Grassmannian* (or *Grassmann manifold*).

Any k-dimensional subspace U of $\mathbb{R}^n$ is spanned by k linearly independent vectors $u_1, \ldots, u_k$ in $\mathbb{R}^n$; write $U = \mathrm{span}(u_1, \ldots, u_k)$. We can define an action $\cdot\colon \mathbf{O}(n) \times G(k,n) \to G(k,n)$ as follows: For any $R \in \mathbf{O}(n)$, for any $U = \mathrm{span}(u_1, \ldots, u_k)$, let

$$R \cdot U = \mathrm{span}(Ru_1, \ldots, Ru_k).$$

We have to check that the above is well defined. If $U = \mathrm{span}(v_1, \ldots, v_k)$ for any other k linearly independent vectors $v_1, \ldots, v_k$, we have

$$v_i = \sum_{j=1}^{k} a_{ij} u_j, \quad 1 \leq i \leq k,$$

for some $a_{ij} \in \mathbb{R}$, and so

$$Rv_i = \sum_{j=1}^{k} a_{ij} Ru_j, \quad 1 \leq i \leq k,$$

which shows that

$$\mathrm{span}(Ru_1, \ldots, Ru_k) = \mathrm{span}(Rv_1, \ldots, Rv_k);$$

that is, the above action is well defined.

We claim this action is transitive. This is because if U and V are any two k-planes, we may assume that $U = \mathrm{span}(u_1, \ldots, u_k)$ and

$V = \text{span}(v_1, \ldots, v_k)$, where the u_i's form an orthonormal family and similarly for the v_i's. Then we can extend these families to orthonormal bases $(u_1, \ldots, u_n)$ and $(v_1, \ldots, v_n)$ on $\mathbb{R}^n$, and w.r.t. the orthonormal basis $(u_1, \ldots, u_n)$, the matrix of the linear map sending u_i to v_i is orthogonal. Hence $G(k, n)$ is a homogeneous space.

In order to represent $G(k, n)$ as a quotient space, Proposition C.14 implies it is enough to find the stabilizer of any k-plane. Pick $U = \text{span}(e_1, \ldots, e_k)$, where $(e_1, \ldots, e_n)$ is the canonical basis of $\mathbb{R}^n$ (i.e., $e_i = (0, \ldots, 0, 1, 0, \ldots, 0)$, with the 1 in the ith position). Any $R \in \mathbf{O}(n)$ stabilizes U iff R maps $e_1, \ldots, e_k$ to k linearly independent vectors in the subspace $U = \text{span}(e_1, \ldots, e_k)$, i.e., R is of the form

$$R = \begin{pmatrix} S & 0 \\ 0 & T \end{pmatrix},$$

where S is $k \times k$ and T is $(n-k) \times (n-k)$. Moreover, as R is orthogonal, S and T must be orthogonal, that is $S \in \mathbf{O}(k)$ and $T \in \mathbf{O}(n-k)$. We deduce that the stabilizer of U is isomorphic to $\mathbf{O}(k) \times \mathbf{O}(n-k)$ and we find that

$$\mathbf{O}(n)/(\mathbf{O}(k) \times \mathbf{O}(n-k)) \cong G(k, n).$$

It turns out that this makes $G(k, n)$ into a smooth manifold of dimension

$$\frac{n(n-1)}{2} - \frac{k(k-1)}{2} - \frac{(n-k)(n-k-1)}{2} = k(n-k)$$

called a *Grassmannian*.

The restriction of the action of $\mathbf{O}(n)$ on $G(k, n)$ to $\mathbf{SO}(n)$ yields an action $\cdot\colon \mathbf{SO}(n) \times G(k, n) \to G(k, n)$ of $\mathbf{SO}(n)$ on $G(k, n)$. Then it is easy to see that this action is transitive and that the stabilizer of the subspace U is isomorphic to the subgroup $(\mathbf{O}(k) \times \mathbf{O}(n-k))$ of $\mathbf{SO}(n)$ consisting of the rotations of the form

$$R = \begin{pmatrix} S & 0 \\ 0 & T \end{pmatrix},$$

with $S \in \mathbf{O}(k)$, $T \in \mathbf{O}(n-k)$ and $\det(S)\det(T) = 1$. Thus, we also have

$$\mathbf{SO}(n)/S(\mathbf{O}(k) \times \mathbf{O}(n-k)) \cong G(k, n).$$

If we recall the projection map of Example C.12 in Section C.2, namely $pr\colon \mathbb{R}^{n+1} - \{0\} \to \mathbb{RP}^n$, by definition, a k-*plane* in $\mathbb{RP}^n$ is the image under pr of any $(k+1)$-plane in $\mathbb{R}^{n+1}$. So, for example, a line in $\mathbb{RP}^n$ is the image of a 2-plane in $\mathbb{R}^{n+1}$, and a hyperplane in $\mathbb{RP}^n$ is the image of a hyperplane in $\mathbb{R}^{n+1}$. The advantage of this point of view is that the k-planes in $\mathbb{RP}^n$

are arbitrary; that is, they do not have to go through "the origin" (which does not make sense, anyway!). Then we see that we can interpret the Grassmannian, $G(k+1, n+1)$, as a space of "parameters" for the k-planes in $\mathbb{RP}^n$. For example, $G(2, n+1)$ parametrizes the lines in $\mathbb{RP}^n$. In this viewpoint, $G(k+1, n+1)$ is usually denoted $\mathbb{G}(k, n)$.

It can be proved (using some exterior algebra) that $G(k, n)$ can be embedded in $\mathbb{RP}^{\binom{n}{k}-1}$. Much more is true. For example, $G(k, n)$ is a projective variety, which means that it can be defined as a subset of $\mathbb{RP}^{\binom{n}{k}-1}$ equal to the zero locus of a set of homogeneous equations. There is even a set of quadratic equations known as the *Plücker equations* defining $G(k, n)$. In particular, when $n = 4$ and $k = 2$, we have $G(2, 4) \subseteq \mathbb{RP}^5$, and $G(2, 4)$ is defined by a single equation of degree 2. The Grassmannian $G(2, 4) = \mathbb{G}(1, 3)$ is known as the *Klein quadric*. This hypersurface in $\mathbb{RP}^5$ parametrizes the lines in $\mathbb{RP}^3$.

Complex Grassmannians are defined in a similar way, by replacing $\mathbb{R}$ by $\mathbb{C}$ and $\mathbf{O}(n)$ by $\mathbf{U}(n)$ throughout. The complex Grassmannian $G_{\mathbb{C}}(k, n)$ is a complex manifold as well as a real manifold, and we have

$$\mathbf{U}(n)/(\mathbf{U}(k) \times \mathbf{U}(n-k)) \cong G_{\mathbb{C}}(k, n).$$

As in the case of the real Grassmannians, the action of $\mathbf{U}(n)$ on $G_{\mathbb{C}}(k, n)$ yields an action of $\mathbf{SU}(n)$ on $G_{\mathbb{C}}(k, n)$, and we get

$$\mathbf{SU}(n)/S(\mathbf{U}(k) \times \mathbf{U}(n-k)) \cong G_{\mathbb{C}}(k, n),$$

where $S(\mathbf{U}(k) \times \mathbf{U}(n-k))$ is the subgroup of $\mathbf{SU}(n)$ consisting of all matrices $R \in \mathbf{SU}(n)$ of the form

$$R = \begin{pmatrix} S & 0 \\ 0 & T \end{pmatrix},$$

with $S \in \mathbf{U}(k)$, $T \in \mathbf{U}(n-k)$ and $\det(S)\det(T) = 1$.

Closely related to Grassmannians are the *Stiefel manifolds* $S(k, n)$. Again we begin with the real case.

Definition C.19. For any $n \geq 1$ and any k with $1 \leq k \leq n$, the set $S(k, n)$ of all orthonormal k-frames, that is, of k-tuples of orthonormal vectors $(u_1, \ldots, u_k)$ with $u_i \in \mathbb{R}^n$, is called a *Stiefel manifold*.

Obviously, $S(1, n) = S^{n-1}$ and $S(n, n) = \mathbf{O}(n)$, so assume $k \leq n-1$. There is a natural action $\cdot\colon \mathbf{SO}(n) \times S(k, n) \to S(k, n)$ of $\mathbf{SO}(n)$ on $S(k, n)$ given by

$$R \cdot (u_1, \ldots, u_k) = (Ru_1, \ldots, Ru_k).$$

This action is transitive, because if $(u_1, \ldots, u_k)$ and $(v_1, \ldots, v_k)$ are any two orthonormal k-frames, then they can be extended to orthonormal bases (for example, by Gram-Schmidt) $(u_1, \ldots, u_n)$ and $(v_1, \ldots, v_n)$ with the same orientation (since we can pick u_n and v_n so that our bases have the same orientation), and there is a unique orthogonal transformation $R \in \mathbf{SO}(n)$ such that $Ru_i = v_i$ for $i = 1, \ldots, n$.

In order to apply Proposition C.14, we need to find the stabilizer of the orthonormal k-frame $(e_1, \ldots, e_k)$ consisting of the first canonical basis vectors of $\mathbb{R}^n$. A matrix $R \in \mathbf{SO}(n)$ stabilizes $(e_1, \ldots, e_k)$ iff it is of the form

$$R = \begin{pmatrix} I_k & 0 \\ 0 & S \end{pmatrix}$$

where $S \in \mathbf{SO}(n-k)$. Therefore, for $1 \leq k \leq n-1$, we have

$$\mathbf{SO}(n)/\mathbf{SO}(n-k) \cong S(k, n).$$

This makes $S(k, n)$ a smooth manifold of dimension

$$\frac{n(n-1)}{2} - \frac{(n-k)(n-k-1)}{2} = nk - \frac{k(k+1)}{2} = k(n-k) + \frac{k(k-1)}{2}.$$

Remark: It should be noted that we can define another type of Stiefel manifolds, denoted by $V(k, n)$, using linearly independent k-tuples $(u_1, \ldots, u_k)$ that do not necessarily form an orthonormal system. In this case, there is an action $\cdot\colon \mathbf{GL}(n, \mathbb{R}) \times V(k, n) \to V(k, n)$, and the stabilizer H of the first k canonical basis vectors $(e_1, \ldots, e_k)$ is a closed subgroup of $\mathbf{GL}(n, \mathbb{R})$, but it doesn't have a simple description (see Warner [Warner (1983)], Chapter 3). We get an isomorphism

$$V(k, n) \cong \mathrm{GL}(n, \mathbb{R})/H.$$

The version of the Stiefel manifold $S(k, n)$ using orthonormal frames is sometimes denoted by $V^0(k, n)$ (Milnor and Stasheff [Milnor and Stasheff (1974)] use the notation $V^0_k(\mathbb{R}^n)$). Beware that the notation is not standardized. Certain authors use $V(k, n)$ for what we denote by $S(k, n)$!

Complex Stiefel manifolds are defined in a similar way by replacing $\mathbb{R}$ by $\mathbb{C}$ and $\mathbf{SO}(n)$ by $\mathbf{SU}(n)$. For $1 \leq k \leq n-1$, the complex Stiefel manifold $S_{\mathbb{C}}(k, n)$ is isomorphic to the quotient

$$\mathbf{SU}(n)/\mathbf{SU}(n-k) \cong S_{\mathbb{C}}(k, n).$$

If $k = 1$, we have $S_{\mathbb{C}}(1, n) = S^{2n-1}$, and if $k = n$, we have $S_{\mathbb{C}}(n, n) = \mathbf{U}(n)$.

The Grassmannians can also be viewed as quotient spaces of the Stiefel manifolds. Every orthonomal k-frame $(u_1, \ldots, u_k)$ can be represented by an $n \times k$ matrix Y over the canonical basis of $\mathbb{R}^n$, and such a matrix Y satisfies the equation

$$Y^\top Y = I.$$

We have a right action $\cdot \colon S(k,n) \times \mathbf{O}(k) \to S(k,n)$ given by

$$Y \cdot R = YR,$$

for any $R \in \mathbf{O}(k)$. This action is well defined since

$$(YR)^\top YR = R^\top Y^\top YR = I.$$

However, this action is not transitive (unless $k = 1$), but the orbit space $S(k,n)/\mathbf{O}(k)$ is isomorphic to the Grassmannian $G(k,n)$, so we can write

$$G(k,n) \cong S(k,n)/\mathbf{O}(k).$$

Similarly, the complex Grassmannian is isomorphic to the orbit space $S_{\mathbb{C}}(k,n)/\mathbf{U}(k)$:

$$G_{\mathbb{C}}(k,n) \cong S_{\mathbb{C}}(k,n)/\mathbf{U}(k).$$

Appendix D

Hilbert Spaces

D.1 The Projection Lemma, Duality

If E is a complex vector space, recall that a map $\langle -, - \rangle \colon E \times E \to \mathbb{C}$ is a *hermitian form* if it satisfies the following properties for all $x, y, x_1, x_2, y_1, y_2 \in E$ and all $\lambda \in \mathbb{C}$: it is *sesquilinear*, which means that

$$\begin{aligned} \langle x_1 + x_2, y \rangle &= \langle x_1, y \rangle + \langle x_2, y \rangle \\ \langle x, y_1 + y_2 \rangle &= \langle x, y_1 \rangle + \langle x, y_2 \rangle \\ \langle \lambda x, y \rangle &= \lambda \langle x, y \rangle \\ \langle x, \lambda y \rangle &= \overline{\lambda} \langle x, y \rangle, \end{aligned}$$

and satisfies the *hermitian property*,

$$\langle y, x \rangle = \overline{\langle x, y \rangle}.$$

The hermitian property implies that $\langle x, x \rangle \in \mathbb{R}$ for all $x \in E$.

A hermitian form $\langle -, - \rangle \colon E \times E \to \mathbb{C}$ is *positive* if

$$\langle x, x \rangle \geq 0 \qquad \text{for all } x \in E.$$

A positive hermitian form satisfies the *Cauchy-Schwarz inequality*:

$$|\langle x, y \rangle|^2 \leq \langle x, x \rangle \langle y, y \rangle, \qquad \text{for all } x, y \in E.$$

A positive hermitian form is *positive definite* if for all $x \in E$,

$$\langle x, x \rangle = 0 \quad \text{implies that } x = 0,$$

or equivalently,

$$\langle x, x \rangle > 0 \qquad \text{for all } x \neq 0.$$

A positive definite hermitian form on E is often called a *hermitian inner product* on E, and E is called a *hermitian space* (sometimes a *pre-Hilbert space*).

Given a hermitian space $\langle E, \langle -, - \rangle \rangle$, the function $\| \, \| \colon E \to \mathbb{R}$ defined such that $\|u\| = \sqrt{\langle u, u \rangle}$, is a norm on E. Thus, E is a normed vector space. If E is also complete, then it is a very interesting space.

In a hermitian space $\langle E, \langle -, - \rangle \rangle$, the inner product $\langle -, - \rangle$ can be recovered from the norm $\| \, \|$ using the following *polarization identities*: In the complex case,

$$\langle x, y \rangle = \frac{1}{4}(\|x+y\|^2 - \|x-y\|^2 + i\,\|x+iy\|^2 - i\,\|x-iy\|^2),$$

and in the real case,

$$\langle x, y \rangle = \frac{1}{2}(\|x+y\|^2 - \|x\|^2 - \|y\|^2).$$

Recall that completeness has to do with the convergence of Cauchy sequences. A normed vector space $\langle E, \| \, \| \rangle$ is automatically a metric space under the metric d defined such that $d(u, v) = \|v - u\|$ (see Chapter B for the definition of a normed vector space and of a metric space, or Lang [Lang (1996, 1997)], or Dixmier [Dixmier (1984)]). Given a metric space E with metric d, a sequence $(a_n)_{n \geq 1}$ of elements $a_n \in E$ is a *Cauchy sequence* iff for every $\epsilon > 0$, there is some $N \geq 1$ such that

$$d(a_m, a_n) < \epsilon \quad \text{for all} \quad m, n \geq N.$$

We say that E is *complete* iff every Cauchy sequence converges to a limit (which is unique, since a metric space is Hausdorff).

Every finite dimensional vector space over $\mathbb{R}$ or $\mathbb{C}$ is complete. For example, one can show by induction that given any basis $(e_1, \ldots, e_n)$ of E, the linear map $h \colon \mathbb{C}^n \to E$ defined such that

$$h((z_1, \ldots, z_n)) = z_1 e_1 + \cdots + z_n e_n$$

is a homeomorphism (using the *sup*-norm on $\mathbb{C}^n$). One can also use the fact that any two norms on a finite dimensional vector space over $\mathbb{R}$ or $\mathbb{C}$ are equivalent (see Theorem B.3, or Lang [Lang (1997)], Dixmier [Dixmier (1984)], Schwartz [Schwartz (1991)]).

However, if E has infinite dimension, it may not be complete. When a hermitian space is complete, a number of the properties that hold for finite dimensional hermitian spaces also hold for infinite dimensional spaces.

For example, any closed subspace has an orthogonal complement, and in particular, a finite dimensional subspace has an orthogonal complement. hermitian spaces that are also complete play an important role in analysis. Since they were first studied by Hilbert, they are called Hilbert spaces.

Definition D.1. A (complex) hermitian space $\langle E, \langle -, - \rangle \rangle$ which is a complete normed vector space under the norm $\| \ \|$ induced by $\langle -, - \rangle$ is called a *Hilbert space*. A real Euclidean space $\langle E, \langle -, - \rangle \rangle$ which is complete under the norm $\| \ \|$ induced by $\langle -, - \rangle$ is called a *real Hilbert space*.

All the results in this section hold for complex Hilbert spaces as well as for real Hilbert spaces. We state all results for the complex case only, since they also apply to the real case, and since the proofs in the complex case need a little more care.

Example D.1. The space ℓ^2 of all countably infinite sequences $x = (x_i)_{i \in \mathbb{N}}$ of complex numbers such that $\sum_{i=0}^{\infty} |x_i|^2 < \infty$ is a Hilbert space. It will be shown later that the map $\langle -, - \rangle \colon \ell^2 \times \ell^2 \to \mathbb{C}$ defined such that

$$\langle (x_i)_{i \in \mathbb{N}}, (y_i)_{i \in \mathbb{N}} \rangle = \sum_{i=0}^{\infty} x_i \overline{y_i}$$

is well defined, and that ℓ^2 is a Hilbert space under $\langle -, - \rangle$. In fact, we will prove a more general result (Proposition D.14).

Example D.2. The set $\mathcal{C}^{\infty}[a, b]$ of smooth functions $f \colon [a, b] \to \mathbb{C}$ is a hermitian space under the hermitian form

$$\langle f, g \rangle = \int_a^b f(x) \overline{g(x)} dx,$$

but it is not a Hilbert space because it is not complete. It is possible to construct its completion $\mathrm{L}^2([a, b])$, which turns out to be the space of Lebesgue square-integrable functions on $[a, b]$.

A simple adaptation of the completion theorem for normed vector spaces (Theorem A.72) shows that every hermitian space has a Hilbert space completion.

Theorem D.1. *If $(E, \langle -, - \rangle)$ is a hermitian space, then the Banach space $(\widehat{E}, \| \ \|_{\widehat{E}})$, which is the completion of the normed vector space $(E, \| \ \|)$ where $\|x\| = \sqrt{\langle x, x \rangle}$ for all $x \in E$, is Hilbert space with the inner product $\langle -, - \rangle_h$ given by*

$$\langle x, y \rangle_h = \frac{1}{4} (\|x + y\|_{\widehat{E}}^2 - \|x - y\|_{\widehat{E}}^2 + i \, \|x + iy\|_{\widehat{E}}^2 - i \, \|x - iy\|_{\widehat{E}}^2),$$

and in the real case,

$$\langle x, y \rangle_h = \frac{1}{2}(\|x+y\|_{\widehat{E}}^2 - \|x\|_{\widehat{E}}^2 - \|y\|_{\widehat{E}}^2)$$

for all $x, y \in \widehat{E}$. *Furthermore, the linear map* $\varphi\colon E \to \widehat{E}$ *given by Theorem A.72 is inner-product preserving.*

Proof. Since E is dense in E, $E \times E$ is dense $\widehat{E} \times \widehat{E}$, and since the map

$$(x, y) \mapsto \frac{1}{4}(\|x+y\|^2 - \|x-y\|^2 + i\,\|x+iy\|^2 - i\,\|x-iy\|^2),$$

and in the real case,

$$(x, y) = \frac{1}{2}(\|x+y\|^2 - \|x\|^2 - \|y\|^2),$$

is uniformly continuous, by Theorem A.61, these maps have unique continuous extensions, and if we define

$$\langle x, y \rangle_h = \frac{1}{4}(\|x+y\|_{\widehat{E}}^2 - \|x-y\|_{\widehat{E}}^2 + i\,\|x+iy\|_{\widehat{E}}^2 - i\,\|x-iy\|_{\widehat{E}}^2),$$

and in the real case,

$$\langle x, y \rangle_h \mapsto \frac{1}{2}(\|x+y\|_{\widehat{E}}^2 - \|x\|_{\widehat{E}}^2 - \|y\|_{\widehat{E}}^2),$$

for all $x, y \in \widehat{E}$, it is easy to check that we obtain positive definite hermitian forms with associated norm $\|\ \|_{\widehat{E}}$, so $\widehat{E}$ a Hilbert space with this inner product. For another proof, see Bourbaki [Bourbaki (1981)]. ☐

One of the most important facts about finite-dimensional hermitian (and Euclidean) spaces is that they have orthonormal bases. This implies that, up to isomorphism, every finite-dimensional hermitian space is isomorphic to $\mathbb{C}^n$ (for some $n \in \mathbb{N}$) and that the inner product is given by

$$\langle (x_1, \ldots, x_n), (y_1, \ldots, y_n) \rangle = \sum_{i=1}^{n} x_i \overline{y_i}.$$

Furthermore, every subspace W has an orthogonal complement $W^\perp$, and the inner product induces a natural duality between E and E^*, where E^* is the space of linear forms on E.

When E is a Hilbert space, E may be infinite dimensional, often of uncountable dimension. Thus, we can't expect that E always have an orthonormal basis. However, if we modify the notion of basis so that a "Hilbert basis" is an orthogonal family that is also dense in E, i.e., every $v \in E$ is the limit of a sequence of finite combinations of vectors from the

Hilbert basis, then we can recover most of the "nice" properties of finite-dimensional hermitian spaces. For instance, if $(u_k)_{k\in K}$ is a Hilbert basis, for every $v \in E$, we can define the Fourier coefficients $c_k = \langle v, u_k\rangle/\|u_k\|$, and then, v is the "sum" of its Fourier series $\sum_{k\in K} c_k u_k$. However, the cardinality of the index set K can be very large, and it is necessary to define what it means for a family of vectors indexed by K to be summable. We will do this in Section D.2. It turns out that every Hilbert space is isomorphic to a space of the form $\ell^2(K)$, where $\ell^2(K)$ is a generalization of the space of Example D.1 (see Theorem D.19, usually called the Riesz–Fischer theorem).

Our first goal is to prove that a closed subspace of a Hilbert space has an orthogonal complement. We also show that duality holds if we redefine the dual E' of E to be the space of *continuous* linear maps on E. Our presentation closely follows Bourbaki [Bourbaki (1981)]. We also were inspired by Rudin [Rudin (1987)], Lang [Lang (1996, 1997)], Schwartz [Schwartz (1991, 1980)], and Dixmier [Dixmier (1984)]. In fact, we highly recommend Dixmier [Dixmier (1984)] as a clear and simple text on the basics of topology and analysis. We first prove the so-called projection lemma.

Recall that in a metric space E, a subset X of E is *closed* iff for every convergent sequence (x_n) of points $x_n \in X$, the limit $x = \lim_{n\to\infty} x_n$ also belongs to X. The *closure* $\overline{X}$ *of* X is the set of all limits of convergent sequences (x_n) of points $x_n \in X$. Obviously, $X \subseteq \overline{X}$. We say that the subset X of E is *dense in* E iff $E = \overline{X}$, the closure of X, which means that every $a \in E$ is the limit of some sequence (x_n) of points $x_n \in X$. Convex sets will again play a crucial role.

First, we state the following easy "parallelogram inequality", whose proof is left as an exercise.

Proposition D.2. *If E is a hermitian space, for any two vectors $u, v \in E$, we have*

$$\|u+v\|^2 + \|u-v\|^2 = 2(\|u\|^2 + \|v\|^2).$$

From the above, we get the following proposition:

Proposition D.3. *If E is a hermitian space, given any $d, \delta \in \mathbb{R}$ such that $0 \le \delta < d$, let*

$$B = \{u \in E \mid \|u\| < d\} \quad \textit{and} \quad C = \{u \in E \mid \|u\| \le d + \delta\}.$$

For any convex set such A that $A \subseteq C - B$, we have

$$\|v - u\| \leq \sqrt{12d\delta},$$

for all $u, v \in A$ (see Figure D.1).

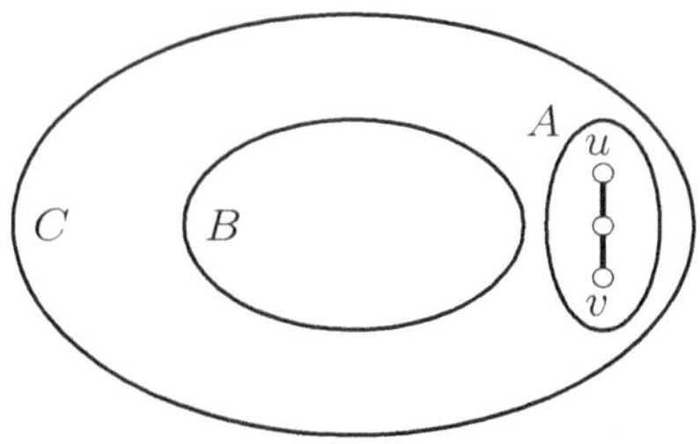

Fig. D.1 Inequality of Proposition D.3.

Proof. Since A is convex, $\frac{1}{2}(u+v) \in A$ if $u, v \in A$, and thus, $\|\frac{1}{2}(u+v)\| \geq d$. From the parallelogram inequality written in the form

$$\left\|\frac{1}{2}(u+v)\right\|^2 + \left\|\frac{1}{2}(u-v)\right\|^2 = \frac{1}{2}\left(\|u\|^2 + \|v\|^2\right),$$

since $\delta < d$, we get

$$\begin{aligned}\left\|\frac{1}{2}(u-v)\right\|^2 &= \frac{1}{2}\left(\|u\|^2 + \|v\|^2\right) - \left\|\frac{1}{2}(u+v)\right\|^2 \\ &\leq (d+\delta)^2 - d^2 = 2d\delta + \delta^2 \leq 3d\delta,\end{aligned}$$

from which

$$\|v - u\| \leq \sqrt{12d\delta},$$

as claimed. ☐

If X is a nonempty subset of a metric space (E, d), for any $a \in E$, recall that we define the *distance $d(a, X)$ of a to X* as

$$d(a, X) = \inf_{b \in X} d(a, b).$$

Also, the *diameter $\delta(X)$ of X* is defined by

$$\delta(X) = \sup\{d(a, b) \mid a, b \in X\}.$$

It is possible that $\delta(X) = \infty$. We leave the following standard two facts as an exercise (see Dixmier [Dixmier (1984)]):

Proposition D.4. *Let E be a metric space.*

(1) For every subset $X \subseteq E$, $\delta(X) = \delta(\overline{X})$.
(2) If E is a complete metric space, for every sequence (F_n) of closed nonempty subsets of E such that $F_{n+1} \subseteq F_n$, if $\lim_{n\to\infty} \delta(F_n) = 0$, then $\bigcap_{n=1}^{\infty} F_n$ consists of a single point.

We are now ready to prove the crucial projection lemma.

Proposition D.5. *(Projection lemma) Let E be a Hilbert space.*

(1) For any nonempty convex and closed subset $X \subseteq E$, for any $u \in E$, there is a unique vector $p_X(u) \in X$ such that
$$\|u - p_X(u)\| = \inf_{v\in X} \|u - v\| = d(u, X).$$
See Figure D.2.
(2) The vector $p_X(u)$ is the unique vector $w \in E$ satisfying the following property (see Figure D.3):
$$w \in X \quad \text{and} \quad \Re\langle u - w, z - w\rangle \leq 0 \quad \text{for all } z \in X. \qquad (*)$$

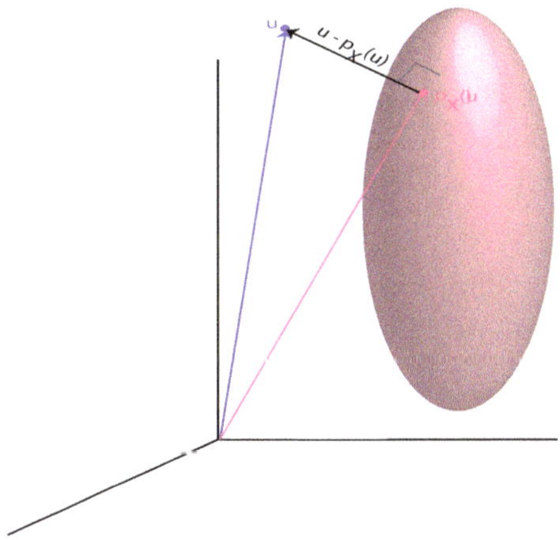

Fig. D.2 Let X be the solid pink ellipsoid. The projection of the purple point u onto X is the magenta point $p_X(u)$.

Proof. (1) Let $d = \inf_{v\in X} \|u - v\| = d(u, X)$. We define a sequence X_n of subsets of X as follows: for every $n \geq 1$,
$$X_n = \left\{ v \in X \mid \|u - v\| \leq d + \frac{1}{n} \right\}.$$

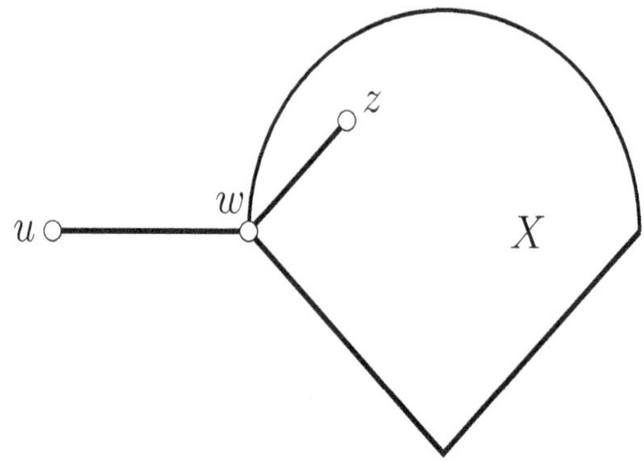

Fig. D.3 Inequality of Proposition D.5.

It is immediately verified that each X_n is nonempty (by definition of d), convex, and that $X_{n+1} \subseteq X_n$. Also, by Proposition D.3, we have

$$\sup\{\|w - v\| \mid v, w \in X_n\} \leq \sqrt{12d/n},$$

and thus, $\bigcap_{n\geq 1} X_n$ contains at most one point. We will prove that $\bigcap_{n\geq 1} X_n$ contains exactly one point, namely, $p_X(u)$. For this, define a sequence $(w_n)_{n\geq 1}$ by picking some $w_n \in X_n$ for every $n \geq 1$. We claim that $(w_n)_{n\geq 1}$ is a Cauchy sequence. Given any $\epsilon > 0$, if we pick N such that

$$N > \frac{12d}{\epsilon^2},$$

since $(X_n)_{n\geq 1}$ is a monotonic decreasing sequence, for all $m, n \geq N$, we have

$$\|w_m - w_n\| \leq \sqrt{12d/N} < \epsilon,$$

as desired. Since E is complete, the sequence $(w_n)_{n\geq 1}$ has a limit w, and since $w_n \in X$ and X is closed, we must have $w \in X$. Also observe that

$$\|u - w\| \leq \|u - w_n\| + \|w_n - w\|,$$

and since w is the limit of $(w_n)_{n\geq 1}$ and

$$\|u - w_n\| \leq d + \frac{1}{n},$$

given any $\epsilon > 0$, there is some n large enough so that

$$\frac{1}{n} < \frac{\epsilon}{2} \quad \text{and} \quad \|w_n - w\| \leq \frac{\epsilon}{2},$$

and thus

$$\|u - w\| \leq d + \epsilon.$$

Since the above holds for every $\epsilon > 0$, we have $\|u - w\| = d$. Thus, $w \in X_n$ for all $n \geq 1$, which proves that $\bigcap_{n \geq 1} X_n = \{w\}$. Now, any $z \in X$ such that $\|u - z\| = d(u, X) = d$ also belongs to every X_n, and thus $z = w$, proving the uniqueness of w, which we denote as $p_X(u)$. See Figure D.4.

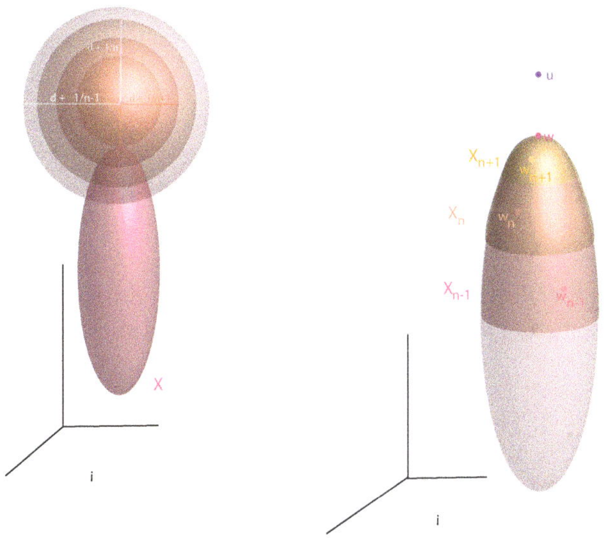

Fig. D.4 Let X be the solid pink ellipsoid with $p_X(u) = w$ at its apex. Each X_n is the intersection of X and a solid sphere centered at u with radius $d + 1/n$. These intersections are the colored "caps" of Figure ii. The Cauchy sequence $(w_n)_{n \geq 1}$ is obtained by selecting a point in each colored X_n.

(2) Let $w \in X$. Since X is convex, $z = (1 - \lambda)p_X(u) + \lambda w \in X$ for every λ, $0 \leq \lambda \leq 1$. Then, we have

$$\|u - z\| \geq \|u - p_X(u)\|$$

for all λ, $0 \leq \lambda \leq 1$, and since

$$\begin{aligned}\|u - z\|^2 &= \|u - p_X(u) - \lambda(w - p_X(u))\|^2 \\ &= \|u - p_X(u)\|^2 + \lambda^2\|w - p_X(u)\|^2 - 2\lambda\Re\langle u - p_X(u), w - p_X(u)\rangle\end{aligned}$$

for all λ, $0 < \lambda \leq 1$, we get

$$\begin{aligned}\Re\langle u - p_X(u), w - p_X(u)\rangle &= \frac{1}{2\lambda}\left(\|u - p_X(u)\|^2 - \|u - z\|^2\right) \\ &\quad + \frac{\lambda}{2}\|w - p_X(u)\|^2,\end{aligned}$$

and since this holds for every λ, $0 < \lambda \leq 1$ and

$$\|u - z\| \geq \|u - p_X(u)\|,$$

we have

$$\Re \langle u - p_X(u), w - p_X(u) \rangle \leq 0.$$

Conversely, assume that $w \in X$ satisfies the condition

$$\Re \langle u - w, z - w \rangle \leq 0$$

for all $z \in X$. For all $z \in X$, we have

$$\|u - z\|^2 = \|u - w\|^2 + \|z - w\|^2 - 2\Re \langle u - w, z - w \rangle \geq \|u - w\|^2,$$

which implies that $\|u - w\| = d(u, X) = d$, and from (1), $w = p_X(u)$. ☐

The vector $p_X(u)$ is called the *projection of u onto X*, and the map $p_X \colon E \to X$ is called the *projection of E onto X*. In the case of a real Hilbert space, there is an intuitive geometric interpretation of the condition

$$\langle u - p_X(u), z - p_X(u) \rangle \leq 0$$

for all $z \in X$. If we restate the condition as

$$\langle u - p_X(u), p_X(u) - z \rangle \geq 0$$

for all $z \in X$, this says that the absolute value of the measure of the angle between the vectors $u - p_X(u)$ and $p_X(u) - z$ is at most $\pi/2$. See Figure D.5. This makes sense, since X is convex, and points in X must be on the side opposite to the "tangent space" to X at $p_X(u)$, which is orthogonal to $u - p_X(u)$. Of course, this is only an intuitive description, since the notion of tangent space has not been defined!

If X is a closed subspace of E, then Condition $(**)$ says that the vector $u - p_X(u)$ is orthogonal to X, in the sense that $u - p_X(u)$ is orthogonal to every vector $z \in X$.

The map $p_X \colon E \to X$ is continuous, as shown below.

Proposition D.6. *Let E be a Hilbert space. For any nonempty convex and closed subset $X \subseteq E$, the map $p_X \colon E \to X$ is continuous.*

Proof. For any two vectors $u, v \in E$, let $x = p_X(u) - u$, $y = p_X(v) - p_X(u)$, and $z = v - p_X(v)$. Clearly,

$$v - u = x + y + z,$$

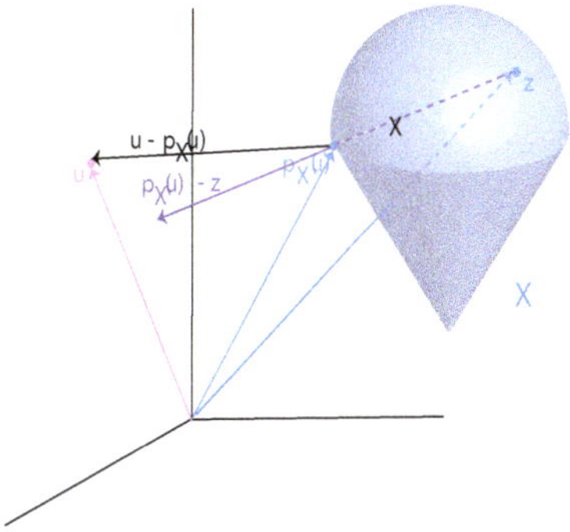

Fig. D.5 Let X be the solid blue ice cream cone. The acute angle between the black vector $u - p_X(u)$ and the purple vector $p_X(u) - z$ is less than $\pi/2$.

(see Figure D.6), and from Proposition D.5(2), we also have

$$\Re\langle x, y\rangle \geq 0 \quad \text{and} \quad \Re\langle z, y\rangle \geq 0,$$

from which we get

$$\begin{aligned}\|v - u\|^2 &= \|x + y + z\|^2 = \|x + z + y\|^2 \\ &= \|x + z\|^2 + \|y\|^2 + 2\Re\langle x, y\rangle + 2\Re\langle z, y\rangle \\ &\geq \|y\|^2 = \|p_X(v) - p_X(u)\|^2.\end{aligned}$$

However, $\|p_X(v) - p_X(u)\| \leq \|v - u\|$ obviously implies that p_X is continuous. ☐

We can now prove the following important proposition.

Proposition D.7. *Let E be a Hilbert space.*

(1) For any closed subspace $V \subseteq E$, we have $E = V \oplus V^{\perp}$, and the map $p_V : E \to V$ is linear and continuous.
(2) For any $u \in E$, the projection $p_V(u)$ is the unique vector $w \in E$ such that

$$w \in V \quad \textit{and} \quad \langle u - w, z\rangle = 0 \quad \textit{for all } z \in V.$$

Proof. (1) First, we prove that $u - p_V(u) \in V^{\perp}$ for all $u \in E$. For any $v \in V$, since V is a subspace, $z = p_V(u) + \lambda v \in V$ for all $\lambda \in \mathbb{C}$, and since V

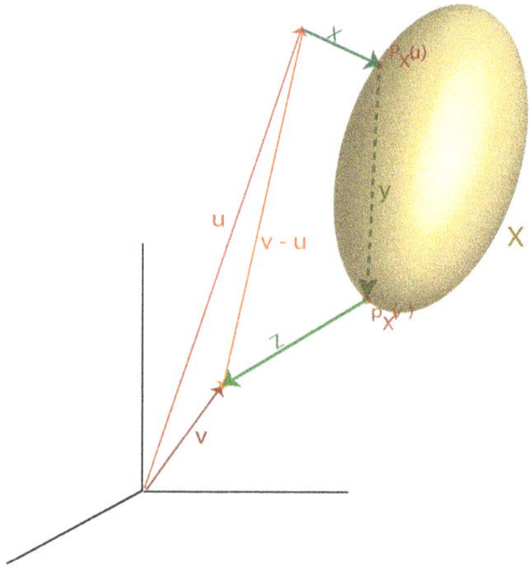

Fig. D.6 Let X be the solid gold ellipsoid. The vector $v-u$ is the sum of the three green vectors, each of which is determined by the appropriate projections.

is convex and nonempty (since it is a subspace), and closed by hypothesis, by Proposition D.5 (2), we have

$$\Re(\overline{\lambda}\,\langle u-p_V(u),v\rangle) = \Re(\langle u-p_V(u),\lambda v\rangle = \Re\,\langle u-p_V(u),z-p_V(u)\rangle \leq 0$$

for all $\lambda \in \mathbb{C}$. In particular, the above holds for $\lambda = \langle u-p_V(u),v\rangle$, which yields

$$|\,\langle u-p_V(u),v\rangle\,| \leq 0,$$

and thus, $\langle u-p_V(u),v\rangle = 0$. See Figure D.7. As a consequence, $u-p_V(u) \in V^{\perp}$ for all $u \in E$. Since $u = p_V(u)+u-p_V(u)$ for every $u \in E$, we have $E = V+V^{\perp}$. On the other hand, since $\langle -,-\rangle$ is positive definite, $V \cap V^{\perp} = \{0\}$, and thus $E = V \oplus V^{\perp}$.

We already proved in Proposition D.6 that $p_V \colon E \to V$ is continuous. Also, since

$$\begin{aligned}p_V(\lambda u+\mu v)-(\lambda p_V(u)+\mu p_V(v)) &= p_V(\lambda u+\mu v)-(\lambda u+\mu v)\\ &\quad +\lambda(u-p_V(u))+\mu(v-p_V(v)),\end{aligned}$$

for all $u,v \in E$, and since the left-hand side term belongs to V, and from what we just showed, the right-hand side term belongs to $V^{\perp}$, we have

$$p_V(\lambda u+\mu v)-(\lambda p_V(u)+\mu p_V(v)) = 0,$$

showing that p_V is linear.

(2) This is basically obvious from (1). We proved in (1) that $u - p_V(u) \in V^{\perp}$, which is exactly the condition

$$\langle u - p_V(u), z\rangle = 0$$

for all $z \in V$. Conversely, if $w \in V$ satisfies the condition

$$\langle u - w, z\rangle = 0$$

for all $z \in V$, since $w \in V$, every vector $z \in V$ is of the form $y - w$, with $y = z + w \in V$, and thus, we have

$$\langle u - w, y - w\rangle = 0$$

for all $y \in V$, which implies the condition of Proposition D.5(2):

$$\Re \langle u - w, y - w\rangle \leq 0$$

for all $y \in V$. By Proposition D.5, $w = p_V(u)$ is the projection of u onto V. $\square$

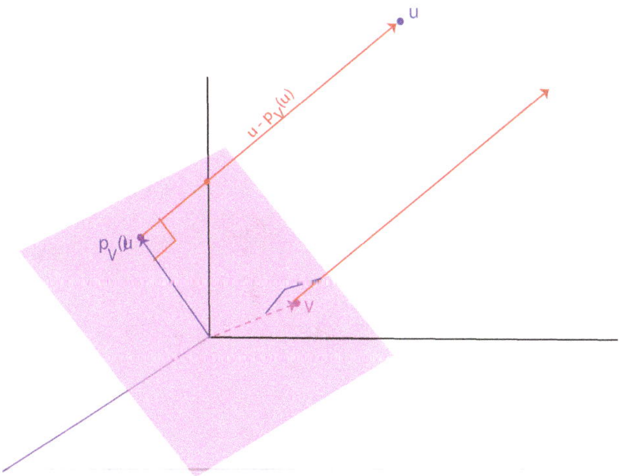

Fig. D.7 Let V be the pink plane. The vector $u - p_V(u)$ is perpendicular to any $v \in V$.

Definition D.2. Let E be a Hilbert space. For any closed subspace $V \subseteq E$, the linear and continuous map $p_V \colon E \to V$ given by Proposition D.7 is called the *orthogonal projection* of E onto V (recall that $E = V \oplus V^{\perp}$).

Let us illustrate the power of Proposition D.7 on the following "least squares" problem. Given a real $m \times n$-matrix A and some vector $b \in \mathbb{R}^m$, we would like to solve the linear system

$$Ax = b$$

in the least-squares sense, which means that we would like to find some solution $x \in \mathbb{R}^n$ that minimizes the Euclidean norm $\|Ax - b\|$ of the error $Ax - b$. It is actually not clear that the problem has a solution, but it does! The problem can be restated as follows: Is there some $x \in \mathbb{R}^n$ such that

$$\|Ax - b\| = \inf_{y \in \mathbb{R}^n} \|Ay - b\|,$$

or equivalently, is there some $z \in \mathrm{Im}\,(A)$ such that

$$\|z - b\| = d(b, \mathrm{Im}\,(A)),$$

where $\mathrm{Im}\,(A) = \{Ay \in \mathbb{R}^m \mid y \in \mathbb{R}^n\}$, the image of the linear map induced by A. Since $\mathrm{Im}\,(A)$ is a closed subspace of $\mathbb{R}^m$, because we are in finite dimension, Proposition D.7 tells us that there is a unique $z \in \mathrm{Im}\,(A)$ such that

$$\|z - b\| = \inf_{y \in \mathbb{R}^n} \|Ay - b\|,$$

and thus, the problem always has a solution since $z \in \mathrm{Im}\,(A)$, and since there is at least some $x \in \mathbb{R}^n$ such that $Ax = z$ (by definition of $\mathrm{Im}\,(A)$). Note that such an x is not necessarily unique. Furthermore, Proposition D.7 also tells us that $z \in \mathrm{Im}\,(A)$ is the solution of the equation

$$\langle z - b, w \rangle = 0 \quad \text{for all } w \in \mathrm{Im}\,(A),$$

or equivalently, that $x \in \mathbb{R}^n$ is the solution of

$$\langle Ax - b, Ay \rangle = 0 \quad \text{for all } y \in \mathbb{R}^n,$$

which is equivalent to

$$\langle A^\top (Ax - b), y \rangle = 0 \quad \text{for all } y \in \mathbb{R}^n,$$

and thus, since the inner product is positive definite, to $A^\top (Ax - b) = 0$, i.e.,

$$A^\top Ax = A^\top b.$$

Therefore, the solutions of the original least-squares problem are precisely the solutions of the the so-called *normal equations*

$$A^\top Ax = A^\top b,$$

discovered by Gauss and Legendre around 1800. We also proved that the normal equations always have a solution.

Computationally, it is best not to solve the normal equations directly, and instead, to use methods such as the QR-decomposition (applied to A) or the SVD-decomposition (in the form of the pseudo-inverse). We will come back to this point later on.

As another corollary of Proposition D.7, for any continuous nonnull linear map $h\colon E \to \mathbb{C}$, the null space

$$H = \operatorname{Ker} h = \{u \in E \mid h(u) = 0\} = h^{-1}(0)$$

is a closed hyperplane H, and thus, $H^{\perp}$ is a subspace of dimension one such that $E = H \oplus H^{\perp}$. This suggests defining the dual space of E as the set of all continuous maps $h\colon E \to \mathbb{C}$.

Remark: If $h\colon E \to \mathbb{C}$ is a linear map which is **not** continuous, then it can be shown that the hyperplane $H = \operatorname{Ker} h$ is dense in E! Thus, $H^{\perp}$ is reduced to the trivial subspace $\{0\}$. This goes against our intuition of what a hyperplane in $\mathbb{R}^n$ (or $\mathbb{C}^n$) is, and warns us not to trust our "physical" intuition too much when dealing with infinite dimensions.

Definition D.3. Given two vector spaces E and F over the complex field $\mathbb{C}$, a function $f\colon E \to F$ is *semilinear* if

$$\begin{aligned} f(u + v) &= f(u) + f(v), \\ f(\lambda u) &= \overline{\lambda} f(u), \end{aligned}$$

for all $u, v \in E$ and all $\lambda \in \mathbb{C}$.

Instead of defining semilinear maps, we can define the vector space $\overline{E}$ as the vector space with the same carrier set E whose addition is the same as that of E, but whose multiplication by a complex number λ is given by

$$(\lambda, u) \mapsto \overline{\lambda} u.$$

Then it is easy to check that a function $f\colon E \to \mathbb{C}$ is semilinear iff $f\colon \overline{E} \to \mathbb{C}$ is linear.

A fundamental fact about a *finite-dimensional* hermitian space is that the hermitian inner product induces a bijection (i.e., independent of the choice of bases) between the vector space E and its dual space E^*.

Given a hermitian space E, for any vector $u \in E$, let $\varphi^l_u\colon E \to \mathbb{C}$ be the map defined such that

$$\varphi^l_u(v) = \overline{\langle u, v \rangle}, \quad \text{for all } v \in E.$$

Similarly, for any vector $v \in E$, let $\varphi_v^r \colon E \to \mathbb{C}$ be the map defined such that

$$\varphi_v^r(u) = \langle u, v \rangle, \quad \text{for all } u \in E.$$

Since the hermitian product is linear in its first argument u, the map φ_v^r is a linear form in E^*, and since it is semilinear in its second argument v, the map φ_u^l is also a linear form in E^*. Thus, we have two maps $\flat^l \colon E \to E^*$ and $\flat^r \colon E \to E^*$, defined such that

$$\flat^l(u) = \varphi_u^l, \quad \text{and} \quad \flat^r(v) = \varphi_v^r.$$

Actually, $\varphi_u^l = \varphi_u^r$ and $\flat^l = \flat^r$. Indeed, for all $u, v \in E$, we have

$$\flat^l(u)(v) = \varphi_u^l(v) = \overline{\langle u, v \rangle} = \langle v, u \rangle = \varphi_u^r(v) = \flat^r(u)(v).$$

Therefore, we use the notation φ_u for both φ_u^l and φ_u^r, and $\flat$ for both $\flat^l$ and $\flat^r$.

Theorem D.8. *Let E be a hermitian space E. The map $\flat \colon E \to E^*$ defined such that*

$$\flat(u) = \varphi_u \quad \textit{for all } u \in E$$

is semilinear and injective. When E is also of finite dimension, the map $\flat \colon \overline{E} \to E^*$ is a canonical isomorphism.

Proof. That $\flat \colon E \to E^*$ is a semilinear map follows immediately from the fact that $\flat = \flat^r$, and that the hermitian product is semilinear in its second argument. If $\varphi_u = \varphi_v$, then $\varphi_u(w) = \varphi_v(w)$ for all $w \in E$, which by definition of φ_u and φ_v means that

$$\langle w, u \rangle = \langle w, v \rangle$$

for all $w \in E$, which by semilinearity on the right is equivalent to

$$\langle w, v - u \rangle = 0 \quad \text{for all } w \in E,$$

which implies that $u = v$, since the hermitian product is positive definite. Thus, $\flat \colon E \to E^*$ is injective. Finally, when E is of finite dimension n, E^* is also of dimension n, and then $\flat \colon E \to E^*$ is bijective. Since $\flat$ is semilinear, the map $\flat \colon \overline{E} \to E^*$ is an isomorphism. $\square$

However, if E is infinite dimensional, the map $\flat \colon E \to E^*$ is not surjective, since the linear forms of the form $u \mapsto \langle u, v \rangle$ (for some fixed vector $v \in E$) are continuous (the inner product is continuous), but there are linear forms that are not continuous.

We now show that by redefining the dual space of a Hilbert space as the set of continuous linear forms on E, we recover Theorem D.8.

Definition D.4. Given a Hilbert space E, we define the *dual space* E' *of* E as the vector space of all continuous linear forms $h\colon E \to \mathbb{C}$. Maps in E' are also called *bounded linear operators, bounded linear functionals, or simply, operators or functionals.*

Theorem D.8 is generalized to Hilbert spaces as follows.

Theorem D.9. *(Riesz representation theorem) Let E be a Hilbert space. Then, the map $\flat\colon E \to E'$ defined such that*

$$\flat(v) = \varphi_v,$$

is semilinear, continuous, and bijective.

Proof. The proof is basically identical to the proof of Theorem D.8, except that a different argument is required for the surjectivity of $\flat\colon E \to E'$, since E may not be finite dimensional. For any nonnull linear operator $h \in E'$, the hyperplane $H = \operatorname{Ker} h = h^{-1}(0)$ is a closed subspace of E, and by Proposition D.7, $H^{\perp}$ is a subspace of dimension one such that $E = H \oplus H^{\perp}$. Then, picking any nonnull vector $w \in H^{\perp}$, observe that H is also the kernel of the linear operator φ_w, with

$$\varphi_w(u) = \langle u, w\rangle\,,$$

and thus, since any two nonzero linear forms defining the same hyperplane must be proportional, there is some nonzero scalar $\lambda \in \mathbb{C}$ such that $h = \lambda\varphi_w$. But then, $h = \varphi_{\overline{\lambda}w}$, proving that $\flat\colon E \to E'$ is surjective. ☐

Theorem D.9 is known as the *Riesz representation theorem, or "Little Riesz Theorem."* It shows that the inner product on a Hilbert space induces a natural linear isomorphism between E and its dual E'.

Remarks:

(1) Actually, the map $\flat\colon E \to E'$ turns out to be an isometry. To show this, we need to recall the notion of norm of a linear map, which we do not want to do right now.
(2) Many books on quantum mechanics use the so-called Dirac notation to denote objects in the Hilbert space E and operators in its dual space E'. In the Dirac notation, an element of E is denoted as $|x\rangle$, and an element of E' is denoted as $\langle t|$. The scalar product is denoted as $\langle t|\cdot|x\rangle$. This uses the isomorphism between E and E', except that the inner product is assumed to be semi-linear on the left, rather than on the right.

Theorem D.9 allows us to define the adjoint of a continuous linear map, as in the hermitian case.

Proposition D.10. *Given a Hilbert space E, for every continuous linear map $f\colon E \to E$, there is a unique linear map $f^*\colon E \to E$, such that*

$$\langle f^*(u), v\rangle = \langle u, f(v)\rangle$$

for all $u, v \in E$. The map f^* is called the adjoint of f.

Proposition D.11 will show that if f is continuous, then f^* is also continuous. As in the hermitian case, given two Hilbert spaces E and F, for any continuous linear map $f\colon E \to F$, such that

$$\langle f(u), v\rangle_2 = \langle u, f^*(v)\rangle_1$$

for all $u \in E$ and all $v \in F$. The linear map f^* is also called the adjoint of f.

The following results will be needed in Vol II, Section 2.5.

Proposition D.11. *Let E a Hilbert space. For every continuous linear map $f\colon E \to E$, we have*

$$\begin{aligned}
\|f^*\| &= \|f\| \\
\|f^* \circ f\| &= \|f\|^2 \\
\|f \circ f^*\| &= \|f^* \circ f\| .
\end{aligned}$$

In the above equations, we use the operator norm induced by the inner product on E. The first equation implies that f^* is continuous.

Proof. Since f^* is the adjoint of f we have

$$\langle f(x), y\rangle = \langle x, f^*(y)\rangle \quad \text{for all } x, y \in E.$$

By the Cauchy-Schwarz inequality and properties of the operator norm,

$$|\langle x, f^*(y)\rangle| = |\langle f(x), y\rangle| \leq \|f(x)\|\,\|y\| \leq \|f\|\,\|x\|\,\|y\| .$$

If we let $x = f^*(y)$, we obtain

$$\|f^*(y)\|^2 \leq \|f\|\,\|f^*(y)\|\,\|y\| ,$$

which implies that

$$\|f^*(y)\| \leq \|f\|\,\|y\| , \quad \text{for all } y \in E,$$

so by definition of the operator norm $\|f^*\|$,

$$\|f^*\| \leq \|f\| .$$

Repeating the same argument with f^* substituted for f and the fact that $(f^*)^* = f$ we get $\|f\| \leq \|f^*\|$, and so $\|f^*\| = \|f\|$.

Since $(f^*)^* = f$, the map f is the adjoint of f^* and we have

$$\langle f^*(f(x)), x\rangle = \langle f(x), f(x)\rangle = \|f(x)\|^2 \quad \text{for all } x \in E,$$

so by the Cauchy-Schwarz inequality,

$$\|f(x)\|^2 \leq \|f^*(f(x))\| \, \|x\| .$$

Since we are using the operator norm,

$$\|f(x)\|^2 \leq \|f^*(f(x))\| \, \|x\| \leq \|f^* \circ f\| \, \|x\|^2 \quad \text{for all } x \in E,$$

which implies (first take square roots) that

$$\|f\|^2 \leq \|f^* \circ f\| .$$

However, by a well-known property of the operator norm and the fact that $\|f^*\| = \|f\|$, we have

$$\|f^* \circ f\| \leq \|f^*\| \, \|f\| = \|f\|^2 .$$

Therefore, $\|f^* \circ f\| = \|f\|^2$.

The above equation with f replaced by f^* yields $\|f \circ f^*\| = \|f^*\|^2$, and since $\|f^*\| = \|f\|$, we obtain $\|f \circ f^*\| = \|f^*\|^2 = \|f\|^2 = \|f^* \circ f\|$, which is the third equation. ☐

As a corollary of Proposition D.11, if f is self-adjoint, that is, $f^* = f$, then

$$\|f \circ f\| = \|f\| .$$

D.2 Total Orthogonal Families (Hilbert Bases), Fourier Coefficients

We conclude our quick tour of Hilbert spaces by showing that the notion of orthogonal basis can be generalized to Hilbert spaces. However, the useful notion is not the usual notion of a basis, but a notion which is an abstraction of the concept of Fourier series. Every element of a Hilbert space is the "sum" of its Fourier series.

Definition D.5. Given a Hilbert space E, a family $(u_k)_{k\in K}$ of nonnull vectors is an *orthogonal family* iff the u_k are pairwise orthogonal, i.e., $\langle u_i, u_j\rangle = 0$ for all $i \neq j$ $(i, j \in K)$, and an *orthonormal family* iff $\langle u_i, u_j\rangle = \delta_{i,j}$, for all $i, j \in K$. A *total orthogonal family (or system)* or *Hilbert basis* is an orthogonal family that is dense in E. This means that for every $v \in E$, for every $\epsilon > 0$, there is some finite subset $I \subseteq K$ and some family $(\lambda_i)_{i\in I}$ of complex numbers, such that

$$\left\| v - \sum_{i\in I} \lambda_i u_i \right\| < \epsilon.$$

Given an orthogonal family $(u_k)_{k\in K}$, for every $v \in E$, for every $k \in K$, the scalar $c_k = \langle v, u_k\rangle / \|u_k\|^2$ is called the *k-th Fourier coefficient of v over* $(u_k)_{k\in K}$.

Remark: The terminology Hilbert basis is misleading, because a Hilbert basis $(u_k)_{k\in K}$ is not necessarily a basis in the algebraic sense. Indeed, in general, $(u_k)_{k\in K}$ does not span E. Intuitively, it takes linear combinations of the u_k's with infinitely many nonnull coefficients to span E. Technically, this is achieved in terms of limits. In order to avoid the confusion between bases in the algebraic sense and Hilbert bases, some authors refer to algebraic bases as *Hamel bases* and to total orthogonal families (or Hilbert bases) as *Schauder bases*.

Given an orthogonal family $(u_k)_{k\in K}$, for any finite subset I of K, we often call sums of the form $\sum_{i\in I} \lambda_i u_i$ *partial sums of Fourier series*, and if these partial sums converge to a limit denoted as $\sum_{k\in K} c_k u_k$, we call $\sum_{k\in K} c_k u_k$ a *Fourier series*.

However, we have to make sense of such sums! Indeed, when K is unordered or uncountable, the notion of limit or sum has not been defined. This can be done as follows (for more details, see Dixmier [Dixmier (1984)]):

Definition D.6. Given a normed vector space E (say, a Hilbert space), for any nonempty index set K, we say that a family $(u_k)_{k\in K}$ of vectors in E is *summable with sum* $v \in E$ iff for every $\epsilon > 0$, there is some finite subset I of K, such that,

$$\left\| v - \sum_{j\in J} u_j \right\| < \epsilon$$

for every finite subset J with $I \subseteq J \subseteq K$. We say that the family $(u_k)_{k\in K}$ is *summable* iff there is some $v \in E$ such that $(u_k)_{k\in K}$ is summable with sum v. A family $(u_k)_{k\in K}$ is a *Cauchy family* iff for every $\epsilon > 0$, there is a finite subset I of K, such that,

$$\left\|\sum_{j\in J} u_j\right\| < \epsilon$$

for every finite subset J of K with $I \cap J = \emptyset$,

If $(u_k)_{k\in K}$ is summable with sum v, we usually denote v as $\sum_{k\in K} u_k$. The following technical proposition will be needed:

Proposition D.12. *Let E be a complete normed vector space (say, a Hilbert space).*

(1) For any nonempty index set K, a family $(u_k)_{k\in K}$ is summable iff it is a Cauchy family.
(2) Given a family $(r_k)_{k\in K}$ of nonnegative reals $r_k \geq 0$, if there is some real number $B > 0$ such that $\sum_{i\in I} r_i < B$ for every finite subset I of K, then $(r_k)_{k\in K}$ is summable and $\sum_{k\in K} r_k = r$, where r is least upper bound of the set of finite sums $\sum_{i\in I} r_i$ $(I \subseteq K)$.

Proof. (1) If $(u_k)_{k\in K}$ is summable, for every finite subset I of K, let

$$u_I = \sum_{i\in I} u_i \quad \text{and} \quad u = \sum_{k\in K} u_k.$$

For every $\epsilon > 0$, there is some finite subset I of K such that

$$\|u - u_L\| < \epsilon/2$$

for all finite subsets L such that $I \subseteq L \subseteq K$. For every finite subset J of K such that $I \cap J = \emptyset$, since $I \subseteq I \cup J \subseteq K$ and $I \cup J$ is finite, we have

$$\|u - u_{I\cup J}\| < \epsilon/2 \quad \text{and} \quad \|u - u_I\| < \epsilon/2,$$

and since

$$\|u_{I\cup J} - u_I\| \leq \|u_{I\cup J} - u\| + \|u - u_I\|$$

and $u_{I\cup J} - u_I = u_J$ since $I \cap J = \emptyset$, we get

$$\|u_J\| = \|u_{I\cup J} - u_I\| < \epsilon,$$

which is the condition for $(u_k)_{k\in K}$ to be a Cauchy family.

Conversely, assume that $(u_k)_{k\in K}$ is a Cauchy family. We define inductively a decreasing sequence (X_n) of subsets of E, each of diameter at most

$1/n$, as follows: For $n = 1$, since $(u_k)_{k \in K}$ is a Cauchy family, there is some finite subset J_1 of K such that

$$\|u_J\| < 1/2$$

for every finite subset J of K with $J_1 \cap J = \emptyset$. We pick some finite subset J_1 with the above property, and we let $I_1 = J_1$ and

$$X_1 = \{u_I \mid I_1 \subseteq I \subseteq K,\ I \text{ finite}\}.$$

For $n \geq 1$, there is some finite subset J_{n+1} of K such that

$$\|u_J\| < 1/(2n+2)$$

for every finite subset J of K with $J_{n+1} \cap J = \emptyset$. We pick some finite subset J_{n+1} with the above property, and we let $I_{n+1} = I_n \cup J_{n+1}$ and

$$X_{n+1} = \{u_I \mid I_{n+1} \subseteq I \subseteq K,\ I \text{ finite}\}.$$

Since $I_n \subseteq I_{n+1}$, it is obvious that $X_{n+1} \subseteq X_n$ for all $n \geq 1$. We need to prove that each X_n has diameter at most $1/n$. Since J_n was chosen such that

$$\|u_J\| < 1/(2n)$$

for every finite subset J of K with $J_n \cap J = \emptyset$, and since $J_n \subseteq I_n$, it is also true that

$$\|u_J\| < 1/(2n)$$

for every finite subset J of K with $I_n \cap J = \emptyset$ (since $I_n \cap J = \emptyset$ and $J_n \subseteq I_n$ implies that $J_n \cap J = \emptyset$). Then, for every two finite subsets J, L such that $I_n \subseteq J, L \subseteq K$, we have

$$\|u_{J-I_n}\| < 1/(2n) \quad \text{and} \quad \|u_{L-I_n}\| < 1/(2n),$$

and since

$$\|u_J - u_L\| \leq \|u_J - u_{I_n}\| + \|u_{I_n} - u_L\| = \|u_{J-I_n}\| + \|u_{L-I_n}\|,$$

we get

$$\|u_J - u_L\| < 1/n,$$

which proves that $\delta(X_n) \leq 1/n$. Now, if we consider the sequence of closed sets $(\overline{X_n})$, we still have $\overline{X_{n+1}} \subseteq \overline{X_n}$, and by Proposition D.4, $\delta(\overline{X_n}) = \delta(X_n) \leq 1/n$, which means that $\lim_{n\to\infty} \delta(\overline{X_n}) = 0$, and by Proposition D.4, $\bigcap_{n=1}^{\infty} \overline{X_n}$ consists of a single element u. We claim that u is the sum of the family $(u_k)_{k \in K}$.

For every $\epsilon > 0$, there is some $n \geq 1$ such that $n > 2/\epsilon$, and since $u \in \overline{X_m}$ for all $m \geq 1$, there is some finite subset J_0 of K such that $I_n \subseteq J_0$ and

$$\|u - u_{J_0}\| < \epsilon/2,$$

where I_n is the finite subset of K involved in the definition of X_n. However, since $\delta(X_n) \leq 1/n$, for every finite subset J of K such that $I_n \subseteq J$, we have

$$\|u_J - u_{J_0}\| \leq 1/n < \epsilon/2,$$

and since

$$\|u - u_J\| \leq \|u - u_{J_0}\| + \|u_{J_0} - u_J\|,$$

we get

$$\|u - u_J\| < \epsilon$$

for every finite subset J of K with $I_n \subseteq J$, which proves that u is the sum of the family $(u_k)_{k \in K}$.

(2) Since every finite sum $\sum_{i \in I} r_i$ is bounded by the uniform bound B, the set of these finite sums has a least upper bound $r \leq B$. For every $\epsilon > 0$, since r is the least upper bound of the finite sums $\sum_{i \in I} r_i$ (where I finite, $I \subseteq K$), there is some finite $I \subseteq K$ such that

$$\left| r - \sum_{i \in I} r_i \right| < \epsilon,$$

and since $r_k \geq 0$ for all $k \in K$, we have

$$\sum_{i \in I} r_i \leq \sum_{j \in J} r_j$$

whenever $I \subseteq J$, which shows that

$$\left| r - \sum_{j \in J} r_j \right| \leq \left| r - \sum_{i \in I} r_i \right| < \epsilon$$

for every finite subset J such that $I \subseteq J \subseteq K$, proving that $(r_k)_{k \in K}$ is summable with sum $\sum_{k \in K} r_k = r$. $\square$

The notion of summability implies that the sum of a family $(u_k)_{k \in K}$ is independent of any order on K. In this sense, it is a kind of "commutative summability". More precisely, it is easy to show that for every bijection $\varphi\colon K \to K$ (intuitively, a reordering of K), the family $(u_k)_{k \in K}$ is summable iff the family $(u_l)_{l \in \varphi(K)}$ is summable, and if so, they have the same sum.

The following proposition gives some of the main properties of Fourier coefficients. Among other things, at most countably many of the Fourier coefficient may be nonnull, and the partial sums of a Fourier series converge. Given an orthogonal family $(u_k)_{k\in K}$, we let $U_k = \mathbb{C}u_k$, and $p_{U_k}\colon E \to U_k$ is the projection of E onto U_k.

Proposition D.13. *Let E be a Hilbert space, $(u_k)_{k\in K}$ an orthogonal family in E, and V the closure of the subspace generated by $(u_k)_{k\in K}$. The following properties hold:*

(1) For every $v \in E$, for every finite subset $I \subseteq K$, we have

$$\sum_{i\in I} |c_i|^2 \leq \|v\|^2,$$

where the c_k are the Fourier coefficients of v.

(2) For every vector $v \in E$, if $(c_k)_{k\in K}$ are the Fourier coefficients of v, the following conditions are equivalent:

(2a) $v \in V$.

(2b) The family $(c_k u_k)_{k\in K}$ is summable and $v = \sum_{k\in K} c_k u_k$.

(2c) The family $(|c_k|^2)_{k\in K}$ is summable and $\|v\|^2 = \sum_{k\in K} |c_k|^2$.

(3) The family $(|c_k|^2)_{k\in K}$ is summable, and we have the Bessel inequality:

$$\sum_{k\in K} |c_k|^2 \leq \|v\|^2.$$

As a consequence, at most countably many of the c_k may be nonzero. The family $(c_k u_k)_{k\in K}$ forms a Cauchy family, and thus, the Fourier series $\sum_{k\in K} c_k u_k$ converges in E to some vector $u = \sum_{k\in K} c_k u_k$. Furthermore, $u = p_V(v)$.

Proof. (1) Let

$$u_I = \sum_{i\in I} c_i u_i$$

for any finite subset I of K. We claim that $v - u_I$ is orthogonal to u_i for every $i \in I$. Indeed,

$$\begin{aligned}
\langle v - u_I, u_i\rangle &= \left\langle v - \sum_{j\in I} c_j u_j, u_i\right\rangle \\
&= \langle v, u_i\rangle - \sum_{j\in I} c_j \langle u_j, u_i\rangle \\
&= \langle v, u_i\rangle - c_i\|u_i\|^2 \\
&= \langle v, u_i\rangle - \langle v, u_i\rangle = 0,
\end{aligned}$$

since $\langle u_j, u_i \rangle = 0$ for all $i \neq j$ and $c_i = \langle v, u_i \rangle / \|u_i\|^2$. As a consequence, we have

$$\begin{aligned}\|v\|^2 &= \left\| v - \sum_{i \in I} c_i u_i + \sum_{i \in I} c_i u_i \right\|^2 \\ &= \left\| v - \sum_{i \in I} c_i u_i \right\|^2 + \left\| \sum_{i \in I} c_i u_i \right\|^2 \\ &= \left\| v - \sum_{i \in I} c_i u_i \right\|^2 + \sum_{i \in I} |c_i|^2,\end{aligned}$$

since the u_i are pairwise orthogonal, that is,

$$\|v\|^2 = \left\| v - \sum_{i \in I} c_i u_i \right\|^2 + \sum_{i \in I} |c_i|^2.$$

Thus,

$$\sum_{i \in I} |c_i|^2 \leq \|v\|^2,$$

as claimed.

(2) We prove the chain of implications $(a) \Rightarrow (b) \Rightarrow (c) \Rightarrow (a)$.

$(a) \Rightarrow (b)$: If $v \in V$, since V is the closure of the subspace spanned by $(u_k)_{k \in K}$, for every $\epsilon > 0$, there is some finite subset I of K and some family $(\lambda_i)_{i \in I}$ of complex numbers, such that

$$\left\| v - \sum_{i \in I} \lambda_i u_i \right\| < \epsilon.$$

Now, for every finite subset J of K such that $I \subseteq J$, we have

$$\begin{aligned}\left\| v - \sum_{i \in I} \lambda_i u_i \right\|^2 &= \left\| v - \sum_{j \in J} c_j u_j + \sum_{j \in J} c_j u_j - \sum_{i \in I} \lambda_i u_i \right\|^2 \\ &= \left\| v - \sum_{j \in J} c_j u_j \right\|^2 + \left\| \sum_{j \in J} c_j u_j - \sum_{i \in I} \lambda_i u_i \right\|^2,\end{aligned}$$

since $I \subseteq J$ and the u_j (with $j \in J$) are orthogonal to $v - \sum_{j \in J} c_j u_j$ by the argument in (1), which shows that

$$\left\| v - \sum_{j \in J} c_j u_j \right\| \leq \left\| v - \sum_{i \in I} \lambda_i u_i \right\| < \epsilon,$$

and thus, that the family $(c_k u_k)_{k \in K}$ is summable with sum v, so that

$$v = \sum_{k \in K} c_k u_k.$$

$(b) \Rightarrow (c)$: If $v = \sum_{k \in K} c_k u_k$, then for every $\epsilon > 0$, there some finite subset I of K, such that

$$\left\| v - \sum_{j \in J} c_j u_j \right\| < \sqrt{\epsilon},$$

for every finite subset J of K such that $I \subseteq J$, and since we proved in (1) that

$$\|v\|^2 = \left\| v - \sum_{j \in J} c_j u_j \right\|^2 + \sum_{j \in J} |c_j|^2,$$

we get

$$\|v\|^2 - \sum_{j \in J} |c_j|^2 < \epsilon,$$

which proves that $(|c_k|^2)_{k \in K}$ is summable with sum $\|v\|^2$.

$(c) \Rightarrow (a)$: Finally, if $(|c_k|^2)_{k \in K}$ is summable with sum $\|v\|^2$, for every $\epsilon > 0$, there is some finite subset I of K such that

$$\|v\|^2 - \sum_{j \in J} |c_j|^2 < \epsilon^2$$

for every finite subset J of K such that $I \subseteq J$, and again, using the fact that

$$\|v\|^2 = \left\| v - \sum_{j \in J} c_j u_j \right\|^2 + \sum_{j \in J} |c_j|^2,$$

we get

$$\left\| v - \sum_{j \in J} c_j u_j \right\| < \epsilon,$$

which proves that $(c_k u_k)_{k \in K}$ is summable with sum $\sum_{k \in K} c_k u_k = v$, and $v \in V$.

(3) Since $\sum_{i \in I} |c_i|^2 \leq \|v\|^2$ for every finite subset I of K, by Proposition D.12, the family $(|c_k|^2)_{k \in K}$ is summable. The Bessel inequality

$$\sum_{k \in K} |c_k|^2 \leq \|v\|^2$$

is an obvious consequence of the inequality $\sum_{i \in I} |c_i|^2 \leq \|v\|^2$ (for every finite $I \subseteq K$). Now, for every natural number $n \geq 1$, if K_n is the subset of K consisting of all c_k such that $|c_k| \geq 1/n$, the number of elements in K_n is at most

$$\sum_{k \in K_n} |nc_k|^2 \leq n^2 \sum_{k \in K} |c_k|^2 \leq n^2 \|v\|^2,$$

which is finite, and thus, at most a countable number of the c_k may be nonzero.

Since $(|c_k|^2)_{k\in K}$ is summable with sum c, for every $\epsilon > 0$, there is some finite subset I of K such that

$$\sum_{j\in J} |c_j|^2 < \epsilon^2$$

for every finite subset J of K such that $I \cap J = \emptyset$. Since

$$\left\|\sum_{j\in J} c_j u_j\right\|^2 = \sum_{j\in J} |c_j|^2,$$

we get

$$\left\|\sum_{j\in J} c_j u_j\right\| < \epsilon.$$

This proves that $(c_k u_k)_{k\in K}$ is a Cauchy family, which, by Proposition D.12, implies that $(c_k u_k)_{k\in K}$ is summable, since E is complete. Thus, the Fourier series $\sum_{k\in K} c_k u_k$ is summable, with its sum denoted $u \in V$.

Since $\sum_{k\in K} c_k u_k$ is summable with sum u, for every $\epsilon > 0$, there is some finite subset I_1 of K such that

$$\left\|u - \sum_{j\in J} c_j u_j\right\| < \epsilon$$

for every finite subset J of K such that $I_1 \subseteq J$. By the triangle inequality, for every finite subset I of K,

$$\left\|u - v\right\| \leq \left\|u - \sum_{i\in I} c_i u_i\right\| + \left\|\sum_{i\in I} c_i u_i - v\right\|.$$

By (2), every $w \in V$ is the sum of its Fourier series $\sum_{k\in K} \lambda_k u_k$, and for every $\epsilon > 0$, there is some finite subset I_2 of K such that

$$\left\|w - \sum_{j\in J} \lambda_j u_j\right\| < \epsilon$$

for every finite subset J of K such that $I_2 \subseteq J$. By the triangle inequality, for every finite subset I of K,

$$\left\|v - \sum_{i\in I} \lambda_i u_i\right\| \leq \|v - w\| + \left\|w - \sum_{i\in I} \lambda_i u_i\right\|.$$

Letting $I = I_1 \cup I_2$, since we showed in (2) that

$$\left\|v - \sum_{i\in I} c_i u_i\right\| \leq \left\|v - \sum_{i\in I} \lambda_i u_i\right\|$$

for every finite subset I of K, we get

$$\begin{aligned}\|u - v\| &\leq \Big\|u - \sum_{i\in I} c_i u_i\Big\| + \Big\|\sum_{i\in I} c_i u_i - v\Big\| \\ &\leq \Big\|u - \sum_{i\in I} c_i u_i\Big\| + \Big\|\sum_{i\in I} \lambda_i u_i - v\Big\| \\ &\leq \Big\|u - \sum_{i\in I} c_i u_i\Big\| + \|v - w\| + \Big\|w - \sum_{i\in I} \lambda_i u_i\Big\|,\end{aligned}$$

and thus

$$\|u - v\| \leq \|v - w\| + 2\epsilon.$$

Since this holds for every $\epsilon > 0$, we have

$$\|u - v\| \leq \|v - w\|$$

for all $w \in V$, i.e. $\|v - u\| = d(v, V)$, with $u \in V$, which proves that $u = p_V(v)$. ☐

D.3 The Hilbert Space $\ell^2(K)$ and the Riesz–Fischer Theorem

Proposition D.13 suggests looking at the space of sequences $(z_k)_{k\in K}$ (where $z_k \in \mathbb{C}$) such that $(|z_k|^2)_{k\in K}$ is summable. Indeed, such spaces are Hilbert spaces, and it turns out that every Hilbert space is isomorphic to one of those. Such spaces are the infinite-dimensional version of the spaces $\mathbb{C}^n$ under the usual Euclidean norm.

Definition D.7. Given any nonempty index set K, the space $\ell^2(K)$ is the set of all sequences $(z_k)_{k\in K}$, where $z_k \in \mathbb{C}$, such that $(|z_k|^2)_{k\in K}$ is summable, i.e., $\sum_{k\in K} |z_k|^2 < \infty$.

Remarks:

(1) When K is a finite set of cardinality n, $\ell^2(K)$ is isomorphic to $\mathbb{C}^n$.
(2) When $K = \mathbb{N}$, the space $\ell^2(\mathbb{N})$ is the space ℓ^2 from Example D.1. It is a Hilbert space, and we now prove this fact for any index set K.

Proposition D.14. *Given any nonempty index set K, the space $\ell^2(K)$ is a Hilbert space under the hermitian product*

$$\langle (x_k)_{k\in K}, (y_k)_{k\in K}\rangle = \sum_{k\in K} x_k \overline{y_k}.$$

The subspace consisting of sequences $(z_k)_{k\in K}$ such that $z_k = 0$, except perhaps for finitely many k, is a dense subspace of $\ell^2(K)$.

Proof. First, we need to prove that $\ell^2(K)$ is a vector space. Assume that $(x_k)_{k\in K}$ and $(y_k)_{k\in K}$ are in $\ell^2(K)$. This means that $(|x_k|^2)_{k\in K}$ and $(|y_k|^2)_{k\in K}$ are summable, which, in view of Proposition D.12, is equivalent to the existence of some positive bounds A and B such that $\sum_{i\in I}|x_i|^2 < A$ and $\sum_{i\in I}|y_i|^2 < B$, for every finite subset I of K. To prove that $(|x_k+y_k|^2)_{k\in K}$ is summable, it is sufficient to prove that there is some $C > 0$ such that $\sum_{i\in I}|x_i+y_i|^2 < C$ for every finite subset I of K. However, the parallelogram inequality implies that

$$\sum_{i\in I}|x_i+y_i|^2 \leq \sum_{i\in I}2(|x_i|^2+|y_i|^2) \leq 2(A+B),$$

for every finite subset I of K, and we conclude by Proposition D.12. Similarly, for every $\lambda \in \mathbb{C}$,

$$\sum_{i\in I}|\lambda x_i|^2 \leq \sum_{i\in I}|\lambda|^2|x_i|^2 \leq |\lambda|^2 A,$$

and $(\lambda_k x_k)_{k\in K}$ is summable. Therefore, $\ell^2(K)$ is a vector space.

By the Cauchy-Schwarz inequality,

$$\begin{aligned}\sum_{i\in I}|x_i\overline{y_i}| \leq \sum_{i\in I}|x_i||y_i| &\leq \Big(\sum_{i\in I}|x_i|^2\Big)^{1/2}\Big(\sum_{i\in I}|y_i|^2\Big)^{1/2}\\ &\leq \sum_{i\in I}(|x_i|^2+|y_i|^2)/2 \leq (A+B)/2,\end{aligned}$$

for every finite subset I of K. Here, we used the fact that

$$4CD \leq (C+D)^2,$$

which is equivalent to

$$(C-D)^2 \geq 0.$$

By Proposition D.12, $(|x_k\overline{y_k}|)_{k\in K}$ is summable. The customary language is that $(x_k\overline{y_k})_{k\in K}$ is absolutely summable. However, it is a standard fact that this implies that $(x_k\overline{y_k})_{k\in K}$ is summable (For every $\epsilon > 0$, there is some finite subset I of K such that

$$\sum_{j\in J}|x_j\overline{y_j}| < \epsilon$$

for every finite subset J of K such that $I \cap J = \emptyset$, and thus

$$\Big|\sum_{j\in J}x_j\overline{y_j}\Big| \leq \sum_{i\in J}|x_j\overline{y_j}| < \epsilon,$$

proving that $(x_k\overline{y_k})_{k\in K}$ is a Cauchy family, and thus summable). We still have to prove that $\ell^2(K)$ is complete.

Consider a sequence $((\lambda_k^n)_{k\in K})_{n\geq 1}$ of sequences $(\lambda_k^n)_{k\in K} \in \ell^2(K)$, and assume that it is a Cauchy sequence. This means that for every $\epsilon > 0$, there is some $N \geq 1$ such that

$$\sum_{k\in K} |\lambda_k^m - \lambda_k^n|^2 < \epsilon^2$$

for all $m, n \geq N$. For every fixed $k \in K$, this implies that

$$|\lambda_k^m - \lambda_k^n| < \epsilon$$

for all $m, n \geq N$, which shows that $(\lambda_k^n)_{n\geq 1}$ is a Cauchy sequence in $\mathbb{C}$. Since $\mathbb{C}$ is complete, the sequence $(\lambda_k^n)_{n\geq 1}$ has a limit $\lambda_k \in \mathbb{C}$. We claim that $(\lambda_k)_{k\in K} \in \ell^2(K)$ and that this is the limit of $((\lambda_k^n)_{k\in K})_{n\geq 1}$.

Given any $\epsilon > 0$, the fact that $((\lambda_k^n)_{k\in K})_{n\geq 1}$ is a Cauchy sequence implies that there is some $N \geq 1$ such that for every finite subset I of K, we have

$$\sum_{i\in I} |\lambda_i^m - \lambda_i^n|^2 < \epsilon/4$$

for all $m, n \geq N$. Let $p = |I|$. Then,

$$|\lambda_i^m - \lambda_i^n| < \frac{\sqrt{\epsilon}}{2\sqrt{p}}$$

for every $i \in I$. Since λ_i is the limit of $(\lambda_i^n)_{n\geq 1}$, we can find some n large enough so that

$$|\lambda_i^n - \lambda_i| < \frac{\sqrt{\epsilon}}{2\sqrt{p}}$$

for every $i \in I$. Since

$$|\lambda_i^m - \lambda_i| \leq |\lambda_i^m - \lambda_i^n| + |\lambda_i^n - \lambda_i|,$$

we get

$$|\lambda_i^m - \lambda_i| < \frac{\sqrt{\epsilon}}{\sqrt{p}},$$

and thus,

$$\sum_{i\in I} |\lambda_i^m - \lambda_i|^2 < \epsilon,$$

for all $m \geq N$. Since the above holds for every finite subset I of K, by Proposition D.12, we get

$$\sum_{k \in K} |\lambda_k^m - \lambda_k|^2 < \epsilon,$$

for all $m \geq N$. This proves that $(\lambda_k^m - \lambda_k)_{k \in K} \in \ell^2(K)$ for all $m \geq N$, and since $\ell^2(K)$ is a vector space and $(\lambda_k^m)_{k \in K} \in \ell^2(K)$ for all $m \geq 1$, we get $(\lambda_k)_{k \in K} \in \ell^2(K)$. However,

$$\sum_{k \in K} |\lambda_k^m - \lambda_k|^2 < \epsilon$$

for all $m \geq N$, means that the sequence $(\lambda_k^m)_{k \in K}$ converges to $(\lambda_k)_{k \in K} \in \ell^2(K)$. The fact that the subspace consisting of sequences $(z_k)_{k \in K}$ such that $z_k = 0$ except perhaps for finitely many k is a dense subspace of $\ell^2(K)$ is left as an easy exercise. ☐

Remark: The subspace consisting of all sequences $(z_k)_{k \in K}$ such that $z_k = 0$, except perhaps for finitely many k, provides an example of a subspace which is not closed in $\ell^2(K)$. Indeed, this space is strictly contained in $\ell^2(K)$, since there are countable sequences of nonnull elements in $\ell^2(K)$ (why?).

We just need two more propositions before being able to prove that every Hilbert space is isomorphic to some $\ell^2(K)$.

Proposition D.15. *Let E be a Hilbert space, and $(u_k)_{k \in K}$ an orthogonal family in E. The following properties hold:*

(1) For every family $(\lambda_k)_{k \in K} \in \ell^2(K)$, the family $(\lambda_k u_k)_{k \in K}$ is summable. Furthermore, $v = \sum_{k \in K} \lambda_k u_k$ is the only vector such that $c_k = \lambda_k$ for all $k \in K$, where the c_k are the Fourier coefficients of v.
(2) For any two families $(\lambda_k)_{k \in K} \in \ell^2(K)$ and $(\mu_k)_{k \in K} \in \ell^2(K)$, if $v = \sum_{k \in K} \lambda_k u_k$ and $w = \sum_{k \in K} \mu_k u_k$, we have the following equation, also called Parseval identity:

$$\langle v, w \rangle = \sum_{k \in K} \lambda_k \overline{\mu_k}.$$

Proof. (1) The fact that $(\lambda_k)_{k \in K} \in \ell^2(K)$ means that $(|\lambda_k|^2)_{k \in K}$ is summable. The proof given in Proposition D.13(3) applies to the family $(|\lambda_k|^2)_{k \in K}$ (instead of $(|c_k|^2)_{k \in K}$), and yields the fact that $(\lambda_k u_k)_{k \in K}$ is summable. Letting $v = \sum_{k \in K} \lambda_k u_k$, recall that $c_k = \langle v, u_k \rangle / \|u_k\|^2$. Pick some $k \in K$. Since $\langle -, - \rangle$ is continuous, for every $\epsilon > 0$, there is some $\eta > 0$ such that

$$|\langle v, u_k \rangle - \langle w, u_k \rangle| < \epsilon \|u_k\|^2$$

whenever

$$\|v - w\| < \eta.$$

However, since for every $\eta > 0$, there is some finite subset I of K such that

$$\left\| v - \sum_{j \in J} \lambda_j u_j \right\| < \eta$$

for every finite subset J of K such that $I \subseteq J$, we can pick $J = I \cup \{k\}$, and letting $w = \sum_{j \in J} \lambda_j u_j$, we get

$$\left| \langle v, u_k \rangle - \left\langle \sum_{j \in J} \lambda_j u_j, u_k \right\rangle \right| < \epsilon \|u_k\|^2.$$

However,

$$\langle v, u_k \rangle = c_k \|u_k\|^2 \quad \text{and} \quad \left\langle \sum_{j \in J} \lambda_j u_j, u_k \right\rangle = \lambda_k \|u_k\|^2,$$

and thus, the above proves that $|c_k - \lambda_k| < \epsilon$ for every $\epsilon > 0$, and thus, that $c_k = \lambda_k$.

(2) Since $\langle -, - \rangle$ is continuous, for every $\epsilon > 0$, there are some $\eta_1 > 0$ and $\eta_2 > 0$, such that

$$|\langle x, y \rangle| < \epsilon$$

whenever $\|x\| < \eta_1$ and $\|y\| < \eta_2$. Since $v = \sum_{k \in K} \lambda_k u_k$ and $w = \sum_{k \in K} \mu_k u_k$, there is some finite subset I_1 of K such that

$$\left\| v - \sum_{j \in J} \lambda_j u_j \right\| < \eta_1$$

for every finite subset J of K such that $I_1 \subseteq J$, and there is some finite subset I_2 of K such that

$$\left\| w - \sum_{j \in J} \mu_j u_j \right\| < \eta_2$$

for every finite subset J of K such that $I_2 \subseteq J$. Letting $I = I_1 \cup I_2$, we get

$$\left| \left\langle v - \sum_{i \in I} \lambda_i u_i, w - \sum_{i \in I} \mu_i u_i \right\rangle \right| < \epsilon.$$

Furthermore,

$$\begin{aligned}\langle v, w\rangle &= \left\langle v - \sum_{i\in I}\lambda_i u_i + \sum_{i\in I}\lambda_i u_i, w - \sum_{i\in I}\mu_i u_i + \sum_{i\in I}\mu_i u_i\right\rangle \\ &= \left\langle v - \sum_{i\in I}\lambda_i u_i, w - \sum_{i\in I}\mu_i u_i\right\rangle + \sum_{i\in I}\lambda_i\overline{\mu_i},\end{aligned}$$

since the u_i are orthogonal to $v - \sum_{i\in I}\lambda_i u_i$ and $w - \sum_{i\in I}\mu_i u_i$ for all $i \in I$. This proves that for every $\epsilon > 0$, there is some finite subset I of K such that

$$\left|\langle v, w\rangle - \sum_{i\in I}\lambda_i\overline{\mu_i}\right| < \epsilon.$$

We already know from Proposition D.14 that $(\lambda_k\overline{\mu_k})_{k\in K}$ is summable, and since $\epsilon > 0$ is arbitrary, we get

$$\langle v, w\rangle = \sum_{k\in K}\lambda_k\overline{\mu_k},$$

as claimed. □

The next proposition states properties characterizing Hilbert bases (total orthogonal families).

Proposition D.16. *Let E be a Hilbert space, and let $(u_k)_{k\in K}$ be an orthogonal family in E. The following properties are equivalent:*

(1) The family $(u_k)_{k\in K}$ is a total orthogonal family.
(2) For every vector $v \in E$, if $(c_k)_{k\in K}$ are the Fourier coefficients of v, then the family $(c_k u_k)_{k\in K}$ is summable and $v = \sum_{k\in K} c_k u_k$.
(3) For every vector $v \in E$, we have the Parseval identity:

$$\|v\|^2 = \sum_{k\in K}|c_k|^2.$$

(4) For every vector $u \in E$, if $\langle u, u_k\rangle = 0$ for all $k \in K$, then $u = 0$.

Proof. The equivalence of (1), (2), and (3), is an immediate consequence of Proposition D.13 and Proposition D.15.

(4) If $(u_k)_{k\in K}$ is a total orthogonal family and $\langle u, u_k\rangle = 0$ for all $k \in K$, since $u = \sum_{k\in K} c_k u_k$ where $c_k = \langle u, u_k\rangle/\|u_k\|^2$, we have $c_k = 0$ for all $k \in K$, and $u = 0$.

Conversely, assume that the closure V of $(u_k)_{k\in K}$ is different from E. Then, by Proposition D.7, we have $E = V \oplus V^{\perp}$, where $V^{\perp}$ is the orthogonal complement of V, and $V^{\perp}$ is nontrivial since $V \neq E$. As a consequence, there is some nonnull vector $u \in V^{\perp}$. But then, u is orthogonal to every vector in V, and in particular,

$$\langle u, u_k \rangle = 0$$

for all $k \in K$, which, by assumption, implies that $u = 0$, contradicting the fact that $u \neq 0$. □

Remarks:

(1) If E is a Hilbert space and $(u_k)_{k\in K}$ is a total orthogonal family in E, there is a simpler argument to prove that $u = 0$ if $\langle u, u_k \rangle = 0$ for all $k \in K$, based on the continuity of $\langle -, - \rangle$. The argument is to prove that the assumption implies that $\langle v, u \rangle = 0$ for all $v \in E$. Since $\langle -, - \rangle$ is positive definite, this implies that $u = 0$. By continuity of $\langle -, - \rangle$, for every $\epsilon > 0$, there is some $\eta > 0$ such that for every finite subset I of K, for every family $(\lambda_i)_{i\in I}$, for every $v \in E$,

$$\left| \langle v, u \rangle - \left\langle \sum_{i\in I} \lambda_i u_i, u \right\rangle \right| < \epsilon$$

whenever

$$\left\| v - \sum_{i\in I} \lambda_i u_i \right\| < \eta.$$

Since $(u_k)_{k\in K}$ is dense in E, for every $v \in E$, there is some finite subset I of K and some family $(\lambda_i)_{i\in I}$ such that

$$\left\| v - \sum_{i\in I} \lambda_i u_i \right\| < \eta,$$

and since by assumption, $\left\langle \sum_{i\in I} \lambda_i u_i, u \right\rangle = 0$, we get

$$|\langle v, u \rangle| < \epsilon.$$

Since this holds for every $\epsilon > 0$, we must have $\langle v, u \rangle = 0$

(2) If V is any nonempty subset of E, the kind of argument used in the previous remark can be used to prove that $V^{\perp}$ is closed (even if V is not), and that $V^{\perp\perp}$ is the closure of V.

We will now prove that every Hilbert space has some Hilbert basis. This requires using a fundamental theorem from set theory known as *Zorn's lemma*, which we quickly review.

Given any set X with a partial ordering $\leq$, recall that a nonempty subset C of X is a *chain* if it is totally ordered (i.e., for all $x, y \in C$, either $x \leq y$ or $y \leq x$). A nonempty subset Y of X is *bounded* iff there is some $b \in X$ such that $y \leq b$ for all $y \in Y$. Some $m \in X$ is *maximal* iff for every $x \in X$, $m \leq x$ implies that $x = m$. We can now state Zorn's lemma. For more details, see Rudin [Rudin (1987)], Lang [Lang (1993)], or Artin [Artin (1991)].

Proposition D.17. *Given any nonempty partially ordered set X, if every (nonempty) chain in X is bounded, then X has some maximal element.*

We can now prove the existence of Hilbert bases. We define a partial order on families $(u_k)_{k\in K}$ as follows: For any two families $(u_k)_{k\in K_1}$ and $(v_k)_{k\in K_2}$, we say that

$$(u_k)_{k\in K_1} \leq (v_k)_{k\in K_2}$$

iff $K_1 \subseteq K_2$ and $u_k = v_k$ for all $k \in K_1$. This is clearly a partial order.

Proposition D.18. *Let E be a Hilbert space. Given any orthogonal family $(u_k)_{k\in K}$ in E, there is a total orthogonal family $(u_l)_{l\in L}$ containing $(u_k)_{k\in K}$.*

Proof. Consider the set $\mathcal{S}$ of all orthogonal families greater than or equal to the family $B = (u_k)_{k\in K}$. We claim that every chain in $\mathcal{S}$ is bounded. Indeed, if $C = (C_l)_{l\in L}$ is a chain in $\mathcal{S}$, where $C_l = (u_{k,l})_{k\in K_l}$, the union family

$$(u_k)_{k\in \bigcup_{l\in L} K_l}, \text{ where } u_k = u_{k,l} \text{ whenever } k \in K_l,$$

is clearly an upper bound for C, and it is immediately verified that it is an orthogonal family. By Zorn's lemma D.17, there is a maximal family $(u_l)_{l\in L}$ containing $(u_k)_{k\in K}$. If $(u_l)_{l\in L}$ is not dense in E, then its closure V is strictly contained in E, and by Proposition D.7, the orthogonal complement $V^\perp$ of V is nontrivial since $V \neq E$. As a consequence, there is some nonnull vector $u \in V^\perp$. But then, u is orthogonal to every vector in $(u_l)_{l\in L}$, and we can form an orthogonal family strictly greater than $(u_l)_{l\in L}$ by adding u to this family, contradicting the maximality of $(u_l)_{l\in L}$. Therefore, $(u_l)_{l\in L}$ is dense in E, and thus, it is a Hilbert basis. $\square$

Remark: It is possible to prove that all Hilbert bases for a Hilbert space E have index sets K of the same cardinality. For a proof, see Bourbaki [Bourbaki (1981)].

At last, we can prove that every Hilbert space is isomorphic to some Hilbert space $\ell^2(K)$ for some suitable K.

Theorem D.19. *(Riesz–Fischer) For every Hilbert space E, there is some nonempty set K such that E is isomorphic to the Hilbert space $\ell^2(K)$. More specifically, for any Hilbert basis $(u_k)_{k\in K}$ of E, the maps $f\colon \ell^2(K) \to E$ and $g\colon E \to \ell^2(K)$ defined such that*

$$f\left((\lambda_k)_{k\in K}\right) = \sum_{k\in K} \lambda_k u_k \quad \textit{and} \quad g(u) = \left(\langle u, u_k\rangle / \|u_k\|^2\right)_{k\in K} = (c_k)_{k\in K},$$

are bijective linear isometries such that $g \circ f = \mathrm{id}$ and $f \circ g = \mathrm{id}$.

Proof. By Proposition D.15(1), the map f is well defined, and it it clearly linear. By Proposition D.13(3), the map g is well defined, and it is also clearly linear. By Proposition D.13(2b), we have

$$f(g(u)) = u = \sum_{k\in K} c_k u_k,$$

and by Proposition D.15(1), we have

$$g(f\left((\lambda_k)_{k\in K}\right)) = (\lambda_k)_{k\in K},$$

and thus $g \circ f = \mathrm{id}$ and $f \circ g = \mathrm{id}$. By Proposition D.15(2), the linear map g is an isometry. Therefore, f is a linear bijection and an isometry between $\ell^2(K)$ and E, with inverse g. ☐

Remark: The surjectivity of the map $g\colon E \to \ell^2(K)$ is known as the *Riesz–Fischer* theorem.

Having done all this hard work, we sketch how these results apply to Fourier series. Again, we refer the readers to Rudin [Rudin (1987)] or Lang [Lang (1996, 1997)] for a comprehensive exposition.

Let $\mathcal{C}(T)$ denote the set of all periodic continuous functions $f\colon [-\pi, \pi] \to \mathbb{C}$ with period 2π. There is a Hilbert space $\mathrm{L}^2(T)$ containing $\mathcal{C}(T)$ and such that $\mathcal{C}(T)$ is dense in $\mathrm{L}^2(T)$, whose inner product is given by

$$\langle f, g\rangle = \int_{-\pi}^{\pi} f(x)\overline{g(x)}dx.$$

The Hilbert space $\mathrm{L}^2(T)$ is the space of *Lebesgue square-integrable periodic functions* (of period 2π).

It turns out that the family $(e^{ikx})_{k\in\mathbb{Z}}$ is a total orthogonal family in $\mathrm{L}^2(T)$, because it is already dense in $\mathcal{C}(T)$ (for instance, see Rudin [Rudin (1987)]). Then, the Riesz–Fischer theorem says that for every family $(c_k)_{k\in\mathbb{Z}}$ of complex numbers such that

$$\sum_{k\in\mathbb{Z}} |c_k|^2 < \infty,$$

there is a unique function $f \in \mathrm{L}^2(T)$ such that f is equal to its Fourier series

$$f(x) = \sum_{k\in\mathbb{Z}} c_k e^{ikx},$$

where the Fourier coefficients c_k of f are given by the formula

$$c_k = \frac{1}{2\pi}\int_{-\pi}^{\pi} f(t)e^{-ikt}dt.$$

The Parseval theorem says that

$$\sum_{k=-\infty}^{+\infty} c_k\overline{d_k} = \frac{1}{2\pi}\int_{-\pi}^{\pi} f(t)\overline{g(t)}dt$$

for all $f, g \in \mathrm{L}^2(T)$, where c_k and d_k are the Fourier coefficients of f and g.

Thus, there is an isomorphism between the two Hilbert spaces $\mathrm{L}^2(T)$ and $\ell^2(\mathbb{Z})$, which is the deep reason why the Fourier coefficients "work". Theorem D.19 implies that the Fourier series $\sum_{k\in\mathbb{Z}} c_k e^{ikx}$ of a function $f \in \mathrm{L}^2(T)$ converges to f in the L^2-sense, i.e., in the mean-square sense. This does not necessarily imply that the Fourier series converges to f pointwise! This is a subtle issue, and for more on this subject, the reader is referred to Lang [Lang (1996, 1997)] or Schwartz [Schwartz (1993a,b)].

We can also consider the set $\mathcal{C}([-1,1])$ of continuous functions $f\colon [-1,1] \to \mathbb{C}$. There is a Hilbert space $\mathrm{L}^2([-1,1])$ containing $\mathcal{C}([-1,1])$ and such that $\mathcal{C}([-1,1])$ is dense in $\mathrm{L}^2([-1,1])$, whose inner product is given by

$$\langle f, g\rangle = \int_{-1}^{1} f(x)\overline{g(x)}dx.$$

The Hilbert space $\mathrm{L}^2([-1,1])$ is the space of *Lebesgue square-integrable functions* over $[-1,1]$. The Legendre polynomials $P_n(x)$ form a Hilbert basis of $\mathrm{L}^2([-1,1])$.

Recall that if we let f_n be the function

$$f_n(x) = (x^2 - 1)^n,$$

$P_n(x)$ is defined as follows:

$$P_0(x) = 1, \quad \text{and} \quad P_n(x) = \frac{1}{2^n n!} f_n^{(n)}(x),$$

where $f_n^{(n)}$ is the nth derivative of f_n. The reason for the leading coefficient is to get $P_n(1) = 1$. It can be shown with much efforts that

$$P_n(x) = \sum_{0 \le k \le n/2} (-1)^k \frac{(2(n-k))!}{2^n (n-k)! k! (n-2k)!} x^{n-2k}.$$

Appendix E

Well-Ordered Sets, Ordinals, Cardinals, Alephs

The purpose of this chapter is to define the notions of ordinal, cardinal and alephs, and to review some of their main properties. Intuitively the ordinals are the equivalence classes of well-ordered sets under the equivalence relation of order-isomorphism (the order-types). This idea goes back to Cantor; see Levy [Levy (1979)] for a thorough discussion of this approach. However, such a definition does not make sense because the collection of well-ordered sets is not a set. To circumvent this difficulty, following Von Neumann, we can define an ordinal as a certain special type of set.

E.1 Well-Ordered Sets

We begin by reviewing the notions of partial orders, total orders, strict partial orders, and strict total orders. Given a set X and a binary relation $\preceq\, \subseteq X \times X$ on X, we write $x \preceq y$ for $(x, y) \in\, \preceq$ and $x \not\preceq y$ for $\neg(x \preceq y)$.

Definition E.1. Given a set X, a binary relation $\leq$ on X is a *partial order* if it satisfies the following properties:

(1) The relation $\leq$ is *reflexive*, which means that for all x, if $x \in X$, then $x \leq x$.
(2) The relation $\leq$ is *transitive*, which means that for all x, y, z, if $x, y, z \in X$, $x \leq y$ and $y \leq z$, then $x \leq z$.
(3) The relation $\leq$ is *antisymmetric*, which means that for all x, y, if $x, y \in X$, $x \leq y$ and $y \leq x$, then $x = y$. The pair $(X, \leq)$ is called a *partially ordered set*.

A binary relation $\leq$ on X is a *total order (or simple order)* if it is a partial order and if it is *strongly connected*, which means that for all x, y, if $x, y \in X$, then either $x \leq y$ or $y \leq x$. The pair $(X, \leq)$ is called a *totally ordered set*.

The empty set (with the empty relation) is trivially a partially and a totally ordered set.

Example E.1.

(1) Given any nonempty set X, the inclusion relation $Y \subseteq Z$ on subsets Y and Z of X is a partial order which is not a total order if X has at least two elements.
(2) The set $\mathbb{N}$ of natural numbers with its usual ordering is a totally ordered set.
(3) The set $\mathbb{Z}$ of integers with its usual ordering is a totally ordered set.
(4) The relation $\ll$ on $\mathbb{N} \times \mathbb{N}$ defined such that for all $(m_1, n_1), (m_2, n_2) \in \mathbb{N} \times \mathbb{N}$,

$$(m_1, n_1) \ll (m_2, n_2) \quad \text{iff} \quad \begin{cases} m_1 = m_2 \text{ and } n_1 = n_2, \text{ or} \\ m_1 < m_2, \text{ or} \\ m_1 = m_2 \text{ and } n_1 < n_2 \end{cases}$$

is a total order.

Definition E.2. Given a set X, a binary relation $\leq$ on X is a *strict partial order* if it satisfies the following properties:

(1) The relation $\leq$ is *asymmetric*, which means that for all x, y, if $x, y \in X$, then either $x \not\leq y$ or $y \not\leq x$, equivalently $\neg((x \leq y) \wedge (y \leq x))$.
(2) The relation $\leq$ is *transitive*, which means that for all x, y, z, if $x, y, z \in X$, $x \leq y$ and $y \leq z$, then $x \leq z$. The pair $(X, \leq)$ is called a *strictly partially ordered set*.

A binary relation $\leq$ on X is a *strict total order (or strict simple order)* if it is a strict partial order and if it is *connected*, which means that for all x, y, if $x, y \in X$ and $x \neq y$, then $x \leq y$ or $y \leq x$. The pair $(X, \leq)$ is called a *strictly totally ordered set*.

The empty set (with the empty relation) is trivially a strictly partially and a strictly totally ordered set.

Example E.2.

(1) Given any nonempty set X, the strict inclusion relation $Y \subseteq Z$ and $Y \neq Z$ on subsets Y and Z of X is a strict partial order which is not a strict total order if X has at least two elements.

(2) The set $\mathbb{N}$ of natural numbers with the strict ordering $m < n$ (namely $m \leq n$ and $m \neq n$) is a strictly totally ordered set.
(3) The set $\mathbb{Z}$ of integers with its strict ordering $m < n$ (namely $m \leq n$ and $m \neq n$) is a strictly totally ordered set.
(4) The relation $\ll$ on $\mathbb{N} \times \mathbb{N}$ defined such that for all $(m_1, n_1), (m_2, n_2) \in \mathbb{N} \times \mathbb{N}$,

$$(m_1, n_1) \ll (m_2, n_2) \quad \text{iff} \quad \begin{cases} m_1 < n_1, \text{ or} \\ m_1 = n_1 \text{ and } m_2 < n_2 \end{cases}$$

is a strict total order.

Definition E.3. Given a set X, a partial order $\leq$ on X is a *well-order* if every nonempty subset Y of X has a smallest element, which can be expressed as follows: for all Y, if $Y \neq \emptyset$ and $Y \subseteq X$, then there is some $x \in Y$ such that for y, if $y \in Y$, then $x \leq y$. The pair $(X, \leq)$ is called a *well-ordered set*.

A strict partial order $\leq$ on X is a *strictly well-order* if every nonempty subset Y of X has a smallest element. The pair $(X, \leq)$ is called a *strictly well-ordered set*.

The empty set (with the empty relation) is trivially a well-ordered set and strictly well-ordered set. If a well-ordered set is nonempty, then by picking $Y = \{x, y\}$ for any $x, y \in X$, since Y must have a smallest element, we see that either $x \leq y$ or $y \leq x$, that is, a well-ordered set is totally ordered. The same reasoning shows that a strictly well-ordered set is strictly totally ordered.

Example E.3.

(1) The partial order of Example E.1 is not a well-order (in fact, it is not a total order).
(2) The set $\mathbb{N}$ is well-ordered under its natural ordering.
(3) The set $\mathbb{Z}$ is not well-ordered under its natural ordering. For example, the subset $\{n \in \mathbb{Z} \mid n \leq 0\}$ does not have a smallest element.
(4) The set $\mathbb{N} \times \mathbb{N}$ under the total order of Example E.1 is well-ordered.

Proposition E.1. *Let $(X, \leq)$ be a partially ordered set. The relation $<$ on X given by*

$$x < y \quad \textit{iff} \quad x \leq y \textit{ and } x \neq y$$

is a strict partial order on X. If $(X, \leq)$ is a totally ordered set, then the relation $<$ on X defined above is a strict total order. If $(X, \leq)$ is a well-ordered set, then the relation $<$ on X defined above is a strict well-order.

Proof. Assume that $(X, \leq)$ is a partially ordered set. The relation $<$ is transitive because if $x < y$ and $y < z$, then $x \leq y$, $y \leq z$, $x \neq y$ and $y \neq z$, so by transitivity of $\leq$ we have $x \leq z$. If $x = z$, then $y \leq z$ is equivalent to $y \leq x$, and since $x \leq y$, and $\leq$ is antisymmetric, we get $x = y$, a contradiction. The relation $<$ is asymmetric, because if $x < y$ and $y < x$, then $x \leq y$, $y \leq x$ and $x \neq y$, but since $\leq$ is antisymmetric, $x = y$, a contradiction.

The other statements are left as exercises to the reader. □

We say that $(X, <)$ is the strictly partially ordered set associated with the partially ordered set $(X, \leq)$, *etc.*

A detailed exposition of the above results and much more can be found in Suppes [Suppes (1972)].

The importance of well-orders has to do with the fact that they support a powerful induction principle.

Definition E.4. For any partially ordered set $(E, \leq)$, for any $x \in E$, the subset $s(x) = \{y \in E \mid y < x\} = \{y \in E \mid y \leq x,\ y \neq x\}$ is called an *initial segment* of E.

Theorem E.2. *Let $(E, \leq)$ be a well-ordered set. For any subset A of E, if for every $a \in E$,*

$$\textit{if } a \in A \textit{ whenever } b \in A \textit{ for all } b \in E \textit{ such that } b < a,$$

then $A = E$. Equivalently, for all $a \in E$, if $s(a) \subseteq A$ implies that $a \in A$, then $A = E$.

Proof. Suppose by contradiction that $A \neq E$. Then the subset $E - A$ is nonempty, and since E is well-ordered, it has a least element $b \notin A$. We claim that $s(b) \subseteq A$. Indeed, $y \in s(b)$ iff $y < b$, but then we can't have $y \in E - A$, because this would contradict the fact that b is the smallest element of $E - A$, so $y \in A$. Since $s(b) \subseteq A$, by hypothesis $b \in A$, a contradiction. □

Theorem E.2 immediately implies the following induction principle.

Theorem E.3. *Let $(E, \leq)$ be a well-ordered set and let $P(x)$ be a first-order formula with free variable x. For every $a \in E$, if $P(a)$ holds whenever $P(b)$ holds for all $b \in E$ such that $b < a$, then $P(x)$ holds for all $x \in E$.*

Theorem E.3 follows immediately from Theorem E.2 by setting $A = \{a \in E \mid P(a) = \mathbf{true}\}$. The induction principle in Theorem E.3 is sometimes called *transfinite induction on a well-ordered set*. It is a generalization of complete induction on $\mathbb{N}$.

Definition E.5. Let $(X_1, \leq_1)$ and $(X_2, \leq_2)$ be two partially ordered sets. A function $f\colon X_1 \to X_2$ is an *(order) isomorphism* if it is a bijection and if

$$x \leq_1 y \quad \text{iff} \quad f(x) \leq_2 f(y), \qquad \text{for all } x, y \in X_1.$$

The same definition applies if $(X_1, \leq_1)$ and $(X_2, \leq_2)$ are two strictly partially ordered sets, and if the orderings are total or well-orders.

Note that a well-ordered set may be isomorphic to a proper subset of itself. For example, $(\mathbb{N}, \leq)$ is isomorphic to $(2\mathbb{N}, \leq)$ (where $2\mathbb{N} = \{2n \mid n \in \mathbb{N}\}$). However, we have the following important results.

Proposition E.4. *Let $(E, \leq)$ be a well-ordered set. If $f\colon E \to E$ is a function such that for all $x, y \in E$, if $x \neq y$ and $x \leq y$ implies that $f(x) \neq f(y)$ and $f(x) \leq f(y)$, then*

$$x \leq f(x) \quad \textit{for all } x \in E.$$

Using Proposition E.4 we can prove the following result.

Proposition E.5. *Let $(E_1, \leq_1)$ and $(E_2, \leq_2)$ be two well-ordered sets. If $f\colon E_1 \to E_2$ and $g\colon E_1 \to E_2$ are isomorphisms, then $f = g$.*

As a corollary of Proposition E.5 it can be shown that if $(E, \leq)$ is a well-ordered set, then there is no isomorphism between E and any initial segment $s(x) = \{y \in E \mid y < x\}$, for any $x \in E$.

E.2 Ordinals

Technically, the definition of an ordinal depends on the precise axiomatic definition chosen for set theory (in first-order logic), specifically whether individual constants other than the symbol $\emptyset$ (the empty set) are allowed. Suppes [Suppes (1972)] allows such individual symbols. For simplicity we follow Krivine [Krivine (1969)] who does not allow such symbols. What this means is that the sets under consideration only have other sets as members, building up from the empty set.

Definition E.6. An *ordinal* is a set α such that

(1) The membership relation $x \in y$ on α (with $x, y \in \alpha$) is a strict well-order.
(2) For every x, if $x \in \alpha$, then $x \subseteq \alpha$. By definition of the inclusion relation, this means that for all x, y, if $y \in x$ and $x \in \alpha$, then $y \in \alpha$. Sometimes it is said that α is a *transitive set*.[4]

Remark: One of the axioms of set theory, the *sum axiom, also called the union axiom*, states that for every set X, the collection of all y such that $y \in x$ for some $x \in X$ is a set, denoted $\bigcup X$ or $\bigcup_{x \in X} x$. Then Condition (2) of Definition E.6 is equivalent to the condition

$$\bigcup \alpha \subseteq \alpha.$$

This condition is used in Suppes [Suppes (1972)] and Levy [Levy (1979)].

We see that Definition E.6 implies that an ordinal is a set of sets of sets, *etc.*

For example, $\emptyset$, $\{\emptyset\}$, $\{\emptyset, \{\emptyset\}\}$, $\{\emptyset, \{\emptyset\}, \{\emptyset, \{\emptyset\}\}\}$ are ordinals, and more generally, if α is an ordinal, then $\alpha \cup \{\alpha\}$ is also an ordinal denoted α^+. The ordinal $\emptyset$ is also denoted by 0.

This is the method used by Von Neumann to define the natural numbers. The number 0 is represented by the empty set, 1 is represented by $\{\emptyset\} = \{0\}$, 2 is represented by $\{\emptyset, \{\emptyset\}\} = \{0, 1\}$, 3 is represented by $\{\emptyset, \{\emptyset\}, \{\emptyset, \{\emptyset\}\}\} = \{0, 1, 2\}$, and if α represents a natural number, then $\alpha^+ = \alpha \cup \{\alpha\}$ represents the natural number $\alpha + 1$. For this reason, we also denote α^+ as $\alpha + 1$.

We now list (mostly) without proof the most important properties of ordinals. Proofs can be found in Suppes [Suppes (1972)] and Krivine [Krivine (1969)]. A more advanced, rigorous and very thorough presentation can be found in Levy [Levy (1979)].

Proposition E.6. *Let α be an ordinal.*

(1) For any $\xi \in \alpha$, we have $s(\xi) = \{\eta \in \alpha \mid \eta \in \xi\} = \xi$.
(2) If $\xi \in \alpha$, then ξ is an ordinal.

Proposition E.7. *For every ordinal α, we have $\alpha \notin \alpha$.*

Proof. For any $\xi \in \alpha$, since the membership relation $\in$ on α is a strict order, we have $\xi \notin \xi$. Then if $\alpha \in \alpha$, we also have $\alpha \notin \alpha$, a contradiction. ☐

[4]This use of the word transitive is unfortunate since it differs from its meaning in Definition E.1(2).

Using Theorem E.3 the following result can be shown.

Proposition E.8. *For any two ordinals α, β, if there is an isomorphism between α and β (each equipped with the strict order of membership), then $\alpha = \beta$.*

Proposition E.9. *For any two ordinals α, β, either $\alpha = \beta$, $\alpha \in \beta$, or $\beta \in \alpha$, and these three cases are mutually exclusive.*

Proposition E.9 implies that for any two ordinals α, β, we have $\alpha \subseteq \beta$ iff $\alpha = \beta$ or $\alpha \in \beta$. It follows that the relation $\alpha \subseteq \beta$ is a total order on the ordinals, and we also write $\alpha \leq \beta$ instead of $\alpha \subseteq \beta$ and $\alpha < \beta$ for $\alpha \in \beta$. Observe that the relation $\alpha \in \beta$ is the strict total order associated with the total order $\subseteq$.

Proposition E.10. *For any ordinal α, the ordinal $\alpha^+ = \alpha \cup \{\alpha\}$ is the smallest ordinal strictly greater than α.*

Proposition E.11. *For any set S of ordinals, the set $\beta = \bigcup_{\alpha \in S} \alpha = \bigcup S$ is an ordinal which is the least upper bound of the set S.*

Proposition E.12. *For any set S of ordinals, the membership relation on S is a strict well-order. As a consequence, for any ordinal α, the ordinals $\beta < \alpha$ form a strictly well-ordered set (under inclusion).*

Proposition E.13. *(Burali–Forti paradox) The collection of all ordinals is not a set.*

Proof. Assume that the collection of all ordinals is a set α. Then by Proposition E.12, the set α is strictly well-ordered. Also, by definition of the set α, if $\beta \in \alpha$, then β is an ordinal, and since by Proposition E.6(2), every $\xi \in \beta$ is an ordinal, we have $\xi \in \alpha$ (since α is the set of all ordinals), so $\beta \subseteq \alpha$. Then by definition of an ordinal, α is an ordinal, and since α is the set of all ordinals, $\alpha \in \alpha$, contradicting Proposition E.7. $\square$

Proposition E.14 confirms that the concept of ordinal captures the idea that the ordinals are the "order-types" of well-ordered sets.

Proposition E.14. *For every well-ordered set $(S, \leq)$, there is a unique ordinal α and a unique isomorphism between $(S, <)$ and α (where $(S, <)$ is the strictly well-ordered set associated with the well-ordered set $(S, \leq)$ and α is strictly well-ordered by the membership relation).*

Proposition E.14 is proven using Theorem E.3.

Finite and infinite ordinals are defined as follows.

Definition E.7. An ordinal α is *finite* if either $\alpha = \emptyset$ or for every $\beta \subseteq \alpha$ with $\beta \neq \emptyset$, there is some ordinal ξ such that $\beta = \xi + 1$. An *infinite ordinal* is an ordinal that is not finite.

Remark: Definition E.7 is the definition found in Levy [Levy (1979)] and Krivine [Krivine (1969)]. A different definition is used in Suppes [Suppes (1972)].

So far we don't know if infinite ordinals exist! The axiom of infinity asserts that infinite ordinals exist.

Axiom of Infinity. There exists an infinite ordinal.

It can be shown that the axiom of infinity is equivalent to the fact that the collection of finite ordinals is a set (which is an ordinal), denoted ω; see Krivine [Krivine (1969)].

Remark: In an axiomatic presentation of the axioms of Zermelo–Frankel set theory it is customary to state a version of the axiom of infinity which does not involve the notion of ordinal. It can be shown that this version of the axiom of infinity is equivalent to the above version about ordinals. For this classical approach, see Suppes [Suppes (1972)] and Levy [Levy (1979)]. Since it is not our intention to give an axiomatic presentation of Zermelo–Frankel set theory, the above version of the axiom of infinity is preferable.

Definition E.8. Under the axiom of infinity, the set of all finite ordinals is an ordinal denoted ω.

In the Von Neumann approach, the natural numbers are identified with the finite ordinals. Thus ω is the set of natural numbers and it is also denoted $\mathbb{N}$ by most mathematicians. The ordinal ω is not a finite ordinal. It is the smallest infinite ordinal because if ξ is an infinite ordinal such that $\xi \in \omega$, then ξ is a finite ordinal (ω is the set of all finite ordinals), a contradiction.

Definition E.9. An ordinal $\alpha \neq \emptyset$ is a *limit ordinal* if for all $\beta \in \alpha$, we also have $\beta + 1 \in \alpha$.

It is easy to see that an ordinal $\alpha \neq \emptyset$ is a limit ordinal iff there is no ordinal β such that $\alpha = \beta + 1$ iff

$$\alpha = \bigcup \alpha = \bigcup_{\beta \in \alpha} \beta$$

Furthermore, it can be shown that every limit ordinal is infinite and that the axiom of infinity is equivalent to the existence of a limit ordinal; see Krivine [Krivine (1969)].

E.3 Cardinals, Alephs ($\aleph_\alpha$) and Beths ($\beth_\alpha$)

Having defined the ordinals, we can define cardinals and the cardinality of a set. This is where the axiom of choice shows its nose.

Definition E.10. A *cardinal* is an ordinal $\mathfrak{a}$ such that if β is any ordinal in bijection with $\mathfrak{a}$, then $\mathfrak{a} \subseteq \beta$.

A cardinal is often referred to as an *initial ordinal*. It appears that the universal notation adopted to denote cardinals is to use lower case German letters ("Fraktur" font), $\mathfrak{a}, \mathfrak{b}$, *etc.* This convention is convenient since if we denote ordinals by lower case Greek letters (as it is customary), then we have a visual mechanism to distinguish between ordinals and cardinals. As we will see shortly, cardinals are also denoted using the Hebrew letter aleph with an ordinal subscript ($\aleph_\alpha$).

Proposition E.15. *Every finite ordinal is a cardinal.*

Definition E.11. The smallest infinite ordinal ω is a cardinal, which is also denoted $\aleph_0$.

As we will see later, there is no largest cardinal, but this is not easy to prove; see Suppes [Suppes (1972)] (Section 7.3, Theorem 60).

Assume that the **axiom of choice holds**. An easy-going version of the axiom of choice is that for any two nonempty sets X and Y, for any surjection $f\colon X \to Y$, there is some injection $g\colon Y \to X$ such that $f \circ g = \mathrm{id}_Y$.

Theorem E.16. *(Zermelo) Every set has some well-ordering.*

A proof of Theorem E.16 can be found in all set theory texts, in particular Suppes [Suppes (1972)] Krivine [Krivine (1969)]. Theorem E.16 is one of many results equivalent to the famous axiom of choice. If you think

Theorem E.16 is obvious, try finding a well-ordering on the power set $\mathcal{P}(\mathbb{N})$ of the set $\mathbb{N}$ of natural numbers.

Now, *if we accept the axiom of choice*, since by Theorem E.16 every set X has some well-order (not unique if X has at least two elements), by Proposition E.14, there is a bijection between X and some ordinal α. Then it is not hard to show that the ordinals β that are in bijection with X form a set (because if γ is an ordinal in bijection with the power set $\mathcal{P}(X)$, then $\beta \in \gamma$), so by Proposition E.12, there is a smallest ordinal, denoted $|X|$, among the ordinals in bijection with X.

Definition E.12. Given any set X, the smallest ordinal $|X|$ (also denoted $\mathrm{card}(X)$) in bijection with X is a cardinal called the *cardinal number (or cardinality)* of X.

It can be shown that the collection of cardinals numbers is *not* a set.

Remark: It is possible to define the notion of cardinality of a set even if we do not assume the axiom of choice. But then the cardinal $|X|$ of set X is a certain kind of set that may not be an ordinal. In fact, the cardinal $|X|$ is an ordinal iff the set $|X|$ is well-orderable. See Levy [Levy (1979)] (Chapter III, Section 2).

Definition E.13. The cardinality of the set $\mathbb{R}$ of real numbers is denoted by $\mathfrak{c}$ and is called the *cardinality of the continuum* (or *power of the continuum*).

It is a standard theorem of set theory that there is a bijection between $\mathcal{P}(\mathbb{N})$, the power set of the set $\mathbb{N}$ of natural numbers, and the set $\mathbb{R}$ of real numbers; see Section 6.7 of Suppes [Suppes (1972)].

Definition E.14. For any cardinal $\mathfrak{a}$, the cardinality of the power set $\mathcal{P}(\mathfrak{a})$ of $\mathfrak{a}$ is denoted $2^{\mathfrak{a}}$.

Using the above definition, the fact that there is a bijection between $\mathcal{P}(N)$ and $\mathbb{R}$ is restated as $\mathfrak{c} = 2^{\aleph_0}$. Cantor's theorem (which says that there is no surjection of a set X onto its power set $\mathcal{P}(X)$) stated in terms of cardinals says that for any cardinal $\mathfrak{a}$, we have

$$\mathfrak{a} < 2^{\mathfrak{a}}.$$

Our next goal is to show that it is possible to provide an enumeration of the infinite cardinals indexed by the ordinals. We first proceed informally. The idea is to define the infinite cardinal $\aleph_\alpha$ for every ordinal α as follows: the cardinal $\aleph_\alpha$ is the infinite cardinal β such that the set

$\{\xi \mid \xi \in \beta, \xi \text{ is an infinite cardinal}\}$ is isomorphic (as a strictly well-ordered set under the membership relation) to α (also with the strict order of membership). Intuitively, $\aleph_\alpha$ is the α's infinite cardinal. So $\aleph_1$ is the smallest cardinal of cardinality strictly greater than $\aleph_0$, then $\aleph_2$ is the smallest cardinal of cardinality strictly greater than $\aleph_1$, and more generally $\aleph_{\alpha+1}$ is the smallest cardinal of cardinality strictly greater than $\aleph_\alpha$. See Definition E.17 for a rigorous approach (which needs to deal with the case where α is a limit ordinal).

Then Cantor's theorem implies that

$$\aleph_{\alpha+1} \subseteq 2^{\aleph_\alpha}.$$

Whether or not the above inequality is actually an equality is a famous problem called the *generalized continuum hypothesis*. For $\alpha = 0$, famous results of Gödel and Cohen show that the statement $\aleph_1 = \mathfrak{c} = 2^{\aleph_0}$ is independent of Zermelo–Frankel set theory (with the axiom of choice).

Our definition of the *alephs* (denoted $\aleph_\alpha$) is not rigorous. It is actually possible to define rigorously the alephs *without assuming the axiom of choice*, as explained in Suppes [Suppes (1972)].

We need to recall the following notation.

Definition E.15. Let A and B be any two sets.

(1) We write $A \approx B$ if there is a bijection from A to B. In this case we say that A and B are *equipollent*.
(2) We write $A \preceq B$ if there is a subset C of B such that $A \approx C$.
(3) We write $A \prec B$ if $A \preceq B$ and $\neg(B \preceq A)$, and we write $A \succ B$ if $B \preceq A$ and $\neg(A \preceq B)$.

Two ordinals may be equipollent and yet be very different in terms of they order structure. A simple example consists of the two ordinals ω and $\omega + 1 = \omega \cup \{\omega\}$. We can define the bijection f from $\omega + 1$ to ω given by $f(\{\omega\}) = 0$ and $f(n) = n + 1$, for any $n \in \omega$.

If $\varphi(\alpha)$ is a first-order formula in which α ranges over the ordinals, it can be shown that if there is some ordinal α such that $\varphi(\alpha)$ holds, then there is a smallest ordinal β such that $\varphi(\beta)$ holds; see Suppes [Suppes (1972)] (Section 7.1, Theorem 5). The above fact suggests the definition of the smallest ordinal $\mu_\alpha(\varphi(\alpha))$ satisfying a first-order formula $\varphi(\alpha)$ (where α denotes an ordinal). If $\forall\alpha\neg\varphi(\alpha)$, that is, $\varphi(\alpha)$ is not satisfied by any ordinal, then we set $\mu_\alpha(\varphi(\alpha)) = 0$.

Definition E.16. Given a first-order formula $\varphi(\alpha)$ where α denotes an ordinal, the ordinal $\mu_\alpha(\varphi(\alpha))$ is defined such that for every ordinal β, we have the equivalence

$$\mu_\alpha(\varphi(\alpha)) = \beta \quad \text{iff} \quad [\varphi(\beta) \land \forall\gamma(\varphi(\gamma) \Longrightarrow (\beta \subseteq \gamma))] \lor [\forall\alpha\neg\varphi(\alpha) \land (\beta = 0)].$$

Then it can be shown that

(1) If $\varphi(\beta)$ holds for some ordinal β, then $\mu_\alpha(\varphi(\alpha)) \subseteq \beta$.
(2) If $\exists\alpha\varphi(\alpha)$ holds, then $\varphi(\mu_\alpha(\varphi(\alpha)))$ holds.

The alephs are then defined by transfinite recursion as follows.

Definition E.17. The ordinals $\aleph_\alpha$ (the *alephs*) are defined as follows:

(1) $\aleph_0 = \omega$.
(2) For any successor ordinal $\alpha + 1$,
$$\aleph_{\alpha+1} = \mu_\beta(\beta \succ \aleph_\alpha).$$
(3) For any limit ordinal α,
$$\aleph_\alpha = \bigcup_{\beta\in\alpha} \aleph_\beta.$$

Actually, we really have to justify why a recursive definition as in Definition E.17 is legitimate. To do so requires delving into axiomatic set theory more than we want to for the purpose of this appendix. Let us just say that the *axiom schema of replacement* (due to Zermelo) is required. Intuitively, this axiom says that if $\varphi(x, y)$ is a functional relation, which means that for all x, y_1, y_2, $\varphi(x, y_1)$ and $\varphi(x, y_2)$ implies that $y_1 = y_2$, then for any set A, the image of A by φ, that is, the collection of y such that $\varphi(x, y)$ for some $x \in A$, is also a set. Then a powerful version of definition by *transfinite recursion* can be established. For details, see Suppes [Suppes (1972)] (Chapter 7). Incidentally, this version of transfinite recursion is also used to define addition, multiplication, and exponentiation of ordinals.

Returning to the alephs, the following properties can be shown; see Suppes [Suppes (1972)] (Chapter 7).

Actually, it is not obvious at all that for every $\aleph_\alpha$, there is some ordinal β such that $\beta \succ \aleph_\alpha$, so that in Clause (2) of Definition E.17, some nonzero ordinal is returned. The next proposition shows that this is indeed the case; see Suppes [Suppes (1972)] (Section 7.3, Theorem 63).

Proposition E.17. *For any ordinal α, there is some ordinal β such that for every ordinal $\gamma \in \alpha$ we have $\beta \succ \aleph_\gamma$.*

Proposition E.17 implies the following result which, together with the equation $\aleph_0 = \omega$, can be used as a definition of $\aleph_\alpha$.

Proposition E.18. *If α is a nonzero ordinal, then*

$$\aleph_\alpha = \mu_\beta(\forall\gamma((\gamma \in \alpha) \Longrightarrow (\beta \succ \aleph_\gamma))).$$

Proposition E.19. *For every ordinal α, the ordinal $\aleph_\alpha$ is an infinite cardinal.*

Proposition E.20. *For any two ordinals α, β, if $\alpha \in \beta$, then $\aleph_\alpha \in \aleph_\beta$.*

Proposition E.20 implies that there is no largest aleph.

Proposition E.21. *For any ordinal α, there is no infinite cardinal β such that $\aleph_\alpha \in \beta \in \aleph_{\alpha+1}$.*

It can also be shown that every cardinal $\aleph_\alpha$ is a limit ordinal; see Levy [Levy (1979)]. Finally, every infinite cardinal arises as some aleph, which means that there is an "enumeration" of the infinite cardinals by the ordinals.

Theorem E.22. *For every infinite cardinal $\mathfrak{a}$, there is an ordinal α (necessarily unique by Proposition E.20) such that $\mathfrak{a} = \aleph_\alpha$.*

All the above results do *not* rely on the axiom of choice. However, the axiom of choice is needed to show that every set has a cardinal (is in bijection with a cardinal).

Remark: Let us again assume that the axiom of choice holds. Then we can restate the generalized continuum hypothesis by introducing cardinals known as the *beth's*.

Definition E.18. We define by transfinite recursion the cardinals *beth α*, denoted $\beth_\alpha$, as follows: for every ordinal α,

$$\begin{aligned}
\beth_0 &= \aleph_0 \\
\beth_{\alpha+1} &= \mathrm{card}(\mathcal{P}(\beth_\alpha)) \\
\beth_\alpha &= \bigcup_{\beta<\alpha} \beth_\beta \quad \text{if } \alpha \text{ is a limit ordinal.}
\end{aligned}$$

Observe that

$$\beth_1 = \mathfrak{c},$$

the cardinality of the continuum. We can show by transfinite induction that

$$\aleph_\alpha \leq \beth_\alpha$$

for every ordinal α, and the generalized continuum hypothesis is restated as

$$\aleph_\alpha = \beth_\alpha$$

for every ordinal α. The continuum hypothesis is restated as

$$\aleph_1 = \beth_1.$$

Infinite ordinals beyond ω are very hard to understand. A way to get a better grasp of the infinite ordinals is to generalize the operations of addition, multiplication, and exponentiation defined on the natural numbers (the finite ordinals) to infinite ordinals. This is done by generalizing the familiar recursive definitions to infinite ordinals, and the trick for doing so is to extends these recursive definitions to limit ordinals. However such developments are too peripheral to harmonic analysis to be covered here.

Bibliography

Ahlfors, L. V. and Sario, L. (1960). *Riemann Surfaces*, Princeton Math. Series, No. 2 (Princeton University Press).

Apostol, T. M. (1976). *Introduction to Analytic Number Theory*, 1st edn., Undergraduate Texts in Mathematics (Springer).

Artin, M. (1991). *Algebra*, 1st edn. (Prentice Hall).

Berger, M. (1990). *Géométrie 1* (Nathan), english edition: Geometry 1, Universitext, Springer Verlag.

Bourbaki, N. (1965a). *Elements of Mathematics. Intégration, Chapters 1–4*, 2nd edn., Eléments de Mathématiques (Hermann).

Bourbaki, N. (1965b). *Elements of Mathematics. Intégration, Chapters 7–8. Mesure de Haar, Convolution et Représentations*, 2nd edn., Eléments de Mathématiques (Hermann).

Bourbaki, N. (1967a). *Elements of Mathematics. Intégration, Chapter 5. Intégration des mesures*, 2nd edn., Eléments de Mathématiques (Hermann).

Bourbaki, N. (1967b). *Elements of Mathematics. Théories Spectrales, Chapters 1–2*, 1st edn., Eléments de Mathématiques (Hermann).

Bourbaki, N. (1968a). *Algèbre, Chapitre 8*, Eléments de Mathématiques (Hermann).

Bourbaki, N. (1968b). *Elements of Mathematics. Intégration, Chapter 6. Intégration vectorielle*, 1st edn., Eléments de Mathématiques (Hermann).

Bourbaki, N. (1981). *Espaces Vectoriels Topologiques*, Eléments de Mathématiques (Masson).

Bourbaki, N. (1990a). *Topologie Générale, Chapitres 1-4*, Eléments de Mathématiques (Masson).

Bourbaki, N. (1990b). *Topologie Générale, Chapitres 5-10*, Eléments de Mathématiques (CCLS).

Bredon, G. E. (1993). *Topology and Geometry*, 1st edn., GTM No. 139 (Springer Verlag).

Carter, R., Segal, G., and Macdonald, I. (1995). *Lectures on Lie Groups and Lie Algebras*, 1st edn. (Cambridge University Press).

Chevalley, C. (1946). *Theory of Lie Groups I*, 1st edn., Princeton Mathematical Series, No. 8 (Princeton University Press), eighth printing.

Deitmar, A. (2002). *A First Course in Harmonic Analysis*, 1st edn., UTM (Springer Verlag).

Dieudonné, J. (2003a). *Éléments d'Analyse, Tome V. Groupes de Lie Compacts, Groupes de Lie Semi-Simples*, 1st edn. (Edition Jacques Gabay).

Dieudonné, J. (2003b). *Éléments d'Analyse, Tome VI. Analyse Harmonique*, 1st edn. (Edition Jacques Gabay).

Dieudonné, J. (2005). *Éléments d'Analyse, Tome II. Chapitres XII à XV*, 1st edn. (Edition Jacques Gabay).

Dieudonné, J. (2010). *Éléments d'Analyse, Tome I. Chapitres I à XI*, 3rd edn. (Edition Jacques Gabay).

Dixmier, J. (1984). *General Topology*, 1st edn., UTM (Springer Verlag).

Dubrovin, B., Fomenko, A., and Novikov, S. (1985). *Modern Geometry–Methods and Applications. Part I*, 2nd edn., GTM No. 93 (Springer Verlag).

Duistermaat, J. and Kolk, J. (2000). *Lie Groups*, 1st edn., Universitext (Springer Verlag).

Dunford, N. and Schwartz, J. T. (1988). *Linear Operators, Part I: General Theory*, 1st edn. (Wiley).

Edgar, G. A. (1992). *Measure, Topology, and Fractal Geometry*, 1st edn., Undergraduate Texts in Mathematics (Springer Verlag).

Folland, G. B. (1992). *Fourier Analysis and its Applications*, 1st edn. (Brooks/Cole Publishing Company).

Folland, G. B. (1995). *A Course in Abstract Harmonic Analysis*, 1st edn. (CRC Press).

Folland, G. B. (1999). *Real Analysis. Modern Techniques and Their Applications*, 2nd edn. (Wiley-Interscience).

Fulton, W. (1995). *Algebraic Topology, A first course*, 1st edn., GTM No. 153 (Springer Verlag).

Gallier, J. H. (2011). *Geometric Methods and Applications, For Computer Science and Engineering*, 2nd edn., TAM, Vol. 38 (Springer).

Gallier, J. H. and Quaintance, J. (2020a). *Differential Geometry and Lie Groups. A Computational Perspective*, 1st edn., Geometry and Computing 12 (Springer).

Gallier, J. H. and Quaintance, J. (2020b). *Differential Geometry and Lie Groups. A Second Course*, 1st edn., Geometry and Computing 13 (Springer).

Gallier, J. H. and Quaintance, J. (2020c). *Linear Algebra and Optimization with Applications to Machine Learning, Volume I*, 1st edn. (World Scientific).

Gallot, S., Hulin, D., and Lafontaine, J. (1993). *Riemannian Geometry*, 2nd edn., Universitext (Springer Verlag).

Halmos, P. R. (1978). *Measure Theory*, 1st edn., GTM, Vol. No 18 (Springer-Verlag).

Hewitt, E. and Ross, K. A. (1979). *Abstract Harmonic Analysis I*, 2nd edn., Grundlehren der Mathematischen Wissenschaften (Springer-Verlag).

Horn, R. A. and Johnson, C. R. (1990). *Matrix Analysis*, 1st edn. (Cambridge University Press).

Kahane, J.-P. and Lemaré-Rieusset, P. G. (1998). *Séries de Fourier et Ondelettes*, 1st edn. (Cassini).

Kolmogorov, A. and Fomin, S. (1975). *Introductory Real Analysis*, 1st edn. (Dover).

Krivine, J.-L. (1969). *Théorie Axiomatique des Ensembles*, 1st edn. (PUF).

Lang, S. (1993). *Algebra*, 3rd edn. (Addison Wesley).

Lang, S. (1996). *Real and Functional Analysis*, 3rd edn., GTM 142 (Springer Verlag).

Lang, S. (1997). *Undergraduate Analysis*, 2nd edn., UTM (Springer Verlag).

Lax, P. (2002). *Functional Analysis*, 1st edn. (Wiley).

Levy, A. (1979). *Basic Set Theory*, 1st edn., Perspectives in Mathematical Logic (Springer-Verlag).

Malliavian, P., Airault, H., Kay, L., and Letac, G. (1995). *Integration and Probability*, 1st edn., GTM, Vol. No. 157 (Springer-Verlag).

Marle, C.-M. (1974). *Mesures et Probabilités*, 1st edn. (Hermann).

Massey, W. S. (1987). *Algebraic Topology: An Introduction*, 2nd edn., GTM No. 56 (Springer Verlag).

Massey, W. S. (1991). *A Basic Course in Algebraic Topology*, 1st edn., GTM No. 127 (Springer Verlag).

Milnor, J. W. and Stasheff, J. D. (1974). *Characteristic Classes*, 1st edn., Annals of Math. Series, No. 76 (Princeton University Press).

Mneimné, R. and Testard, F. (1997). *Introduction à la Théorie des Groupes de Lie Classiques*, 1st edn. (Hermann).

Munkres, J. R. (1991). *Analysis on Manifolds* (Addison Wesley).

Munkres, J. R. (2000). *Topology*, 2nd edn. (Prentice Hall).

Pontryagin, L. (1939). *Topological Groups*, 1st edn. (Princeton University Press).

Pontryagin, L. (1960). *Topological Groups*, 2nd edn. (Gordon and Breach).

Rudin, W. (1987). *Real and Complex Analysis*, 3rd edn. (McGraw Hill).

Rudin, W. (1991). *Functional Analysis*, 2nd edn. (McGraw Hill).

Sagle, A. A. and Walde, R. E. (1973). *Introduction to Lie Groups and Lie Algebras*, 1st edn. (Academic Press).

Schwartz, L. (1980). *Topologie Générale et Analyse Fonctionnelle*, Collection Enseignement des Sciences (Hermann).

Schwartz, L. (1991). *Analyse I. Théorie des Ensembles et Topologie*, Collection Enseignement des Sciences (Hermann).

Schwartz, L. (1992). *Analyse II. Calcul Différentiel et Equations Différentielles*, Collection Enseignement des Sciences (Hermann).

Schwartz, L. (1993a). *Analyse III. Calcul Intégral*, Collection Enseignement des Sciences (Hermann).

Schwartz, L. (1993b). *Analyse IV. Applications à la Théorie de la Mesure*, Collection Enseignement des Sciences (Hermann).

Serre, D. (2010). *Matrices, Theory and Applications*, 2nd edn., GTM No. 216 (Springer Verlag).

Serre, J.-P. (1973). *A Course in Arithmetic*, 1st edn., Graduate Text in Mathematics, No. 7 (Springer).

Stein, E. M. and Shakarchi, R. (2003). *Fourier Analysis. An Introduction*, Princeton Lectures in Analysis, I (Princeton University Press).

Suppes, P. (1972). *Axiomatic Set Theory* (Dover).

Terras, A. (1999). *Fourier Analysis on Finite Groups and Applications*, 1st edn., Student Text 43 (Cambridge University Press).

Warner, F. (1983). *Foundations of Differentiable Manifolds and Lie Groups*, 1st edn., GTM No. 94 (Springer Verlag).

Weil, A. (1979). *L'Intégration dans les Groupes Topologiques et ses Applications*, 2nd edn. (Hermann).

Index

Symbol Index

www.ingramcontent.com/pod-product-compliance
Lightning Source LLC
LaVergne TN
LVHW010321220926
848484LV00004BB/823
9789811291715